XIV Mexican School on Particles and Fields 2010

Journal of Physics: Conference Series Volume 287

Morelia, Michoacan, Mexico
8 – 12 November 2010

ISBN: 978-1-61782-909-3
ISSN: 1742-6588

Printed from e-media with permission by:

Curran Associates, Inc.
57 Morehouse Lane
Red Hook, NY 12571

Some format issues inherent in the e-media version may also appear in this print version.

Printed by Curran Associates, Inc. (2011)

For permission requests, please contact the Institute of Physics
at the address below.

Institute of Physics
Dirac House, Temple Back
Bristol BS1 6BE UK

Phone: 44 1 17 929 7481
Fax: 44 1 17 920 0979

techtracking@iop.org

Additional copies of this publication are available from:

Curran Associates, Inc.
57 Morehouse Lane
Red Hook, NY 12571 USA
Phone: 845-758-0400
Fax: 845-758-2634
Email: curran@proceedings.com
Web: www.proceedings.com

TABLE OF CONTENTS

MINI COURSES

PLENARY SESSIONS

PARALLEL SESSIONS

POSTER SESSIONS

PREFACE

The XIV Mexican School on Particles and Fields took place from 8–12 November, 2010, in the colonial city of Morelia, Michoacan, Mexico. The format of the school was such that the morning sessions were devoted to theoretical and experimental reviews, whereas parallel thematic sessions were held in the afternoons. All the reviews and seminars were delivered by experts of international prestige on subjects which are of current interest to the global scientific community and are also actively pursued within Mexico.

In order to equip the attending graduate students and post docs with the necessary introductory tools to allow them to benefit substantially from the specialized seminars, a series of mini-courses were offered prior to the event from 4–7 November 2010, in the Auditorium of the Faculty of Science of the University of Michoacan (UMSNH). The length of each course was about 5 hours, English being the language of instruction. An informal and friendly atmosphere was encouraged during the courses so that the students could overcome their inhibitions and actively participate in the discussions. A novel feature of this event was a colloquium aimed at the general public and younger students of pre-undergraduate level, which allowed the expert scientists to reach out to a wider community and raise their awareness and interest in one of the most fascinating and vital fields of knowledge.

The XIV-MSPF was organized by the Division of Particles and Fields of the Mexican Physical Society. It was generously sponsored by several institutions: *Consejo Estatal de Ciencia y Tecnológico (COECyT) del Estado de Michoacán, Universidad Michoacana de San Nicolás de Hidalgo, Universidad de Sonora, Universidad Nacional Autónoma de México, Universidad de Guanajuato, Universidad de Sinaloa, Centro de Investigaciones de Estudios Avanzados del IPN (CINVESTAV), Consejo Nacional de Ciencia y Tecnología (CONACyT), la Academia Mexicana de Ciencias* and, most importantly, the *Red Nacional de Física de Altas Energías*.

At a personal level, we are very grateful to Dr Juan Carlos D'Olivo (President of the *Red Nacional de Física de Altas Energías*), Dr Pedro Mata Vázquez (Director of COECyT), Dr Ricardo Becerril Bárcenas (Director of the Institute of Physics and Mathematics, UMSNH), Dr Rigoberto Vera Mendoza (Director of the Faculty of Science, UMSNH) and Dr José Napoleón Guzmán Ávila (Coordinator of Scientific Research, UMSNH) for their invaluable support in all organizational matters, which enabled the school to become a reality.

We gratefully acknowledge the help of our colleagues in the organizing committee: Alexis Aguilar, Alejandro Ayala, Wolfgang Bietenholz, Alberto Güijosa, Gabriela Murguía, Sarira Sahu (UNAM), Eduard de la Cruz Burelo, Abdel Pérez-Lorenzana (CINVESTAV), Elena Cáceres (UCOL), David Delepine (UG), Mariana Kirchbach (UASLP), Ildefonso León (UAS), Juan Carlos Arteaga-Velázquez (for his impeccable work in managing the web page of the school) and Víctor Villanueva (UMSNH). Most of them contributed to the extra work involved in refereeing the contributions submitted for this publication. Many thanks also go to all the student volunteers for the efficiency and dedication with which they carried out their duties. At the registration desk, we relied on the hard work of Xiomara Gutiérrez, Enif Gutiérrez (UMSNH) and Mara Diaz Pancardo. Several post docs and PhD students provided invaluable support in all organizational matters: Adolfo Huet, Cliffor Compeán, Rocío Bermúdez, Saúl Sánchez, Anabel Trejo, Iraís Rubalcava, Khépani Raya, José Juan González, Saúl Hernández Ortiz (UMSNH), Alfredo Galaviz, and Alan Aganza (USON). Their help in carrying out the organization of the school was essential and without their collaboration, this school would not have been the same. We also acknowledge the help of the administrative secretary Maria Esperanza Jaramillo of IFM (UMSNH).

We would like to take this opportunity to thank all the speakers for delivering excellent lectures and seminars which made this event a success. We are grateful to all the participants for providing their write-ups in time, including the notes of the mini-courses, the review articles, the contributions stemming from the parallel talks and the summarized versions of the posters presented by students.

In conclusion, we cannot resist the temptation to comment that to our utmost delight, the students participated very enthusiastically and we hope that this school will contribute considerably towards their academic development. The future of scientific endeavour always rests upon the students.

Adnan Bashir (UMSNH, Morelia)
Guillermo Contreras (CINVESTAV, Mérida)
Alfredo Raya (UMSNH, Morelia)
Maria Elena Tejeda-Yeomans (UNISON, Hermosillo)

Journal of Physics: Conference Series **287** (2011) 012001 doi:10.1088/1742-6596/287/1/012001

A Minicourse on Supersymmetry

Liliana Velasco-Sevilla
CINVESTAV, Ave. IPN 2508, Col. San Pedro Zacatenco CP 07360, México D.F., México
E-mail: lvelasco@fis.cinvestav.mx

Abstract. An immediate non-comprehensive and fast reading three hours immersion in supersymmetry. Get yourself prepared for the LHC.

1. Introduction

The series of minicourses of the Mexican Schools of Particles and Fields are thought to be an immediate, non-comprehensive, introduction to a particular field for graduate students, typically comprising three hours of lectures. Given the complexity of supersymmetry, the task seems undoable. There is a real benefit in presenting a minicourse if one at the very least attempts to make clear the main motivation of studying supersymmetry, the main ingredients necessary for it and the current status of the field, providing a damn good set of references! This is what I attempt in this contribution.

To begin with, I cite the references which I find really useful. First, the classic "A Supersymmetry Primer" [1] by S. Martin. Then the book by M. Drees, R. M. Godbole and P. Roy [2], which I think is an excellent comprehensive book for the serious supersymmetric student. There are three other references that I like. The first one of these is the very recent "Cambridge Lecures on Supersymmetry & Extra Dimensions" by F. Quevedo, S. Krippendorf and O. Schlotterer, [3]. The second one is the classic book by J. Wess and J. Bagger [4]. This one is a bit cryptic, but may be more useful for the more mathematically inclined mind. The third one is the report by M. Sohnius [5].

The real motivation of the particle physics community for studying supersymmetry is the elegant solution to the *Hierarchy Problem.* That is, if we think the Standard Model (SM) is an effective theory, broken at the Electroweak scale, and the Planck scale is a fundamental scale, why are there more than 16 orders of magnitude, in units of GeV, difference among the two scales? What would cancel the huge corrections to the Higgs mass? What fills the huge desert between the two scales? I present this motivation in Section 2, which is now, more than ever, relevant to review giving the exciting time for the possibility of discovering supersymmetric particles at the LHC (Large Hadron Collider).

The actual mathematical formalism to define entities used in supersymmetry, Superalgebra, Grassmann variables, superspace and finally supermultiplets are introduced in Sections 3, 4 and 5, respectively. These topics are superbly covered in the book by Wess and Bagger [4]. A guideline for constructing supersymmetric Lagrangians follows in Section 6. This is done in a very cryptic way, but hopefully useful enough to equip the reader for understanding the last Sections. In Section 7 I introduce the minimal supersymmetric extension of the SM (MSSM). In Section 8, I talk about mechanisms to break supersymmetry in general and in particular in

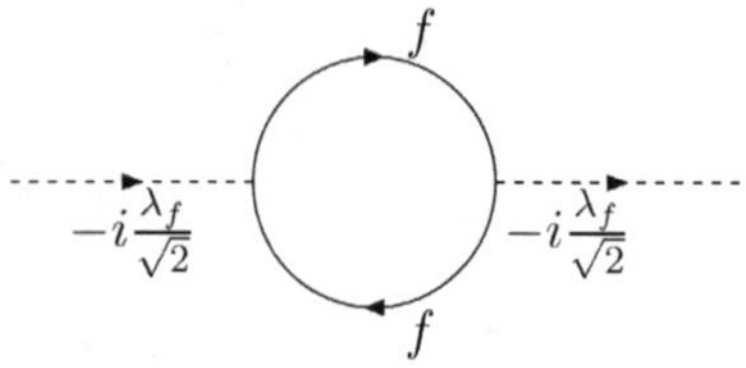

Figure 1: One loop contribution from a fermion f coupled to the Higgs field h.

the MSSM. In Section 9, I introduce the general Lagrangian that the MSSM could have after supersymmetry breaking. Here I also talk briefly about how to obtain realistic masses in the MSSM and state the current status on limits for such masses. Finally in Section 10, I mention possible extensions of the MSSM. In each section I mention further references in addition to the ones above.

2. Motivation: Hierarchy Problem & Quadratic Divergences

Let us consider the Higgs of the Standard Model (SM) [1], Φ, near its vacuum, v, where we can make the expansion $\Phi = \frac{1}{\sqrt{2}}(h+v)$. We know that $v^2 = (\sqrt{2}G_F)^{-1}$, $G_F = 1.166371 \times 10^{-5}$ GeV^{-2} and so $v \simeq 246$ GeV. The interaction with a SM fermion f is given by the Yukawa Lagrangian

$$\mathcal{L}_{\bar{f}f\Phi} = -\frac{\lambda_f}{\sqrt{2}} h\bar{f}f - \frac{\lambda_f}{\sqrt{2}} v\bar{f}f. \tag{1}$$

Via the Higgs mechanism [2], the fermion f acquires the tree level mass $m_f(0) = \lambda_f v/\sqrt{2}$ and the Higgs field obtains the tree level mass squared $m_h^2(0) = v^2/2$.

The fermionic one loop (1L) contribution to the scalar two point function reads

$$\begin{aligned}
\Pi^f_{hh}(0) &= -1\int \frac{d^4\kappa}{(2\pi)^4}\mathrm{Tr}\left[\left(-i\frac{\lambda_f}{\sqrt{2}}\right)\frac{i}{\not{\kappa}-m_f}\left(-i\frac{\lambda_f}{\sqrt{2}}\right)\frac{i}{\not{\kappa}-m_f}\right] \\
&= -2\lambda_f^2\int\frac{d^4\kappa}{(2\pi)^4}\frac{\kappa^2+m_f^2}{\left(\kappa^2-m_f^2\right)^2} = -2\lambda_f^2\int\frac{d^4\kappa}{(2\pi)^4}\left[\frac{\kappa^2-m_f^2}{\left(\kappa^2-m_f^2\right)^2}+\frac{2m_f^2}{\left(\kappa^2-m_f^2\right)^2}\right] \\
&= -2\lambda_f^2\int\frac{d^4\kappa}{(2\pi)^4}\left[\frac{1}{\kappa^2-m_f^2}+\frac{2m_f^2}{\left(\kappa^2-m_f^2\right)^2}\right].
\end{aligned} \tag{2}$$

The first term in Eq. (2) is quadratically divergent. It is particularly bad because the divergence does not depend on the mass of h. So what comes to mind is whether or not we can introduce a cut-off scale just as it is done with the logarithmic divergences. Performing integral of Eq. (2) with the substitution

$$\int\frac{d^4\kappa}{(2\pi)^4} \longrightarrow \int d\Omega\int_0^\Lambda d_{\kappa_E}\frac{\kappa^3}{8\pi^2}\frac{\kappa_E^3}{\kappa_E^2+m_f^2}, \tag{3}$$

[1] For the most part of this contribution I adopt the notation of [2].
[2] See the notes of the lectures by Poul Henrik Damgaard for an enlightening discussion of the term *spontaneous symmetry breaking*.

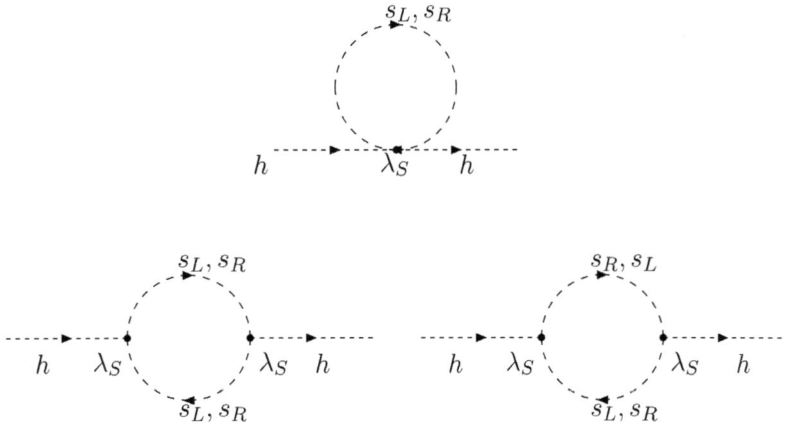

Figure 2: One loop contribution from a pair of scalars s_L and s_R coupled to the Higgs field h as in Eq. (5).

we have

$$\Pi^f_{hh}(0) = 2\lambda_f^2 \frac{\Lambda^2}{8\pi^2} i \;\Rightarrow\; m_h^2 = m_h^2(0) + 2\lambda_f^2 \frac{\Lambda^2}{8\pi^2}. \tag{4}$$

If the SM was the complete fundamental theory of particle physics, then the cut-off scale Λ could be anything. For $\Lambda = O(100)$ GeV the correction would not be orders of magnitude larger than the tree level value. However if the SM is not regarded as such and we expect to include somehow gravitational effects, we should then consider $\Lambda = O(M_{\rm P}) \simeq 2.4 \times 10^{18}$ GeV, the highest scale known in particle physics. This would be a disaster: the 1L correction to $m_h^2(0)$ would be 32 orders of magnitude larger than the tree level value. One can think also on that instead of using a momentum cut-off, Λ, we could use dimensional regularization where there would be no λ^2 piece. However, in this scheme there would enter contributions that cannot be made small if one makes an ultraviolet completion of the SM. We introduce two additional complex scalar fields s_L and s_R, with the coupling to the Higgs field

$$\begin{aligned} \mathcal{L}_{s_L s_R \Phi} &= \lambda_S |\Phi|^2 \left(|s_L|^2 + |s_R|^2\right) + \left(\lambda_f A_f \Phi s_L s_R^* + h.c.\right) \\ \xrightarrow{\Phi=\frac{1}{\sqrt{2}}(h+v)} &= \frac{1}{2}\lambda_S h^2 \left(|s_L|^2 + |s_R|^2\right) + v\lambda_S h \left(|s_L|^2 + |s_R|^2\right). \end{aligned} \tag{5}$$

Then we can calculate for this example the scalar contributions to the self energy of h, produced by the 1L diagrams of Fig. 2. These are

$$\begin{aligned} \Pi^{s_L,s_R}_{hh}(0) &= -i\lambda_S \int \frac{d^4\kappa}{(2\pi)^4} \left[\frac{i}{\kappa^2 - m^2_{s_L}} + \frac{i}{\kappa^2 - m^2_{s_R}}\right] \\ &+ (\lambda_S v)^2 \int \frac{d^4\kappa}{(2\pi)^4} \left[\frac{1}{\left(\kappa^2 - m^2_{s_L}\right)^2} + \frac{1}{\left(\kappa^2 - m^2_{s_R}\right)^2}\right] \\ &+ |\lambda_f A_f|^2 \int \frac{d^4\kappa}{(2\pi)^4} \frac{1}{\kappa^2 - m^2_{s_L}} \frac{1}{\kappa^2 - m^2_{s_R}}. \end{aligned} \tag{6}$$

We cancel the first two terms of Eq. (6) with the first term of the last expression in Eq. (2), provided that $-\lambda_S = \lambda_f^2$, $m_{s_L} = m_{s_R} = m_f$. Furthermore, using that $\int \frac{d^4\kappa}{(2\pi)^4} \frac{i}{\kappa^2 - m^2_{s_L}}$

$= 2\pi^2 \int_0^\Lambda \frac{d\kappa_E}{16\pi^2} \frac{\kappa_E^3}{\kappa_E^2+m_f^2}$, we can cancel all the quadratic divergences with the sum of expressions Eq. (2) and Eq. (6). The remaining logarithmic divergences in $\Pi^f_{hh}(0) + \Pi^s_{hh}(0)$ can be cancelled by replacing the logarithmic divergences by the logarithm of the square of the renormalization scale μ [3]. However by performing the integrals in the second row of Eq. (6), using the regularization of the divergent integrals as $\int \frac{d^4\kappa}{(i\pi)^2} \frac{1}{(\kappa^2-m^2)^2} \longmapsto -\ln\left[\frac{m^2}{\mu^2}\right]$,

$$\int \frac{d^4\kappa}{(2\pi)^4} \left[\frac{1}{\kappa^2 - m_1^2} \frac{1}{\kappa^2 - m_2^2}\right]$$
$$\longmapsto (m_1^2 - m_2^2) B_0(0, m_1^2, m_2^2) = m_1^2 \left(1 - \ln \frac{m_1^2}{\mu^2}\right) - m_2^2 \left(1 - \ln \frac{m_2^2}{\mu^2}\right)$$

and summing up the fermionic and scalar contributions, where from now own we call

$$s_{L,R} \doteq \tilde{f}_{L,R}, \tag{7}$$

we obtain the total 1L correction to $m_h^2(0)$

$$\begin{aligned}
\Pi^f_{hh}(0) + \Pi^{s_L,s_R}_{hh}(0) &= -2\lambda_f^2 \int \frac{d^4\kappa}{(2\pi)^4} \left[\frac{1}{\kappa^2 - m_f^2} + \frac{2m_f^2}{\left(\kappa^2 - m_f^2\right)^2}\right] \\
&- \lambda_{\tilde{f}} \int \frac{d^4\kappa}{(2\pi)^4} \left[\frac{1}{\left(\kappa^2 - m^2_{\tilde{f}_L}\right)^2} + \frac{1}{\left(\kappa^2 - m^2_{\tilde{f}_R}\right)^2}\right] \\
&+ (\tilde{\lambda}_f v)^2 \int \frac{d^4\kappa}{(2\pi)^4} \left[\frac{1}{\kappa^2 - m^2_{\tilde{f}_L}} + \frac{1}{\kappa^2 - m^2_{\tilde{f}_R}}\right] \\
&= i\frac{\lambda_f^2}{16\pi^2} \left[-2m_f^2 \left(1 - \ln \frac{m_f^2}{\mu^2}\right) + 4m_f^2 \ln \frac{m_f^2}{\mu^2}\right. \\
&+ \left. 2m_{\tilde{f}}^2 \left(1 - \ln \frac{m_{\tilde{f}}^2}{\mu^2}\right) - 4m_{\tilde{f}}^2 \ln \frac{m_{\tilde{f}}^2}{\mu^2}\right].
\end{aligned} \tag{8}$$

We can see that indeed we need

$$\lambda_{\tilde{f}} = -\lambda_f^2, \quad m_{\tilde{f}_L} = m_{\tilde{f}_R} = m_f \tag{9}$$

to cancel exactly the 1L correction to $m_h^2(0)$. We have seen that introducing scalar particles with the same masses to those of the fermions and whose couplings are related to the couplings of the latter, Eq. (9), quadratic divergences can be cancelled at the 1L level. To cancel these kind of divergences at all orders in perturbation theory would be remarkable. Well, this is precisely what supersymmetry does. Despite the fact that we have not yet seen any experimental evidence of it, in the next sections we will explore why is so fascinating.

3. Supersymmetry Algebra

Since supersymmetric transformations change a particle with integer spin to a particle with half-integer spin, Q must be a spinor. Then we introduce properly its two spinorial indices

$$Q_A, \quad A = 1, 2. \tag{10}$$

[3] Check references [2, 1] for further comments about this.

However Q should be identified with a four dimensional (4D) quantity and it is indeed a 4D Majorana spinor, which can be decomposed as 2D Weyl spinors

$$Q_a = \begin{pmatrix} Q_A \\ \overline{Q}_{\dot{A}} \end{pmatrix}. \tag{11}$$

The natural question to ask immediately is whether or not Q_a can be a generator of the symmetries of a 4D theory. We know that the Poincaré Group generates the fullest continuous spacetime symmetries of particle interactions. We recall that its generators can be expressed as unitary operators

$$U(a) = e^{ia^\mu P_\mu}, \quad U(\Lambda) = e^{-i/2\omega^{\mu\nu} M_{\mu\nu}}, \tag{12}$$

for translations and for homogeneous Lorentz transformations, respectively. For an infinitesimal transformation on the 4D coordinates we have $x^\mu \longrightarrow x^\mu = (\delta^\mu_\nu + w^\mu_\nu)x^\nu + a^\mu$.

What do we know about the properties of the generators? the Algebra of the generators which is expressed in terms of commutators, for example

$$\begin{aligned} [P_\mu, P_\nu] &= 0, \quad [M_{\mu,\nu}, P_\rho] = i\left(\eta_{\nu\rho} P_\mu - \eta_{\mu\rho} P_\nu\right), \\ [M_{\mu,\nu}, P_{\rho\sigma}] &= -i\left(\eta_{\mu\rho} M_{\nu\sigma} - \eta_{\mu\sigma} M_{\nu\rho} - \eta_{\nu\rho} M_{\mu\sigma} + \eta_{\nu\sigma} M_{\mu\rho}\right). \end{aligned} \tag{13}$$

The Coleman-Mandula theorem [6] states that the full Lie Algebra of the S-Matrix in 4D, described by $\left[T^a, T^b\right] = it^{ab}_c T^c$, is such that the generators T^a are a direct sum of P_μ and $M_{\mu,\nu}$, that is to say

$$[T^a, P_\mu] = 0 = [T^a, M_{\mu\nu}]. \tag{14}$$

However a supersymmetric transformation described by Q_a is transforming internal degrees of freedom from a boson into a fermion, or vice versa $Q|b, f\rangle = |f, b\rangle$, for b a boson with spin n and f a fermion with spin $m/2$. Specifically, we are making a transformation of particles of spin n to particles with spin $(n \pm 1)/2$ and particles with spin $m/2$ to particles with spin m. Thus if we want Q_a to be indeed a generator of a definite group describing the symmetries of our 4D theory then we need to specify the Algebra which is satisfied by these generators. From the Coleman-Mandula theorem we know that it cannot be a simple Lie Algebra and has the following properties: it cannot commute with rotations but commutes with translations. That the generator Q_a of these transformations cannot commute with rotations can be easily seen thinking of a rotation about an axis, described by $U(2\pi)$: $U(2\pi)|b\rangle = |b\rangle$ and $U(2\pi)|f\rangle = -|f\rangle$, then we have

$$\begin{aligned} &U(2\pi)QU(2\pi)^{-1}|f\rangle = -U(2\pi)QU(2\pi)^{-1}U(2\pi)|f\rangle = -U(2\pi)|b\rangle = -Q|f\rangle, \\ &U(2\pi)QU(2\pi)^{-1}|b\rangle = -Q|b\rangle, \end{aligned} \tag{15}$$

that is $U(2\pi)QU(2\pi)^{-1} = -Q$. But it is easy to see that it commutes with translations because the generator Q is not affecting the position of the field describing the particle. In order to get around with the use of Lie Algebras, defined in terms of commutators, then one can think about using Algebras defined also in terms of anti-commutators. Such mathematical constructions enter into the classification of the *Graded Lie Algebras* [4]. For our case, we need to expand the set of commutator relations followed by P_μ and $M_{\mu\nu}$ to include Q_a in such a way that the Poincaré Algebra is recovered and the properties described above (that it commutes with translations and

[4] For a physics perspective on Graded Lie Algebras and specifically on the Superalgebra, the book by J.F. Cornwell [7] is great.

Journal of Physics: Conference Series **287** (2011) 012001 doi:10.1088/1742-6596/287/1/012001

that it cannot commute with rotations) are recovered, by means of anti-commutators. In the two component notation of Eq. (10), we have the $N = 1$ Supersymmetry Algebra in 4D

$$\begin{aligned}
&\{Q_A, \overline{Q}_{\dot{B}}\} = 2\sigma^{\mu}_{A\dot{B}}P_{\mu}, \quad \left\{\overline{Q}^{\dot{A}}, Q^{B}\right\} = 2\overline{\sigma}^{\mu}_{\dot{A}B}P_{\mu}, \\
&\{Q_A, Q_B\} = \left\{\overline{Q}^{\dot{A}}, \overline{Q}^{\dot{B}}\right\} = 0, \\
&[Q_A, P_\mu] = [\overline{Q}_{\dot{A}}, P_\mu] = 0, \\
&[M_{\mu\nu}, Q_A] = -\left(\sigma_{\mu\nu}\right)_A^B Q_B, \quad \left[M_{\mu\nu}, \overline{Q}^{\dot{A}}\right] = -\left(\sigma_{\mu\nu}\right)^{\dot{A}}_{\dot{B}} Q^{\dot{B}}, \\
&[Q_A, R] = Q_A, \quad \left[\overline{Q}^{\dot{A}}, R\right] = -Q^{\dot{A}}.
\end{aligned} \tag{16}$$

The last expression in Eq. (16) reflects an invariance under a *chiral* rotation, that can be thought of being generated by a $U(1)$ generator R. This will satisfy $[Q_a, R] = (\gamma_5)_{ab}Q_b$. $\sigma_{\mu\nu} = \sigma^{\mu}\overline{\sigma}^{\nu} - \sigma^{\nu}\overline{\sigma}^{\mu}$.

The Haag, Łopuszánski, Sohnius theorem [8] states that the $N = 1$ Supersymmetry (SUSY) Algebra in 4D is a unique extension of the Poincaré Algebra in 4D relativistic Quantum Field Theory (QFT) with one generator Q_A. This is really just the statement of saying, if there is another generator Q_A of a 4D QFT, apart from the translations and homogeneous Lorentz transformations, then that will define a unique Algebra, the supersymmetry algebra, $N = 1$ in 4D. But of course this theorem allows for theories with more supersymmetries, that is, there could be more than one generator Q_B in a 4D QFT.

4. Grassmann Variables and Superspace

Supersymmetric theories can be defined in terms of anti-commuting or Grassmann variables. These are such that

$$\{\theta, \overline{\theta}\} \doteq \theta\overline{\theta} + \overline{\theta}\theta = 0 \text{ and } \theta = 0^2 = \overline{\theta}^2. \tag{17}$$

From these basic rules we can develop the properties of calculus on these variables. For example, a function can be *Taylor-Grassmann* expanded as

$$f(\theta) = f_0 + f_1\theta, \tag{18}$$

where $f_0, f_1 \in \mathbb{C}$. Then we have the nice properties

$$\int d\theta\ \theta = 1, \quad \frac{df(\theta)}{d\theta} = f_1, \quad \theta f(\theta) = \theta f_0 + \cancelto{0}{f_1\theta^2}. \tag{19}$$

We can see why Grassmann Variables could be so useful, the expansion Eq. (18) is finite and simple differentiation or integration can project out the components of the functions defined on θ. It happens that supersymmetric Lagrangians and their particle interactions can be written in a space which is spanned by the coordinates of our 4D space plus two Grassmann coordinates

$$z \doteq (x^{\mu}, \theta, \overline{\theta}), \tag{20}$$

this space is called the *Superspace*. However the Grassmann variables in Eq. (20) are not as simple as in Eq. (17), because they will contain two indices, A and B. The indices A and B will label two spinor indices in the $(\frac{1}{2},0)$ representation of the Lorenz group, while the indices $\dot{A}$ and $\dot{B}$ will label two spinor indices in the $(0,\frac{1}{2})$ representation. Accordingly, the tensors $\epsilon^{A,B}$ and $\epsilon_{A,B} = \mathrm{Inv}[\epsilon^{A,B}]$ represent the metric in the space of the $(\frac{1}{2},0)$ representations, while $\epsilon^{\dot{A},\dot{B}} = \epsilon_{AB}$ and $\epsilon_{\dot{A},\dot{B}} = \mathrm{Inv}[\epsilon^{\dot{A},\dot{B}}]$ represent the metric in the space of the $(0,\frac{1}{2})$ representations [5] .

We can then rise and lower the indices as

$$\theta^A = \epsilon^{A,B}\theta_B, \quad \theta_A = \epsilon_{A,B}\theta^B, \quad \overline{\theta}^{\dot{A}} = \epsilon^{\dot{A},\dot{B}}\overline{\theta}_{\dot{B}}, \quad \overline{\theta}_{\dot{A}} = \epsilon_{\dot{A},\dot{B}}\overline{\theta}^{\dot{B}}. \tag{21}$$

The supersymmetry generators can be represented in a differential form in terms of the Grassmann variables $Q_A = -i\left(\partial_A + i\sigma^{\mu}_{A\dot{B}}\overline{\theta}^{\dot{B}}\partial_\mu\right)$ and $\overline{Q}^{\dot{A}} = i\left(\overline{\partial}^{\dot{A}} - i\sigma^{\mu\dot{A}B}\theta_B\partial_\mu\right)$.

Here we note that for the supersymmetry algebra, Eq. (16), to be satisfied, we need $\overline{\epsilon Q} = (\epsilon Q)^\dagger$. Also we define $\sigma^{\mu\dot{A}B} = (1_{2\times 2}, -\vec{\sigma})$ and $\sigma^{\mu}_{A\dot{B}} = (1_{2\times 2}, \vec{\sigma})$. Then we define the chiral covariant and anti-chiral covariant derivatives

$$\begin{aligned} \mathcal{D}_A &= \partial_A - i\sigma^{\mu}_{A\dot{B}}\overline{\theta}^{\dot{B}}\partial_\mu \rightarrow \text{ in the } (\frac{1}{2},0) \text{ representation,} \\ \overline{\mathcal{D}}^{\dot{A}} &= \overline{\partial}^{\dot{A}} + i\sigma^{\mu\dot{A}B}\theta_B\partial_\mu \rightarrow \text{ in the } (0,\frac{1}{2}) \text{ representation.} \end{aligned} \tag{22}$$

In the superspace defined by Eq. (20), the infinitesimal supersymmetric transformations act as

$$\left(x^\mu, \theta, \overline{\theta}\right) \rightarrow \left(x^\mu - i\theta\sigma^\mu\overline{\epsilon} + i\epsilon\sigma^\mu\overline{\theta}, \theta + \epsilon, \overline{\theta} + \overline{\epsilon}\right), \tag{23}$$

where ϵ is an infinitesimal Grassmann parameter.

5. Particle Supermultiplets aka Superfields

Being the case that we have an Algebra, we need to find representations, i.e. multiplets, that satisfy the properties of the Algebra. In the case of supersymmetry, it is necessary to define them in the superspace. The two classes of such supermultiplets or superfields used in particle physics are the *Chiral superfields* and the *Vector superfields.*

Before describing each kind of superfields, we just state their general definition and properties. The definition is a generalization of Eq. (18), that is a function $\mathcal{F}$ defined in superspace [6]

$$\begin{aligned} \mathcal{F}\left(x^\mu, \theta, \overline{\theta}\right) &= f_0(x) + f_1^A(x)\theta_A + \overline{f}_{2\dot{A}}(x)\overline{\theta}^{\dot{A}} \\ &+ f_3(x)\theta\theta + \overline{f}_4(x)\overline{\theta\theta} + f_5^A(x)\theta_A\overline{\theta\theta} + \overline{f}_{6\dot{A}}(x)\overline{\theta}^{\dot{A}}\theta\theta + f_7(x)\theta\theta\overline{\theta\theta}. \end{aligned} \tag{24}$$

There are various things to note. First, since there are two indices for θ, the naive commutation relations of Eq. (17) will not hold. Instead we have

$$\theta\theta = -2\theta^A\theta^B\epsilon_{AB}, \quad \overline{\theta\theta} = -2\overline{\theta}_{\dot{A}}\overline{\theta}_{\dot{B}}\epsilon^{\dot{A}\dot{B}}. \tag{25}$$

Second, the functions $f_i(x)$ are not all of the same type, some carry spinor indices and some do not. In Eq. (24) f_0, f_3, f_4 and f_7 are scalars, f_1 and f_2 must be spinors in the $(1/2,0)$ (left) representation, while f_2 and f_6 must be spinors in the $(0,1/2)$ (right) representation.

5 ϵ_{AB} it is the anti-symmetric tensor in two dimensions.

6 The review by S. Martin [1] is a great introduction to these topics and in general for supersymmetric phenomenology.

Last, the θ's with no apparent indices have been left so because we have contracted explicitly all the spinor indices. Just as in § 4, we can perform simple differentiations or integrations to project out the components of the superfields, e.g.

$$\int d^4\theta \mathcal{F}\left(x^\mu,\theta,\overline{\theta}\right) = f_7(x), \quad \int d^4\theta[\theta\theta\overline{\theta}\overline{\theta}] = 1. \tag{26}$$

Last, there are two terms in Eq. (24) that receive special names: the F term, which is the term that appears multiplied by $\theta\theta$, and the D term, which appears multiplied by $\theta\theta\overline{\theta}\overline{\theta}$.

The supersymmetry transformations act on $\mathcal{F}$ as

$$\delta\mathcal{F} \;=\; \mathcal{F}\left(x^\mu - i\theta\sigma^\mu\overline{\epsilon} + i\epsilon\sigma^\mu\overline{\theta}, \theta+\epsilon, \overline{\theta}+\overline{\epsilon}\right) - \mathcal{F}(x,\theta,\overline{\theta}) \;=\; i(\epsilon Q + \overline{\epsilon}\overline{Q})\mathcal{F}. \tag{27}$$

From this last expression we can work out the supersymmetry transformations for each of the components of $\mathcal{F}$. We will explicitly state them in the next subsections. Finally we note that the superspace is a fascinating research field on its own. This only for its mathematical structure

5.1. *Chiral superfields*

A left chiral superfield is defined such that it depends only on y^μ and θ and a right chiral superfield such that it depends only on $\overline{y}^\mu$ and $\overline{\theta}$, where

$$y^\mu \;\doteq\; x^\mu - i\theta\sigma^\mu\overline{\theta} \quad , \overline{y}^\mu \;\doteq\; x^\mu + i\theta\sigma^\mu\overline{\theta}. \tag{28}$$

According to this definition we can write

$$\Phi(y,\theta) = \phi(y) + \sqrt{2}\theta\xi(y) + \theta\theta F(y), \quad \Phi(\overline{y},\overline{\theta})^\dagger = \phi^*(y) + \sqrt{2}\overline{\theta}\overline{\xi}(\overline{y}) + \overline{\theta}\overline{\theta}F^*(\overline{y}), \tag{29}$$

where the factor $\sqrt{2}$ is introduced for convenience. Both expressions can be put in terms of x

$$\begin{aligned}
\Phi(x^\mu - i\theta\sigma^\mu\overline{\theta},\theta) &= \phi(x) - i\theta\sigma^\mu\overline{\theta}\partial_\mu\phi(x) - \frac{1}{4}\theta\theta\overline{\theta}\overline{\theta}\partial^\mu\partial_\mu\phi(x) \\
&+ \sqrt{2}\theta\xi(x) + \frac{i}{\sqrt{2}}\theta\theta\partial_\mu\xi(x)\sigma^\mu\overline{\theta}, \\
\Phi(x^\mu + i\theta\sigma^\mu\overline{\theta})^\dagger &= \phi(x)^* + i\theta\sigma^\mu\overline{\theta}\partial_\mu\phi^*(x) - \frac{1}{4}\theta\theta\overline{\theta}\overline{\theta}\partial^\mu\partial_\mu\phi^*(x) + \sqrt{2}\overline{\theta}\overline{\xi}(x) \\
&- \frac{i}{\sqrt{2}}\overline{\theta}\overline{\theta}\theta\sigma^\mu\partial_\mu\overline{\xi}(x) + \overline{\theta}\overline{\theta}F^*(x).
\end{aligned} \tag{30}$$

What is the important difference between the expressions of Eq. (29) and Eq. (30)? They are two different representations of the supersymmetry Algebra, Eq. (29) is the *chiral representation* for supersymmetric generators, Eq. (30) is the so called *symmetric* representation, where $\overline{Q} = Q^\dagger$. They are completely equivalent. We can project the components of $\Phi(y,\theta)$, using equally Eq. (29) or Eq. (30) for the latter we have

$$\phi(x) = \Phi(x,\theta,\overline{\theta})|_{\theta=0=\overline{\theta}}, \quad \sqrt{2}\xi_A(x) = \mathcal{D}_A\Phi(x,\theta,\overline{\theta})|_{\theta=0=\overline{\theta}}, \quad F(x) = \frac{1}{4}\mathcal{D}\mathcal{D}\Phi(x,\theta,\overline{\theta})|_{\theta=0=\overline{\theta}}. \tag{31}$$

What is the physics that we can extract from Eq. (29)? There are four real scalar components, contained in $F(x)$ and $\phi(x)$. The spinor ξ_A contains four real fermionic fields. Thus there is an equal number of scalars and fermions. This the *off-shell* description, that means that not all the degrees of freedom are physical. F are called auxiliary fields and have trivial equations of motion, that is $\frac{\partial\mathcal{L}}{\partial F} = 0$, which can be used to eliminate them from the Lagrangian. After this, we have two physical scalar degrees of freedom and two physical fermionic degrees of freedom, both with the same mass. This sets the minimum basis for understanding § 6. To end this section we state the properties, that can be easily derived using the exposition above, of the chiral and anti-chiral superfields:

- $\overline{\mathcal{D}}_{\dot{A}}\Phi = 0, \quad \mathcal{D}^{A}\Phi^{\dagger} = 0,$
- products and linear combinations of chiral superfields are chiral superfields.

5.2. Vector superfields

A vector superfield in superspace is defined such that

$$\begin{aligned} V(x,\theta,\overline{\theta}) &= C(x) + \sqrt{2}\theta\xi(x) + \sqrt{2}\overline{\theta}\overline{\xi}(x) + \theta\theta M(x) + \overline{\theta\theta}M^{*}(x) \\ &+ \theta\sigma^{\mu}\overline{\theta}A_{\mu}(x) + \theta\theta\overline{\theta}\overline{\lambda}(x) + \overline{\theta\theta}\theta\lambda(x) + \frac{1}{2}\theta\theta\overline{\theta\theta}D(x), \end{aligned} \tag{32}$$

where $M(x)$ is a complex scalar. Here ξ and ζ are left-handed Weyl spinors and $\overline{\chi}$ and $\overline{\lambda}$ are right-handed Weyl spinors.

From the definition of a vector superfield, Eq. (32), we can easily obtain its most important property:

- $V = V^{\dagger}$.

6. Construction of a Supersymmetric Langrangian

6.1. General considerations

From the properties of the chiral superfields, Eq. (5.1) and Eq. (5.2), we can extract a very useful recipe for the construction of supersymmetric Lagrangians:

(i) Take all the possible products of chiral superfields which satisfy the symmetries of the proposed model, the F term it will be always part of the Lagrangian, as it transforms as a total derivative under supersymmetric transformations.

(ii) The D term of the product $\Phi\Phi^{\dagger}$ also transforms as a total derivative under supersymmetric transformations.

From the observations above enumerated, we see that we need to put a limit on the product of chiral superfields, because of renormalizability, obviously then we have that the quantity

$$W(\Phi_i) = h^{i}\Phi_i + \frac{1}{2}m^{ij}\Phi_i\Phi_j + \frac{1}{3!}y^{ijk}\Phi_i\Phi_j\Phi_k. \tag{33}$$

called *superpotential*, it is the most general combination of products of superfields that we can have. We say that the superpotential is *holomorphic* because, since each term is a chiral superfield, W can be entirely defined in terms of two different coordinates in superspace: y and θ Eq. (28).

The *holomorphicity* property of the superpotential, excludes explicitly then combinations of the form $B\overline{B}$, where $\overline{B}$ is the supermultiplet defined as $\overline{B} = B^{\dagger}\gamma^{0}$ so it not only carries the opposite quantum numbers of B but the supermultiplet $\overline{B}$ obviously contains the anti-particles of the supermultiplet B! I emphasize this because of the extended vicious notation in the supersymmetric literature where there are expressions of the form

$$W = \frac{1}{2}m^{ij}B_i\overline{B}_j. \tag{34}$$

Well the authors here abuse the notation and the harm done here it is that *we* are introducing a chiral superfield $\overline{B}$ with opposite quantum numbers to those of B but that it does not contain their anti-particle!, i.e. in these cases $\overline{B} \neq B^{\dagger}\gamma^{0}$. The careful reader will note that anyway... If we have terms in the superpotential such as

$$ABC \in W, \tag{35}$$

for A, B, C chiral superfields, then terms of the form

$$A\overline{B}C, \tag{36}$$

are also excluded because for $\overline{B} = B^\dagger\gamma^0$ this is an anti-chiral superfield and destroys explicitly the holomorphicity of W, that is why the emphasis in the notation $W(\Phi_i)$. The $\mathcal{L}$ of the theory can at most change under a SUSY transformations into itself plus a total space-time derivative. The D term of the product $\Phi_i^\dagger\Phi_j$ is

$$\left[\Phi_i^\dagger\Phi_j\right]_D = \theta\theta\overline{\theta\theta}\left(F_i^*F_j + \frac{1}{2}\partial_\mu\Phi_i^*\partial^\mu\Phi_J - \frac{1}{2}\Phi_i^*\partial_\mu\partial^\mu\Phi_j + i\xi_j\sigma^\mu\delta_\mu\overline{\xi}_i\right), \tag{37}$$

which is so special because after the supersymmetric transformations it transforms indeed as a total derivative: $\delta D = i\delta_\mu\left(\zeta\sigma^\mu\overline{\epsilon} + \overline{\lambda}\overline{\sigma}^\mu\epsilon\right)$.

We can show that $[W(\Phi_i) + h.c.]_F$ transforms as a total derivative, then we can construct the following supersymmetric Lagrangian

$$\mathcal{L} = \left[\Phi_i^\dagger\Phi_i\right]_D + \left[W(\Phi_i) + \overline{W}(\Phi_i^\dagger)\right]_F. \tag{38}$$

Would it be possible to define other quantity rather than W, such that its D or F part could supersymmetrically transform exactly into itself or up to a total derivative? The answer is no and this has profound consequences, in particular it restricts the matter content of supersymmetric theories and the way they are coupled, as we will see in § 7. Let us state this in another way. If a given Lagrangian is supersymmetric, then it can be decomposed in the form of Eq. (38) and $W(\Phi_i)$ is uniquely defined.

6.2. Interaction Free Lagrangian

The simplest supersymmetric model is the one proposed by Wess and Zumino (WZ). The action of this model can be written as

$$S = \int d^4z\,(\mathcal{L}_S + \mathcal{L}_F + \mathcal{L}_A), \quad \mathcal{L}_S = -\partial^\mu\phi^*\partial_\mu\phi,\ \mathcal{L}_F = i\psi^\dagger\overline{\sigma}^\mu\partial_\mu\psi,\ \mathcal{L}_A = F^*F. \tag{39}$$

The transformations of bosons into fermions and vice versa, are given by

$$\begin{aligned}
\delta\phi &= \epsilon\psi,\\
\delta\psi_A &= -i\left(\sigma^\mu\epsilon^\dagger\right)_A\partial_\mu\phi + \epsilon_A F,\\
\delta\psi^\dagger_{\dot{A}} &= i\left(\epsilon\sigma^\mu\right)_{\dot{A}}\partial_\mu\phi^* + \epsilon^\dagger_{\dot{A}}F^*,\\
\delta F &= -i\epsilon^\dagger\overline{\sigma}^\mu\partial_\mu\psi,\\
\delta F^* &= i\partial_\mu\psi^\dagger\overline{\sigma}^\mu\epsilon.
\end{aligned} \tag{40}$$

Recall that the auxiliary fields F should disappear in the expression for the *physical* Lagrangian. We can calculate the supersymmetric transformations of $\mathcal{L}_S$ and $\mathcal{L}_F$, ignoring at the moment the auxiliary fields

$$\begin{aligned}
\delta\mathcal{L}_S &= -\delta\left(\partial^\mu\phi^*\partial_\mu\phi\right),\\
&= -\left(\partial^\mu(\delta\phi^*)\partial_\mu\phi^\dagger + \partial^\mu\phi^*\partial_\mu(\delta\phi)\right),\\
&= -\epsilon^\dagger\partial^\mu\psi^\dagger\partial_\mu\phi - \epsilon\partial^\mu\psi\partial_\mu\phi^*,\\
\delta\mathcal{L}_F &= -\epsilon\sigma^\mu\overline{\sigma}^\mu\partial_\mu\psi\partial_\mu\phi^* + \psi^\dagger\overline{\sigma}^\nu\sigma^\mu\epsilon^\dagger\partial_\mu\partial_\nu\phi,
\end{aligned} \tag{41}$$

rearranging terms:

$$\delta\mathcal{L}_S + \delta\mathcal{L}_F = \epsilon\partial^\mu\Phi_\mu\phi^* + \epsilon^\dagger\partial^\mu - \partial_\mu\left(\epsilon\sigma^\nu\overline{\sigma}^\mu\psi\partial_\mu\phi^* + \epsilon\psi\partial^\mu\phi^* + \epsilon^\dagger\psi^\dagger\partial^\mu\phi\right). \quad (42)$$

It can be shown that including the auxiliary fields, the variation can also be written as a total derivative $\delta S = \int d^4x\,(\delta\mathcal{L}_S + \delta\mathcal{L}_F + \delta\mathcal{L}_A) = 0$.

6.3. Interacting Lagrangian

We need to check how we can construct an interacting Lagrangian. The easiest example are the $U(1)$ gauge transformations, under which the superfields transform as

$$\Phi_i \longrightarrow e^{-2igt_i\Lambda(z)}\Phi_i, \quad \Phi_i^\dagger \longrightarrow \Phi_i^\dagger e^{2igt_i\Lambda^\dagger(z)}, \quad (43)$$

where g is the coupling constant, t_i are $U(1)$ charges and Λ is a complex function specifying the local gauge transformations, defined in superspace, $z = (x, \theta, \overline{\theta})$. Now we need to get a term of the sort $\Phi^\dagger V\Phi$, where V is a vector superfield, in order to have an interaction term. Then we can show that

$$\Phi_i^\dagger e^{2gt_iV}\Phi \quad (44)$$

is gauge invariant. To preserve the chirality of the gauge transformations, we need $\mathcal{D}_A\Lambda^\dagger = 0$, $\overline{\mathcal{D}}^{\dot{A}}\Lambda = 0$, that is Λ and $\Lambda^\dagger$ are chiral and anti-chiral respectively. Now defining

$$T_A \doteq -\frac{1}{4}\overline{\mathcal{D}}\overline{\mathcal{D}}\mathcal{D}_AV, \quad \overline{T}_{\dot{A}} \doteq -\frac{1}{4}\mathcal{D}\mathcal{D}\overline{\mathcal{D}}_AV. \quad (45)$$

Note that T^AT_A is a left chiral superfield, then we can construct the following interacting Lagrangian

$$\mathcal{L} = \frac{1}{4}\left[T^AT_A + \overline{T}_{\dot{A}}\overline{T}^{\dot{A}}\right]_F + \left[\Phi_i^\dagger e^{2gt_iV}\Phi_i + 2\eta V\right]_D + \left[W(\phi_i) + h.c.\right]_F, \quad (46)$$

in component form, in the Wess-Zumino (WZ) gauge, once D and F are projected out, we have:

$$\begin{aligned}\mathcal{L} &= \frac{1}{2}D^2 + \eta D + gt_i|\phi_i|^2D - \frac{1}{4}F_{\mu\nu}F^\mu\nu + i\lambda\sigma^\mu\partial_\mu\overline{\lambda} + i\xi_i\sigma^\mu\mathcal{D}_{\mu i}^\dagger\overline{\xi_i}\\ &+ \mathcal{D}_i^\mu\xi_i\mathcal{D}_{\mu i}^\dagger\overline{\xi_i} + F_i^*F_i + \left[\frac{1}{2}\xi_i\xi_jW_{ij}(\phi) + h.c.\right] - \sqrt{2}\,gt_i\left(\overline{\lambda\xi}_i + h.c.\right),\end{aligned} \quad (47)$$

where of course the superpotential $W(\phi_i)$ needs to respect gauge invariance and η is a real constant [7]. In expression Eq. (47) we have used

$$F_i^* = -W_i = \frac{\delta W}{\delta\phi_i}, \quad F_i = -\overline{W}_i = \frac{\delta W^\dagger}{\delta\phi_i^\dagger}, \quad W^i = \frac{\delta W}{\delta\Phi_i}, \quad W^{ij} = m^{ij} + y^{ijk}\Phi_i. \quad (48)$$

[7] The corresponding action to Eq. (46) is

$$\mathcal{S} = \int d^8z\left[\Phi_i^\dagger e^{2gt_iV}\Phi_i + 2\eta V\right] + \int d^6z\left[\frac{1}{4}T^AT_A + W(\Phi)\right] + \int d^6\overline{z}\left[\frac{1}{4}\overline{T}^{\dot{A}}\overline{T}_{\dot{A}} + W^\dagger(\overline{\Phi})\right].$$

Journal of Physics: Conference Series 287 (2011) 012001 doi:10.1088/1742-6596/287/1/012001

We note that the D component of V is real and its gauge invariant:

$$[V]_D = \int d^2\theta V(x,\theta,\overline{\theta}) = \frac{1}{2}D(x), \tag{49}$$

exactly because V must transform under the adjoint representation of $U(1)$, which is scalar and thus $D(x)$ must be a scalar left invariant under $U(1)$. Note that this only happens in a $U(1)$ gauge theory because the expression Eq. (49) is not gauge invariant for a non-Abelian vectors superfield, since they have to be in the adjoint and so the vector field carries indices of the adjoint, $V_{IJ} = 2gV^a T^a_{ij}$.

We have that $\frac{\partial \mathcal{L}}{\partial D} = 0$ then D is another auxiliary field and it can be expressed as

$$D = -g\phi_i^* t_i \phi_i - \eta. \tag{50}$$

Then we can write the scalar potential as

$$U = W_i(\Phi_i)\overline{W}^i(\overline{\Phi}_i) + \frac{1}{2}D^2, = F_i^* F_i + \frac{1}{2}D^2, \tag{51}$$

and it is gauge invariant. When projecting the D term we have

$$\mathcal{L}_{int} = \left(-\frac{1}{2}W^{ij}\psi_i\psi_j + W^i F_i + x^{ij}F_iF_j + c.c.\right) - U, \tag{52}$$

where we can see that the left-hand side is not a symmetric quantity because the interchange of indices i, j with respect to the second term of the right-hand side, where we need to vary all possible indices k. The first part in the right-hand side of Eq. (48) is a symmetric quantity as it is explicitly given by Eq. (33).

7. The Minimal Supersymmetric Standard Model (MSSM)

The Minimal Supersymmetric Standard Model (MSSM) is as its name states the minimal possible way in one the SM can be supersymmetrized. The recipe seems obvious: just embed each of the SM particles in a supermultiplet. Since supersymmetry does commute with the gauge symmetries of the theory, the quantum numbers associated to them will be the same in the whole supermultiplet, except that the supersymmetric particles will have different spins. The consequences are then that:

- For each SM fermionic multiplet with spin 1/2, there will appear a scalar (sfermion) with spin 0.
- For each SM gauge boson, with spin 1, it will appear a fermion, better known as *gaugino*, with spin 1/2.
- Instead of having just one Higgs supermultiplet, there will be two! These supermultiplets will contain each one scalar boson of spin 0 and their fermionic superpartner called *higgsino* with spin 1/2 .

O.k. the last consequence is the non obvious part of the extension. But it is very easy to understand. From Eq. (33) we can see that the only possible way that could enter in a supersymmetric Lagrangian, Eq. (38), is to be defined as a holomorphic function of the superfields. We need to obtain from the Lagrangian of Eq. (38) the couplings of the Higgs field to both kinds of quarks: up and down. How do we do it?

We have said that each of these fields are embedded in different chiral supermultiplets

$$Q = \begin{pmatrix} u_L \\ d_L \end{pmatrix} \in \mathbf{Q}, \quad u_R^\dagger \in \overline{\mathbf{u}}, \quad d_R^\dagger \in \overline{\mathbf{d}} \quad \text{and} \quad H = \begin{pmatrix} H_u^+ \\ H^0 \end{pmatrix} \in \mathbf{H}, \tag{53}$$

Names		spin 0	spin 1/2	$SU(3)_C$, $SU(2)_L$, $U(1)_Y$
squarks, quarks	$\mathbf{Q}$	$(\widetilde{u}_L\ \widetilde{d}_L)$	$(u_L\ d_L)$	$(\mathbf{3},\ \mathbf{2},\ \frac{1}{6})$
(×3 families)	$\overline{\mathbf{u}}$	$\widetilde{u}^*_R$	$u^\dagger_R$	$(\overline{\mathbf{3}},\ \mathbf{1},\ -\frac{2}{3})$
	$\overline{\mathbf{d}}$	$\widetilde{d}^*_R$	$d^\dagger_R$	$(\overline{\mathbf{3}},\ \mathbf{1},\ \frac{1}{3})$
sleptons, leptons	$\mathbf{L}$	$(\widetilde{\nu}\ \widetilde{e}_L)$	$(\nu\ e_L)$	$(\mathbf{1},\ \mathbf{2},\ -\frac{1}{2})$
(×3 families)	$\overline{e}$	$\widetilde{e}^*_R$	$e^\dagger_R$	$(\mathbf{1},\ \mathbf{1},\ 1)$
Higgs, higgsinos	$\mathbf{H}_u$	$(H_u^+\ H_u^0)$	$(\widetilde{H}_u^+\ \widetilde{H}_u^0)$	$(\mathbf{1},\ \mathbf{2},\ +\frac{1}{2})$
	$\mathbf{H}_d$	$(H_d^0\ H_d^-)$	$(\widetilde{H}_d^0\ \widetilde{H}_d^-)$	$(\mathbf{1},\ \mathbf{2},\ -\frac{1}{2})$
		spin 1/2	spin 1	$SU(3)_C$, $SU(2)_L$, $U(1)_Y$
gluino, gluon	$\mathbf{g}$	$\widetilde{g}$	g	$(\mathbf{8},\ \mathbf{1},\ 0)$
winos, W bosons	$\mathbf{W}$	$\widetilde{W}^\pm\ \widetilde{W}^0$	$W^\pm\ W^0$	$(\mathbf{1},\ \mathbf{3},\ 0)$
bino, B boson	$\mathbf{B}$	$\widetilde{B}^0$	B^0	$(\mathbf{1},\ \mathbf{1},\ 0)$

Table 1: Chiral and gauge supermultiplets in the Minimal Supersymmetric Standard Model. For the chiral superfields, the spin-0 fields are complex scalars, and the spin-1/2 fields are left-handed two-component Weyl fermions. The gauginos are spin-1/2 fields which are also left-handed two-component Weyl fermions, note however that they sit in the adjoint representations of the gauge groups, as opposed the SM fermions which sit in the fundamental or trivial representations of each gauge group.

note the name $\overline{\mathbf{u}}$, this is a left-chiral superfield because it contains $u^\dagger_R$. Obviously also Q is chiral and H is defined to be so, that means in superspace we have: $\mathbf{H}(y) \doteq H(y) + \sqrt{2}\theta\xi\tilde{H}(y) + \theta\theta F_H$,

where H is our well known Higgs scalar from the SM and $\tilde{H}$ is a left-chiral fermion. Then the combination

$$\epsilon_{\alpha\beta}\mathbf{Q}^\alpha\mathbf{H}^\beta\overline{\mathbf{u}} \in W \tag{54}$$

it is indeed allowed in W. Here $\epsilon_{\alpha\beta}$ is the anti-symmetric tensor in two dimensions, so the $SU(2)_L$ indices of $\mathbf{Q}$ and $\mathbf{H}$ are properly contracted. Then as we have said § Eq. (6), we cannot have at the same time a term $\epsilon_{\alpha\beta}\mathbf{Q}^\alpha\left(i\sigma^2\mathbf{H}^{*\beta}\right)\overline{\mathbf{d}}$ because $\mathbf{H}^*(y) = H^*(y) + \sqrt{2}\theta\tilde{H}^\dagger(y) + \theta\theta F^*_H(y)$ spoils the holomorphicity that W is required to have, i.e. such that $[W+\overline{W}]_F$ could supersymmetrically transform as a total derivative. Then it happens that we need indeed another Higgs field from which eventually masses to d quarks could be obtained

$$-\epsilon_{\alpha\beta}\mathbf{Q}^\alpha\mathbf{H}^\beta_d\overline{\mathbf{d}} \in W. \tag{55}$$

This time for everything to work out properly, the hypercharge, Y, of $\mathbf{H}_d$ must be $-1/2$, since we need to preserve $U(1)_Y$ transformations and so $Y_{H_d} + Y_Q + Y_{\bar{d}} = \frac{-1}{2} + \frac{1}{6} + \frac{1}{3} = 0$.

Furthermore this is indeed the requirement that we must impose for the cancellation of the triangular anomalies in the theory. The ones relevant to the hypercharge are proportional to $\mathrm{Tr}[T_3^2Y]$ and $\mathrm{Tr}[Y^3]$, where T_3 is the third component of weak isospin. The trace is a sum over all the fermionic degrees of freedom. We know that the SM is anomaly free, that is $\mathrm{Tr}[T_3^2Y] = \mathrm{Tr}[Y^3] = 0$. Then with only one Higgs superfield, we would have only one *higgsino* with hypercharge 1/2, and since already the SM fermionic fields have zero contribution to the traces above, we would have anomalies such that $\mathrm{Tr}[T_3^2Y] \propto 1/2$ and $\mathrm{Tr}[Y^3] \propto 8$. We can see that the problem is solved with the superfield Higgs H_d of exactly $-1/2$ hypercharge.

We have identified the MSSM and for concreteness we borrow the idea from [1] of encoding its content and most commonly used notation in a Table 1. Then the most general superpotential that can be written with these MSSM multiplets is:

$$\begin{aligned} W &= Y_l^{ij}\epsilon_{\alpha\beta}\mathbf{H}_d^\alpha\mathbf{E}_i^c\mathbf{L}_j^\beta + Y_d^{ij}\epsilon_{\alpha\beta}\mathbf{H}_d^\alpha\mathbf{D}_i^c\mathbf{Q}_j^\beta - Y_u^{ij}\epsilon_{\alpha\beta}\mathbf{H}_u^\alpha\mathbf{U}_i^c\mathbf{Q}_j^\beta \\ &\quad + \mu\epsilon_{\alpha\beta}\mathbf{H}_u^\alpha\mathbf{H}_d^\beta \\ &\quad + \frac{1}{2}\lambda^{ijk}\mathbf{L}_i\mathbf{L}_j\overline{e}_k + \frac{1}{2}\lambda'^{ijk}\mathbf{L}_i\mathbf{Q}_j\overline{\mathbf{d}}_k + \mu'^i\mathbf{L}_i\mathbf{H}_u \\ &\quad + \frac{1}{2}\lambda''^{ijk}\overline{\mathbf{u}}_i\overline{\mathbf{d}}_j\overline{\mathbf{d}}_k, \end{aligned} \tag{56}$$

Note that we are using the so called *Left-Right* notation for the Yukawa couplings. The previous line to the last violates the lepton number L by one unit, while the last one violates the baryon number B also by one unit. These are the famous undesired couplings that make possible proton decay $p^+ \rightarrow e^+\pi^0,\ e^+K^0,\ \mu^+\pi^0,\ \mu^+K^0$, etc. Then it was invented a symmetry that could forbid the last two lines of Eq. (56), *R parity* [9] or *matter parity* [10, 11, 12, 13] such that we could remain only with the desired part of W. Models where *R parity* is violated have been some what less studied.

8. Mechanisms to Break a Supersymmetric Lagrangian

8.1. General Considerations and Spontaneous Breaking

Just as in a normal field theory, in supersymmetric field theories matter and gauge fields cannot acquire vacuum expectation values (VEV) different from zero. But as it happens in the SM, scalar fields can do the job. As usual then the supersymmetry breaking part must be coming from the scalar potential. From Eq. (51) we can see that it is in general written by D and F contributions, hence effectively one needs to make one of these non-zero when the scalars acquire VEVs different from zero.

The bad news are that for the MSSM or any of its extensions there is not a single working model where one breaks the $N = 1$ supersymmetry and generates at the same time, the mass terms and trilinear parameters for all the fields of the MSSM. The good news is that there is a way to account for all the possible terms that may appear after the breaking of supersymmetry. This is because these terms can be restricted by imposing the absence of quadratic divergences of the effective theory obtained after supersymmetry breaking. We will check out this in the next section, § 9. Here we will just state how D and F terms can be made non zero.

Spontaneous breaking of the supersymmetry occurs by definition when the vacuum is not invariant under superymmetric transformations, that is $Q_A|0\rangle \neq 0$ and $Q_A^\dagger|0\rangle \neq 0$. In supersymmetry this happens when the ground state energy is lifted from zero. This is very easy to see because for any state $|\alpha\rangle$,

$$\begin{aligned} \mathcal{H} &= \frac{1}{4}\left\{Q_1,\overline{Q}_{\dot{1}}\right\} + \frac{1}{4}\left\{Q_2,\overline{Q}_{\dot{2}}\right\} \implies \\ \langle\alpha|\mathcal{H}|\alpha\rangle &= \frac{1}{4}\sum_{A=1}\sum_{n}\left[|\langle\alpha|Q_A|n\rangle|^2\right] + \left[|\langle\alpha|\overline{Q}_{\dot{A}}|n\rangle|^2\right], \end{aligned} \tag{57}$$

which is a positive definite quantity except for $|\alpha\rangle = |0\rangle$, which is zero.

Is nice to break supersymmetry this way since the Lagrangian does not change its form. Once the supersymmetry is broken there is a split in the supermultiplets, generating a difference in mass between superpartners. In this case the D or F terms do not vanish in the vacuum state. Recall that the $N = 1$ supersymmetry, that we have introduced in the first sections and the one that the MSSM satisfies, is a global supersymmetry. According to Goldstone's theorem when it is broken the resulting theory will contain a Nambu-Goldstone massless particle with the same quantum numbers of the broken symmetry generator [8], Q_A. This is a two component Weyl spinor, therefore the Nambu-Goldstone *goldstino* will be a massless two component Weyl fermion.

8.2. D Term Supersymmetry Breaking

D term supersymmetry breaking happens when a D term in the scalar potential is lifted from zero. An example of this is the Fayet-Iliopoulos Mechanism, which assumes the existence of a $U(1)$ gauge factor in the theory. As we have seen from Eq. (47), in a supersymmetric $U(1)$ gauge Lagrangian, the following terms contain the scalar D

$$\mathcal{L}_{U(1)} \supset \frac{1}{2}D^2 + \eta D + g t_i |\phi_i|^2 D, \quad \mathcal{L}_{FI} \doteq \eta D, \tag{58}$$

where exaclty ηD is the Fayet-Iliopoulos part of the Lagrangian. Obviously D should be a gauge singlet. The D term itself is given by Eq. (50). In this mechanism it is assumed that there are n scalars with charges $t_i = q_i$ under $U(1)$, which may or may not have masses, m_i, coming from the superpotential. Then in general we can write

$$U = -\frac{1}{2}D^2 - \eta D - g\sum_{i=1}^{i=n} q_i|\phi_i|^2 D + m_i^2|\phi_i|^2 \;=\; \frac{1}{2}\left[-\eta - g\sum_{i=1}^{i=n} q_i|\phi_i|^2\right]^2 + m_i^2|\phi_i|^2, \tag{59}$$

where for the last expression of U we have used Eq. (50) and Eq. (51). This cannot vanish and therefore supersymmetry must be broken. Note that the scalars will have masses $m_i^2 + g q_i \eta$ while their fermion partners will have masses m_i^2.

Unfortunately the $U(1)_Y$ gauge factor of the MSSM cannot help to break supersymmetry, in this case because its fermions do not have superpotential mass terms, the m_i's. We see that even in the absence of m_i's, U (Eq. (59)), one may think to use the effective U to break supersymmetry, but it has been shown that does not and instead it does break color and electromagnetism.

However this mechanism can be used to break realistic extensions of the MSSM, which include, other than $U(1)_Y$, $U(1)$ gauge factors. It is also very useful in supersymmetry family symmetry theories where we have different $U(1)$ charges for each family of fermions and at least two scalars θ and θ' with charges 1 and -1. They can be coupled to $QH_u u_R^\dagger$ and $QH_d d_R^\dagger$, generating the hierarchical Yukawa couplings and by means of the U of Eq. (59) fix the order of the VEVs of θ and θ'.

8.3. F Term Supersymmetry Breaking

F term supersymmetry breaking is easier to achieve because one only requires that there exist at least one $F*_i = \frac{\delta W}{\delta \phi_i} \neq 0$, among n possible F_i terms. If supersymmetry needs to be broken in the global minimum of the potential [9] then $F_j = 0\ \forall j \neq$ cannot simultaneously be solved for $F_i = 0$. In this case supersymmetry is indeed broken because $U = \sum_{k=1}^{n} F_k^* F_k \neq 0$. A consequence of $F_i = 0$ is that any superpotential without a linear term in Φ, first term in Eq. (33), cannot produce an F type symmetry breaking because given that $F_k^* = m_{kj}\phi_j + \frac{1}{3} f_{kst}\phi_s\phi_t$ can be made zero, once we take VEVs, $\langle F_k\rangle = 0$, $\forall j$ for $\langle \phi_j \rangle = 0$, then we will have $U = 0$

An example of a successful F type supersymmetry breaking is the O'Raifeartaigh model. In this model one needs a minimum of three chiral fields, Φ_i with the superpotential of the form

$$W = m\Phi_2\Phi_3 + \lambda\Phi_1(\Phi_3^2 - \mu^2), \tag{60}$$

[8] We are used to bosonic symmetry generators because of the standard $U(1)$ bosonic example of a complex scalar field. After symmetry breaking the Nambu-Goldstone massless particle is a boson.

[9] Local minima can generate the so called *metastable* vacua which can have very large lifetimes and therefore under some conditions are suitable for candidates of the "observed" vacua of a given theory.

for μ a given mass parameter. If m, μ, and λ are real then $F_1^* = \lambda(\phi_3^2 - \mu^2)$, $F_2^* = -m\phi_3$ and $F_3^* = -m\phi_2 - 2\lambda\phi_1\phi_3$. Then as O'Raifeartaigh showed, the system of equations $F_i = 0$, $i = 1, 2, 3$, has no simultaneous solution. The minimum of the potential U in this model has two different behaviours when $\mu^2 < m^2/2\lambda^2$ and $\mu^2 > m^2/2\lambda^2$. It is often easy to find global minima that break supersymmetry in models which are characterized by the first case.

8.4. Mechanisms for Supersymmetry Breaking in the MSSM

It so happens that just with the superfields of the MSSM it is not possible to achieve neither D nor F term breakings. Then additional superfields that break supersymmetry must be included, but since many fermionic components can induce dangerous couplings [10] with the MSSM fields, these couplings should be really small. The set of fields that break the MSSM supersymmetry in this way, is called the *Hidden Sector*. Now, it is not enough to break the supersymmetry, it is also necessary to obtain a realistic mass spectrum for the supersymmetric particles. There are two popular mechanisms to attempt it, which we describe in what it follows.

In the **Gravity Mediated Supersymmetry Breaking** mechanism, the hidden sector is communicated only through gravitational interactions to the MSSM fields. That is, the breaking or any coupling to MSSM fields can only be felt via gravity.

Another mechanism to attempt supersymmetry breaking in the MSSM is the **Gauge Mediated Supersymmetry Breaking**, this is just as the gravity mediated supersymmetry breaking but this time the interactions are the same gauge interactions of the MSSM. In order construct realistic models in this scenario, the couplings of the hidden sector must come from loop effects.

Is sad to say it, but up to date there is not a single full working model for supersymmetry Breaking in the MSSM.

9. Soft SUSY Breaking Lagrangian and Mass Spectrum of the MSSM

9.1. Soft SUSY Breaking

As we have mentioned in the previous section, with the MSSM fields alone, it is not possible to break the $N = 1$ supersymmety. However, given the particle content of the MSSM, it is possible to write down all the possible terms that, after symmetry breaking, could exist. These terms can be restricted by imposing the absence of quadratic divergences of the effective theory. Having said so, in the MSSM with R parity, the most general soft supersymmetric Lagrangian that can be written is

$$\begin{aligned} -\mathcal{L}_{\text{soft}} &= \tilde{q}_{Li}^\dagger (m_{\tilde{Q}}^2)^{ij} \tilde{q}_{Lj} + \tilde{u}_{Rj}(m_{\tilde{u}}^2)^{ji} \tilde{u}_{Ri}^* + \tilde{d}_{Rj}(m_{\tilde{d}}^2)^{ji} \tilde{d}_{Ri}^* \\ &\quad + \tilde{l}_{Li}^\dagger (m_{\tilde{L}}^2)^{ij} \tilde{l}_{Lj} + \tilde{e}_{Rj}(m_{\tilde{e}}^2)^{ji} \tilde{e}_{Ri}^* + \tilde{\nu}_{Rj}(m_{\tilde{\nu}}^2)^{ji} \tilde{\nu}_{Ri}^* + m_{h_d}^2 h_d^\dagger h_d + m_{h_u}^2 h_u^\dagger h_u \\ &\quad + (b\, h_d h_u + \text{h.c.}) + \Big(-a_d^{ij} h_d \tilde{d}_{Ri}^* \tilde{q}_{Lj} + a_u^{ij} h_u \tilde{u}_{Ri}^* \tilde{q}_{Lj} - a_l^{ij} h_d \tilde{e}_{Ri}^* \tilde{l}_{Lj} + a_\nu^{ij} h_u \tilde{\nu}_{Ri}^* \tilde{l}_{Lj} \\ &\quad + \frac{1}{2} M_1 \widetilde{B}\widetilde{B} + \frac{1}{2} M_2 \widetilde{W}^a \widetilde{W}^a + \frac{1}{2} M_3 \widetilde{G}^a \widetilde{G}^a + \text{h.c.} \Big), \end{aligned} \tag{61}$$

where SU(2) indices are not written explicitly. They are contracted by $\epsilon_{\alpha\beta}$ and $\delta_{\alpha\beta}$, respectively, i.e. $AB := \epsilon_{\alpha\beta} A^\alpha B^\beta$ and $A^\dagger A := A_\alpha^\dagger A^\alpha$ for SU(2) doublet fields A, B. As we see it is an incredible task to study constraints on all the parameter space of the general MSSM, counting the number of parameters we arrive at an incredible number of 109! Not only this but also the fact that we need to set up typically three main scales

$$M_{\text{UV}} > M_{\not{S}} > M_{\text{EW}}, \tag{62}$$

[10] That, for example, could mediate decay of particles at a rate that is not observed, etc.

which describe the scales at which one should put boundary conditions of all the parameters in Eqs. (56) and (61). $M_{\rm UV}$ represents the *Ultra Violet completion* (UV) of the model, coming from a definite supersymmetric model, $M_{\not{S}}$ is the scale at which supersymmetric particles decouple where we need to match to SM parameters and $M_{\rm EW}$ the Electroweak (EW) scale. On then should employ the appropriate Renormalization Group Equations (RGE) between two different scales and work out appropriate matching conditions [11]. This is a formidable numerical task and there are several computer codes that have been developed for addressing this enterprise [15, 16, 17] and several people managing their own code.

Even when the RGE evolution can be done numerically, there are still formidable hitches to worry about. The strategy followed in the last more than twenty years to tackle this problem is to simplify the number of parameters based on the following guidelines.

(i) *Minimal MSSM.* General couplings in Eq. (61) will easily induce flavour violating processes [12] at a rate that is much above the experimental bounds or measured values. Then we can think a set up where

$$\begin{aligned} m^2_{\tilde{Q}_{ij}} &= m^2_{\tilde{u}ij} = m^2_{\tilde{d}ij} = m^2_{\tilde{e}ij} = \delta_{ij} m_0^2, \\ m^2_{H_d} &= m^2_{H_u} = m_0^2, \quad b = B\mu, \\ a_u^{ij} &= AY_u^{ij},\ a_d^{ij} = AY_d^{ij},\ a_e^{ij} = AY_e^{ij}, \\ M_1 &= M_2 = M_3 = M_{1/2}. \end{aligned} \tag{63}$$

The Constrained MSSM (**CMSSM**) adopts this strategy [13]. The conditions Eq. (63) are set up as a boundary condition at $M_{\rm UV}$ identified with the GUT scale, $M_{\rm G}$. Then the bread and butter free parameters of the CMSSM community are

$$\mathrm{sign}[\mu],\ A,\ m_0,\ M_{1/2} \text{ and } \tan\beta. \tag{64}$$

The parameters B and μ are obtained by the requirement of the Electroweak symmetry breaking conditions. That is to say, they are fitted such that with the spectra obtained from the running of the CMSSM from $M_{\rm G}$ to $M_{\rm EW}$ (passing through $M_{\not{S}}$) it so happens that there is a minimum for the scalar potential of the theory such that the EW symmetry is broken. The only degrees of freedom that are run from $M_{\rm G}$ down to $M_{\rm EW}$ are those of the MSSM, any theory that was above $M_{\rm G}$ is assumed to be decoupling exactly at that scale.

(ii) Minimal Supergravity **mSUGRA**
Here we are entering into a fascinating realm of supersymmetry that we have not yet mentioned, that of local supersymmetric transformations ([14]). An $N = 1$ local supersymmetry transformation turns out to describe a theory which contains a particle of spin 2 that can be identified with a *graviton* and its superpartner, of spin 3/2, the *gravitino.* This local supersymmetry is called *Supergravity* (SUGRA) [].
Supergravity theories apart from having a superpotential, have two other unique functions that are relevant. First, the so called *Kähler potential*, K, which it is a real function of scalar

[11] Two loop beta functions of the MSSM can be found in [14].

[12] Among the most pressing ones are the mixing in the $K^0 - \overline{K^0}$ system whose CP violation parameter, ϵ_K, has been precisely measured, the leptonic decays $\ell_i^- \to \ell_j^- \gamma$, the decay $b \to s\gamma$, known to a good precision.

[13] In practice, since the Yukawa matrices cannot be uniquely determined, when one comes across a *CMSSM* analysis, it is common that the analysis is assuming particular forms of the Yukawa matrices. The most used is that all Yukawa couplings, except those of the heaviest families, are zero. That is, it just takes into account the RGE running of the heaviest fermionic families, i.e. instead of running the 27 parameters from a_u^{ij}, a_d^{ij} and a_e^{ij} and the 27 from Y_u^{ij}, Y_d^{ij} and Y_e^{ij}, it just runs the 3 parameters $A_{u,d,e}$ and the three Yukawa couplings $y_{t,b,\tau}$.

[14] Check out [18] for a great introduction.

fields of the theory. Its *metric*, $K_{ij^*} = \partial^2 K/(\partial\phi_i \partial\phi_j^*)$, appears in terms like $K_{ij^*}\overline{\psi}^{i*}\gamma^\mu\partial_\mu\psi^j$. Second, the *gauge kinetic functions*, f_{ab}, which appear in the supergravity Lagrangian, for example in the terms: $\mathcal{L}_{SUGRA} \propto \mathrm{Re}[f_{ab}]F^a_{\mu\nu}F^{b\mu\nu} + i\mathrm{Im}[f_{ab}]F^a_{\mu\nu}\tilde{F}^{b\mu\nu}$. The gauge kinetic functions are arbitrary analytic functions of the scalar fields of the theory.
There are two remarkable features of this theory. The first is that it is not renormalizable! The second is that a limit, where $M_P \to \infty$ and the mass of the gravitino ($m_{3/2} \propto 1/M_P$) is fixed, produces a supersymmetric theory, free of quadratic divergences and such that those parameters which were free in the soft-supersymmetry breaking Lagrangian of Eq. (61), can be calculated in terms of scalar fields of the hidden sector of the theory, the same on which K_{ij} and f_{ab} depend on.
Minimal Supergravity (mSUGRA) really refers to the choice

$$\begin{aligned} K_{ij} &= \delta_{ij}, \quad \forall \ \text{fields in the theory}, \\ f_{ab} &= \delta_{ab}, \quad \forall \ \text{gauge fields in the theory}. \end{aligned} \tag{65}$$

Independently of the choice of a particular superpotential W, there will be simplifications among the quantities of Eq. (61), in fact those of Eq. (63) plus relations among A, $M_{1/2}$, B and $m_{3/2}$.
Now, the CMSSM is a framework to study what happens when parameters in Eq. (64) are varied, while in mSUGRA those parameters should be identified with some underlying supergravity model. Explicitly in the context of the MSSM, mSUGRA, encompasses all the models which propose a definite superpotential for the hidden sector of the theory. An example of these is the **Polonyi Model** [15]. The defining characteristic of these models is obviously the superpotential of the hidden sector. For example, the Polonyi superpotential is $W = m^2(\Phi + \beta)$, both m and β are mass terms. In particular the choice $\beta = (2 - \sqrt{3})M_P$ produces the following simplifications

$$A = (3 - \sqrt{3})m_{3/2}, \quad m_0^2 = m_{3/2}^2, \quad M_{1/2} = O(m_{3/2}). \tag{66}$$

For the different approaches mentioned here, a good state of the art can be found in [20, 21, 22] and references therein.

(iii) Supersymmetric Family Symmetries They put specific relations among all the parameters in Eqs. (56) and (61) such that apart from reproducing the hierarchy of masses and mixing in the quark and lepton sector, they minimize the processes that change flavour. Generically they do not produce simple relations as Eq. (63), but the couplings of the supersymmetric theory are controlled also by the Family Symmetry [23].

9.2. MSSM Mass Spectrum

Having a look at Table 1 we can classify the following sets of supersymmetric particles: [i] scalar particles ($\widetilde{u}_R^*$, $\widetilde{d}_R^*$, $\widetilde{e}_R^*$, $\widetilde{Q}_L^*$ and $\widetilde{Q}_R^*$), [ii] charged, under $U(1)_{em}$, fermions ($\widetilde{H}_u^+$, $\widetilde{H}_d^-$, $\widetilde{W}^\pm$), [iii] neutral fermions ($\widetilde{H}_d^0$, $\widetilde{H}_u^0$, $\widetilde{B}^0$, $\widetilde{W}^0$) and [iv] the very special gluino, $\widetilde{g}$. Due to the couplings that can be derived from the superpotential of Eq. (56), once the MSSM Lagrangian is constructed via a generalization of Eq. (46) and the contribution from the soft breaking Lagrangian, Eq. (61), there will be mixings among some particles in each set.

Once the mass matrices form the mixing states and we obtain the mass eigenstates of these sets are called *Charginos* ($\chi_i^\pm$) for set [ii] and *Neutralinos* (χ_i^0) for set [iii]. For the set [i], there are indeed 6 $\widetilde{u}_i^*$ (coming from $\widetilde{u}_R^*$ and $\widetilde{Q}_L$) different particles which will mix, the same for $\widetilde{d}_i^*$ and analogously for $\widetilde{e}_i^*$. The mass eigenstates of these different particles are called respectively *up-type squarks* (aka *s-ups*) , *down-type squarks* (aka *s-downs*) and *s-electrons*.

[15] For more examples, check out [19].

Journal of Physics: Conference Series **287** (2011) 012001 doi:10.1088/1742-6596/287/1/012001

The masses of these particles must be RGE evolved from $M_{\rm UV}$ down to $M_{\rm EW}$. After this is achieved we can obtain tree level masses of the *physical* mass-eigenstates mentioned above. That is not all, we have to calculate loop corrections to these masses, fortunately for some cases, there are computer programmes to do so.

Note that a crucial part of obtaining a realistic MSSM mas spectrum is to obtain EW symmetry breaking. The MSSM scalar potential is a bit more complicated than that of the SM. Assuming that at the minimum of the potential, the charged parts of the Higgs scalars are zero, as in the SM, $H_u^+ = 0$, at the minimum of the potential ($\partial U_H/\partial H_u^+ = 0$), we must also have $H_d^- = 0$. Then at the minimum of the potential electromagnetism is unbroken, since the charged components of the Higgs scalars do not get VEVs. With these constraints the MSSM Higgs scalar potential can be written as

$$U_H = (|\mu|^2 + m_{H_u}^2)|H_u^0|^2 + (|\mu|^2 + m_{H_d}^2)|H_d^0|^2 - (bH_u^0 H_d^0 + c.c.) + \frac{(g^2+g'^2)}{8}(|H_u^0|^2 - |H_d^0|^2)^2. \quad (67)$$

It can be shown that a minimum of this potential requires that b, H_u^0 and H_d^0 to be real and positive, so $\langle H_u^0\rangle$ and $\langle H_d^0\rangle$ must have opposite phases. We can therefore use a $U(1)_Y$ gauge transformation to make them both real and positive without loss of generality, since H_u and H_d have opposite weak hypercharges ($\pm 1/2$). Then CP cannot be spontaneously broken by the Higgs scalar potential, since the VEVs and b can be simultaneously chosen real. This means that the Higgs scalar mass eigenstates can be assigned well-defined eigenvalues of CP, at least at tree-level. CP-violating phases in other couplings can induce loop-suppressed CP violation in the Higgs sector. After the spontaneous symmetry breaking the VEVs of the CP even Higgs, $v_u = \langle H_u^0\rangle$, $v_d = \langle H_d^0\rangle$, need to satisfy $v_u^2 + v_d^2 = 246^2/2$ GeV^2. The ratio of these VEVs is known as $\tan\beta = v_u/v_d$. In the SM we will have just one real Higgs, h after spontaneous symmetry breaking, in the MSSM there are four other physical Higgs particles: the other CP even Higgs, H^0, a CP odd neutral Higgs, A^0 and two charged Higgs fields, $H^\pm$. The phenomenology of this zoo on its own is fascinating, for a great book check out [24, 25].

As we have said in the previous section it is a phenomenal task to analyse the full parameter space of the MSSM. As the LHC is running, we would like to have an idea of some reasonable assumptions about the parameters defining the MSSM. The following assumptions are a good starting point: (i) χ_i^0 are the lightest supersymmetric particles, (ii) R-parity is conserved, (iii) with the exception of $\tilde t$ and $\tilde b$ all squarks are degenerate in mass and the masses of the superpartners of left-handed quarks and right-handed quarks are equal at $M_{\rm G}$ and (iv) there is gaugino mass unification at $M_{\rm G}$.

Then some bounds [26], all at $CL = 95\%$, can be obtained on the masses for the sparticles:

(i) for the lightest neutralino χ_1^0 (coming from mixtures of $\widetilde B^0$, $\tilde Z^0$, and $\tilde H_i^0$), $m_{\chi_1^0} > 46$ GeV;

(ii) for $\widetilde\chi_i^\pm$ (mixtures of $\widetilde W^\pm$ and $\widetilde H^\pm$ and $\tan\beta < 40$, $m_{\widetilde\chi_1^\pm} - m_{\widetilde\chi_1^0} > 3$ GeV), $m_{\widetilde\chi_i^\pm} > 94$ GeV;

(iii) $m_{\widetilde e} > 107$ GeV, $m_{\widetilde\mu} > 94$ GeV (here the bounds are assuming that the left parts of the sleptons do not play a significant rôle and that $1 \le \tan\beta \le 40$, $m_{\widetilde\mu} - m_{\widetilde\chi_1^0} > 10$ GeV) and $m_{\widetilde\mu} > 82$ GeV ($m_{\widetilde\tau} - m_{\widetilde\chi_1^0} > 15$ GeV);

(iv) $m_{\widetilde b} > 89$ GeV ($m_{\widetilde b_1} - m_{\widetilde\chi_1^0} > 8$ GeV);

(v) $m_{\widetilde t} > 96$ GeV (obtained from the decay $\widetilde t \to c\chi_i^0$, $m_{\widetilde t} - m_{\widetilde\chi_1^0} > 10$ GeV) and

(vi) $m_{\widetilde g} > 308$ GeV in general (assuming gauge coupling unification at $M_{\rm G}$). There are more specific bounds for particular relations among other sparticles.

10. Glimpse of Extensions of the MSSM

During the last decade neutrino physics, after the observation of their oscillations [27], has become a central arena for studies beyond the SM. They are massive, though we do not yet know their overall mass, so any model BSM should worry about them. There are ways to think that they cannot affect the basic behaviour of the MSSM. For example in scenarios where the see-saw mechanism explains through heavy right-handed neutrino masses, the small masses of oscillating energy neutrinos, the right-handed neutrinos could decouple before M_{G}. There are other models in which it is interesting to decouple right-handed neutrinos after M_{G}. Then the running of the mass spectrum of the MSSM is altered and all the sector which produces neutrino masses should also be taken into account in the running. I find cute the generic name νMSSM for this kind of models.

11. Acknowledgments

I wish to acknowledge the Faculty of Sciences of the University of Guanuajuato, Campus Leon, for their kind support that enabled me to participate in the XIV Mexican School of Particles and Fields. I also thank all the audience of my lectures, specially to those who asked questions and helped me by correcting some typos in my notes. Finally, I specially thank Jörn Kersten for the reading of the manuscript and his helpful comments.

References

[1] Martin S P 1997 (*Preprint* hep-ph/9709356)
[2] M Drees R Godbole P R 2004 **1** 555
[3] Krippendorf S, Quevedo F and Schlotterer O 2010 (*Preprint* 1011.1491)
[4] Wess J and Bagger J 1992
[5] Sohnius M 1985 *Phys.Rept.* **128** 39–204
[6] Coleman S R and Mandula J 1967 *Phys.Rev.* **159** 1251–1256
[7] Cornwell J 1989
[8] Haag R, Lopuszanski J T and Sohnius M 1975 *Nucl.Phys.* **B88** 257
[9] Farrar G R and Fayet P 1978 *Phys.Lett.* **B76** 575–579
[10] Dimopoulos S and Georgi H 1981 *Nucl.Phys.* **B193** 150
[11] Weinberg S 1982 *Phys.Rev.* **D26** 287
[12] Sakai N and Yanagida T 1982 *Nucl.Phys.* **B197** 533
[13] Dimopoulos S, Raby S and Wilczek F 1982 *Phys.Lett.* **B112** 133
[14] Martin S P and Vaughn M T 1994 *Phys.Rev.* **D50** 2282 (*Preprint* hep-ph/9311340)
[15] Allanach B 2002 *Comput.Phys.Commun.* **143** 305–331 (*Preprint* hep-ph/0104145)
[16] Porod W 2003 *Comput.Phys.Commun.* **153** 275–315 (*Preprint* hep-ph/0301101)
[17] Djouadi A, Kneur J L and Moultaka G 2007 *Comput.Phys.Commun.* **176** 426–455 (*Preprint* hep-ph/0211331)
[18] Cerdeno D G, Munoz C and Seto O 2009 *Phys.Rev.* **D79** 023510 (*Preprint* 0807.3029)
[19] Nilles H P 1984 *Phys.Rept.* **110** 1–162
[20] Ellis J R, Heinemeyer S, Olive K, Weber A and Weiglein G 2007 *JHEP* **0708** 083 (*Preprint* 0706.0652)
[21] AbdusSalam S S, Allanach B C, Quevedo F, Feroz F and Hobson M 2010 *Phys.Rev.* **D81** 095012 (*Preprint* 0904.2548)
[22] Ellwanger U, Hugonie C and Teixeira A M 2010 *Phys.Rept.* **496** 1–77 (*Preprint* 0910.1785)
[23] Velasco-Sevilla L 2011 Contribution to the XIV Mexican School of Particles and Fields
[24] Gunion J F, Haber H E, Kane G L and Dawson S 1989
[25] Gunion J F, Haber H E, Kane G L and Dawson S 1992 (*Preprint* hep-ph/9302272)
[26] Amsler C *et al.* (Particle Data Group) 2008 *Phys.Lett.* **B667** 1–1340
[27] Fukuda Y *et al.* (Super-Kamiokande Collaboration) 1998 *Phys.Rev.Lett.* **81** 1562–1567 (*Preprint* hep-ex/9807003)

Journal of Physics: Conference Series **287** (2011) 012002 doi:10.1088/1742-6596/287/1/012002

Pseudo-Hermitian quantum mechanics

Ashok Das

Department of Physics, University of Rochester, Rochester, New York, NY 14627, USA
Saha Institute of Nuclear Physics, 1/AF Bidhannagar, Calcutta 700064, India

E-mail: das@pas.rochester.edu

Abstract. In these lectures, we will describe a systematic procedure for constructing an inner product in a pseudo-Hermitian quantum mechanical system.

In recent years there is a lot of interest in $\mathcal{PT}$ symmetric quantum mechanical theories [1] where $\mathcal{P}$ represents parity or space reflection while $\mathcal{T}$ denotes time reversal. As we know $\mathcal{P}$ represents a linear operator while $\mathcal{T}$ corresponds to an anti-linear operator. As a result, in these theories the Hamiltonian is not Hermitian in the Dirac sense, rather it satisfies

$$H = \mathcal{P}^{-1} H^{\dagger} \mathcal{P} = \mathcal{P} H^{\dagger} \mathcal{P}, \tag{1}$$

where we have used the properties of the parity operator in the last step, namely,

$$\mathcal{P}^{\dagger} = \mathcal{P}, \quad \mathcal{P}^2 = \mathbb{1}. \tag{2}$$

The interest in these theories arises from the fact than even though the Hamiltonians in these theories are not Hermitian, the eigenvalues are real if the ground state of the theory is $\mathcal{PT}$ symmetric. We note that the reality of eigenvalues does not require that the Hamiltonian be Hermitian, namely, Hermiticity is not a necessary condition for eigenvalues to be real. A familiar example of a matrix Hamiltonian which is not Hermitian and yet has real eigenvalues is given by

$$H = \begin{pmatrix} \alpha & \beta \\ 0 & \gamma \end{pmatrix} \neq H^{\dagger}, \qquad \alpha, \beta, \gamma \text{ real},$$

which is not Hermitian, but the eigenvalues of the Hamiltonian are given by $E = \alpha, \gamma$ which are real. In this case, we note that if we define

$$S = \begin{pmatrix} a & \frac{\beta}{\alpha-\gamma}\, a \\ \frac{\beta}{\alpha-\gamma}\, a & d \end{pmatrix}, \quad S^{-1} = \frac{1}{\det S} \begin{pmatrix} d & -\frac{\beta}{\alpha-\gamma}\, a \\ -\frac{\beta}{\alpha-\gamma}\, a & a \end{pmatrix},$$

where a, d are arbitrary constants, we can check that

$$H = S^{-1} H^{\dagger} S, \tag{3}$$

which is quite similar to (1).

Any quantum mechanical system whose Hamiltonian can be related to its adjoint (in the Dirac sense) through a similarity transformation as in (3),

$$H = S^{-1}H^{\dagger}S, \qquad \Rightarrow \qquad SH = H^{\dagger}S, \tag{4}$$

is called a pseudo-Hermitian Hamiltonian [2]. It is worth emphasizing that by definition, the similarity transformation is assumed to be bounded so that it does not take us out of the Hilbert space. We note from the definition in (4) that, in particular, when $S = \mathbb{1}$, the pseudo-Hermitian Hamiltonian coincides with a Hermitian Hamiltonian. Therefore, Hermitian Hamiltonians define a subset of pseudo-Hermitian Hamiltonians. By taking the (Dirac) adjoint of the defining relation (4), we obtain

$$H^{\dagger} = S^{\dagger}H(S^{\dagger})^{-1}, \qquad \Rightarrow \qquad H = (S^{\dagger})^{-1}H^{\dagger}S^{\dagger}, \tag{5}$$

which shows (compare with (3)) that the similarity transformation S can be chosen to be Hermitian

$$S = S^{\dagger}. \tag{6}$$

It is worth emphasizing here that to define the adjoint of an operator, we need a Hilbert space $\mathcal{H}$ with an inner product such that if $|\psi\rangle$,$|\phi\rangle \in \mathcal{H}$

$$\langle\phi|H\psi\rangle = \langle H^{\dagger}\phi|\psi\rangle, \tag{7}$$

and a Hamiltonian is said to be Hermitian (self-adjoint) if

$$H = H^{\dagger}, \tag{8}$$

besides technical questions such as domains of the operators etc. Normally, the inner product in quantum mechanics is understood in the Dirac sense, namely, if $|\psi\rangle, |\phi\rangle \in \mathcal{H}$, then the inner product of the two states is defined to be

$$\langle\phi|\psi\rangle = \int_{-\infty}^{\infty} \mathrm{d}x\, \langle\phi|x\rangle\langle x|\psi\rangle = \int_{-\infty}^{\infty} \mathrm{d}x\, \phi^{*}(x)\psi(x). \tag{9}$$

This inner product is defined independent of the quantum mechanical system under consideration and has the added advantage that it leads to a positive norm for any quantum mechanical state and we choose physical states to satisfy

$$\langle\psi|\psi\rangle = \int_{-\infty}^{\infty} \mathrm{d}x\, |\psi(x)|^2 < \infty, \tag{10}$$

which allows us to give a probabilistic description for the system.

We are, of course, quite familiar with quantum mechanical systems where the Hamiltonian is Hermitian. In these lectures we would address the question of whether we can also have a satisfactory quantum mechanical description of a system whose Hamiltonian is pseudo-Hermitian [3]

$$H = S^{-1}H^{\dagger}S. \tag{11}$$

We note that, in general, the Hamiltonian in this case can have real or complex eigenvalues. The immediate difficulty that arises when H is not Hermitian is that the time evolution operator ($\hbar = 1$)

$$U(t) = e^{-itH}, \tag{12}$$

Journal of Physics: Conference Series **287** (2011) 012002 doi:10.1088/1742-6596/287/1/012002

is not unitary with the inner product (9)

$$\langle\phi(t)|\psi(t)\rangle = \langle\phi(0)|e^{itH^\dagger}e^{-itH}|\psi(0)\rangle \neq \langle\phi(0)|\psi(0)\rangle, \tag{13}$$

since $H \neq H^\dagger$. This would result in nonconservation of probability under time evolution. This is, of course, the standard interpretation for non-Hermitian Hamiltonians and has been used in the past to describe systems with decay of particles. Here we would like to ask if a pseudo-Hermitian system can be defined in such a way that it leads to a probabilistic description together with a unitary evolution.

As we have already mentioned, concepts such as Hermiticity and unitarity are defined with respect to a given inner product. Therefore, the difficulty of nonunitary time evolution can possibly be avoided and we can define a unitary quantum theory if we can have a modified inner product with respect to which the Hamiltonian H will be Hermitian

$$H = H^{\#\#}, \tag{14}$$

where $H^{\#\#}$ denotes the adjoint with respect to the modified inner product.

Before showing that this is possible let us study the quadratic form

$$\langle\phi|\psi\rangle_S = \langle\phi|S|\psi\rangle = \langle\phi|S\psi\rangle = \langle S\phi|\psi\rangle, \tag{15}$$

where we have used (6) in the last step. We note that this quadratic form reduces to the standard Dirac inner product when $S = \mathbb{1}$ as we would like, since in that case the system is described by a Hermitian Hamiltonian. The adjoint with respect to this quadratic form, $H^\#$, would satisfy (see (7) as well as (15))

$$\begin{aligned}\langle H^\#\phi|\psi\rangle_S &= \langle\phi|H\psi\rangle_S = \langle\phi|SH\psi\rangle \\ &= \langle H^\dagger S\phi|\psi\rangle = \langle SS^{-1}H^\dagger S\phi|\psi\rangle \\ &= \langle S^{-1}H^\dagger S\phi|\psi\rangle_S.\end{aligned} \tag{16}$$

Comparing both sides we conclude that

$$H^\# = S^{-1}H^\dagger S, \tag{17}$$

which reduces to $H^\# = H^\dagger$ when $S = \mathbb{1}$ as it should (namely, it should correspond to the Dirac adjoint in this case). We note here that the relation (17) is quite general in the sense that if we define a quadratic form with respect to an operator q, then the adjoint $H^{\#\#}$ with respect to this product would be related to the Dirac adjoint as

$$H^{\#\#} = q^{-1}H^\dagger q. \tag{18}$$

Using (4) (or (11)) we note from (17) that with respect to the quadratic form (15), the Hamiltonian, in fact, becomes Hermitian (self-adjoint)

$$H^\# = H. \tag{19}$$

Since S defines a product with respect to which the Hamiltonian is Hermitian, we expect that time evolution would be unitary with respect to this product

$$\begin{aligned}\langle\phi(t)|\psi(t)\rangle_S &= \langle\phi(t)|S|\psi(t)\rangle = \langle\phi(0)|e^{itH^\dagger}Se^{-itH}|\psi(0)\rangle \\ &= \langle\phi(0)|S(S^{-1}e^{itH^\dagger}S)e^{-itH}|\psi(0)\rangle \\ &= \langle\phi(0)|Se^{itH}e^{-itH}|\psi(0)\rangle \\ &= \langle\phi(0)|S|\psi(0)\rangle = \langle\phi(0)|\psi(0)\rangle_S,\end{aligned} \tag{20}$$

where we have used (4) in the intermediate step. This shows that time evolution is indeed unitary with respect to this product. However, this modified product does not yet lead to a quantum mechanical description.

To better understand the problem, let us note that if $|\psi_E$ and $|\psi_{E'}$ denote two eigenstates of the Hamiltonian (E, E' are in general complex and we denote the complex conjugate with a "bar") satisfying

$$H|\psi_E\rangle = E|\psi_E\rangle, \quad H|\psi_{E'}\rangle = E'|\psi_{E'}\rangle, \quad \langle\psi_{E'}|H^\dagger = \bar{E}'\langle\psi_{E'}|\,, E, E' \in \mathbb{C}, \tag{21}$$

then it follows that (we use (4) in the intermediate step)

$$\begin{aligned} E\langle\psi_{E'}|\psi_E\rangle_S = E\langle\psi_{E'}|S|\psi_E\rangle = \langle\psi_{E'}|SH|\psi_E\rangle = \langle\psi_{E'}|H^\dagger S|\psi_E\rangle \\ = \bar{E}'\langle\psi_{E'}|S|\psi_E\rangle = \bar{E}'\langle\psi_{E'}|\psi_E\rangle_S, \end{aligned} \tag{22}$$

so that we can write

$$(E - \bar{E}')\langle\psi_{E'}|\psi_E\rangle_S = 0. \tag{23}$$

Therefore, this determines

$$\langle\psi_{E'}|\psi_E\rangle_S = e^{-iF(E)}\delta_{E\bar{E}'}. \tag{24}$$

The exponential factor and, therefore, the product can become negative (for example, if S has negative eigenvalues as in the case of the parity operator $\mathcal{P}$) and, therefore, cannot define an inner product. So, although time evolution is unitary with this modified product, the product cannot lead to a probabilistic description. As a result, a quantum mechanical description would not follow.

We note that if S is positive, then the norm of the state is automatically positive. This is because in this case we can write

$$S = g^\dagger g, \tag{25}$$

so that we can write

$$\begin{aligned} & H = g^{-1}(g^{-1})^\dagger H^\dagger g^\dagger g, \\ \text{or,} \quad & gHg^{-1} = (g^{-1})^\dagger H^\dagger g^\dagger = (gHg^{-1})^\dagger, \\ \text{or,} \quad & h = h^\dagger, \end{aligned} \tag{26}$$

where we have identified

$$h = gHg^{-1}, \qquad h^\dagger = (gHg^{-1})^\dagger. \tag{27}$$

Thus, the energy eigenvalue equation can also be written as

$$\begin{aligned} & H|\psi_E\rangle = E|\psi_E\rangle, \\ \text{or,} \quad & gHg^{-1}g|\psi_E\rangle = Eg|\psi_E\rangle, \\ \text{or,} \quad & h(g|\psi_E\rangle) = E(g|\psi_E\rangle). \end{aligned} \tag{28}$$

As we have seen in (26) h is Hermitian so that we have

$$E = \bar{E}, \qquad \langle\psi_E|S|\psi_E\rangle = \langle\psi_E|g^\dagger g|\psi_E\rangle = |g|\psi_E\rangle|^2 \geq 0. \tag{29}$$

However, we are interested in the more general case where S is not necessarily positive. We note here that the exponential factor in (24) cannot vanish since this would imply that the state $|\psi_{\bar{E}}\rangle$ would be orthogonal to every state (pseudo eigenstate) and would reduce the dimensionality of the space. This will contradict the theorem from linear algebra that if S is a bounded operator and $\{|\psi_E\rangle\}$ defines a complete basis, then so does $\{|S|\psi_E\rangle\}$.

We can make the product (15) an inner product if we can remove the exponential factor in (24). This should, of course, be done maintaining the unitary evolution which can be guaranted by defining a new operator and a new product as follows. Let

$$q = SA, \qquad [A, H] = 0, \tag{30}$$

which implies that

$$A = f(H), \qquad q^{\dagger} = A^{\dagger}S = SS^{-1}f(H^{\dagger})S = Sf(H) = SA = q, \tag{31}$$

much like the similarity transformation S. (We note here that (31) can also be shown without assuming that $A = f(H)$.) Let us next define a product with respect to the operator q as

$$\langle\phi|\psi\rangle_q = \langle\phi|q|\psi\rangle = \langle\phi|q\psi\rangle = \langle q\phi|\psi\rangle, \tag{32}$$

which reduces to the Dirac inner product when $q = \mathbb{1}$. From (18) we note that with respect to this product, the adjoint will be given by

$$H^{\#\#} = q^{-1}H^{\dagger}q = A^{-1}S^{-1}H^{\dagger}SA = A^{-1}HA = H, \tag{33}$$

so that the Hamiltonian will be Hermitian with respect to this product and time evolution will be unitary. As a result, for any operator A commuting with the Hamiltonian, the operator q in (30) would define a product (32) which would maintain the unitary evolution.

Let us next show that we can choose the operator A as well as a suitable Hilbert space carefully so that the product (32) becomes an inner product. Let

$$\mathcal{H} = \{|\psi_E\rangle \text{ such that } H|\psi_E\rangle = E|\psi_E\rangle \text{ and } 0 < |\langle\psi_{\bar{E}}|\psi_E\rangle_S| < \infty\}. \tag{34}$$

Let P_E denote projection operators for states in this space such that

$$\begin{aligned} P_E|\psi_{E'}\rangle &= \delta_{EE'}|\psi_E\rangle, \\ \sum_E P_E &= \mathbb{1}. \end{aligned} \tag{35}$$

Therefore, since $[A, H] = 0$ we can express

$$A = \sum_E c_E P_E, \qquad [A, H]|\psi_E\rangle = 0, \tag{36}$$

which implies that we can write

$$A|\psi_E\rangle = c_E|\psi_E\rangle. \tag{37}$$

Namely, the states $|\psi_E\rangle$ are simultaneous eigenstates of A and H and the coefficients of expansion c_E in (36) can be identified with the eigenvalues of A.

Using (36) in (30) we can write

$$q = SA = S\sum_E c_E P_E, \tag{38}$$

and it follows from (32) that

$$\begin{aligned} \langle\psi_{E'}|\psi_E\rangle_q &= \langle\psi_{E'}|S\sum_{E''} c_{E''}P_{E''}|\psi_E\rangle \\ &= c_E\langle\psi_{E'}|S|\psi_E\rangle = c_E e^{-iF(F)}\delta_{E\bar{E}'}. \end{aligned} \tag{39}$$

Here we have used (24) in the last step. It is clear now that if we choose

$$c_E = e^{iF(E)}, \qquad A = \sum_E e^{iF(E)}\, P_E, \tag{40}$$

then we have

$$\langle \psi_{E'} | \psi_E \rangle = \delta_{E\bar{E}'}, \tag{41}$$

so that the product, in fact, becomes an inner product and a quantum mechanical description for the system is possible.

This shows that, while for any arbitrary A commuting with the Hamiltonian H the product (32) leads to a unitary evolution, for the particular choice of A in (40)

$$q = S \sum_E c_E P_E = S \sum_E e^{iF(H)} P_E = S e^{iF(H)} \sum_E P_E = S e^{iF(H)}, \tag{42}$$

the product with respect to q (32) is indeed an inner product. Here we have used the completeness relation in (35). This allows us to identify

$$A = e^{iF(H)} = (\langle \psi_{\bar{E}} | \psi_E \rangle)^{-1}, \tag{43}$$

with $E \to H$ on the right.

While (43), in principle, identifies the operator A and determines q in (42), in practice the construction of q in complicated theories is nontrivial. So, let us describe a systematic method for constructing q using conventional ideas from quantum mechanics. Let us consider the operator equation

$$H\sigma_E = E\sigma_E + \sigma_E k_E, \tag{44}$$

and its adjoint (by multiplying with appropriate inverse operators)

$$H^\dagger (\sigma_E^{-1})^\dagger = \bar{E}(\sigma_E^{-1})^\dagger + (\sigma_E^{-1})^\dagger k_E^\dagger. \tag{45}$$

σ_E and $(\sigma_E^{-1})^\dagger$ are called generators for the eigenstates of H and $H^\dagger$ respectively if

(i) there exists at least one state $|\psi\rangle$ such that

$$k_E|\psi\rangle = 0, \qquad \sigma_E|\psi\rangle \neq 0, \quad \forall E \in \text{Spect}(H). \tag{46}$$

(ii) there exists at least one state $|\phi\rangle$ satisfying

$$k_E^\dagger|\phi\rangle = 0, \qquad \text{with } (\sigma_E^{-1})^\dagger|\phi\rangle \neq 0, \quad \text{and } \langle\psi|\phi\rangle \neq 0, \quad \forall E \in \text{Spect}(H). \tag{47}$$

(iii) σ_E has an inverse σ_E^{-1}, at least on the reference state $|\psi\rangle$, such that

$$\sigma_E^{-1}\sigma_E|\psi\rangle = |\psi\rangle, \tag{48}$$

and an adjoint inverse $(\sigma_E^{-1})^\dagger$ is well defined at least on the reference state $|\phi\rangle$ This requirement is so that $|\psi_E\rangle = \sigma_E|\psi\rangle$ is invertible in the sense that $|\psi\rangle = \sigma_E^{-1}|\psi_E\rangle$).

It follows now from (44) as well as (46) that

$$H\sigma_E|\psi\rangle = E\sigma_E|\psi\rangle + \sigma_E k_E|\psi\rangle = E\sigma_E|\psi\rangle, \tag{49}$$

which allows us to identify the eigenstates of H as

$$|\psi_E\rangle = \sigma_E|\psi\rangle, \quad H|\psi_E\rangle = E|\psi_E\rangle. \tag{50}$$

This is reminiscent of the harmonic oscillator where we can identify the reference state with the ground state with the generators corresponding to $(a^\dagger)^n$ up to normalization factors. Similarly, from (45) as well as (47) we can identify the eigenstates of $H^\dagger$ to correspond to

$$|\phi_E\rangle = (\sigma_{\bar{E}}^{-1})^\dagger|\phi\rangle, \quad H^\dagger|\phi_E\rangle = \bar{E}|\phi_E\rangle, \quad H^\dagger|\phi_{\bar{E}}\rangle = E|\phi_{\bar{E}}\rangle. \tag{51}$$

Some comments are in order here. Since H is not Hermitian, we have two sets of eigenstates

$$H|\psi_E\rangle = E|\psi_E\rangle, \quad H^\dagger|\phi_{\bar{E}}\rangle = E|\phi_{\bar{E}}\rangle, \tag{52}$$

and we have to construct generators for both these sets of states. If $H = H^\dagger$, then the two sets of states would coincide, but when H is pseudo-Hermitian, the two sets would be related by some operator which we will see shortly to correspond to q.

From the definition of a pseudo-Hermitian system (4) we note that

$$\begin{aligned} H|\psi_E\rangle &= E|\psi_E\rangle, \\ \text{or,} \quad SH|\psi_E\rangle &= H^\dagger(S|\psi_E\rangle) = E(S|\psi_E\rangle), \end{aligned} \tag{53}$$

so that comparing with (51) and (52) we can write

$$|\phi_{\bar{E}}\rangle = (\sigma_{\bar{E}}^{-1})^\dagger|\phi\rangle = \alpha_E S|\psi_E\rangle, \tag{54}$$

where α_E is a constant. Taking the inner product with $\langle\psi_{\bar{E}}|$ we obtain

$$\begin{aligned} \langle\psi_{\bar{E}}|\phi_{\bar{E}}\rangle &= \langle\psi|\sigma_{\bar{E}}^\dagger(\sigma_{\bar{E}}^{-1})^\dagger|\phi\rangle = \alpha_E\langle\psi_{\bar{E}}|\psi_E\rangle_S \\ \text{or,} \quad \alpha_E e^{-iF(E)} &= \langle\psi|\phi\rangle. \end{aligned} \tag{55}$$

If we choose the normalization

$$\langle\psi|\phi\rangle = 1, \tag{56}$$

then (54) and (55) determine

$$\alpha_E = e^{iF(E)}, \quad |\phi_{\bar{E}}\rangle = Se^{iF(E)}|\psi_E\rangle = SA|\psi_E\rangle = q|\psi_E\rangle. \tag{57}$$

Furthermore, following (57), let us define

$$|\phi\rangle = q_0|\psi\rangle, \quad \text{such that} \quad \langle\psi|\phi\rangle = \langle\psi|q_0|\psi\rangle = 1. \tag{58}$$

From this analysis we see that for every E we can write

$$\begin{aligned} |\phi_{\bar{E}}\rangle &= q|\psi_E\rangle = (\sigma_{\bar{E}}^{-1})^\dagger|\phi\rangle \\ \text{or,} \quad q\sigma_E|\psi\rangle &= (\sigma_{\bar{E}}^{-1})^\dagger q_0|\psi\rangle. \end{aligned} \tag{59}$$

Therefore, we determine

$$q = \sum_E (\sigma_{\bar{E}}^{-1})^\dagger q_0 \sigma_E^{-1} P_E, \tag{60}$$

so that

$$\begin{aligned} q|\psi_E\rangle &= \sum_{E'} (\sigma_{\bar{E}'}^{-1})^\dagger q_0 \sigma_{E'}^{-1}) P_{E'}|\psi_E\rangle \\ &= \sum_{E'} (\sigma_{\bar{E}}^{-1})^\dagger q_0 \sigma_{E'}^{-1} \delta_{E'E}|\psi_E\rangle \\ &= (\sigma_{\bar{E}}^{-1})^\dagger q_0 \sigma_E^{-1}\sigma_E|\psi\rangle = (\sigma_{\bar{E}}^{-1})^\dagger q_0|\psi\rangle \\ &= (\sigma_{\bar{E}}^{-1})^\dagger|\phi\rangle = |\phi_{\bar{E}}\rangle, \end{aligned} \tag{61}$$

Journal of Physics: Conference Series **287** (2011) 012002 doi:10.1088/1742-6596/287/1/012002

and that

$$\langle\psi_{E'}|\psi_E\rangle_q = \langle\psi_{E'}|SA|\psi_E\rangle = e^{iF(E)}\langle\psi_{E'}|\psi_E\rangle_S = e^{iF(E)}e^{-iF(E)}\delta_{E\bar{E}'} = \delta_{E\bar{E}'}, \tag{62}$$

as we would expect.

This construction of the operator q and the inner product works for any pseudo-Hermitian Hamiltonian with real or complex eigenvalues. Furthermore, since Hermitian Hamiltonians define a subset of pseudo-Hermitian Hamiltonians, it should work for such systems as well leading to $q = \mathbb{1}$ and reducing the inner product (32) to the standard Dirac inner product. Let us check this in the case of the harmonic oscillator where the Hamiltonian has the form ($\hbar = m = \omega = 1$ and we add a constant to remove the zero point energy for simplicity)

$$H = \frac{1}{2}(p^2 + x^2 - 1) = a^\dagger a = H^\dagger. \tag{63}$$

This is a Hermitian Hamiltonian with real energy eigenvalues

$$\begin{aligned} &H|\psi_{E_n}\rangle = H|\psi_n\rangle = E_n|\psi_n\rangle = n|\psi_n\rangle, \quad E_n = n = 0, 1, 2, \cdots, \\ &|\psi_n\rangle = \frac{(a^\dagger)^n}{\sqrt{n!}}|\psi_0\rangle, \quad a|\psi_0\rangle = 0, \\ &\langle\psi_m|\psi_n\rangle = \delta_{mn}. \end{aligned} \tag{64}$$

In this case, we can identify the reference state with the ground state which leads to

$$|\psi\rangle = |\psi_0\rangle, \quad \sigma_{E_n} = \sigma_n = \frac{(a^\dagger)^n}{\sqrt{n!}}. \tag{65}$$

Using the standard commutation relations between the annihilation and creation operators we obtain

$$\begin{aligned} [H, \sigma_n] &= [a^\dagger a\,,\, \frac{(a^\dagger)^n}{\sqrt{n!}}] = n\frac{(a^\dagger)^n}{\sqrt{n!}}, \\ \text{or,} \quad H\sigma_n &= n\sigma_n + \sigma_n H. \end{aligned} \tag{66}$$

Comparing with (44) we now determine

$$k_n = H, \quad k_n|\psi\rangle = H|\psi_0\rangle = 0. \tag{67}$$

Since the ground state is already normalized (see (64)), it follows from (58) that in this Hermitian case,

$$q_0 = \mathbb{1}, \quad \text{leading to} \quad |\phi\rangle = |\psi\rangle = |\psi_0\rangle. \tag{68}$$

Furthermore, the inverses of the generators are easily seen to correspond to

$$\sigma_n^{-1} = \frac{a^n}{\sqrt{n!}}, \quad \text{so that} \quad \sigma_n^{-1}\sigma_n|\psi\rangle = |\psi\rangle. \tag{69}$$

Furthermore, the projection operators on to the energy states are given by

$$P_n = |\psi_n\rangle\langle\psi_n|, \quad \sum_n P_n = \mathbb{1}, \tag{70}$$

which follows from the completeness of the harmonic oscillator states. Furthermore, P_n can be checked to satisfy

$$(\sigma_n^{-1})^\dagger(\sigma_n)^{-1}P_n = \frac{(a^\dagger)^n}{\sqrt{n!}}\frac{(a)^n}{\sqrt{n!}}P_n = P_n. \tag{71}$$

It follows now from (60) that

$$\begin{aligned} q &= \sum_n (\sigma_n^{-1})^\dagger q_0 \sigma_n^{-1} P_n = \sum_n \frac{(a^\dagger)^n}{\sqrt{n!}}\frac{(a)^n}{\sqrt{n!}}P_n = \sum_n P_n = \mathbb{1}, \\ \langle\psi_m|\psi_n\rangle_q &= \langle\psi_m|q|\psi_n\rangle = \langle\psi_m|\psi_n\rangle = \delta_{mn}. \end{aligned} \tag{72}$$

Our construction indeed reduces to the standard inner product for a Hermitian Hamiltonian.

Let us next apply our construction to a simple 2×2 matrix Hamiltonian which is pseudo-Hermitian. Let us consider the Hamiltonian

$$H = \begin{pmatrix} re^{i\theta} & se^{i\phi} \\ te^{-i\phi} & re^{-i\theta} \end{pmatrix} \neq H^\dagger, \quad r,s,t,\theta,\phi \text{ real}.$$

This Hamiltonian is not Hermitian. However, if we define a matrix

$$S = \begin{pmatrix} 0 & e^{i\phi} \\ e^{-i\phi} & 0 \end{pmatrix} = S^{-1} = S^\dagger,$$

we can check that

$$H = S^{-1}H^\dagger S, \tag{73}$$

so that the Hamiltonian is pseudo-Hermitian. Note that this Hamiltonian is not $\mathcal{PT}$ symmetric (unless $s = t, \phi = 0$) where the parity operator is defined by

$$\mathcal{P} = \begin{pmatrix} 0 & 1 \\ 1 & 0 \end{pmatrix},$$

and time reversal $\mathcal{T}$ corresponds to complex conjugation. However, we note that we can define a modified parity operation

$$\overline{\mathcal{P}} = \begin{pmatrix} 0 & \sqrt{\frac{s}{t}} \\ \sqrt{\frac{t}{s}} & 0 \end{pmatrix} = \overline{\mathcal{P}}^{-1},$$

and the model is $\overline{\mathcal{P}}\mathcal{T}$ symmetric ($\mathcal{T}$ is complex conjugation). For $s = t$, $\overline{\mathcal{P}} \to \mathcal{P}$, but note that although $\overline{\mathcal{P}}^2 = \mathbb{1}$, unless $s = t$

$$\overline{\mathcal{P}}^\dagger \neq \overline{\mathcal{P}}, \quad \overline{\mathcal{P}}H\overline{\mathcal{P}} = H^\star \neq H^\dagger, \tag{74}$$

as in (1).

The eigenvalues of H are determined to be

$$\det\begin{pmatrix} re^{i\theta} - E & se^{i\phi} \\ te^{-i\phi} & re^{-i\theta} - E \end{pmatrix} = 0,$$

whose roots are given by

$$\begin{aligned} & E^2 - 2Er\cos\theta + r^2 - st = 0, \\ \text{or,}\quad & E_\pm = r\cos\theta \pm \sqrt{st - r^2\sin^2\theta}. \end{aligned} \tag{75}$$

Therefore, for $st > r^2\sin^2\theta$, the energy eigenvalues are real while for $st < r^2\sin^2\theta$, the two energy eigenvalues are complex conjugates of each other (for $st = r^2\sin^2\theta$ the Hamiltonian cannot be diagonalized). Let us discuss only the case with real eigenvalues for simplicity, the complex case is also straightforward.

Let us define

$$Q = \sqrt{st - r^2\sin^2\theta}, \quad st = Q^2 + r^2\sin^2 = (Q + ir\sin\theta)(Q - ir\sin\theta), \tag{76}$$

so that we can write

$$E_\pm = r\cos\theta \pm Q = \bar{E}_\pm, \text{ real.} \tag{77}$$

We can work out the the eigenstates corresponding to the two eigenvalues easily and they have the forms

$$\begin{aligned} |\psi_{E_+}\rangle &= \frac{1}{\sqrt{s+t}}\begin{pmatrix} \left(\frac{s}{t}\right)^{\frac{1}{4}}\sqrt{Q+ir\sin\theta}\,e^{\frac{i\phi}{2}} \\ \left(\frac{t}{s}\right)^{\frac{1}{4}}\sqrt{Q-ir\sin\theta}\,e^{-\frac{i\phi}{2}} \end{pmatrix}, \\ |\psi_{E_-}\rangle &= \frac{i}{\sqrt{s+t}}\begin{pmatrix} \left(\frac{s}{t}\right)^{\frac{1}{4}}\sqrt{Q-ir\sin\theta}\,e^{\frac{i\phi}{2}} \\ -\left(\frac{t}{s}\right)^{\frac{1}{4}}\sqrt{Q+ir\sin\theta}\,e^{-\frac{i\phi}{2}} \end{pmatrix}. \end{aligned} \tag{78}$$

The projection operators in this case can be determined to correspond to

$$\begin{aligned} P_{E_+} &= \frac{1}{2Q}(H - E_-) = \frac{1}{2Q}\begin{pmatrix} Q + ir\sin\theta & se^{i\phi} \\ te^{-i\phi} & Q - ir\sin\theta \end{pmatrix}, \\ P_{E_-} &= -\frac{1}{2Q}(H - E_+) = \frac{1}{2Q}\begin{pmatrix} Q - ir\sin\theta & -se^{i\phi} \\ -te^{-i\phi} & Q + ir\sin\theta \end{pmatrix}, \end{aligned} \tag{79}$$

which satisfy

$$P_{E_+} + P_{E_-} = \frac{1}{2Q}(E_+ - E_-)\mathbb{1} = \frac{1}{2Q} \times 2Q\mathbb{1} = \mathbb{1}. \tag{80}$$

Therefore, the projection operators project on to complementary spaces. Furthermore, they satisfy the relations

$$P_{E_+} = S^{-1}P_{E_+}^\dagger S, \quad P_{E_-} = S^{-1}P_{E_-}^\dagger S. \tag{81}$$

Let us choose the reference state to be

$$|\psi\rangle = |\psi_{E_-}\rangle, \quad \Rightarrow \quad \sigma_{E_-} = \mathbb{1} = \sigma_{E_-}^{-1}. \tag{82}$$

Furthermore, from the form of the eigenstates in (78) and the definition

$$|\psi_{E_+}\rangle = \sigma_{E_+}|\psi\rangle = \sigma_{E_+}|\psi_{E_-}\rangle, \tag{83}$$

we determine that

$$\sigma_{E_+} = \begin{pmatrix} 0 & i\sqrt{\frac{s}{t}}\,e^{i\phi} \\ -i\sqrt{\frac{t}{s}}\,e^{-i\phi} & 0 \end{pmatrix} = \sigma_{E_+}^{-1}, \quad S\sigma_{E_+} = -\sigma_{E_+}^\dagger S. \tag{84}$$

The defining relation (44) leads to

$$\begin{aligned} H\sigma_{E_-} &= E_-\sigma_{E_-} + \sigma_{E_-}k_{E_-}, \quad \Rightarrow \quad k_{E_-} = \sigma_{E_-}^{-1}(H - E_-)\sigma_{E_-} = 2QP_{E_+}, \\ H\sigma_{E_+} &= E_+\sigma_{E_+} + \sigma_{E_+}k_{E_+}, \quad \Rightarrow \quad k_{E_+} = \sigma_{E_+}^{-1}(H - E_+)\sigma_{E_+} = -2QP_{E_+}. \end{aligned} \tag{85}$$

Journal of Physics: Conference Series **287** (2011) 012002 doi:10.1088/1742-6596/287/1/012002

Similarly, analyzing the eigenstates of $H^\dagger$, we determine that

$$|\phi\rangle = -\frac{s+t}{2Q} S|\psi\rangle = q_0|\psi\rangle, \quad \Rightarrow \quad q_0 = -\frac{s+t}{2Q} S, \tag{86}$$

where the normalization is chosen so that $\langle\psi|\phi\rangle = 1$ (see (56). With all this information, we can now determine q from (60)

$$\begin{aligned} q &= (\sigma_{E_+}^{-1})^\dagger q_0 \sigma_{E_+}^{-1} P_{E_+} + (\sigma_{E_-}^{-1})^\dagger q_0 \sigma_{E_-}^{-1} P_{E_-} \\ &= -\frac{s+t}{2Q} [\sigma_{E_+}^\dagger S \sigma_{E_+} P_{E_+} + S P_{E_-}] = \frac{s+t}{2Q} S(P_{E_+} - P_{E_-}) \\ &= \frac{s+t}{2Q} \begin{pmatrix} 0 & e^{i\phi} \\ e^{-i\phi} & 0 \end{pmatrix} \frac{1}{Q} \begin{pmatrix} ir\sin\theta & se^{i\phi} \\ te^{-i\phi} & -ir\sin\theta \end{pmatrix} \\ &= \frac{s+t}{2Q} \begin{pmatrix} t & -ir\sin\theta\, e^{i\phi} \\ ir\sin\theta\, e^{-i\phi} & s \end{pmatrix}. \end{aligned} \tag{87}$$

It can be checked now that

$$\langle\psi_{E_i}|\psi_{E_j}\rangle_q = \delta_{E_i E_j}, \qquad i, j = \pm. \tag{88}$$

This analysis can be extended to the case of complex eigenvalues in a straightforward manner. (This can also be extended to $n \times n$ matrix systems.)

The construction that we have described seems to require a knowledge of the energy eigenstates of the theory (so that we can construct the generators of energy eigenstates). However, we cannot solve most quantum mechanical systems. In this case, we determine the energy eigenstates perturbatively. Therefore, following our ideas we can construct the q-operator or the inner product perturbatively. The pseudo-Hermitian systems differ from Hermitian systems in this respect. Namely, the inner product in a Hermitian system is given independent of the dynamics whereas in a pseudo-Hermitian system it depends on the dynamics of the system.

Let us consider a pseudo-Hermitian quantum system (with real eigenvalues) whose Hamiltonian can be writen as

$$H = H_0 + \epsilon V(x), \tag{89}$$

where H_0 can be solved exactly and ϵ is a small perturbation parameter. Of course, H_0 can be Hermitian or non-Hermitian. However, in most examples, it can be chosen to be Hermitian which we consider for simplicity. Let us denote the generators of eigenstates of H_0 by $\sigma_E^{(0)}$ so that we can write

$$|\psi_E^{(0)}\rangle = \sigma_E^{(0)}|\psi\rangle, \quad P_E^{(0)} = |\psi_E^{(0)}\rangle\langle\psi_E^{(0)}|, \tag{90}$$

where $|\psi\rangle$ denotes the reference state. Since H_0 is assumed to be Hermitian, as we have shown in the case of the harmonic oscillator (see (68) and (72))

$$q_0 = \mathbb{1}, \quad |\phi\rangle = |\psi\rangle, \quad q^{(0)} = \sum_E (\sigma_E^{(0)-1})^\dagger q_0 \sigma_E^{(0)-1} P_E^{(0)} = \mathbb{1}. \tag{91}$$

The perturbative expansion of the eigenstates of the total Hamiltonian is given by

$$|\psi_E\rangle = \sum_i \epsilon^i\, |\psi_E^{(i)}\rangle, \tag{92}$$

where $|\psi_E^{(i)}\rangle$ denotes the ith order correction to the state $|\psi_E^{(0)}\rangle$. As we know in standard quantum mechanics, the corrections at any order can be chosen to be orthogonal to the state

$$\langle\psi_E^{(0)}|\psi_E^{(i)}\rangle = 0, \quad i > 0. \tag{93}$$

Journal of Physics: Conference Series **287** (2011) 012002 doi:10.1088/1742-6596/287/1/012002

With this, the correction to the energy eigenstate at any order can be determined recursively as $(i > 0)$

$$
\begin{aligned}
|\psi_E^{(i)}\rangle &= \sum_{E'\neq E} \frac{1}{E^{(0)} - E'^{(0)}} \left[\langle \psi_E'^{(0)} | V_E^{(i-1)} \rangle - \sum_{j=1}^{i-1} E^{(j)} \langle \psi_E'^{(0)} | \psi_E^{(i-j)} \rangle \right] \\
&= R^{(i)} |\psi_E^{(0)}\rangle = R^{(i)} \sigma_E^{(0)} |\psi\rangle = \sigma_E^{(i)} |\psi\rangle, \\
E^{(i)} &= \langle \psi_E^{(0)} | V | \psi_E^{(i-1)} \rangle,
\end{aligned} \tag{94}
$$

where we have identified

$$
\sigma_E^{(i)} = R^{(i)} \sigma_E^{(0)}. \tag{95}
$$

To any order we can write

$$
\begin{aligned}
R &= \sum_{i=0} \epsilon^i R^{(i)}, \qquad R^{(0)} = \mathbb{1}, \\
\sigma_E &= \sum_{i=0} \epsilon^i \sigma_E^{(i)} = \sum_{i=0} \epsilon^i R^{(i)} \sigma_E^{(0)} = R \sigma_E^{(0)}.
\end{aligned} \tag{96}
$$

By taking the adjoint we can determine the corrections to the eigenstates of $H^\dagger$ to any order to be

$$
\begin{aligned}
|\phi_E\rangle &= (R^{-1})^\dagger |\phi_E^{(0)}\rangle = (R^{-1})^\dagger (\sigma_E^{(0)-1})^\dagger |\phi\rangle = (R^{-1})^\dagger (\sigma_E^{(0)-1})^\dagger |\psi\rangle \\
&= ((R\sigma_E^{(0)})^{-1})^\dagger |\psi\rangle = (\sigma_E^{-1})^\dagger |\psi\rangle.
\end{aligned} \tag{97}
$$

As a result, we determine

$$
\begin{aligned}
q &= \sum_E (\sigma_E^{-1})^\dagger q_0 \sigma_E^{-1} P_E \\
&= \sum_E (R^{-1})^\dagger (\sigma_E^{(0)-1})^\dagger \sigma_E^{(0)-1} R^{-1} P_E \\
&= (R^{-1})^\dagger \sum_E (\sigma_E^{(0)-1})^\dagger \sigma_E^{(0)-1} P_E^{(0)} R^{-1} \\
&= (R^{-1})^\dagger q^{(0)} R^{-1} = (R^{-1})^\dagger R^{-1}.
\end{aligned} \tag{98}
$$

Thus q depends only on the perturbation operator R for systems with real eigenvalues. If H_0 were not Hermitian, then $q^{(0)}$ would be nontrivial which would have to be determined.

As an application of this formalism, let us consider the $\mathcal{PT}$ symmetric theory

$$
H = \frac{1}{2}(p^2 + x^2 - 1) + i\epsilon x^3 = H_0 + \epsilon V(x). \tag{99}
$$

In this case, H_0 is the harmonic oscillator Hamiltonian which we have already studied.

$$
|\psi_n^{(0)}\rangle = \sigma_n^{(0)} |\psi_0\rangle = \sigma_n^{(0)} |\psi\rangle, \quad E_n^{(0)} = n. \tag{100}
$$

To first order in perturbation, we have

$$
\begin{aligned}
|\psi_n^{(1)}\rangle &= \sum_{m\neq n} \frac{1}{E_n^{(0)} - E_m^{(0)}} \langle \psi_m^{(0)} | V | \psi_n^{(0)} \rangle \, |\psi_m^{(0)}\rangle \\
&= \sum_{m\neq n} \frac{1}{n-m} \langle \psi_m^{(0)} | i x^3 | \psi_n^{(0)} \rangle \, |\psi_m^{(0)}\rangle.
\end{aligned} \tag{101}
$$

Using the definitions of the annihilation and creation operators,

$$x = \frac{1}{\sqrt{2}}(a + a^\dagger), \quad p = -\frac{i}{\sqrt{2}}(a - a^\dagger), \tag{102}$$

we can evaluate the first order correction to the energy eigenstate to be

$$|\psi_n^{(1)}\rangle = \frac{i}{2^{3/2}}\left[-\frac{2}{3}(a - a^\dagger)^3 + \{(a + a^\dagger)^2(a - a^\dagger)\}\right]|\psi_n^{(0)}\rangle, \tag{103}$$

where we have defined

$$\{A^2B\} = 1/3(A^2 + ABA + BA^2). \tag{104}$$

Using (102) we can rewrite (103) also as

$$|\psi_n^{(1)}\rangle = -\left(\frac{2}{3}p^3 + x^2p - ix\right)|\psi_n^{(0)}\rangle, \tag{105}$$

so that we can identify

$$R^{(1)} = -\left(\frac{2}{3}p^3 + x^2p - ix\right). \tag{106}$$

To this order, therefore, we can determine

$$\begin{aligned}
R &= \left(\mathbb{1} - \epsilon\left(\frac{2}{3}p^3 + x^2p - ix\right)\right), \\
q &= (R^{-1})^\dagger R^{-1} = \left(\mathbb{1} - \epsilon\left(\frac{2}{3}p^3 + x^2p - ix\right)\right)^\dagger \left(\mathbb{1} + \epsilon\left(\frac{2}{3}p^3 + x^2p - ix\right)\right) \\
&= \mathbb{1} + \epsilon\left(\frac{2}{3}p^3 + x^2p + ix\right) + \epsilon\left(\frac{2}{3}p^3 + x^2p - ix\right) + O(\epsilon^2) \\
&= \mathbb{1} + \epsilon\left(\frac{4}{3}p^3 + px^2 + x^2p\right) + O(\epsilon^2).
\end{aligned} \tag{107}$$

This perturbative construction of q (and, therefore, the inner product) can be carried out systematically to any order in perturbation.

Acknowledgment

It is a pleasure to thank the organizers of the Mexican school of particles and fields for the invitation to give these lectures. I would also like to thank Mr. Saul Hernandez Otiz and Mr. Khepani Raya for help in typing these lectures. This work was supported in part by US DOE Grant number DE-FG 02-91ER40685.

References

[1] It is impossible to refer to the large number of papers on this subject. Therefore, we only refer to three main reviews where references to other papers can be found.
Bender, C. M, Contemp. Phys. **46**, 277 (2005).
Bender, C. M, Reports Prog. Phys. **70**, 947 (2007).
[2] Mostafazadeh, A, arXiv:0805.1651 [quant-ph].
[3] Das, A. and L. Greenwood, Phys. Lett. **B678**, 504 (2009); A. Das and L. Greenwood, J. Math. Phys. **51**, 042103 (2010).

Journal of Physics: Conference Series **287** (2011) 012003 doi:10.1088/1742-6596/287/1/012003

String Theory: a mini-course

C. Damian and O. Loaiza-Brito[1]

Departamento de Física, DCI, Campus León, Universidad de Guanajuato, C.P. 37150, Guanuajuato, Mexico

E-mail: cesaredas@fisica.ugto.mx, oloaiza@fisica.ugto.mx

Abstract. We provide a brief introduction to string theory. We review some basic aspects of the theory as the quantization of the bosonic and the supersymmetric string. D-branes are studied in some detail as well as some aspects of T-duality. The course is intended for students with some previous knowledge in Quantum Field Theory and General Relativity.

1. Introduction

String theory is our best candidate for a model beyond the Standard Model (SM) in the sense that it provides a quantum mechanically well-defined theory underlying gauge and gravitational interactions. As a theory of quantum gravity, it has the potential to answer questions so far unknown as the nature of space-time and the black hole information paradox. As a gauge theory, it has the potential to unify all forces. More over, it offers a way to relate aspects of gauge theories with gravity and vice versa.

The basic feature of string theory proposes that elementary particles are not point-like, but rather they are small 1-dimensional extended objects. By this simple fact, we shall see that, every symmetry we know is relevant in the real world, can be deduced from the supersymmetric quantum string. More over, we shall review how important aspects of the non-perturbative physics of strings are easily achieved.

From the phenomenological point of view, we are interested in constructing a theory which goes far beyond the Standard Model of Particles (SM) since there are basic questions that can not be answered in the context of quantum field theories. For instance, how is gravity coupled to other forces at the quantum level? Is there an unification of all forces? Why are there three families in the SM? How is supersymmetry broken? Why are there so many free parameters in the SM and even more in the Supersymmetric Standard Model? Why is the cosmological constant so small? Why and what is the dark energy? The presence of singularities, as the Big Bang and the black hole singularity, tells us that the theory breaks down in such limits. What is the extension to General Relativity at the quantum level? and so on. All these questions might be answered in the context of string theory and for this reason, it is worth to study it.

String theory is however, a theory under construction, meaning that we do not know the basic physical principles underlying it. Although a lot of progress has been done in the past years,

[1] Speaker.

Journal of Physics: Conference Series **287** (2011) 012003 doi:10.1088/1742-6596/287/1/012003

still there is not a concrete prediction so far. This is a crucial point we must face in the years to come. Meanwhile, let us begin studying the basics of this beautiful model called *string theory*.

Our review is organized as follows. In section 1 we study the bosonic string and its quantization. Section 2 is devoted to the superstring. Special emphasis is given to the construction of the spectrum of Type IIA and Type IIB theories. In section 4 we concentrate on the description of T-duality and the physics of supersymmetric D-branes. At the end, we briefly comment on some recent studies in string theory. References are restricted to some text books on string theory [1]-[9] and to some review articles written by one of the authors [11]-[12] . We apologize for not given an extensive reference list about the hundreds of excellent reviews on string theory. However, we encourage the interested reader to consult the bibliography reported in the text books we refer to.

2. The Bosonic String

It is well known that the Lagrangian of a relativistic point particle is proportional to the corresponding invariant length, the world-line described by the particle while moving in the space-time. Our first attempt to describe the dynamics of a string is to construct its Lagrangian, which based on the relativistic particle case, we expect it to be proportional to the invariant area described by the string as it moves in a D-dimensional space-time[2]. This is called the string world-sheet and it is a 2-dimensional object with an induced metric denoted by $G_{\alpha\beta}$. The action is given by

$$S = -T \int d^2\sigma \sqrt{-det(G_{\alpha\beta})}, \tag{1}$$

where the world-sheet is parametrized by $\sigma_1 = \sigma$ (a space-like coordinate), and $\sigma_0 = \tau$ (a time-like coordinate). After some direct calculations, one can see that the above action can be written in terms of the D-dimensional coordinates X^M which characterize the position of the world-sheet in the space-time[3] This is the *Nambu-Goto* action and it is given by

$$S_{NG} = -T \int d\sigma d\tau \sqrt{\left(\dot{X} \cdot X'\right)^2 + \dot{X}^2 X'^2}, \tag{2}$$

where

$$\dot{X}^M = \frac{\partial X^M}{\partial \tau} \quad X'^M = \frac{\partial X^M}{\partial \sigma}. \tag{3}$$

However this action is difficult to quantize. An easiest way to do it, requires an action without the square root of the field derivatives[4]. The desired action is known as the *Polyakov action* and it is equivalent to the Nambu-Goto action at the classical level in the sense that it give rise to the same equations of motion for the fields X^M. The Polyakov action is

$$S_P = -\frac{1}{2} \int d^2\sigma \sqrt{-h} h^{\alpha\beta} \partial_\alpha X \cdot \partial_\beta X, \tag{4}$$

where $h = det(h_{\alpha\beta})$, $h_{\alpha\beta}$ is the induced metric on the world-sheet, and α and β runs on 0 and 1. This action posses the following symmetries.

[2] Notice that we start from a space-time with an arbitrary dimensionality $D = d + 1$. One of the most interesting things of string theory, is that it is possible to fix such number. This constitutes an exceptional case in theoretical models as Quantum Field Theory, where the number of dimensions are fixed by hand. In String Theory, such number is fixed by the theory itself under certain circumstance we shall comment later on.

[3] This is formally known as a sigma model.

[4] Here we refer to the path integral method of quantization. In this mini-course we shall take the canonical quantization procedure. However, using the Polyakov action represents as well an easy way to realize the full set of symmetries involved in the bosonic string.

(i) Space-time Poincaré symmetry. The action is invariant under Poincaré transformations. It is a global symmetry.

(ii) Reparametrizations. This is a non-trivial gauge symmetry and represent a large extra symmetry on the world sheet. Essentially tells us that we can select different coordinate frames on the world-sheet.

(iii) Weyl transformation. This symmetry represents the invariance of the action under rescaling of the metric. By transforming the induced metric $h_{\alpha\beta}$ to $\Omega^2 h_{\alpha\beta}$ the equations of motion are the same.

Since we have three constraints given by these symmetries, it is possible to fix the corresponding three parameters in $h_{\alpha\beta}$. By reparametrization invariance we can choose two components of $h_{\alpha\beta}$, such that only one component remains free. But Weyl invariance fixes it. Therefore, the induced metric $h_{\alpha\beta}$ is fixed as $h_{\alpha\beta} = \eta_{\alpha\beta}$, where $\eta_{\alpha\beta}$ is the flat 2-dimensional Minkowskian metric. With this gauge, the Polyakov action reduces to

$$S = \frac{T}{2}\int d^2\sigma(\dot{X}^2 - X'^2). \tag{5}$$

The corresponding equation of motion with respect to the fields X^M, up to some boundary conditions, are given by

$$\partial_\alpha \partial^\alpha X = 0, \tag{6}$$

which is a differential wave equation. Their solutions are then a linear combination of linear terms in τ and σ and periodic functions. Notice that such solutions represents undulating modes of the string. Later on we shall see that such modes can be quantized, and they will represent different massive and massless fields. If some of the massless fields have the properties of already known fields in the context of quantum field theory, we shall conclude that quantum oscillations of a string produces the particles we are familiar with. For instance, let us say that we find that one of the modes is massless, transforms under the Lorentz group as a vector boson, has a $U(1)$ internal symmetry and has a spin 1. We would assure that such mode is indeed a photon. Things are nevertheless, a little bit more complicated, but we shall see actually that some modes have all the desired properties with respect to the particles we already know.

Fixing the gauge as before, implies that each component of the energy-momentum tensor vanishes,

$$T_{\alpha\beta} = \partial_\alpha X \partial_\beta X - \frac{1}{2} h_{\alpha\beta} h^{\gamma\delta} \partial_\gamma X \, \partial_\delta X = 0, \tag{7}$$

and by using the flat induced metric, the components read

$$T_{\alpha\beta} \doteq \begin{bmatrix} \frac{1}{2}\left(\dot{X}^2 + X'^2\right) & \dot{X}X' \\ \dot{X}X' & \frac{1}{2}\left(\dot{X}^2 + X'^2\right). \end{bmatrix} = 0 \tag{8}$$

This expression has a geometrical interpretation. The gauge selection we have performed implies that the coordinate framework on the world-sheet is orthogonal. This allows us to choose a particular useful system of coordinates: the light-cone coordinates. We shall return to this selection later on.

2.1. Boundary Conditions and D-branes

The stationary points are chosen by demanding the invariance of the action under the shifts $X^M \to X^M + \delta X^M$. The variation of the action reads

$$\delta S = \frac{T}{2}\int d^2\sigma(\partial_\alpha\partial^\alpha X_M)\delta X^M - T\int d\tau \left[X'_M \delta X^M \,|_{\sigma=\pi} - X'_M \delta X^M \,|_{\sigma=0}\right], \tag{9}$$

Table 1. Boundary conditions on the bosonic string.

Type	Boundary Condition
Closed	$X^M(\sigma,\tau) = X^M(\sigma+\pi,\tau)$
Open/Neumann	$X'_M = 0, \quad \sigma = 0, \pi$
Open/Dirichlet	$\delta X^M = 0, \quad \sigma = 0, \pi$

from which, the vanishing of the boundary terms leads to different conditions on the string (see Table 1). Strings which endpoints are identified, are called *closed strings*, while *open strings* presents two variations. For Neumann boundary conditions we can see that there is not momentum flowing away from the string endpoints. These strings are free to move over the whole D-dimensional space-time, meaning that we can impose Nuemann conditions on all field coordinates X^M. We can also consider an open string with all its field coordinates fulfilling Dirichlet boundary conditions. This means that each of the endpoints of the strings, are fixed on a point in a D-dimensional space-time. Notice as well that such point is localized in time. Further analysis shows that this is indeed an instanton of the theory. However, let us consider the richest case, in which we impose Dirichlet boundary conditions on some coordinates and Neumann on others. Let us say that for $M = 0, 1, \ldots, p$, X^M satisfies Neumann boundary conditions, while for $M = p+1, \ldots, d$, it satisfies Dirichlet conditions. Henceforth, the endpoints of the string will be attached to move freely on a $p+1$-dimensional hyperplane. Notice that they can not leave this object unless their endpoints coincide. If this happens, the open string becomes closed and it is free to escape away from the D-brane. In this sense we say that D-branes emit closed strings[5].

It turns out that such objects are very important in string theory. They are named *Dp*-branes: *p*- dimensional branes (generalization of membranes) on which the endpoints of an open string are attached to. Orthogonal coordinates of the endpoints satisfy Dirichlet boundary conditions (the "D" of *D*-brane refers to Dirtichlet boundary conditions). For the bosonic case, it is possible to have different *Dp*-branes of all dimensionalities, from $p = -1$ to $p = d$.

2.2. Solution to the Equations of Motion

We are now ready to explicitly show the solutions of the equations of motion. However, we shall take a short-way. We are going to solve the equations of motion in a particular coordinate-frame on the string world-sheet, known as the light-cone coordinates. The advantages of this selection are that we shall arrive to the solutions in a faster and easier way. The price to pay is however, that we have lost covariance in our description of the string dynamics. Also, by selecting this frame, one can check that there are two unphysical degrees of freedom, corresponding to stretching and rotating a string. In a covariant formalism, one is able to arrive at the same conclusion, but in a richer way, i.e., by analyzing the anomalies, ghosts fields and by using the powerful conformal field theory in the string world-sheet.

The light-cone coordinates are given by $\sigma^{\pm} = \tau \pm \sigma$, from which the equations of motion read

$$\partial_+ \partial_- X^M = 0. \qquad (10)$$

[5] We shall see that gravitons are associated to closed strings. Therefore, a D-brane emits and absorbs gravitons, implying that these objects have mass and tension and mainly, that they are dynamical.

For the closed string, the corresponding solution can be written in terms of two waves by

$$X^M(\sigma,\tau) = X_R^M(\sigma^-) + X_L^M(\sigma^+), \tag{11}$$

which represent left and right movers. The most general solution for the closed string given by

$$X_R^M = \frac{1}{2}x^M + \alpha' p^M \sigma^- + \frac{i}{2}l_s \sum \frac{1}{n}\alpha_n^M e^{-2in\sigma^-}, \tag{12}$$

$$X_L^M = \frac{1}{2}x^M + \alpha' p^M \sigma^+ + \frac{i}{2}l_s \sum \frac{1}{n}\tilde{\alpha}_n^M e^{-2in\sigma^+}. \tag{13}$$

Since $X^M(\sigma,\tau)$ is required to be real, the left and right movers X_L^M and X_R^M must be real function as well. This implies that x^M (center of mass) and p^M (momentum of the center of mass) are also real and in consequence, the modes of the string must obey that $\alpha_{-n}^M = (\alpha_n^M)^*$, $\tilde{\alpha}_{-n}^M = (\tilde{\alpha}_n^M)^*$ for all n and where we have used $\alpha_0^M = \sqrt{\alpha'/2}\, p^M$. Introducing X_L and X_R into the energy-momentum tensor we obtain that

$$T_{--} = \alpha' \sum L_m e^{-2im\sigma^-}, \quad T_{++} = \alpha' \sum \tilde{L}_m e^{-2im\sigma^+}, \tag{14}$$

where the Fourier coefficients are actually the Virasoro generators given by

$$L_m = \frac{1}{2}\sum \alpha_{m-n}\alpha_n \quad \tilde{L}_m = \frac{1}{2}\sum \tilde{\alpha}_{m-n}\tilde{\alpha}_n. \tag{15}$$

So the requirement of $T_{++} = T_{--} = 0$ is reflected in the fact that all the modes must vanish. This constitutes a classical constraint for all modes, $L_m = \tilde{L}_m = 0$. An interesting and special case concerns the zero mode, since it offers a way to compute the mass associated to different Fourier modes. Hence, for $m = 0$,

$$\begin{aligned} L_0 + \tilde{L}_0 &= \frac{1}{2}(\alpha_0^2 + \alpha_0^2) + \sum_{n\neq 0}(\alpha_{-n}\cdot\alpha_n, +\tilde{\alpha}_{-n}\cdot\tilde{\alpha}_n) \\ &= \alpha' p^M p_M + \sum_{n\neq 0}(\alpha_{-n}\cdot\alpha_n + \tilde{\alpha}_{-n}\cdot\tilde{\alpha}_n). \end{aligned} \tag{16}$$

Since in a relativistic theory, $p^2 = -m^2$, we arrive at the mass expression for the closed string modes,

$$M^2 = \frac{2}{\alpha'}\sum_{n=1}^{\infty}(\alpha_{-n}\cdot\alpha_n + \tilde{\alpha}_n\cdot\tilde{\alpha}_n). \tag{17}$$

By thinking that a closed string is made of two open strings glued together, then it is easy to see that the solution of the equations of motion and the expansion in Fourier modes, corresponds to just one of the movers. Therefore, for the open string theory, the mass of the modes is given by

$$M^2 = \frac{1}{\alpha}\sum_{n=1}^{\infty}\alpha_{-n}\cdot\alpha_n, \tag{18}$$

where the most general solution of the equations of motion is given by

$$X^M(\tau,\sigma) = x^M + \frac{\alpha'}{2}p^M\tau + \frac{i\alpha'}{2}\sum\frac{1}{m}\alpha_m^M e^{-im\tau}cos(m\sigma). \tag{19}$$

Journal of Physics: Conference Series **287** (2011) 012003 doi:10.1088/1742-6596/287/1/012003

2.3. Canonical Quantization

Our next step consists in quantizing the vibrations modes of the string. We would like to explore the possibility that some modes be representations of quantum fields. There are many ways to quantize the string, being the most popular the covariant quantization. However, since we want to study the quantum string features in a short way, we shall take the canonical quantization procedure. Essentially, this consists in promoting the Fourier modes α_n to operators which satisfy the algebra[6]

$$\left[\alpha_m^M, \alpha_n^N\right] = \left[\tilde{\alpha}_m^M, \tilde{\alpha}_n^N\right] = \eta^{MN}\delta_{m+n,0}, \quad \left[\alpha_m^M, \tilde{\alpha}_n^N\right] = 0. \tag{20}$$

Notice that this is the algebra that satisfies every rising and annihilator operator in the usual quantum harmonic oscillator, meaning that each quantum mode, now promoted to be an operator, creates or destroys quantum states. However, since now the modes are non-commuting operators for $m = 0$, we must order the product appearing on the Virasoro generator operators.

Therefore, it is possible to order the products $\alpha_{-n} \cdot \alpha_n$ (and similarly for left movers) up to the addition of a constant a. In this context, the mass term for the open string is written as

$$M^2 = \frac{1}{\alpha'}\left(\sum \alpha_{-n} \cdot \alpha_n - a\right), \tag{21}$$

while for the closed string, it reads

$$M^2 = \frac{2}{\alpha'}\left(\sum_{n=1}^{\infty} (\alpha_{-n} \cdot \alpha_n + \tilde{\alpha}_{-n} \cdot \tilde{\alpha}_n) - 2a\right), \tag{22}$$

where a is the zero point energy. Similarly as the quantum harmonic oscillator, the product of a lowering and a rising operators denotes another operator called the number operator N given by $N = \sum \alpha_{-n}\alpha_n$ for the left movers and $\tilde{N} = \sum \tilde{\alpha}_{-n}\tilde{\alpha}_n$ for the right ones. In this way, the normal ordering constant a cancels out for the difference $\left(L_0 - \tilde{L}_0\right) \mid \phi\rangle = 0$, which implies $N = \tilde{N}$. This is called the level-matching condition of the bosonic string.

The corresponding Hilbert space is constructed by acting with the modes α_n^i on a vacuum $\mid \Omega\rangle$ annihilated by the lowering operators α_{-n} and $\tilde{\alpha}_{-n}$. It is important to notice that, although not explicitly shown in this notes, the use of the light-cone coordinates forces some coordinates to be non-dynamical. For that reason, the index i in the operators modes runs over $D-2$ values.

Before constructing the quantum states, let us comment on the value of the ordering constant a, appearing in the mass expressions for the string. The value of such constant can be computed as follows. At first sight, it seems we are dwelling with a divergent quantity, since

$$a = \frac{1}{2}(D-2)\sum_{n=1}^{\infty} n. \tag{23}$$

However this sum has an analytic continuation and it is computed by using the ζ-function, which is defined as

$$\zeta(s) = \sum_{n=1}^{\infty} n^{-s}. \tag{24}$$

[6] In the classical picture, the Fourier modes α_n satisfy a Poincaré algebra.

Table 2. Closed string spectrum. For $D = 26$, the second state is masless.

State	Level	M^2/α
$\mid \Omega\rangle_L \otimes \mid \Omega\rangle_R$	$N = \tilde{N} = 0$	$-\frac{D-2}{6}$
$\alpha^i_{-1}\tilde{\alpha}^i_{-1} \mid \Omega\rangle_L \otimes \mid \Omega\rangle_R$	$N = \tilde{N} = 1$	$\frac{26-D}{12}$

This complex function has as unique analytic continuation at $s = -1$, where it takes the value $\zeta(-1) = -1/12$. Therefore by comparison we obtain that,

$$a = -\frac{D-2}{24}. \tag{25}$$

Quantum states are constructed by acting the rising operator modes and by restricting to the level match conditions. This means that for every left mover operator, there must be the corresponding right one. Therefore, there are an infinite tower of string states, generically given by the application of rising operators on the closed vacuum $\mid \Omega\rangle_L \otimes \mid \Omega\rangle_R$ or on the open vacuum $\mid \Omega\rangle$. As the reader might expect, one way to construct gauge bosons, is by considering open strings.

2.4. The Bosonic String Spectrum

Let us firstly, consider the closed string spectrum shown in Table 2. The first state corresponds to vanishing number operators, implying that such state is a tachyon, i.e., a quantum state with negative squared mass. Observe that this not depend on the value of D. The presence of tachyons, as it is now understood, reflects the fact that we are constructing a perturbative theory on a false vacuum. This means that there are some variable which must run down to a stable point on which the corresponding perturbative theory would not have tachyons in its spectrum. Closed string tachyons have been considered in literature, mainly in cosmological models.

The next state is given by a left and right number operators $N = \tilde{N} = 1$. Its mass is then $(26-D)/12$. It is a state with $(D-2)^2$ degrees of freedom. However, since the theory is Lorentz invariant, we expect that physical states belong to representations of the little group $SO(D-2)$. Recall that the little group is the subgroup of the Lorentz group which leaves invariant the D-dimensional momentum of a particle. Therefore, only a massless state would have the necessary degrees of freedom of a little group representation (a massive one would transform with respect to the little group $SO(D-1)$). This fixes completely the dimension number D to 26. Some comments are in order: first, the above argument, although not formal, yields the same result as the obtained by a covariant quantization[7]. Second, by requiring the theory to be Lorentz invariant at the perturbative level, we have fixed the number of dimensions. These theories are called *critical*[8].

The critical dimension 26 can be as well obtained from the open string spectrum, shown in Table 3. The first excited state has $(D-2)$ degrees of freedom and must transform as a vector

[7] There, the dimension is fixed by canceling the Weyl anomaly. If this anomaly is present, there would be inconsistencies as the presence of non-unitary states and unphysical degrees of freedom.

[8] There are models in which the Lorentz invariance is preserved and Weyl anomalies are cancelled non-perturbatively with $D \neq 26$. They are called non-critical string theories.

Table 3. Open string spectrum for $D = 26$

State	Level	M^2/α
$\mid\Omega\rangle$	$N = 0$	-1
$\alpha^i_{-1}\mid\Omega\rangle$	$N = 1$	0

under the little group $SO(D-2)$. The number of degrees of freedom match only for a masless state, i.e., for $D = 26$.

Therefore, in the critical bosonic theory, the massless first excited state in the closed string, transforms as a two-index object under $SO(24)$. It contains 24^2 degrees of freedom, which can be decomposed on irreducible representations of $SO(24)$. Actually

$$\mathbf{24^2} = \mathbf{1} \oplus \mathbf{276} \oplus \mathbf{299}, \tag{26}$$

corresponding to (all of them being massless) a scalar field ϕ called *the dilaton*, an antisymmetric two-index field B_{MN} and a symmetric two-index field G_{MN}. This last field is the cause that so many people around the world are interested in string theory and that they have been working on it for the last 30 years. G_{MN} has all the properties we expect from a quantum particle of gravity, *the graviton*!

3. The Superstring

Bosonic string theory opened up a new and huge area of research. Quantum gravity, an old dream, was now a tangible subject. However, it failed by the first and the simplest feature we must have in any physical theory: the presence of fermions. Bosonic string theory has only bosonic fields as quantum vibrations, and that is the main reason that renders it to be unphysical.

The simplest way to incorporate fermions into the theory considers the presence of an extra symmetry at the level of the Polyakov action. A symmetry that relates a fermionic field for each bosonic field present in the two-dimensional string world-sheet. This is called supersymmetry. According with the Ramond-Neveu-Schwarz (R-NS) formalism[9], the bosonic fields $X^M(\sigma,\tau)$ are paired with fermionic partners $\psi^M(\sigma,\tau)$. In this case, the action of the bosonic string is modified by including the standard Dirac action. Therefore, the supersymmetric string action is

$$S_{R-NS} = -\frac{1}{2\pi}\int d^2\sigma\, (\partial_\alpha X_M \partial^\alpha X^M + \overline{\psi}^M \rho^\alpha \partial_\alpha \psi_M) \tag{27}$$

where α is a world-sheet index and ρ^α are 2×2 matrices which obey the Dirac algebra. We shall use the representation given by

$$\rho^0 \doteq \begin{bmatrix} 0 & -1 \\ 1 & 0 \end{bmatrix} \quad \rho^1 \doteq \begin{bmatrix} 0 & 1 \\ 1 & 0 \end{bmatrix}, \tag{28}$$

The two-dimensional spinor ψ^M has two Weyl components according to

$$\psi^M = \begin{pmatrix} \psi^M_- \\ \psi^M_+ \end{pmatrix}. \tag{29}$$

[9] There exists other formalisms as the Green-Schwarz which consider supersymmetry on the space-time, and recently and developed by Berkovits, the pure spinor formalism.

In the same way as the bosonic action, using the ligth-cone coordinates the fermionic part of the string action is rewriten as

$$S_f = \frac{i}{\pi} \int d^2\sigma \; (\psi_-^M \partial_+ \psi_{M-} + \psi_+^M \partial_- \psi_{M-}). \tag{30}$$

The whole action is invariant under supersymmetric transformations of the form

$$\begin{aligned} \delta X^M &= i\left(\epsilon_+ \psi_-^M - \epsilon_- \psi^M\right), \\ \delta\psi_-^M &= -2\partial_- X^M \epsilon_+ \\ \delta\psi_+^M &= 2\partial_+ X^M \epsilon_- \end{aligned} \tag{31}$$

where $\epsilon_\pm$ are infinitesimal supersymmetric parameters corresponding to constant two-dimensional Majorana spinors.

3.1. Equations of Motion and Boundary Conditions

The equations of motion with respect to X^i and ψ^i (notice that there is again just $(D-2)$ degrees of freedom) are given respectively by the wave equation in (10) and the Dirac equations

$$\partial_\pm \psi_\mp^M = 0. \tag{32}$$

The corresponding boundary terms for the fermionic part is given by[10]

$$\delta S = -T \int d\tau \left(\psi_+^M \delta\psi_{M+} - \psi_-^M \delta\psi_{M-}\right)|_{\sigma=\pi} - \left(\psi_+^M \delta\psi_{M+} - \psi_-^M \delta\psi_{M-}\right)|_{\sigma=0}, \tag{33}$$

which vanishes in several different ways. Each way establishes the presence of different uncorrelated sectors of the theory as we shall see shortly.

3.1.1. Boundary Conditions for the Open String Let us consider an open string, i.e., a string which endpoints are not coincident, and let us analyze the boundary conditions on the fermionic component of the action. Being an open string, both boundary terms in the variation of the action, must vanish separately. This is achieved if

$$\psi_+^M = \pm\psi_-^M, \tag{34}$$

wich just reflects the very well known fact that fermions need a double rotation to recover their initial sign[11]. Let us say that for two of the endpoints at $\sigma = 0$, both Weyl spinors are coincident., i.e., that $\psi_+^M\,|_{\sigma=0} = \psi_-^M\,|_{\sigma=0}$. For the other endpoint there are two possible cases.

Periodic boundary conditions, $\psi_+^M\,|_{\sigma=\pi} = \psi_-^M\,|_{\sigma=\pi}$, are commonly called *Ramond (R) boundary conditions.* A solution satisfying this condition is said to belong to the R-sector, and it reads

$$\begin{aligned} \psi_-^M(\sigma,\tau) &= \frac{1}{\sqrt{2}} \sum_{n\in\mathbf{Z}} d_n^M e^{-in\sigma^-}, \\ \psi_+^M(\sigma,\tau) &= \frac{1}{\sqrt{2}} \sum_{n\in\mathbf{Z}} d_n^M e^{-in\sigma^+}. \end{aligned} \tag{35}$$

[10] Boundary conditions for the bosonic part are the same as in the bosonic string. This means that *Dp*-branes are also present in the superstring theory.

[11] Formally speaking, two-dimensional Weyl fermions are representations of the Lie group $SU(2)$, a double covering of the rotational group $SO(3)$.

Journal of Physics: Conference Series **287** (2011) 012003 doi:10.1088/1742-6596/287/1/012003

The Majorana conditions requires these expansions to be real and hence $d^M_{-n} = d^{M\dagger}_n$. Notice that the index runs over integer numbers.

Anti-periodic boundary conditions, $\psi^M_+ \mid_{\sigma=\pi} = -\psi^M_- \mid_{\sigma=\pi}$, are referred as *Neveu-Schwarz (NS)* boundary conditions. The corresponding solutions belong to the NS-sector, and are of the form

$$\begin{aligned} \psi^M_-(\sigma,\tau) &= \frac{1}{2}\sum_{r\in\mathbf{Z}+\frac{1}{2}} d^M_r e^{-ir\sigma^-}, \\ \psi^M_+(\sigma,\tau) &= \frac{1}{2}\sum_{r\in\mathbf{Z}+\frac{1}{2}} d^M_r e^{-ir\sigma^+}. \end{aligned} \tag{36}$$

Therefore, the supersymmetric open string solutions are separated into two different sectors. We shall see that space-time bosons are constructed in the NS-sector, while space-time fermions are constructed in the R-sector.

3.1.2. Boundary Conditions for the Closed String. As in the bosonic string, the closed superstring solution can be represented in terms of left and right movers. On each of them is possible to impose periodicity or anti-periodicity boundary conditions, $\psi_\pm(\sigma) = \pm\, \psi_\pm(\sigma+\pi)$. This establishes four ways to determine a solution of the equations of motion. For instance, if both left and right movers satisfy NS boundary conditions, we say that the corresponding solution belongs to the Neveu-Schwartz-Neveu-Schwartz (NS-NS) sector. Hence, the four sectors are NS-NS, NS-R, R-NS and R-R. We shall see that bosonic fields arise from the NS-NS and R-R sector, while fermions belong to the R-NS and NS-R sectors. The solutions are denoted as

$$\psi^M_-(\sigma,\tau) = \frac{1}{2}\sum_{n\in\mathbf{Z}} d^M_n e^{-2in\sigma^-}, \quad \psi^M_-(\sigma,\tau) = \frac{1}{2}\sum_{r\in\mathbf{Z}+\frac{1}{2}} b^M_r e^{-2ir\sigma^-}, \tag{37}$$

for the right movers with R and NS boundary conditions respectively, and

$$\psi^M_+(\sigma,\tau) = \frac{1}{2}\sum_{n\in\mathbf{Z}} \tilde{d}^M_n e^{-2in\sigma^-}, \quad \psi^M_+(\sigma,\tau) = \frac{1}{2}\sum_{r\in\mathbf{Z}+\frac{1}{2}} \tilde{b}^M_r e^{-2ir\sigma^-}, \tag{38}$$

for the left movers. The specific fields arising as quantum modes of these solutions, will depend on which type of spinor we consider in the the supersymmetric action. There are two possibilities: chiral or non-chiral. So, there are at least two different superstring theories consisting on closed strings.

3.2. Canonical Quantization of the Supersymmetric String

As in the bosonic string case, we shall take a short-way and we will proceed to quantize the superstring by using the canonical quantization procedure, which means that Fourier modes are promoted to be operators satisfying some algebra. For the bosonic modes, the commutation relations are the same as in the bosonic string. For the fermionic modes we have that

$$\{b^M_r, b^N_s\} = \eta^{MN}\delta_{r+s,0} \quad \{d^M_m, d^N_n\} = \eta^{MN}\delta_{m+n,0}, \tag{39}$$

and similar expressions for the left movers. The reader should observe that such relations are precisely the algebra of raising and lowering operators in a harmonic oscillator satisfying the fermi statistics. Therefore, we can assure that such modes act as creator and annihilator operators. By acting them on a suitable defined background state, we can describe the quantized string vibrations. Let us then define the background state $|\,\Omega\rangle_R$ for the right movers. For $n \geq 0$, $|\,\Omega\rangle_R$ is annihilated by the operators α_n^M and d_n^M in the R sector, while for the NS-sector, the annihilation comes from operator b_r^M, for $r > 0$. Similarly, for the left movers, we have that for the R sector, $\tilde{\alpha}_n^M \mid 0\rangle_L = \tilde{d}_n^M \mid 0\rangle_L = 0$, while for the NS-sector, $\tilde{b}_r^M \mid \Omega\rangle_L = 0$, for positive indices.

Therefore, excited states of the closed string are constructed by acting negative modes operators on the ground state $|\,\Omega\rangle_R \otimes |\,\Omega\rangle_L$. It turns out that the ground state for left or right movers is degenerated in the R-sector and unique in the NS-sector. In order to construct the respective excited states for the closed string, let us start by analyzing the construction on some left or right movers, which is equivalent to study the spectrum of the open string.

In the NS sector, the ground state is unique, since all rising operators increment its mass. This follows, as in the bosonic case, from the energy momentum tensor. Classically, the energy-momentum tensor nonzero components read

$$\begin{aligned} T_{++} &= \partial_+ X_M \partial_+ X^M + \frac{1}{2}\psi_+^M \partial_+ \psi_{+M}, \\ T_{--} &= \partial_- X_M \partial_- X^M + \frac{1}{2}\psi_-^M \partial_- \psi_{-M}, \end{aligned} \tag{40}$$

which vanishes by the super Weyl symmetry. Hence, by substituting the solutions of the equations of motion, these energy-momentum tensor components can be written in terms of super Virasoro generators L_m and G_r satisfying some algebra. The generators L_m, with m an integer, are given by[12]

$$L_m = \frac{1}{2}\sum_m \alpha_{m-n}\cdot\alpha_m + \frac{1}{4}\sum_r (2r-m)\psi_{m-r}\cdot\psi_r + a\delta_{m,0}, \tag{41}$$

where the product of modes operators are already ordered. The constant a is, as in the bosonic case, a measure of the zero point energy. For the R-sector, fermionic fields contribute with a same quantity than bosons, but with an opposite sign, i.e., $a = a_b + a_f = -\frac{1}{12}(D-2)(1-1) = 0$.

For the NS-sector, the situation is different. Being the modes labeled by fractional indices, the zero point energy contribution reads $a_f = -\frac{1}{48}(D-2)$, from which the total contribution to the zero point energy is $-1/2$.

Therefore, the mass is computed from the zero mode generator L_0, which in the NS-sector, is given by

$$\alpha' M^2 = \sum_{n\in\mathbf{Z}} \alpha_{-n}\cdot\alpha_n + \sum_{r\in\mathbf{Z}+1/2} r b_{-r}\cdot b_r - \frac{1}{2}, \tag{42}$$

Notice that a generic excited state is constructed by arbitrary values of N_B and N_F and that since all the oscillators transform as space-time vectors then all excited sates are bosons. In the R sector the ground state is degenerated since the operators d_0^M can act without changing the mass, which is given by

$$\alpha' M^2 = \sum_{n\in\mathbf{Z}} \alpha_{-n}\cdot\alpha_n + \sum_{n\in\mathbf{Z}} n d_{-n}\cdot d_n. \tag{43}$$

[12] We encourage the interested reader to work out the details behind these calculations following standard text books as the ones shown in the reference list.

This follows from the fact that zero modes d_0 satisfy the algebra $\{d_0^M, d_0^N\} = \eta^{MN}$ yielding the non-existence of a unique solution for $d_0^M \mid \Omega\rangle = 0$. Therefore, the set of ground states in the R sector must transform as a representation of Dirac algebra,

$$d_0^M \mid a\rangle = \frac{1}{\sqrt{2}}\Gamma_{ab}^M \mid b\rangle, \tag{44}$$

where Γ is a Dirac matrix. This implies that the R-sector ground state is actually a space-time fermion.

3.3. *The Superstring Spectrum*

For the NS sector we have seen that there is a negative-quadratic-mass state which is identified as the tachyon. Consistency requires the tachyon to be projected out. This can be achieved if we can project some states out, and the remaining states are self-contained, i.e., that by the action of any operator, the resulting state is contained among that set of states. It turns out that such projection exists and it is called the *the GSO projection* (Gliozzi, Scherk and Olive). Basically the action of the GSO projector on the NS and R sector is to take out a half of the states arising at different energy values of the string.

Being more explicit, the action of the GSO projector on NS states is given by $(-1)^F$, with F being the world-sheet fermion number. Negative states under the GSO projector are removed. Then, the tachyonic state is projected out (the background state has fermion number $F = 1$) and the first state which survives GSO projection is $\psi_{1/2}^M|\Omega\rangle$, with $M = 1, \cdots, 8$. The coordinates $0, 9$ represent non-physical degrees of freedom[13]. It is possible to show that this state is a massless gauge boson transforming in the adjoint of $U(1)$ gauge group, and it is expressed as the vectorial representation $\mathbf{8_V}$ of the little Lorentz group $SO(8)$.

For the R sector, the energy of ground state is zero, implying there is a degeneration in the energy levels on states given by the zero mode ψ_0^M. In the light-cone gauge (or after considering the effect of ghosts) we have eight zero modes which can be used to construct 4 raising and 4 lowering operators by a linear combination of two zero modes. Applying the raising operator to a bosonic vacuum $|\Omega\rangle$ we generate $2^4 = 16$ states.

Now, on the one hand, there are at least 16 supersymmetry generators, corresponding to the Weyl components of the full spinor, which has 32 degrees of freedom. This spinor is a Majorana-Weyl spinor, which splits it into two real spinors with opposite chirality, i.e., $\mathbf{32} = \mathbf{16} \oplus \mathbf{16'}$. Then, under spinorial representations of the Lorentz group $SO(1,9)$, we have the following decomposition

$$\begin{aligned} SO(1,9) &\longrightarrow SO(1,1) \times SO(8) \\ \mathbf{16} &\longrightarrow (\frac{\mathbf{1}}{\mathbf{2}}, \mathbf{8_s}) \oplus (-\frac{\mathbf{1}}{\mathbf{2}}, \mathbf{8_c})\,. \end{aligned} \tag{45}$$

On the other hand, there is one constraint (Dirac equation) that threw out representations with quantum number $-\frac{1}{2}$ of $SO(1,1)$. That is why we kept just 16 states that correspond to those constructed by raising operators described above.

[13] Remember we are using the light-cone gauge. In the context of covariant quantization, this is an effect of ghosts on the string dynamics.

Table 4. Open superstring spectrum

Sector	State	Level	M^2/α	$SO(8)$ rep.	Field	GSO
NS	$\mid\Omega\rangle_{NS}$	$N_b = N_f = 0$	$-1/2$	-	Tachyon	$-$
NS	$\psi^M_{-1}\mid\Omega\rangle_{NS}$	$N_b = 0,\ N_f = 1$	0	$\mathbf{8_v}$	Gauge boson	$+$
R	$\mid\Omega\rangle_R$	$N_b = N_f = 0$	0	$\mathbf{8_c}$	Gauginos	$-$
R	$\mid\Omega\rangle_R$	$N_b = N_f = 0$	0	$\mathbf{8_s}$	Gauginos	$+$

Table 5. Type IIA superstring spectrum

Sector	State	$SO(8)$ rep.	Fields
NS-NS	$\psi^M_{-1}\tilde{\psi}^N_{-1}\mid\Omega\rangle_L\otimes\mid\Omega\rangle_R$	$\mathbf{8_v}\otimes\mathbf{8_v} = \mathbf{1}\oplus\mathbf{26}\oplus\mathbf{35}$	G_{MN}, B_{MN}, ϕ
R-R	$\mid+\rangle_L\otimes\mid-\rangle_R$	$\mathbf{8_s}\otimes\mathbf{8_c} = \mathbf{8_v}\oplus\mathbf{56}_t$	$C_{(1)}, C_{(3)}$
NS-R	$\psi^M_{-1}\mid\Omega\rangle_L\otimes\mid-\rangle$	$\mathbf{8_v}\otimes\mathbf{8_c} = \mathbf{8_s}\oplus\mathbf{56_c}$	$\lambda^\alpha,\ \psi^\alpha_M$
R-NS	$\mid+\rangle_L\otimes\tilde{\psi}^M_{-1}\mid\Omega\rangle_R$	$\mathbf{8_s}\otimes\mathbf{8_v} = \mathbf{8_c}\oplus\mathbf{56_s}$	$\lambda^{\dot{\alpha}},\ \psi^{\dot{\alpha}}_M$

Table 6. Type IIB superstring spectrum

Sector	State	$SO(8)$ rep.	Fields
NS-NS	$\psi^M_{-1}\tilde{\psi}^N_{-1}\mid\Omega\rangle_L\otimes\mid\Omega\rangle_R$	$\mathbf{8_v}\otimes\mathbf{8_v} = \mathbf{1}\oplus\mathbf{26}\oplus\mathbf{35}$	G_{MN}, B_{MN}, ϕ
R-R	$\mid+\rangle_L\otimes\mid+\rangle_R$	$\mathbf{8_s}\otimes\mathbf{8_s} = \mathbf{1}\oplus\mathbf{26}\oplus\mathbf{35}$	$C_{(0)}, C_{(2)}$
NS-R	$\psi^M_{-1}\mid\Omega\rangle_L\otimes\mid+\rangle$	$\mathbf{8_v}\otimes\mathbf{8_s} = \mathbf{8_s}\oplus\mathbf{56_s}$	$\lambda^\alpha,\ \psi^\alpha_M$
R-NS	$\mid+\rangle_L\otimes\tilde{\psi}^M_{-1}\mid\Omega\rangle_R$	same as NS-R	

Therefore, GSO projection in RR sector is basically the application of the chiral matrix $\overline{\Gamma} = i\Gamma^0\Gamma^1\cdots\Gamma^9$ to the spinor states. Because the states in the **16** of $SO(8)$ are spinors, the action of $\overline{\Gamma}$ split up them into two Weyl fermions, which are representations of $SO(8)$ given by $\mathbf{8_s}$ and $\mathbf{8_c}$ with different chirality. After GSO projection, the R-sector contains 16 spinorial states in the $\mathbf{8_s}$ of $SO(8)$ that represents the super-partners of $\mathbf{8_v}$ in the NS sector (gauginos). The spectrum of the open string is shown in Table 4.

3.3.1. Closed superstring spectrum Once the spectrum of the open string has been computed, it is straightforward to compute the closed string spectrum. The resulting fields are constructed by the tensorial product of the corresponding states for the left and right movers. However, there are two different ways to do it.

The theory is called *type IIA* if left and right-moving R-sector ground states are chosen to have the opposite chirality (which reflects in the fact that the two R-sectors have opposite parity. In the *type IIB* left and right-moving sector ground states have the same chirality. These two consistent ways to construct a superstring theory spectrum, are summarized in Tables 5 and 6.

Journal of Physics: Conference Series **287** (2011) 012003 doi:10.1088/1742-6596/287/1/012003

3.3.2. The Type IIA Superstring spectrum As mentioned, the representation of the complete states is obtained by tensoring the representation of the left and right movers. The NS-NS sector contains a symmetric tensor of two index corresponding to a graviton (G_{MN}), an antisymmetric tensor or 2-form (B_{MN}) and a dilaton (Φ). The R-R sector contains a set of completely anti-symmetric tensors (p-forms). In particular a 1-form(C_M) and a 3-form (C_3). It is sometimes convenient to introduce the Hodge duals of these, which are a 5-form (C_5) and a 7-form (C_7). Finally, it is also useful to introduce a 9-form C_9, with no dynamics. The space-time fermions arise from the R-NS and NS-R sectors which contain the gravitino (56 states) and a dilatino (8 states) with opposite chirality. The R-NS and NS-R sectors contain fermions, which are a spin 3/2 gravitino and a spin 1/2 dilatino, each of these fermions with opposite chirality.

3.3.3. The Type IIB Superstring spectrum For this case, we choose left and right movers in R-sector to have the same chirality (in the $\mathbf{8_s}$ and $\mathbf{8_s}$ of $SO(8)$). The NS-NS sector contains the same fields as in the Type IIA case. The R-NS and NS-R sectors contain fermions with the same chirality. Bassically they are two copies of fermions in the NS-R sector in type IIA. The RR sector contains a 0-form C_0 (scalar), a 2-form C_2 and a 4-form $C_4^\dagger$ with a self-dual field strength. It is sometimes convenient to introduce the Hodge duals of these, which are a 6-form B_6, and 8-form C_8. Finally, it is also useful to introduce a 10-form C_{10}, which does not have any propagating degrees of freedom, since it has no space-time kinetic term.

The Type IIA and IIB superstring theories have as low energy description (when just massless modes are considered) the Types IIA and IIB supergravity (SUGRA) versions. In other words, II refers to supersymmetry $\mathcal{N} = 2$ and we have two gravitinos in each supergravity multiplet with $D = 10$ [14].

4. Non-perturbative aspects of Superstring Theory

Until now we have described the perturbative sector of type II superstring theories[15] in a flat space-time. However, it is interesting and important to compactify some of the spatial extra dimensions that the theory predicts. The first step is wrapping one such dimension on a circle. Here we find a big difference by comparing strings to quantum fields. While for fields on a compact dimension we get only Kaluza-Klein (KK) modes arising by periodicity, strings can also be wrapped on compact coordinates. In this form, for the string there are two values the energy depends on. One is the energy required to stretch the string in a space with less volume because of the compactification. Essentially these are the KK modes; smaller the radius R of the compact dimension, bigger the energy required to stretch the string. KK modes goes as $\frac{n}{R}$ with $n \in \mathbf{Z}$. When the string is wrapped on the compact dimension, then we need energy to wrap it, i.e, the string has a strength tension in order to be wrapped around the circle. Then, smaller the radius of the compact dimension, smaller the energy needed to wrap it. The number of times a string winds a compact dimension is called the winding number, and the energy is

[14] It is surprising that at SUGRA level, space-time anomalies due to gauge and gravitational fields, coupled to chiral fermions, and cancel each other exactly. This is reflected in string theory (or viceversa) as the absence of tadpole diagramas at the fermionic sector. However, if we couple IIB theory (strings or SUGRA) to super Yang-Mills (SYM) gauge bosons, there is an extra term in the anomaly, which makes the theory inconsistent. On the string theory side, this means we are adding a D9-brane to the background which introduces an open string sector and breaks supersymmetry. This theory is not consistent and one can see that by including open strings and SYM gauge theories, more details must be considered. In Type IIA theory such anomalies are not present because the theory is not chiral.

[15] There are other three types, we shall briefly mention at the end of this note.

proportional to mR with $m \in \mathbf{Z}$ being the winding number.

The mass of a string with KK modes and winding number is given by

$$M^2 = \frac{n^2}{R^2} + \frac{m^2R^2}{\alpha'^2} + \frac{2}{\alpha'}(N_B + \tilde{N_B} r N_F + r\tilde{N_F} - E_0), \tag{46}$$

with $N_B, \tilde{N_B}$ being the bosonic energy number level for the right and left movers respectively, $N_F, \tilde{N_F}$ the fermionic ones and E_0 the ground energy level for each sector.

Now, consider the symmetric two-order tensor G_{MN} from the NS-NS sector. Upon compactification on a circle, one obtains a nine-dimensional two-tensor G_{ij}, a gauge field $A_i^L = G_{i9}$ and a scalar $\phi = G_{99}$. In the same way, the two-form B_{MN} gives rise to a nine-dimensional two-form B_{ij} and a gauge field $A_i^R = B_{i9}$. Stringy states carrying KK modes or winding numbers are charged under such gauge fields. So, by compactification there is an extra gauge symmetry $U(1) \times U(1)$ under which massive states are charged. In other words, there are states that are charged under NS-NS fields. However, at perturbative level, there is not evidence that there exist massive objects charged under RR fields. This is in some sense paradoxical because compactification also affects RR fields. In fact, the p-forms are reduced to $(p-1)$- and p-forms on a nine-dimensional space-time, so we expect that $(p-1)$-forms must be the fields under which some objects are charged. For the NS-R and R-NS sector, under compactification into a circle, we obtain a nine-dimensional theory in which fermionic fields have no chirality (there is no such chiral Dirac matrix $\overline{\Gamma}$). Gravitinos and gauginos in Type IIA theory are mapped according to

$$\begin{array}{lccc} \text{R-NS:} & \psi_\mu^\alpha, \lambda^\alpha & \longrightarrow & \lambda^{\alpha'}, \lambda^\beta, \psi_i^{\alpha'} \\ & \mathbf{8_s} \otimes \mathbf{8_v} = \mathbf{8_c} \oplus \mathbf{56_s} & \longrightarrow & \mathbf{8} + \mathbf{8} + \mathbf{48} \\ \text{NS-R:} & \psi_\mu^{\dot\alpha}, \lambda^{\dot\alpha} & \longrightarrow & \lambda^{\alpha'}, \lambda^\beta, \psi_i^{\alpha'} \\ & \mathbf{8_c} \otimes \mathbf{8_v} = \mathbf{8_s} \oplus \mathbf{56_s} & \longrightarrow & \mathbf{8} + \mathbf{8} + \mathbf{48} \end{array} \tag{47}$$

where $i = 0, \cdots, 8$. Notice the two extra gravitinos $\mathbf{8}$ are the superpartners of the gauge fields in the NS-NS sector, A_i^R and A_i^L and of the $(p-1)$-forms in the RR-sector. For Type IIB we have the same fields in the nine-dimensional spacetime.

4.1. T-duality and D-branes

Consider the mass expression given by Eq. (46). We see that mass of states, i.e. their energy, is unchanged under the identification

$$\begin{array}{ccc} m & \longleftrightarrow & n \\ \frac{\alpha'}{R} & \longleftrightarrow & R\,. \end{array} \tag{48}$$

Then it follows that under such change of parameters, the theory has the same spectrum and it is indistinguishable of the original one. We said that two such theories are dual one to each other and the transformation is called *T-duality.*

This identification also has an effect on the bosonic and fermionic modes. T-duality maps X_L^μ into $-X_L^\mu$ and by superconformal symmetry, the fermionic modes are changed in a similar way, $\psi_L^\mu \rightarrow -\psi_L^\mu$. The latter map has a profound and very important consequence in the theory. Mapping the left fermionic mode into its negative one, implies a change in the chirality of the spinors constructed with them. Let us explain it with more detail. A basis of states in the RR sector is given by $|s_1s_2s_3s_4\rangle$ with $s_i = \pm\frac{1}{2}$ for all i values. They give us the 16 states we studied previously. We set the ground state as the element given by $|\frac{1}{2}\frac{1}{2}\frac{1}{2}\frac{1}{2}\rangle$ and then the rest of the basis is generated by the raising operator $S_a^\dagger = \frac{1}{\sqrt{2}}(\psi_0^a + i\psi^{a+1})$ which raises the value of s_i by a half for each i.

When ψ^a is mapped to its negative mode by performing T-duality on the $a+1$-coordinate, $S_a^\dagger$ transforms into S_a, which corresponds to a lowering operator. For instance, take $a = 9$; this means that the state $|s_1s_2s_3 - \frac{1}{2}\rangle$ will be projected out by the lowering operator. On the other hand, if initially we have the state $|s_2s_2s_3\frac{1}{2}\rangle$, then under T-duality this state is replaced by $|s_1s_2s_3 - \frac{1}{2}\rangle$, i.e, under T-duality, the number of negative or positive one-half factors are interchanged. But this is the definition of the action of GSO projection on RR states. Then T-duality changes the chirality. The most important consequence of this is that Type IIA and Type IIB theories are T-dual.

Under toroidal compactifications (taking many wrapped dimensions on circles) it is possible to elucidate the existence of certain planes in which the endpoints of open strings are fixed. Although open strings has not been considered, we shall see it is possible to include them under some constraints. Take for instance the ninth-coordinate wrapped into a circle, and consider a closed and an open string in the bulk. It is important to point out an open string is not able to wind a circle because always can be unwinded; this is a big difference between closed and open strings. The latter has no winding number.

Now let the radius R tends to zero. What does happen to the strings? At the limit, the open string looks like being fixed by its endpoint, in a eight-dimensional plane. This is the first evidence of such hyperplanes. It turns out that the y are actually the D-branes we have studied previously. For the closed string the story is more surprising: at the limit $R \to 0$, the closed string is flatten, identifying the points under $\sigma \to 2\pi - \sigma$. This gives two fixed points which are identified to the endpoints of an unoriented open string. The next step is asking, what are the T-dual version for these cases? The last one gives rise to the construction of another consistent string theory called *Type I*, which contains closed an open unortiented strings.

Let us perform T-duality for an open string on a compact dimension. Under T-duality we know that $X_L^\mu \to -X_L^\mu$ and then we can define a T-dual bosonic field $\overline{X}^\mu = X_R^\mu - X_L^\mu$. Working on the dual coordinates the Neumann boundary condition, which an open string satisfies, is transformed into a Dirichlet condition $\partial_\tau \overline{X}^\mu = 0$. In consequence the endpoints are fixed on a D-brane. But, is there any constraint to include open strings to Type II theories? To include oriented open strings we require that D9-branes do not exist because their existence implies also the existence of SYM degrees of freedom in the ten-dimensional space-time, and as we know, the coupling between Type II SUGRA and SYM is anomalous, meaning that R tadpoles are not cancelled.

However, D-branes are more than just hyperplanes. Actually, they are also BPS states (Bogomol'nyi-Prasad-Sommerfeld), i.e., states that break down a half of supersymmetry on the bulk, and their mass were determined by their charge under some fields. Because of that, supersymmetry protects them from radiative corrections. Their mass and charge are exact. The relation between mass and charge is given by the supersymmetry algebra in ten dimensions with $\mathcal{N} = 2$. In the same way, it was found that some solitonic objects described in the context of supergravity (generalizations of black-holes) known as p-branes, are the low energy limit of Dp-branes. In this form, Dp-branes are the solitonic objects (with tension $T \sim \frac{1}{g_s}$) that allow us to study the non-perturbative sector in string theory; we are able to study their excited states by the perturbative description in terms of open strings. This is a crucial point in our knowledge of the theory.

As we said, D-branes must be charged by some fields because of the BPS condition $M = |Q|$. These fields are precisely the RR fields. Then the objects we require to be charged under RR fields are Dp-branes. In general, a $(p+1)$-form C_{p+1} couples to a Dp-brane by

$$\mu_p \int_{W_{p+1}} C_{p+1} \,, \tag{49}$$

where μ_p is the charge of the Dp-brane and W_{p+1} is its worldvolume. Then, by the RR spectrum for Type IIA and IIB theories, we can deduce which values of p are allowed in each theory. For Type IIA we have Dp-branes for p even, and for Type IIB, p must be odd.

The picture we should have in mind is that open strings has their endpoints attached to Dp-branes and supersymmetry is broken to one half of the bulk. This means that for Type II theories we have supersymmetry $\mathcal{N} = 1$ on the brane and $\mathcal{N} = 2$ in the bulk. Far away from the brane, we have locally a closed string theory. This can be regarded also as follows: take an open string on the Dp-brane, and consider the process in which the two endpoints are coming forward to the same point on the D-brane worldvolume. When two endpoints are glued into a single point, we have essentially a closed string (the charge at the endpoints cancel each other) and it is said that a Dp-brane emits a closed string into the bulk.

Emission of closed string has a relevant role in D-brane physics. Consider two Dp-branes parallel to each other and interchanging a closed string and take into account the tree level amplitude of such a process. There is a well studied duality between the 1-loop open string amplitude and the tree level closed one. We can consider an open string connecting the two Dp-branes and forming a cylinder (1-loop) or a closed string interchanged by the branes at tree-level amplitude. At the open-sector we see that the amplitude is zero because the theory is supersymmetric, then there is not net force between the branes. But if one return to the closed-sector is easy to see there must be a force due to NS-NS and RR fields. It is concluded that NS-NS contribution cancels the RR one. In other words, a Dp-brane has tension (mass) and then could emit gravitons. Also it is able to emit B-fields and dilatons, and C_{p+1}-forms. All of them contribute to a non-zero amplitude but all at once give a zero net force.

The bosonic part of the action which describes a Dp-brane classically is,

$$S = \int d^p x \, e^{-\phi} \sqrt{det(G_{\mu\nu} + \frac{1}{2\pi\alpha'}(B_{\mu\nu} + F_{\mu\nu}))} + \int G_{p+2} \wedge *G_{p+2} + \mu_p \int C_{p+1} \,,$$

where G_{p+2} is the strength field given by $G_{p+2} = dC_{p+1}$, $G_{\mu\nu}$ and $B_{\mu\nu}$ are the induced metric and B-field on the D-brane and $F_{\mu\nu}$ is the magnetic flux on the Dp-brane (μ and ν run over the Dp-brane coordinates excepting the light-gauge cone ones). The magnetic flux arises as a background field on the worldvolume of the brane. It is the classical version of the strength two-form $F_2 = dA$ whit A being the gauge field obtained in the NS sector of the open string $\psi^{\mu}_{\frac{1}{2}}|0; k\rangle$.

The presence of open strings in Type II theories are due to the existence of D-branes. Under this context there is an enhancement of gauge symmetry given by the NS gauge bosons. Consider N Dp-branes at the same position. For each open string there are N possible states each endpoint has. It can be attached to the 'first' brane, or to the jth-brane with $j = 1, \cdots, N$. The same holds for the other endpoint. Then for stringy states we have to label the states by ij.

These extra non-dynamical degrees of freedom are called *Chan-Paton factors*. It can be seen that they introduce a $U(N)$ symmetry and NS gauge bosons turn out into NS non-abelian gauge bosons which transform in the adjoint representation of $U(N)$. The endpoints are charged by these 1-form gauge fields because there is a natural topological coupling between point-particles and 1-forms. By this, there is an enhancement of gauge symmetry $U(1) \to U(N)$. We say that there is a gauge field in the worldvolume of a Dp-brane and if this field is non-trivial, there is also a field strength given by the two-form F which we refer previously as a magnetic flux.

T-duality also plays an important role in the non-perturbative regime. Under it a Dp-brane turns into a D-brane with a higher or lower dimension depending in which coordinates T-duality is taken. For instance, if we take T-duality on transverse coordinates to the brane, its dimension will grow up according the number of coordinates we have considered. When T-duality is performed on longitudinal coordinates the brane dimension decreases. For example, if we have a Dp-brane on coordinates $0, 1, \cdots, p-1, p$ and we take T-duality transformation on one of these coordinates we obtain a D$(p-1)$-brane. This because the Neumann boundary conditions are transformed into Dirichlet ones.

Returning to our original picture (before taking T-dual transformation) the existence of a magnetic flux establishes an extra coupling on the classical action. Every $(p+1)$-form will we couple to the worldvolume W_p of a Dp-brane. Then if we have a two-form given by the strength gauge field $\mathcal{F}$, this together with a $(p-1)$-form establish a $(p+1)$-form that couples to W_p. This kind of coupling is known as Chern-Simons term and is given generally by

$$\int_{\Sigma} \Sigma_p C_{p+1} \wedge Tr\, e^{\mathcal{F}} \,, \tag{50}$$

Now, if this kind of couplings are allowed then the presence of RR charges related to low-dimensional D-branes is implied when a magnetic flux is turned on over the Dp-brane. Considering just the first non-trivial term in the expansion of $e^{\mathcal{F}}$ we induce a RR charge of a D$(p-2)$-brane. Then we hope to obtain a similar spectrum of open strings attached to D-branes as the obtained by analyzing D-branes with magnetic fluxes.

The spectrum of open strings attached to N coincident Dp-branes is as follows. For the NS-sector the state is $\psi^i_{1/2}|0;k\rangle$ (i labels longitudinal D-brane coordinates) which gives us N^2 $(p-1)$-dimensional gauge vectors of $SO(p-1)$ in the adjoint of $U(N)$. Also we have $(8-p)$ scalars $\psi^a_{1/2}|0;k\rangle$ of the Lorentz group $SO(p-1)$. In the R-sector, $\mathbf{8_s}$ flips out under $SO(8) \to SO(p-1) \times SO(8-p)$ into suitable representations of the Lorentz group $SO(p-1)$ in the adjoint of $U(N)$. The values of p for which Dp-branes generates a supersymmetric Yang-Mills theory are $p = 9, 5, 4, 2$. For instance, a D9-brane gives a SYM theory $D = 10$ and $\mathcal{N} = 1$. Also, for a D5-brane, we have a SYM theory $D = 6$ and $\mathcal{N} = 2$.

The RR fields of Type IIA theory are differential forms C_{p+1} with p even and for Type IIB p is odd. Denote by G_{p+2} the field strength given by $G_{p+2} = dC_{p+1}$. The field strength is invariant under gauge transformations $C_{p+1} \to C_{p+1} + d\Lambda_p$, that implies a gauge invariant action given by

$$S \sim \int_{\mathcal{M}_{10}} dC_{p+1} \wedge *dC_{p+1} + \int_{W_p} C_{p+1} \,. \tag{51}$$

The RR charge is given by

$$Q_E = \int_{S^{8-p}} *dC_{p+1} \,. \tag{52}$$

Also we can built magnetic dual objects to the Dp-brane, which corresponds to an object coupling to a $(7-p)$-form. This object is a D$(6-p)$-brane with magnetic charge given by

$$Q_M = \int_{S^{p+2}} dC_{p+1} \,. \tag{53}$$

By this, for instance a D5-brane in Type IIB is the magnetic dual to a D1-brane and so on. In the same way the magnetic dual to a string under NS-NS charge is a five-brane namely NS5-brane.

4.1.1. M-theory Together with type II superstring theories, there are three more supersymmetric string theories. The first one, we already have sai a little about it. It is called type I theory, and consists of a ten-dimensional background in which un-oriented open and closed strings are allowed to exist. Since it contains open string, it also have ten-dimensional gauge bosons transforming in the adjoint of $SO(32)$. Essentially this follows from the fact that there is a gravitational and gauge anomalies which cancel each other only for gauge groups with 496 generators. For type I theory, it turns out that we require the presence of the gauge group $SO(32)$. An alternative way to construct type I theory is by means of type IIB. Type I is constructed by adding 32 D9-branes to type IIB and by projecting out all oriented strings. This is the action of another non-perturbative object called *Orientifold.*

Another way to construct consistent supersymmetric theories involves the mixture between bosonic and fermionic strings. Take a closed string and let us say that left-movers are bosonic, and impose supersymmetry on the right movers. "Half of the string" propagates on a ten-dimensional space-time, while the other, being bosonic, propagates on a space-time with 26 dimensions. The apparent contradiction is solved when one realizes that the extra 16 coordinates on the bosonic side behave as gauge degrees of freedom. Once again, the fact that we have chiral gauged fermions, implies the presence of gauge anomalies, which cancel for gauge groups $SO(32)$ and $E_8 \times E_8$. These are precisely the extra pair of superstring theories we have not mentioned. They are called Heterotic $SO(32)$ (Het($SO(32)$)) and Heterotic $E_8 \times E_8$ (Het($E_8 \times E_8$)).

Five different and consistent string theories seem to much for an unified description of nature. This issue was softened once people realized that these five string theories and actually related by a set of mappings, called dualities. We have already discussed one of the, T-duality. The other one is called S-duality, and essentially relates a weak coupled string theory with a strong coupled one. Nowadays, we believed that the five different theories are just different limits on the moduli space of a bigger 11-dimensional theory, so far called *M-theory.* No much is known about it, but since the theory is still under progress, we expect to develop more powerful techniques which allow us to explore the fundamentals of M-theory.

5. Recent developments

String theory has become a huge framework on which different topics can be addressed. Enumerating all of them is a task far beyond the scope of this course. However, we shall mention some of them according to our interests. First of all, big progress has been performed in the so called string phenomenology branch. There is a lot of people working on effective four-dimensional theories constructed from string theory. See [10] (and references therein) for a more extensive treatment. One of the main topics covered includes the construction of Standard-Model-like scenarios and their corresponding supersymmetric extensions from Type II superstrings and from

Heterotic strings as well.

Historically, people used Type I and Heterotic theories to construct phenomenologically viable models. The main reason for that, was that such theories already had gauge groups, necessary to incorporate the Standard Model gauge groups. This vision changed once D-branes were introduced in Type II theories.

The construction of vacua from string scenarios, established an starting point for some very nice researches. First of all, we wanted to know the conditions, physical or mathematical, that the extra six dimensions must satisfy in order to obtain a four-dimensional effective physics, close enough to what we daily observe. Soon, people realized that the extra six dimensions must be enrolled in a six-dimensional mathematical object known as *Calabi-Yau manifold.* Some properties of the field content in the effective theory, as the number of families, was related to topological properties of such manifold. In the last decade, there was a huge progress towards the construction of realistic vacua, as the flux compactification and the generalization of Calabi-Yau manifolds, a topic that has enriched the develop of new mathematics [13]. People have also explored the possibility that our universe is just one choice among thousands of options in the so called String Landscape.

Other applications of String theory involve the construction of cosmological models, the AdS/CFT correspondence (a correspondence between gauge theories and gravity which has become a huge area of research) and its implications on physical predictions on the quark-gluon plasma. Also, new methods of calculations on amplitude scatterings by the use of twistors have been developed and the always fascinating study of black holes.

String theory is doubtless a very rich scenario in which many open questions of theoretical physics might be answered. The theory is still under construction and we hope that in the near future we realize whether or not, string theory plays a role in our world.

Acknowledgments
O. L.-B. thanks the organizers of the XIV Mexican School of Particles and Fields for inviting him to give this mini-course. O.L.-B. is partially supported by a CONACyT grant with contract number 60209. C. D.-A. is supported by a PhD CONACyT grant.

References
[1] K. Becker, M. Becker and J. H. Schwarz, *Cambridge, UK: Cambridge Univ. Pr. (2007) 739 p*
[2] E. Kiritsis, *Princeton, USA: Univ. Pr. (2007) 588 p*
[3] J. Polchinski, *Cambridge, UK: Univ. Pr. (1998) 402 p*
[4] J. Polchinski, *Cambridge, UK: Univ. Pr. (1998) 531 p*
[5] B. Zwiebach, *Cambridge, UK: Univ. Pr. (2004) 558 p*
[6] M. B. Green, J. H. Schwarz and E. Witten, *Cambridge, Uk: Univ. Pr. (1987) 596 P. (Cambridge Monographs On Mathematical Physics)*
[7] M. B. Green, J. H. Schwarz and E. Witten, *Cambridge, Uk: Univ. Pr. (1987) 469 P. (Cambridge Monographs On Mathematical Physics)*
[8] D. Lust and S. Theisen, Lect. Notes Phys. **346**, 1 (1989).
[9] C. V. Johnson, *Cambridge, USA: Univ. Pr. (2003) 548 p*
[10] M. Dine, *Cambridge, UK: Cambridge Univ. Pr. (2007) 515 p*
[11] H. Garcia-Compean and O. Loaiza-Brito, arXiv:hep-th/0003019.
[12] H. Garcia-Compean and O. Loaiza-Brito, AIP Conf. Proc. **562**, 86 (2001) [arXiv:hep-th/0010046].
[13] O. Loaiza-Brito and L. Vazquez-Mercado, AIP Conf. Proc. **1287**, 52 (2010).

Journal of Physics: Conference Series **287** (2011) 012004 doi:10.1088/1742-6596/287/1/012004

Chiral Random Matrix Theory and Chiral Perturbation Theory

Poul H. Damgaard

The Niels Bohr International Academy and Discovery Center, The Niels Bohr Institute, Blegdamsvej 17, DK-2100 Copenhagen, Denmark

E-mail: phdamg@nbi.dk

Abstract. Spontaneous breaking of chiral symmetry in QCD has traditionally been inferred indirectly through low-energy theorems and comparison with experiments. Thanks to the understanding of an unexpected connection between chiral Random Matrix Theory and chiral Perturbation Theory, the spontaneous breaking of chiral symmetry in QCD can now be shown unequivocally from first principles and lattice simulations. In these lectures I give an introduction to the subject, starting with an elementary discussion of spontaneous breaking of global symmetries.

1. Introduction

Many tend to think of spontaneous breaking of global symmetries in quantum field theory as something rather simple, almost trivial. To illustrate that this is certainly not so, let us first briefly review the standard textbook treatment of this phenomenon. Typically, one considers a field theory of a real scalar field with a potential that, appropriately for this school, can be dubbed a "sombrero potential",

$$V(\phi) \; = \; -\frac{1}{2}m^2\phi(x)^2 + \frac{\lambda}{4!}\phi(x)^4 \; . \tag{1}$$

From a perspective of classical physics this looks unstable. Certainly a constant field configuration of $\phi(x) = 0$ cannot possibly be a good starting point for a perturbative expansion. Instead, one considers the two minima of the potential at

$$\phi \; = \; \pm v \; = \; \pm\sqrt{\frac{6}{\lambda}}m \tag{2}$$

and says that a consistent treatment of such a field theory must be based on one vacuum that is *either* at $\phi = v$ or at $\phi = -v$, which one being undecidable. Here a comparison is often made to a classical picture, like a pencil standing exactly on its tip: such a system is rotationally symmetric, but any small perturbation will evidently make the pencil fall in the direction of the perturbation. Spontaneous symmetry breaking! Indeed, the pencil will now lie flat on the table, and it will have picked just one random direction, the one that was induced by the perturbation, a perturbation we can make as small as we like. What about the potential energy of the pencil? It was first converted to kinetic energy, and then, on impact with the table, to heat. Such a system is dissipative: If it were not, the pencil would simply bounce back up. There may be issues such as accuracy of initial conditions and so on, but this last point illustrates the difficulty with a simple picture of spontaneous symmetry breaking.

To investigate this a little more closely, let us proceed to quantum mechanics and glue two harmonic oscillator potentials together, separated by a distance $2a$[1]. The potential is thus

$$V(x) \;=\; \frac{1}{2}m\omega^2\left(|x|-a\right)^2 \tag{3}$$

The ground state of this quantum mechanical system must have no nodes, and we can almost trivially identify the correct ground state by drawing two Gaussian bumps centered at $x = \pm a$,

$$\psi_{\pm}(x) \;\sim\; \exp\left[-\frac{1}{2}m\omega(x\mp a)^2\right] \tag{4}$$

which are, individually, ground states of either of the two harmonic oscillators. If we smoothly join the two at $x = 0$ so that we construct, roughly,

$$\psi_0(x) \;\sim\; \psi_+(x) + \psi_-(x) \tag{5}$$

this will be a very good approximation to the true ground state. What about the 1st excited state? It should have one node, and again we can easily guess its form. Suppose we glue the two wave functions $\psi_{\pm}(x)$ together, but with a twist: we flip the sign of one of them to get

$$\psi_1(x) \;\sim\; \psi_+(x) - \psi_-(x) \tag{6}$$

This state will have a *slightly* higher energy than $\psi_0(x)$, but it is clear from this simple construction that $\psi_0(x)$ and $\psi_1(x)$ are almost degenerate. The higher excited states can be visualized similarly, by gluing two wave functions together around $x = 0$, more and more excited states, and alternatingly adding and subtracting combinations.

Does such a system display spontaneous symmetry breaking? It is well known that in this quantum mechanical case one of the two apparently distinct vacua will never be selected. Although the first excited state is almost degenerate with the ground state, there will always be a finite energy difference $E_1 - E_0$. If we prepare a state that initially is localized in only one of the two wells, it will tunnel to the other side. The tunneling time goes like $(E_1 - E_0)^{-1}$. Only when the two minima are infinitely far from each other does that tunneling time go to infinity. So our naive picture of spontaneous symmetry breaking can clearly not be valid for such a simple quantum mechanical system of just one degree of freedom. One cannot contain the probability density on just one side of $x = 0$.

These two examples, one classical and one quantum mechanical, illustrate that the phenomenon of spontaneous symmetry breaking in quantum field theory cannot be quite as simple as the intuition based on the sombrero potential seems to suggest. Nevertheless, as is so often the case in physics: the argument may not be quite right, but the answer is correct. There is no spontaneous breaking of symmetry in quantum mechanical systems of a finite number of degrees of freedom. But there *can* be spontaneous breaking of symmetries in quantum field theory. The distinction here comes precisely from the infinite number of degrees of freedom, the infinite number of quantum field theory modes. Although each of them individually can prevent spontaneous breaking of symmetry by tunneling, this mechanism can be blocked when an infinite number of modes have to tunnel. Subtleties in the arguments have been discussed on and off in the literature, see for instance refs. [2, 3]. For a more recent discussion, with plenty of references to earlier literature, see ref. [4].

[1] This is also a standard textbook example, see *e.g.*, ref. [1] for a nice and clear discussion.

To make the confusion complete, let us also point out another related phenomenon in quantum field theory: tunneling from one vacuum to another. We have just argued that in quantum field theory the tunneling between degenerate vacua can be prohibited due to the fact that an infinite number of degrees of freedom have to simultaneously align. Yet, if we tilt the sombrero potential slightly so that one vacuum really becomes energetically favored it is also common knowledge that in quantum field theory we *do* have tunneling to the true vacuum. How can this be? It turns out to be another example of a superficial argument that is basically correct nevetheless, and there is an endless amount of literature on this subject. Suffice it to say here that when there is a genuine non-degeneracy of vacua, one can have bubble formation and a first order phase transition where the true vacuum gobbles up the false vacuum. For a nice discussion of the essential physics of this situation see, *e.g.*, ref. [5].

Of particular importance is the phenomenon of spontaneous breaking of continuous global symmetries because it is linked to what is known as Goldstone's Theorem. This theorem implies the appearance of one massless mode associated with each broken generator of the continuous group of symmetries. Again there are caveats, and there are several. One is space-time dimensionality: a massless scalar in two space-time dimensions is not well defined in an infinite volume due to infrared singularities, and indeed there is no Goldstone phenomenon in two space-time dimensions [6]. In higher dimensions the theorem is incredibly strong, as it says that there is no mass gap, an otherwise forbiddingly difficult question to tackle. The Goldstone Theorem is most concisely stated in the operator language. Since the symmetry is continuous, let us consider the associated conserved current operator. Ordinarily the vacuum $|0\rangle$ is assumed to be annihilated by the corresponding conserved charge operator Q (the vacuum is not charged under Q). If the vacuum is not annihilated by Q, the symmetry is spontaneously broken. Remarkably, this implies the existence of an associated massless mode. This is the content of Goldstone's Theorem.

In QCD, the spontaneous breaking of chiral symmetry is a profound phenomenon that has been inferred indirectly over a long period of both theoretical and experimental development, starting from the days when the fundamental theory of strong interactions was not even known. In hindsight, it is incredible that it could be gotten at without a correct understanding of the underlying microscopic mechanism. Today we're much better off: we know the fundamental Lagrangian of QCD and it is "just" a matter of checking whether chiral symmetries are broken spontaneously or not. In practice, this is not simple *at all*. In fact, the question of showing spontaneous breaking of chiral symmetry in QCD and QCD-like theories is a challenge at the level of showing confinement and the existence of a mass gap in pure Yang-Mills theory. It is one of the remarkable successes of the numerical non-perturbative approach to QCD – lattice QCD – that this now has been shown beyond any doubt. An essential ingredient in this comes from the new understanding of chiral symmetry breaking through Random Matrix Theory.

The purpose of these very elementary lectures was to acquaint students with the new developments in our understanding of chiral symmetry breaking in QCD. The amount of time was insufficient to give a complete review and this is reflected in these notes which have focused on only a few of the many interesting aspects. That chiral Random Matrix Theory can say something *exact* about a complicated quantum field theory such as QCD with light quarks sounds like a wild idea, and it was initially met by a lot of skepticism in the lattice QCD community. All of this skepticism turned out to be unwarranted. There is now complete understanding of how and why chiral Random Matrix Theory represents an exact limit of the light quark partition function of QCD in a specific finite-volume regime. A precise mapping can be made between observables computed in QCD in that regime, the standard chiral Lagrangian of light-quark QCD and chiral Random Matrix Theory. This is a fantastic achievement, a most surprising and deep relation between a quantum field theory as complicated as QCD and a sequence of universal phenomena that follow from the spontaneous breaking of chiral symmetry alone. One

Journal of Physics: Conference Series **287** (2011) 012004 doi:10.1088/1742-6596/287/1/012004

example suffices to illustrate the enormous progress that has been achieved. Let us consider the eigenvalues of the QCD Dirac operator D:

$$D\psi_n = \lambda_n\psi_n \tag{7}$$

To make the discussing well-defined, let us consider this eigenvalue problem in a theory with a finite ultraviolet cut-off Λ and a finite infrared cut-off L. As will be discussed further down in these lectures, spontaneous breaking of chiral symmetry requires that the *smallest* non-zero eigenvalues λ_n accumulate towards zero at a rate proportional to $1/L^4$ in four dimensions. This is thus a simple test: do they or do they not? Before the developments described here, the strongest statement that could be made on basic principles was a proof that the λ_n's accumulate *at least* as fast as $1/L$ [7]. But this is a very weak condition. Even in a free theory, eigenvalues accumulate as $1/L$. All the theorem says, then, is that Dirac eigenvalues in QCD do not accumulate with a rate that is *slower* than that of free fermions, *i.e.*, as if there were no gauge interactions whatsoever. Can one not do better? With the advent of chiral Random Matrix Theory this problem has been solved, and it is now known that in QCD with two light flavors (the u and d quarks), Dirac operator eigenvalues *do* accumulate towards the origin at a much faster rate that goes exactly as $1/L^4$. Not only that, the precise probability distributions of single individual Dirac eigenvalues, $\lambda_i, i = 1, 2, 3, \ldots$ ordered according to their distance from the origin can be computed exactly. These distributions, as well as all spectral correlation functions of these smallest Dirac operator eigenvalues follow universal scaling laws whose exact analytical forms are known.

2. Chiral symmetry in QCD

If we look at the Lagrangian density of QCD with massless quarks, we are struck by the fact that the left-handed and right-handed fields decouple. If we had exactly N_f massless quarks in QCD, the global symmetry group would be $U(N_f) \times U(N_f)$. Of this a subgroup is $U(1)$ baryon number, which remains unbroken in QCD. The singlet chiral $U(1)$ symmetry is apparent only: it is broken by the chiral anomaly, and is therefore not a symmetry of the quantum theory. What remain are two independent flavor rotations $SU(N_f)$ for the left and right handed quarks. If this symmetry is broken as expected, $SU(N_f)_L \times SU(N_F)_R \to SU(N_f)$ due to the formation of a flavor-independent chiral condensate $\langle\bar{\psi}\psi\rangle$, we have $N_f^2 - 1$ broken generators and hence, by Goldstone's Theorem, $N_f^2 - 1$ massless bosons. This is a profound statement! How this slowly came to be realized before quarks had even been introduced is a story in itself. One consequence of this is that a low-energy representation of the QCD partition function had been inferred long before QCD was considered. In the next subsection we will give a lightning review of the moderne viewpoint on this.

2.1. The Chiral Lagrangian

Because of Goldstone's Theorem, the low-energy degrees of freedom of massless QCD are those of the Nambu-Goldstone bosons. In reality, even the u and d quarks are not exactly massless. But on the QCD scale of ~ 1 GeV they look nearly massless; their masses are on the order a few MeV. At some point it must be possible to treat the quark masses as small perturbations on top of a theory of genuine Goldstone bosons. The chiral Lagrangian does exactly this. Consider first a truly massless N_f-flavor theory where chiral symmetry is spontaneously broken as discussed above. To describe the coset of chiral symmetry breaking it is convenient to introduce an unusual non-linear field representation in terms of group elements $U(x) = \exp[i\sqrt{2}\Phi(x)/F]$, where F is the pion decay constant and $\Phi(x) = \lambda^a\phi^a(x)/\sqrt{2}$ represents the collection of $N_f^2 - 1$ pion fields. A field theory based on this should have the following ingredients: (1) kinetic energy terms for the pions with canonical normalization, (2) vanishing interactions at zero energy (as follows from Goldstone's Theorem) and (3) invariance under the chiral rotations. In the absence of any other principle that could fix terms in the Lagrangian the only solution will be to write down *all* terms that comply with the two conditions above. This gives us an endless series of invariants:

$$\mathcal{L} = \frac{F^2}{4}Tr[\partial_\mu U^\dagger \partial^\mu U] + L_1(\mathrm{Tr}[\partial_\mu U^\dagger \partial^\mu U])^2 + L_2\mathrm{Tr}[\partial_\mu U^\dagger \partial_\nu U]\mathrm{Tr}[\partial^\nu U^\dagger \partial^\mu U] + \ldots \tag{8}$$

with an infinite series of couplings (we display only the first two, L_1 and L_2 here). Values of all these couplings are left totally unspecified. They must depend on details of the dynamics (which gauge group is responsible for the symmetry breaking and so on). This is the chiral Lagrangian.

This cannot be the full story because we also want to treat quark masses, at least as a specified perturbation to the above Lagrangian. There is a beautifully simple way to do this, which goes under the name of the spurion technique. The chiral Lagrangian (8) is by construction invariant under global transformations $U \to U_R U U_L^\dagger$. A mass term in the QCD Lagrangian,

$$\bar{\psi}_L m \psi_R + \bar{\psi}_R m^\dagger \psi_L \tag{9}$$

is of course *not* invariant under $\psi_L \to U_L\psi_L, \psi_R \to U_R\psi_R$. But if the mass (matrix) were a field transforming like $m \to U_L m U_R^\dagger$ then the QCD Lagrangian with this mass term *would* be invariant. Then we know what the corresponding terms must look like in the chiral Lagrangian. The leading term will be

$$\mathcal{L} \;=\; \Sigma\mathrm{Tr}[mU + m^\dagger U^\dagger] \tag{10}$$

and there will of course be higher order terms involving higher powers of m and derivatives. Such a chiral Lagrangian will be invariant under the combined transformation

$$m \to U_L m U_R^\dagger \;, \qquad U \to U_R U U_L^\dagger \;. \tag{11}$$

A Lagrangian constructed in this way will thus explicitly break chiral symmetries in precisely the same way as the underlying theory of QCD.

2.2. Different counting schemes

So far we have defined the chiral Lagrangian by the totality of all terms that transform properly under chiral transformations. This looks rather hopeless, since there are infinitely many terms. Does such a theory have any predictive power at all? Weinberg [8] showed that it indeed does. First, since all possible terms are already included in the Lagrangian, it is by construction *renormalizable* since quantum corrections cannot produce terms not already present in the tree-level Lagrangian. This goes counter to everybody's intuition about renormalizable field theories, but it is correct. The chiral Lagrangian is an *effective field theory* and as such perfectly renormalizable. The theory, however, is not well-defined at all energy scales, and this is what we should consider next. The trouble is the appearance of higher and higher derivatives in the Lagrangian. Normally, this indicates that theory will not be a truly local theory and this may hold here as well. Certainly, the more derivatives, the more there will be contributions from the high energy scale. But the theory is not even supposed to hold at the high energy scale – it is precisely a low-energy effective theory. The way to deal with this was explained by Weinberg: consider an ordering of the Lagrangian where one expands in the number of derivatives. If this is to make sense, the Lagrangian must necessarily be viewed as a theory with a cut-off. In QCD, one would expect the cut-off Λ to be around 1 GeV; in detail this shows up in the form of the combination $4\pi F$, which indeed is of that order. Dependence on this cut-off can be removed for observables below this energy scale.

The expansion in terms of the number of derivatives can be made precise, see *e.g.* ref. [9]. A counting is introduced which makes it systematic. Since it is in terms of derivatives (or momenta), this is referred to as the p-expansion. The pion mass will also enter explicitly in the chiral Lagrangian, due to non-vanishing quark masses that explicitly break chiral symmetry in QCD. So an ordering with respect to m_π^2 needs to be introduced as well. The natural expectation works: one can systematicall treat m_π as being of order momentum p. Then chiral perturbation has only one expansion parameter, p. The scale is given by the expected ultraviolet cut-off $4\pi F$ so that the dimensionless expansion parameter is $\sim p/(4\pi F)$. This expansion works quite well phenomenologically, at least in the sector of the two lightest quarks (and we shall only be concerned with this light quark sector here).

The p-expansion breaks down eventually if one takes the massless limit at finite four-volume V. This is most easily seen by considering the pion propagator

$$\Delta(p^2) \;=\; \frac{1}{V}\frac{1}{p^2+m_\pi^2} \; . \tag{12}$$

For simplicity, let us take the finite volume V to be a symmetric four-torus of length L. The smallest (quantized) momentum is thus of order $1/L$. In a massless theory this makes the propagator $\Delta(p^2)$ vanish for large L like L^{-2}. But for the momentum *zero mode* the cut-off is entirely given by the mass. Clearly, for $m_\pi \gg 1/L$ the propagator is still protected by the mass, but as $m_\pi L \sim 1$ or much smaller we enter a new regime. From this point and onward the usual perturbative expansion cannot make sense in the zero-mode sector. Is this of importance? Can we not just ignore the zero mode? The answer is unfortunately no. Although it concerns only one mode, this one mode potentially overwhelms all other contributions from the perturbative expansion. The resolution of this problem is to treat the zero-mode sector exactly in a sense that will be more clear below [10, 11]. Gasser and Leutwyler developed a modified perturbation theory that includes this feature, while retaining all other properties of the usual pertubative expansion of chiral Lagrangians. This has become know as an ϵ-expansion (not to be confused with the expansion of similar name that expands in dimensionality $4-d$). The non-zero momentum modes still retain their usual pertubative expansion since the propagator for these modes will go at least as L^{-2}, even in the chiral limit.

A systematic chiral counting for this so-called ϵ-regime is as follows [11]. Let $\epsilon \sim 1/L$. Then

$$m_\pi \sim \epsilon^2 \; , \qquad p \sim \epsilon \; . \tag{13}$$

It is more instructive to think of this in terms of the microscopic degrees of freedom. Let us restrict ourselves to two degenerate light quarks of mass m. Due to the Gell-Mann–Oakes–Renner relation

$$F^2 m_\pi^2 \;=\; 2m\Sigma \tag{14}$$

where Σ is the chiral condensate, we see that $m \sim \epsilon^4$. In the ϵ-regime the quark masses scale as inverse powers of the volume. In fact, we can think of this as a regime where the extreme chiral limit $m \to 0$ is taken in a way that correlates with the way the four-volume V is sent to infinity. The relevant proportionality factor must have dimension 3, and indeed the right way to think of it is that the combination $m\Sigma V$ is kept of order unity. Here the condensate Σ provides the missing constant of proportionality, which is directly related to the fact that the combination $m\bar{\psi}\psi$ is a renormalization group invariant, in fact it is the explicit chiral symmetry breaking term in the QCD Lagrangian.

There is another way in which we can understand the appearance of the scale $m\Sigma V$. In two-flavor QCD, the chiral condensate $\langle\bar{\psi}\psi\rangle$ at finite quark mass m and finite volume V is the trace of the quark propagator,

$$\langle\bar{\psi}\psi\rangle \;=\; \frac{1}{2V}\partial_m \ln Z$$

$$
\begin{aligned}
&= \frac{1}{V}\left\langle \mathrm{Tr}\frac{1}{D+m}\right\rangle \\
&= \frac{1}{V}\left\langle \sum_j \frac{1}{i\lambda_j + m}\right\rangle \\
&= \frac{1}{V}\left\langle \sum_{j>0}\left(\frac{1}{i\lambda_j + m} + \frac{1}{-i\lambda_j + m}\right)\right\rangle \\
&= \frac{1}{V}\left\langle \sum_{j>0}\frac{2m}{\lambda_j^2 + m^2}\right\rangle
\end{aligned} \tag{15}
$$

where we have used that each quark contributes to $\langle\bar{\psi}\psi\rangle$ with the same amount. Use has also been made of the fact that every non-vanishing eigenvalue λ_j can be matched by its opposite-sign counterpart. This follows directly from the spectrum being chiral, $\{D,\gamma^5\} = 0$. The above spectral representation can be used to deduce the so-called Banks-Casher relation for the infinite-volume chiral condensate:

$$
\Sigma = \lim_{m\to 0}\lim_{V\to\infty} 2m\int_0^\infty d\lambda\, \frac{\rho(\lambda)}{\lambda^2 + m^2} = \pi\rho(0) \tag{16}
$$

This is a formal expression that should be understood only in terms of a cut-off theory. The chiral condensate $\langle\bar{\psi}\psi\rangle$ is ill-defined in the ultraviolet, and indeed one expects the spectral density $\rho(\lambda)$ to behave like $\sim \lambda^3$ (as in a free theory) for very large momenta, and this would lead to a quadratic divergence $\langle\bar{\psi}\psi\rangle \sim m\Lambda^2$. A term like $\sim m\ln(\Lambda)$ is also expected. However, the important point here is not the behavior at the ultraviolet end of the spectrum, but near the origin. Whether chiral symmetry is spontaneously broken or not is determined by whether or not the spectral density $\rho(\lambda)$ vanishes there. If $\rho(0)$ is to be non-vanishing, the discrete eigenvalues λ_j must accumulate there at a rate inversely proportional to the volume [12]:

$$
\Delta\lambda \sim 1/V \qquad \text{near } \lambda \sim 0 \tag{17}
$$

so as to yield a finite density at the origin. The inverse of the proportionality factor is precisely the condensate Σ on account of the Banks-Casher relation. We thus find that $\lambda\Sigma V$ will be of order unity for the smallest eigenvalues if we are to generate a chiral condensate. As will become clear shortly, there is an ϵ-regime of a *partially quenched* chiral Lagrangian hidden in this statement.

Before defining the partially quenched theory, let us return to the integral representation of $\langle\bar{\psi}\psi\rangle$. If we know how this condensate behaves as a function of the mass m, can we not invert the relation to get $\rho(\lambda)$? The trouble is that $\rho(\lambda)$ is also a function of quark masses m. If it were not, we could indeed simply invert the relation. The trick to achieve this is to do partial quenching: one introduces new unphysical quarks that are used to produce an intermediate generating function, but which are removed at the end. The spectral density will then be independent of the masses of these physical quarks, but if we know the analogue of their condensates $\langle\bar{\psi}\psi\rangle$, we can invert the integral representation and compute the spectral density. There are two ways of introducing such partially quenched quarks: one is by means of a graded structure [13], the other by means of replicas [14, 15]. Both methods work beautifully in perturbation theory, and for the same reason: observables computed in perturbation theory will depend polynomially on the number of additional fermionic N_f and bosonic N_b species. If both fermions and bosons are kept, the anticommutation sign from fermions just cancel the corresponding contribution from the fermions. So an equal number of degenerate fermions and bosons do not contribute to the partition function, a statement that looks entirely trivial if phrased formally in terms of functional determinants. In the replica method, one keeps only the fermions (or, equivalently,

Journal of Physics: Conference Series **287** (2011) 012004 doi:10.1088/1742-6596/287/1/012004

only the bosons) and instead removes them from the partition function by taking the replica limit $N_f \to 0$. Since the behavior is polynomial in perturbation theory, it is unambiguous how to do so in that setting.

In the ϵ-regime, integrals have to be done exactly, beyond perturbation theory. Curiously, then, both the graded method and the replica method run into technical difficulties. In the graded method it turns out that the naive extension of the chiral Lagrangian to a graded coset fails to yield convergent integrals, a problem that never surfaces in perturbation theory. This problem can be overcome by a judicious choice of bosonic integration [16]: The graded integration domain must be a combination of a compact and non-compact regions. In the replica framework the difficulty arises from the precise specification of how to analytically continue in the integer N_f. In perturbation theory this is trivial (because dependence on N_f is polynomial only), but beyond perturbation theory one needs more structure as guidance – knowing a function at all integers is of course not sufficient for defining an unambiguous analytic continuation into the real line. Fortunately, the required structure is precisely present in the cases of interest [17], in fact there is a remarkable connection to the theory of integrable systems and the so-called Toda lattice hierarchy. It is beyond the scope of these simple lectures to go into that fascinating subject, to which we here only refer to the literature [18]. It was based on this structure, and its relation to the graded analog, that the partially quenched partition function of the chiral Lagrangian was first computed in the ϵ-regime [17, 19].

3. What is Random Matrix Theory?

What is a "random matrix"? It is perhaps not obvious, but with the conventional definition we must fix its size. Let us first, for simplicity, consider square matrices of size $N \times N$. Elements in this matrix can now be chosen at random, according to our chosen distribution. But this cannot be the whole story, because some matrices have particular symmetries and particular transformation properties under conjugation. Indeed, a sensible definition of Random Matrix Theory is based on a division into certain classes of random matrices, a classification that dates back to Dyson [20]. The three main classes are labelled by an index β, which can take values 1,2 or 4. Here we will focus the $\beta = 2$ class where one picks the random matrices to be Hermitian, and thus having real eigenvalues.

To produce a Random Matrix Theory one does as in statistical mechanics and *sums over an ensemble* of a given set of random matrices M. This produces a partition function,

$$Z = \int dM \, \exp\left[-N\text{Tr}(M^2)\right] , \tag{18}$$

where, for simplicity, we have here restricted ourselves to a Gaussian distribution. We can associate with this a probability distribution

$$P(M) = \frac{1}{Z}\exp\left[-N\text{Tr}(M^2)\right] \tag{19}$$

and for Hermitian matrices the measure is

$$dM = \prod_{i=1}^{N} dM_{ii} \prod_{i<j} d(\text{Re}M_{ij})d(\text{Im}M_{ij}) . \tag{20}$$

We are not interested in the matrices themselves, but rather in their real eigenvalues λ_i. It is therefore advantageous to go to an eigenvalue representation of the partition function by means of diagonalization. This is achieved by a unitary transformation:

$$M = U^{\dagger}DU , \quad D = \text{diag}(\lambda_1, \ldots, \lambda_N) \tag{21}$$

Journal of Physics: Conference Series **287** (2011) 012004 doi:10.1088/1742-6596/287/1/012004

with $U^\dagger U = 1$.

We note that the Hermitian matrix M is described by N^2 coefficients, while the diagonal matrix D has N coefficients and U has $N^2 - N$. So we can indeed make the change of variables

$$M \quad \rightarrow \quad D, U \; .$$

Since the Haar measure on $U(N)$ is left and right invariant, the asociated Jacobian J of this change of variables can only depend on D, *i.e.*, $J = J(\lambda_1, \ldots, \lambda_N)$ and

$$dM \;=\; \left(\prod_{i=1}^{N} d\lambda_i\right) J(\lambda_1, \ldots, \lambda_N) dU \; . \tag{22}$$

To compute J, it suffices to go near the identity in $U(N)$:

$$U \;=\; 1 + i\varepsilon \tag{23}$$

where, for unitarity, $\varepsilon^\dagger = \varepsilon$. Then

$$M \;=\; U^\dagger D U \;=\; D - i[\varepsilon, D] \tag{24}$$

or, in components,

$$M_{ij} \;=\; \lambda_i \delta_{ij} + i\varepsilon_{ij}(\lambda_i - \lambda_j) \;\; [\text{no sum on } i,j] \; . \tag{25}$$

In other words,

$$\begin{aligned} i = j \;\; &: \;\; dM_{ii} = d\lambda_i \\ i < j \;\; &: \;\; d(\mathrm{Re}M_{ij}) = (\lambda_i - \lambda_j) d(\mathrm{Re}(i\varepsilon_{ij})) \\ & \;\;\;\;\; d(\mathrm{Im}M_{ij}) = (\lambda_i - \lambda_j) d(\mathrm{Im}(i\varepsilon_{ij})) \end{aligned} \tag{26}$$

so that

$$\begin{aligned} dM \;&=\; \left(\prod_{i=1}^{N} dM_{ii}\right) \prod_{i<j} d(\mathrm{Re}M_{ij}) d(\mathrm{Im}M_{ij} \\ &=\; \left(\prod_{i=1}^{N} d\lambda_i\right) \prod_{i<j} (\lambda_i - \lambda_j)^2 \end{aligned} \tag{27}$$

up to the group volume of $U(N)$. Note also that

$$\mathrm{Tr}(M^2) \;=\; \mathrm{Tr}(U^\dagger D U U^\dagger D U) \;=\; \mathrm{Tr}(D^2) \;=\; \sum_{i=1}^{N} \lambda_i^2 \tag{28}$$

so that also the probability measure can written entirely in terms of eigenvalues and the partition is then

$$Z \;=\; \int \left(\prod_{i=1}^{N} d\lambda_i\right) \prod_{i<j} (\lambda_i - \lambda_j)^2 e^{-N \sum_{i=1}^{N} \lambda_i^2} \; . \tag{29}$$

There is an easy mnemonic to see that this Random Matrix Theory is in the $\beta = 2$ class since it can be read off from the power "2" in

$$\prod_{i<j} (\lambda_j - \lambda_i)^2 \; .$$

This particular combination of eigenvalues can be written in the form of a so-called Vandermonde determinant raised to the second power:

$$\prod_{i<j}(\lambda_j - \lambda_i)^2 \;=\; \Delta(\{\lambda_i\})^2 \tag{30}$$

where

$$\Delta(\{\lambda_i\}) \;=\; \det(\lambda_j^i) \qquad j = 1,\ldots n,\; i = 0\ldots n-1\;. \tag{31}$$

The Random Matrix Theory (29) is of course well defined for all N. When N is finite, the N eigenvalues will jump around in a rather random manner. But the presence of the Vandermonde determinant (squared) in front of the Gaussian damping factor has an important consequence: *level repulsion.* The probability of two eigenvalues approaching each other goes to zero. For finite N, there could be two ways to get around this: 1) all eigenvalues could spread all over the real line or 2) eigenvalues could be confined to a finite interval, but level repulsion would force their distributions to lock in an essentially equidistant form. This will clearly depend on how we rescale eigenvalues. It turns out that the Gaussian damping factor in the probability distribution (29) is so strong that eigenvalues are forced to lie on a finite interval, and it is thus the 2nd realization of eigenvalue distributions we are seeing with such measures. The factor of N in the exponent is precisely inserted so as to ensure this compact distribution without further rescalings of the eigenvalues.

The distribution of eigenvalues becomes beautifully simple in the limit of very large N: we get the famous Wigner semicircle distribution.

3.1. Chiral Random Matrix Theory

We are now ready to consider a Random Matrix Theory that is chiral. The identification of this class of theories to be the one relevant for low-energy QCD was first made by Verbaarschot in a remarkable series of papers about 15 years ago [21, 22]. The idea was to base the Random Matrix Theory on transformation properties of the Dirac operator D in QCD. In particular, since D anticommutes with γ^5, $\{D, \gamma^5\} = 0$, a random matrix M is introduced that has a similar structure. Moreover, gauge field configurations can be assigned to different classes depending on their topology, which is classified according to the invariant

$$\nu \;=\; \frac{g^2}{16\pi^2}\int d^4x \mathrm{Tr}(\tilde{F}_{\mu\nu}F^{\mu\nu}) \tag{32}$$

where $\tilde{F}_{\mu\nu} = (1/2)\epsilon_{\mu\nu\alpha\beta}F^{\mu\nu}$ is the dual QCD field strength tensor. In the QCD partition function one sums over all these different classes that have $\nu = 0, \pm1, \pm2, \pm3, \ldots$, but it can be useful to consider also sectors of fixed ν. A well known mathematical theorem ensures that in such sectors there are exact zero modes of the Dirac operator, all of definite chirality. The number of positive minus negative chirality zero modes is precisely the index ν. Leutwyler and Smilga [12] suggested to study properties of the Dirac operator in such sectors of fixed topology by means of the relationship to the chiral Lagrangian. This gives an exciting new opportunity for cross-checking how low-energy observables in the full theory (QCD) are in one-to-one correspondence with observables in the effective large-distance theory of the chiral Lagrangian.

Verbaarschot and Shuryak [21] introduced the following chiral Random Matrix Theory:

$$Z \;=\; \int dM \,(\det(M))^{N_f} \exp\left[-(N/2)\mathrm{Tr}V(M^2)\right] \tag{33}$$

where $V(M^2)$ will be discussed below, and where

$$M \;\equiv\; \begin{pmatrix} 0 & W^\dagger \\ -W & 0 \end{pmatrix}$$

Journal of Physics: Conference Series **287** (2011) 012004 doi:10.1088/1742-6596/287/1/012004

is a $(2N+\nu)\times(2N+\nu)$ block Hermitian matrix composed of the complex $N\times(n+\nu)$ rectangular matrix W. The integral is over the elements of this matrix. We note that both measure and integrand is then invariant under

$$W \to V^{\dagger}WU \ , \qquad U \in U(N) \ , \qquad V \in U(N+\nu) \tag{34}$$

and for this reason this Random Matrix Theory is called chiral unitary. But why is it 'chiral'? Consider the $(2N+\nu)\times(2N+\nu)$ matrix

$$\gamma^5 \equiv \begin{pmatrix} \mathbf{1}_N & 0 \\ 0 & -\mathbf{1}_{N+\nu} \end{pmatrix}$$

This matrix clearly anticommutes with the matrix M,

$$\{M, \gamma^5\} = 0 \ , \tag{35}$$

which implies that all non-zero eigenvalues of M come paired $\pm\lambda_j$. So the matrix M shares the $U(1)$ chirality property of the full Dirac opertator of QCD. Moreover, due to its rectangular block decomposition, there are ν zero modes. The spectrum of M therefore has two parts: there are $2N$ non-zero eigenvalues that are chirally paired and ν zero modes.

Actually, the chiral Random Matrix Theory (33) was originally introduced with just a Gaussian measure, but it was believed from the beginning that the so-called 'microscopic' results that could be derived from it would be *universal*, *i.e.* to a very large extent independent of the chosen potential. Indeed, it was proven in ref. [23] that all results to be discussed below hold for an arbitrary potential

$$V(M^2) = \sum_{k\geq 1} \frac{g_k}{k} M^{2k} \tag{36}$$

with, essentially, only one single constraint: the spectrum of eigenvalues of M must have support at the origin: $\rho(0) \neq 0$. This is in beautiful accord with the intuition, to be made more precise below, that a non-vanishing chRMT spectral density at the origin is required for this theory to describe aspects of chiral symmetry breaking – in heuristic analogy with the Banks-Casher relation in QCD. The universality and robustness of this result is crucial for the understanding of why chiral Random Matrix Theory can describe chiral symmetry breaking in a full-fledged quantum field theory such as QCD.

Interestingly, one can also consider a chiral Random Matrix Theory where the spectral density at the origin $\rho(0)$ is carefully tuned so as to produce a zero: $\rho(0) = 0$ [24]. One might hope that this could represent chiral symmetry restoration, one way or another. This, however, has never been established. There are still universality classes in this case, one for each order of zero of the spectral density. Each class can be reached, successively, by tuning more and more of the parameters g_k in (36). This is an example of universality in what can be dubbed multicriticality (while the more general condition $\rho(0) \neq$ corresponds to the non-critical case). A pity that there seems to be few instances of quantum field theory where these more sophisticated universality classes can be of relevance! One possibility could be two space-time dimensions, where spontaneous symmetry breaking is prohibited, and the spectral properties of the Dirac operator can be studied in, say, the Schwinger model [25]. In any case, the mere existence of these extended classes shows that the universality domain of the non-critical ensemble is restricted and by no means trivially implied for any chiral Random Matrix Theory potential $V(M^2)$.

Very similar to the way we could introduce an eigenvalue representation (29) for the non-chiral (Gaussian) ensemble, one can also find an eigenvalue representation

$$Z = \int_{-\infty}^{\infty} \prod_{i=1}^{N} \left(dz_i^2 \, z_i^{2(N_f+\nu)} e^{-NV(z_i^2)} \right) \left(\det_{ij} z_j^{2(i-1)} \right)^2$$

Journal of Physics: Conference Series **287** (2011) 012004 doi:10.1088/1742-6596/287/1/012004

$$= \int_0^\infty \prod_{i=1}^N \left(d\lambda_i\ \lambda_i^{N_f+\nu} e^{-NV(\lambda_i)}\right)\left(\det_{ij} \lambda_j^{i-1}\right)^2 \tag{37}$$

where the z_i^2's are the real (and positive) non-vanishing eigenvalues of $W^\dagger W$, and where in the last line we have just transformed to more convenient variables λ_i. We have also used the fact that diagonalization of $W^\dagger W$ automatically guarantees factorization of the potential part in the manner shown. It is interesting to note how the dependence on N_f and ν has merged into the combination $N_f+\nu$ only. This is a simplifying property that holds only in the massless case. The case of massive quarks will be discussed below. The square of the Vandermonde determinant has appeared just as in the non-chiral case. Also here it will result in an eigenvalue repulsion.

There are standard methods for computing the spectral correlation functions of a Random Matrix Theory with an eigenvalue representation such as (37). A convenient method is based on orthogonal polynomials, using a beautiful formalism that dates back to Dyson. Limitations of space does not permit a detailed discussion of this, and here we will restrict ourselves to explaining the main results that follow from that kind of analysis.

Let us from now on work entirely in rescaled variables $\zeta \equiv 2\pi N\rho(0)\lambda$. For fixed ζ this implies smaller and smaller eigenvalues – those near the origin. A central object is the *kernel* $K(\zeta_1, \zeta_2)$. If we know this kernel we know all spectral correlation functions

$$\rho(\zeta_1, \ldots, \zeta_k) = \det_{ij} K(\zeta_i, \zeta_j) \tag{38}$$

and in particular the spectral density itself,

$$\rho(\zeta) \;=\; K(\zeta, \zeta)\ . \tag{39}$$

The kernel appropriate for QCD is the *Bessel kernel* derived by Verbaarschot and Zahed [22],

$$K(\zeta_1, \zeta_2) = \sqrt{\zeta_1\zeta_2}\frac{\zeta_1 J_{\alpha+1}(\zeta_1)J_\alpha(\zeta_2) - \zeta_2 J_\alpha(\zeta_1)J_{\alpha+1}(\zeta_2)}{\zeta_1^2 - \zeta_2^2}\ , \tag{40}$$

where $\alpha = N_f + \nu$ (we consider only positive ν here). Despite appearance, this is regular at coincident points where it gives the microscopic spectral density

$$\rho(\zeta) = \frac{\zeta}{2}\left(J_\alpha(\zeta)^2 - J_{\alpha+1}(\zeta)J_{\alpha-1}(\zeta)\right)\ . \tag{41}$$

This is the most simple and remarkable result of the chiral Random Matrix Theory approach. The claim is that this gives the microscopic spectral density of QCD with N_f massless quarks in the finite-volume regime we consider. Below we shall see why this is so, and how to include massive quarks as well. Amazingly, if we boldly set $N_f = 0$ this should give the spectral density of the Dirac operator in pure Yang-Mills theory! When we see the connection with the partially quenched chiral Lagrangian this makes it clear why this should be possible (with some caveats). Another immediate check on the above spectral density comes from the fact that it reproduces exact spectral sum rules derived by Leutwyler and Smilga [12] on the basis of the chiral Lagrangian alone.

3.2. Spectral properties of the Dirac operator

The pioneering paper that showed how these detailed predictions for the spectral density of the Dirac operator can be tested in lattice gauge theory simulation was ref. [26]. Quenched staggered fermions with gauge group $SU(2)$ was used there. Because the chiral symmetries of staggered fermions away from the continuum limit are different from those of continuum fermions when the fermions carry real or pseudo-real representations of the gauge group (pseudo-real in the case of fundamental fermions and gauge group $SU(2)$), the detailed predictions were not those of eq. (41). But also for this case the analytical prediction had been worked out [27] and although it differs in detail, it shares many of the properties with that of (41). The analytical expressions for continuum fermions based on this gauge group and fermions in the fundamental representation had been worked out as well [28]. We shall not go into great detail with these different predictions, but it is an amazing fact, first recognized by Verbaarschot [29], that there is a one-to-one correspondence between the Dyson classification of (chiral) Random Matrix Theory and the three chiral symmetry breaking patterns in vector-like gauge theories:

- The fermion representation r is pseudo-real: Chiral symmetries are enhanced from $SU(N_f)\times SU(N_f)$ to $SU(2N_f)$, and the expected symmetry breaking pattern is $SU(2N_f) \to Sp(2N_f)$.
- The fermion representation r is complex: The expected symmetry breaking pattern is $SU(N_f) \times SU(N_f) \to SU(N_f)$.
- The fermion representation r is real: Chiral symmetries are again enhanced to $SU(2N_f)$, and the expected chiral symmetry breaking pattern is $SU(2N_f) \to SO(2N_f)$.

That there is indeed a one-to-one matching with the three Dyson classes, and that the chiral symmetry breaking patterns really do follow the above classification, has been tested on the lattice in great detail for a variety of different fermion representations and it works in all cases [30].

There is one further general peculiarity of simulations with staggered fermions on the lattice which in the beginning caused some confusion, but which really is just one further confirmation of the above classification scheme. This arises from the fact that staggered fermions away from the continuum limit have an additional $U(1)$ factor in the coset of chiral symmetry breaking (this holds for all three classes). As we will show below, this means that all results for staggered fermions away from the continuum are *equivalent* to those we would get if we had projected on the $\nu = 0$ sector. Any attempt at a further projection on sectors with $\nu \neq 0$ leads to the same partition function. So for staggered fermions away from the continuum it will *look* as if only the sector of $\nu = 0$ appears. This has been confirmed by a high-statistics study where gauge field configurations were even selected according to their (approximative) topological indices: all spectral data of the staggered Dirac operator look as if they have been obtained from the $\nu = 0$ sector [31]. This should of course change as the continuum limit is reached. By using highly improved actions as well as improved Dirac operators that smear both configurations and observables it has been checked that predictions of continuum fermions do match those of staggered fermions eventually [32].

While most initial studies of the spectral properties of the Dirac operator were restricted to gauge group $SU(2)$, results were soon available for the QCD gauge group of $SU(3)$ [33], nicely confirming the prediction (41). A far more exciting possibility was to check all of these predictions with 'chirally good' fermions based on the overlap operator. This was first done in ref. [34]. Here the effect of topology can be clearly seen, and this works well even for approximate realizations of the overlap operator [35, 36]. A high-statistics study with several eigenvalues included has been made in ref. [37], showing clear agreement with the predictions. It is also interesting to see how, with an infinite number of colors, chiral symmetry can break even at finite four-volume. Results still fall right on the chiral Random Matrix Theory predictions, now as a result of scaling in the number of colors N_c [38]. Here one sees clearly how color and volume acquire complementary roles at large-N_c, the corresponding scaling variable being $\lambda\Sigma N_c V$.

So far our discussion has been restricted to the massless limit, which is easily taken in lattice gauge theory if one restrict oneself to the quenched theory. An interesting question is how to include quark masses in all of these preditions. If we consider the chiral Random Matrix Theory

(33), it is clear that masses m must be rescaled at the same rate as eigenvalues λ_i. Let us re-emphasize that m in (33) is simply a dimensionless parameter in the matrix integral. It is *not* a quark mass. However, just as the universal predictions for rescaled eigenvalues $N\rho(0)\lambda$ correspond to volume-rescaled eigenvalues $V\Sigma\lambda$ on the QCD side, so $N\rho(0)m$ will correspond to the scaling variable $V\Sigma m$ in QCD. The precise predictions were worked out in ref. [40]. The clue to the solution was there found in the theory of orthogonal polynomials, which is known to solve the spectral problem of eigenvalue representations like those of (33). Intuitively, it comes about as follows: if we know a set of orthogonal polynomials corresponding to (33) *without* the determinant in the measure, we can construct the set of orthogonal polynomials with *one determinant* in the measure. Basically, one expands the new set of orthogonal polynomials in the basis of the old set. The coefficient in this expansion can be fixed by what is known as Christoffel's Theorem. This solves the $N_f = 1$ case in terms of the quenched case. To illustrate, one finds the spectral density in sector of topological [40] charge $\nu = 0$,

$$\rho(\zeta,\mu) = \frac{\zeta}{2}\left(J_0(\zeta)^2 + J_1(\zeta)^2\right) - \zeta\frac{J_0(\zeta)[I_1(\mu)\mu J_0(\zeta) + I_0(\mu)\zeta J_1(\zeta)]}{(\zeta^2+\mu^2)I_0(\mu)} \tag{42}$$

In this way of writing it, we nicely illustrate an essential feature:

$$\lim_{\mu\to\infty} \rho(\zeta,\mu) = \rho(\zeta) \tag{43}$$

in other words: when we take the (rescaled) quark mass to infinity we recover the quenched spectral density. This is as we would intuitively expect, since the quark should then decouple and the eigenvalues should distribute themselves according to the quenched distributions.

All of this was generalized in ref. [40] to any number of flavors, and the universality proof was there extended to this generalized case as well. Moreover, the explicit expressions for the massive kernels were provided there. From these, all spectral correlation functions with massive quarks follow. However, the expressions can become unwieldy, and there is actually a much more compact way of understanding these results. It turns out that, quite surprisingly, one can express all microscopic spectral correlators compactly in terms of finite volume QCD effective partition functions. The crucial ingredient is a powerful identity derived by Zinn-Justin [41] in the case of the ordinary unitary ensemble, and which is readily generalized to the chiral case. Let us consider the $\nu = 0$ case in detail [42]. The kernel can then be expressed as

$$K_N(z,z';m_1,\ldots,m_{N_f}) = \frac{e^{-\frac{N}{2}(V(z^2)+V(z'^2))}\sqrt{zz'}\prod_f\sqrt{(z^2+m_f^2)(z'^2+m_f^2)}}{\tilde{\mathcal{Z}}_\nu^{(N_f)}(m_1,\ldots,m_{N_f})} \times \int_0^\infty \prod_{i=1}^{N-1}\left(d\lambda_i\lambda_i^\nu(\lambda_i-z^2)(\lambda_i-z'^2)\prod_{f=1}^{N_f}(\lambda_i+m_f^2)e^{-NV(\lambda_i)}\right)\left|\det\lambda_j^{i-1}\right|^2 \tag{44}$$

The last integral is over $(N-1)$ eigenvalues only. However, in the large-N limit we shall consider below, this distinction can be ignored. Thus, in the large-N limit we have

$$K_N(z,z';m_1,\ldots,m_{N_f}) = e^{-\frac{N}{2}(V(z^2)+V(z'^2))}\sqrt{zz'}\prod_f\sqrt{(z^2+m_f^2)(z'^2+m_f^2)} \times\frac{\tilde{\mathcal{Z}}_\nu^{(N_f+2)}(m_1,\ldots,m_{N_f},iz,iz')}{\tilde{\mathcal{Z}}_\nu^{(N_f)}(m_1,\ldots,m_{N_f})}, \tag{45}$$

where the matrix model partition function in the numerator is evaluated for a theory corresponding to (N_f+2) fermions, of which two have imaginary mass. By means of the usual factorization property, all higher n-point spectral correlation functions are then also explicitly expressed in terms of the two matrix model partition functions $\tilde{\mathcal{Z}}_\nu^{(N_f)}$ and $\tilde{\mathcal{Z}}_\nu^{(N_f+2)}$. The spectral density corresponds to the two additional (imaginary) masses being equal:

$$\rho^{(N_f)}(z;m_1,\ldots,m_{N_f}) \;=\; \lim_{N\to\infty} K_N(z,z;m_1,\ldots,m_{N_f})\;. \tag{46}$$

All of this is exact, and at the level of unrescaled variables. We can now take the microscopic scaling limit in which $\zeta \equiv zN2\pi\rho(0)$ and $\mu_i \equiv m_iN2\pi\rho(0)$ are kept fixed as $N\to\infty$. In this limit the pre-factor $\exp[-(N/2)(V(z^2)+V(z'^2))]$ becomes replaced by unity. Identifying $\Sigma = 2\pi\rho(0)$, this is the limit in which we can compare with the finite-volume partition function of QCD. Proceeding in this way, we get, with C being an overall constant,

$$\rho^{(N_f,\nu)}(\zeta;\mu_1,\ldots,\mu_{N_f}) \;=\; C|\zeta|\prod_f(\zeta^2+\mu_f^2)\,\frac{\mathcal{Z}_\nu^{(N_f+2)}(\mu_1,\ldots,\mu_{N_f},i\zeta,i\zeta)}{\mathcal{Z}_\nu^{(N_f)}(\mu_1,\ldots,\mu_{N_f})}\;, \tag{47}$$

and all n-point correlation functions are given by one magical Master Formula [42]

$$\rho^{(N_f,\nu)}(\zeta_1,\ldots,\zeta_n;\mu_1,\ldots,\mu_{N_f}) \;=\; \det_{a,b} K(\zeta_a,\zeta_b;\mu_1,\ldots,\mu_{N_f})\;. \tag{48}$$

This generalizes to any ν [43] and in fact leads to some surprising identities among the partition functions involved, identities that express the connection to an associated integrable hierarchy. The advantage is that the finite-volume partition functions are known in exact analytical forms [44]. Results have also been generalized to the other chiral symmetry breaking classes [45, 46].

One immediate check on these massive spectral densities is that they reproduce exact *massive* spectral sum rules [47] just as the massless spectral densities reproduce the massless spectral sum rules of Leutwyler and Smilga.

Amazingly, the same procedure generalizes to the computation of *individual eigenvalue distributions*. Such a notion makes clear sense in the chiral Random Matrix Theory framework since there is a natural starting point from where to start counting eigenvalues (the origin). After the lattice gauge theory community had accepted that chiral Random Matrix Theory gives exact results for the spectral density of the Dirac operator, and all higher-point spectral correlation functions, there remained a curious belief that even the very notion of ordered Dirac operator eigenvalues and their precise distributions could not be given field theoretic meaning. This is of course not so: ordered Dirac operator eigenvalues have precisely as much meaning in the field theory framework as the spectral density and all spectral correlation functions. This will become clear below. But as first step towards this, we should describe how individual eigenvalue distributions of the chiral Random Matrix Theory can be given explicitly in terms of finite volume partition functions for first the smallest Dirac operator eigenvalue [48] and then the general distribution for the kth Dirac operator eigenvalue [49], as counted from the origin. Let us denote the joint probability distribution of the smallest k ordered eigenvalues by $\omega_k(\zeta_1,\ldots,\zeta_{k-1},\zeta_k)$, where $\zeta_1\le\ldots\le\zeta_k$. By manipulations very similar to the ones used above, one can show that it can be written in terms of finite volume partition functions [49]

$$\begin{aligned}\omega_k(\zeta_1,\ldots,\zeta_{k-1},\zeta_k) \;=\;& Ce^{-\zeta_k^2/4}\zeta_k\prod_{i=1}^{k-1}\left(\zeta_i^{2\nu+1}\prod_{j=1}^{N_f}(\zeta_i^2+\mu_j^2)\right)\prod_{i>j}^{k-1}(\zeta_i^2-\zeta_j^2)^2\prod_{j=1}^{N_f}\mu_j^\nu\\ &\times\frac{Z_2(\{\sqrt{\mu_i^2+\zeta_k^2}\},\{\sqrt{\zeta_k^2-\zeta_i^2}\},\{\zeta_k\})}{Z_\nu(\{\mu_i\})}\end{aligned} \tag{49}$$

where the "masses" $\sqrt{\zeta_k^2 - \zeta_i^2}$ are doubly degenerate (and $1 \le i \le k-1$). Similarly, the set of "masses" $\{\zeta_k\}$ is ν-fold degenerate.

A very high-accurate Monte Carlo study of these individual eigenvalue distributions has been done in ref. [50], for both the quenched case and for theories with dynamical finite-mass fermions. The agreement is quite spectacular, with parameter-free fits that lie right on the analytical predictions. Of course, if one for some reason would fear these predictions, one could also just look at the chiral condensate which follows a very precise scaling law based on these analytical predictions for the chiral condensate [51, 52].

One important question is how individual eigenvalue distributions, ordered in the way we have discussed above, can be understood and derived from the quantum field theory point of view. How do we even define such distributions in quantum field theory? It turns out that there is a simple answer [54]. For simplicity, let us here restrict ourselves to the case of just the smallest Dirac operator eigenvalue, the distributions of the larger eigenvalues follow analogously [54]. Here one can show that the so-called gap probability has a convergent expansion of the form

$$E_0(s) = 1 - \int_0^s d\lambda_1\ \rho(\lambda_1) + \frac{1}{2}\int_0^s d\lambda_1 d\lambda_2\ \rho_2(\lambda_1,\lambda_2) + \ldots \tag{50}$$

involving higher and higher spectral correlation functions. All of these spectral correlation functions can be computed in quantum field theory by means of suitable sources. The distribution of the smallest eigenvaluie now follows by differentiation:

$$p_1(s) \ = \ -\frac{\partial}{\partial s}E_0(s)\ . \tag{51}$$

For the kth eigenvalue this follows by a simple iterative scheme that involves increasingly higher order correlation function as the first term in the expansion [54]. But the convergence is fast in the appropriate region of λ, and this scheme is in fact a quite practical way of computing these distributions if one is only interested in high numerical accuracy.

Another issue concerns gauge field topology. It is clear from the chiral Random Matrix Theory construction that topology is intimately tied together with the whole construction. While a summation over topology is straightforward in QCD, there seems to be no way to peform such a summation in the context of chiral Random Matrix Theory. Every sector corresponds to matrices of different sizes. However, all physical observables can be decomposed into their counterparts in distinct topological sectors, and a summation over topology is therefore indirectly possible through [53]

$$\langle \mathcal{O}\rangle \ = \ \frac{1}{Z}\sum_\nu Z_\nu \langle \mathcal{O}\rangle_\nu \tag{52}$$

where Z is the full partition function and subscript ν indicates when the average is taken in a sector of fixed topological charge. Note how averages are weighted by the partition function in fixed sectors. Using this relationship and the Master Formula derived above, it is possible to perform the sum analytically and derive an analogous formula for the spectral density of the full theory [53]. It is also particularly illuminating to see analytically how the triviality of results in the $N_f = 1$ theory are recovered from the highly non-trivial predictions in sectors of fixed topological charge.

3.3. Matching chiral Random Matrix Theory with the chiral Lagrangian

The Master Formula (48) gives a strong indication that all these chiral Random Matrix Theory results can be derived entirely from chiral Lagrangian framework. This is indeed the case [55, 16]. The framework is partially quenched chiral perturbation theory, which we have already discussed above. Through it, one can derive a one-to-one correspondence between the spectrum derived

Journal of Physics: Conference Series **287** (2011) 012004 doi:10.1088/1742-6596/287/1/012004

from the chiral Lagrangian and the spectrum derived from chiral Random matrix Theory. The final steps showing that this holds for all spectral correlation functions were taken by Akemann and Basile in ref. [56].

What has been achieved is thus a precise and detailed understanding of how chiral Random Matrix Theory can yield exact results. It provides an exact representation of the leading-order chiral Lagrangian in the extreme finite-volume ϵ-regime. The proof goes through the partially quenched theory which here serves two purposes at one time: (1) it provides predictions for genuine partially quenched lattice gauge theories in this finite-volume regime, and (2) it serves as a generating function of spectral correlation functions. In chiral Random Matrix Theory one does often not make use of a partially quenched (graded) version of the theory because other simpler technqiues (such as the one based on orthogonal polynomials) are available. It is precisely by going through the graded version that the full equivalence has finally been established for all correlation functions.

An interesting question is whether one can extract information about the other low-energy constants F as well as all higher-order coupling constants of the chiral Lagrangian by means of spectral properties of the Dirac operator. In principle, this should be possible since the influence of these couplings on the distributions of the Dirac operator eigenvalues can computed. By a careful comparison to numerical lattice data, one should be able to extract these other low-energy parameters of QCD by considering only a few of the lowest Dirac operator eigenvalue – a remarkable thought. However, in practice it is not that easy. The simplest example is the leading correction to Σ, which we already discussed above. In principle, by considering a variety of different volumes V, and perhaps different geometries as well, one can extract the pion decay constant F from this.

Other chiral Random Matrix Theory approaches, aimed specifically at determining F, have also been invented. One consists in introducing a new source in the chiral Lagrangian that couples directly to F to leading order in the chiral counting, while retaining anti-Hermiticity of the Dirac operator in that context. Although it is to be considered only as a source, it actually has a physical interpretation as imaginary isospin potential. With an external isospin potential the QCD measure is free of the sign problem that plagues QCD with baryon chemical potential. Making the potential purely imaginary ensures that the Dirac operator eigenvalues with this type of chemical potential remain anti-Hermitian. This approach was first developed in the chiral Lagrangian framework [57], but inspired by a construction due to Osborn [58] who introduced a two-matrix chiral theory do describe baryon chemical potential, a chiral two-matrix theory was set up to describe imaginary isospin chemical potential [59]. Indeed, results from this theory, which we will describe next, agree precisely with those obtained from the chiral Lagrangian [57]. Because all the powerful machinery of the more usual chiral Random Matrix Theory is preserved, this formulation allows for an explicit calculation of all spectral correlation function.

The first step consists in considering the eigenvalues of two different Dirac operators in QCD,

$$\begin{aligned} D_1\psi_1^{(n)} &\equiv [D(A)+i\mu_1\gamma_0]\psi_1^{(n)} = i\lambda_1^{(n)}\psi_1^{(n)} \\ D_2\psi_2^{(n)} &\equiv [D(A)+i\mu_2\gamma_0]\psi_2^{(n)} = i\lambda_2^{(n)}\psi_2^{(n)} \ . \end{aligned} \tag{53}$$

Imaginary isospin potential corresponds to $\mu \equiv \mu_1 = -\mu_2$. Such a source couples directly to F is the chiral Lagrangian for which the leading-order terms in a suitably defined ϵ-regime read:

$$Z_\nu^{(N_f)} = \int_{U(N_f)} dU\,(\det U)^\nu e^{\frac{1}{4}VF_\pi^2\mathrm{Tr}[U,B][U^\dagger,B]+\frac{1}{2}\Sigma V\mathrm{Tr}(\mathcal{M}^\dagger U+\mathcal{M}U^\dagger)} \ . \tag{54}$$

where the matrix $B = \mathrm{diag}(\{\mu_1\},\{\mu_2\})$ contains the two chemical potentials introduced above, and $\mathcal{M} = \mathrm{diag}(m_1,\ldots,m_{N_f})$ is the mass matrix. Chemical potential μ_1 has been assigned to N_1 of the quarks and chemical potential μ_2 to the remaining $N_f - N_1 = N_2$ quarks.

A two-matrix chiral theory that corresponds to this is

$$\mathcal{Z}_\nu^{(N_f)} \;=\; \int d\Phi d\Psi\; e^{-N\mathrm{Tr}(\Phi^\dagger\Phi+\Psi^\dagger\Psi)} \prod_{f1=1}^{N_1} \det[\mathcal{D}_1 + m_{f1}] \prod_{f2=1}^{N_2} \det[\mathcal{D}_2 + m_{f2}] \tag{55}$$

where $\mathcal{D}$ is defined by

$$\mathcal{D}_f = \begin{pmatrix} 0 & i\Phi + i\mu_f\Psi \\ i\Phi^\dagger + i\mu_f\Psi^\dagger & 0 \end{pmatrix} \;,\quad f = 1,2 \;.$$

The matrices Φ and Ψ are complex rectangular matrices of size $N\times(N+\nu)$ in the same manner as in the usual chiral ensemble.

Without going into the technical details, an eigenvalue representation can also be found for this theory. Let us define

$$\begin{aligned} c_1 &= (1+\mu_2^2)/\delta^2 \;, \quad c_2 \;=\; (1+\mu_1^2)/\delta^2 \\ d &= (1+\mu_1\mu_2)/\delta^2 \;, \quad 1-\tau \;=\; d^2/(c_1c_2) \\ \delta &= \mu_2 - \mu_1 \;, \end{aligned} \tag{56}$$

Then,

$$\begin{aligned} \mathcal{Z}_\nu^{(N_f)} &= \int_0^\infty \prod_i^N \left(dx_i dy_i (x_i y_i)^{\nu+1} \prod_{f1=1}^{N_1} (x_i^2 + m_{f1}^2) \prod_{f2=1}^{N_2} (y_i^2 + m_{f2}^2) \right) \\ &\times \;\Delta(\{x^2\})\Delta(\{y^2\}) \det\left[I_\nu(2dN x_i y_j)\right] e^{-N\sum_i^N c_1 x_i^2 + c_2 y_i^2} \;. \end{aligned} \tag{57}$$

The two sets of eigenvalues x_i and y_i are obtained after diagonalizing

$$\begin{aligned} \Phi_1 &\equiv \Phi + \mu_1\Psi \\ \Phi_2 &\equiv \Phi + \mu_2\Psi \;, \end{aligned} \tag{58}$$

and due to this redefinition the original matrices now become coupled in the exponent. This looks horribly complicated, but it actually has some recognizable structure. First of all, the parameter δ is what measures the strength of the external source, the imaginary chemical potential. Next, by a stroke of luck, a formalism of bi-orthogonal polymials had recently been invented for precisely this kind of problem by Eynard and Mehta [60]. Analytical miracles occur, and one ends with very simple analytical expressions for all spectral correlation functions. Particularly interesting spectral observables are those that *vanish* if $\delta = 0$. This leads to a strong signal with which the pion decay constant can be measured (since δ couples to F in the chiral Lagrangian). All results derived from the chiral Lagrangian in [57] are reproduced in this Random Matrix Theory approach. Moreover, all spectral correlation functions can now be expressed in closed analytical forms [59]. Also individual eigenvalue distributions can be computed analytically in this two-matrix theory [61]. Detailed comparisons with lattice gauge theory data, for a variety of different volumes and including the finite-volume corrections [62], have been made recently [63].

3.4. Beyond chiral Random Matrix Theory

Once one has realized the connection between the leading-order chiral Lagrangian in the ϵ-regime and chiral Random Matrix Theory, it becomes obvious how to combine the two. The most difficult part of the ϵ-expansion is precisely the non-perturbative contribution from the momentum zero modes of the pseudo-Goldstone bosons. We have seen how a partially quenched chiral Lagrangian had to be understood in this context just in order to derive properties of the spectral density. This is extremely useful for lattice gauge simulations since there one is often interested in doing partial quenching as an *approximation* to a real simulation. For example, one might wish to scan more parameter values with the same lattice configurations, or one has difficulty simulating with light enough fermions. Here partial quenching can provide additional information.

The first issue concerns pion-loop contributions to the leading-order chiral Lagrangian. This was investigated in the original paper by Gasser and Leutwyler [11]. Considering the contribution to the chiral condensate, they found that the Lagrangian remains form invariant except for a rescaling of the Σ-parameter,

$$\Sigma \;\rightarrow\; \Sigma_{eff} \;=\; \Sigma\left[1 - \frac{N_f^2-1}{N_f}\frac{\beta_1}{\sqrt{V}} + \mathcal{O}(1/V)\right] \tag{59}$$

for a four-dimensional torus of volume V. The coefficient is one of an infinite sequence of shape coefficients that appear in the ϵ-expansion [64]. This rescaled low-energy constant Σ_{eff} appears naturally in the expansion of any observable in the ϵ-regime. We see that the correction term disappears for large volume as $1/\sqrt{V}$, which is of order ϵ^2, as expected. The expansion thus seems to work well. In the fully quenched case this is not the case. There a new scale, set by the η'-mass in physical terms, enters and there is no longer a simple one-parameter expansion. As usual, when things can go wrong, they will go wrong. In this case one ends up with a new term that grows logarithmically with volume – a "quenched finite volume logarithm" [65]. Although such a correction term can be kept under control if one only compares two different volumes (at least to that order), it is just one more indication of the fundamental difficulty of quenched theories. In a physical theory with dynamical quarks such a logarithm still shows up at higher orders, but always suppressed by additional inverse powers of L, and therefore harmless.

Hansen [66] was the first to explore the ϵ-regime beyond leading order for mesonic correlation functions in chiral perturbation theory. This was before Leutwyler and Smilga [12] had shown the usefulness of working in sectors of fixed topological charge ν, and a re-analysis of Hansen's results in such sectors has been done [67]. The generalization of Hansen's results to the quenched case was also done there. What is interesting about mesonic correlation functions in the ϵ-regime is their *polynomial* dependence on the euclidean length scale. In the p-regime correlators have the conventional exponential fall-off, but in the ϵ-regime, where one is essentially "inside the pion", the correlation function is almost flat. One can understand this very simply. Consider the zero-momentum projection of the standard massive pion propagator,

$$\int d^3x\ \Delta(x) \;=\; \frac{\cosh\left(m_\pi(T/2-t)\right)}{2m_\pi\sinh(m_\pi T/2)}\ , \tag{60}$$

where T is the (arbitrarily chosen) euclidean time extent. Apart from the overall factor of $1/m_\pi^2$, we can Taylor expand this in powers of m_π^2 to find

$$\int d^3x\ \Delta(x) = \frac{1}{Tm_\pi^2} + \frac{T}{2}\left[\left(\tau-\frac{1}{2}\right)^2 - \frac{1}{12}\right] + \frac{T^3}{24}\left[\tau^2(\tau-1)^2 - \frac{1}{30}\right]m_{\pi^2} + \ldots \tag{61}$$

The finite-order polynomials of this expansion are precisely what in the ϵ-regime replace the ordinarily exponentially decaying correlation functions. The pole-term at $m_\pi = 0$ is the contribution from the momentum zero-mode, which indeed blows up in the simple expansion above (but which is treated exactly in the ϵ-expansion). This is the way in which the ϵ-regime expansion of correlation functions merge smoothly with the standard p-regime correlation functions. Such an approach has also been developed further, see, for instance, ref. [68].

An interesting question is whether one can combine the ϵ-regime with the p-regime, either partially quenched or not. Similarly, one can consider the question of so-called 'mixed actions' in this context (partial quenching where the light quark limit is reached by using configurations generated with different types of fermions, for example). Ideally, one would like to be able to do only one calculation which would be valid in a domain that stretches between the two different counting regimes. This is an active area of research at the moment [69, 70, 71] and results are currently being compared to large-scale simulations based on overlap fermions [72].

4. Outlook

There is a fascinating interplay between the chiral Lagrangian and chiral Random Matrix Theory. Here we have reviewed some of the exact relations that have been established. The fact that chiral Random Matrix Theory provides the leading contribution to the ϵ-expansion of chiral perturbation theory has turned out to give an analytically very powerful new view on low-energy QCD.

In lattice gauge theory, where most of the results discussed have very direct applicability, one is dealing with a discretized version of the continuum field theory. This induces lattice artifacts that are not physical. In a first approximation such discretization errors can be neglected, but in a more careful analysis they must be taken into account. In fact, measuring the quantitative effects of such lattice artifacts can be done in a systematic way following what is known as the Symanzik program of effective lattice field theory. There is then a modified chiral Lagrangian that includes discretization effects, but phrased in the continuum language. One can then also ask for the effect of discretization errors on the Dirac operator spectrum [73]. Remarkably, chiral Random Matrix Theory can be extended to include such effects as well [74, 75]. It is particular interesting to see how the special case of $N_f = 1$, where there is no spontaneous breaking of chiral symmetries, looks in this context. The full partition function is, to leading order, a trivial exponential of the free energy since there is a mass gap even in the chiral limit. But it contains now two terms: one proportional to the quark mass m, and one proportional to the leading lattice-spacing artifact a^2. These two terms compete trivially in the full theory for which the leading term in the effective partition function simply reads

$$Z = \exp\left[m\Sigma V - 2W_8 V a^2\right] \tag{62}$$

where W_8 is a new low-energy constant of Wilson fermions in this curious $N_f = 1$ theory (the constant of 2 in front is a convention inherited from the general-N_f theory where one has a genuine chiral Lagrangian). If one splits the partition function into sectors of fixed index ν, there is a non-trivial interplay between the two terms above due to the fixing of the index [76]. It is quite amazing how much carries over from the case of spontaneous chiral symmetry breaking once the index ν has been fixed. For the more physical case of a larger number of dynamical quarks also other observables, such as space-time correlation functions, can be computed for lattice gauge theory observables in the ϵ-regime with explicit a^2 corections. A variety of different scaling regimes can be considered. Such a program is presently being carried out for the case of Wilson fermions [77]. This is one of the clearly identifiable directions in which one can envision future research within this field.

ACKNOWLEDGMENT: The author is most grateful to Wolfgang Bietenholz for a careful reading of the manuscript and for many useful discussions.

References

[1] E. Merzbacher, *Quantum Mechanics*, John Wiley & Sons, Inc. (New York) 1998.
[2] T. Banks, C. M. Bender and T. T. Wu, Phys. Rev. D **8** (1973) 3346.
[3] M. M. Nieto, V. P. Gutschick, F. Cooper, D. Strottman and C. M. Bender, Phys. Lett. B **163** (1985) 336.
[4] A. Perez and D. Sudarsky, arXiv:0811.3181 [hep-th].

[5] M. B. Voloshin, "False vacuum decay,", *in* Erice School of Subnuclear Physics (1995) pp. 88-124.
[6] S. R. Coleman, Commun. Math. Phys. **31** (1973) 259.
[7] C. Vafa and E. Witten, Commun. Math. Phys. **95** (1984) 257.
[8] S. Weinberg, Physica A **96** (1979) 327.
[9] J. Gasser and H. Leutwyler, Annals Phys. **158** (1984) 142.
[10] H. Neuberger, Phys. Rev. Lett. **60** (1988) 889.
[11] J. Gasser and H. Leutwyler, Phys. Lett. B **184** (1987) 83.
[12] H. Leutwyler and A. V. Smilga, Phys. Rev. D **46** (1992) 5607.
[13] C. W. Bernard and M. F. L. Golterman, Phys. Rev. D **49** (1994) 486 [arXiv:hep-lat/9306005].
[14] P. H. Damgaard and K. Splittorff, Nucl. Phys. B **572** (2000) 478 [arXiv:hep-th/9912146]. P. H. Damgaard, Phys. Lett. B **476** (2000) 465 [arXiv:hep-lat/0001002]. G. Akemann, D. Dalmazi, P. H. Damgaard and J. J. M. Verbaarschot, Nucl. Phys. B **601** (2001) 77 [arXiv:hep-th/0011072].
[15] P. H. Damgaard and K. Splittorff, Phys. Rev. D **62** (2000) 054509 [arXiv:hep-lat/0003017].
[16] P. H. Damgaard, J. C. Osborn, D. Toublan and J. J. M. Verbaarschot, Nucl. Phys. B **547** (1999) 305 [arXiv:hep-th/9811212].
[17] K. Splittorff and J. J. M. Verbaarschot, Phys. Rev. Lett. **90** (2003) 041601 [arXiv:cond-mat/0209594].
[18] K. Splittorff and J. J. M. Verbaarschot, Nucl. Phys. B **683** (2004) 467 [arXiv:hep-th/0310271]; Nucl. Phys. B **695** (2004) 84 [arXiv:hep-th/0402177].
[19] Y. V. Fyodorov and G. Akemann, JETP Lett. **77** (2003) 438 [Pisma Zh. Eksp. Teor. Fiz. **77** (2003) 513] [arXiv:cond-mat/0210647].
[20] F. J. Dyson, J. Math. Phys. **3** (1962) 140.
[21] E. V. Shuryak and J. J. M. Verbaarschot, Nucl. Phys. A **560** (1993) 306 [arXiv:hep-th/9212088].
[22] J. J. M. Verbaarschot and I. Zahed, Phys. Rev. Lett. **70** (1993) 3852 [arXiv:hep-th/9303012]; J. J. M. Verbaarschot, Phys. Rev. Lett. **72** (1994) 2531 [arXiv:hep-th/9401059]; J. J. M. Verbaarschot, Phys. Lett. B **329** (1994) 351 [arXiv:hep-th/9402008]; J. J. M. Verbaarschot, Nucl. Phys. B **426** (1994) 559 [arXiv:hep-th/9401092].
[23] G. Akemann, P. H. Damgaard, U. Magnea and S. Nishigaki, Nucl. Phys. B **487** (1997) 721 [arXiv:hep-th/9609174].
[24] G. Akemann, P. H. Damgaard, U. Magnea and S. M. Nishigaki, Nucl. Phys. B **519** (1998) 682 [arXiv:hep-th/9712006].
[25] W. Bietenholz and I. Hip, PoS **LAT2009** (2009) 086 [arXiv:0909.2241 [hep-lat]].
[26] M. E. Berbenni-Bitsch, S. Meyer, A. Schafer, J. J. M. Verbaarschot and T. Wettig, Phys. Rev. Lett. **80** (1998) 1146 [arXiv:hep-lat/9704018].
[27] T. Nagao and P.J. Forrester, Nucl. Phys. B **435** (1995) 401.
[28] J. J. M. Verbaarschot, Nucl. Phys. B **426** (1994) 559 [arXiv:hep-th/9401092].
[29] J. J. M. Verbaarschot, Phys. Rev. Lett. **72** (1994) 2531 [arXiv:hep-th/9401059].
[30] P. H. Damgaard, U. M. Heller, R. Niclasen and B. Svetitsky, Nucl. Phys. B **633** (2002) 97 [arXiv:hep-lat/0110028].
[31] P. H. Damgaard, U. M. Heller, R. Niclasen and K. Rummukainen, Phys. Rev. D **61** (1999) 014501 [arXiv:hep-lat/9907019].
[32] E. Follana, A. Hart and C. T. H. Davies [HPQCD Collaboration and UKQCD Collaboration], Phys. Rev. Lett. **93** (2004) 241601 [arXiv:hep-lat/0406010].
[33] P. H. Damgaard, U. M. Heller and A. Krasnitz, Phys. Lett. B **445** (1999) 366 [arXiv:hep-lat/9810060].
[34] R. G. Edwards, U. M. Heller, J. E. Kiskis and R. Narayanan, Phys. Rev. Lett. **82** (1999) 4188 [arXiv:hep-th/9902117].
[35] F. Farchioni, I. Hip, C. B. Lang and M. Wohlgenannt, Nucl. Phys. B **549** (1999) 364 [arXiv:hep-lat/9812018]; W. Bietenholz and S. Shcheredin, Nucl. Phys. B **754** (2006) 17 [arXiv:hep-lat/0605013].
[36] W. Bietenholz, K. Jansen and S. Shcheredin, JHEP **0307** (2003) 033 [arXiv:hep-lat/0306022].
[37] L. Giusti, M. Luscher, P. Weisz and H. Wittig, JHEP **0311** (2003) 023 [arXiv:hep-lat/0309189].
[38] R. Narayanan and H. Neuberger, Nucl. Phys. B **696** (2004) 107 [arXiv:hep-lat/0405025].
[39] T. DeGrand and S. Schaefer, Phys. Rev. D **76** (2007) 094509 [arXiv:0708.1731 [hep-lat]]; A. Hasenfratz, R. Hoffmann, S. Schaefer, Phys. Rev. **D78** (2008) 054511. [arXiv:0806.4586 [hep-lat]].
[40] P. H. Damgaard and S. M. Nishigaki, Nucl. Phys. B **518** (1998) 495 [arXiv:hep-th/9711023]; P. H. Damgaard and S. M. Nishigaki, Phys. Rev. D **57** (1998) 5299 [arXiv:hep-th/9711096].
[41] P. Zinn-Justin, Nucl. Phys. B **497** (1997) 725 [arXiv:cond-mat/9703033].
[42] P. H. Damgaard, Phys. Lett. B **424** (1998) 322 [arXiv:hep-th/9711110].
[43] G. Akemann and P. H. Damgaard, Nucl. Phys. B **528** (1998) 411 [arXiv:hep-th/9801133]. G. Akemann and P. H. Damgaard, Phys. Lett. B **432** (1998) 390 [arXiv:hep-th/9802174].
[44] R. Brower, P. Rossi and C. I. Tan, Nucl. Phys. B **190** (1981) 699; A. D. Jackson, M. K. Sener and

J. J. M. Verbaarschot, Phys. Lett. B **387** (1996) 355 [arXiv:hep-th/9605183].
[45] G. Akemann and E. Kanzieper, Phys. Rev. Lett. **85** (2000) 1174 [arXiv:hep-th/0001188].
[46] T. Nagao and S. M. Nishigaki, Phys. Rev. D **62** (2000) 065006 [arXiv:hep-th/0001137]; Phys. Rev. D **62** (2000) 065007 [arXiv:hep-th/0003009].
[47] P. H. Damgaard, Phys. Lett. B **425** (1998) 151 [arXiv:hep-th/9711047].
[48] S. M. Nishigaki, P. H. Damgaard and T. Wettig, Phys. Rev. D **58** (1998) 087704 [arXiv:hep-th/9803007].
[49] P. H. Damgaard and S. M. Nishigaki, Phys. Rev. D **63** (2001) 045012 [arXiv:hep-th/0006111].
[50] P. H. Damgaard, U. M. Heller, R. Niclasen and K. Rummukainen, Phys. Lett. B **495** (2000) 263 [arXiv:hep-lat/0007041].
[51] J. J. M. Verbaarschot, Phys. Lett. B **368** (1996) 137 [arXiv:hep-ph/9509369].
[52] P. H. Damgaard, R. G. Edwards, U. M. Heller and R. Narayanan, Phys. Rev. D **61** (2000) 094503 [arXiv:hep-lat/9907016].
[53] P. H. Damgaard, Nucl. Phys. B **556** (1999) 327 [arXiv:hep-th/9903096].
[54] G. Akemann and P. H. Damgaard, Phys. Lett. B **583** (2004) 199 [arXiv:hep-th/0311171].
[55] J. C. Osborn, D. Toublan and J. J. M. Verbaarschot, Nucl. Phys. B **540** (1999) 317 [arXiv:hep-th/9806110].
[56] F. Basile and G. Akemann, JHEP **0712** (2007) 043 [arXiv:0710.0376 [hep-th]].
[57] P. H. Damgaard, U. M. Heller, K. Splittorff and B. Svetitsky, Phys. Rev. D **72** (2005) 091501 [arXiv:hep-lat/0508029]; Phys. Rev. D **73** (2006) 074023 [arXiv:hep-lat/0602030]. P. H. Damgaard, U. M. Heller, K. Splittorff, B. Svetitsky and D. Toublan, Phys. Rev. D **73** (2006) 105016 [arXiv:hep-th/0604054].
[58] J. C. Osborn, Phys. Rev. Lett. **93** (2004) 222001 [arXiv:hep-th/0403131].
[59] G. Akemann, P. H. Damgaard, J. C. Osborn and K. Splittorff, Nucl. Phys. B **766** (2007) 34 [Erratum-ibid. B **800** (2008) 406] [arXiv:hep-th/0609059].
[60] B. Eynard and M.L. Mehta, J. Phys. A **31** (1998) 4449 [cond-mat/9710230].
[61] G. Akemann and P. H. Damgaard, JHEP **0803** (2008) 073 [arXiv:0803.1171 [hep-th]].
[62] G. Akemann, F. Basile and L. Lellouch, JHEP **0812** (2008) 069 [arXiv:0804.3809 [hep-lat]]; P. H. Damgaard, T. DeGrand and H. Fukaya, JHEP **0712** (2007) 060 [arXiv:0711.0167 [hep-lat]]; C. Lehner and T. Wettig, JHEP **0911** (2009) 005 [arXiv:0909.1489 [hep-lat]].
[63] C. Lehner, S. Hashimoto and T. Wettig, JHEP **1006** (2010) 028 [arXiv:1004.5584 [hep-lat]]; C. Lehner, J. Bloch, S. Hashimoto and T. Wettig, arXiv:1101.5576 [hep-lat].
[64] P. Hasenfratz and H. Leutwyler, Nucl. Phys. B **343** (1990) 241.
[65] P. H. Damgaard, Nucl. Phys. B **608** (2001) 162 [arXiv:hep-lat/0105010].
[66] F. C. Hansen, Nucl. Phys. B **345** (1990) 685. F. C. Hansen and H. Leutwyler, Nucl. Phys. B **350** (1991) 201.
[67] P. H. Damgaard, M. C. Diamantini, P. Hernandez and K. Jansen, Nucl. Phys. B **629** (2002) 445 [arXiv:hep-lat/0112016]; P. H. Damgaard, P. Hernandez, K. Jansen, M. Laine and L. Lellouch, Nucl. Phys. B **656** (2003) 226 [arXiv:hep-lat/0211020].
[68] L. Giusti, P. Hernandez, M. Laine, P. Weisz and H. Wittig, JHEP **0401** (2004) 003 [arXiv:hep-lat/0312012]. L. Giusti, P. Hernandez, M. Laine, P. Weisz and H. Wittig, JHEP **0404** (2004) 013 [arXiv:hep-lat/0402002]. W. Bietenholz, T. Chiarappa, K. Jansen, K. I. Nagai and S. Shcheredin, JHEP **0402** (2004) 023 [arXiv:hep-lat/0311012].
[69] P. H. Damgaard and H. Fukaya, Nucl. Phys. B **793** (2008) 160 [arXiv:0707.3740 [hep-lat]].
[70] F. Bernardoni and P. Hernandez, JHEP **0710** (2007) 033 [arXiv:0707.3887 [hep-lat]]; F. Bernardoni, P. H. Damgaard, H. Fukaya and P. Hernandez, JHEP **0810** (2008) 008 [arXiv:0808.1986 [hep-lat]]; F. Bernardoni, P. Hernandez and S. Necco, JHEP **1001** (2010) 070 [arXiv:0910.2537 [hep-lat]]; F. Bernardoni, P. Hernandez, N. Garron, S. Necco and C. Pena, arXiv:1008.1870 [hep-lat].
[71] P. H. Damgaard and H. Fukaya, JHEP **0901** (2009) 052 [arXiv:0812.2797 [hep-lat]].
[72] H. Fukaya, S. Aoki, S. Hashimoto, T. Kaneko, J. Noaki, T. Onogi and N. Yamada [JLQCD collaboration], Phys. Rev. Lett. **104** (2010) 122002 [Erratum-ibid. **105** (2010) 159901] [arXiv:0911.5555 [hep-lat]]; H. Fukaya *et al.* [JLQCD and TWQCD collaborations], arXiv:1012.4052 [hep-lat].
[73] S. R. Sharpe, Phys. Rev. D **74** (2006) 014512 [arXiv:hep-lat/0606002].
[74] P. H. Damgaard, K. Splittorff and J. J. M. Verbaarschot, Phys. Rev. Lett. **105** (2010) 162002 [arXiv:1001.2937 [hep-th]]; G. Akemann, P. H. Damgaard, K. Splittorff and J. J. M. Verbaarschot, arXiv:1012.0752 [hep-lat]; arXiv:1012.0752 [hep-lat].
[75] J. C. Osborn, arXiv:1012.4837 [hep-lat].
[76] G. Akemann, P. H. Damgaard, K. Splittorff and J. J. M. Verbaarschot, PoS **LATTICE2010** (2010) 079 [arXiv:1011.5121 [hep-lat]].
[77] A. Shindler, Phys. Lett. B **672** (2009) 82 [arXiv:0812.2251 [hep-lat]]. O. Bar, S. Necco and S. Schaefer, JHEP **0903** (2009) 006 [arXiv:0812.2403 [hep-lat]]. O. Bar, S. Necco and A. Shindler, JHEP **1004** (2010) 053 [arXiv:1002.1582 [hep-lat]]; S. Necco and A. Shindler, arXiv:1101.1778 [hep-lat].

Collider Physics an Experimental Introduction
Journal of Physics: Conference Series

Carmine Elvezio Pagliarone[1]
Università di Cassino & INFN Trieste, Italy

E-mail: *pagliarone@fnal.gov*

Abstract. This paper reviews shortly a small part of the contents of a set of lectures, presented at the XIV International School of Particles and Fields in Morelia, state of Michoacán, Mexico, during November 2010. The main goal of those lectures was to introduce students to some of the basic ideas and tools required for experimental and phenomenological analysis of collider data. In particular, after an introduction to the scientific motivations, that drives the construction of powerful accelerator complexes, and the need of reaching high center of mass energies and luminosities, some basic concept about collider particle detectors will be discussed. A status about the present running colliders and collider experiments as well as future plans and research and development is also given.

1. Introduction
The need of understanding the nature and its laws in the sub-nuclear domain has always been the driving force behind the development of particle accelerators and particle detectors and related techniques. Particle physics experiments, at the highest possible energies, with their unique ability to explore the uncharted territory at the energy frontier have been providing and will provide exciting opportunities for great scientific advances. In the past four decades, hadron colliders such as the Intersecting Storage Rings (ISR, CERN), that on January 27th 1971 delivered the first pp collision [1], the Tevatron Collider, the Fermilab $p\bar{p}$ machine, and at the present the Large Hadron Collider (LHC, CERN) have held the high-energy frontier resulting in the discovery of the W and Z-bosons [2], the top quark [3, 4, 5], together with other particles and other important measurements. Very important test about the Standard Model of Particle Physics (SM) have been performed using lepton colliders such as the Large Electron Positron (LEP) at CERN, the Positron Electron Project (PEP) at SLAC (the first e^+e^-linear collider in the history) and the Hadron Electron Ring Accelerator (HERA) an ep collider at DESY. Other collider facilities, specifically, designed to produce a large number of B mesons, the so-called B-factories, such as the PEP-II at SLAC or KEKB at KEK in Japan have been used to intensively and widely studying the CP violation. In the next years, LHC will provide, excellent chances to discover the Higgs boson and new physics at the TeV scale, while the Tevatron collider experiments will end collecting data by September 2011, continuing the process of analyzing the collected data for other years, contributing to the process of searching for new physics.

[1] To whom any correspondence should be addressed.

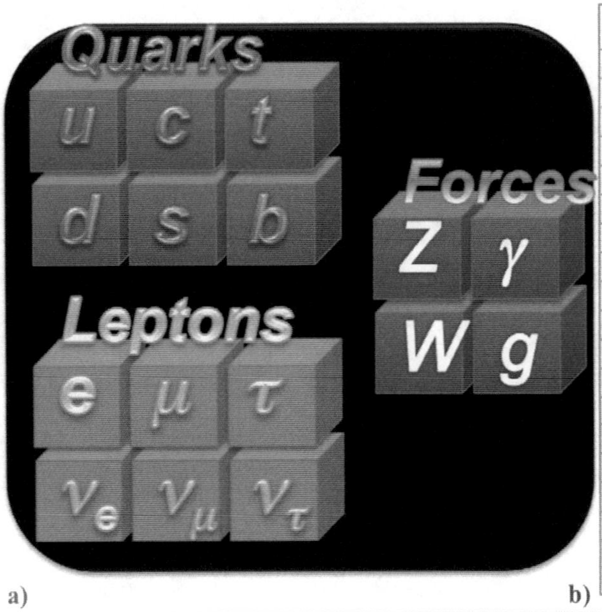

a)

b)

Symbol	Description	Renormalization scheme (point)	Value
m_e	Electron mass		511 keV
m_μ	Muon mass		105.7 MeV
m_τ	Tau mass		1.78 GeV
m_u	Up quark mass	$\mu_{\overline{MS}} = 2$ GeV	1.9 MeV
m_d	Down quark mass	$\mu_{\overline{MS}} = 2$ GeV	4.4 MeV
m_s	Strange quark mass	$\mu_{\overline{MS}} = 2$ GeV	87 MeV
m_c	Charm quark mass	$\mu_{\overline{MS}} = m_c$	1.32 GeV
m_b	Bottom quark mass	$\mu_{\overline{MS}} = m_b$	4.24 GeV
m_t	Top quark mass	On-shell scheme	172.7 GeV
θ_{12}	CKM 12-mixing angle		13.1°
θ_{23}	CKM 23-mixing angle		2.4°
θ_{13}	CKM 13-mixing angle		0.2°
δ	CKM CP-violating Phase		0.995
g_1	U(1) gauge coupling	$\mu_{\overline{MS}} = m_Z$	0.357
g_2	SU(2) gauge coupling	$\mu_{\overline{MS}} = m_Z$	0.652
g_3	SU(3) gauge coupling	$\mu_{\overline{MS}} = m_Z$	1.221
θ_{QCD}	QCD vacuum angle		~0
μ	Higgs quadratic coupling		Unknown
λ	Higgs self-coupling strength		Unknown

Figure 1: a) The "Periodic Table" of Particle Physics; b) list of the Standard Model free parameters and their values, as measured up today [6].

2. The Physics challenge

The Standard Model of Particle Physics represents the simplest and most economical theory, which is able to describe jointly weak and electromagnetic interactions. At present it continues to survive all experimental tests, providing a remarkably successful description of known phenomena; some precision observable tests it at 10^{-3} (see Fig. 1.a and Fig. 1.b) [7]. In spite of that, there are plenty of aspects that we do not understand yet and that may suggest the SM to be most likely, a low energy effective theory of spin-1/2 matter fermions interacting via spin-1gauge bosons. As matter of fact, the SM leaves many important questions unanswered. It is also widely acknowledged that, from the theoretical standpoint, the SM must be part of a larger theory, "beyond" the SM (BSM), which is yet to be experimentally "confirmed". Within the kinematic regime and the viable physical signatures, which have been possible to explore so far, none, of the ample spectrum of particles, predicted by models beyond the SM [8], as well as the Higgs boson, that is a key building block of the SM, has been found so far. Most of these particles may have, depending on the theoretical scenario under consideration, production cross-sections that lies well below the most common QCD production cross sections, as shown in Fig. 2.a and in the case of LHC in Fig. 8.a. The basic formula for evaluating the cross section and kinematic distributions for a 2 → N scattering process is the following:

$$d\sigma = \frac{1}{2s}\left(\prod_{i=1}^{N}\frac{d^3p_i}{(2\pi)^3}\frac{1}{2E_i}\right)\cdot(2\pi)^4\delta^4\left(p_A+p_B-\sum p_i\right)\cdot\left|M\left(p_A,p_B\rightarrow\{p_i\}\right)\right|^2 \qquad [1]$$

where M is the invariant matrix element, also know as scattering amplitude, and $p_i \equiv (E_i, \vec{p}_i)$ are the 4-momenta of the final-state particles. Note that M contains the entire information specific for the process under consideration, information such as coupling constant dependence and so on, whereas all other ingredients are simply kinematic factors common for any 2 → N process. Equation [1] is written in the center of mass (cms) frame of the colliding particles, it is in fact invariant under boosts parallel to the collision axis. This feature is important when we consider, in particular, hadron collisions. If the colliding beams are unpolarized, one needs to average the quantity $|M|^2$ over all possible initial-states

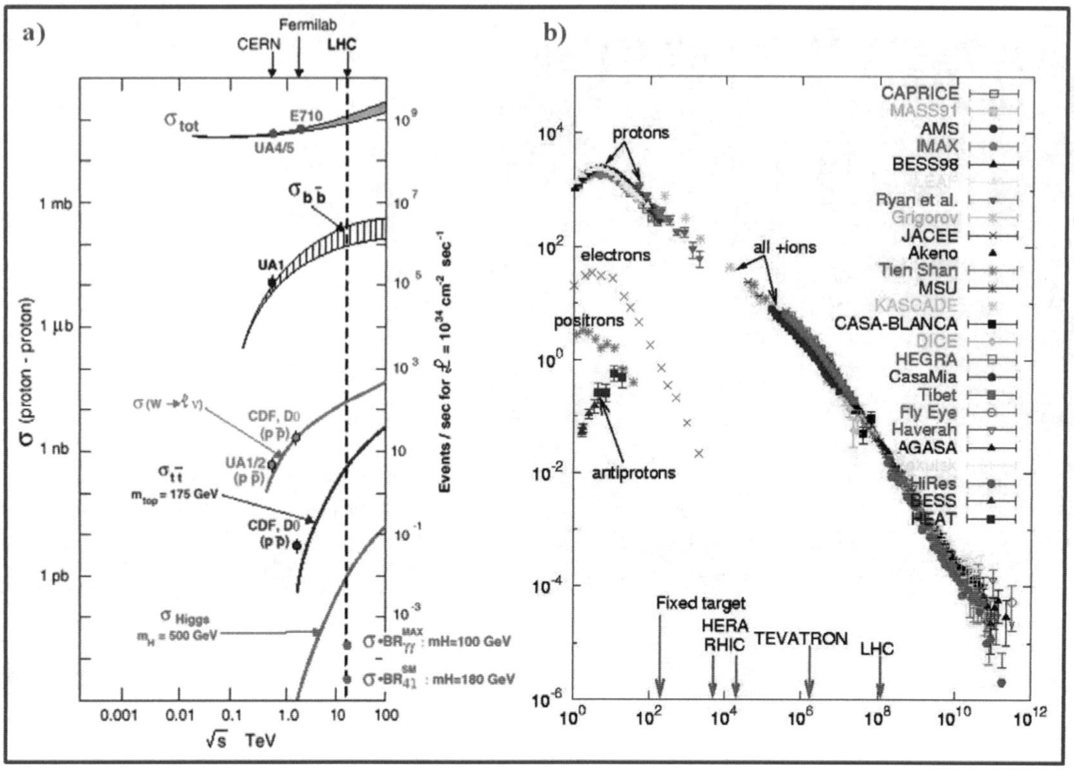

Figure 2: a) pp and $p\bar{p}$ cross-sections as a function of the center of mass energy $\sqrt{s}$ [9]; b) global view of the full measured cosmic ray spectrum as performed by several different experiments [10].

polarizations. If the beams are polarized (this was the case of the Stanford Linear Collider, SLC, and may be implemented at the ILC), an appropriately weighted average should be computed instead. In addition, if the final-state particles have spin, $|M|^2$ should typically be summed over all possible spin states, since no collider detector is capable of detecting spins of individual particles. (Exception occurs when the final-state particles decay promptly, in which case the angular distribution of their decay products may carry information about their polarization state.) The appropriately averaged and/or summed squared scattering amplitude will be denoted by $|M|^2$. A summary of the cross sections, for the most important SM processes and not, in the case of pp and $p\bar{p}$ collisions, as a function of $\sqrt{s}$, is given in Fig. 2.a and in Fig. 8.a for the LHC case. The total scattering cross section $\sigma_{Tot} \approx 0.1\ bn$ is completely dominated by low-$p_\perp$ pure QCD scattering, and depends only weakly (logarithmically) on the center of mass energy ($\sqrt{s}$). The inclusive cross section for b-quark production is 2÷3 orders of magnitude lower, reflecting the effective $p_\perp$ cutoff provided by the b-quark mass. The strongest electroweak processes, inclusive W and Z production, have cross-sections of order 10÷100 nb; in other words, there is one electroweak boson produced per 1÷10 million of pure QCD events. Of course, the Data Acquisition System (DAQ) of each experiment does not collect the vast majority of these purely QCD events, as they are not relevant for most of the searches for physics beyond the SM or for the SM Higgs searches or for other specific SM channels under investigation. For this reason is common to use specific "trigger" systems (triggers), that may be hardware or software or both of them, with single or multiple trigger levels, able to select, with high efficiency and large background rejection, those events that match specific final state topologies. Even if triggers are necessary, they anyhow introduce unavoidable bias. Generally speaking, experiments running on hadron colliders are more sensitive to the need of having more complex trigger systems.

3. Particle Accelerators

When J. J. Thomson discovered the electron, he did not call the instrument, he was using, an accelerator, but in reality it was an accelerator as electron was accelerated between two electrodes to which he had applied a difference in electric potential. He manipulated the resulting beam with electric and magnetic fields to determine the ratio between the charge and the mass of the so-called cathode rays (q_e/m_e). Thomson achieved his discovery by studying the properties of the beam itself not its impact on a target or on another beam, as we do today. Accelerators have, since that time, become indispensable in the quest to understand Nature at smaller and smaller scales. And although they are much bigger and far more complex, they still operate on much the same physical principles as Thomson's device. It took another half century, however, before accelerators became entrenched as the key tools in the search for subatomic particles. The energies and techniques have changed, but the scientific methods have not. These machines, often little known, in the world outside particle physics, are extensively used for general industrial use (~1500 machines for sterilization, imaging, etc.), ion implantation and surface implantation (~6000 machines for controlled semiconductor doping, change of surface properties, etc.), radioisotope production (~200 machines for cancer treatment, imaging organs for medical use, etc.), radiotherapy (~4500 machines for cancer treatment with X-rays, protons, neutrons and ions) and research (~1200 machines for producing synchrotron light for biomedical, physics, chemistry, biology, material research) [11].

3.1. The Cosmic Accelerator

Before starting to describe as a collider work we want to remind that the most energetic collisions, available for experimental analysis, do not occur in the particle colliders rather they are produced during the collision of cosmic ray primaries (CRs) with nuclei in the Earth atmosphere [12]. The spectrum of those particles has been measured, extensively, over more than ten orders of magnitude in energy ($E < 10^{11}$) and about 30 orders of magnitude in flux. Figure 2.b shows a global view of the full measured cosmic ray spectrum as performed by several different experiments [13]. It is believed that most CR primaries are accelerated in shock fronts, surrounding celestial bodies [14]. The basic requirement, for such acceleration mechanisms to work, is that there must be a sufficiently strong magnetic field B extending over a sufficient length L giving rise to a sufficiently large energy CR primaries. In this scenario, the neutral primaries would have to be produced when charged particles collide with the matter, but these would require even higher energies for the charged particles than what is commonly observed on the Earth. It is possible, in principle, to use high energy cosmic rays to make studies similar to the one we do using particle colliders. Of course these studies are performed but, because of the very low particle flux, the number of events, expected, is very small even if we construct very large surface detectors. In many cases, anyhow, the study of Ultra High-energy Cosmic Rays (UHCRs) is very useful and complementary to the study performed using particle colliders [15].

3.2. Colliding beams and the Energy frontier

Colliders are only the last product of the evolution of devices built in order to perform violent collisions of particles on other particles. Earlier version used accelerated beams impinging on fixed targets. The focus is on colliding beam machines (colliders) because they provide the highest center of mass energies. As matter of fact if $p_\mu(j) \equiv (E_j, \vec{p}_j)$, with j=1,2, are the 4-momentuma of the two colliding particles of mass m_1 and m_2 we have that the center of mass energy, during the collisions is:

$$E_{CM} = (p_\mu(1) + p_\mu(2)) = (p^\mu(1) + p^\mu(2)) = \sqrt{m_1 + m_2 + 2E_1E_2 + 2p_1p_2} \approx \sqrt{4E_1E_2} \quad [2]$$

In the case of a fixed target experiment, in the laboratory frame, assuming that the particle 1 is the fixed target particle of mass m_T and the particle 2 is the colliding beam of mass m_B with a beam

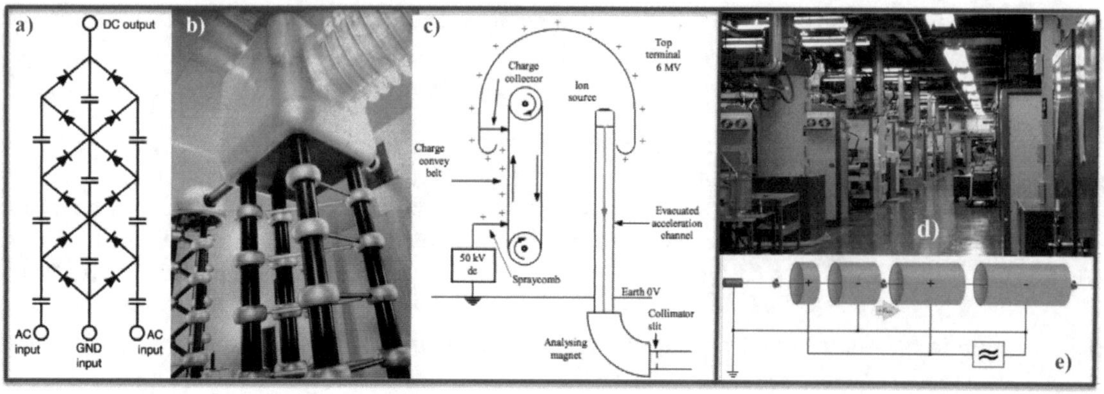

Figure 3: a) Schematic view of a Cockcroft-Walton generator; b) the Fermilab Cockcroft-Walton accelerator [16]; c) a schematic view of a generic Van de Graaff generator with an accelerator inside; d) the Fermilab Linac Klystrons and Klystron modulators; e) schematic description of the linac working principle [16].

energy of E_B, we have: $p_\mu(1) \equiv (m_T, 0)$ and $p_\mu(B) \equiv (E_B, \vec{p}_B)$, then, for a fixed target collision, the center of mass energy will be lower:

$$E_{CM} = \sqrt{m_T + m_B + 2m_T E_B} \approx \sqrt{2m_T E_B} \qquad [3]$$

R. Widerow laid down, for the first time in the history of physics, the concept of a particle collider in a German patent that was registered in 1943 but not published until 1952. It proposed storing beams and allowing them to collide repeatedly so to obtain a high energy in the center of mass. This result has been achieved, after several years of fixed target experiments, only the 27 January 1971, 40 years ago, when two beams of protons collided for the first time in the CERN Intersecting Storage Rings (ISR) [1]. It was the world's first hadron collider, and ran from 1971 to 1984, achieving the maximum center of mass energy of 62 GeV. From its initial startup, the collider itself had the capability to produce particles like the J/ψ and the upsilon, as well as to observe jets; however, the particle detector experiments were not configured to observe events with large momentum transverse to the beamline, leaving these discoveries to be made at other experiments in the mid-1970s. Nevertheless, the construction of the ISR involved many advances in accelerator physics, including the first use of stochastic cooling, and it held the record for luminosity at a hadron collider until surpassed by the Fermilab Tevatron collider in 2004.

In order to achieve very high-energy collisions a set of different kinds of particle accelerators is needed. Each specific accelerator operate to step-up the beam energy up to some level and deliver the beam to the next machine until the beam will not reach the energy needed to be injected into the collider. Later on we will give an example of this, describing how the two presently running colliders, the Tevatron collider and the Large Hadron Collider work.

Cockcroft-Walton and Van de Graaff Accelerators

Common to all accelerators is the use of electric fields for the acceleration of charged particles; however, the manner in which the fields are applied varies widely. The most straightforward type of accelerators results from the application of a potential difference between two terminals. To obtain more than about 200 kV of accelerating voltage, it is necessary to use one or more stages of voltage-doubling circuits. J. D. Cockcroft and E. T. S. Walton built the first of such a device in 1932. Cockcroft-Walton accelerators are still widely used nowadays as injectors for much larger accelerators (see Fig. 3.a and Fig. 3.b). Another kind, pioneered at beginning of 1929 by R. J. Van de Graaff is the Van de Graaff accelerator, in which a high potential difference is built up and maintained on a smooth conducting surface by the continuous transfer of positive static charges from a moving belt to the

Figure 4: a) Schematic description of the acceleration process inside a RF cavity; the particles fell, always, a force in the forward direction; b) detail of a LEP RF cavity [17]; c) a LHC Four-cavity module during the assembly [18]; d) ILC 1.3 GHz nine-cell superconducting RF cavity [19]. These niobium cavities are supposed to operate at an accelerating gradient of 31.5 MV/m.

surface (see Fig 3.c). The limitation of these types of accelerators arises from the maximum practical potential difference that can be held by the charged surfaces.

Linear Accelerator

In a linear accelerator (Linac), an ion is injected into an accelerating tube containing a number of electrodes. A high-frequency alternating voltage, from an oscillator, is applied between groups of electrodes. An ion traveling down the tube will be accelerated in the gap between the electrodes if the voltage is in the proper phase. The distance between electrodes increases along the length of the tube so that the particle stays in phase with the voltage (see Fig. 3.d and Fig. 3.e). R. Wideroe built the first linear accelerator in 1928. Since that time, a sizable number of linear accelerators, also called linacs, have come into operation, both for electron and proton acceleration, as well as several heavy-ion linacs. In the next sections we will discuss the evolution of linacs with the use of superconducting Radio Frequencies (RF) cavities (SCRF).

Cyclotron

The best known and one of the most successful devices for acceleration of ions to millions of electron volts is the cyclotron, which was invented by E. O. Lawrence in 1929. The first working model produced 80 keV protons in 1930. A cyclotron, as well as a linac, uses multiple acceleration by a RF electrical field. However, the ions in a cyclotron are constrained by a magnetic field to move in a spiral path. The ions are injected at the center of the magnet between two semicircular electrodes called dees. As the particle spirals outward it gets accelerated each time it crosses the gap between the dees. The time it takes a particle to complete an orbit is constant, since the distance it travels increases at the same rate as its velocity, allowing it to stay in phase with the RF. As relativistic energies are approached, this condition breaks down, limiting cyclotrons in energy. However, cyclotrons are still in use all over the world for nuclear science studies, radioisotope production, and medical therapy.

Journal of Physics: Conference Series **287** (2011) 012005 doi:10.1088/1742-6596/287/1/012005

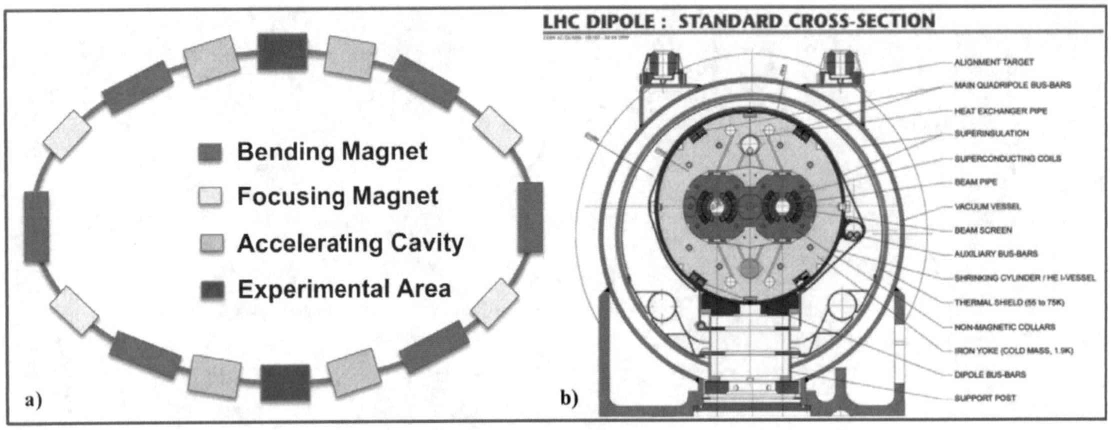

Figure 5: a) A schematic view of a generic circular collider; b) cutaway of a LHC dipole [18]. Two opposite sign dipolar fields are used to make the proton bunches circulating in opposite directions.

Synchrotron
The synchrotron was developed to overcome the energy limitations of cyclotrons imposed by special relativity. In a synchrotron, the radius of the orbit is kept constant by a magnetic field that increases with time as the momentum of a particle increases. The acceleration is provided by a RF oscillator that supplies an energy increment every time the particle beams crosses an accelerating gap.

Main collider components
In a circular accelerator, particles circulate in a vacuum tube and are manipulated using electromagnetic devices: dipole magnets keep the particles in their nearly circular orbits, quadrupole magnets focus the beam, and accelerating cavities are electromagnetic resonators that accelerate particles and then keep them at a constant energy by compensating the Larmor energy losses, described in the next section. Fig. 4.a gives a schematic description of the acceleration process inside a RF cavity, where the particles fell, always, a force in the forward direction. In Fig. 4.b, Fig. 4.c and Fig. 4.d we show a LEP RF cavities, a LHC SCRF cavity and an example of a 1.3 GHz nine-cell SCRF cavity for the International Linear Collider (ILC). The LHC and ILC colliders will be described in some detail later. A schematic description of a generic circular collider is given in Fig. 5.a together with a cutaway of a LHC dipole (see Fig 5.b).

3.3. The Luminosity Frontier
Luminosity of a high-energy collider is a key parameter that is second, in importance, only to the collider energy reach. Luminosity measures the flux of particles capable of creating a reaction of interest. The reaction rate (R), that is the number of events produced in a collision per unit time, is directly proportional to the luminosity and is given by the following equation:

$$R_X(t) = \sigma(AB \to X + \sum_j Y_j; E) L(t) \quad [4]$$

where $\sigma(AB \to X; E)$ is defined to be the total X particle production cross section. Though the units for the cross section are conventionally taken as cm^2, these units are too big in order to be used for sub-atomic particle scattering, and thus more suitable units, called a barn, are introduced as follow: 1 cm^2 =10^{24} barn= 10^{27} mb= 10^{36} pb= 10^{39} fb= 10^{42} ab. It may also be convenient to use these units for the luminosity accordingly like 1 cm^{-2}s^{-1} = 10^{-33}nb^{-1}s^{-1}. If we call N , the total number of event produced in a collider, for a given process, after an "effective" running time T , we will have:

$$N_X = \sigma(AB \to X + \sum_j Y_j; E) \int_0^T L(t) dt \quad [5]$$

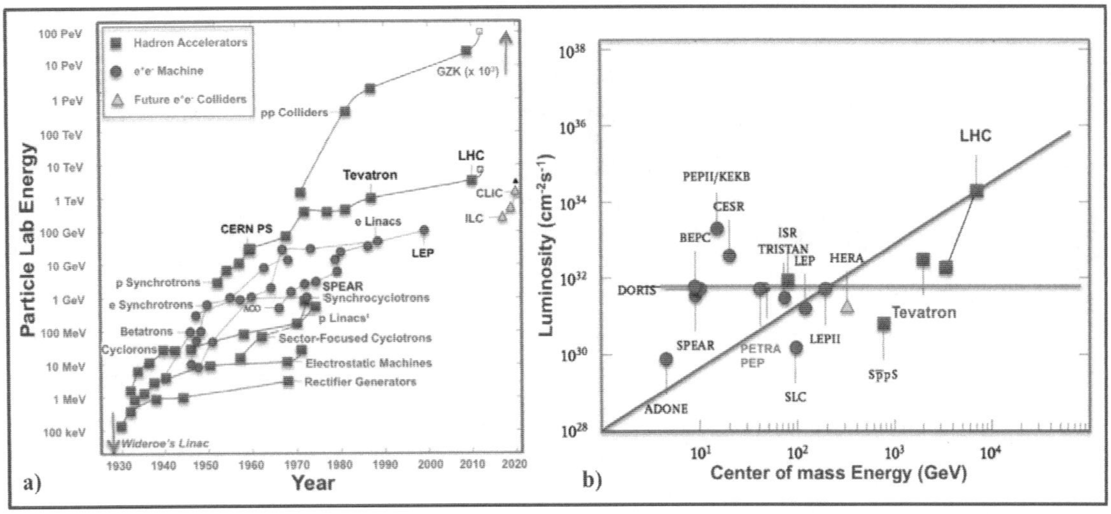

Figure 6: a) Livingston plot showing the historical exponential growth with time in the energy reach from Wideröe's Linac to the GZK limit (solid symbols are for existing machines, open symbols are for nonexistent machines); b) Panofsky plot showing the collider luminosity versus $\sqrt{s}$ [20].

For a given integrated luminosity the number of events observed in an experiment is:

$$N_{obs} = \sigma(AB \to X + \sum_j Y_j; \sqrt{s}) \times \varepsilon_{Detection} \times \int_0^T L\, dt + N_{Background} \quad [6]$$

where $\varepsilon_{Detection}$ is the probability that a signal event will be observed in a given detector and depends from the overall detector and trigger efficiency and acceptance and $N_{Background}$ is the number of events produced from other processes that got counted incorrectly between the signal events. At colliders, the luminosity depends on both the beam intensities and the beam densities and it is generally defined as 4-dimensional overlap integral of two colliding bunches in space and time and in the ultrarelativistic limit given by:

$$L = f_c \int\int\int\int_{-\infty}^{+\infty} \rho^A(x,y,s+ct)\rho^B(x,y,s+ct) 2c\, dt\, ds\, dx\, dy \quad [7]$$

where ρ^A and ρ^B are the particle distributions within the bunches. Note that the relative velocity of the colliding bunches in laboratory space is $2c$ and $f_c = c/b$ is the collision frequency with b the distance between successive bunches in the beam, as shown below.

If we assume a Gaussian distributions for ρ^A and ρ^B and, for simplification, we also assume

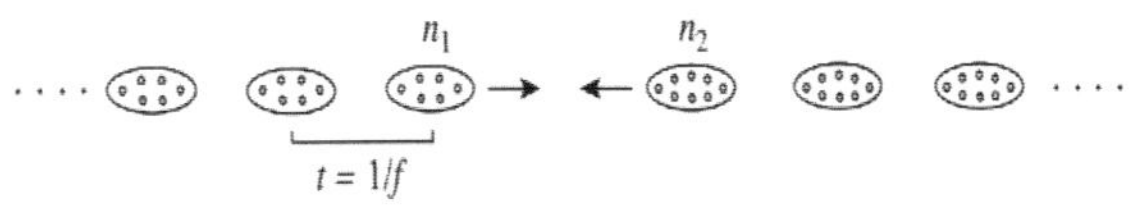

identical *rms* beam parameters for both beams and we also allow the beams to collide not head-on but at a small crossing angle where $2\theta << 1$ we may write:

$$L = \frac{f_c N_A N_B}{4\pi\sigma_x^*\sigma_y^*} S \quad [8]$$

where σ_x^* and σ_y^* are the *rms* beam radii, at collision point. The beam radii before and after the interaction point are expressed as follow:

$$\sigma_j = \sigma_j^* \sqrt{1 + \left(\frac{s}{\beta_j^*}\right)^2}, \qquad j = x, y \quad [9]$$

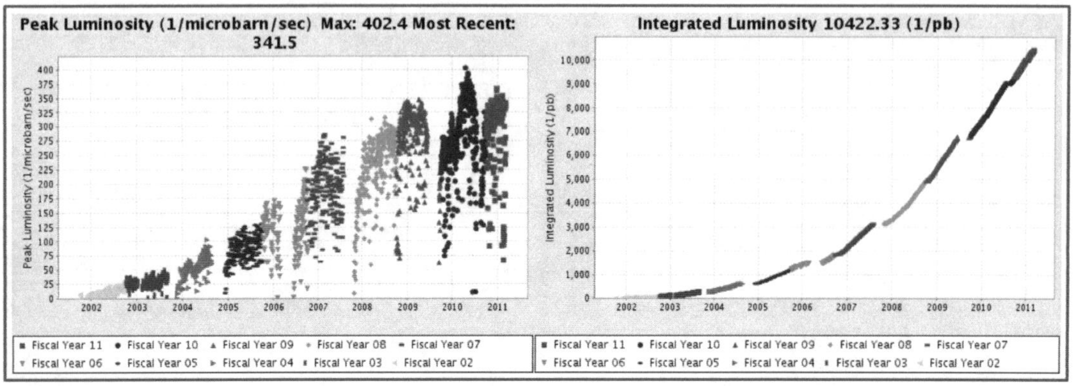

Figure 7: a) Tevatron collider peak luminosity as a function of time; the grow, in the machine performances, is clearly visible [21]; b) Tevatron integrated luminosity as a function of time [21].

The luminosity suppression factor S describes the increase of effective interaction area due to the divergence from the formula and to the crossing angle between the beams. Equation 8 can be recast in terms of emittances and amplitude functions as

$$L = f\frac{n_1 n_2}{\sqrt{\varepsilon_x \beta_x^* \varepsilon_y \beta_y^*}} \qquad [10]$$

Then, in order to achieve high luminosities, all one has to do is make high population bunches of low emittance to collide at high frequency at locations where the beam optics provides as low values of the amplitude functions as possible. M. Stanley Livingston, that was the first student of Ernest Lawrence to build the first cyclotrons, plotted the laboratory energy versus calendar year. The Livingston chart shows, in a very striking way, how the succession of new ideas and new technologies has relentlessly pushed up accelerator beam energies over five decades at the rate of over one and a half orders of magnitude per decade (see Fig. 6.a). One repeatedly sees a new idea, which rapidly increases the available beam energy, but only to be surpassed by yet another new idea. Meanwhile the first idea continues into saturation and possibly into quasi oblivion. Figure 6.b shows the Panofsky plot, which gives collider luminosity versus center of mass energy. In Fig. 7.a and in Fig. 7.b are plotted the Tevatron collider peak and integrated luminosities. As it is possible to see, colliders are very complex and sophisticated machine that in order to be brought to the edge need time, upgrades and continues tuning of their performances.

3.4. Synchrotron radiation energy losses

Charged particles emit light when accelerated. These energy losses are often a large problem when designing a very high-energy accelerator and represent an important limiting factor in the maximum attainable energy. An accelerated charged particle emits photons according to the Larmor law:

$$P = -\frac{2}{3}\frac{e^2}{m^2c^3}\left(\frac{dp_\mu}{d\tau}\frac{dp^\mu}{d\tau}\right) \qquad [11]$$

where p_μ is the charged particle's 4-momentum. For a given applied force the radiated power depends inversely on the square of the mass of the particle involved. Consequently these effects are larger in the case of electron beams. In a linear accelerator the motion is one-dimensional consequently:

$$P = \frac{2}{3}\frac{e^2}{m^2c^3}\left(\frac{dE}{dx}\right)^2 \qquad [12]$$

showing that for linear motion the power radiated depends only on the external forces which determine the rate of change of particle energy with distance. The ratio of power radiated to power supplied by the external sources (i.e. a RF cavity) is expressed by following formula:

$$G = \frac{P}{dE/dx} = \frac{2}{3}\frac{e^2}{m^2c^3}\frac{1}{v}\frac{dE}{dx} \rightarrow \frac{2}{3}\frac{(e^2/mc^2)}{mc^2}\frac{dE}{dx} \qquad [13]$$

This equation shows that the radiation loss will be unimportant unless the gain in energy is of the order $mc^2 = 0.511$ MeV in a distance of $e^2/mc^2 = 2.82 \cdot 10^{-15}$ m, that means of the order of $2 \cdot 10^{14}$ MeV/m. In a linear collider such as ILC, assuming a design cavity gradient of about 31.5 MV/m the expected synchrotron radiation energy losses can be considered negligible. The situation changes drastically in the case of a circular accelerator. The radiated power in this case becomes:

$$P = \frac{2}{3}\frac{e^2}{m^2c^3}\gamma^2\omega^2|p|^2 = \frac{2}{3}\frac{e^2c}{\rho^2}\beta^4\gamma^4 \qquad [14]$$

where $\omega = (c\beta/\rho)$ and ρ is the accelerator radius. The energy loss, per evolution, can be then written as follow:

$$\delta E = \frac{2\pi\rho}{c\beta}P = \frac{4\pi}{3}\frac{e^2}{\rho}\beta^3\gamma^4 \qquad [15]$$

that becomes, in the case of an ultrarelativistic accelerator $(\beta \approx 1)$, as follow:

$$\delta E(MeV) = 8.85 \cdot 10^{-2}\frac{E^4([GeV])}{\rho([m])} \qquad [16]$$

the radiated power for a circular accelerators can be then expressed as follow:

$$P[W] = 10^6 \cdot \delta E([MeV]) \cdot J([A]) \qquad [17]$$

If synchrotron radiation is problem when designing a very high-energy accelerator, however, it is an excellent probe of the surface structures of matter and a tool in producing modern microelectronics. As should be expected, synchrotron radiation plays a major role in the design and optimization of the e^+e^- colliders. While vacuum stability and electron clouds can be of concern in the positron rings, synchrotron radiation along with the restoration of longitudinal momentum by the RF system have the positive effect of generating very small transverse beam sizes and small momentum spread. Further reduction of beam size at the interaction points using standard beam optics techniques and successfully contending with high beam currents has led to record luminosities in these rings, far exceeding those of hadron colliders. To maximize integrated luminosity the beam can be "topped off" by injecting new particles without removing existing ones – a feature difficult to imitate in hadron colliders.

3.5. Hadron and Lepton Colliders

The different experimental conditions and, particularly, the different interacting projectiles of hadron and lepton colliders will generally lead to different sensitivities for specific processes at the energy scale under consideration, so the two types of colliders are also complementary to a certain extent and there are advantages to operating both types of machines. This complementarity also applies to the to types of lepton colliders e^+e^- and $\mu^+\mu^-$, but a lesser extent. Protons are composite particles, made of "partons" of quark and gluons. The quarks and gluons are the fundamental degrees of freedom to participate in strong reactions at high energies according to QCD. The proton is much heavier than the electron. These lead to important differences between a hadron collider and an e^+e^- and $\mu^+\mu$ collider. Due to the heavier mass of the proton, hadron colliders can provide much higher center of mass energies in head-on collisions. Lepton colliders can provide on the other hand precise measurements and perform studies using polarized beams. We do not want here to use, anyhow, the common point of view according to which the hadron collider are discovery machines and lepton colliders provides only precise measurements as particles like J/ψ, Υ, the jets, the gluons, etc. have been for example discovered in $e+e^-$ colliders and, on the other hand, a hadron collider like Tevatron performed and is performing the most precise measurement of M_W mass. A summary of most relevant collider parameters of past, present and future machines both for lepton-lepton, hadron-hadron and e-p colliders can be found here [12].

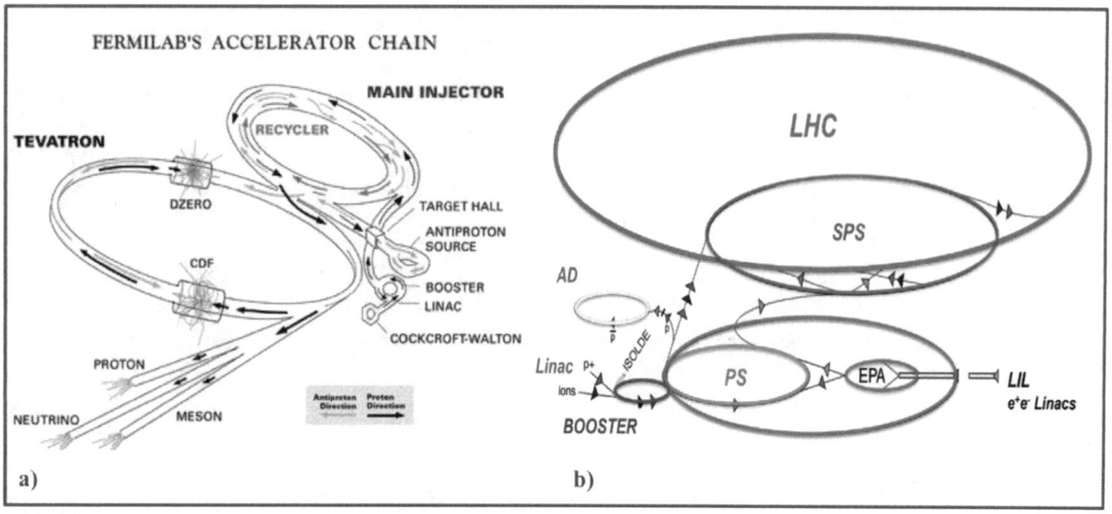

Figure 8: a) Fermilab Tevatron Collider complex [22]; b) the CERN Large Hadron Collider [18].

3.5.1. The Fermilab Tevatron Collider

In order to offer a more detailed description of how a collider works we will describe, with more details, the Tevatron collider and accelerator complex. The Tevatron is a circular particle accelerator at the Fermi National Accelerator Laboratory (FNAL) close to Batavia, Illinois (USA). The Tevatron Collider has been built in order to accelerate beams of protons and anti-protons and make them collide at center of mass energy of 1.80, during the Run I (1992-1996) and 1.98 TeV, during the Run II (2001-2011). Around 2:30 a.m. on October 13, 1985, scientists, working in the Collider Detector at Fermilab experiment (CDF-II), observed the Tevatron collider's first antiproton–proton collision. The collision's center-of-mass energy was of 1.6 TeV, three times higher than the previous world record set by the Super Proton Synchrotron collider (SPS) at CERN. Over the next 20 years, a steady stream of improvements to the accelerator complex and the collider detectors, some big and some small, kept the Tevatron at the leading edge of particle physics. Researchers with the CDF-II and DØ collaborations continue to exploit the growing Tevatron luminosity, which is now 400 times the 1985 design goal, and have produced a magnificent collection of scientific results. During this initial operating period, called Run I, the top quark was discovered in 1995 [3, 4, 5]. After an upgrade period of almost 5 years, the Tevatron resumed operations in early 2001, at a higher energy of $\sqrt{s}$ = 1.96 TeV and higher intensities. The Tevatron has continued to operate at this higher capacity during what is called Run II and is scheduled to be turned off by the end of September 2011, as announced on January 10, 2011. Achieving energy of 980 GeV for each beam requires several acceleration steps, which can be seen in Fig. 8.a. These steps are described in the following sections.

Cockcroft-Walton

The starting point, in the accelerator chain, begins with the Cockcroft-Walton, which is an accelerator that provides a continuous beam of H^- ions at 750 keV. These ions come from the Proton Source, which contains pure hydrogen gas. The hydrogen gas is placed in an electric field, which strips the hydrogen atoms of an electron to become H^+. These protons (H^+) are then attracted to a cesium anode where they attach and acquire two electrons to become H^-, which is then repelled, by the anode. The Cockcroft-Walton then accelerates these H^- ions to 750 keV.

Linac

A Linear Accelerator, called Linac accepts H^- ions from the Cockcroft-Walton at 750 keV. These ions are accelerated by AC electric fields between successive drift tubes of increasing length. This field is varied at high frequency such that when exiting a drift tube some of these negative ions will see an electric field pointing in the opposite direction of their velocity in each gap region, which increases their energy. The ions which arrive out of phase with the RF system will be decelerated, and not make it through the entire Linac. The result of this is a beam consisting of bunched H^- ions in the Linac. These bunches are accelerated to an energy of 400 MeV by the Linac. At the end of this 130 m accelerator the H^- ions pass through a thin carbon foil where the electrons are stripped away from H^- leaving H^+ (protons).

Booster
The Booster is the first of several synchrotrons, in the accelerator chain, which accelerates 400 MeV bunches of protons from the Linac to an energy of 8 GeV. This is done in 16,000 revolutions where each revolution is about 475m. This is on average in increase of 475 keV per revolution.

Main Injector
The Main Injector is another synchrotron, with a circumference of approximately 3 km, and serves multiple purposes in the accelerator chain. It accelerates 8 GeV protons and anti-protons to 150 GeV for injection into the Tevatron ring. The MI also accelerates 8 GeV protons to 120 GeV, after which they are sent to and collided with a nickel alloy target producing antiprotons through the interaction $p+Ni \rightarrow p+\bar{p}+p+Ni$, with an efficiency of ~ 20×10^{-6}. Thus, it takes about 1 million protons to produce 20 antiprotons, due to the low efficiency, which must be selected among the resultant particles. This is done using a lithium lens and pulsing dipole magnet; only negatively charged particles with the mass of a proton will be bent at the angle necessary to continue in the accelerator. The antiprotons are stochastically cooled before they are sent to the Accumulator where they are stored and cooled further.

Debuncher, Accumulator, and Recycler
After the anti-protons are created and filtered out they enter the Debuncher in bunches which tend to have a large energy spread. The debuncher stochastically cools this beam translating the large energy spread into a large time spread such that the anti-protons have a more uniform energy distribution. Anti-protons are sent from the debuncher to the Accumulator roughly every 1.2 s. The accumulator is an 8 GeV storage ring that is used to collect successive injections of anti-protons from the debuncher. After the accumulator has collected a sufficient amount of anti-protons they will be transferred to the Recycler which is an 8 GeV storage ring made of permanent magnets. Anti-protons will be transferred from the recycler to the main injector before they are put into the Tevatron.

Tevatron Collider
The Tevatron Collider is a superconducting synchrotron with a circumference of 6.3 km that accelerates both protons and antiprotons from 150 GeV to their final energy of 980 GeV. The two beams are steered in opposite directions in the same beam-pipe by a set of niobium-titanium superconducting magnets; these 774 dipole magnets, 240 quadrupole magnets with a maximum magnetic field of 4.2 T, and dozens of other types of magnets are cryogenically cooled by liquid helium at a temperature of 4.2 K. Electrostatic separators keep the beams from colliding except at two points along the ring – these points are where the detectors, CDF-II and DØ, are located (see Fig. 13.a and Fig. 13.b). Within these detectors, the beams are focused using quadrupole magnets; the width of each bunch is reduced to approximately 35 μm, although the length increases to 30 cm during the focusing. The beams are then crossed to induce collisions at the center of each detector, at a collision rate of approximately 1.7 MHz. The peak luminosity reached by this collider is $4.02\times10^{32}cm^{-2}s^{-1}$ the 16th of April 2010 (see Fig. 7.a). Both CDF-II and DØ will be collecting data until the Tevatron collider will be shut down. The total amount of data collected so far is shown in Fig. 7.b.

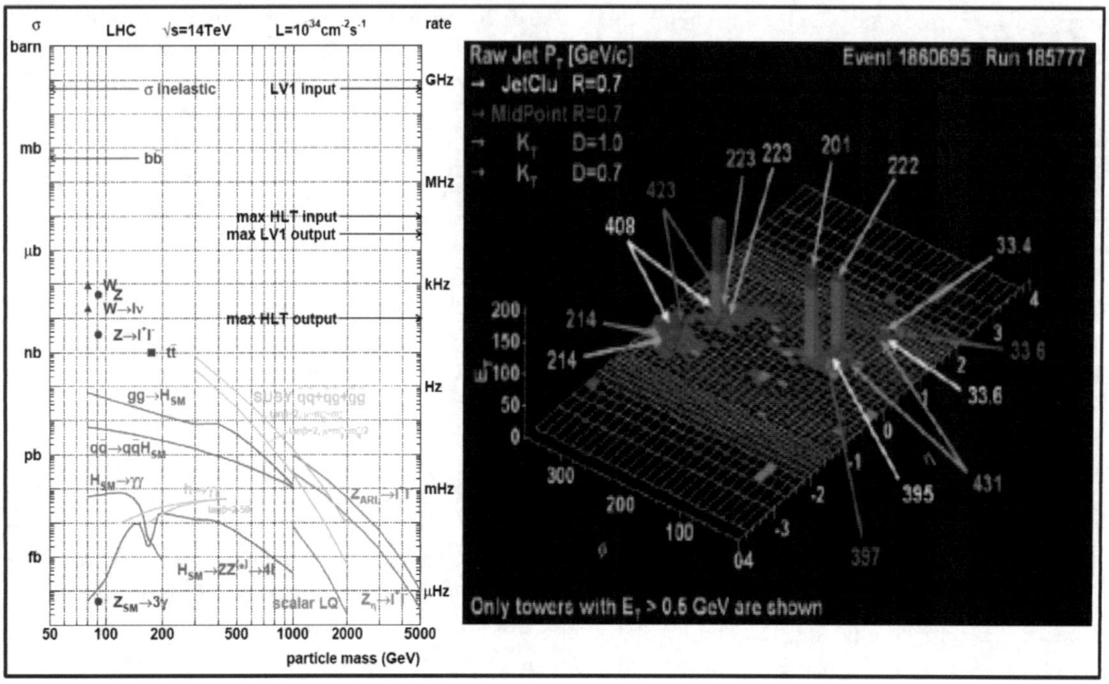

Figure 9: a) Expected production cross-section and rate for several physics processes, assuming a cms energy of $\sqrt{s} = 14\,TeV$ and a luminosity of $10^{34}\,cm^{-2}\,s^{-2}$; b) A CDF-II QCD candidate event [23].

3.5.2. The CERN Large Hadron Collider

The Large Hadron Collider (LHC), starting from December 2009, is the world's largest and highest-energy particle accelerator complex presently taking data. The collider physical reach for a design luminosity of $\sqrt{s} = 14\,TeV$ is reported in Fig. 9.a [24]. LHC has been built by the European Organization for Nuclear Research (CERN), and lies underneath the Franco-Swiss border near Geneva, Switzerland. The collider is contained in a circular tunnel, with a circumference of 27 kilometers, at a depth ranging from 50 to 175 meters underground. The 3.8 m diameter, concrete-lined tunnel, constructed between 1983 and 1988, was formerly used to house the Large Electron-Positron Collider (LEP). The collider tunnel contains two adjacent parallel beam pipes, each containing a proton beam; beam pipes travel in opposite directions around the ring and they intersect at four points. Some 1,232 dipole magnets keep the beams on their circular path, while an additional 392 quadrupole magnets are used to keep the beams focused, in order to maximize the chances of interaction between the particles in the four intersection points, where the two beams will cross. In total, over 1,600 superconducting magnets are installed, with most weighing over 27 tons. Approximately 96 tons of liquid helium is needed to keep the magnets at their operating temperature of 1.9 K, making the LHC the largest cryogenic facility in the world at liquid helium temperature.

The Accelerator complex

The CERN accelerator complex is a succession of machines with increasingly higher energies. Each machine injects the beam into the next one, which takes over to bring the beam to an even higher energy. A schematic overview of the Large Hadron collider accelerator complex is given in Fig. 8.b. The design, LHC center of mass energy is $\sqrt{s} = 14\,TeV$. At the present, the CERN hadron collider is running at the lower energy of $\sqrt{s} = 7\,TeV$, reached March 30th 2010, after a previous run at 2.36 TeV. The proton acceleration chain can be summarized as follows:

- Protons are produced by H atoms, stripping off the orbiting electrons (H^+), and sent to a linac that brings their energies up to 50 MeV;
- Protons are then injected into a Proton Synchrotron Booster (PSB) that accelerates them to energy of 1.4 GeV;
- The beam is then fed to the Proton Synchrotron (PS) where it is accelerated to 25 GeV;
- Protons are then sent into the Super Proton Synchrotron (SPS) where they reach energy of 450 GeV.

At this point p can be, finally, injected into the Large Hadron Collider. As this accelerator collides protons with protons, the particles are injected into two different accelerator lines; one accelerates beams clockwise the other anticlockwise by using an opposite sign dipole fields.

The beam time structure

Protons circulate in well defined bunches. As we saw, the bunch structure is a direct consequence of RF acceleration scheme. In the LHC, under nominal operating conditions, each proton beam has 2808 bunches, each bunch containing about 10^{11} protons. The bunch size is not constant around the accelerator, as circulating, it gets squeezed and expanded. When they are far from the collision points, bunches of particles measure a few centimeters long and a millimeter wide. However, as they approach the collision points, they are squeezed to about 16 μm in order to get a higher probability of having a proton colliding another proton. Increasing the number of bunches is one of the ways to increase luminosity in a machine. The LHC has opted for a bunch spacing of 25 ns (about 7 m), which introduces many technical challenges. A beam might circulate for 10 hours, travelling more that 10 billion kilometers; a proton, in the LHC, makes 11 245 circuits per second.

The main accelerator components

In an accelerator, particles circulate in a vacuum tube and are manipulated using electromagnetic devices: dipole magnets keep the particles in their nearly circular orbits, quadrupole magnets focus the beam, and accelerating cavities are electromagnetic resonators that accelerate particles and then keep them at a constant energy by compensating the Larmor energy losses.

- The Vacuum systems: the internal pressure at the LHC is 10^{-13} atm (ultrahigh vacuum), because we want to avoid collisions with gas molecules. There is ~6500 m^3 of pumped volume in the LHC that is like pumping down a cathedral.
- Magnets: there is a large variety of magnets in the LHC, including dipoles (see Fig. 5.b), quadrupoles, sextupoles, octupoles, decapoles, etc. for a total number of about 9300 magnets. Each type of magnet contributes to optimizing a particle's trajectory. Most of the correction magnets are embedded in the cold mass of the main dipoles and quadrupoles. The LHC magnets either have a twin aperture (for example, the main dipoles), or a single aperture (for example, some of the insertion quadrupoles). Insertion quadrupoles are special magnets used to focus the beam down to the smallest possible size at the collision points, thereby maximizing the chances of two protons smashing head-on into each other. The biggest magnets are the 1232 dipoles.
- RF Cavities: the main role of the LHC cavities is to keep the 2808 proton bunches tightly bunched to ensure high luminosity at the collision points and consequently, maximize the number of collisions. They also deliver radiofrequency power to the beam during acceleration to the top energy.

Superconducting cavities with small energy losses and large stored energy are the best solution. The LHC uses eight cavities per beam, each delivering 2 MV (an accelerating field of 5 MV/m) at 400 MHz (see Fig. 4.c). Cavities operate at 4.5 K (the LHC magnets will use super-fluid helium at 1.8 K). For the LHC they are grouped in fours in cryomodules, with two cryomodules per beam, and installed in a long straight section of the machine where the transverse inter-beam distance has been increased from the normal 195 mm to 420 mm. The precision level of LHC machine is so extreme that phenomenon like tides in the ocean can alter its activities. At the new Moon and when the Moon is

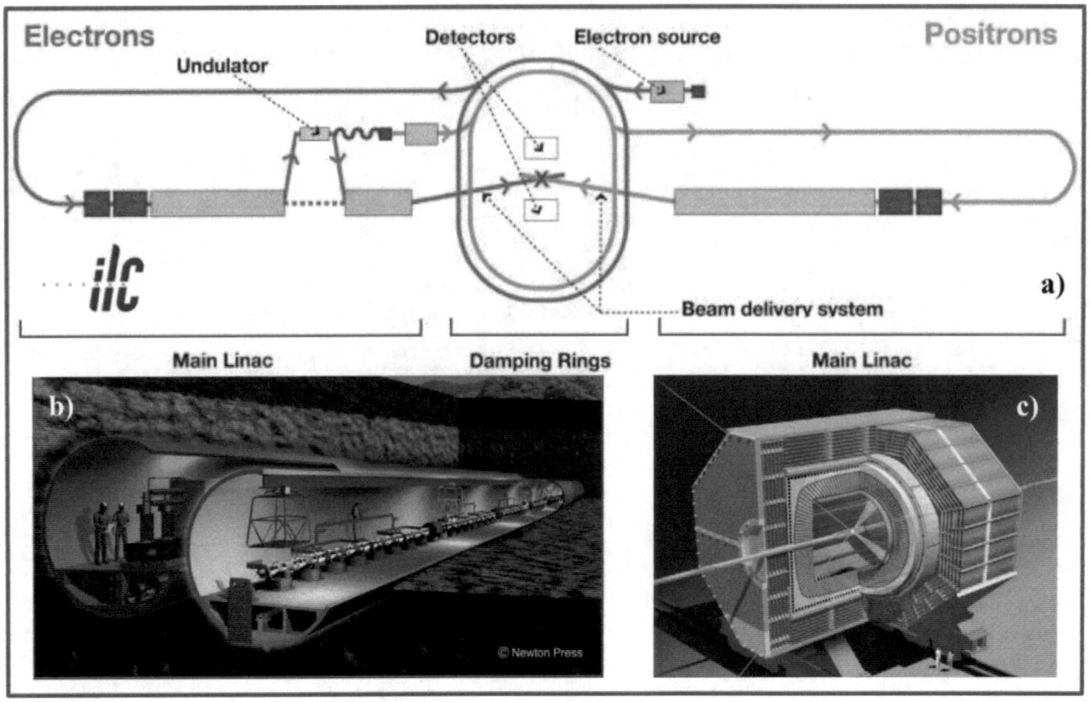

Figure 10: a) ILC general layout [19]; b) schematic view of the serving tunnel containing the klystrons and the modulators [19]; c) a schematic drawing of a possible ILC detector [19, 25].

full, the Earth's crust rises by some 25 cm in the Geneva area under the effect of these 'ground tides'. This movement causes a variation of 1 mm in the circumference of the LHC.

The LHC dipoles

The dipoles of the LHC represented the most important technological challenge for the LHC design. In a proton accelerator like the LHC, the maximum energy that can be achieved is directly proportional to the strength of the dipole field, given a specific acceleration circumference. At the LHC the dipole magnets are superconducting and able to provide the very high field of 8.3 T over their length. No practical solution could have been designed using 'warm' magnets instead of superconducting ones. The LHC dipoles are realized using niobium-titanium (NbTi) multiwire cables, which become superconducting below a temperature of 10 K.

The LHC Cryogenic System

LHC superconducting magnets sit in a 1.9 K bath of superfluid helium at atmospheric pressure. This bath has been cooled by low-pressure liquid helium flowing in heat-exchanger tubes threaded along the string of magnets. The reliability and effectiveness of this sophisticated cryoloop are key factors in achieving the required magnet performance. Liquid helium is used because it can keep things cool over long distances. In all, LHC cryogenics needed 4×10^4 leak-tight pipe junctions, and 96 t of helium have been required by the LHC machine to keep the magnets at their operating temperature. 60% of the helium is in the magnet cold masses while the remaining 40% is shared between the distribution system and the refrigerators. Eight refrigerators operating at 18 kW and 4.5 K have been used to cool down the LHC machine (36 800 t of mass) in about 2 weeks. There were three main phases: first cool down to 4.5 K, second filling with liquid helium of the magnet cold masses and at last the final cool down to 1.9 K.

A layout of the Compact Solenoid Detector (CMS) and of the ATLAS experiment is given in Fig. 13.c. and Fig. 13.d. A candidate event produced by a pp collision at LHC is given, for the ATLAS detector in Fig. 9.b.

3.6. Future Colliders

Here we want to summarize some of the R&Ds under consideration of the scientific community. Present design activity emphasizes a lepton collider as the next major HEP project following initial results from the LHC. Synchrotron radiation precludes a higher energy successor to LEP. Three alternatives are noted in this section: two approaches to an electron-positron linear collider, the International Linear Collider and the Compact Linear Collider (CLIC) and a muon ring collider.
We will discuss shortly also the possibility of realizing a Very Large Hadron Collider.

3.6.1. The International Linear Collider

At present the highest energy accelerators are all circular colliders, but it is likely that limits have been reached in respect of compensating for synchrotron radiation losses for electron accelerators, and the next generation will probably be linear accelerators 10 times the current length. An example of such a next generation electron accelerator could be the ~40 km long International Linear Collider. The maximal center of mass energy is designed to be $\sqrt{s}=500\,GeV$, and a peak luminosity of $2\times10^{34}\text{cm}^{-2}\text{s}^{-1}$, with a possible upgrade to 1TeV, where physics runs must be possible for every energy above $\sqrt{s}=200\,GeV$ and some luminosity for calibration runs is needed at $\sqrt{s}=91\,GeV$ [25]. For mass measurements threshold scans are required so that it must be possible to change the beam energy fast in small steps. The total luminosity is required to be around 500 fb^{-1} within the first four years and about 1000 fb^{-1} during the first phase of operation. For the electron beam, polarization with a degree of larger than ±80% is mandatory. For the positron beam, a polarization of more than ±50% is useful which should be relatively easy to achieve with the undulator positron source in the present ILC design. To reduce systematic uncertainties, the polarization direction has to be switchable on the train-by-train basis. Beam energy and polarization have to be stable and measurable at a level of about 0.1%. The ILC is based on 1.3 GHz nine-cell superconducting radio-frequency accelerating cavities that should operate with a nominal accelerating gradient of 31.5 MV/m, a pulse rate of 5 Hz and a beam length of 1 ms with an average beam current in pulse of about 9.0 mA. For a general layout of the International Linear collider, the related tunnel and hypothetical collider experiment see Fig. 10.a, Fig. 10.b and Fig. 10.c. In ILC, the Main Linacs (ML) accelerate the beam from 15 GeV to a maximum energy of 250 GeV at an average accelerating gradient of 31.5 MV/m. The linacs are composed of RF units, each of which are formed by three contiguous SCRF cryomodules containing 26 nine-cell cavities. A picture of a typical 9-cell SCRF cavity is given in Figure 4.d. The positron linac contains 278 RF units, and the electron linac has 282 RF units. Each RF unit has a stand-alone RF source, which includes a conventional or other type pulse-transformer type high-voltage (120 kV) modulator, a 10 MW multi-beam klystron, and a waveguide system that distributes the RF power to the cavities. It also includes the low-level RF (LLRF) system to regulate the cavity field levels, interlock systems to protect the source components, and the power supplies and support electronics associated with the operation of the source. To operate the cavities at 2 K, they are immersed in a saturated He II bath, and helium gas-cooled shields intercept thermal radiation and thermal conduction at 5-8 K and at 40-80 K. The estimated static and dynamic cryogenic heat loads per RF unit at 2 K are 5.1 W and 29 W, respectively. Liquid helium for the main linacs and the RTML is supplied from 10 large cryogenic plants, each of which has an installed equivalent cooling power of ~20 kW at 4.5 K. The main linacs follow the average Earth's curvature to simplify the liquid helium transport. The cryomodule design is a modification of the Type-3 version (Figure 2.2) developed and used at DESY. The Main Linac components are housed in two tunnels, an accelerator tunnel and a service tunnel, each of which has an interior diameter of 4.5 meters. To facilitate maintenance and limit radiation exposure, the RF source is housed mainly in the service tunnel as illustrated in Figure xx.

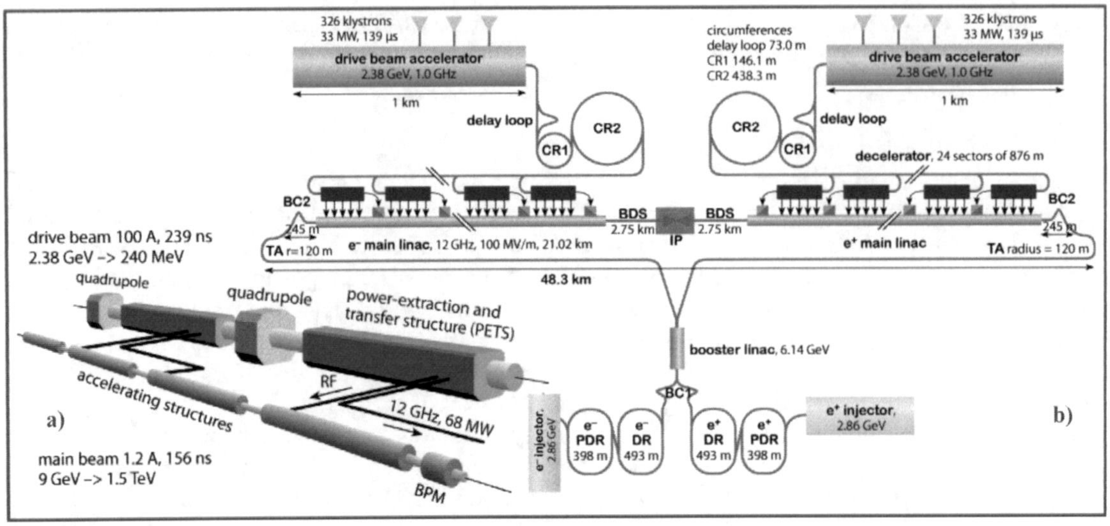

Figure 11: a) Compact Linear Collider (CLIC) module layout; b) the CLIC general layout [26].

The Americas sample site lies in Northern Illinois near Fermilab. The site provides a range of locations to position the ILC in a north-south orientation. The site chosen has approximately one-quarter of the machine on the Fermilab site. The surface is primarily flat. The long tunnels are bored in a contiguous dolomite rock strata (Galena Platteville), at a typical depth of 30-100 m below the surface. The Asian site has been chosen from several possible ILC candidate sites in Japan. The sample site has a uniform terrain located along a mountain range, with a tunnel depth ranging from 40 m to 600 m.

3.6.2. The Compact Linear Collider

The Compact Linear Collider (CLIC) is an under study possible future electron-positron collider that would allow physicists to explore a new energy region beyond the capabilities of today's particle accelerators [27]. It would provide significant fundamental physics information even beyond that available from the LHC and a lower-energy linear e^+/e^- collider, as a result of its unique combination of high energy and experimental precision. Within the framework of a worldwide collaboration on Linear Colliders, the Compact Linear Collider study aims at a center-of-mass energy range for electron-positron collisions of 0.5 to 5 TeV, optimized for a nominal center-of-mass energy of 3 TeV (3 TeV CLIC). In order to reach this energy in a realistic and cost efficient scenario, the accelerating gradient has to be very high - CLIC aims at an acceleration of 100 MV/m. Superconducting technology being fundamentally limited to lower gradients, only room temperature travelling wave structures at high frequency (12 GHz) are likely to achieve this gradient.In order to optimize the production of sufficient RF power for this high gradient, and CLIC relies upon a two-beam-acceleration concept: The 12 GHz RF power is generated by a high current electron beam (drive beam) running parallel to the main beam. This drive beam is decelerated in special power extraction structures (PETS) and the generated RF power is transferred to the main beam. This leads to a very simple tunnel layout without any active RF components (i.e. klystrons). Both beams can be generated in a central injector complex and are transported along the linac. A schematic general layout of the CLIC collider together with the layout of an accelerating module is given in Fig. 11.a and in Fig. 11.b.

3.6.3. The Muon collider

All of the problems with TeV-scale e^+e^- colliders are associated with the relative smallness of the electron mass. The proposed technology of a muon collider aims to solve, or at least greatly reduce, these problems by instead colliding muons, which are leptons that are 207 times heavier than

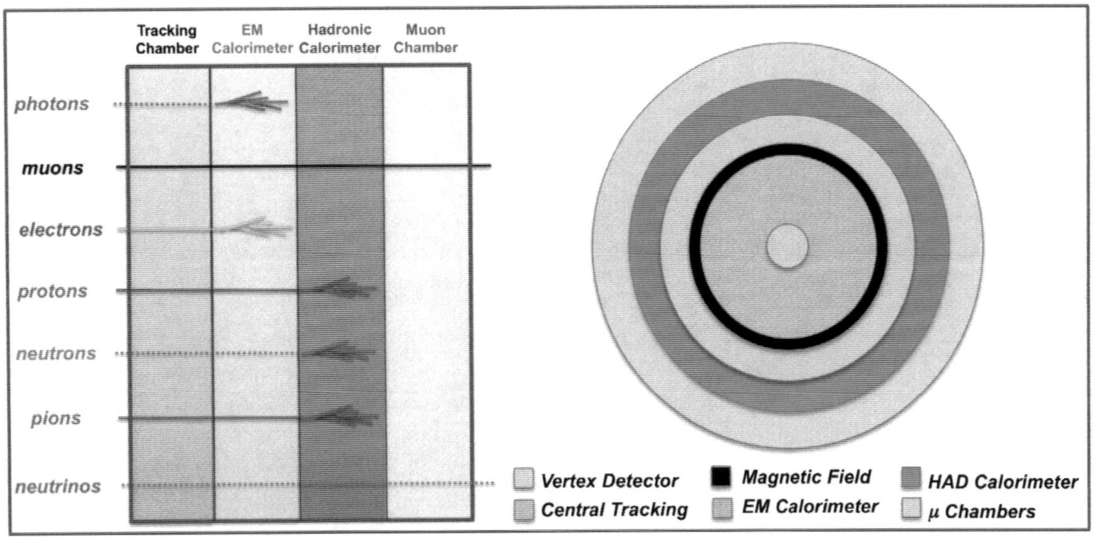

Figure 12: Schematic layout of the central part of a general-purpose collider detector [28].

electrons. Replacing the mass-related problems of e^+e^- colliders, the main problems at muon colliders arise because muons are unstable particles, with an average lifetime of approximately 2.197 μs in their rest frame. The preparation, acceleration and collision of the muon beams must all be done quickly and the supply of muons must be replenished often. The products of the muon decays also cause problems: the decay electrons deposit energy all along the path of the muon beams and create backgrounds in the detectors and, more surprisingly, the neutrinos can cause a radiation hazard in the surroundings of the collider ring.

3.6.4. The Very Large Hadron Collider

The Very Large Hadron Collider (VLHC) is a proposed hadron collider with performances significantly beyond the CERN Large Hadron Collider. There is no planned location or time schedule for this machine, as the project has not been approved, so far. In this paragraph we want to discuss, mainly, the technological feasibility of such a collider and the ways that it might be designed. The basic idea of the Design Study [29] consists in constructing a low field (1.9 T) collider (VLHC stage 1) design in a tunnel, which also can be used for a future high field (9.8 T) collider (VLHC stage 2) [30]. While the second collider is seen as far in the future, one needs to anticipate its requirements as best one can before laying out the components of the first machine since they ultimately must reside in same tunnel. This entails anticipating the geometric requirements, beam optics requirements, and physical space requirements for future technical equipment – much of which may not be known yet. The fundamental lattice parameters, common to both low and high field rings are the following. By design, the collider circumference is 233 km with an average arc radius of 35.0 km and two interaction points. The number of buckets should be 41280 and the bunch spacing (53.1 MHz) 5.645 m. In the stage one, the collider is supposed to have 3440 dipoles with a dipole field at collision energy of 1.9 T and delivers proton-proton collision at a center of mass energy of 40 TeV, with a peak luminosity of $1\times10^{34}cm^{-2}s^{-1}$ and a luminosity lifetime or about 24 hours. In this regime the average number of interactions, for beam crossing, is expected to be around 21. In the phase 2 the collider should work at center of mass energy of 175 TeV, with a peak luminosity of $2\times10^{34}cm^{-2}s^{-1}$, and a luminosity lifetime or about 8 hours. In these conditions, an average number of 58 interactions, per beam crossing, is expected. The stage 2 can be reached by using 9728 dipoles with a dipole field, at collision energy, of 9.8 T [31].

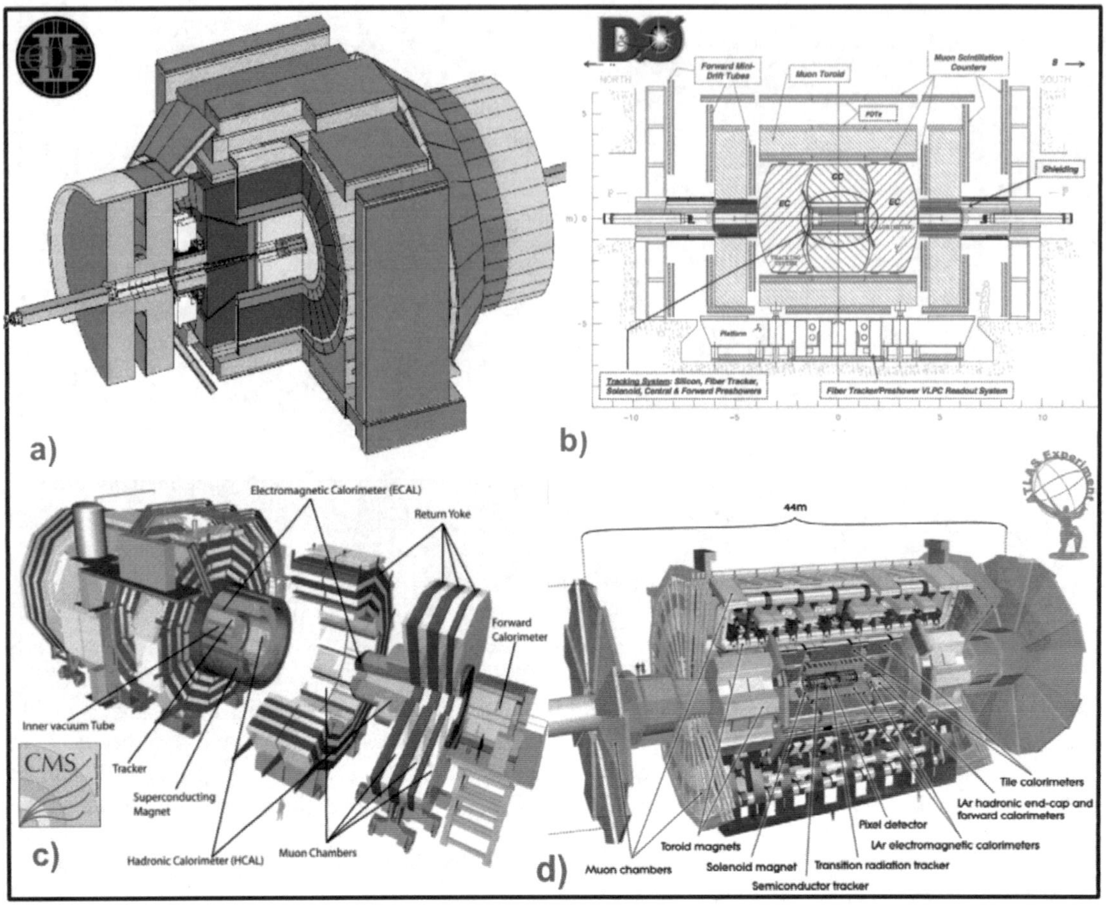

Figure 13: A schematic view of four hadron collider detectors: a) the CDF-II Experiment [32, 23], b) the DØ Experiment at Tevatron Collider [33], c) the Compact Muon Solenoid Experiment (CMS) [34] and d) the ATLAS Detector [35].

4. Collider Particle Detectors

In order to perform physical measurements or particle searches we need to collect a "proper" number of signal events over the possible sources of backgrounds. As we saw, the importance of reaching the energy frontier and to achieve the highest possible beam intensity is needed in order to get a rate, for a given physical process, that is high enough to make possible to perform a measurement or to hunt for new particles, predicted in models beyond the SM. Besides of these aspects, high precision experiments, with the highest possible overall acceptances and efficiencies, are needed. The detection (identification) of the products of each high-energy collision (very often several in the same event) is crucial in order to be able to reconstruct, topologically, the initial states that produced them. In order to perform such a task it is necessary to identify the particles present in the event. This implies the experimental ability to reconstruct the 4-momentum (m and $\vec{p}_{Lab}$), the charge and the spin of the collision products. From a experimental point of view, in most of the case, it's not possible to fully reconstruct all these information because for example part of the energy is unseen along the longitudinal direction or because our detectors cannot measure directly the spin and so on. In elementary particle physics most particles produced have unit charge. Both the charge sign, of non-neutral particles and their momentum can be measured using a magnetic field. The deflection of a charged particle in a magnetic field determines its momentum p; the radius of curvature ρ is given by:

$$\rho = \frac{p}{ZeB} \propto \frac{p}{Z} = \frac{\gamma m_0 \beta c}{Z} \qquad [18]$$

where m_0 is the particle mass in the rest frame and Z its charge, B the magnetic bending field and as usual $\beta = v/c$ and $\gamma = \left(1-\beta^2\right)^{-1/2}$ is the Lorentz factor. The particle velocity can be determined, e.g., by a time–of–flight method yielding: $\beta = 1/\tau$. A calorimetric measurement provides a determination of the kinetic energy:

$$E^{kin} = (\gamma - 1) m_0 c^2 \qquad [19]$$

that is achieved making, the particles entering the calorimetric apparatus, initiating a particle shower that is then deposited in the calorimeter, collected, and measured. Particle detectors at lepton or hadron colliders have, then, evolved to be pretty similar. The technologies used, in each component, differ but they all have the same basic layout as shown in Fig. 12. Starting from the interaction point moving to the outside, there is a tracking volume with almost no material and a high magnetic field. This is used to measure the momentum of charged particles with high precision. It normally has an inner, high resolution section, built of silicon detectors to recognize the decays of short lived particles and an outer tracker made of less expensive materials and optimized for momentum and higher multiplicities measurements. Moving outside of the tracking system, we find the so-called calorimeters that are detectors made of very heavy materials, which absorbs and detects almost all strongly and electromagnetically interacting particles. It is normally divided into a high Z electromagnetic part (EM) and a cheaper outside hadronic part (HAD). A muon detection system, which measures the momentum of any muons, which passed through the calorimeters, follows the outside of those detectors. For massless particles, which is a good approximation for the decay products of almost any particle in a collider, the rapidity function *y* reduces to a new variable called pseudo-rapidity:

$$\eta = -\ln\left(tg(\theta/2)\right) \qquad [20]$$

The pseudorapidity is commonly used as a spatial coordinate describing the angle of a particle relative to the beam axis. Detectors are then segmented in η and designed so that each detector element covers the same area in the $\eta - \phi$ space. In a particle detector is possible to identify photons, neutrons, π, K, protons, muons, b-quarks/c-quarks, light-quarks, neutrinos or any weakly interacting neutral particles like i.e. the Lightest Supersymmetric Particle (LSP). A μ in a collider experiments is a minimum ionizing particle (MIP). It will be characterized by having a charge track originated in the interaction point (IP), matching the "stub" reconstructed in the muon detectors (the multiple coulombian scattering is taken into account) and releasing an amount of energy, in both EM and HAD calorimeters, compatible with a MIP. An electron will be characterized as a charged track, originated in the IP, completely absorbed in the EM calorimeter. A photon will only appear to be as a trackless object releasing all its energy in the EM calorimeter and so on. Fig. 12 summarizes the signatures for the most common collision products. Quarks are identified as a cluster of tracks called jets that are reconstructed by using special clustering algorithms. Between the quarks, in particular, b-quarks are identified by their unique feature of hadronizing in B-hadrons. B-hadrons have sufficient lifetime to travel some distance before decaying. This distance can be measured by using high precision dedicated silicon detectors, then b-jets can be tagged. The detector cannot directly observe neutrinos but its transverse momentum can be inferred by the unbalancing of the event total observed momenta:

$$\vec{\not{p}}_T = -\sum \vec{p}_T^{\,observed} \qquad [21]$$

This quantity is called missing transverse momentum and is identified with the neutrino momentum: $\not{p}_T = p_T^{\nu}$. The Missing Transverse Energy ($\not{E}_T$ or MET) is similarly defined: $\not{E}_T = E_T^{\nu}$. This variable works only if the $p_\perp$ and direction of all scattered particles, except the missing particle, are detected. This requires a hermetic detector, a detector that covers almost all of 4π solid angle with active components. Such coverage is very difficult to achieve and it is expensive. Detectors such as DØ and

CDF-II, for example, have an active calorimetry down to a pseudorapidity of $|\eta|< 4$ and then are they are able to achieve $\not{p}_T$ resolution of the order of 5 GeV/c. However there is a potential for very large $\not{E}_T$ fluctuations that have to be taken into account in the data analysis. For this and for other similar reasons any result obtained by a Montecarlo simulation, of the physical process, have to be passed trough a detailed Detector simulation before being able to make a comparison between data and models. The detector and trigger efficiencies and acceptances have also to be carefully evaluated.
Figures 13 gives some example of collider detectors such as the CDF-II and DØ presently taking data at Tevatron collider and the ATLAS and CMS detectors working on LHC.

5. Conclusions

In the present paper we reviewed some of the contents of a set of lectures, presented at the XIV International School of Particles and Fields in Morelia, Mexico. Those lectures were intended for students and the main goal have been to introduce some basic concept about particle accelerators, energy and luminosity. We described for this reason some of colliders presently running. A small description of the future collider R&Ds was also given together with a description of how common detectors work and are used for data analysis.

6. Acknowledgments

I would like to heartily thank the organizers of the XIV School of Particles and Fields in Morelia, for inviting me to give this set of lectures in such beautiful land. I would like to thank in particular A. Ayala, A. Bashir, G. Contreras, I. León, G. Murguía, A. Raya and M. E. Tejeda for their very kind hospitality and for providing a greatly stimulating and lively environment in which have been possible to wonderfully work, exchange ideas and relax.

7. Bibliography

[1] CERN Courier, vol. 51, number 1, January/February 2011.
[2] C. Rubbia, Phys.Rept. 239 (1994) 241-284.
[3] F. Abe, *et al.* [CDF collaboration] Phys.Rev. D**50** (1994) 2966-3026; Phys.Rev.Lett. **73** (1994) 225-231.
[4] F. Abe, *et al.* [CDF collaboration], *Phys. Rev. Lett.* **74,** 2626 (1995).
[5] S. Abachi, *et al.* [DØ collaboration], *Phys. Rev. Lett.* **74,** 2632 (1995).
[6] ALEPH, CDF, D0, DELPHI, L3, OPAL, SLD, the LEP EWK Group, the Tevatron EWK Group, and the SLD EWK and heavy flavour groups Collaboration, *Precision electroweak measurements and constraints on the Standard Model*, Fermilab-tm-2480-ppd.
[7] G. Altarelli, CERN-PH-TH-2010-249. Oct 2010. 17 pp.
[8] C. Pagliarone, *Searching for Physics beyond the SM with High Energy Colliders*, to appear on the Proceedings of the XIV Mexican School on Particle & Fields, Morelia, Mexico, November 2010.
[9] G. Flügge, *Future Research in High Energy Physics*. N. Ellis and M. B. Gavela, editors, *1993 European School of High Energy Physics*, Yellow reports. CERN 94-04, 1994.
[10] T. K. Gaisser, *The cosmic-ray spectrum: from the knee to the ankle*, J. Phys. Conf. Ser. **47**, 15-20 (2006).
[11] H. Wiedemann, Particle Accelerator Physics, Springer-Verlag, 2007.
[12] N. Nakamura *et al.* [Particle Data Group], J. Phys. G **37**, 075021 (2010).
[13] T. K. Gaisser, J. Phys. Conf. Ser. 47, 15 (2006).
[14] V. Tatischeff, arXiv:0804.1004 [astro-ph], E. G. Berezhko, Adv. Space Res. 41, 429 (2008).
[15] C. Pagliarone, *From Accelerators to Cosmic Rays Physics*, published on the Conference Proceedings of the XII Mexican Workshop on Particles and Fields, Mazatlan, 2009.
[16] The Fermilab chain of accelerators can be found here: http://www-bd.fnal.gov/public/.
[17] This picture have been taken close to the CERN cafeteria by the author, see also http://livefromcern.web.cern.ch/livefromcern/antimatter/factory/AM-factory02.html.

Journal of Physics: Conference Series **287** (2011) 012005 doi:10.1088/1742-6596/287/1/012005

[18] L. Evans and P. Bryant, (editors), LHC Machine, JINST **3**, S08001 (2008).
[19] J. Brau, *et al.* [ILC Collaboration], *ILC Reference Design Report: ILC Global Design Effort and World Wide Study*, [arXiv:0712.1950 [physics.acc-ph]].
[20] W. K. H. Panofsky, Evolution of Particle Accelerators & Colliders, Spring 1997.
[21] http://www-bd.fnal.gov/pplot/today/DataSummaryTables.html.
[22] See also http://www-bd.fnal.gov/.
[23] C. Pagliarone, *Results on QCD Physics from the CDF-II Experiment*, Published on the Proceedings of New trends in High energy physics, Edited by P.N. Bogolyubov, L. Jankovszky, V.K. Magas, Z.I. Vakhnenko. Kiev, Bogolyubov Inst. Theor. Phys., 2006. 287p.
[24] K. Jacobs, *Physics at the LHC and sLHC*, to be published on NIM A, ISBN 0168-9002.
[25] T.Behnke, *et al.* [ILC Collaboration], ILC Reference Design Report Volume 4 - Detectors, [arXiv:0712.2356 [physics.ins-det]].
[26] J. P. Delahaye, *Towards CLIC Feasibility*, IPAC-2010-FRXCMH01. May 2010; see also http://clic-study.web.cern.ch/CLIC-Study/intro.html.
[27] J. P. Delahaye, Towards CLIC Feasibility, Proceeding of IPAC'10, Kyoto, 2010.
[28] C. Grupen, *Physics of Particle Detection*, Proceedings of the ICFA School, 2001.
[29] P. Limon et al., *Design Study for a Staged VLHC*, Fermilab-TM-2149, May 2001.
[30] J. Johnstone et al., *Lattice Design of the VLHC Rings*, PAC 01, Chicago, 2001.
[31] M. Blaskiewiczt et al., *VLHC accelerator physics*, Fermilab-TM-2158, November 2001.
[32] C. Pagliarone, *Recent results from CDF and status of CDF-II*, published on the Proceedings of the XI International School on Particles and Cosmology, Baksan, FERMILAB-CONF-01-363-E.
[33] V. M. Abazov *et al.* [D0 Collaboration], *The Upgraded D0 Detector*, Fermilab-Pub-05/341-E.
[34] G. Tonelli [CMS Collaboration], *The CMS experiment: Status and highlights*, PoS **ICHEP2010**, 565 (2010).
[35] G. Aad *et al.* [ATLAS Collaboration], *The ATLAS Experiment at the CERN Large Hadron Collider*, JINST **3**, S08003 (2008).

Journal of Physics: Conference Series **287** (2011) 012006 doi:10.1088/1742-6596/287/1/012006

Finite temperature effective actions

Ashok Das

Department of Physics, University of Rochester, Rochester, New York, NY 14627, USA
Saha Institute of Nuclear Physics, 1/AF Bidhannagar, Calcutta 700064, India

E-mail: das@pas.rochester.edu

Abstract. In this lecture we will describe a systematic method for evaluating effective actions at finite temperature. The method will be illustrated with the effective actions for the $0+1$ dimensional massive QED as well as for the $1+1$ dimensional QED with massless or massive fermions.

The calculation of the effective action at finite temperatures has been an interesting question for quite sometime, particularly because the imaginary part of the effective action is expected to give the vacuum decay rate and one would like to understand the influence of temperature on this rate. In this talk we will describe a systematic method for such a calculation and will illustrate the method by calculating the temperature dependent effective actions for the $0+1$ dimensional massive QED as well as for the $1+1$ dimensional QED both for massless and massive fermions. We will also comment on the imaginary parts of these effective actions as well as their physical meanings. The talk would be based primarily on our recent papers [1, 2, 3] (other references to the subject can be found there).

To begin with, let us quickly review how effective actions are constructed at zero temperature. For example, let us consider a system of massive fermions interacting with a background electromagnetic field described by the Lagrangian density (at this point we leave the dimension arbitrary and also emphasize that the discussion applies to other field theories as well)

$$\mathcal{L} = \overline{\psi}\left(\gamma^{\mu}(i\partial_{\mu} - eA_{\mu}) - m\right)\psi = \overline{\psi}(i\not{D} - m)\psi, \tag{1}$$

where m denotes the fermion mass and the gauge covariant derivative is defined to be

$$D_{\mu}\psi = (\partial_{\mu} + iA_{\mu})\psi. \tag{2}$$

When fermions are integrated out in the path integral, the effective action (unnormalized) for such a theory is obtained to be

$$\Gamma_{\text{eff}}[A] = -i\text{Tr}\,\ln\left(i\not{D} - m\right) = -i\text{Tr}\,\ln H, \tag{3}$$

where for reasons that will be clear shortly, we have defined the operator

$$H = i\not{D} - m, \tag{4}$$

and the "Tr" in (3) denotes the trace over Dirac indices as well as the trace over a complete set of basis states. (For completeness we note here that the normalized effective action is defined such that $\Gamma_{\text{eff}}[A=0] = 0$.)

Journal of Physics: Conference Series **287** (2011) 012006 doi:10.1088/1742-6596/287/1/012006

The effective action in (3) is (ultraviolet) divergent at zero temperature and needs to be regularized. Schwinger gave a beautiful method for evaluating this at zero temperature with a gauge invariant regularization known as the proper time method. Basically, the method involves giving an integral representation for the effective action (3) in the form

$$\Gamma_{\rm eff}[A] = \lim_{\nu\to 0} -i \int_0^\infty \frac{\mathrm{d}\tau}{\tau^{1-\nu}}\, \mathrm{Tr}\, e^{-\tau H}, \tag{5}$$

where the integration parameter τ is known as the proper time parameter. It is clear now from the structure of the integrand in (5) that H can be thought of as the generator of the (Euclidean) time evolution in τ. Therefore, Schwinger had mapped the complicated loop evaluation to a problem of the proper time evolution of a quantum mechanical particle interacting with a background. If the proper time evolution equations for the motion of the particle can be solved in a closed form, we can determine a coordinate (or momentum) basis in which the trace in (5) can be evaluated leading to a closed form expression for the effective action. On the other hand, if the proper time evolution equations cannot be solved in a closed form, rather only perturbatively, then the effective action also will have a perturbative determination coinciding with that determined by evaluating Feynman diagrams with a gauge invariant regularization. The proper time method is a precursor to the ζ-function regularization. Schwinger used this method to evaluate the complete effective action for fermions interacting with a constant electromagnetic field (Euler-Heisenberg action as well as other cases) and the vacuum decay rate at zero temperature.

There have been several attempts to generalize the method of Schwinger to finite temperature in order to determine the temperature dependent effective actions and these have led to conflicting results. The subtleties behind these results have to do with the imposition of the anti-periodicity condition required at at finite temperature as well as on the finite temperature formalism used to carry out the calculation. We know that there are various formalisms available to describe a finite temperature field theory. Very broadly speaking, they can be divided into two classes, namely, the imaginary time formalism due to Matsubara and the real time formalism. Imaginary time formalism, where time is rotated to the negative imaginary axis with a finite interval, is simpler to use and through analytic continuation to Minkowski time leads naturally to retarded and advanced Green's functions (amplitudes), but not to the (time ordered) Feynman amplitudes which is of relevance in evaluating the effective action. In contrast, in the real time formalism, where we do not rotate time to the imaginary axis and introduce temperature through periodicity/anti-periodicity conditions of the amplitudes, can lead to Feynman amplitudes as well as the retarded/advanced amplitudes in a straightforward manner. We believe this is, therefore, the natural formalism to use in evaluating the effective action at finite temperature.

The price one has to pay in using the real time formalisms is that the field degrees of freedom double. As a result, it is slightly more complicated in practice than the imaginary time formalism, nonetheless this is necessary for a proper evaluation of the effective action at finite temperature. For example, we use the real time formalism known as Schwinger's closed time path formalism where the contour in the complex time plane has the form shown in Fig. 1. The two field degrees of freedom correspond to fields on the two real branches of the contour (see [4] for details) and the branch along the imaginary axis decouples and does not contribute to any amplitude. As a result, in the real time formalism the propagator becomes a 2×2 matrix at finite temperature. Furthermore, calculationally, it is much simpler to use the real time formalism in the mixed space where the space coordinates have been transformed to the Fourier space.

Let us further from the definition of the effective action in (3) that

$$\frac{\partial \Gamma_{\rm eff}[A]}{\partial m} = \mathrm{Tr}\, S(x,x) = \int \mathrm{d}x\, \mathrm{tr}\, S(x,x), \tag{6}$$

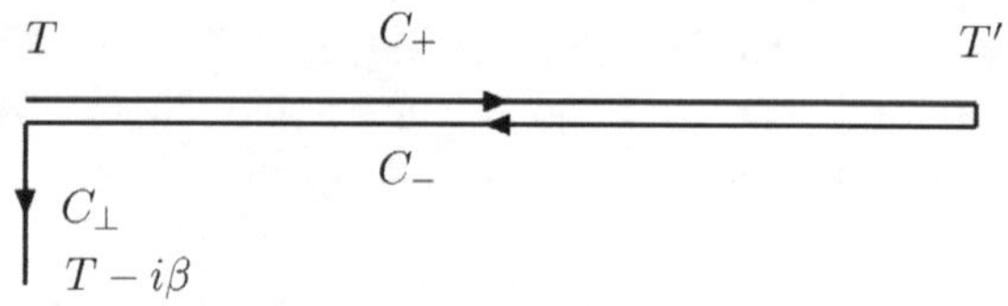

Figure 1. The closed time path contour in the complex t-plane. Here $T \to -\infty$, while $T' \to \infty$ and β denotes the inverse temperature (in units of the Boltzmann constant k) [4].

where "tr" denotes the trace over spinor indices and $S(x, x)$ represents the complete fermion propagator in the background at coincident points. Therefore, if we know the complete fermion propagator at coincident points $S(x, x)$, then integrating (6), we can determine the effective action. This is, however, not useful if the fermion is massless. Alternatively, we note from (3) that the current in this theory can be written as

$$J^\mu(x) = \frac{\delta \Gamma_{\text{eff}}[A]}{\delta A_\mu(x)} = e \,\text{tr}\, \gamma^\mu S(x, x), \tag{7}$$

independent of whether the mass vanishes or not. Once again we note that if the complete fermion propagator at the coincident points we can obtain the effective action by functionally integrating (7) which is particularly useful when $m = 0$.

It is well known that there is no ultraviolet divergence at finite temperature. Since Schwinger's method was designed to provide a gauge invariant regularization (for the ultraviolet divergences) at zero temperature, we believe that it is not necessary to generalize the proper time method of Schwinger to finite temperature. Rather, it is the complete fermion propagator that is important in evaluating the effective action at finite temperature. Therefore, we propose that in order to calculate the effective action at finite temperature, it is sufficient to determine the complete fermion propagator of the theory such that

(i) it satisfies the equations of motion.

(ii) it satisfies all the relevant symmetry properties such as the Ward identities of the theory *etc.*

(iii) most importantly, it satisfies the anti-periodicity appropriate for a finite temperature fermion propagator.

It is the last condition that is missing at zero temperature and is the most powerful in determining the propagator at finite temperature. Once the propagator is determined, then using (6) or (7), as the case may be, the effective action can be determined. Some comments are in order here. Clearly, the fermion propagator that we are interested in corresponds to the tree level propagator and, consequently, does not have any ultraviolet divergence that needs to be regularized. However, in taking the coincident point and integration over the coincident coordinate would introduce the ultraviolet divergence which is present at zero temperature. However, we are only interested in the finite temperature part of the effective action which is free from ultraviolet divergence. Therefore, this method of determining the effective action may or may not lead to the correct zero temperature effective (which is not of interest to us). If the theory does not have any ultraviolet divergence at zero temperature, our effective action would contain the correct zero temperature effective action as well. However, when ultraviolet divergences are present at zero temperature, the method will yield only the correct temperature dependent effective action which is our goal (the zero temperature part can presumably be obtained from other methods such as Schwinger's proper time method).

Let us illustrate our proposal with a few examples. First, let us consider the 0+1 dimensional massive QED described by the Lagrangian

$$L = \overline{\psi}(t)(i\partial_t - m - eA(t))\psi(t), \tag{8}$$

where m denotes the mass of the fermion and $A(t)$ represents the external background electromagnetic field. In 0+1 dimensions, the electromagnetic potential has only one component (the time component) which we are not indicating for simplicity. At finite temperature, the field degrees of freedom would have to be doubled on the complex contour in Fig. 1 and the contour ordered propagator $S_c(t,t')$ of the theory satisfies the equations

$$\begin{aligned} &(i\partial_t - m - eA_c(t))S_c(t,t') = i\delta_c(t-t'), \\ &S_c(t,t')(i\overleftarrow{\partial}_{t'} + m + eA_c(t')) = -i\delta_c(t-t'), \end{aligned} \tag{9}$$

where the subscript "c" characterizes a function on the contour. On the contour, the step function is defined naturally as [4]

$$\theta_c(t-t') = \begin{cases} \theta(t-t') & \text{if both } t,t' \in C_+, \\ \theta(t'-t) & \text{if both } t,t' \in C_- \text{ or } \in C_\perp, \\ 1 & \text{if } t \in C_- \text{ (or } \in C_\perp\text{)and } t' \in C_+, \\ 0 & \text{if } t \in C_+ \text{and } t' \in C_- \text{ (or } \in C_\perp\text{)}, \end{cases} \tag{10}$$

and the delta function on the contour is defined in the standard manner as

$$\delta_c(t-t') = \partial_t\theta_c(t-t'). \tag{11}$$

The Ward identity of the theory requires that

$$i\partial_t S_c(t,t) = 0. \tag{12}$$

Together with the anti-periodicity, (9) and (12) determine the complete propagator of the theory to be

$$S_c(t,t') = e^{-imt-ie\int d\bar{t}\,\theta_c(t-\bar{t})A_c(\bar{t})}\left(\theta_c(t-t') - n_{\rm F}(m+\frac{iea}{\beta})\right)e^{imt'+ie\int d\bar{t}\,\theta_c(t'-\bar{t})A_c(\bar{t})}. \tag{13}$$

where

$$a = a_+ - a_-, \quad a_\pm = \int_{-\infty}^{\infty} dt\, A_\pm(t). \tag{14}$$

This propagator can be checked to satisfy the Lippmann-Schwinger equation to all orders, namely, it agrees with the perturbative calculations order by order to all orders. We can integrate this propagator at the coincident point ($t = t'$) to obtain the complete (normalized) effective action

$$\Gamma_{\rm eff}[a_+,a_-] = -i\ln\left[\cos\frac{e(a_+-a_-)}{2} + i\tanh\frac{\beta m}{2}\sin\frac{e(a_+-a_-)}{2}\right]. \tag{15}$$

Since the $0+1$ dimensional theory is free of ultraviolet divergences, the effective action (15) contains the correct zero temperature effective action as well. If we restrict this effective action to the C_+ branch of the contour in Fig. 1 (by setting $A_- = 0$), this action coincides with the earlier studies on the effective action of this theory in connection with large gauge invariance. However, being the complete effective action of the theory, we can show that all the retarded and advanced amplitudes vanish in this theory. This is basically a consequence of the fact that the effective action is a functional of the difference $(a_+ - a_-)$. (Since the earlier studies were restricted only to the C_+ branch of the contour in Fig. 1, this result is impossible to derive within that framework.)

As a second example, let us consider massless QED in $1+1$ dimensions, also known as the Schwinger model. The theory is described by the Lagrangian density

$$\mathcal{L} = \overline{\psi}\gamma^\mu \left(i\partial_\mu - eA_\mu\right)\psi. \tag{16}$$

Since the fermions are massless, the theory naturally separates into left and right handed spinors as

$$\mathcal{L} = \psi_{\rm R}^\dagger \left(i\partial_+ - eA_+\right)\psi_{\rm R} + \psi_{\rm L}^\dagger \left(i\partial_- - eA_-\right)\psi_{\rm L}, \tag{17}$$

where we have defined the light-cone components

$$x^\pm = \frac{x^0 \pm x^1}{2}, \quad \partial_\pm = \partial_0 \pm \partial_1, \quad A_\pm = A_0 \pm A_1. \tag{18}$$

The Schwinger model is ultraviolet divergent at zero temperature and a gauge invariant regularization mixes the left and the right handed sectors at the two point level which manifests as the chiral anomaly in the theory. At finite temperature, however, as we have argued, there is no ultraviolet divergence. Consequently, we do not expect any mixing between the two sectors at finite temperature so that we can study the temperature dependent effective actions coming from each sector separately. The total temperature dependent effective action will be a sum of the two contributions arising from the left and the right handed fermion sectors.

Let us first concentrate on the right handed sector of the theory. If we consider x^+ to correspond to the light-cone time variable in this sector, we can Fourier transform x^- and introduce the following operator notation in this mixed space,

$$\begin{aligned}
&A_+(x^+, p-k) = \langle p|\hat{A}_+(x^+)|k\rangle,\\
&S(x^+, x'^+; p, k) = \langle p|\hat{S}(x^+, x'^+)|k\rangle, \quad S(x^+, x'^+; p) = \int \frac{\mathrm{d}k}{2\pi}\langle p+k|\hat{S}(x^+, x'^+|k\rangle,
\end{aligned} \tag{19}$$

where S denotes the fermion propagator in the right handed sector. With this operator notation, the contour ordered propagator satisfies the simple equations

$$\begin{aligned}
(i\partial_+ - e\hat{A}_c(x^+))\hat{S}_c(x^+, x'^+) &= \frac{i}{2}\,\delta_c(x^+ - x'^+),\\
\hat{S}_c(x^+, x'^+)(i\overleftarrow{\partial}'_+ + e\hat{A}_c(x'^+)) &= -\frac{i}{2}\,\delta_c(x^+ - x'^+).
\end{aligned} \tag{20}$$

The Ward identity in the right handed sector requires (we note that since there is no ultraviolet divergence at finite temperature, the Ward identity is not expected to be violated)

$$\partial_+\hat{S}_c(x^+, x'^+) = 0. \tag{21}$$

Furthermore, the charge conjugation of the theory implies that the complete fermion propagator is odd under

$$\hat{A}_c(x^+) \to -\hat{A}_c(x^+), \tag{22}$$

as well as the reflection of the momentum of the fermion propgator $S_c(x^+, x'^+; p)$.

The propagator satisfying (20)-(22) as well as the anti-periodicity condition can be determined to have the form

$$\hat{S}_c(x^+, x'^+) = \frac{1}{4}\, e^{-ie\int \mathrm{d}\bar{x}^+\, \theta_c(x^+-\bar{x}^+)\hat{A}_{+c}(\bar{x}^+)} \left(\mathrm{sgn}_c(x^+ - x'^+) + 1 - 2(\hat{\mathcal{O}}_+ + 1)^{-1}\right) \times e^{ie\int \mathrm{d}\bar{x}^+\, \theta_c(x'^+-\bar{x}^+)\hat{A}_{+c}(\bar{x}^+)}, \tag{23}$$

where we have defined

$$\hat{\mathcal{O}}_+ = e^{\frac{ie\hat{a}_+}{2}}\, e^{\frac{\beta\hat{K}}{2}}\, e^{\frac{ie\hat{a}_+}{2}}, \quad \hat{a}_+ = \hat{a}_{+(+)} - \hat{a}_{+(-)}, \quad \hat{a}_{+(\pm)} = \int \mathrm{d}x^+\, \hat{A}_{+(\pm)}(x^+), \tag{24}$$

and $\hat{K}$ denotes the momentum operator. Here "$(\pm)$" denote the thermal indices corresponding to fields on the $C_\pm$ branches in the contour in Fig. 1. This is the complete propagator in the right handed sector of the theory in an arbitrary gauge background at finite temperature. It can be checked that this propagator satisfies the Lippmann-Schwinger equation to all orders so that it leads to the correct perturbative expansion of the propagator.

In this case, the fermion is massless and, therefore, taking the propagator at the coincident point, we can use (7) (in the mixed space) to determine the effective action arising from the right handed sector of the theory. Since at zero temperature there is an ultraviolet divergence in the theory (the mixing of the two sectors at the two point level), we do not expect our method to reproduce this. However, the temperature dependent part of the effective action can be determined to be contained in

$$\Gamma_{\mathrm{R,\,eff}}[a_+] = -\frac{i}{2}\int \frac{\mathrm{d}k}{2\pi}\, \langle k|\ln\cosh\frac{\ln\hat{\mathcal{O}}_+}{2} - \ln\cosh\frac{\beta\hat{K}}{4}|k\rangle. \tag{25}$$

The thermal part of the effective action coming from the right handed sector coincides order by order with the perturbative determination of the effective action (for the case $A_{+(-)} = 0$) in [5] and has the right structure already noted there. Similarly, the temperature dependent effective action arising from the left handed sector of the theory can be determined to be contained in the form

$$\Gamma_{\mathrm{L,\,eff}}[a_-] = -\frac{i}{2}\int \frac{\mathrm{d}k}{2\pi}\, \langle k|\ln\cosh\frac{\ln\hat{\mathcal{O}}_-}{2} - \ln\cosh\frac{\beta\hat{K}}{4}|k\rangle, \tag{26}$$

where the momentum k is the momentum conjugate to the light-cone variable x^+ and

$$\hat{\mathcal{O}}_- = e^{\frac{ie\hat{a}_-}{2}}\, e^{\frac{\beta\hat{K}}{2}}\, e^{\frac{ie\hat{a}_-}{2}}, \quad \hat{a}_- = \hat{a}_{-(+)} - \hat{a}_{-(-)}, \quad \hat{a}_{-(\pm)} = \int \mathrm{d}x^-\, \hat{A}_{-(\pm)}(x^-). \tag{27}$$

The complete temperature dependent effective action, therefore, can be identified to be contained in

$$\Gamma_{\mathrm{eff}}^{(m=0)}[a_+, a_-] = \Gamma_{\mathrm{R,\,eff}}[a_+] + \Gamma_{\mathrm{L,\,eff}}[a_-], \tag{28}$$

whose perturbative expansion coincides order by order with the explicit determination (from Feynman diagram calculations) in [5] for the case when the fields on the branch C_- (in Fig. 1) are set to zero. However, since (28) represents the closed form expression for the complete real time effective action at finite temperature, we can now show, as in the $0+1$ dimensional case, that all the retarded and advanced amplitudes vanish in this theory. This has to be contrasted with the fact that such a result had only been observed up to the four point amplitude through an explicit calculation.

The massless case in $1+1$ dimensions has the simplicity that the theory separates into left handed and right handed sectors. The massive Schwinger model where the fermions have a mass m is described by the Lagrangian density

$$\mathcal{L} = \overline{\psi}\gamma^\mu\left(i\partial_\mu - eA_\mu\right)\psi - m\overline{\psi}\psi. \tag{29}$$

In this case, we can no longer separate the theory into decoupled left and right handed sectors. Nonetheless, following our proposal, we can determine the complete propagator of the theory. Let us only describe the results here and the reader is refereed to [3] for more details. From the form of the complete fermion propagator of the theory, the temperature dependent effective action can be determined to be contained in the normalized action

$$\Gamma_{\text{eff}} = \Gamma_{\text{eff}}^{(0)} + \Gamma_{\text{eff}}^{(m)} - \Gamma_{\text{eff}}^{(m)}(e=0), \tag{30}$$

such that

$$\Gamma_{\text{eff}}(e=0) = 0, \quad \Gamma_{\text{eff}}(m=0) = \Gamma_{\text{eff}}^{(m=0)}, \tag{31}$$

where

$$\Gamma_{\text{eff}}^{(m)} = -\frac{i}{2}\,\text{Tr}\,\ln\left(\mathbb{1} + m^2(\hat{S}^{(0)}\tau)^2\right). \tag{32}$$

Here $\hat{S}^{(0)}$ denotes the complete propagator for the massless ($m=0$) Schwinger model (including both left handed and right handed components) and

$$\tau = \begin{pmatrix} 1 & 0 \\ 0 & -1 \end{pmatrix}, \tag{33}$$

denotes the thermal matrix describing the mass insertions on the two thermal branches $C_{\pm}$ in Fig. 1.

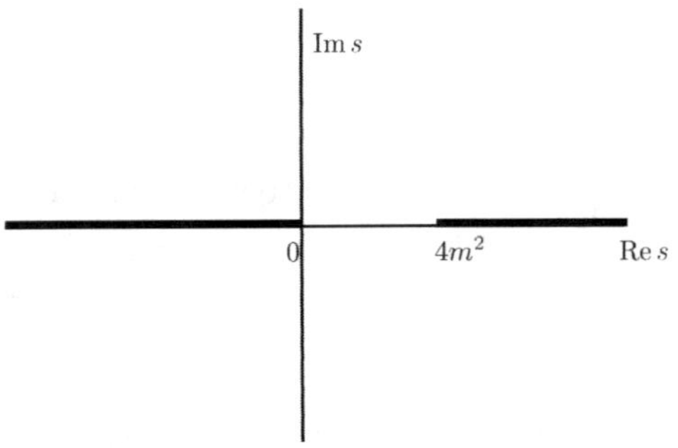

Figure 2. The two branch cuts in the two point function in massive QED at finite temperature. Here $s \equiv p^2$ where p^μ denotes the external momentum. The branch cut for $s \geq 4m^2$ corresponds to pair production while the other for $s \leq 0$ represents the effects due to thermal scattering.

Let us now discuss the imaginary parts of the thermal effective action in all these examples. We note from (15) that the thermal effective action in general is complex and it has a nontrivial temperature dependent imaginary part. In the case of the Schwinger model ($m=0$), it can be checked from (26)-(28) that the temperature dependent part of the effective action is purely imaginary (this has to be contrasted with the fact that the zero temperature part of the effective action which gives the photon a mass is purely real). In the case of the massive Schwinger model, (30)-(32) lead to the fact that the temperature dependent part of the effective action is in general complex and has a nontrivial imaginary part. Thus, we see that temperature does contribute to the imaginary part of the effective action and, therefore, the vacuum "decay rate". At zero temperature, the decay of the vacuum takes place through pair production. At finite temperature, on the other hand, there are other channels of reactions possible (see [4] for details) since a particle can scatter from the real (physical) particles present in the medium.

These additional channels of reaction give rise to new branch cuts in the amplitudes. For example, in the case of massive QED (massive Schwinger model), the two point function at finite temperature has two branch cuts shown in Fig. 2. Correspondingly, the imaginary part of the temperature dependent effective action should be thought of as the "vacuum decay" rate arising from the contributions of all these channels (and not just from pair production).

In conclusion, we have proposed a very powerful method for evaluating temperature dependent effective actions for particles (bosons/fermions) interacting with a background field. Even though our discussion has been within the context of $0+1$ and $1+1$ dimensional theories, this indeed covers all the (known) solvable models of fermions interacting with electric fields in $3+1$ dimensions.

Acknowledgment

It is a pleasure to thank the organizers of the Mexican school of particles and fields for the invitation to give this lecture at the conference. This work was supported in part by US DOE Grant number DE-FG 02-91ER40685.

References

[1] Das, A. and J. Frenkel, Phys. Lett. **B680**, 195 (2009).
[2] Das, A. and J. Frenkel, Phys. Rev. **D80**, 125039 (2009).
[3] Das, A. and J. Frenkel, Phys. Rev. **D82**, 125002 (2010).
[4] Das, A, *Finite Temperature Field Theory*, World Scientific, Singapore (1997).
[5] Das, A. and A. J. da Silva, Phys. Rev. **D59**, 105011 (1999).

Journal of Physics: Conference Series **287** (2011) 012007 doi:10.1088/1742-6596/287/1/012007

Light-Front Quantization and AdS/QCD: An overview

Guy F. de Téramond[1] **and Stanley J. Brodsky**[2]

[1] Universidad de Costa Rica, San José, Costa Rica

[2] SLAC National Accelerator Laboratory, Stanford University, Stanford, CA 94309

Abstract. We give an overview of the light front holographic approach to strongly coupled QCD, whereby a confining gauge theory quantized on the light front is mapped to a higher-dimensional anti de Sitter (AdS) space incorporating the AdS/CFT correspondence as a useful guide. One can start from the Hamiltonian equation of motion in physical space time by studying the off-shell dynamics of the bound state wavefunctions as a function of the invariant mass of the constituents. To a first semiclassical approximation, where quantum loops and quark masses are not included, this leads to a light-front Hamiltonian equation which describes the bound state dynamics of light hadrons in terms of an invariant impact variable ζ, which measures the separation of the partons within the hadron at equal light-front time. Alternatively, one can start from the gravity side by studying the propagation of hadronic modes in a fixed effective gravitational background which encodes salient properties of the QCD dual theory, such as the ultraviolet conformal limit at the AdS boundary at $z \to 0$, as well as modifications of the background geometry in the large z infrared region to describe confinement. In the semiclassical approximation both approaches are equivalent. This allows us to identify the holographic variable z in AdS space with the impact variable ζ. Light-front holography also allows a precise mapping of transition amplitudes from AdS to physical space-time. In contrast with the usual AdS/QCD framework, in light front holography the internal structure of hadrons is explicitly introduced and the angular momentum of the constituents plays a key role.

1. Introduction

The AdS/CFT correspondence between gravity or string theory on a higher dimensional anti de Sitter (AdS) space and conformal field theories in physical space time [1] has led to a semiclassical approximation for strongly-coupled QCD, which provides physical insights into its non-perturbative dynamics. The correspondence is holographic in the sense that it determines a duality between theories in different number of space time dimensions. In practice, the duality provides an effective gravity description in a $(d+1)$-dimensional AdS space-time in terms of a flat d-dimensional conformally-invariant quantum field theory defined at the AdS asymptotic boundary [2, 3]

$$e^{iW_{\rm CFT}[J=\Phi_0]} = Z_{\rm grav}[\Phi] = \int \mathcal{D} e^{iS_{eff}[\Phi]}, \tag{1}$$

where $W_{\rm CFT}$ is the generating functional of the conformal theory and Z the string partition function given in terms of fields Φ in AdS space. Thus, in principle, one can compute physical observables in a strongly coupled gauge theory in terms of a classical gravity theory.

Anti-de Sitter AdS_5 space is the maximally symmetric space-time with negative curvature and a four dimensional space-time boundary. The most general group of transformations that

Journal of Physics: Conference Series **287** (2011) 012007 doi:10.1088/1742-6596/287/1/012007

leave the AdS_{d+1} differential line element

$$ds^2 = \frac{R^2}{z^2}\left(\eta_{\mu\nu}dx^\mu dx^\nu - dz^2\right), \tag{2}$$

invariant, the isometry group, has dimensions $(d+1)(d+2)/2$. Five dimensional Anti-de Sitter space AdS_5 has 15 isometries, in agreement with the number of generators of the conformal group in four dimensions.

Since the AdS metric (2) is invariant under a dilatation of all coordinates $x^\mu \to \lambda x^\mu$ and $z \to \lambda z$, it follows that the additional dimension, the holographic variable z, acts like a scaling variable in Minkowski space: different values of z correspond to different energy scales at which the hadron is examined. As a result, a short space-like or time-like invariant interval near the light-cone, $x_\mu x^\mu \to 0$, maps to the conformal AdS boundary near $z \to 0$. This also corresponds to the $Q \to \infty$ ultraviolet (UV) zero separation distance. On the other hand, a large invariant four-dimensional interval of confinement dimensions $x_\mu x^\mu \sim 1/\Lambda^2_{\text{QCD}}$ maps to the large infrared (IR) region of AdS space $z \sim 1/\Lambda_{\text{QCD}}$.

QCD is fundamentally different from conformal theories since its scale invariance is broken by quantum effects. A gravity dual to QCD is not known, but the mechanisms of confinement can be incorporated in the gauge/gravity correspondence by modifying the AdS geometry near a large infrared value $z \sim 1/\Lambda_{\text{QCD}}$, which sets the scale of the strong interactions. In this simplified approach hadronic modes propagate in a fixed effective gravitational background which encodes salient properties of the QCD dual theory, such as the ultraviolet conformal limit at the AdS boundary at $z \to 0$, as well as modifications of the background geometry for large z to describe confinement. The modified theory generates the point-like hard behavior expected from QCD, instead of the soft behavior characteristic of extended objects. [4] It is striking that the QCD dimensional counting rules [5, 6] are also a key feature of nonperturbative models [4] based on the gauge/gravity duality. Although the mechanisms are different, both the perturbative QCD and the AdS/QCD approaches depend on the leading twist (dimension minus spin) interpolating operators of the hadron and their structure at short distances.

Incorporating the AdS/CFT correspondence as a useful guide, light-front holographic methods were originally introduced [7] by matching the expression for electromagnetic (EM) current matrix elements in AdS space [8] with the corresponding matrix elements using light-front theory in physical space time. This allows us to identify the holographic variable z in AdS space with an impact variable ζ. [7] It was also shown that one obtains identical holographic mapping using the matrix elements of the gravitational or energy-momentum tensor [9] by perturbing the AdS metric [10] around its static solution (2).

Light-front quantization is the ideal framework to describe the structure of hadrons in terms of their quark and gluon degrees of freedom. The simple structure of the light-front (LF) vacuum allows an unambiguous definition of the partonic content of a hadron in QCD and of hadronic light-front wavefunctions (LFWFs), the underlying link between large distance hadronic states and the constituent degrees of freedom at short distances. The LFWFs of relativistic bound states in QCD provide a description of the structure and internal dynamics of hadronic states in terms of their constituent quark and gluons at the same LF time $\tau = x^0 + x^3$, the time marked by the front of a light wave, [11] instead of the ordinary instant time $t = x^0$. The constituent spin and orbital angular momentum properties of the hadrons are also encoded in the LFWFs. In fact, the definition of quark and gluon angular momentum is unambiguous in Dirac's front form in light-cone gauge $A^+ = 0$, and the gluons have physical polarization $S^z_g = \pm 1$. Unlike instant time quantization, the Hamiltonian equation of motion in the light-front is frame independent and has a structure similar to the eigenmode equations in AdS space. This makes a direct connection of QCD with AdS/CFT methods possible. The identification of orbital angular momentum of the constituents is a key element in our description of the

Journal of Physics: Conference Series **287** (2011) 012007 doi:10.1088/1742-6596/287/1/012007

internal structure of hadrons using holographic principles, since hadrons with the same quark content but different orbital angular momentum have different masses. In contrast with the usual AdS/QCD approach [12, 13] where fields in the bulk geometry are introduced to match the chiral symmetries of QCD and axial and vector currents become the primary entities as in effective chiral theory, in light-front holography a direct connection with the internal constituent structure of hadrons is established using light front quantization. [7, 9, 14]

A physical hadron in four-dimensional Minkowski space has four-momentum P_μ and invariant hadronic mass states determined by the light-front Lorentz-invariant Hamiltonian equation for the relativistic bound-state system $P_\mu P^\mu|\psi(P)\rangle = \mathcal{M}^2|\psi(P)\rangle$, where the operator $P_\mu P^\mu$ is determined canonically from the QCD Lagrangian. On AdS space the physical states are represented by normalizable modes $\Phi_P(x,z) = e^{-iP\cdot x}\Phi(z)$, with plane waves along Minkowski coordinates x^μ and a profile function $\Phi(z)$ along the holographic coordinate z. The hadronic invariant mass $P_\mu P^\mu = \mathcal{M}^2$ is found by solving the eigenvalue problem for the AdS wave equation. Each light-front hadronic state $|\psi(P)\rangle$ is dual to a normalizable string mode $\Phi_P(x,z)$. For fields near the AdS boundary the behavior of $\Phi(z)$ depends on the scaling dimension of the corresponding interpolating operators.

We have shown recently a remarkable connection between the description of hadronic modes in AdS space and the Hamiltonian formulation of QCD in physical space-time quantized on the light-front at equal light-front time τ. [14] Indeed, one may take the light-front Hamiltonian equation of motion in physical space time as a starting point to derive relativistic wave equations by studying the off-shell dynamics of the bound state wavefunctions. [14] To a first semiclassical approximation, where quantum loops and quark masses are not included, this leads to a light-front Hamiltonian equation which describes the bound state dynamics of light hadrons in terms of an invariant impact variable ζ, which measures the separation of the partons within the hadron at equal light-front time. This procedure leads to relativistic light-front wave equations which are equivalent to the equations of motion which describe the propagation of hadronic modes in AdS space. [14] Remarkably, the AdS equations correspond to the kinetic energy terms of the partons inside a hadron, whereas the interaction terms build confinement and correspond to the truncation of AdS space in an effective dual gravity approximation. [14] Early attempts to derive effective one-body equations in light-front QCD are described in reference. [15] We should also mention previous work by 't Hooft, who obtained the spectrum of two-dimensional QCD in the large N_C limit in terms of a Schrödinger equation as a function of the parton x-variable. [16] In the scale invariant limit, this equation is equivalent to the equation of motion for a scalar field in AdS_3 space. [17] In this case, there is a mapping between the variable x and the radial coordinate in AdS_3.

2. Higher Spin Hadronic Modes in AdS Space

The description of higher spin modes in AdS space is a notoriously difficult problem. [18, 19, 20] A spin-J field in AdS_{d+1} is represented by a rank J tensor field $\Phi(x^M)_{M_1\cdots M_J}$, which is totally symmetric in all its indices. Such a tensor contains also lower spins, which can be eliminated by imposing gauge conditions. The action for a spin-J field in AdS_{d+1} space time in presence of a dilaton background field $\varphi(z)$ is given by

$$S = \frac{1}{2}\int d^dx\, dz\, \sqrt{g}\, e^{\varphi(z)}\Big(g^{NN'}g^{M_1M_1'}\cdots g^{M_JM_J'}D_N\Phi_{M_1\cdots M_J}D_{N'}\Phi_{M_1'\cdots M_J'} - \mu^2 g^{M_1M_1'}\cdots g^{M_JM_J'}\Phi_{M_1\cdots M_J}\Phi_{M_1'\cdots M_J'} + \cdots\Big), \tag{3}$$

where D_M is the covariant derivative which includes parallel transport

$$[D_N, D_K]\Phi_{M_1\cdots M_J} = -R^L_{M_1NK}\Phi_{L\cdots M_J} - \cdots - R^L_{M_JNK}\Phi_{M_1\cdots L}, \tag{4}$$

Journal of Physics: Conference Series **287** (2011) 012007 doi:10.1088/1742-6596/287/1/012007

and the omitted terms refer to terms with different contractions. Conformal invariance in (3) is broken by the dilaton profile $\varphi(z)$ which is a function of the holographic coordinate z and vanishes in the conformal limit $z \to 0$. The coordinates of AdS are the Minkowski coordinates x^μ and the holographic variable z labeled $x^M = (x^\mu, z)$, with $M, N = 1, \cdots, d+1$

A physical hadron has plane-wave solutions and polarization indices μ_i, $i = 1 \cdots J$, along the $3+1$ physical coordinates $\Phi_P(x,z)_{\mu_1 \cdots \mu_J} = e^{-iP\cdot x}\Phi(z)_{\mu_1\cdots\mu_J}$, with four-momentum P_μ and invariant hadronic mass $P_\mu P^\mu = \mathcal{M}^2$. All other components vanish identically: $\Phi_{z\mu_2\cdots\mu_J} = \cdots = \Phi_{\mu_1\mu_2\cdots z} = 0$. One can then construct an effective action in terms of high spin modes $\Phi_{\mu_1\mu_2\cdots\mu_J}$, with only the physical degrees of freedom. [21, 22] In this case the system of coupled differential equations which follow from (3) reduce to a homogeneous equation in terms of the field $\Phi_P(x,z)_{\mu_1\cdots\mu_J}$ with all the polarization indices along the physical $3+1$ coordinates. In terms of fields with tangent indices

$$\hat{\Phi}_{A_1 A_2 \cdots A_J} = e^{M_1}_{A_1} e^{M_2}_{A_2} \cdots e^{M_J}_{A_J} \Phi_{M_1 M_2 \cdots M_J} = \left(\frac{z}{R}\right)^J \Phi_{A_1 A_2 \cdots A_J}, \tag{5}$$

we find the effective action [22]

$$S = \frac{1}{2}\int d^d x\, dz\, \sqrt{g}\, e^{\varphi(z)} \Big(g^{NN'} \eta^{\mu_1 \mu_1'} \cdots \eta^{\mu_J \mu_J'} \partial_N \hat{\Phi}_{\mu_1 \cdots \mu_J} \partial_{N'} \hat{\Phi}_{\mu_1' \cdots \mu_J'} - \mu^2 \eta^{\mu_1 \mu_1'} \cdots \eta^{\mu_J \mu_J'} \hat{\Phi}_{\mu_1 \cdots \mu_J} \hat{\Phi}_{\mu_1' \cdots \mu_J'} \Big), \tag{6}$$

upon rescaling the fifth dimensional mass μ. The vielbein e^A_M is defined by $g_{MN} = e^A_M e^B_N \eta_{AB}$, where $A, B = 1, \cdots, d+1$ are tangent AdS space indices.

Variation of the action gives the AdS wave equation for the spin-J mode $\Phi_{\mu_1\cdots\mu_J}$

$$\left[-\frac{z^{d-1-2J}}{e^{\varphi(z)}} \partial_z \left(\frac{e^{\varphi}(z)}{z^{d-1-2J}} \partial_z \right) + \left(\frac{\mu R}{z} \right)^2 \right] \Phi(z)_{\mu_1 \cdots \mu_J} = \mathcal{M}^2 \Phi(z)_{\mu_1 \cdots \mu_J}, \tag{7}$$

where $\hat{\Phi}(z)_{\mu_1\cdots\mu_J} = (z/R)^J \Phi(z)_{\mu_1\cdots\mu_J}$ has scaling behavior $\hat{\Phi}_J(z \to 0) \sim z^\tau$ and scaling dimension τ given by the relation $(\mu R)^2 = (\tau - J)(\tau - d + J)$. Eq. (7) is the basic equation we shall use here to describe the propagation of spin-J hadronic modes in a fixed gravitational background, asymptotic to AdS space. Its eigenvalues determine the invariant mass $\mathcal{M}^2$ of hadronic states $|\psi(P)\rangle$ dual to the hadronic modes $\Phi_P(x,z)_{\mu_1\cdots\mu_J}$. We describe below how Eq. (7) is mapped to the QCD light-front wave equations in physical space-time. A detailed derivation of (7) and comparison with the scaling described in Ref. [21] will be given in Ref. [22]. The wave equation (7) also follows from considering the wave equation for a scalar mode Φ in AdS and rescaling the solution by shifting dimensions according to $\Phi_J(z) = (z/R)^{-J}\Phi(z)$. [14, 23]

3. Light-Front Quantization of QCD

One can express the hadron four-momentum generator $P = (P^+, P^-, \mathbf{P}_\perp)$, $P^\pm = P^0 \pm P^3$, in terms of the dynamical fields, the Dirac field ψ_+, where $\psi_\pm = \Lambda_\pm \psi$, $\Lambda_\pm = \gamma^0\gamma^\pm$, and the transverse field $\mathbf{A}_\perp$ in the $A^+ = 0$ gauge [24] quantized on the light-front at fixed light-cone time x^+, $x^\pm = x^0 \pm x^3$

$$P^- = \frac{1}{2}\int dx^- d^2\mathbf{x}_\perp \overline{\psi}_+ \gamma^+ \frac{(i\nabla_\perp)^2 + m^2}{i\partial^+} \psi_+ + \text{(interactions)}, \tag{8}$$

$$P^+ = \int dx^- d^2\mathbf{x}_\perp \overline{\psi}_+ \gamma^+ i\partial^+ \psi_+, \tag{9}$$

$$\mathbf{P}_\perp = \frac{1}{2}\int dx^- d^2\mathbf{x}_\perp \overline{\psi}_+ \gamma^+ i\nabla_\perp \psi_+, \tag{10}$$

where the integrals are over the null plane $x^+ = 0$, the hyper-plane tangent to the light cone. This is the initial-value surface for the fields where the commutation relations are fixed. The LF Hamiltonian P^- generates LF time translations $[\psi_+(x), P^-] = i\frac{\partial}{\partial x^+}\psi_+(x)$, to evolve the initial conditions to all space-time, whereas the LF longitudinal P^+ and transverse momentum $\mathbf{P}_\perp$ are kinematical generators. For simplicity we have omitted from (8 - 10) the contributions from the gluon field $\mathbf{A}_\perp$.

According to Dirac's classification of the forms of relativistic dynamics, [11] the fundamental generators of the Poincaré group can be separated into kinematical and dynamical generators. The kinematical generators act along the initial surface and leave the light-front plane invariant: they are thus independent of the dynamics and therefore contain no interactions. The dynamical generators change the light-front position and depend consequently on the interactions. In addition to P^+ and $\mathbf{P}_\perp$, the kinematical generators in the light-front frame are the z-component of the angular momentum J^z and the boost operator $\mathbf{K}$. In addition to the Hamiltonian P^-, J^z and J^y are also dynamical generators. The light-front frame has the maximal number of kinematical generators.

A physical hadron in four-dimensional Minkowski space has four-momentum P_μ and invariant hadronic mass states $P_\mu P^\mu = \mathcal{M}^2$ determined by the Lorentz-invariant Hamiltonian equation for the relativistic bound-state system

$$P_\mu P^\mu|\psi(P)\rangle = \left(P^-P^+ - \mathbf{P}_\perp^2\right)|\psi(P)\rangle = \mathcal{M}^2|\psi(P)\rangle, \tag{11}$$

where the hadronic state $|\psi\rangle$ is an expansion in multiparticle Fock eigenstates $|n\rangle$ of the free light-front Hamiltonian: $|\psi\rangle = \sum_n \psi_n|\psi\rangle$. The state $|\psi(P^+, \mathbf{P}_\perp, J^z)\rangle$ is an eigenstate of the total momentum P^+ and $\mathbf{P}_\perp$ and the total spin J^z. Quark and gluons appear from the light-front quantization of the excitations of the dynamical fields ψ_+ and $\mathbf{A}_\perp$, expanded in terms of creation and annihilation operators at fixed LF time τ. The Fock components $\psi_n(x_i, \mathbf{k}_{\perp i}, \lambda_i^z)$ are independent of P^+ and $\mathbf{P}_\perp$ and depend only on relative partonic coordinates: the momentum fraction $x_i = k_i^+/P^+$, the transverse momentum $\mathbf{k}_{\perp i}$ and spin component λ_i^z. Momentum conservation requires $\sum_{i=1}^n x_i = 1$ and $\sum_{i=1}^n \mathbf{k}_{\perp i} = 0$. The LFWFs ψ_n provide a *frame-independent* representation of a hadron which relates its quark and gluon degrees of freedom to their asymptotic hadronic state.

4. A Semiclassical Approximation to QCD

We can compute $\mathcal{M}^2$ from the hadronic matrix element

$$\langle\psi(P')|P_\mu P^\mu|\psi(P)\rangle = \mathcal{M}^2\langle\psi(P')|\psi(P)\rangle, \tag{12}$$

expanding the initial and final hadronic states in terms of its Fock components. The computation is simplified in the frame $P = \left(P^+, \mathcal{M}^2/P^+, \vec{0}_\perp\right)$ where $P^2 = P^+P^-$. We find

$$\mathcal{M}^2 = \sum_n \int [dx_i]\left[d^2\mathbf{k}_{\perp i}\right] \sum_q \left(\frac{\mathbf{k}_{\perp q}^2 + m_q^2}{x_q}\right) |\psi_n(x_i, \mathbf{k}_{\perp i})|^2 + (\text{interactions}), \tag{13}$$

plus similar terms for antiquarks and gluons ($m_g = 0$). The integrals in (13) are over the internal coordinates of the n constituents for each Fock state

$$\int [dx_i] \equiv \prod_{i=1}^n \int dx_i\, \delta\Big(1 - \sum_{j=1}^n x_j\Big), \quad \int \left[d^2\mathbf{k}_{\perp i}\right] \equiv \prod_{i=1}^n \int \frac{d^2\mathbf{k}_{\perp i}}{2(2\pi)^3}\, 16\pi^3\, \delta^{(2)}\Big(\sum_{j=1}^n \mathbf{k}_{\perp j}\Big), \tag{14}$$

with phase space normalization $\sum_n \int [dx_i]\left[d^2\mathbf{k}_{\perp i}\right] |\psi_n(x_i, \mathbf{k}_{\perp i})|^2 = 1$.

Journal of Physics: Conference Series **287** (2011) 012007 doi:10.1088/1742-6596/287/1/012007

The LFWF $\psi_n(x_i, \mathbf{k}_{\perp i})$ can be expanded in terms of $n-1$ independent position coordinates $\mathbf{b}_{\perp j}$, $j = 1, 2, \ldots, n-1$, conjugate to the relative coordinates $\mathbf{k}_{\perp i}$, with $\sum_{i=1}^{n} \mathbf{b}_{\perp i} = 0$. We can also express (13) in terms of the internal impact coordinates $\mathbf{b}_{\perp j}$ with the result

$$\mathcal{M}^2 = \sum_n \prod_{j=1}^{n-1} \int dx_j \, d^2\mathbf{b}_{\perp j} \, \psi_n^*(x_j, \mathbf{b}_{\perp j}) \sum_q \left(\frac{-\nabla^2_{\mathbf{b}_{\perp q}} + m_q^2}{x_q} \right) \psi_n(x_j, \mathbf{b}_{\perp j}) + (\text{interactions}). \quad (15)$$

The normalization is defined by $\sum_n \prod_{j=1}^{n-1} \int dx_j d^2\mathbf{b}_{\perp j} \left|\psi_n(x_j, \mathbf{b}_{\perp j})\right|^2 = 1$. To simplify the discussion we will consider a two-parton hadronic bound state. In the limit of zero quark mass $m_q \to 0$

$$\mathcal{M}^2 = \int_0^1 \frac{dx}{x(1-x)} \int d^2\mathbf{b}_\perp \, \psi^*(x, \mathbf{b}_\perp) \left(-\nabla^2_{\mathbf{b}_\perp}\right) \psi(x, \mathbf{b}_\perp) + (\text{interactions}). \quad (16)$$

The functional dependence for a given Fock state is given in terms of the invariant mass

$$\mathcal{M}_n^2 = \Big(\sum_{a=1}^{n} k_a^\mu \Big)^2 = \sum_a \frac{\mathbf{k}^2_{\perp a} + m_a^2}{x_a} \to \frac{\mathbf{k}^2_\perp}{x(1-x)}, \quad (17)$$

giving the measure of the off-energy shell of the bound state, $\mathcal{M}^2 - \mathcal{M}_n^2$. Similarly, in impact space the relevant variable for a two-parton state is $\zeta^2 = x(1-x)\mathbf{b}^2_\perp$. Thus, to first approximation LF dynamics depend only on the boost invariant variable $\mathcal{M}_n$ or ζ, and hadronic properties are encoded in the hadronic mode $\phi(\zeta)$ from the relation

$$\psi(x, \zeta, \varphi) = e^{iL\varphi} X(x) \frac{\phi(\zeta)}{\sqrt{2\pi\zeta}}, \quad (18)$$

thus factoring out the angular dependence φ and the longitudinal, $X(x)$, and transverse mode $\phi(\zeta)$. This is a natural factorization in the light front since the corresponding canonical generators, the longitudinal and transverse generators P^+ and $\mathbf{P}_\perp$ and the z-component of the orbital angular momentum J^z are kinematical generators which commute with the LF Hamiltonian generator P^-. We choose the normalization of the LF mode $\phi(\zeta) = \langle \zeta | \psi \rangle$ as $\langle \phi | \phi \rangle = \int d\zeta \, |\langle \zeta | \phi \rangle|^2 = 1$.

We can write the Laplacian operator in (16) in circular cylindrical coordinates (ζ, φ) and factor out the angular dependence of the modes in terms of the $SO(2)$ Casimir representation L^2 of orbital angular momentum in the transverse plane. Using (18) we find [14]

$$\mathcal{M}^2 = \int d\zeta \, \phi^*(\zeta) \sqrt{\zeta} \left(-\frac{d^2}{d\zeta^2} - \frac{1}{\zeta}\frac{d}{d\zeta} + \frac{L^2}{\zeta^2} \right) \frac{\phi(\zeta)}{\sqrt{\zeta}} + \int d\zeta \, \phi^*(\zeta) U(\zeta) \phi(\zeta), \quad (19)$$

where $L = L^z$. In writing the above equation we have summed up the complexity of the interaction terms in the QCD Lagrangian in the addition of the effective potential $U(\zeta)$, which is then modeled to enforce confinement at some IR scale. The LF eigenvalue equation $P_\mu P^\mu |\phi\rangle = \mathcal{M}^2 |\phi\rangle$ is thus a light-front wave equation for ϕ

$$\left(-\frac{d^2}{d\zeta^2} - \frac{1 - 4L^2}{4\zeta^2} + U(\zeta) \right) \phi(\zeta) = \mathcal{M}^2 \phi(\zeta), \quad (20)$$

a relativistic single-variable LF Schrödinger equation. Its eigenmodes $\phi(\zeta)$ determine the hadronic mass spectrum and represent the probability amplitude to find n-partons at transverse impact separation ζ, the invariant separation between pointlike constituents within the hadron [7] at equal LF time. Extension of the results to arbitrary n follows from the x-weighted definition of the transverse impact variable of the $n-1$ spectator system [7]: $\zeta = \sqrt{\frac{x}{1-x}} \left| \sum_{j=1}^{n-1} x_j \mathbf{b}_{\perp j} \right|$, where $x = x_n$ is the longitudinal momentum fraction of the active quark. One can also generalize the equations to allow for the kinetic energy of massive quarks using Eqs. (13) or (15). In this case, however, the longitudinal mode $X(x)$ does not decouple from the effective LF bound-state equations.

5. Light-Front Holographic Mapping and Hadronic Spectrum

The structure of the QCD Hamiltonian equation (11) for the state $|\psi(P)\rangle$ is similar to the structure of the wave equation (7) for the J-mode $\Phi_{\mu_1 \cdots \mu_J}$ in AdS; they are both frame-independent and have identical eigenvalues $\mathcal{M}^2$, the mass spectrum of the color-singlet states of QCD, a possible indication of a more profound connection between physical QCD and the physics of hadronic modes in AdS space. However, important differences are also apparent: Eq. (11) is a linear quantum-mechanical equation of states in Hilbert space, whereas Eq. (7) is a classical gravity equation; its solutions describe spin-J modes propagating in a higher dimensional warped space. Physical hadrons are composite and thus inexorably endowed of orbital angular momentum. Thus, the identification of orbital angular momentum is of primary interest in finding a connection between both approaches.

As shown in Sect. 4, one can indeed systematically reduce the LF Hamiltonian eigenvalue Eq. (11) to an effective relativistic wave equation (20), analogous to the AdS equations, by observing that each n-particle Fock state has an essential dependence on the invariant mass of the system and thus, to a first approximation, LF dynamics depend only on $\mathcal{M}_n^2$. In impact space the relevant variable is the boost invariant variable ζ, which measures the separation of the constituents and which also allows one to separate the dynamics of quark and gluon binding from the kinematics of the constituent internal angular momentum.

Upon the substitution $z \to \zeta$ and $\phi_J(\zeta) = (\zeta/R)^{-3/2+J} \, e^{\varphi(z)/2} \, \Phi_J(\zeta)$, in (7), we find for $d = 4$ the QCD light-front wave equation (20) with effective potential [23]

$$U(\zeta) = \frac{1}{2}\varphi''(z) + \frac{1}{4}\varphi'(z)^2 + \frac{2J-3}{2z}\varphi'(z). \tag{21}$$

The fifth dimensional mass μ is not a free parameter but scales as $(\mu R)^2 = -(2-J)^2 + L^2$.

If $L^2 < 0$ the LF Hamiltonian is unbounded from below $\langle \phi | P_\mu P^\mu | \phi \rangle < 0$ and the spectrum contains an infinite number of negative values of $\mathcal{M}^2$, which can be arbitrarily large. As $\mathcal{M}^2$ increases in absolute value the particle becomes localized within a very small region near $\zeta = 0$, since the effective potential is conformal at small ζ. For $\mathcal{M}^2 \to -\infty$ the particle is localized at $\zeta = 0$, the particle "falls towards the center". [25] The critical value $L = 0$ corresponds to the lowest possible stable solution, the ground state of the light-front Hamiltonian. For $J = 0$ the five dimensional mass μ is related to the orbital momentum of the hadronic bound state by $(\mu R)^2 = -4 + L^2$ and thus $(\mu R)^2 \geq -4$. The quantum mechanical stability condition $L^2 \geq 0$ is thus equivalent to the Breitenlohner-Freedman stability bound in AdS. [26] The scaling dimensions are $2 + L$ independent of J, in agreement with the twist-scaling dimension of a two-parton bound state in QCD. It is important to notice that in the light-front the $SO(2)$ Casimir for orbital angular momentum L^2 is a kinematical quantity, in contrast with the usual $SO(3)$ Casimir $L(L+1)$ from non-relativistic physics which is rotational, but not boost invariant. The $SO(2)$ Casimir form L^2 corresponds to the group of rotations in the transverse LF plane. Indeed, the Casimir operator for $SO(N) \sim S^{N-1}$ is $L(L+N-2)$.

We consider here the positive-sign dilaton profile $\exp(+\kappa^2 z^2)$ which explicitly confines the constituents to distances $\langle z \rangle \sim 1/\kappa$ [27, 28]. It is also the solution compatible with (7), which naturally encodes the internal structure of hadrons and their orbital angular momentum. From (21) we obtain the effective potential [27] $U(\zeta) = \kappa^4\zeta^2 + 2\kappa^2(L + S - 1)$, where $J^z = L^z + S^z$, which corresponds to a transverse oscillator in the light-front. Equation (20) has eigenfunctions

$$\phi_{n,L}(\zeta) = \kappa^{1+L}\sqrt{\frac{2n!}{(n+L)!}}\,\zeta^{1/2+L}e^{-\kappa^2\zeta^2/2}L_n^L(\kappa^2\zeta^2), \tag{22}$$

and eigenvalues

$$\mathcal{M}^2_{n,L,S} = 4\kappa^2\left(n + L + \frac{S}{2}\right). \tag{23}$$

The meson spectrum has a string-theory Regge form: the square of the masses are linear in both the internal orbital angular momentum L and radial quantum number n, where n counts the number of nodes of the wavefunction in the radial variable ζ. The spectrum also depends on the internal spin S. The lowest possible solution for $n = L = S = 0$ has eigenvalue $\mathcal{M}^2 = 0$. This is a chiral symmetric bound state of two massless quarks with scaling dimension 2 and size $\langle\zeta^2\rangle \sim 1/\kappa^2$, which we identify with the lowest state, the pion. Thus one can compute the hadron spectrum by simply adding $4\kappa^2$ for a unit change in the radial quantum number, $4\kappa^2$ for a change in one unit in the orbital quantum number and $2\kappa^2$ for a change of one unit of spin to the ground state value of $\mathcal{M}^2$. Remarkably, the same rule holds for baryons. [29] This is an important feature of light-front holography, which predicts the same multiplicity of states for mesons and baryons as it is observed experimentally. [30]

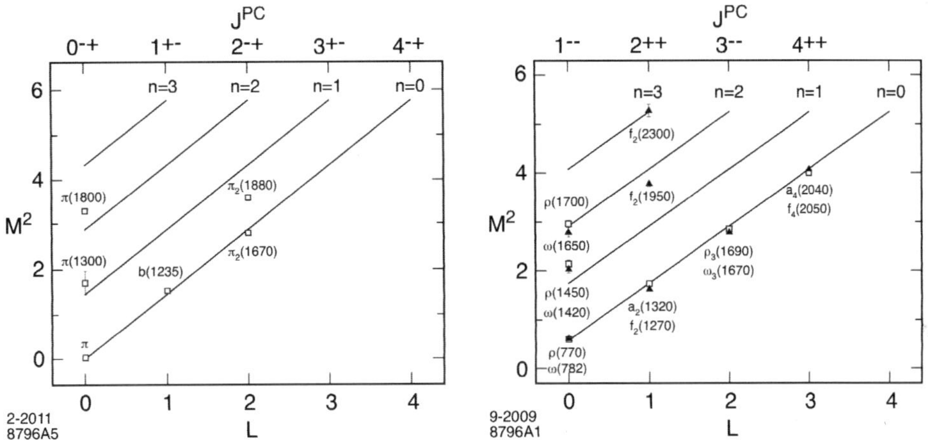

Figure 1. Regge trajectories for π-meson family with $\kappa = 0.6$ GeV (left); $I=1$ ρ-meson and $I=0$ ω-meson families with $\kappa = 0.54$ GeV (right). Only confirmed PDG states [31] are shown.

Individual hadron states are identified by their interpolating operators at $z \to 0$. Pion interpolating operators are constructed by examining the behavior of bilinear covariants $\overline{\psi}\Gamma\psi$ under charge conjugation and parity transformation. Thus, for example, a pion interpolating operator $\overline{q}\gamma_5 q$ create a state with quantum numbers $J^{PC} = 0^{-+}$, and a vector meson interpolating operator $\overline{q}\gamma_\mu q$ a state 1^{--}. Likewise the operator $\overline{q}\gamma_\mu\gamma_5 q$ creates a state with 1^{++} quantum numbers, the $a_1(1260)$ positive parity meson. If we include orbital excitations the pion interpolating operator is $\mathcal{O}_{2+L} = \overline{q}\gamma_5 D_{\{\ell_1} \cdots D_{\ell_m\}}q$. This is an operator with total internal orbital momentum $L = \sum_{i=1}^m \ell_i$, twist $\tau = 2 + L$ and canonical dimension $\Delta = 3 + L$. The scaling of the AdS field $\Phi(z) \sim z^\tau$ at $z \to 0$ is precisely the scaling required to match the scaling dimension of the local meson interpolating operators. The spectral predictions for light meson and vector meson states are compared with experimental data in Fig. 1 for the positive sign dilaton model discussed here.

The baryon interpolating operator $\mathcal{O}_{3+L} = \psi D_{\{\ell_1} \dots D_{\ell_q} \psi D_{\ell_{q+1}} \dots D_{\ell_m\}} \psi$, $L = \sum_{i=1}^{m} \ell_i$, is a twist 3, dimension $9/2 + L$ operator with scaling behavior given by its twist-dimension $3 + L$. As in the meson sector, the increase in the mass squared for higher baryonic states is $\Delta n = 4\kappa^2$, $\Delta L = 4\kappa^2$ and $\Delta S = 2\kappa^2$, relative to the lowest ground state, the proton. [29]

The predictions for the positive parity light baryons are shown in Fig. 2. As for the predictions for mesons in Fig. 1, only confirmed PDG [31] states are shown. The Roper state $N(1440)$ and the $N(1710)$ are well accounted for in this model as the first and second radial states. Likewise the $\Delta(1660)$ corresponds to the first radial state of the Δ family. The model is successful in explaining the important parity degeneracy observed in the light baryon spectrum, such as the $L=2$, $N(1680)$–$N(1720)$ pair and the $\Delta(1905), \Delta(1910), \Delta(1920), \Delta(1950)$ states which are degenerate within error bars. The parity degeneracy of baryons is also a property of the hard wall model, but radial states are not well described by this model. [32] For other recent calculations of the hadronic spectrum in the framework of AdS/QCD, see Refs. [33, 34, 35, 36, 37, 38, 39, 40, 41, 42, 43, 44, 45, 46, 47, 48, 49, 50].

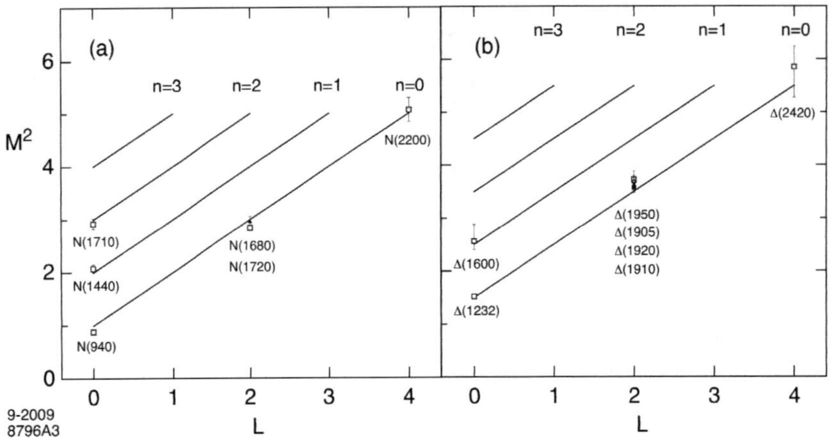

Figure 2. Positive parity Regge trajectories for the N and Δ baryon families for $\kappa = 0.5$ GeV. Data from [31].

The proton in light-front holography is described by a two component wave function $\psi(\zeta) = \psi_+(\zeta)u_+ + \psi_-(\zeta)u_-$, where $u_\pm$ are four-dimensional spinors. The $L^z = 0$ and $L^z = +1$ orbital components are combined with spin components $S^z = +1/2$ and $S^z = -1/2$ respectively. An interesting feature of light front holography for baryons and massless quarks is the fact that the lowest valence Fock states with $L^z = 0$ and $L^z = \pm 1$ have the same probability $\int d\zeta\, |\psi_+(\zeta)|^2 = \int d\zeta\, |\psi_-(\zeta)|^2$, a manifestation of the chiral invariance of the theory for massless quarks. [51] This implies that the quarks carry zero angular momentum $\langle S_z = 0 \rangle$ in the proton with $J^z = \pm 1/2$ and $\langle L^z = 1/2 \rangle$.

6. Light Front Holographic Mapping of Current Matrix Elements

The great advantage of the front form – as emphasized by Dirac – is that boost operators are kinematic. Unlike the instant form, the boost operators in the front form have no interaction terms. The calculation of a current matrix element $\langle P+q|J^\mu|P\rangle$ requires boosting the hadronic eigenstate from $|P\rangle$ to $|P+q\rangle$, a task which becomes hopelessly complicated in the instant form.

The form factor is computed in the light front from the matrix elements of the plus-component of the current J^+, in order to avoid coupling to Fock states with different numbers of constituents

$$\langle \psi(P')|J^+(0)|\psi(P)\rangle = (P + P')^+ F(q^2), \tag{24}$$

where $Q = P' - P$ and $J^+(x) = \sum_q e_q(x)\overline{q}\gamma^+ q(x)$ is the quark current which couples locally to the quarks in the hadron. Expanding the initial and final Fock states in terms of Fock components, we obtain Drell-Yan-West (DYW) expression [52, 53] upon integration over the intermediate variables in the q^+ frame:

$$F(q^2) = \sum_n \int [dx_i]\, [d^2\mathbf{k}_{\perp i}] \sum_j e_j \psi_n^*(x_i, \mathbf{k}'_{\perp i}, \lambda_i)\psi_n(x_i, \mathbf{k}_{\perp i}, \lambda_i), \tag{25}$$

where the variables of the light cone Fock components in the final-state are given by $\mathbf{k}'_{\perp i} = \mathbf{k}_{\perp i} + (1 - x_i)\,\mathbf{q}_\perp$ for a struck constituent quark and $\mathbf{k}'_{\perp i} = \mathbf{k}_{\perp i} - x_i\,\mathbf{q}_\perp$ for each spectator. The formula is exact if the sum is over all Fock states n. The form factor can also be conveniently written in impact space

$$F(q^2) = \sum_n \prod_{j=1}^{n-1} \int dx_j d^2\mathbf{b}_{\perp j} \exp\Big(i\mathbf{q}_\perp \cdot \sum_{j=1}^{n-1} x_j \mathbf{b}_{\perp j}\Big) \left|\psi_n(x_j, \mathbf{b}_{\perp j})\right|^2, \tag{26}$$

corresponding to a change of transverse momentum $x_j\mathbf{q}_\perp$ for each of the $n-1$ spectators.

On the higher dimensional gravity theory, the hadronic matrix element corresponds to the non-local coupling of an external electromagnetic field $A^M(x, z)$ propagating in AdS with the extended mode $\Phi(x, z)$ [8]

$$\int d^4x\, dz\, \sqrt{g}\, e^{\varphi(z)} A^M(x, z)\Phi^*_{P'}(x, z) \overleftrightarrow{\partial}_M \Phi_P(x, z). \tag{27}$$

Can the transition amplitudes be related? How can we recover hard pointlike scattering at large Q out of soft collision of extended objects? [8] Although the expressions for the transition amplitudes look very different, one can show that a precise mapping of the J^+ elements at fixed light-front time can be carried out in the semiclassical approximation discussed in Sect. 4. In fact, if both expressions for the form factor are identical for arbitrary values of q, then the factorization (18) is required. For a two parton state we find $\phi(\zeta) = (\zeta/R)^{-3/2}\, e^{\varphi(z)/2}\, \Phi(\zeta)$ as before, and $X(x) = \sqrt{x(1-x)}$. In this case the longitudinal mode does not decouple, but instead a specific form for $X(x)$ is obtained. [7] A precise mapping can be established for an arbitrary number of partons. [7] Identical results are obtained for the mapping of the matrix elements of the energy momentum tensor. [9]

The results described above correspond to a "free" current propagating on AdS space and dual to the EM pointlike current in the DYW LF formula, which allow us to map state-by-state. [1] This mapping has the shortcoming that the pole structure of the form factor is not built on the timelike region. Furthermore, the moments of the form factor at $Q^2 = 0$ diverge, giving for example an infinite charge radius. The pole structure is generated when the EM current is confined, this means, when the EM current propagates on a IR deformed AdS space to mimic confinement. This also leads to finite moments at $Q^2 = 0$ as illustrated on Fig. 3. Since the computation of the form factor corresponds to a twist-3 current J^+, the poles do not correspond to the physical poles of the twist-2 transverse current $\mathbf{J}_\perp$ present in the annihilation channel. Consequently the location of the poles in the final result should be rescaled to their physical positions. When this is done the results agree extremely well with available data. The non-perturbative effects from the "dressed" current correspond to an infinite sum of diagrams. One should however be careful to avoid a double counting of terms. This important point deserves further investigation.

[1] In general the mapping relates the AdS density $\Phi^2(z)$ to an effective LF single particle transverse density. [7]

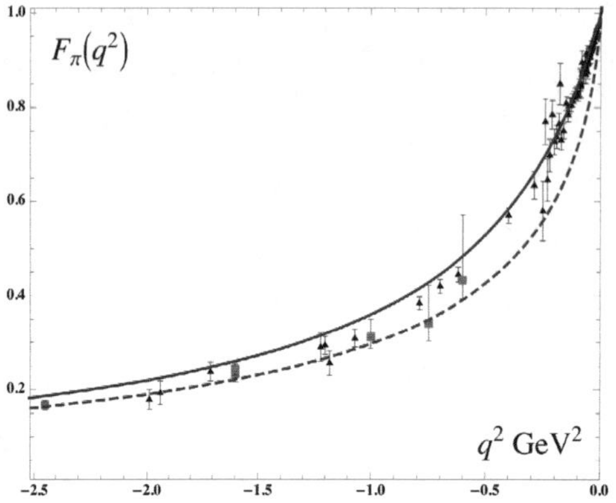

Figure 3. Space-like pion form factor $F_\pi(q^2)$ as a function of q^2. Continuous line: confined current, dashed line free current. Triangles are the data compilation from Baldini [54], boxes are JLAB data. [55]

7. Conclusions

The light front holographic approach described in this paper provides a direct correspondence between an effective gravity theory defined in a fifth-dimensional warped space, which encodes the salient features of QCD, and a semiclassical approximation to strongly coupled QCD quantized on the light-front. This duality leads to a Lorentz-invariant relativistic Schrödinger wave equation [14] which provides a successful prediction for the light-quark meson and baryon spectra as a function of hadron spin, quark angular momentum, and radial quantum numbers. It also predicts the same multiplicity of states for mesons and baryons which is observed experimentally. We originally derived this correspondence using the identity between electromagnetic and gravitational form factors computed in AdS and light-front theory. [7, 9] The results for hadronic form factors are also successful, and the predicted power law fall-off agrees with dimensional counting rules as required by conformal invariance at small z. In particular we have shown the importance of including non-perturbative effects corresponding to the infinite sum of diagrams contained in the structure of the currents propagating in the confining geometry. This "dressed" current is required to reproduce the time-like pole structure and finite moments near $Q^2 = 0$, effects which are not seen in the semiclassical approximation when comparing state by state. As in the Schrödinger equation, the semiclassical approximation to light-front QCD described in this paper does not account for particle creation and absorption; it is thus expected to break down at short distances where hard gluon exchange and quantum corrections become important. Other shortcomings of holographic methods are discussed in Ref. [56].

Acknowledgements

Invited talk presented by GdT at XIV School of Particles and Fields, Morelia, México, November 8-12, 2010. GdT is grateful to the organizers for their outstanding hospitality. We thank H. G. Dosch and J. Erlich, for helpful comments and collaborations. This research was supported by the Department of Energy contract DE–AC02–76SF00515. SLAC-PUB-14393.

[1] J. M. Maldacena, Int. J. Theor. Phys. **38**, 1113 (1999) [arXiv:hep-th/9711200].
[2] S. S. Gubser, I. R. Klebanov and A. M. Polyakov, Phys. Lett. B **428**, 105 (1998) [arXiv:hep-th/9802109].
[3] E. Witten, Adv. Theor. Math. Phys. **2**, 253 (1998) [arXiv:hep-th/9802150].
[4] J. Polchinski and M. J. Strassler, Phys. Rev. Lett. **88**, 031601 (2002) [arXiv:hep-th/0109174].
[5] S. J. Brodsky and G. R. Farrar, Phys. Rev. Lett. **31**, 1153 (1973).
[6] V. A. Matveev, R. M. Muradian and A. N. Tavkhelidze, Lett. Nuovo Cim. **7**, 719 (1973).
[7] S. J. Brodsky and G. F. de Teramond, Phys. Rev. Lett. **96**, 201601 (2006) [arXiv:hep-ph/0602252]; Phys. Rev. D **77**, 056007 (2008) [arXiv:0707.3859 [hep-ph]].
[8] J. Polchinski and M. J. Strassler, JHEP **0305**, 012 (2003) [arXiv:hep-th/0209211].
[9] S. J. Brodsky and G. F. de Teramond, Phys. Rev. D **78**, 025032 (2008) [arXiv:0804.0452 [hep-ph]].
[10] Z. Abidin and C. E. Carlson, Phys. Rev. D **77**, 095007 (2008), [arXiv:0801.3839 [hep-ph]].
[11] P. A. M. Dirac, Rev. Mod. Phys. **21**, 392 (1949).
[12] J. Erlich *et al.*, Phys. Rev. Lett. **95**, 261602 (2005) [arXiv:hep-ph/0501128].
[13] L. Da Rold and A. Pomarol, Nucl. Phys. B **721**, 79 (2005) [arXiv:hep-ph/0501218].
[14] G. F. de Teramond and S. J. Brodsky, Phys. Rev. Lett. **102**, 081601 (2009) [arXiv:0809.4899 [hep-ph]].
[15] H. C. Pauli, Nucl. Phys. Proc. Suppl. **108**, 273 (2002) [arXiv:hep-ph/0202179].
[16] G. 't Hooft, Nucl. Phys. B **75**, 461 (1974).
[17] E. Katz and T. Okui, JHEP **0901**, 113 (2009) [arXiv:0710.3402 [hep-th]].
[18] C. Fronsdal, Phys. Rev. **D20**, 848 (1979).
[19] E. S. Fradkin, M. A. Vasiliev, Nucl. Phys. **B291**, 141 (1987).
[20] R. R. Metsaev, Phys. Rev. D **78**, 106010 (2008) [arXiv:0805.3472 [hep-th]].
[21] A. Karch *et al.* Phys. Rev. D **74**, 015005 (2006) [arXiv:hep-ph/0602229].
[22] S. J. Brodsky, G. F. de Teramond, H. G. Dosch and J. Erlich, *in preparation.*
[23] G. F. de Teramond and S. J. Brodsky, AIP Conf. Proc. **1296**, 128 (2010) [arXiv:1006.2431 [hep-ph]].
[24] S. J. Brodsky, H. C. Pauli and S. S. Pinsky, Phys. Rept. **301**, 299 (1998) [arXiv:hep-ph/9705477].
[25] L. D. Landau and E. M. Lifshitz, *Quantum Mechanics*, Pergamon, New York, 1958, Sec. 35.
[26] P. Breitenlohner and D. Z. Freedman, Annals Phys. **144**, 249 (1982).
[27] G. F. de Teramond and S. J. Brodsky, Nucl. Phys. B, Proc. Suppl. **199**, 89 (2010) [arXiv:0909.3900 [hep-ph]].
[28] O. Andreev and V. I. Zakharov, Phys. Rev. D **74**, 025023 (2006) [arXiv:hep-ph/0604204].
[29] G. F. de Teramond, S. J. Brodsky, AIP Conf. Proc. **1257**, 59 (2010) [arXiv:1001.5193 [hep-ph]].
[30] E. Klempt and A. Zaitsev, Phys. Rept. **454**, 1 (2007) [arXiv:0708.4016 [hep-ph]].
[31] C. Amsler *et al.* (Particle Data Group), Phys. Lett. B **667**, 1 (2008).
[32] G. F. de Teramond, S. J. Brodsky, Phys. Rev. Lett. **94**, 201601 (2005) [hep-th/0501022].
[33] H. Boschi-Filho, N. R. F. Braga and H. L. Carrion, Phys. Rev. D **73**, 047901 (2006) [arXiv:hep-th/0507063].
[34] N. Evans and A. Tedder, Phys. Lett. B **642**, 546 (2006) [arXiv:hep-ph/0609112].
[35] D. K. Hong, T. Inami and H. U. Yee, Phys. Lett. B **646**, 165 (2007) [arXiv:hep-ph/0609270].
[36] P. Colangelo *et al.* Phys. Lett. B **652**, 73 (2007) [arXiv:hep-ph/0703316].
[37] H. Forkel, Phys. Rev. D **78**, 025001 (2008) [arXiv:0711.1179 [hep-ph]].
[38] A. Vega and I. Schmidt, Phys. Rev. D **78**, 017703 (2008) [arXiv:0806.2267 [hep-ph]].
[39] K. Nawa, H. Suganuma and T. Kojo, Mod. Phys. Lett. A **23**, 2364 (2008) [arXiv:0806.3040 [hep-th].
[40] W. de Paula *et al.*, Phys. Rev. D **79**, 075019 (2009) [arXiv:0806.3830 [hep-ph]].
[41] P. Colangelo *et al.*, Phys. Rev. D **78**, 055009 (2008) [arXiv:0807.1054 [hep-ph]].
[42] H. Forkel and E. Klempt, Phys. Lett. B **679**, 77 (2009) [arXiv:0810.2959 [hep-ph]].
[43] H. C. Ahn *et al.*, Phys. Rev. D **80**, 054001 (2009) [arXiv:0904.3731 [hep-ph]].
[44] Y. Q. Sui *et al.*, Phys. Rev. D **81**, 014024 (2010) [arXiv:0909.3887 [hep-ph]].
[45] J. I. Kapusta and T. Springer, Phys. Rev. D **81**, 086009 (2010) [arXiv:1001.4799 [hep-ph]].
[46] P. Zhang, Phys. Rev. D **81** (2010) 114029 [arXiv:1002.4352 [hep-ph]]; JHEP **1005** (2010) 039 [arXiv:1003.0558 [hep-ph]]; Phys. Rev. **D82**, 094013 (2010) arXiv:1007.2163 [hep-ph].
[47] I. Iatrakis, E. Kiritsis and A. Paredes, Phys. Rev. D **81**, 115004 (2010) [arXiv:1003.2377 [hep-ph]].
[48] T. Branz *et al.*, Phys. Rev. **D82**, 074022 (2010) arXiv:1008.0268 [hep-ph].
[49] M. Kirchbach, C. B. Compean, Phys. Rev. **D82**, 034008 (2010) [arXiv:1003.1747 [hep-ph]].
[50] Y. -Q. Sui, Y. -L. Wu, Y. -B. Yang, [arXiv:1012.3518 [hep-ph]].
[51] S. J. Brodsky, G. F. de Teramond, Acta Phys. Pol. B 41, 2605 (2010) [arXiv:1009.4232 [hep-ph]].
[52] S. D. Drell and T. M. Yan, Phys. Rev. Lett. **24**, 181 (1970).
[53] G. B. West, Phys. Rev. Lett. **24**, 1206 (1970).
[54] R. Baldini *et al.*, Eur. Phys. J. C **11**, 709 (1999).
[55] V. Tadevosyan *et al.*, Phys. Rev. C **75**, 055205 (2007) [arXiv:nucl-ex/0607007]; T. Horn *et al.*, Phys. Rev. Lett. **97**, 192001 (2006) [arXiv:nucl-ex/0607005].
[56] C. Csaki, M. Reece and J. Terning, JHEP **0905**, 067 (2009) [arXiv:0811.3001 [hep-ph]].

Journal of Physics: Conference Series **287** (2011) 012008 doi:10.1088/1742-6596/287/1/012008

Prospects for an Accelerator Program in Mexico Focused on Photon Science

C. Hernandez-Garcia*, and M. Napsuciale**

*Jefferson Lab, 12000 Jefferson Ave, Suite 19, Newport News, VA 23606

chgarcia@jlab.org

** Departamento de Física, Universidad de Guanajuato Campus León, Lomas del Bosque 103, Fraccionamiento Lomas del Campestre, 37150, León, Guanajuato, México.

Abstract. Recent interest in developing Mexican expertise in Accelerator Science and Technology has resulted in several actions by the Division of Particles and Fields in Mexico, and by the electron accelerator community in the United States. We report on the very encouraging activities over the past two years which were aimed at developing a light source as the most effective starting point. We present a number of possibilities to initiate and grow an accelerator science program and present a path that would lead to building, commissioning and operating a third or fourth generation light source in Mexico.[1]

1. Background

In March 2009, representatives of Mexican high-energy physics groups (theorists and experimenters) attended a meeting, hosted by the National Council for Science and Technology of México (CONACyT), to discuss the perspectives and prospects for their disciplines and to propose plans to develop them. Most of the participants were members of the Division of Particles and Fields of the Mexican Physical Society (DPyC-SMF). Among the conclusions of this meeting was the need to train people in "accelerator physics". At present, the Mexican high energy physics community is involved in theoretical and experimental research (LHC, AUGER, FERMILAB) with expertise is in data analysis and detector development. There are no Mexican physicists or engineers with accelerator science expertise.

In the summer of 2009 one graduate student from Universidad Autonoma de Sinaloa spent 8 weeks at Jefferson Lab's Free Electron Laser (FEL) learning the basics of accelerator physics, and developing

[1] Notice: Authored by Jefferson Science Associates, LLC under U.S. DOE Contract No. DE-AC05-06OR23177.

Journal of Physics: Conference Series **287** (2011) 012008 doi:10.1088/1742-6596/287/1/012008

practical skills working with high voltage DC photoemission electron guns. Later in November of that year, the corresponding author presented a talk about the JLab FEL program at the XII Mexican Workshop in Particles and Fields, hosted by Universidad Autónoma de Sinaloa in Mazatlan, Mexico. In a meeting with members of the DPyC-SMF, the intentions amongst the particle physics groups to develop accelerator physics programs in Mexico were discussed, with the ultimate goal of building and operating a particle accelerator.

2. Some initial considerations

2.1. Why would México want to embark on developing an accelerator program?

Accelerators are very important tools in many scientific fields. These fields increasingly include medical, biological, material sciences, to name just a few, in addition to the traditional nuclear and high energy physics areas. This is exemplified by the growth in accelerator installations world-wide. One third of the 15,000 accelerators in operation worldwide are used in radiotherapy and almost two-thirds are used for industrial applications such as semiconductor production, sterilization, and materials preparation or in processes such as x-ray micro-machining. These few examples show the economic benefits of accelerators.

The development of an accelerator physics program trains a new generation of students, scientists, engineers, and technicians in this inherently interdisciplinary field and has the potential for economic stimulus for indigenous industry. For example, Klystrons, (used to accelerate particles and are expensive to develop), can be replaced by solid-state Radio Frequency (RF) amplifiers that can be built in Mexico by bringing expertise from the microwave telecommunications industry to a new technical field. Cooling water systems, control electronics and software and vacuum systems can be developed and constructed as a multi-institution and multi-disciplinary effort. Direct economic benefit comes also from manufacturing specialized equipment such as magnets and DC power supplies, from training and employing technicians, engineers, surveyors, etc. not to mention the civil construction of the accelerator enclosure and associated laboratory space and office buildings. All the research done by Mexican institutions is an additional stimulus along with the influx of international researchers looking for light sources.

2.2. What is the ultimate goal?

A Light Source, based electron accelerator technology, would be an ideal entry machine for Mexico to eventually do forefront photon science and become part of a worldwide accelerator-based light source community. A light source benefits a broader research community (e.g., materials, solid state, medical, biology, chemistry). In addition, light sources tend to be smaller installations than typical particle accelerators which are primarily nuclear and high-energy physics machines.

To place the photon science relevance in perspective, during the past five years, the United States Department of Energy has organized 10 Workshops with more than 1,500 participants from academia, industry and national laboratories through its Basic Energy Sciences division to make a case for the century's Science Grand Challenges. The report, "Directing Matter and Energy: 5 Challenges for Science & the imagination" by Graham Fleming and Mark Ratner (chairs) [1], presents the following challenges:

1. How do we control materials processes at the level of the electrons? *Pump-probe and time dependent dynamics*
2. How do we design and perfect atom- and energy-efficient synthesis of new forms of matter with tailored properties? *Pulsed Laser Deposition, photochemistry, X-Ray Spectroscopy.*

3. How do remarkable properties of matter emerge from the complex correlations of atomic and electronic constituents and how can we control these properties? *Pump-probe time dependent dynamics and X-Ray Spectroscopy*
4. How can we master energy and information on the nanoscale to create new technologies with capabilities rivaling those of living things? *Pump-probe time dependent dynamics and X-Ray Spectroscopy*
5. How do we characterize and control matter away – especially very far away – from equilibrium? *Non-linear dynamics, ultra-bright photon sources.*

These challenges are not unique to one nation and provide the impetus for the construction of third generation light sources in Australia, Spain, Brazil and France and for the construction of fourth generation light sources by the U.S., China, Japan, and a collaboration of European countries. The fundamental physics and technological challenges involved in developing accelerators capable of meeting the stringent requirements for producing X-ray beams with unprecedented brightness, fluence and pulse length, while maintaining temporal and special coherence are just being explored by the international accelerator and photon science communities. There are many opportunities for a new generation of scientist and engineers to contribute to the field.

3. Accelerator R&D

We focus on electron beams due to the fact that electron beams are best suited to generate light. Some of the multi-disciplinary areas involved in accelerator physics are briefly described below.

Beam physics studies the behavior of a group of charged particles, known as a charged particle bunch, subject to internal forces of the bunch itself, as well as applied external electric and magnetic fields. The fundamentals of beam physics can be studied with an understanding of relativity, electromagnetism and calculus. Often the beam physics design of an accelerator, from beam generation through beam acceleration and transport, requires no hardware other than a computer for simulations since the collective effects of all particles composing the electron bunch are far too complex to be treated analytically.

Beam generation. The beam is generated under vacuum conditions at the cathode, and typically accelerated in the range of 10keV to 2 MeV in what is known as an electron gun or injector. The most common type is the thermionic gun similar to that found in cathode ray tubes (CRTs). An electrically heated element such as oxide coated metal filaments or Hexaborides, emit electrons that have overcome the potential barrier (work function) by thermal energy. Electrons gain kinetic energy by means of a suitably biased the cathode. In photoemission electron guns, the electrons are generated in a photosensitive material, the photocathode. When a laser pulse illuminates the photocathode surface electrons are emitted due to photoelectric effect. The actual photoemission process consists of three main steps: photon absorption, electron transport to the surface, and emission into the vacuum. Once the electron bunch is generated, it is quickly accelerated to relativistic energies in order to reduce the transverse beam size spread due to the Coulomb repulsion of electrons within the bunch. In DC guns, an electrostatic accelerating field is generated with the cathode biased at 10-500 kV with respect to grounded anode which is placed a few centimeters from the cathode. An aperture in the anode allows the electron beam to exit the gun. In RF guns, the longitudinal component of an alternating electromagnetic field inside a specially shaped cavity accelerates the beam to ~2 MeV. If the cavity is made of copper, the gun is known as normal conducting radio frequency (NCRF) gun. If the cavity is made of cryogenically cooled Niobium, the gun is known as superconducting radio frequency (SRF) gun. Technical challenges such as ultra-high vacuum, very high voltage, cryogenics and RF high power are intertwined with the physics of beam generation for all types of electron guns, offering a multitude of training opportunities, R&D and career specialization.

Beam transport. The generated electron beam, composed of a train of electron bunches, exits the gun and travels in evacuated beam pipes. The transverse size of the beam is manipulated by electromagnetic elements mounted around the beam pipe to prevent beam scraping at the walls, which results in beam loss and generation of undesired radiation in the form of neutrons, x-rays and gamma radiation. Some elements such as solenoids and quadrupoles focus the beam, dipole magnets bend the beam and air-core coils provide vertical and horizontal steering. DC power supplies energize the magnets and high power magnet coils are cooled down with low conductivity water recirculation systems instead of air.

Beam acceleration. Typically, the beam leaving the gun/injector would be further accelerated. The final beam energy can be anywhere from 10 MeV for medical applications and up to tens of GeV for nuclear physics, high-energy physics, and light sources. The acceleration of electron bunches is achieved in either NCRF or SRF cavities, usually assembled in groups of 2-10 in larger, modular accelerator components. The traveling electron bunch is accelerated in a cavity by the longitudinal electric field (in the direction of beam velocity) component of an electromagnetic wave oscillating at radio-frequencies in the order of 500 to 1500 MHz coupled to the cavity. The RF power needed to accelerate the beam is provided by velocity modulation in electron tubes called Klystrons, and is transported to the cavities by waveguides or specially shielded cables, depending on the frequency and power. The module assembly is then transported to the accelerator enclosure for installation in the beam line, and then connected to the RF power systems, and in the case of SRF technology, it is also connected to cryogenic lines. There are a multitude of disciplines involved in the physics and engineering designs and in the construction of these modules, such as particle beam physics, material science, mechanical, electrical and cryogenics engineering. Also wide ranges of degrees and skills ranging from bachelors to doctoral degrees in science and engineering fields are needed to realize such accelerating modules from concept to the actual product.

Beam diagnostics. This is an area that offers a range of topics in disciplines such as electrical engineering, computer science, mechanical engineering, instrumentation & control and beam physics. There are two main branches in beam diagnostics: one provides information about the beam (bunch) properties and the other provides information about the beam position and timing as it travels through the accelerator. In beam position monitors, opposite and electrically isolated plates coaxial to the beam path register the image charge, the difference in charge between the plates gives a relative position of the beam. In pulsed mode, the electron beam transverse profile can be viewed indirectly by the coherent transition radiation when the beam collides with a thin foil inserted in its path. The beam energy spread can be inferred from synchrotron radiation imaged on CCD cameras as the beam path is bent in dipole magnets. Other devices such as moving wires and multi-slits are inserted into the beam path for measuring transverse emittance, helping characterize the transverse phase space properties at a particular location along the accelerator.

Other technical areas such as vacuum systems and beam loss detection by radiation monitoring, essential for personal safety and for machine protection systems, also present opportunities at various levels of proficiency. If SRF acceleration technology is chosen for the machine, scientists, engineers and technicians can be trained in cryogenic systems. Following the Light Source in Brazil example [2], insertion devices (undulators/wigglers) can be designed by a new generation of scientists and engineers, and built by specialized companies such as STI. A new generation of scientists and engineers could also develop the X-ray beam transport lines and user end-stations.

Mexican engineers engaging undergraduate and graduate students in specific projects can develop the control system for all of the machine sub-systems.

4. Developing a Mexican accelerator program

Successful projects have often been envisioned, planned and executed by a core group of people with a clearly defined goal. The formation of a "steering committee" would be essential to take the endeavor from the early stages to the ultimate goal. This is something that needs to be driven by Mexican institutions interested in both photon science and accelerator physics- necessarily multi-disciplinary and multi-institutional. The committee will need to remain focused despite unavoidable hurdles (politics, resources, etc). A paramount task would be the creation of a nation-wide case for photon science towards building a national user community, which could potentially expand to the international photon user community and drive the development and construction of the first light source in Mexico. In parallel with lobbying at the government level for a science case, development of a foundation of accelerator scientists, students, engineers and technicians could be carried out by the means described below. The formation of the committee should not preclude the start of a training program, which in fact, has already started.

4.1. *The photon science case*: Organization of workshops will be essential in attracting potential users among diverse disciplines such as material and life sciences, applied R&D for industry and medicine towards developing a photon science case for building a Light Source. The idea is the establishment of a truly National Facility offering unprecedented research resources not only to Mexican institutions but also to U.S. and Latin American counterparts. Multi-national sources of funding can be explored if credible proposals with strong science are presented.

4.2. *The student foundation*: Training a new generation of students in beam physics theory and simulations can be started immediately by encouraging undergraduate students to seek universities with graduate programs specialized in accelerator physics such as Old Dominion University, Cornell, Berkeley, University of California Los Angeles, Stony Brook, Idaho State, and Northern Illinois University. These activities should be handled skilfully to foster a strong motivation for graduated students to return to Mexico for becoming the professors and the scientists who will lead the Accelerator Program.

4.3. *The scientist and engineer foundation*: Establish collaborations between Mexican and International Institutions to send recent PhD graduates to become post-doctoral fellows, visiting professors and visiting researchers to institutions with well established accelerator machines and programs such as Thomas Jefferson National Accelerator Facility, Cornell light source (CHESS), Lawrence Berkeley National Lab, UCLA Particle Beam Physics Laboratory, Brookhaven National Laboratory, Argonne National Light Source, University of Wisconsin Light Source (Aladdin)and Idaho Accelerator Center.

4.4. *The technical expertise foundation*: Establish collaborations with aforementioned institutions to send engineers and technicians to learn new skills on topics such as magnet design, construction and operation, Radio Frequency systems, Vacuum systems, DC power system, Low conductivity and cooling water systems, Cryogenics, Radiation control, Machine protection systems, Instrumentation and control, etc.

4.5. *Small accelerators as training tools*: There are many decommissioned medical and industrial accelerators that can be acquired as "educational loans" to be assembled and operated in university laboratories with minor infrastructure requirements. New Ph.D. students returning to Mexico could take on the lead in training students and technicians during the assembly and operation of such accelerators. Of special interest would be the formation of a committee to visit Idaho State University (ISU) to learn from their model. ISU has over 20 small accelerators and a very productive educational program. Other example is the Cyclotron Kids project [3] sponsored by Jefferson Lab, in which two high school students have spent the last two summers at Jefferson Lab working and learning from scientists, engineers and technicians who have enthusiastically worked on the project for free.

Materials such as 2 tons of steel for the magnet have been donated by local industries, vacuum equipment has been rebuilt from spare parts donated also by Jefferson Lab. RF amplifiers and electronics have been purchased at very modest prices through e-bay. This could be an ideal approach to train students and technicians in the technical and scientific aspects of an accelerator since the small cyclotron shares many of the same principles of operation as its larger counterparts. In addition, its small size allows it to be developed and operated in a typical university laboratory space, with minimal radiation shielding.

5. Suggested time line

2010-2015:

•Students start Ph.D. programs in Accelerator Physics

•Interested Mexican Universities establish connections with U.S. institutions and industries to acquire equipment by "educational loans" to build small accelerator for training and education

•Develop synergy between physics and light source community in Mexico to draft plans for workshops towards forming a photon science case

2015-2020:

•The first generation of students returns to Mexico and starts assembling and operating the equipment, while teaching basic courses to students and training technicians.

•At the end of the five year period, demonstrate successful operation of the first accelerator for educational and training purposes

•Continue training a second generation of students

•Continue physics and photon science workshops towards elaborating science case

2020-2025:

•Develop formal photon science case based on workshops and white papers. The document should also include a study of the economic benefit that an accelerator-based light source would have in the country, by training a new generation of highly skilled physicists, engineers and technicians who would eventually build, maintain and upgrade such a machine or use their acquired skills in industry or start their own enterprises.

•Establish a PhD program in Mexican Universities focused on accelerator physics, for the continuing training of a new generation of physicists and engineers.

2025-2030:

Begin design and construction of the accelerator-based light source, followed by commissioning and operation. This phase can be accomplished by taking a de-commissioned machine, or by designing and building a brand new one. The latter would be more difficult but will be the ideal case since the machine would be designed and built by the people who have actually being trained. In either case, the accelerator-based light source would be operated and maintained by a new generation of scientist, engineers and technicians that will be subject-matter-experts in many disciplines.

We would like to remark that the suggested time-line can be shortened to at most ten years if an aggressive plan to develop the accelerator program is supported by the Mexican funding agencies.

6. Resources and opportunities:

The United States Particle Accelerator School (USPAS) conducts graduate and undergraduate level courses at U.S. universities, holding two such programs per year, one in June and one in January. These courses, running 2 weeks in duration, take place at leading universities across the United States. By successfully completing the 2-week course requirements, which include forty-five contact hours as well as daily problems and examinations, students earn 3 semester hours of university credit. The course accreditation for students enrolled in Mexican Universities would have to be worked out

Journal of Physics: Conference Series **287** (2011) 012008 doi:10.1088/1742-6596/287/1/012008

directly with USPAS. In Norfolk VA, Old Dominion University (ODU) Center for Accelerator Science grants Ph.D. in accelerator physics where students do their thesis work on projects under the guidance of JLab staff who are also ODU professors. The visit of one of the authors (M. N.) to Jefferson Lab in May 2010 generated genuine interest, willingness and cooperation - which can provide not only theoretical but hands-on experience for particle accelerators. Since JLab is a leading SRF facility and since SRF technology is favored for most future accelerators including the International LINAC Collider and fourth generation X-ray light sources, the lab's encouragement is fortunate. Students can potentially earn Ph.D. degrees not only at ODU but at other universities with accelerator physics programs.

6.1. Some of the steps already taken

In the Fall 2010, one graduate student from Universidad de Guanajuato was accepted into ODU's Ph.D. program in accelerator physics. The student will develop his thesis work at JLab under the guidance of Dr. G. Krafft, Director of the Center for Advanced Studies in Accelerators. ODU's physics department Chair, Dr. G. Dodge is looking forward to receiving more applications from excellent and motivated Mexican student to pursue their Ph.D. degrees. During Dr. Napsuciale's visit to ODU, discussions about developing a joint degree program between ODU took place. Based on ODU's experience with a French University, developing agreements for joint degrees seems extremely difficult. As an alternative, it was proposed that the student can hold dual degree, that is, both his/her academic Mexican institution and the U.S. institution would grant Ph.D. degrees. It is always a viable option for the student to enroll in graduate programs at U.S. Universities with assistantships from CONACyT. This is an area that will have to be resolved case by case by Mexican Universities.

Two undergraduate students from Universidad de Guanajuato and one from Universidad de Puebla spent 8 weeks during the 2010 summer at JLab participating in the Cyclotron Kids project [3]. The students experienced unique hands-on training on vacuum techniques and laboratory safety. They also designed, built and operated an ion source, and received basic concept training on accelerator physics. Imagine what can be accomplished with basic equipment in laboratories at Mexican universities in areas such as instrumentation & control, vacuum systems, electron guns, mechanical design & engineering, beam physics theory and simulations, etc. with modest, perhaps used equipment that could potentially be donated or loaned under special agreements by U.S. laboratories.

The corresponding author has discussed the prospects for training Mexican students in beam physics with USPAS director Dr. W. Barletta and with Dr. H. Winick (SLAC) who point out that Brazil and Australia have developed successful accelerator programs and built light sources and can be used as role models. Dr. Barletta reports that there are fewer students from Mexico attending USPAS compared to other countries in South and Central America. Quoting Dr. Barletta: "It was seem that USPAS could play an extremely valuable role in the enterprise...certainly the enthusiastic support and dedicated engagement of some notable users would push a project forward". Dr. Barletta expressed also that he would be most enthusiastic to assist, both through the USPAS and through the American Physical Society [4].

7. Conclusion

The possibilities for developing an accelerator physics program in Mexico focused on photon science are plentiful and the commitment of talented and enthusiastic leading scientists in the U.S. is real and encouraging. The vision of developing, building and operating the first light source in Mexico as a true national facility has sparked enthusiastic discussions in the high energy physics group and can potentially open a brand new field for graduate students and future scientists.

The formation of a core group or committee should be seriously considered. The committee would have tasks such as establishing collaborations with international institutions, encouraging the

Journal of Physics: Conference Series **287** (2011) 012008 doi:10.1088/1742-6596/287/1/012008

formation of a user group in Mexico which ultimately will drive and define the science case for an accelerator-based light source, organizing photon science and accelerator R&D workshops, organizing visits to other light sources in countries like Brazil, Spain, etc. In this way, a coherent and focused effort can be developed. For reference, it took about 10 years to build the Brazilian Light Source.

The immediate enthusiasm should perhaps be directed in sending students to earn their Ph.D. degrees in accelerator physics and to engage with the photon user community.

As Dr. Herman Winick, Assistant Director and Professor Emeritus SLAC National Accelerator Laboratory pointed out "...I think that lots of help is potentially available, but the lead must be taken by Mexico and Mexicans... "[5].

8. Acknowledgments

The authors wish to express their most sincere gratitude to Dr. Hari Areti (JLab) for his continuing enthusiastic support and guidance.

[1] www.science.doe.gov/bes/reports/list.html

[2] "The Brazilian Synchrotron Light Source", P. F. Tavares and J. A. Brum, Proceedings of the 2005 Particle Accelerator Conference, Knowxville, TN, pp. 325-329.

[3] www.thecyclotronkids.org/archive/

[4] www.aps.org/

[5] H. Winick, private communication.

Journal of Physics: Conference Series **287** (2011) 012009 doi:10.1088/1742-6596/287/1/012009

Family Symmetries in the era of the LHC

Liliana Velasco-Sevilla

CINVESTAV, Ave. IPN 2508, Col. San Pedro Zacatenco CP 07360, México D.F., México

E-mail: lvelasco@fis.cinvestav.mx

Abstract. The LHC is set to achieve a total 14 TeV collision energy for every pair of protons colliding. At this energy it will be possible to study interactions of the SM particles that may contain particles beyond it. We therefore will continue improving our understanding of particle physics, at the very least. Ideally many particles would be discovered and its interactions would be studied. A framework study for understanding the decay, mixing and CP breaking processes of these and the SM particles are the *family symmetries*. Proposed to explain the hierarchy of fermion masses and mixing in the SM but posed for explanations and predictions of what may happen beyond the SM. In this talk I will review the present bounds on flavour changing neutral currents and make a summary of how family symmetries can explain them.

1. Introduction

The *flavour problem* in the Standard Model (SM) is our ignorance in understanding the origin of the hierarchical values of quarks and charged leptons and the structure of the mixing in the quark sector. On the other hand, the *CP violation problem* is the fact that the Yukawa Lagrangian is not CP invariant, that is *Charge* and *Parity* transformations, respectively, are not conserved.

The CP violation problem can be put in very easy terms: if the Yukawa matrices were real, there would be no CP violation. However, since there is indeed CP violation this implies that there is a not trivial phase in one of the Yukawa matrices, either in the up or the down sectors or a combination of both. Specifically there is a not a trivial phase in the allowed redefinitions of the Yukawa matrices, $\mathrm{Im}\left[Y^dY^{d\dagger}Y^uY^{u\dagger}+Y^uY^{u\dagger}Y^dY^{d\dagger}\right]\neq 0$. *Family symmetries* (FS) address the two problems above. However, since the observables that we know are just quark masses and their mixing matrix, the Cabibbo-Kobayashi-Masakawa (CKM) matrix, Yukawa matrices cannot be uniquely determined and hence the way to address the flavour problem in the SM is not unique. Furthermore, we know that neutrino matrices are not zero and their mixing is not trivial. Therefore any unifying attempt to describe the way fermion masses are generated must include right-handed neutrinos. The biggest group which is candidate for a FS group that includes also three right-handed neutrinos is $U(3)^6$. This is simply just because for each kind of quarks, Q_L, u_R, d_R and leptons, L_L, e_R and ν_R, there is a factor $U(3)$. The factor $U(3)$ includes the three generations when no breaking of the FS has taken place. Hence the possible FS subgroups are many and quite varied: discrete or continuous. One should also decide the nature of the groups: global or gauged. We will refer in this contribution just to the supersymmetric FS models that have addressed both problems above mentioned, that is, aim to explain *both*

(i) the hierarchy of fermion masses and mixing *and*

(ii) the CP problem.

Journal of Physics: Conference Series **287** (2011) 012009 doi:10.1088/1742-6596/287/1/012009

In the SM these problems are intimately related to the Glashow-Iliopoulos-Maiani (GIM) mechanism, which is effectively a framework to understand why even though we do not have a way to *explain* flavour and CP problems in the SM, we understand why flavour changing processes are under control. The reason is because there are three species of quarks with a unitary mixing matrix whose off-diagonal CKM elements are really small [1] and because quark masses are so hierarchical.

Enter supersymmetry and the flavour and CP problems are magnified. Not only there are many more sources of *flavour*, that is matrices with a non trivial mixing and not understood eigenvalues, but it is also hard to understand how to control the many sources of CP violation.

In this contribution I point out why Supersymmetric Family Symmetries (SFS) are a relevant feature in the way we think supersymmetric spectra and their CP and flavour sources can be controlled and what we could expect from them in the era of the LHC. Furthermore, I point out the added benefits of constructing an effective framework, a lá GIM mechanism, to be able to relate directly the structures of masses and mixing, *both* of fermions *and* sfermions with the way flavour changing processes are suppressed.

2. Family Symmetries

The ambiguity in determining the Yukawa matrices

$$Y = U_L^\dagger \hat{Y}^f U_R, \quad f = u, d, \tag{1}$$

has to do with the fact that we just know the mass eigenvalues (in the SM we have $\hat{Y}^u = \sqrt{2}\,\text{Diagonal}[m_u, m_c, m_t]/v$, $\hat{Y}^d = \sqrt{2}\,\text{Diagonal}[m_d, m_s, m_b]/v$, $v^2 = (\sqrt{2}G_F)^{-1}$) and the CKM matrix, $V_{CKM} = U_L^u U_L^{d\dagger}$. Hence the assumptions that should be made are reflected in the right-handed diagonalizing matrices U_R^u and U_R^d. In the FS that are the subject of this note, the SM Yukawa Lagrangian, $\mathcal{L} = -Y_{ij}^u \overline{Q}_i H u_j - Y_{ij}^d \overline{Q}_i \left[i\sigma_2 H^*\right] d_j + h.c.$, is understood as an effective infrared completion where the Yukawa matrices are functions of fields, θ,

$$Y^f = Y^f(\theta_i). \tag{2}$$

Usually these θ fields, charged under the FS group, are scalars which are fancily called *flavons*. Alternatively, the Higgs sector of the SM can be extended such that there is not only one Higgs scalar H but many that couple in a non trivial way to the matter fields of the SM and such that the sum of their contributions to the Yukawa couplings produce the observed hierarchy of masses and mixing. If, in addition one attempts to explain the hierarchy of neutrino masses and their non trivial mixing the problem gets more involved.

The supersymmetric version of the approach that Froggatt and Nielsen proposed [1] (FN) is is illustrated in the diagrams of Figure 1. In this example there are two scalars θ and $\overline{\theta}$, with charges 1 and -1, respectively, under the FS group and singlets under the G_{SM} group. They are coupled to the fermions f_R and f_L of the SM in different ways, through heavy fermion fields χ with a mass term M. For the fermion f coupled exactly in the way of Figure 1 (a) to χ, $\overline{\chi}'$, H and θ; after H, θ and $\overline{\theta}$ take vacuum expectation values (VEVs) and assuming $\langle\theta\rangle < M$, we obtain the following mass term

$$m_f = \langle H\rangle \frac{\langle\theta\rangle}{M} \doteq \langle H\rangle \epsilon. \tag{3}$$

Giving $U(1)_{FS}$ charges to the MSSM plus right-handed neutrinos, as they appear in Table 1,

[1] Roughly of order λ, λ^2 and λ^3, where $\lambda \approx 0.226$.

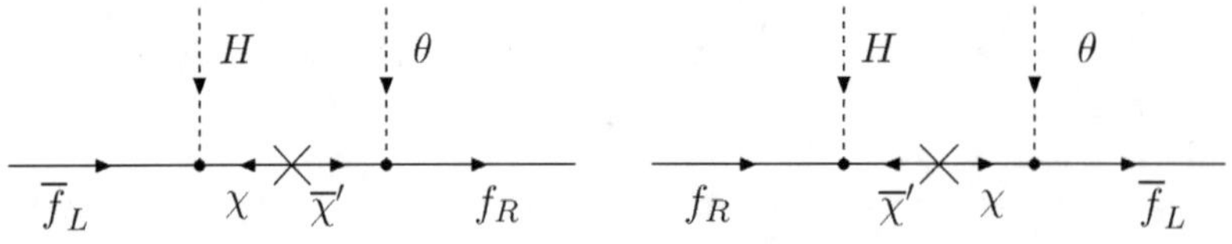

Figure 1: Simple Froggatt-Nielsen mechanism where one scalar θ is coupled to the fermions f_R and f_L of the SM in different ways.

$U(1)_{FS}$ example								
ν MSSM fields	$\mathbf{Q}_i$	$\overline{\mathbf{u}}_i$	$\overline{\mathbf{d}}_i$	$\mathbf{L}_i$	$\overline{\mathbf{e}}_i$	$\overline{\mathbf{N}}_i$	$\mathbf{H}_u$	$\mathbf{H}_d$
$U(1)_{FS}$ Charges	q_i	u_i	d_i	l_i	e_i	n_i	h_u	h_d

Table 1: $U(1)_{FS}$ charges for the ν MSSM fields. I use the notation of

the Yukawa couplings for the different kinds of quarks are then given by

$$\begin{aligned} Y^u_{ij} &= a^u_{ij}\epsilon^{|Q^u_{ij}|}, & Y^d_{ij} &= a^d_{ij}\epsilon^{|Q^d_{ij}|} \\ Q^u_{ij} &= -(q_i - q_3) - (u_j - u_3), & Q^d_{ij} &= -(q_i - q_3) - (d_j - d_3), \\ Y^\nu_{ij} &= a^\nu_{ij}\epsilon^{|Q^\nu_{ij}|}, & Y^e_{ij} &= a^e_{ij}\epsilon^{|Q^e_{ij}|}, \\ Q^\nu_{ij} &= -(l_i - l_3) - (n_j - n_3), & Q^e_{ij} &= -(l_i - l_3) - (n_j - n_3), \end{aligned}$$

where a^f_{ij} are $O(1)$ coefficients. This example illustrates pretty well the way hierarchical Yukawa matrices can be obtained: the smaller the Yukawa coupling, the bigger the power needed and so the bigger the combination of charges that appear in Eq. (4). This corresponds simply to more insertions of the heavy fermion pairs χ and $\overline{\chi}'$ and the θ or $\overline{\theta}$ fields, that can be represented by diagrams like those in Figure 1. Using the Fayet-Iliopoulos term that any supersymmetric $N = 1$ theory containing a $U(1)$ gauge factor can have, we can fix the VEVs of the theta fields and we could even determine the charges of the MSSM fields through cancellation of anomalies (e.g. [2]).

Now, the principal problem with this approach alone is that the coefficients a^f_{ij} cannot be related to each other and if they have to be entirely fixed by phenomenological requirements, then the productivity of such scenarios is in trouble. In combination with a Supersymmetric Grand Unified Theory (SGUT) some of the problems are alleviated and both ingredients blend into a rich Family Symmetry SGUT that can give interesting predictions [3] .

To relate the coefficients a^f_{ij} among the same kind of fermions, one can think on non-Abelian FS groups such as $U(2)$ [4], $SU(3)$ [5] and $SO(3)$ [6]. Their structures are richer because matter fields can be put in their fundamental representations such that when the FS groups are broken the a^f_{ij} coefficients of the fermions in the same representation will be related. This reduces neatly the number of parameters and that is why they are so attractive. All the examples in this paragraph satisfy the conditions mentioned in the Introduction.

The prime example of discrete models are based on the group A_4 but they concentrate in describing very well only the mixing in the neutrino sector, leaving aside the question of fermion hierarchy, except for example [7]. To describe the hierarchy of fermion masses they use a GUT [8].

When we would like to relate the hierarchy of mixing in the quark masses with that of the eigenvalues, we can estimate of how the Yukawa matrices will look like [9]

$$Y_d \propto \begin{bmatrix} \leq \epsilon_d^4 & \epsilon_d^3 & \leq \epsilon_d^3 \\ \leq \epsilon_d^3 & \leq \epsilon_d^2 & \epsilon_d^2 \\ \leq \epsilon_d & \leq 1 & 1 \end{bmatrix}, \quad Y_u \propto \begin{bmatrix} \leq \epsilon_u^4 & \epsilon_u^3 & \leq \epsilon_u^3 \\ \leq \epsilon_u^3 & \leq \epsilon_u^2 & \epsilon_u^2 \\ \leq \epsilon_u & \leq 1 & 1 \end{bmatrix}, \tag{4}$$

for $\epsilon_d = 0.13$ and $\epsilon_u = 0.04$. These Ansatzs are the ones that $SO(3)$ and $SU(3)$ models reproduce.

3. Flavour & CP Problems in SUSY Extensions of the SM

Let us consider the Minimal Supersymmetric Standard Model (MSSM) plus three right-handed neutrinos, it superpotential can be written as

$$\begin{aligned} W_{\nu MSSM} &= Y_l^{ij}\epsilon_{\alpha\beta}\mathbf{H}_d^\alpha \mathbf{E}_i^c \mathbf{L}_j^\beta - Y_\nu^{ij}\epsilon_{\alpha\beta}\mathbf{H}_u^\alpha \mathbf{N}_i^c L_j^\beta + Y_d^{ij}\epsilon_{\alpha\beta}\mathbf{H}_d^\alpha \mathbf{D}_i^c \mathbf{Q}_j^\beta - Y_u^{ij}\epsilon_{\alpha\beta}\mathbf{H}_u^\alpha \mathbf{U}_i^c \mathbf{Q}_j^\beta \\ &+ \mu\epsilon_{\alpha\beta}\mathbf{H}_u^\alpha \mathbf{H}_d^\beta + \frac{1}{2}M_\nu^{ij}\mathbf{N}_i^c\mathbf{N}_j^c, \end{aligned} \tag{5}$$

where $\epsilon_{\alpha\beta}$ is the anti-symmetric tensor in two dimensions and the rest of the notation can be found in [10]. Its soft supersymmetric Lagrangian $\mathcal{L}_{\text{soft}}$ is

$$\begin{aligned} -\mathcal{L}_{\text{soft }\nu MSSM} &= \tilde{q}_{Li}^\dagger (m_{\tilde{Q}}^2)^{ij}\tilde{q}_{Lj} + \tilde{u}_{Rj}(m_{\tilde{u}}^2)^{ji}\tilde{u}_{Ri}^* + \tilde{d}_{Rj}(m_{\tilde{d}}^2)^{ji}\tilde{d}_{Ri}^* \\ &+ \tilde{l}_{Li}^\dagger (m_{\tilde{L}}^2)^{ij}\tilde{l}_{Lj} + \tilde{e}_{Rj}(m_{\tilde{e}}^2)^{ji}\tilde{e}_{Ri}^* + \tilde{\nu}_{Rj}(m_{\tilde{\nu}}^2)^{ji}\tilde{\nu}_{Ri}^* + m_{h_d}^2 h_d^\dagger h_d + m_{h_u}^2 h_u^\dagger h_u \\ &+ (b\, h_d h_u + \frac{1}{2}B_\nu^{ij}M_\nu^{ij}\tilde{\nu}_{Ri}^*\tilde{\nu}_{Rj}^* + \text{h.c.}) \\ &+ \Big(-a_d^{ij}h_d\tilde{d}_{Ri}^*\tilde{q}_{Lj} + a_u^{ij}h_u\tilde{u}_{Ri}^*\tilde{q}_{Lj} - a_l^{ij}h_d\tilde{e}_{Ri}^*\tilde{l}_{Lj} + a_\nu^{ij}h_u\tilde{\nu}_{Ri}^*\tilde{l}_{Lj} \\ &\quad + \frac{1}{2}M_1\widetilde{B}\widetilde{B} + \frac{1}{2}M_2\widetilde{W}^a\widetilde{W}^a + \frac{1}{2}M_3\widetilde{G}^a\widetilde{G}^a + \text{h.c.} \Big), \end{aligned} \tag{6}$$

where SU(2) indices are not written explicitly. Then the flavour and CP supersymmetric problems are easy to identify: arbitrary values of masses and trilinear couplings in the supersymmetric breaking soft Lagrangian, give arbitrary flavour changing neutral currents (FCNC) and can easily exceed CP bounds. From $W_{\nu MSSM}$ a we will not have more flavour or CP sources than those in the SM plus three right-handed neutrinos, except of course those coming from the coupling $\mu\epsilon_{\alpha\beta}\mathbf{H}_u^\alpha\mathbf{H}_d^\beta$.

To be concrete, we can consider the case of the CP asymmetry in the K^0 system, $\bar{K}^0 - K^0$, which are a mixture of $s\bar{d}$ and $\bar{s}d$. In the SM, this mixing can only be mediated by W^- and the up-type quarks, Figure 2. The mixing is small due to the GIM mechanism. That is the VEV of the effective Hamiltonian describing the $\bar{K}^0 - K^0$ mixing, $\langle K^0|\mathcal{H}_{\Delta S=2}^{eff}|\bar{K}^0\rangle$, at M_W,

$$\begin{aligned} \langle K^0|\mathcal{H}_{\Delta S=2}^{eff}|\bar{K}^0\rangle &= \frac{G_F^2 M_W^2}{4\pi^2}X\,\langle K^0|[\bar{d}\gamma^\mu\mathcal{P}_L s][\bar{d}\gamma_\mu\mathcal{P}_L s]|\bar{K}^0\rangle, \\ X|_{M_W} &= \sum_{i,j=1}^{3} V_{js}V_{id}^*V_{jd}V_{is}^*S\left(\frac{m_i^2}{M_W^2}, \frac{m_j^2}{M_W^2}\right), \end{aligned} \tag{7}$$

is small. The GIM mechanism manifest here simply as follows. Firstly due to the unitarity of the $\mathbf{3}\times\mathbf{3}$ V_{CKM} matrix such that combinations of the type $V_{js}V_{id}^*V_{jd}V_{is} = (V_{cs}V_{ud}^*)^2$ are impossible to appear. Secondly because of the value of the function $S(x_i, x_j) = S\left(\frac{m_i^2}{M_W^2}, \frac{m_j^2}{M_W^2}\right)$ for the

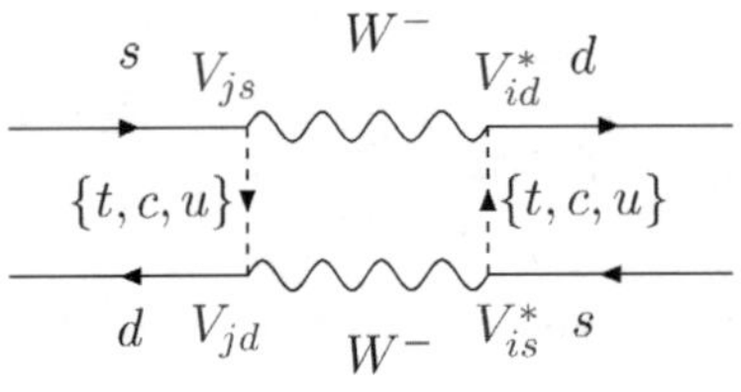

Figure 2: Mixing in the system, $\bar{K}^0 - K^0$, where only W^- and the up-type quarks, mediate it trough the $V_{ij} = V_{\mathrm{CKM}ij}$ elements, for $i = u, c, t$, $j = s, d$.

different relevant quarks (c and t) [2]

$$\begin{aligned}
X &= [V_{cs}V^*_{cd}]^2 S\left(\frac{m_c^2}{M_W^2}\right) + [V_{ts}V^*_{td}]^2 S\left(\frac{m_t^2}{M_W^2}\right) + V_{cs}V^*_{cd}V_{ts}V^*_{td}\, S\left(\frac{m_c^2}{M_W^2}, \frac{m_t^2}{M_W^2}\right) \\
S&\left(\frac{m_c^2}{M_W^2}\right) = 2.49\times 10^{-4},\; S\left(\frac{m_t^2}{M_W^2}\right) = 2.37,\; S\left(\frac{m_c^2}{M_W^2}, \frac{m_t^2}{M_W^2}\right) = 1.97\times 10^{-3}, \\
&[V_{cs}V^*_{cd}]^2 \propto \lambda^2, \qquad [V_{ts}V^*_{td}]^2 \propto \lambda^{10}, \qquad V_{cs}V^*_{cd}V_{ts}V^*_{td} \propto \lambda^7.
\end{aligned} \tag{8}$$

The CP violation parameter

$$\begin{aligned}
\epsilon_K &\doteq \left.\frac{-\mathrm{Im}\left[\langle K^0|\mathcal{H}^{eff}_{\Delta S=2}|\bar{K}^0\rangle\right]}{\mathrm{Re}\left[\langle K^0|\mathcal{H}^{eff}_{\Delta S=2}|\bar{K}^0\rangle\right]}\right|_{\mu=\bar{s}d} = \frac{-2\mathrm{Im}\left[\langle K^0|\mathcal{H}^{eff}_{\Delta S=2}|\bar{K}^0\rangle\right]}{\Delta m_{K^0}}, \\
&= \frac{G_F^2 f_K^2 m_K M_W^2}{6\sqrt{2}\pi^2 \Delta m_K} B_K \mathrm{Im}[X]|_{\mu=\bar{s}d}
\end{aligned}$$

is a great way to check the sensitivity to CP violation. In Eq. (9) m_K is the mass of K^0, f_K and B_K constants of the Kaon system. The QCD NLO (next to leading order) [11] and the experimental values are

$$\epsilon_K{}^{SM}_{NLO} = (0.00178 \pm 0.00025)\,, \quad \epsilon_K^{exp} = (0.00229 \pm 0.00010)\,, \tag{9}$$

so both quantities are compatible within 2σ C.L. While it is truly amazing that within the SM it is possible to achieve this compatibility, we wonder if we have achieved enough precision in the computation of ϵ_K or if there is a hint of something beyond the SM (BSM). Certainly the error in its experimental value is a thought test for theories BSM and in particular for supersymmetry. For the general $\nu MSSM$ in consideration, it is easy to see the many sources of contributions to this observable. The box diagrams associated to $\Delta F = 2$ processes in this theory are given in Figure 3, for the K^0 mixing under consideration, $q_j = s$ and $q_i = d$. We recognize that only the charged Higgs bosons are mediated by the couplings of the SM, the elements of V_{CKM}, Figure 3 (a). In all the other diagrams of Figure 2 ((b),(c),(d)), there appear new mixing elements. In diagram (b) we need to compute the mixing of the quarks and squarks mass eigenstates, identified by the elements of the matrix K^f. This mixing matrix is the unitary matrix diagonalizing the

[2] In Eq. (7) $\mathcal{P}_L$ is the left-handed projection operator and the loop functions are $S(x_i, x_i) = S(x_i) = \frac{x_i^3 - 11x_i^2 + 4x_i}{4(1-x_i)^2} - \frac{3x_i^3\log(x_i)}{2(1-x_i)^3}$, $S(x_i, x_j)|_{i\neq j} = x_i\left(-\frac{3x_ix_j}{4(1-x_j)} + \log\left(\frac{x_j}{x_i}\right) - \frac{3x_j^2\log(x_j)}{4(1-x_j)^2}\right)$.

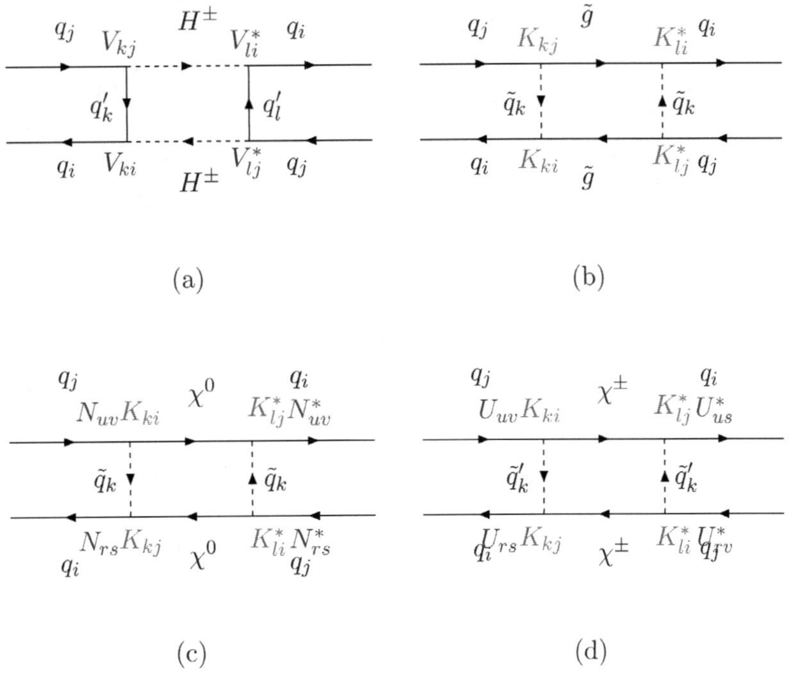

Figure 3: Flavour changing $\Delta F = 2$ box diagrams, q' represents a different kind of quark than q. For the $K^0 - \bar{K}^0$ system $q_j = s$ and $q_i = d$.

effective mass matrix for sfermions $\tilde{f}$ in the basis where Yukawa matrices are diagonal, the super CKM (SCKM) basis, for example for squarks:

$$
\begin{aligned}
(\mathcal{M}_{\tilde{f}}^{\text{SCKM}})^2_{ij} &= \begin{bmatrix} M^{\text{SCKM}}{}^2_{LL} & M^{\text{SCKM}\dagger}{}^2_{LR} \\ M^{\text{SCKM}}{}^2_{LR} & M^{\text{SCKM}}{}^2_{RR} \end{bmatrix}_{ij} \equiv (\widehat{\mathcal{M}}^2_{\tilde{f}})_{ij} \\
&= \begin{bmatrix} (U^f_L M^2_{\tilde{Q}} U^{f\dagger}_L)_{ij} + \hat{m}^2_{f_i}\delta_{ij} + D^f_L & -((U^f_L a_f U^{f\dagger}_R)_{ij} v_f + \mu^* \tan^p \beta \hat{M}_{f_i}\delta_{ij}) \\ -((U^f_R a^\dagger_f U^{f\dagger}_L)_{ij} v_f + \mu \tan^p \beta \hat{M}_{f_i}\delta_{ij}) & (U^f_R M^2_{\tilde{f}_R} U^{f\dagger}_R)_{ij} + \hat{m}^2_{f_i}\delta_{ij} + D^f_R \end{bmatrix},
\end{aligned} \quad (10)
$$

$$
\mathcal{L}^{\text{eff}}_{m_{\tilde{q}}} = -(\tilde{q}'_L, \tilde{q}'_R)_i (\mathcal{M}^2_{\tilde{q}'})_{ij} \left(\tilde{q}'^*_L \tilde{q}'^*_R \right)_j = [\tilde{q}_L, \tilde{q}_R]_{i=1,2,3} K^{f\dagger} \begin{bmatrix} m^2_{\tilde{q}_1} & & \\ & \vdots & \\ & & m^2_{\tilde{q}_6} \end{bmatrix} K^f \begin{bmatrix} \tilde{q}^*_L \\ \tilde{q}^*_R \end{bmatrix}_{j=1,2,3}.
$$

Here $D^f_{L,R}$ remain diagonal, and $\hat{m}_f$ is the diagonal matrix of the f type fermions, the indices i, j run over the different quark generations.

The diagram of Figure 3 (c), brings along the mixing of all the neutral $U(1)_{em}$ sparticles, the neutralinos. Their mixing is described the Lagrangian $\mathcal{L} = -\frac{1}{2}\widetilde{\psi}^0 N N^\dagger \mathcal{M}_{\widetilde{\psi}^0} N^* N^T (\widetilde{\psi}^0)^T + \text{h.c.}$, where the gauge eigenstates are $\widetilde{\psi}^0 = (-i\tilde{b}, -iw, \tilde{h}^0_d, \tilde{h}^0_u)$ and the mass eigenstates need to be computed from

$$\mathcal{M}_{\tilde{\psi}^0} = \begin{bmatrix} M_1 & 0 & -M_Z c_\beta s_W & M_Z s_\beta s_W \\ 0 & M_2 & M_Z c_\beta c_W & -M_Z s_\beta c_W \\ -M_Z c_\beta s_W & M_Z c_\beta c_W & 0 & -\mu \\ M_Z s_\beta s_W & -M_Z s_\beta c_W & -\mu & 0 \end{bmatrix} = N M_{\chi^0} N^T, \qquad (11)$$

where N matrices are diagonalizing matrices and the mass eigenstates, $\tilde{\chi}_i^0$, are given by $\tilde{\chi}_i^0 = (\tilde{\psi}^0 N)_i$. Finally, the diagram of Figure 3 (d) introduces the mixing of all the charged $U(1)_{em}$ sparticles, the charginos. Their effective mass Lagrangian is given by $\mathcal{L} = -\frac{1}{2}\left[\tilde{\psi}^+ V V^+ \mathcal{M}_{\tilde{\psi}^+} U^* U^T (\tilde{\psi}^-)^T + \tilde{\psi}^- \mathcal{M}^T_{\tilde{\psi}^+} (\tilde{\psi}^+)^T\right]$, where $\tilde{\psi}^+ = (-i\tilde{w}^+, \tilde{h}_u^+)$, $\tilde{\psi}^- = (-i\tilde{w}^-, \tilde{h}_d^-)$. The mass matrix

$$\mathcal{M}_{\tilde{\psi}^+} = \begin{pmatrix} M_2 & \sqrt{2}\, M_W c_\beta \\ \sqrt{2}\, M_W s_\beta & \mu \end{pmatrix} \qquad (12)$$

is diagonalized by two different unitary matrices U and U', $M_{\tilde{\chi}^\pm} = U \mathcal{M}_{\tilde{\psi}^+} U'^*$. It is easy to guess that light gluinos can give contributions that easily exceed the observed value of ϵ_K. This is because there is just one type of mixing matrix involved, K. For example, for masses of gluinos and s-downs both less than 400 GeV and $K_{is,id} = V_{\mathrm{CKM}is,id}$ this would give $\epsilon_K \gtrsim O(10^{-3})$! These observations bring along the minimal flavour violating (MFV) scheme [12], where although we do not have the relations $K_{is,id} = V_{\mathrm{CKM}is,id}$, Yukawa matrices are assumed to be the only sources of flavour and CP violation. This is done by imposing as boundary conditions at a determined scale, μ_B, the relations

$$(m^2_{\tilde{Q}})^{ij} = (m^2_{\tilde{u}})^{ij} = (m^2_{\tilde{d}})^{ij} = (m^2_{\tilde{L}})^{ij} = (m^2_{\tilde{e}})^{ij} = \delta_{ij}, \quad a_f^{ij} = Y_{ij}^f A_0, \qquad (13)$$

which in the constrained MSSM (CMSSM) they are set at $\mu_B = M_{\mathrm{G}}$. This will be satisfied at the Electroweak scale only if $A_0 = 0$ [13]. This is a very strong restriction, which the FS do not in general satisfy, we will come back to this issue in the next section.

4. Interplay

4.1. Effective supersymmetric theories

In an effective supersymmetric theory with a FS, both the *effective* superpotential W_{FS} and the soft Lagrangian $\mathcal{L}_{\mathrm{soft\ FS}}$ will be invariant under the action of the FS group G_{FS},

$$\begin{aligned} W_{\nu MSSM} &= W_{FS} \stackrel{G_{FS}}{\longrightarrow} W_{FS} \\ \mathcal{L}_{\mathrm{soft}\ \nu MSSM} &= \mathcal{L}_{\mathrm{soft}\ FS} \stackrel{G_{FS}}{\longrightarrow} \mathcal{L}_{\mathrm{soft}\ FS}, \end{aligned} \qquad (14)$$

where we now assume that Eqs. (5) and (6) are FS invariant. From the rest of this contribution I will assume that flavons are always scalars and that Yukawa couplings, trilinear terms and supersymmetric mass terms are always combinations of flavons, for example for two different flavons ϕ_1 and ϕ_2 we have

$$\frac{\langle \bar{\phi}_1^i \rangle}{M_1} \frac{\langle \bar{\phi}_2^j \rangle}{M_2} \epsilon_{\alpha\beta} \mathbf{H}_d^\alpha \mathbf{E}_i^c \mathbf{L}_j^\beta \quad \rightarrow \quad Y_l^{ij} = \frac{\langle \bar{\phi}_1^i \rangle}{M_1} \frac{\langle \bar{\phi}_2^j \rangle}{M_2}. \qquad (15)$$

For concrete examples I will refer to $SU(3)$ FS models which have been somewhat extensively studied [14, 15, 16] and in particular to the set up in [17]. In these models usually MSSM matter fields, e.g. $\mathbf{E}_i^c$ and $\mathbf{L}_j^\beta$, sit in antitriplet representations of $SU(3)$ and so $\bar{\phi}^i$ and $\bar{\phi}^j$ sit in triplet representations, assuming the Higgs bosons are singlets under $SU(3)$. Once the flavons take VEVs, W_{FS} is an effective renormalizable superpotential. This superpotential should have come indeed from a renormalizable superpotential before the FS is broken, for example

$$\begin{aligned} W_O &= M_1\bar{\chi}^f_{1i}\chi^f_{1i} + M_2\bar{\chi}^f_2\chi^f_2 + F_i H_f \chi^f_{0i} + \bar{\chi}^f_{1i}\bar{\phi}_{1i}\chi^f_2 + \bar{\chi}^f_2\bar{\phi}_{2i} f^c_i + \mu H_u H_d, \\ F &= \mathbf{L},\ \mathbf{Q}, \quad f^c = \mathbf{U}^c,\ \mathbf{D}^c,\ \mathbf{E}^c_i, \end{aligned} \tag{16}$$

then Eq. (5) should be obtained by integrating out the *heavy* fields χ^f_1 and χ^f_2 $\partial W_{FS}/\partial\chi^f_i = \partial W_{FS}/\partial\overline{\chi}^f_i = 0$. These fields χ^f_i are often called *messenger fields*.

The observables to which W_{FS}, Eq. (5), is confronted are again the hierarchical fermion masses ($\hat{Y}^u = \sqrt{2}$ Diagonal$[m_u, m_c, m_t]/(v\sin\beta)$, $\hat{Y}^d = \sqrt{2}$ Diagonal$[m_d, m_s, m_b]/(v\cos\beta)$ [3]) and the CKM matrix $V_{\rm CKM}$. Now of course we have the free parameter $\tan\beta$ but mass ratios among the same kind of quarks are not affected by it.

It is evident that since the FS controls the full superpotential W_{FS} at least from it, we will not have more flavour or CP sources than those in the SM, except of course just coming from the right-handed neutrinos and the coupling $\mu\epsilon_{\alpha\beta}\mathbf{H}^\alpha_u\mathbf{H}^\beta_d$.

Note that in fact the gain is obvious, we reduce the arbitrariness in selecting the possible Yukawa couplings by constraining them through observables directly related to the supersymmetric vertices produced by W_{FS}. What it does introduce much more flavour violating terms and sources of CP violation is the soft-supersymmetric Lagrangian $\mathcal{L}_{\rm soft\ FS}$.

Since $\mathcal{L}_{\rm soft\ FS}$ must be FS invariant, just by FS arguments we can interpret the trilinear terms as functions of the flavons

$$\frac{\langle\bar{\phi}^i_1\rangle}{M_1}\frac{\langle\bar{\phi}^j_2\rangle}{M_2}A_0 h_d \tilde{e}^*_i\tilde{l}_j \quad\rightarrow\quad a^{ij}_l = \frac{\langle\bar{\phi}^i_1\rangle}{M_1}\frac{\langle\bar{\phi}^j_2\rangle}{M_2}A_0, \tag{17}$$

where A_0 is just a mass term like in the CMSSM. Note that in effective supersymmetric theories Eqs. (15) and (17) have many ways to be generated because the only restrictions are the way the flavon fields, ϕ, of the theory can be contracted with matter fields. Note however there is not any way to relate Yukawa couplings, Eq. (15), and trilinear terms Eq. (17) even if by FS arguments they are given by the same combinations of flavons, simply because we cannot determine the exact coefficients in front of them and the relation among them, that is

$$c^{ij}_Y =^? = c^{ij}_a, \quad Y^{ij}_l = c^{ij}_Y\frac{\langle\phi^i\rangle}{M_1}\frac{\langle\phi^j\rangle}{M_2}, \quad a^{ij}_l = c^{ij}_a\frac{\langle\phi^i\rangle}{M_1}\frac{\langle\phi^j\rangle}{M_2}A_0. \tag{18}$$

The ambiguities produced by this underdetermination can be lifted by introducing free parameters but this of course turns in to a question of how predictive the theory would be and in particular how many couplings to the heavy fields χ^f_1 will be allowed.

These theories have been successful to a point where the details of the messenger fields are ignored, all their masses are assumed to be the same and all what it matters to get the right hierarchy of fermion masses are the VEVs of the flavon fields. This is done by studying the scalar potential of the theory, in the example given just as a supersymmetric theory with W_O of Eq. (16) and assuming the VEV of Higgs is set to a much lower scale than the scale of breaking of the FS, we would have

$$U_{FS} = \sum_i\left[W^*_{Oi}W_{Oi} + \sum_a D^a_i D^a_i\right]\Bigg|_{VEVs} = g^2_{FS}\left[\sum_a \langle\bar{\phi}\rangle^\dagger_{1x} T^a_{xy}\langle\bar{\phi}\rangle_{1y} + \langle\bar{\phi}\rangle^\dagger_{2x} T^a_{xy}\langle\bar{\phi}\rangle_{2y}\right]. \tag{19}$$

It is easy to see that since $W^*_i = \delta W_O/\delta\bar{\phi}_i$, $i = 1, 2$, are not proportional to any scalar then they cannot take any role in setting the VEVs of the flavons and all what it matters are the

[3] Here $\tan\beta = \langle H_u\rangle / = \langle H_d\rangle = v_d/v_u$.

D terms. It is usually desirable to set these terms to zero because we do not want to break supersymmetry at the scale of the FS breaking. However in general two flavon fields are not enough to generate the desired hierarchy of Yukawa couplings and at the same time cancel the D terms. Hence many more fields should be introduced in the theory and their F terms usually also play an important role to achieve a possible hierarchy of fermion masses and mixing Eq. (4). It is important to mention that it is also through U_{FS} that the CP symmetry must be broken exactly in the way it is observed at low energies. This is not an easy task, but some attempts have been made [18].

4.2. Effective FS supergravity theories

In effective supergravity theories one really needs to specify the superpotential $W_{\rm O}$, like the one in Eq. (16), and the Kähler potential of the complete MSSM, flavons and messenger fields, for example [17]

$$
\begin{aligned}
K &= F_i^\dagger F_i\left[1+\xi_F+\frac{\lambda_H^2\,(1+\xi_{\bar{\chi}_0^f})}{M^2_{\chi_0^f}}H_f^\dagger H_f\right] + f_i^c f_j^{c\dagger}\left[\delta_{ij}+\xi_{f^c}\delta_{ij}+\frac{\lambda_2^2\,(1+\xi_{\chi_2^f})}{M^2_{\chi_2^f}}\bar{\phi}_{2i}\bar{\phi}^\dagger_{2j}\right] \\
&\quad + H_f^\dagger H_f\left(1+\xi_{H_f}\right)+(Z_H\, H_u H_d+\text{h.c.})+\,\ldots\,+K_{\rm H} \qquad (20)\\
&\equiv \tilde{K}_{F_i^\dagger F_j}\,F_i^\dagger F_j+\tilde{K}_{f_i^c f_j^{c\dagger}}\,f_i^c f_j^{c\dagger}+\tilde{K}_{H_f^\dagger H_f}\,H_f^\dagger H_f+(Z_H\,H_u H_d+\text{h.c.})+\,\ldots\,+K_{\rm H}\,,
\end{aligned}
$$

Where $K_{\rm H}$ represents the hidden-sector including the fields breaking supersymmetery and the family symmetry, that is the flavons. The soft-squared masses and trilinears are then given as functions of the the gravitino mass $m^2_{3/2}=\langle e^{K_{\rm H}/M_{\rm P}^2}|W_{\rm H}|^2\rangle\,/M_{\rm P}^4=$ const. and all the fields of the hidden sector. Assuming that there is no contribution from D terms [19], we have

$$
m'^2_{\bar{\alpha}\beta} = m^2_{3/2}\,\langle\tilde{K}_{\bar{\alpha}\beta}\rangle-\left\langle\mathcal{F}^{*\bar{m}}\left(\partial^*_{\bar{m}}\partial_n\tilde{K}_{\bar{\alpha}\beta}-(\partial^*_{\bar{m}}\tilde{K}_{\bar{\alpha}\gamma})\,\tilde{K}^{\gamma\bar{\delta}}\,\partial_n\tilde{K}_{\bar{\delta}\beta}\right)\mathcal{F}^n\right\rangle, \qquad (21a)
$$

$$
\begin{aligned}
a'_{\alpha\beta\gamma} &= \langle\mathcal{F}^m\rangle\left[\left\langle\frac{\partial_m K_{\rm H}}{M_{\rm P}^2}\right\rangle Y'_{\alpha\beta\gamma}+\frac{\mathcal{N}\partial Y_{\alpha\beta\gamma}}{\partial\langle h_m\rangle}\right] \\
&\quad -\langle\mathcal{F}^m\rangle\left[\left\langle\tilde{K}^{\delta\bar{\rho}}\,(\partial_m\tilde{K}_{\bar{\rho}\alpha})\right\rangle Y'_{\delta\beta\gamma}+(\alpha\leftrightarrow\beta)+(\alpha\leftrightarrow\gamma)\right], \qquad (21b)
\end{aligned}
$$

where $\tilde{K}_{\bar{\alpha}\beta}\equiv\frac{\partial^2 K}{\partial C^\dagger_{\bar{\alpha}}\partial C_\beta}$ with $C_\alpha=(F_i,f_i^{c\dagger},H_f)$ and $\tilde{K}^{\gamma\bar{\delta}}$ denotes the elements of the inverse matrix. Besides, $\partial_m\equiv\partial/\partial h_m$, $\partial^*_{\bar{m}}\equiv\partial/\partial h^*_{\bar{m}}$, and e.g. $\langle\mathcal{F}^{\bar{\phi}_1}\rangle\,\partial/\partial\bar{\phi}_1\equiv\langle\mathcal{F}^{\bar{\phi}_{1i}}\rangle\,\partial/\partial\bar{\phi}_{1i}$. We have expressed the formula for the trilinear couplings in terms of $Y'_{f_i^c F_j H_f}\equiv\mathcal{N}\,Y_{f_i^c F_j H_f}\equiv\mathcal{N}\,\lambda_H\lambda_1\lambda_2\,\frac{\langle\bar{\phi}_2\rangle_i\langle\bar{\phi}_1\rangle_j}{M_{\chi_0^f}M_{\chi_2^f}}$, where $W'_{\rm O}=W_{\rm O}\left\langle\frac{W^*_{\rm H}}{|W_{\rm H}|}\,e^{\frac{1}{2M_{\rm P}^2}\sum_m|h_m|^2}\right\rangle\equiv\mathcal{N}\,W_{\rm O}$. We see then that supergravity give us an exact prescription or how can we relate Yukawa and trilinear couplings to flavons and how the spectra of soft-squared masses can indeed be controlled by the FS.

5. Expectations at the LHC

What can we expect at the LHC from such SFS theories? The first thing to note is that for having a definite answer we must assume something about the way supersymmetry is broken and to set the scale of its breaking. This is the first limitation of this scenario, it would be great if one can combine the breaking of supersymmetry with the breaking of the FS. Putting aside this question, one needs then to assume a definitive value of $m_{3/2}$ and the relation between the fields breaking supersymmetry and the flavon fields. Let us assume that this is trivial, that is that their mixing is zero. As we have said from concreteness we take the example of [17], to which Eqs. (16) and (20) correspond. Then the soft squared masses will have the form, at energies of the order $\langle\bar{\phi}\rangle$,:

$$\hat{m}^2_{\tilde{f}^\dagger\tilde{f}} \sim m^2_{3/2}\begin{pmatrix}1 & \tilde{\epsilon}^2_f\,\epsilon^2_d & \tilde{\epsilon}^2_f\,\epsilon^2_d\\ \cdot & 1+\tilde{\epsilon}^2_f & \tilde{\epsilon}^2_f\\ \cdot & \cdot & 1\end{pmatrix} \quad , \quad f=u,\,d,\,Q,\,e,\,L\,, \tag{22}$$

where $\tilde{\epsilon}$ parameters have the following constraints

$$|\tilde{\epsilon}_Q\,\tilde{\epsilon}_u| \sim |\epsilon_u|^2 \sim \left[\frac{Y'_{f^c_2F_2H_u}}{Y'_{f^c_3F_3H_u}}\right]^2 = 0.04^2, \quad |\tilde{\epsilon}_Q\,\tilde{\epsilon}_d| \sim |\epsilon_d|^2 \sim \left[\frac{Y'_{f^c_2F_2H_d}}{Y'_{f^c_3F_3H_d}}\right]^2 = 0.13^2,$$
$$|\tilde{\epsilon}_L\,\tilde{\epsilon}_e| \sim |\epsilon_d|^2\,. \tag{23}$$

As we can see, they cannot be uniquely determined, but exactly these kind of relations can be tested through FCNC processes, before doing so, there are a couple of observations worth mentioning here. First, the matrices Eq. (22) are not like in the MFV case, that is diagonal at a high scale $\langle\bar{\phi}\rangle \gtrsim M_G$. Second, this is not a problem because the off-diagonal elements are controlled by parameters related to the Yukawa couplings, Eq. (23), even thought not uniquely determined. This is the *weakest* point of the SFS and so a further understanding on how these parameters could be controlled should be addressed. To have an idea of the type of amount of flavour violation that these models have, let us present it in the mass insertion (MI) approximation where the flavour violating terms are given by dimensionless parameters δ defined as

$$(\delta^f_{\rm RR})_{ij} \doteq \frac{(\tilde{m}^2_{\tilde{f},\rm RR})_{ij}}{(\tilde{m}^2_{\tilde{f},\rm RR})_{ii}}, \quad \left(\delta^f_{LR,RL}\right)_{ij} \doteq \frac{(\tilde{m}^2_{\tilde{f},\rm LR,RL})_{ij}}{\sqrt{(\tilde{m}^2_{\tilde{f},\rm LL})_{ii}(\tilde{m}^2_{\tilde{f},\rm RR})_{jj}}}, \tag{24}$$

we do not have to compute the exact mixing matrices K, N and U' of Eqs. (11), (11) and (12) respectively, it is just enough to get Eq. (22) roughly at the EW to have an idea of the flavour violation induced. Analogously we can compute the $(\delta^f_{\rm XY})$ parameters

$$\begin{aligned}\left(\delta^d_{LR,RL}\right)_{ij} &= \frac{v}{\sqrt{1+\tan^2\beta}}\left[-\frac{\widetilde{a}_{\tilde{d}^c_i\tilde{Q}_jH_f}}{30m_0^2} + \frac{\mu^*\tan\beta\,\tilde{Y}^{\rm diag}_{d^c_iQ_j}}{30m_0^2\sqrt{2}}\right],\\ \left(\delta^u_{LR,RL}\right)_{ij} &= \frac{v\tan\beta}{\sqrt{1+\tan^2\beta}}\left[-\frac{\widetilde{a}_{\tilde{u}^c_i\tilde{Q}_jH_f}}{30m_0^2} + \frac{\mu^*\tilde{Y}^{\rm diag}_{u^c_iQ_j}}{30m_0^2\tan\beta\sqrt{2}}\right],\end{aligned} \tag{25}$$

where $\tilde{}$ denotes the quantities in the SCKM basis and the factor 30 is a renormalization running factor. Once this is done on the basis where all the fields have canonically normalised kinetic terms, we can compute the flavour violating parameters δ's and confront them with bounds obtained from several experimental constraints. To illustrate the point, we present these parameters for the example of this section in Table 2.

What do we learn from here? That FS can control flavour violating effects without the need of assuming MFV and through parameters relevant to the Yukawa couplings. Furthermore, FCNC processes really put to test these symmetries because there are parameters ($(\delta^d_{\rm LR,RL})_{12}$ and $(\delta^d_{\rm LL})_{12}$) that are already at the limit of their corresponding bounds.

The next question, it is what could be the real impact on the supersymmetric spectra? This was probed in [13]. The basic results are as follows. When the running of off-diagonal Yukawa

Journal of Physics: Conference Series **287** (2011) 012009 doi:10.1088/1742-6596/287/1/012009

	Example of [17]	Bound
$(\delta^d_{\mathrm{RR}})_{12}$	$\frac{\tilde{\epsilon}^2_d\,\epsilon_d}{30} \sim 7 \cdot 10^{-5}$	$9 \cdot 10^{-3}$
$(\delta^d_{\mathrm{LL}})_{12}$	$\frac{\tilde{\epsilon}^2_Q\,\epsilon_d}{30} \sim 7 \cdot 10^{-5}$	$1 \cdot 10^{-2}$
$(\delta^d_{\mathrm{LR,RL}})_{12}$	$\frac{v}{\sqrt{1+\tan^2\beta}}\left[-\frac{nA_0\epsilon^3_d}{30m_0^2}\right] \sim 4n \cdot 10^{-6}$	$1 \cdot 10^{-5}$
$(\delta^d_{\mathrm{LL}})_{23}$	$\frac{\tilde{\epsilon}^2_Q}{30} \sim 6 \cdot 10^{-4}$	$2 \cdot 10^{-1}$
$(\delta^e_{\mathrm{LL}})_{12}$	$\frac{\tilde{\epsilon}^2_L\,\epsilon_d}{4} \sim 6 \cdot 10^{-4}$	$6 \cdot 10^{-4}$
$(\delta^d_{\mathrm{LR,RL}})_{23}$	$\frac{v}{\sqrt{1+\tan^2\beta}}\left[-\frac{nA_0\epsilon^2_d}{30m_0^2}\right] \sim 4n \cdot 10^{-4}$	$1 \cdot 10^{-3}$

Table 2: An example for the flavour violating parameters δ for the SPS 1a point($m_0 = 100\,\mathrm{GeV}$, $m_{1/2} = 250\,\mathrm{GeV}$, $A_0 = -100$ GeV and $\tan\beta = 10$,), together with the corresponding experimental limit. For a detailed description of the formulas see the text in this section.

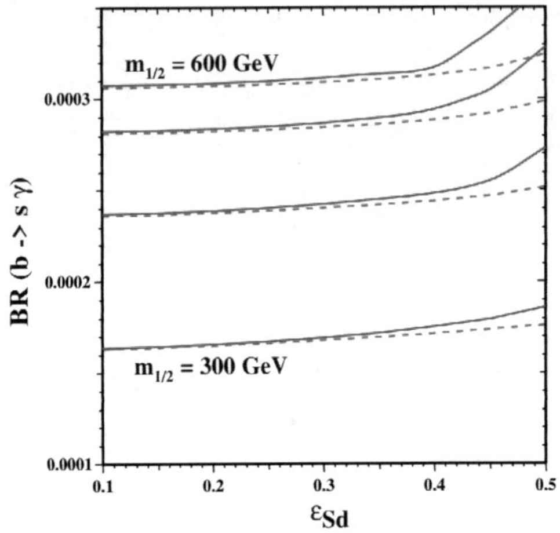

Figure 4: Variation of $BR(b \to s\gamma)$ with respect to $\tilde{\epsilon}_d = \epsilon_{S_d}$. From top to bottom the lines correspond to $(m_{1/2}, m_0, A_0) = (600, 410, 600)$ GeV, $(500, 360, 500)$ GeV, $(400, 340, 400)$ GeV and $(300, 380, 300)$ GeV. At QCD NNLO we have$BR(b \to s\gamma) = (3.15 \pm 0.23) \times 10^{-4}$ [20, 21] while the experimental value is [5] $BR(b \to s\gamma) = (3.55 \pm 0.24 \pm 9) \times 10^{-4}$.

couplings, trilinear terms and soft-squared masses is taken into account according to parameters defined by the SFS, e.g. Eq. (23), we can use precise measurements such as ϵ_k, $BR(b \to s\gamma)$ and $BR(b \to \ell^+\ell^-\gamma)$ to asses their impact on soft masses. It is true that the splitting of the s-fermions will be mainly controlled about what we assume of the way supersymmetry is broken, because Yukawa couplings are small and so the running cannot be greatly affected. However trilinear terms and soft-squared masses are quite sensitive to FCNC. In Figure (4) from [13] we plot $BR(b \to s\gamma)$ as a function of our parameter $\tilde{\epsilon}_d = \epsilon_{S_d}$, there it was let to vary from zero to 0.5. What it is interesting about this plot? The points on the left-hand side of all the curves correspond effectively to the CMSSM predictions of $BR(b \to \ell^+\ell^-\gamma)$ for different $m_{1/2}$ masses, from 300 to 600 GeV. Important departures from what it can be obtained in the CMSSM take place for $\epsilon_{S_d} \geq 0.4$. For example in the CMSSM the particular point shown for $m_{1/2} = 400$ GeV would be experimentally excluded by $BR(b \to s\gamma)$ but for a SFS $\epsilon_{S_d} = 0.44$ would be allowed. $BR(b \to \ell^+\ell^-\gamma)$ it is also quite sensitive to these changes and it has the right sensitivity for the LHC, unfortunately $BR(b \to s\gamma)$ requires photon energies of the order of few GeVs, which could be overseen at the LHC. Apart from the works in [14, 13], interesting analysis of FCNCS of effective supersymmetric $SU(3)$ family symmetries are presented in [11] and constraints from electric dipole moments (EDM) are studied in [23]. They are an important step towards the characterization of this kind of models, but the task is far from complete. It is thus an important and challenging task to determine extensively the changes in the parameter space which are typical of Supersymmetric Flavour Symmetries.

6. Final comments

Giving the exciting opportunity of discovery of the Higgs boson and supersymmetric particles at the LHC, we need to make an extensive catalogue of the possible features of the different supersymmetric models we propose. We have argued here the case of the supersymmetric family symmetries (SFS) that attempt to explain, both the hierarchy of mixing and fermion masses, the CP violation *and* the structure of sfermion masses and mixing.

I would like to emphasize that SFS explored in the literature have not concerned too much about the messenger sector of the theory, which is a decisive point towards a really predictive model.. There have been few attempts, [24, 16, 17], but a serious understanding within a formal supergravity context should be addressed. Nevertheless, some characteristics of the SFS have already gave us information about the way FCNC and EDMs are under control and their impact on the supersymmetric spectra. We have ordered some pieces of the puzzle of the *GIM supersymmetric family symmetry mechanism.* Also as in the SM model, in the MSSM the mixing in FCNC processes needs to be small. The amount of this mixing can be controlled through parameters that are related to the mixing and eigenvalues of the Yukawa matrices. Therefore SFS have found a way to control mixing in both sectors, for fermions and sfermions, however a missing piece is to determine if one could really understand the structure of the supersymmetric eigenvalues from a SFS point of view and the conditions to get the CP violating phases from minimizing an scalar potential.

7. Acknowledgments

I wish to acknowledge the Faculty of Sciences from the University of Guanuajuato, Campus Leon, for their kind support that enabled me to participate in the XIV Mexican School of Particles and Fields. I also thank K. Kadota, J. Kersten, K. Olive and E. de la Cruz Burelo for enlightening discussions of issues related to the ones that appear in this contribution.

References

[1] Froggatt C D and Nielsen H B 1979 *Nucl. Phys.* **B147** 277
[2] Dudas E, Grojean C, Pokorski S and Savoy C A 1996 *Nucl.Phys.* **B481** 85–108 (*Preprint* hep-ph/9606383)
[3] Altarelli G, Feruglio F and Masina I 2000 *JHEP* **0011** 040 (*Preprint* hep-ph/0007254)
[4] Barbieri R, Dvali G and Hall L J 1996 *Phys.Lett.* **B377** 76–82 (*Preprint* hep-ph/9512388)
[5] King S and Ross G G 2001 *Phys.Lett.* **B520** 243–253 (*Preprint* hep-ph/0108112)
[6] King S 2005 *JHEP* **0508** 105 (*Preprint* hep-ph/0506297)
[7] Bazzocchi F, Kaneko S and Morisi S 2008 *JHEP* **0803** 063 (*Preprint* 0707.3032)
[8] Cooper I K, King S F and Luhn C 2010 *Phys.Lett.* **B690** 396–402 (*Preprint* 1004.3243)
[9] King S F, Peddie I N R, Ross G G, Velasco-Sevilla L and Vives O 2005 *JHEP* **07** 049 (*Preprint* hep-ph/0407012)
[10] Velasco-Sevilla L 2011 Contribution to the XIV Mexican School of Particles and Fields
[11] Altmannshofer W, Buras A J, Gori S, Paradisi P and Straub D M 2010 *Nucl. Phys.* **B830** 17–94 (*Preprint* 0909.1333)
[12] D'Ambrosio G, Giudice G, Isidori G and Strumia A 2002 *Nucl.Phys.* **B645** 155–187 (*Preprint* hep-ph/0207036)
[13] Olive K A and Velasco-Sevilla L 2008 *JHEP* **0805** 052 (*Preprint* 0801.0428)
[14] Ross G G, Velasco-Sevilla L and Vives O 2004 *Nucl. Phys.* **B692** 50–82 (*Preprint* hep-ph/0401064)
[15] de Medeiros Varzielas I and Ross G G 2006 *Nucl. Phys.* **B733** 31–47 (*Preprint* hep-ph/0507176)
[16] Calibbi L, Chun E J and Velasco-Sevilla L 2010 *JHEP* **1011** 090 (*Preprint* 1005.5563)
[17] Kadota K, Kersten J and Velasco-Sevilla L 2010 *Phys.Rev.* **D82** 085022 (*Preprint* 1007.1532)

[18] Ross G G and Vives O 2003 *Phys. Rev.* **D67** 095013 (*Preprint* hep-ph/0211279)
[19] Brignole A, Ibañez L E and Muñoz C 1997 (*Preprint* hep-ph/9707209)
[20] Misiak M, Asatrian H, Bieri K, Czakon M, Czarnecki A *et al.* 2007 *Phys.Rev.Lett.* **98** 022002 (*Preprint* hep-ph/0609232)
[21] Misiak M and Steinhauser M 2007 *Nucl.Phys.* **B764** 62–82 (*Preprint* hep-ph/0609241)
[22] Asner D *et al.* (Heavy Flavor Averaging Group) 2010 Long author list - awaiting processing (*Preprint* 1010.1589)
[23] Calibbi L, Jones-Perez J, Masiero A, Park J h, Porod W *et al.* 2010 *Nucl.Phys.* **B831** 26–71 (*Preprint* 0907.4069)
[24] King S F and Malinsky M 2006 *JHEP* **0611** 071 (*Preprint* hep-ph/0608021)

Ultra-High Energy Cosmic Rays and Neutrinos

Shigehiro Nagataki

Yukawa Institute for Theoretical Physics, Kyoto University, Oiwake-cho Kitashirakawa Sakyo-ku, Kyoto 606-8502, Japan

E-mail: nagataki@yukawa.kyoto-u.ac.jp

Abstract. In this paper, simulation of propagation of UHE-protons from nearby galaxies is presented. We found good parameter sets to explain the arrival distribution of UHECRs reported by AGASA and energy spectrum reported by HiRes. Using a good parameter set, we demonstrated how the distribution of arrival direction of UHECRs will be as a function of event numbers. We showed clearly that 1000-10000 events are necessary to see the clear source distribution. We also showed that effects of interactions and trapping of UHE-Nuclei in a galaxy cluster are very important. Especially, when a UHECR source is a bursting source such as GRB/AGN flare, heavy UHE-Nuclei are trapped for a long time in the galaxy cluster, which changes the spectrum and chemical composition of UHECRs coming from the galaxy cluster. We also showed that such effects can be also important when there have been sources of UHE-Nuclei in Milky Way. Since light nuclei escape from Milky Way in a short timescale, the chemical composition of UHECRs observed at the Earth can be heavy at high-energy range. Finally, we showed how much high-energy neutrinos are produced in GRBs. Since GRB neutrinos do not suffer from magnetic field bending, detection of high-energy neutrinos are very important to identify sources of UHECRs. Especially, for the case of GRBs, high-energy neutrinos arrive at the earth with gamma-rays simultaneously, which is very strong feature to identify the sources of UHECRs.

1. Introduction

Many observations on Ultra High-Energy Cosmic Rays (UHECRs) have been done, and there are many interesting information have been reported such as energy spectrum, arrival direction, and chemical composition. However, sources of UHECRs are still unknown. Why is it difficult to hunt the sources of UHECRs? The reason is magnetic fields. The trajectories of UHECRs are not straight by the magnetic field bending effect. This is serious especially for heavy nuclei. Also, the time delay effect due to the magnetic fields is serious. For example, if the source of UHECRs is a Gamma-Ray Burst (GRB), the arrival time difference between GRB and UHECRs can be as large as 1000yrs.

What can we do to identify the sources of UHECRs? More observations for higher statistics are important, of course. This is the matter for observers. From a theoretical point of view, there are two approaches. One is to simulate propagation of UHECRs assuming a lot of source distributions so that we can compare the simulations with observations. This approach will tell us what source distribution is good to explain observations of UHECRs. Another approach is to calculate secondaries (very high energy neutrinos and gamma-rays) from sources of UHECRs. Since the secondary particles are charge-neutral, they do not suffer from magnetic field effects. This is important because VHE neutrinos and VHE gamma-rays (if they do not suffer from

Journal of Physics: Conference Series **287** (2011) 012010 doi:10.1088/1742-6596/287/1/012010

considerable energy-loss during propagation) can arrive at the earth with a bursting source (such as a GRB), which tells us the GRB is the source of VHE-neutrinos/gamma-rays and their parent UHECRs (parent CRs will have about 20 times greater energy than VHE neutrinos/gamma-rays).

In this paper, we would like to show our studies to identify the sources of UHECRs. We show our study on numerical simulation of propagation of UHECRs from nearby galaxies in section 2, propagation of UHE-Nuclei in a galaxy cluster in section 3, propagation of UHE-Nuclei in Milky Way in section 4, and VHE-neutrinos from GRBs in section 5.

2. Numerical Simulations of Propagation of UHECRs

2.1. Formulation

This subsection provides the method of Monte Carlo simulations for propagating protons in intergalactic space. At first, we assume that the composition of UHECRs is proton, and inject an E^{-2} spectrum within the range of $10^{19.5} - 10^{22}$eV. A total of 10,000 protons are injected in each of 26 energy bins, that is, 10 bins per decade of energy for the cases of $l_c = 40$Mpc that is the correlation length of the extragalactic magnetic field. For other cases we inject 5000 protons in each of the bins.

Particles below $\sim 8 \times 10^{19}$eV lose their energies mainly by pair creations and above it by photopion production in collisions with photons of the CMB. The pair production can be treated as a continuous loss process considering its small inelasticity. As for the photopion process, we use the interaction length and the energy distribution of final protons as a function of initial proton energy, which is calculated by simulating the photopion production with the event generator SOPHIA.

Extra Galactic Magnetic Fields (EGMFs) are little known theoretically and observationally. So we studied propagation of UHECRs for various magnetic field models. The mean values of EGMFs are chosen to be 1nG, 10nG, 100nG. The coherent length of magnetic fields are set to be 1Mpc, 10Mpc, 40Mpc. We assume that the magnetic field is represented as the Gaussian random field with zero mean and a power-law (Kolmogorov) spectrum.

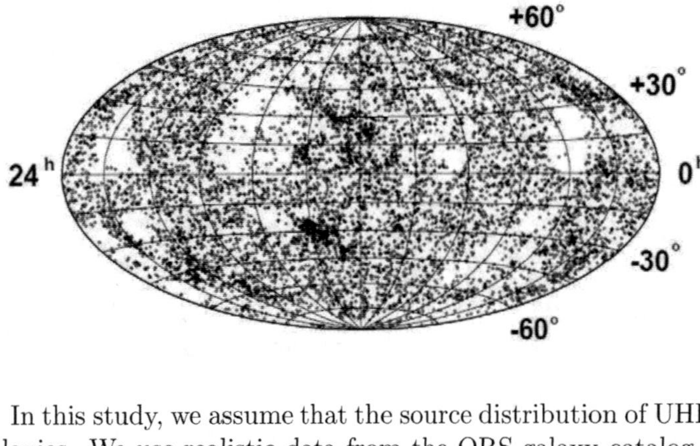

Figure 1. Distribution of galaxies in our ORS sample within 8000 km s^{-1}.

In this study, we assume that the source distribution of UHECRs is proportional to that of the galaxies. We use realistic data from the ORS galaxy catalog, which is a nearly full-sky survey. Compared with IRAS PSCz Survey data, Local Super Cluster members are more included (such as Virgo Cluster). The distribution of galaxies in ORS data is shown in Figure 1.

2.2. Results

By doing a lot of simulations, we found some good parameter sets to explain the observations of UHECRs. We show the results below.

In Figure 2, two-point correlation function predicted by a specific source scenario in the case that the number fraction (NF)$\sim 10^{-1.7}$ of the ORS galaxies more luminous than $M_{\rm lim} = -20.5$ is selected as UHECR sources. The number of simulated events is set to be 57 with energies of $10^{19.6} - 10^{20.3}$eV. The histograms represent the AGASA data in this energy range.

In Figure 3, energy spectra predicted by sources selected from the ORS galaxies more luminous than $M_{\rm lim} = -20.5$ in the case of $(B, l_c,$NF$)=(1,-20.5,10^{-1.7})$. They are fitted to the data of HiRes I detector (squares and error bars). The shaded regions represent 1σ error due to the source selection from our ORS sample.

Using the most favorite parameter sample, we demonstrate arrival directions of UHECRs above 4×10^{19}eV for various event number. Distribution of selected sources within 200Mpc is also shown as circles of radius inversely proportional to their distances. Only the sources within 100Mpc are shown with bold circles. The upper right panel shows the case that event number is 100 that is similar to the event number of PAO in 2010. As the event number increases (1500, 5000, 15000), we can see clear correlation between source distribution and arrival direction of UHECRs. Thus we conclude that about more than 10 times more events than PAO in 2010 are necessary to see the source distribution. South/North Auger, TA, and JEM-EUSO should provide us with such important data in the future. Please see [1, 2] for details.

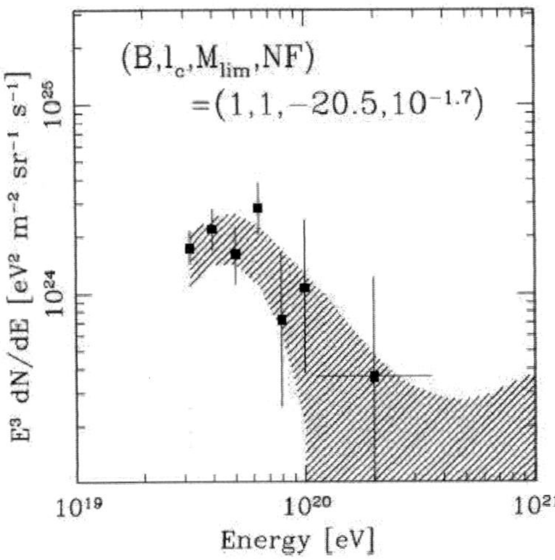

Figure 2. Two-point correlation function predicted by a specific source scenario in the case that the number fraction (NF)$\sim 10^{-1.7}$ of the ORS galaxies more luminous than $M_{\rm lim} = -20.5$ is selected as UHECR sources. The number of simulated events is set to be 57 with energies of $10^{19.6} - 10^{20.3}$eV. The histograms represent the AGASA data in this energy range.

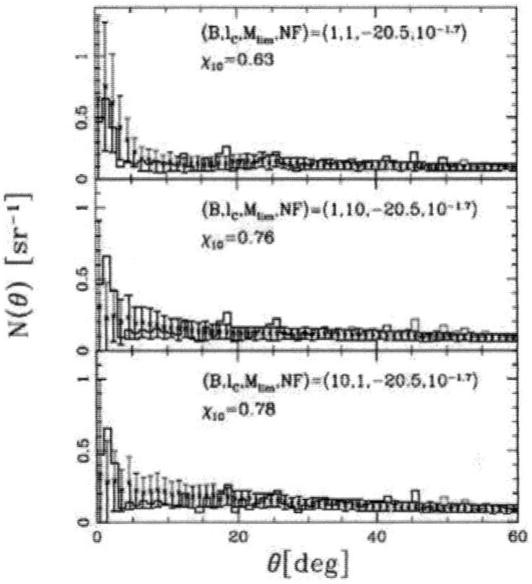

Figure 3. Energy spectra predicted by sources selected from the ORS galaxies more luminous than $M_{\rm lim} = -20.5$ in the case of $(B, l_c,$NF$)=(1,-20.5,10^{-1.7})$. They are fitted to the data of HiRes I detector (squares and error bars). The shaded regions represent 1σ error due to the source selection from our ORS sample.

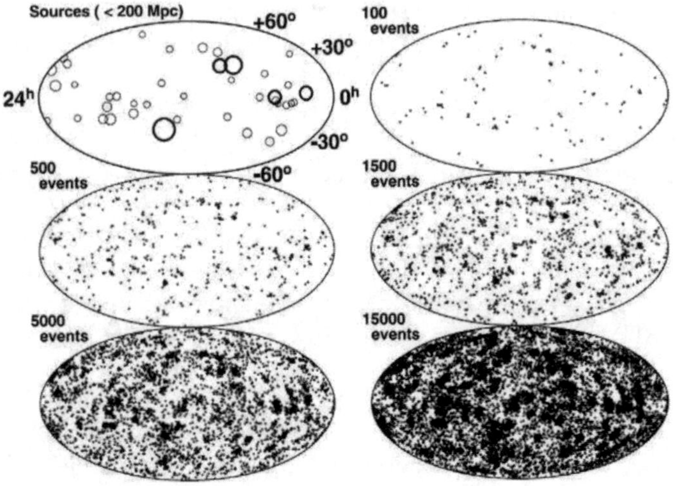

Figure 4. Arrival directions of UHECRs above 4×10^{19}eV predicted by a specific source scenario when 1/50 of the ORS galaxies more luminous than $M_{\rm lim} = -20.5$ are selected as UHECR sources. Distribution of selected sources within 200Mpc is also shown as circles of radius inversely proportional to their distances. Only the sources within 100Mpc are shown with bold circles.

3. Propagation of UHECRs in Galaxy Clusters

3.1. Formulation

In this section, we present our modeling of cluster of galaxies for UHECR propagation. This comprises a three-dimensional modeling of the magnetic field and infrared photon background, a consistent baryonic background profile and an adequate choice of sources for injection. We model our cluster magnetic field using the three dimensional outputs of MHD simulations run by Dubois and Teyssier (2008)[3]. The simulations were run including dark matter, gas, ultraviolet heating, hydrogen and helium cooling, star formation, and magnetic fields with the Adaptive Mesh Refinement code RAMSES.

At the center of a cooled core born in their simulation, we injected UHECRs assuming that the source is AGN. We assumed that the lifetime of $t_{\rm AGN}$=10My s and the chemical composition of UHECRs is same with solar abundances. The transport scheme in the magnetic field was adapted from Kotera and Lemoine (2008)[4]. The interactions of protons and nuclei with CMB, infrared, optical and ultraviolet photon backgrounds were mostly modeled according to the Monte Carlo methods of Allard et al. (2005)[5].

In Figure 5, we show evolution of the cosmic-ray spectrum in time for the case of a cool core cluster of central magnetic field $B_c = 10\mu$G. Each panel presents the spectrum at the time indicated at the top right-hand corner. The injection from the source (AGN) is assumed to begin at $t = 0$. The contribution of the different chemical components are shown. The thick black line is the total spectrum and the thin black line indicates the total flux obtained for an infinite AGN lifetime and an an integration of the flux over a Hubble time (stationary regime). The spectra are normalized to the value of the stationary flux obtained at $E = 10^{19}$eV.

Figure 5. shows that light nuclei such as proton and helium can escape from the cool core first, then heavy nuclei such as iron follow. Especially, the upper right panel shows that the UHECRs are composed of iron mainly. This may be interesting possibility to explain the observations of chemical composition of UHECRs reported by PAO. Please see [6] for details.

4. Propagation of UHECRs in Milky Way

In this section, we consider the possibility that UHECR (but lower than GZK cutoff) spectrum and composition may be explained due to the past GRBs in Milky Way. Stellar explosions in our own Galaxy can accelerate both protons and nuclei, but, while the protons leave the Galaxy promptly, the heavier and less mobile nuclei get trapped in the turbulent magnetic field and

Journal of Physics: Conference Series **287** (2011) 012010 doi:10.1088/1742-6596/287/1/012010

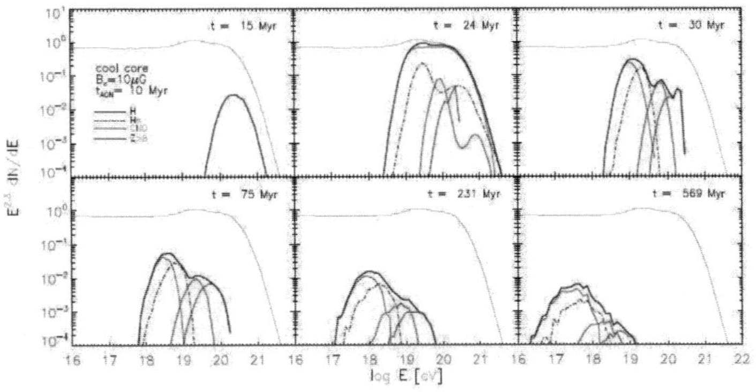

Figure 5. Evolution of the cosmic-ray spectrum in time, assuming a lifetime of $t_{\rm AGN}$=10Myr for the central AGN, for the case of a cool core cluster of central magnetic field $B_c = 10\mu$G. Each panel presents the spectrum at the time indicated at the top right-hand corner. The injection from the source (AGN) is assumed to begin at $t = 0$. The contribution of the different chemical components are shown.

linger longer than protons. As a result, the local density of nuclei is increased, and they bombard Earth in greater numbers, as seen by the Pierre Auger Observatory.

The ultrahigh-energy nuclei observed today have been trapped in the web of Galactic magnetic fields for millions of years, and their arrival directions have been completely randomized by the numerous twists and turns in the tangled field. However, we predict, the protons escaping from other galaxies should still be seen at the highest energies and should point back to their sources.

We did one-dimensional Monte Carlo simulations for the propagation of UHE-protons and Nuclei, assuming that the sources produce 90% protons and 10% iron, with identical spectra $\propto E^{-2.3}$, and that the source distribution traces the distribution of stars in the Galaxy. We used samples of 10^3 GRBs at random locations with time intervals of 10^5 years. The magnetic field was assumed to be 4μG, coherent over $l_0 = 0.2$ kpc domains. The overall power and the iron fraction were adjusted to fit the PAO data points. For each random sample, the fit parameters differ slightly, depending on the location of the latest or closest burst. The result is shown in Figure 6. As for the anisotropy of UHECRs, the result is shown in Figure 7. Please see [7] for details.

5. Ultra-High Energy Neutrinos from Gamma-Ray Bursts

In this section, high energy neutrino emission from GRBs is discussed. In this section, by using GEANT4, we calculate proton cooling efficiency including pion-multiplicity and proton-inelasticity in photomeson production. First, we estimate the maximum energy of accelerated protons in GRBs. Using the obtained results, neutrino flux from diffuse neutrino background is evaluated quantitatively. We also take account of cooling process of pion and muon, which are crucial for resulting neutrino spectra. We introduced the nonthermal baryon-loading factor $\epsilon_{\rm acc}$. We found that the obtained background can be comparable with the prediction of Waxman and Bahcall limit. Detection of high-energy neutrinos from GRBs will be one of the strong evidences that protons are accelerated to very high energy in GRBs. We also considered neutrino flux from X-ray flares and optical flares found by Swift satellite.

In figure 8, we show diffuse neutrino background from GRBs for $z_{\rm max} = 20$ on several SFR models. R-R means the case using Rowan-Robinson SFR. SF1-SF3 correspond to the models of

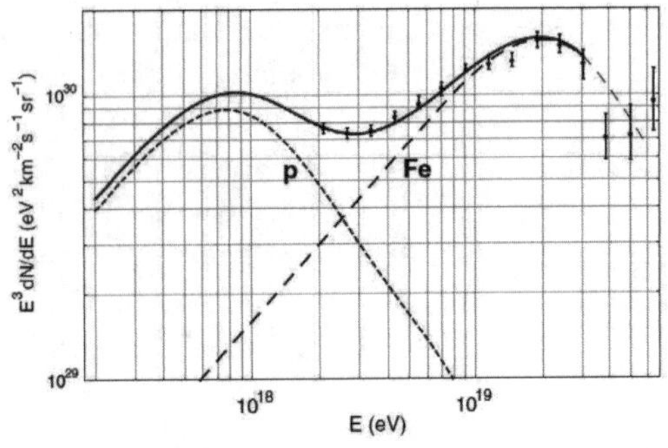

Figure 6. Predicted UHECR spectra, assuming that the sources produce 90% protons and 10% iron, with identical spectra $\propto E^{-2.3}$, and that the source distribution traces the distribution of stars in the Galaxy. We used samples of 10^3 GRBs at random locations with time intervals of 10^5 years. The magnetic field was assumed to be 4μG, coherent over $l_0 = 0.2$ kpc domains. The overall power and the iron fraction were adjusted to fit the PAO data points. For each random sample, the fit parameters differ slightly, depending on the location of the latest or closest burst. To model the spectrum at $E > 3 \times 10^{19}$eV, one has to account for energy losses and (proton) contribution of extragalactic sources, which we leave for future work.

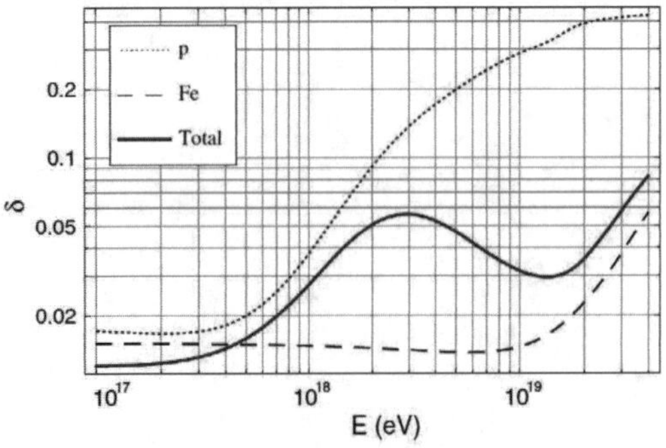

Figure 7. Galactocentric anisotropy for a source distribution that traces the stellar counts in the Milky Way, modeled by random generation of 10^3 bursts separated by time intervals of 10^5yr. The model parameters are the same as in Fig.6. Although the anisotropy in protons is large at high energies, their contribution to the total flux is small, so the total anisotropy < 10%, consistent with the observations. The latest GRBs do not introduce a large degree of anisotropy, as it would be in the case of UHE protons, but they can creat 'hot spots' and clusters of events.

Porciani and Madau. The left lines are for set A, while the right is for set B. For comparison, we show WB bounds (three dashed lines). The upper WB bound is for z-evolution of QSOs. The lower is for no z-evolution. $\epsilon_{\rm acc}$=10 and ϵ_B=1 are assumed. In figure 9, we show diffuse neutrino background from GRBs for $z_{\rm max} = 20$ with $\epsilon_{\rm acc}$=100. ϵ_B is set to be 0.1, 1, 10. All lines use the SF3 model. In figure 10, we show diffuse neutrino background from flares. X-ray flare (the upper dashed line), FUV-ray flare (the dotted line), and GRB are shown respectively. WB is Waxman-Bahcall bounds. Please see [8, 9] for details.

6. Summary and Conclusion

In this paper, simulation of propagation of UHE-protons from nearby galaxies is presented. We found good parameter sets to explain the arrival distribution of UHECRs reported by AGASA and energy spectrum reported by HiRes. Using a good parameter set, we demonstrated how the distribution of arrival direction of UHECRs will be as a function of event numbers. We showed clearly that 1000-10000 events are necessary to see the clear source distribution. We also showed that effects of interactions and trapping of UHE-Nuclei in a galaxy cluster are very important. Especially, when a UHECR source is a bursting source such as GRB/AGN flare, heavy UHE-Nuclei are trapped for a long time in the galaxy cluster, which changes the spectrum and chemical composition of UHECRs coming from the galaxy cluster. We also showed that such effects can be also important when there have been sources of UHE-Nuclei in Milky Way.

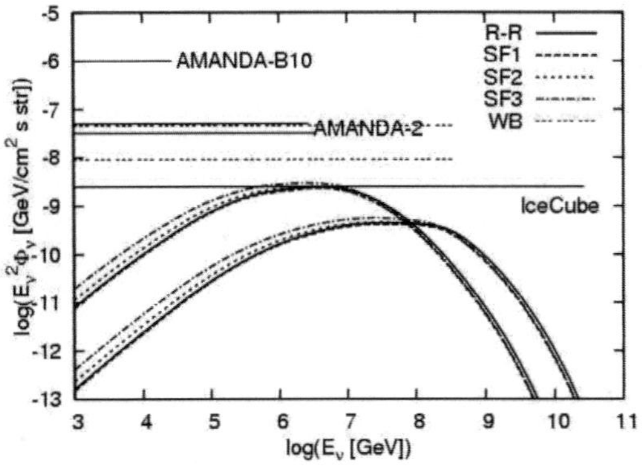

Figure 8. Diffuse neutrino background from GRBs for $z_{\max} = 20$ on several SFR models. R-R means the case using Rowan-Robinson SFR. SF1-SF3 correspond to the models of Porciani and Madau. The left lines are for set A, while the right is for set B. For comparison, we show WB bounds (three dashed lines). The upper WB bound is for z-evolution of QSOs. The lower is for no z-evolution. $\epsilon_{\rm acc}$=10 and ϵ_B=1 are assumed.

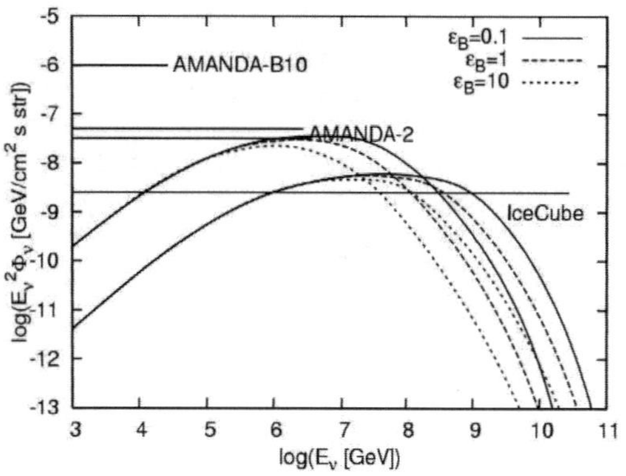

Figure 9. Diffuse neutrino background from GRBs for $z_{\max} = 20$ with $\epsilon_{\rm acc}$=100. ϵ_B is set to be 0.1, 1, 10. All lines use the SF3 model.

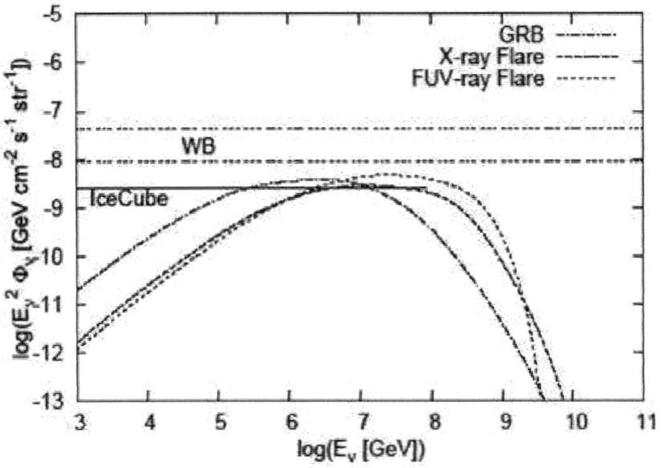

Figure 10. The diffuse neutrino background from flares. X-ray flare (the upper dashed line), FUV-ray flare (the dotted line), and GRB are shown respectively. WB is Waxman-Bahcall bounds.

Since light nuclei escape from Milky Way in a short timescale, the chemical composition of UHECRs observed at the Earth can be heavy at high-energy range. Finally, we showed how much high-energy neutrinos are produced in GRBs. Since GRB neutrinos do not suffer from magnetic field bending, detection of high-energy neutrinos are very important to identify sources of UHECRs. Especially, for the case of GRBs, high-energy neutrinos arrive at the earth with gamma-rays simultaneously, which is very strong feature to identify the sources of UHECRs.

References

[1] H. Yoshiguchi, S. Nagataki, S. Tsubaki, K. Sato, *Astrophys. J.* **586** (2003a) 1211.

[2] H. Yoshiguchi, S. Nagataki, K. Sato, *Astrophys. J.* **592** (2003b) 311.
[3] Y. Dubois, R. Teyssier, *Astron. Astrophys.* **482** (2008) L13.
[4] K. Kotera, M. Lemoine, *Physical Review D* **77** (2008) 023005.
[5] D. Allard, E. Parizot, A.V. Olinto, E. Khan, S. Goriely, *Astron. Astrophys.* **443** (2005) L29.
[6] K. Kotera, D. Allard, K. Murase, J. Aoi, Y. Dubois, T. Pierog, S. Nagataki, *Astrophys. J.* **707** (2009) 370.
[7] A. Calvez, A. Kusenko, S. Nagataki, *Physical Review Letters* **105** (2010) 091101.
[8] K. Murase, S. Nagataki, *Physical D* **73** (2006a) 063002.
[9] K. Murase, S. Nagataki, *Physical Review Letters* **97** (2006b) 051101.

Journal of Physics: Conference Series **287** (2011) 012011 doi:10.1088/1742-6596/287/1/012011

Searching for physics beyond the Standard Model with High Energy Colliders

Carmine Elvezio Pagliarone[1]
Università di Cassino & INFN Trieste, Italy

E-mail: *pagliarone@fnal.gov*

Abstract. In this paper we review a selection of recent results obtained in the area of experimental searches for physics beyond the Standard Model performed at present high-energy physics colliders. In particular we illustrate searches for squarks and gluinos, searches for scalar top and scalar bottom in different scenarios, searches for large extra dimensions, and more in general signature based searches. As no evidence for new physics have been found so far, 95% Confidence Level limits have been set for all the theoretical scenarios that have been investigated by present high energy colliders.

1. Introduction

The Standard Model of Particle Physics (SM) represents the simplest and most economical theory, which is able to describe jointly weak and electromagnetic interactions. At present it continues to survive all experimental tests, providing a remarkably successful description of known phenomena; some precision observable tests it at 10^{-3}. In spite of that, there are plenty of aspects that we do not understand yet and that may suggest the SM to be most likely, a low energy effective theory of spin-1/2 matter fermions interacting via spin-1gauge bosons. As matter of fact, the SM leaves many important questions unanswered. It is also widely acknowledged that, from the theoretical standpoint, the SM must be part of a larger theory, "beyond" the SM (BSM), which is yet to be experimentally confirmed.

2. Supersymmetry

An excellent candidate of a new theory, able to describe physics at arbitrarily high energies, may be the Supersymmetry. Supersymmetry (SUSY) is a larger space-time symmetry, which relates bosons to fermions so that, in the vast space of all viable physics theories, SUSY is not simply a point. Almost any theory can be supersymmetrized and the large array of choices, for spontaneous SUSY breaking (SB), just enhances these possibilities. A comprehensive SUSY search is almost impossible: i.e. the most general Minimal Supersymmetric extension of the SM (MSSM) counts 124 truly independent parameters [1]. The strategy is then to search for signals suggested by particular models in which

[1] To whom any correspondence should be addressed.

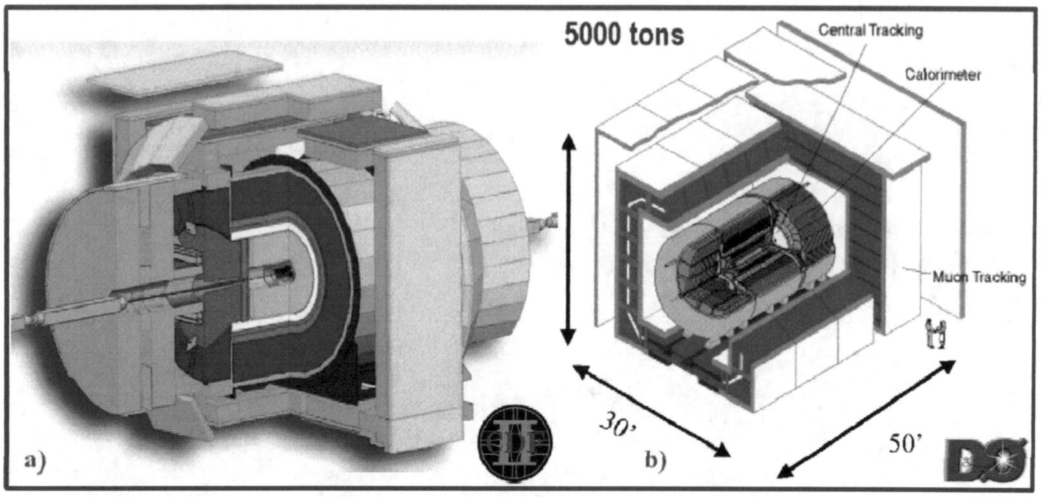

Figure 1: a) Schematic layout of the CDF-II and b) DØ Experiments.

theoretical assumptions are also adopted to reduce the number of free parameters to a few. Even if we don't have direct experimental evidences of SUSY, there are remarkable theoretical properties that provide ample motivation for its study. SUSY describes electroweak data equally well as SM but, in addition, allows the unification of the gauge couplings constants, the unification of the Yukawa couplings and do not require the incredible fine tuning, endemic to the SM Higgs sector. Naturally, SUSY cannot be an exact symmetry of the nature, as none of the predicted spin 0 partners of the quarks or leptons and none of the spin 1/2 partners of the gauge bosons have been observed so far. In Supersymmetry, fermions can couple to a sfermion and a fermion, violating lepton and/or baryon number. To avoid this problem, a discrete multiplicative quantum number, the R-parity was introduced: $R \equiv (-1)^{3B+L+2S}$. SUSY models can be constructed assuming either conservation (RP) or violation of this quantum number (RPV). The assumption of R-parity conservation has, for example, deep phenomenological consequences: SUSY particles can only be pair produced; the Lightest Supersymmetric Particle (LSP) does exist and it is stable and interacts very weakly with the ordinary matter, leading to a robust missing transverse energy signature $\not{E}_T$ [2]; the LSP may be a natural candidate for the dark matter. However, SUSY does not R-parity conservation and viable R-parity violating models can be built, for example, by adding explicitly B-violating and/or L-violating couplings to the SUSY Lagrangian. SUSY signals are of particular interest, as they provide a natural explanation for the Dark Matter, known to pervade our universe, and help us to understand the fundamental connection between particle physics and cosmology. A large amount of theoretical effort has been spent trying to understand the mechanism for soft Supersymmetry Breaking (SB) that produces the desired properties in the superpartner masses and interactions. The three most extensively studied mechanisms are: the Gravity Mediated Supersymmetry Breaking, the Gauge Mediated Supersymmetry Breaking and the Anomaly Mediated Supersymmetry Breaking.

2.1. Gravity Mediated Supersymmetry Breaking Models

Gravity Mediated Supersymmetry Breaking is a method of communicating supersymmetry breaking to the supersymmetric Standard Model through gravitational interactions. It was the first method

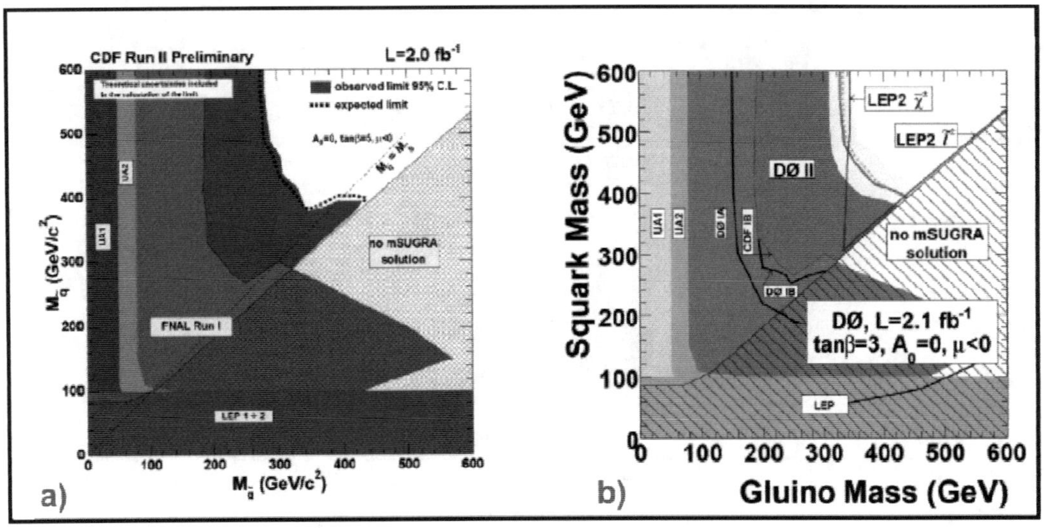

Figure 2: a) Results on the searches for squark and gluinos at Tevatron. a) Results coming from the CDF-II experiment, b) results from the DØ collaboration.

proposed to communicate supersymmetry breaking. In gravity mediated supersymmetry breaking models, there is a part of the theory that only interacts with the MSSM through gravitational interaction. This hidden sector of the theory breaks supersymmetry. Through the supersymmetric version of the Higgs mechanism, the gravitino, the supersymmetric version of the graviton, acquires a mass. After the gravitino has a mass, gravitational radiative corrections to soft masses are incompletely cancelled beneath the gravitino mass. It is currently believed that it is not generic to have a sector completely decoupled from the MSSM and there should be higher dimension operators that couple different sectors together with the higher dimension operators suppressed by the Planck scale. These operators give as large of a contribution to the soft supersymmetry breaking masses as the gravitational loops; therefore, today people usually consider gravity mediation to be gravitational sized direct interactions between the hidden sector and the MSSM. Between these models there are so called minimal supergravity (mSUGRA). This theoretical scenario is one of the most widely investigated models of particle physics due to its predictive power requiring only 4 input parameters and a sign, to determine the low energy phenomenology from the scale of Grand Unification.

2.2. Gauge Mediated Supersymmetry Breaking Models

The theories with gauge-mediated Supersymmetry breaking (GMSB) provide an interesting alternative scenario. In GMSB, for instance, the dynamical Supersymmetry breaking (DSB) is mediated by gauge interactions. In recent years, many mechanisms for DSB have been found and realistic models have been constructed. This class of models assumes that Supersymmetry is broken with a scale $\sqrt{F}$ in a sector of the theory, which contains heavy non-Standard-Model particles. This sector then couples to a set of particles with Standard Model interactions, called messengers, which have a mass of order M. The mass splitting, between the superpartners in the messenger multiplets, depends by $\sqrt{F}$ and the SUSY particles get their masses via gauge interactions, so there is no flavor changing neutral currents (FCNC). These theories have very distinctive phenomenological features. The typical SUSY spectra is

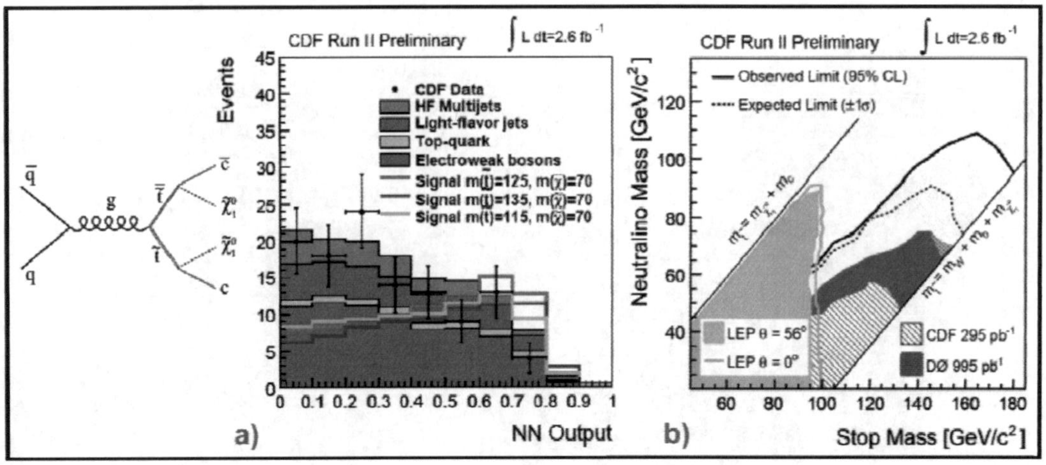

Figure 3: CDF search for scalar top in the case in which $\tilde{t}_1$ is the next-to-LSP: $p\bar{p} \rightarrow \bar{\tilde{t}}\,\tilde{t}_1 \rightarrow c\tilde{\chi}_1^0\bar{c}\tilde{\chi}_1^0$.

different from those predicted in the SUGRA models; the LSP is the gravitino $\tilde{G}$, the next lightest supersymmetric particle (NLSP) has a lifetime that can vary strongly from model to model (1μ< cτ< several km) and decays into a $\tilde{G}$.

2.3. Anomaly Mediated Supersymmetry Breaking Models
The Anomaly Mediated Supersymmetry Breaking is a special type of gravity mediated supersymmetry breaking that results in supersymmetry breaking being communicated to the supersymmetric Standard Model through the conformal anomaly.

2.4. Experimental searches for SUSY at Tevatron
The Tevatron collider is a circular particle accelerator at the Fermi National Accelerator Laboratory (FNAL) close to Batavia, Illinois (USA). This machine has been built in order to accelerate beams of protons and anti-protons and make them collide at center of mass energy of 1.80, during the Run I (1992-1996) and 1.98 TeV, during the Run II (2001-2011) [2]. Two experiments have been equipped in order to study the collision products: the CDF-II detector and the DØ detector. In Fig. 1.a and in Fig. 1.b we give a schematic layout of both these Tevatron experiments. A particle leaving the interaction point (IP) will see first a set of silicon detectors, mainly used to tag jets originated by the b-quarks, and a bigger tracking volume. Both these tracking detectors are immersed in a magnetic field to avoid charge and momentum measurement. An electromagnetic and hadronic calorimeter follows the internal silicon and conventional trackers. Outside the magnetic field there are a set of muon detectors. In order to suppress the strong background, coming from QCD processes, and to be able to collect only those events that may be considered of interest, dedicated trigger systems have been developed in both the experiments, for the different channels under investigation.

2.4.1. Searches for squarks and gluinos
The search for squarks ($\tilde{q}$) and gluinos ($\tilde{g}$) is one of the most relevant SUSY searches at Tevatron. As matter of fact, squarks and gluinos are expected to be copiously produced via strong interaction.

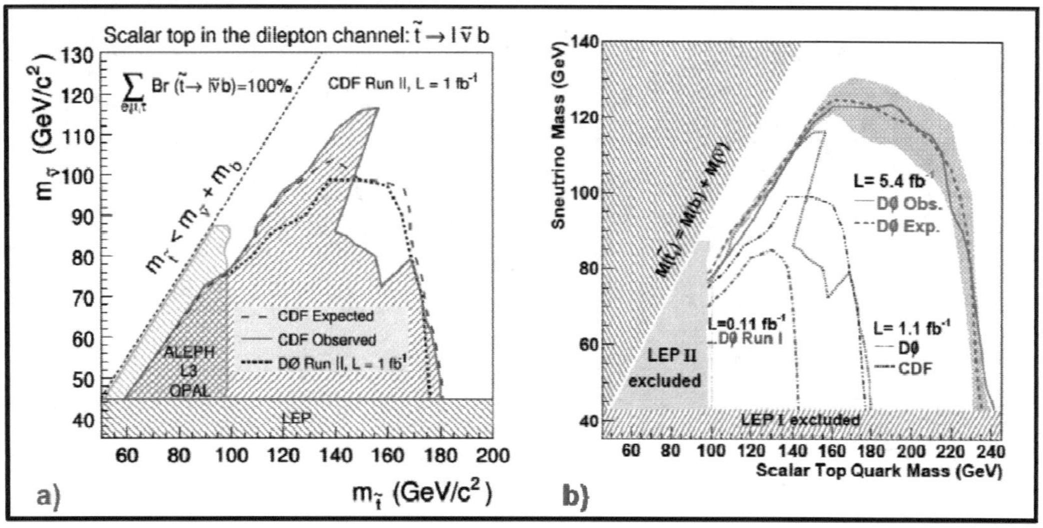

Figure 4: Search for scalar top decaying in two leptons: CDF-II (a) and DØ (b) results.

Due to the large production cross-sections, the inclusive production of squarks and gluinos have always been considered one of the most promising discovery channels for SUSY. Squarks and gluinos are produced at Tevatron trough one of the following three processes: $p\bar{p} \to \tilde{q}\tilde{q} \to q\tilde{\chi}^0 q\tilde{\chi}^0 + X$, $p\bar{p} \to \tilde{g}\tilde{q} \to q\bar{q}\tilde{\chi}^0 q\tilde{\chi}^0 + X$ and $p\bar{p} \to \tilde{g}\tilde{g} \to q\bar{q}\tilde{\chi}^0 q\bar{q}\tilde{\chi}^0 + X$. Because of the presence of two neutralinos ($\tilde{\chi}^0$) in the final states, neutralino that behaves from an experimental point of view as a neutrino, the signature for these searches is 2 or 4 jets plus a large amount of missing transverse energy ($\not{E}_T$, MET) produced by the two neutralinos. The limits obtained by the CDF-II [3] and by the DØ experiment [4] are given in Figure 2.a and in Figure 2.b. The CDF-II experiment, by using a total integrated luminosity of 2 fb^{-1}, have been able to excludes, at 95% Confidence Level, gluinos with masses below 208 GeV/c^2, for any value of the squark masses. DØ performed a similar analysis based on a data set of 2.1 fb^{-1}. In this case gluino masses below 308 GeV/c^2 have been excluded at 95% CL for all squark masses. Both CDF-II and DØ ruled out the existence of squarks with masses below 380 GeV/c^2 for all gluino masses. In mSUGRA, region that was not accessed by LEP experiments have been also excluded by these two experiments.

2.4.2. Searches for scalar top

Search for scalar top ($\tilde{t}$) is particularly relevant since, in many SUSY models the top-squark mass eigenstate $\tilde{t}_1$ is expected to be the lightest squark viable in nature. As matter of fact, the strong Yukawa coupling, between top/stop and Higgs fields gives rise to potentially large mixing effects and large mass splitting between the two scalar to mass eigentates: $\tilde{t}_1$ and $\tilde{t}_2$ returning the following mass hierarchy: $m_{\tilde{t}_1} < m_{\tilde{t}_2}$. The CDF-II collaboration has performed a search for scalar top ($\tilde{t}_1$), in the case in which the stop is next to Lightest Supersymmetric Particle. In this specific scenario the decay channel $p\bar{p} \to \bar{\tilde{t}}\tilde{t}_1 \to c\tilde{\chi}_1^0 \bar{c}\tilde{\chi}_1^0$ is open. The results for such a search are given in Figure 3. The analysis have

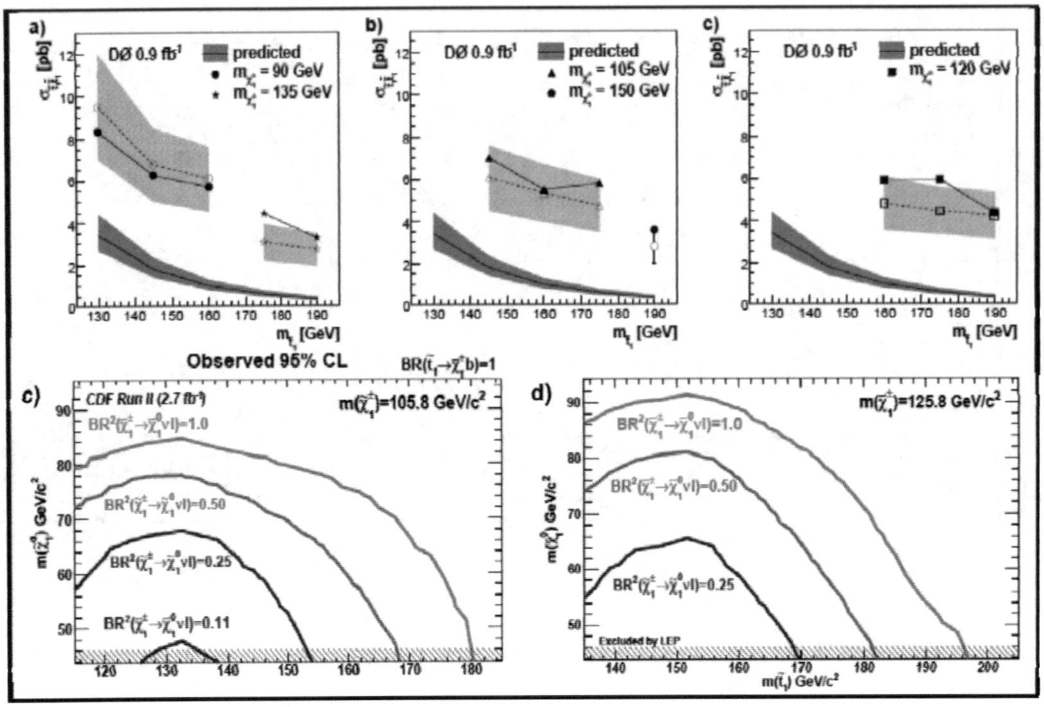

Figure 5: a), b), c) DØ limits in terms of the 3-dimensional space: $M_{\tilde{t}} \times M_{\chi_1^\pm} \times M_{\chi_1^0}$; d) CDF-II limits in the $M_{\tilde{t}_1}$ versus $M_{\chi_1^0}$ plan for two different chargino masses and $Br(\chi_1^\pm \to \chi_1^0 \nu \ell)$.

been performed by using a total integrated luminosity of about 2.6 fb^{-1} and represented as an exclusion region in the stop mass versus neutralino mass plane (see Fig. 3).

2.4.3. Searches for scalar top in the dilepton channel

If sneutrino is lighter than the stop, the dominant decay channel is: $\tilde{t}_1 \to b\ell\bar{\nu}$ with $\bar{\nu} \to \nu\tilde{\chi}_1^0$. In this case the final states are quite similar to the ones that is possible to have in the top dilepton decay. Even if the signature is similar, anyhow the kinematic is different and leptons show up to be softer. The results obtained by searching for the scalar top in the dilepton channel are given in Fig. 4.a [5] and in Fig. 4.b [6].

2.4.4. Searches for scalar top in top-like events

In the case in which chargino is lighter than the stop ($M_{\tilde{\chi}^\pm} < M_{\tilde{t}}$), the dominant stop decay becomes the following: $\tilde{t} \to b\tilde{\chi}^+$ giving rise to a top-like decay scalar top decay: $p\bar{p} \to \tilde{t}\bar{\tilde{t}} \to b\bar{b}\tilde{\chi}^+\tilde{\chi}^-$. As charginos decay in one the two following ways: $\tilde{\chi}^\pm \to \ell^\pm\tilde{\chi}^0$ or $\tilde{\chi}^\pm \to qq'\tilde{\chi}^0$ we end up with the same $t\bar{t}$ finals states: dilepton, lepton + jets, all hadronic. This SUSY scenario has been studied from both the Tevatron experiments. DØ searched for scalar top in the lepton + jets channel [7]. CDF-II studied the stop decay in the dilepton channel. [8] The results of both the analysis are shown from Fig. 5.a to Fig. 5.d.

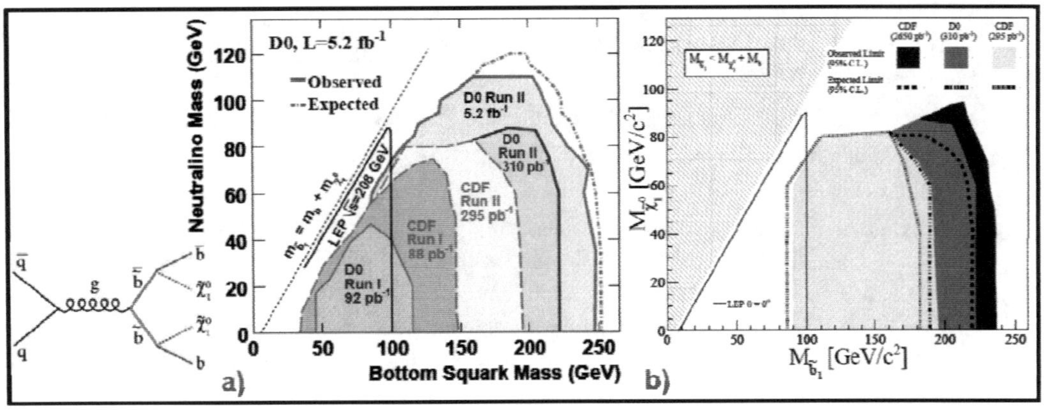

Figure 6: Search for a sbottom decaying in $\tilde{b} \rightarrow b\tilde{\chi}_1^0$ at Tevatron.

DØ extracted the limits in terms of the 3-dimensional space: $M_{\tilde{t}} \times M_{\chi_1^\pm} \times M_{\chi_1^0}$ (see Fig. 5.a, Fig. 5.b and Fig. 5.c). The CDF-II results are expressed in the $M_{\tilde{t}_1}$ versus $M_{\chi_1^0}$ plan, for two different chargino masses and $Br(\chi_1^\pm \rightarrow \chi_1^0 \nu \ell)$.

2.4.5. Searches for scalar bottom

At large $\tan(\beta)$, the scalar bottom ($\tilde{b}$) may be the lightest colored particle. Such sbottom quark can decay as follow: $\tilde{b} \rightarrow b\tilde{\chi}_1^0$ giving rise to final states containing two b-jets and large missing transverse energy ($p\bar{p} \rightarrow \bar{\tilde{b}}\tilde{b} \rightarrow b\bar{b}\tilde{\chi}_1^0\tilde{\chi}_1^0$). The limits, coming from both the Tevatron searches can be seen in Figure 6.a and in Figure 6.b [9, 10]. Both the analysis has excluded sbottom masses up to 250 GeV/c^2.

2.4.6. GMSB Models searches at Tevatron

In the gauge mediated SUSY breaking (GMSB) scenario, the gravitino is the LSP and the next-to-lightest SUSY particle (NLSP) may be the lightest neutralino, which decays to a gravitino and a photon: $\chi_1^0 \rightarrow \gamma\tilde{G}$. At Tevatron, pair production of SUSY particles decaying to a neutralino NLSP with negligible lifetime would therefore lead to final states with two acoplanar photons and missing transverse energy. Both CDF and DØ Collaborations have performed searches for such events. No excess was found in the data [11]. The process is then used to set a limit on the GMSB scale Λ, which also gives the scale of the gaugino masses. For a $\tilde{\chi}_1^0$ lifetime of 0 ns, neutralino masses below 149 GeV/c^2 have been excluded.

2.5. Higgsless Models

Since the Higgs mechanism was first introduced, in early 1960s by Peter W. Higgs, there have been several alternatives proposed. All of the alternative mechanisms use strongly interacting dynamics to produce a vacuum expectation value that breaks electroweak symmetry. A partial list of these alternative mechanisms includes: Technicolor models, Extra-dimensional Higgsless models, Models of composite W and Z vector bosons, Top quark condensate and several other models.

Journal of Physics: Conference Series **287** (2011) 012011 doi:10.1088/1742-6596/287/1/012011

2.6. *Models with Extra Dimensions*

In the Standard Model of we assume, commonly, that effects of gravity can be neglected, because the scale where such effects become large is the Planck Scale. The question of why the 4-dimensional Planck Scale, $G_{Pl}^{-1/2} \approx 10^{19}$ GeV, is much larger than the electroweak (EWK) scale, $G_{Pl}^{-1/2} \approx 10^{19}$ GeV, is an outstanding problem in contemporary physics. The hierarchy problem is in the essence the difficulty to explain the large disparity between these two numbers. Motivated in part by naturalness issues, numerous scenarios have emerged recently, that address the hierarchy problem within the context of the old idea that some part of the physical world (i.e. the SM-world) is confined to a *brane* in a higher dimensional space. Although Supergravity theories were formulated up to 11 dimensions and Superstring theories in 10 dimensions were known since the 70's, the idea to extend this extra spatial dimension paradigm (ESD) to other contexts, received a new impulse in the recent past years. As we don't experience in our everyday world, more then 3 spatial dimensions, we have to assume that any possible ESD is hidden. There is a simple and elegant way to hide possible extra spatial dimensions: the compactification. The result is achieved by assuming, for example, that the extra dimensions form, at each point of the 4-dimensional space, a torus of volume $(2\pi)^D R_1 R_2 \dots R_D$. In this way, it is possible to allow the gravity to live in the D large extra dimensions, the *bulk*, while the SM fields will lie on a $3-D$ surface, the *brane*. In presence of a compactified extra spatial dimension y, a field $\phi(x_\mu, y)$ of mass m_0 is periodic over y and can be Fourier developed as follow:

$$\phi(x_\mu, y) = \sum_{k=-\infty}^{+\infty} e^{\frac{iky}{R}} \phi^{(k)}(x_\mu) \qquad [1]$$

where R is the radius of the compact ESD. The $4D$ terms $\phi^{(k)}(x_\mu)$ are the Kaluza-Klein states,(KK) also called modes or excitations. The mass of each KK mode is then expressed by the formula:

$$m_k^2 = m_0^2 + k^2/R^2 \qquad [2]$$

If M_F is the actual fundamental scale of gravitational interactions and if V_{extra} is the volume of the extra dimensional $D-fold$, then, by Gauss's law, at distances larger than the inverse mass of the lightest KK mode in the theory, the gravitational force will follow an inverse square law with an effective coupling of:

$$M_{Pl}^{-2} = M_F^{-(D+2)} V_{extra}^{-1} \qquad [3]$$

We see indeed that gravity becomes strong in the full $4+D$-dimensional space at a scale M_F of few TeV, which is far below the conventional Planck Scale ($M_{Pl}^2 = M_F^{D+2} V_{extra}$). At the present there are many models, which assume the existence of extra spatial dimensions predicting the appearance of new physics signatures that can be probed at energy scale above 1 TeV. Most of the models fall, between the others, into one of the three following classes: Large Extra Dimension scenario (LED), KK gauge bosons, warped extra spatial dimensions. The large extra dimension scenario (LED) started with the works of Arkani-Hamed, Dimopoulos and Dvali (ADD). In this model the SM particles live on a $3+1$-dimensional space (3-brane) while the gravity is free to propagate in higher- dimensional space, extra dimensions. This model predicts essentially the emission and exchange of large Kaluza-

Journal of Physics: Conference Series **287** (2011) 012011 doi:10.1088/1742-6596/287/1/012011

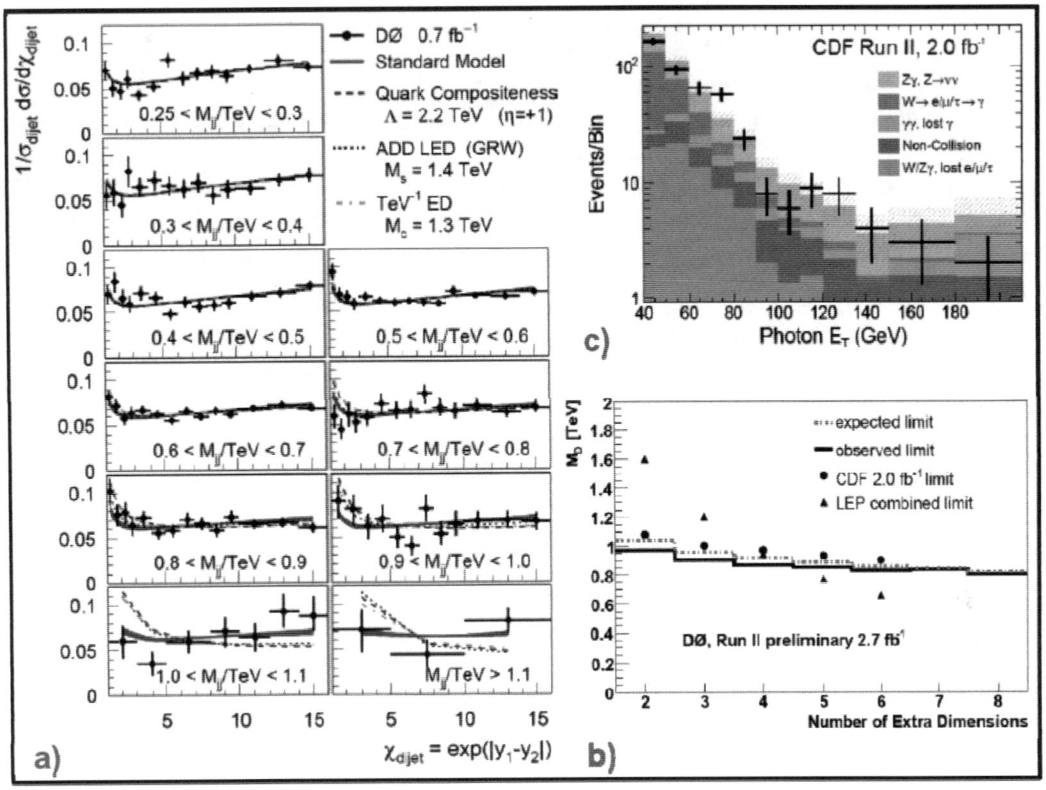

Figure 7: Searches for Large Extra Dimension at Tevatron; a) DØ indirect search, b) and c) results coming from the CDF-II Experiment.

Klein towers of gravitons that are finely spaced in mass. The ADD model was first proposed to solve the hierarchy problem by requiring the compactified dimensions to be of very large size. A second possibility comes from all those models where the extra spatial dimensions are of TeV scale size. In these classes of models there are KK excitations of the SM gauge fields with masses of the order a TeV, which can show up in collider experiments as resonances. Another approach, for extra spatial dimensions, has been proposed by L. Randall and R. Sundrum (RS models or WED). In this scenario two $4D$ branes with tension V and V' are situated in the position, $y=0$ and $y=\pi r_c$, of a $5D$ bulk with cosmological constant λ, where the gravitation lives. One of the interesting consequence of this assumptions is the fact that the fundamental mass scale, on the brane at $y=0$ is then red-shifted by a factor $e^{-2k|y|}$ (warp factor) on the other brane at $y=\pi r_c$. In this way the EWK scale ($O(1\ TeV)$ can be produced from the Planck Scale. In the RS models the $4D$ Planck mass may be expressed as follow:

$$M_{Pl}^2 = \frac{M_5^3}{k}\left[1-e^{-2kr_c\pi}\right] \qquad [4]$$

2.6.1. Experimental searches for Extra Dimension at Tevatron

The experimental limits coming from Tevatron experiments are given both for direct and indirect LED searches in Figure 7 [12, 13].

2.7. *Black Hall production and related signatures*

Black holes (BH) have always been considered, since their introduction, objects of great interest both in theoretical physics both in astrophysics. In the last decade the interest for BH production and observation have been extensively studied in the context of collider experiments. As a matter of fact, BH can be produced in particle collisions if the center of mass energy is above the Planck Scale ($\sqrt{s} > M_{Pl}$). In LED models the Planck Scale can be effectively O(1 TeV), opening up the interesting possibility of producing and studying BH using collider experiments. The observability of such BH at future colliders will depend on the value of the fundamental Planck Scale. At Hadron colliders, where it is possible to reach the highest center of mass energies, compared to other machines, the BH production cross section can be written as follow:

$$\sigma_{pp\to BH}(M_{\min}, s) = \sum_{i,j}\int_{\tau_m}^{1} d\tau \int_{\tau}^{1}\frac{dx}{x} F_i(x)F_j(\tau / x)\sigma_{ij\to BH}(\tau s) \qquad [5]$$

where *i* and *j* are the two colliding partons, *x* and *τ/x* are the momentum fractions of *i* and *j* and *F* are the parton distribution functions. If we assume the Thorne's hoop conjecture, that states that horizons form when and only when a mass M is compacted into a region whose circumference in every directions is less than $2\pi R_{BH}(M)$, we obtain the important result that the black hole production cross section is:

$$\sigma_{ij\to BH}(s) \approx \pi R_{BH}^2(\sqrt{s}) \qquad [6]$$

The precise mass of the BH, formed in a collision, depends on the amount of energy and matter, which becomes trapped behind the event horizon. If the scale of gravity is the TeV scale, BH production could be a dominant process at hadron colliders beyond LHC. For example for $M_{Pl} = 1\ TeV$ and $D = 10$, at a Very Large Hadron Collider (VLHC) where center of mass energy is supposed to be *100 TeV* assuming an integrated luminosity per year of $100\ fb^{-1}yr^{-1}$, black halls, of mass around *10 TeV*, should be produced with a rate of *1 kHz*.

Once produced black holes decay. The decay process is rather complex and occurs in several stages. A *Balding phase*, in which there is emission of gauge and gravitational radiation that will settle down the BH to a symmetrical rotating object with a growing horizon. A *Spin-down phase*, in which the BH Hawking radiates, emitting quanta, having angular moment $\ell \approx 1$. A *Schwarzschild-Hawking Evaporation phase* in which the spin-down phase leaves a Schwarzschild black hall that continues to radiate Hawking radiation. Instant quanta are emitted with a thermal spectrum around Hawking temperature (TH). A *Planck phase* in which, once the BH reach the Planck mass $M \approx M_{Pl}$, the black hall completely decays emitting few quanta with energies $O(M_{Pl})$.

Because of the large cross section, multiplicity and visible energy, black halls, at hadron colliders, give rise to very spectacular events. The BH production and decay is characterized by the following specific signatures: suppression of hard perturbative scattering processes at energy in which the BH production start to dominate; very large production cross sections that grows with the energy; high multiplicity events (for $M_{BH} \approx 10\ TeV$, $D = 10$ the expected multiplicity is around 50) with visible transverse energy of the order of ~1/3 of the total energy; high sphericity events because of the small BH boost: $\langle\gamma\beta\rangle < 1$; a ratio of ~1/5 between leptonic and hadronic activity, because of the

Schwarzschild- Hawking evaporation phase.

2.8. *Technicolor*

Between the theories beyond the SM, another interesting theory as two why matter has mass is the Technicolor theory. In this class of theoretical scenarios, instead of introducing elementary Higgs bosons, Technicolor models hide electroweak symmetry and generate masses for the W and Z bosons through the dynamics of new gauge interactions. Although asymptotically free at very high energies, these interactions must become strong and confining (and hence unobservable) at lower energies that have been experimentally probed. Technicolor differs from Supersymmetry in several ways. It only has two particles, called Techniparticles, or Techniquarks, that are bound together by a new force called Technicolor. Technicolor does not require the multitude of new particles that SUSY does. Technicolor may be simpler then Supersymmetry but it is not without flaws: the theory fails to explain why the forces unit at high energy. Technicolor theories naturally contain dark matter candidates. Almost certainly, models can be built in which the lowest-lying technibaryon, a technicolor-singlet bound state of technifermions, is stable enough to survive the evolution of the universe.

2.9. *Compositeness*

Quarks may not be fundamental particles, but rather an agglomeration of smaller constituents called preons. These features are visible above a characteristic energy scale, the so called compositeness scale Λ, below which, quarks appear to be point like. The parameter Λ characterizes both the strength of the preon coupling and the physical size of the composite scale. If fermions are composite particles made up of more basic constituents, characteristic phenomenological effects could be observable at the particle colliders. If the scale of compositeness is sufficiently low, narrow resonant states of excited fermions could be produced on shell. If, however, the compositeness scale is much larger than the center of mass energy of the colliding partons, the manifestation of compositeness will be an effective 4-fermion contact interaction. In a commonly used model, the contact interaction is described as an interaction between two left-handed quark currents:

$$L_{ffff}(\Lambda) = \frac{g^2}{2(\Lambda_{LL}^{+})^2} \eta \bar{\Psi}_f^L \gamma_\mu \Psi_f^L \bar{\Psi}_f^L \gamma^\gamma \Psi_f^L \qquad [7]$$

where $\eta = \pm 1$ is a constructive or destructive interference sign. Λ is defined in such a way that $g^2/4\pi = 1$. Note that, the possible discovery of contact interaction alone is not enough to prove compositeness because other possible new phenomena can be described by a contact interaction Lagrangian. Early measurements of both jet p_T and dijet mass distributions at the Tevatron showed an excess in the rate above QCD expectations. These effects were eventually explained by larger than expected high-x tail in the gluon Parton Distribution Function (PDF). After these measurements, less PDF-sensitive studies based on ratio of cross sections and angular distributions have been developed by LHC experiments in order to be able to search for contact interactions both in the dijet and in the dimuon channel.

2.10. *Global Model independent searches*

The search for physics beyond the SM can be widely performed by searching for anomalies in the data sets classified by their final states. Then event classes are created according to the number and types of objects detected in the final states. A comparison is then performed between the numbers of events in each class and the expected one coming from purely SM processes. In Figure 8 we give a couple of

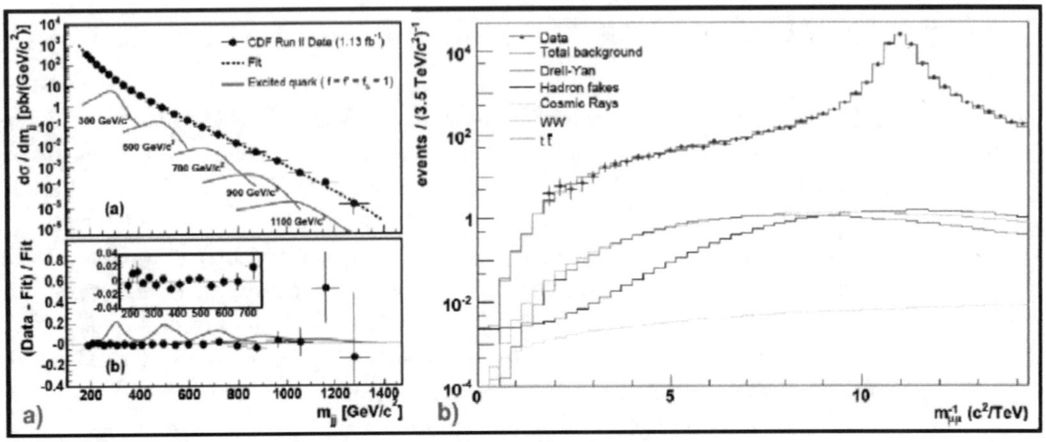

Figure 8: a) Distribution of Dijet mass, b) Distribution of the di-muon inverse mass, in CDF-II data.

example of such a searches where the dijet mass and the inverse di-muon mass ($M^{-1}_{\mu^+\mu^-}$) is compared with the SM expectations [14, 15].

3. Conclusions

This paper reviewed some of the aspects discussed, in the plenary talk, on physics beyond the Standard Model at present colliders. A selection of recent results, obtained in the area of experimental searches, for physics beyond the Standard Model, has been given. Despite of the intensive searches performed on data, there are no evidence yet, no convincing signs, of physics beyond the SM at the energy scale that have been possible to explore so far. Tevatron have been testing, in the last two decades, models with increasing sensitivity. A new era, the LHC era is finally started and very soon we may get closer to the understanding which direction we have to take to further increase our knowledge on the Nature.

4. Acknowledgments

I would like to heartily thank the organizers of the XIV School of Particles and Fields in Morelia, for asking me to give this review talk about recent experimental results on the physics beyond the SM. I would like to thank in particular A. Ayala, A. Bashir, G. Contreras, I. León, G. Murguía, A. Raya and M. E. Tejeda for their very kind hospitality and for providing a greatly stimulating and lively environment in which have been possible to wonderfully work, exchange ideas and relax.

5. Bibliography

[1] N. Nakamura *et al.* [Particle Data Group], J. Phys. G **37**, 075021 (2010).
[2] C. Pagliarone, *Collider Physics an experimental introduction*, to appear on the Proceedings of the XIV Mexican School on Particle & Fields, Morelia, Mexico, November 2010.
[3] A. Abulencia *et al.* [CDF-II Collaboration], Phys.Rev.Lett. **102**, 121801 (2009).
[4] V. M. Abazov *et al.* [DØ Collaboration], Phys.Lett. **B660** 449-527 (2008).
[5] T. Aaltonen *et al.* [CDF-II Collaboration], Phys.Rev. **D82** 092001 (2010).
[6] V. M. Abazov *et al.* [DØ Collaboration], arXiv:1009.5950.
[7] V. M. Abazov *et al.* [DØ Collaboration], Phys.Lett. **B674**, 4-10 (2009).

Journal of Physics: Conference Series **287** (2011) 012011 doi:10.1088/1742-6596/287/1/012011

[8] T. Aaltonen *et al.* [CDF-II Collaboration], Phys.Rev.Lett. **103**, 251801 (2010).
[9] V. M. Abazov *et al.* [DØ Collaboration], Phys.Lett. **B693**, 95-101,2010.
[10] T. Aaltonen *et al.* [CDF-II Collaboration], Phys.Rev.Lett. **105**, 081802 (2010).
[11] V. M. Abazov *et al.* [DØ Collaboration], Phys. Lett. **B659** (2008).
[12] T. Aaltonen *et al.* [CDF-II Collaboration], Phys.Rev.Lett. **101**, 181602 (2009).
[13] V. M. Abazov *et al.* [DØ Collaboration], Phys. Rev. Lett. **102**, 051601 (2009).
[14] T. Aaltonen *et al.* [CDF-II Collaboration], Phys.Rev. **D79**, 112002 (2009).
[15] T. Aaltonen *et al.* [CDF-II Collaboration], Phys.Rev.Lett. **102**, 031801 (2009).

A Review of Noncommutative Field Theories

Victor O. Rivelles

Instituto de Física, Universidade de São Paulo, Rua do Matão Travessa R 187, 05508-090 São Paulo, Brazil

E-mail: rivelles@fma.if.usp.br

Abstract. We present a brief review of selected topics in noncommutative field theories ranging from its revival in string theory, its influence on quantum field theories, its possible experimental signatures and ending with some applications in gravity and emergent gravity.

1. Introduction

The uncertainty principle of quantum mechanics $\Delta x^i \Delta p^j > \delta^{ij}\hbar$ teach us that if we want to probe short distances we need high energies. Now general relativity teach us that if we keep increasing the energy to explore smaller regions a black hole will be formed as soon as we goes beyond the Schwarzschild radius. Since no information can scape from a black hole it seems that there is a lower bound for measuring lengths! This means that there must exist an uncertainty relations for the coordinates

$$\Delta x^\mu \Delta x^\nu > \theta^{\mu\nu},$$

where $\theta^{\mu\nu}$ has dimensions of (lenght)^2, so that coordinates no longer commute. This simple argument was presented some years ago [1] but the idea of noncommutativity is much older. In the late 30s Heisenberg was struggling with the ultraviolet divergences of QED and thought that noncommuting coordinates could provide a cutoff to regulate them. He then wrote to Peierls, who told to Pauli, who told to Oppenheimer, who asked one of his students, Snyder, to think about it. Then in 1947 Snyder wrote a paper where he considered the following commutation relations [2]

$$\begin{aligned}
[x^\mu, x^\nu] &= i\ell^2\hbar^{-1}(x^\mu p^\nu - x^\nu p^\mu), \\
[x^\mu, p_\nu] &= i\hbar\delta^\mu_\nu + i\ell^2\hbar^{-1}p^\mu p_\nu, \\
[p_\mu, p_\nu] &= 0,
\end{aligned} \tag{1}$$

generalizing the usual ones and with a new constant ℓ. However the renormalization program for QED started to become successful so further studies in noncommutativity were discouraged.

Even so we can ask how we can do field theory with noncommutative coordinates satisfying a simpler set of commutation relations

$$\begin{aligned}
[\hat{q}_i, \hat{p}_j] &= i\hbar\delta_{ij}, \\
[\hat{q}_i, \hat{q}_j] &= i\theta_{ij}, \\
[\hat{p}_i, \hat{p}_j] &= 0,
\end{aligned}$$

with θ_{ij} a constant antisymmetric matrix. We can make an association of fields $\phi(x)$ in the usual space with commutative coordinates x^i to operator valued objects $\hat{\Phi}(\hat{q})$ as

$$\hat{\Phi}(\hat{q}) = \int dp\ e^{ip\hat{q}}\tilde{\phi}(p),$$

where $\tilde{\phi}(p)$ is the usual Fourier transform of $\phi(x)$, $\tilde{\phi}(p) = \int dx\ e^{-ipx}\phi(x)$. After using the Hausdorff-Campbell formula for the product of noncommutative fields we get

$$\hat{\Phi}_1(\hat{q})\hat{\Phi}_2(\hat{q}) = \int dp_1\ dp_2\ e^{i(p_1+p_2)\hat{q}-\frac{1}{2}p_1^\mu p_2^\nu\theta_{\mu\nu}}\phi_1(p_1)\phi_2(p_2),$$

so that we can associate to this product a deformed product of commutative fields in ordinary space

$$\hat{\Phi}_1(\hat{q})\hat{\Phi}_2(\hat{q}) \leftrightarrow (\phi_1 \star \phi_2)(x),$$

know as the star or Moyal product

$$(\phi_1 \star \phi_2)(x) \equiv \left[e^{i\frac{1}{2}\theta^{\mu\nu}\frac{\partial}{\partial x^\mu}\frac{\partial}{\partial y^\nu}}\phi_1(x)\phi_2(y)\right]_{y=x} = \phi_1(x)\phi_2(x) + i\frac{1}{2}\theta^{\mu\nu}\partial_\mu\phi_1\partial_\nu\phi_2 + \dots.$$

In this way we can work in the usual space with commutative coordinates and conventional fields but replacing the ordinary product of fields by the Moyal product. We can even consider the quantization of such field theories and one remarkable result is that the structure of ultraviolet divergences is not modified by the Moyal product [3]. This is quite surprising because when the Moyal product is expanded there is an infinite number of higher derivative terms which contribute to the interaction vertices in Feynman diagrams. We would expect that the ultraviolet structure would be spoiled but the Moyal structure works in such a way as to preserve it. At the end, Heisenberg's dream of using noncommutativity to tame the divergences did not come true. But, as we shall see, other effects do appear and in fact are deadly for most field theories.

Besides the importance of Filks work in 1996 it did not call much attention since only a few people were interested in noncommutativity on those days. The revolution in noncommutativity had to wait a few years more for another important discovery, but this time in string theory. Dp-branes are extended objects in p space dimensions in which open strings can end. When a set of N D-branes are nearly coincident the string spectrum includes a $U(N)$ gauge theory in the world-volume of the D-branes. It also includes a matrix model of $N \times N$ matrices for each transverse dimension of the brane. We can consider the situation when the NS-NS antisymmetric field $B_{\mu\nu}$ is turned on in the presence of D3-branes. It was found that there is a low energy limit where the closed strings decouple and the effective theory living on the D3-brane is a noncommutative gauge theory [4]. The noncommutativity is induced by the Moyal product with $\theta_{\mu\nu}$ being related to the NS-NS field $B_{\mu\nu}$. Since these noncommutative theories arouse from a consistent truncation of string theory they must be consistent in some sense and this gave rise to an intense period of research in noncommutative field theories.

Since then the proposal of noncommutativity has been applied to many areas of physics and mathematics and would be impossible to review all advancements happened along these years. Instead I will concentrate on topics which I have been involved, mostly concerned with quantum field theory and gravity. A couple of older reviews are also useful as an introduction to the subject [5, 6].

2. Mixing of Divergences

An important feature of quantum noncommutative field theories is the mixing of ultraviolet and infrared divergences usually known as the UV/IR mixing [7]. Consider the noncommutative scalar field theory

$$S=\int d^4x\,\left(\frac{1}{2}\partial_\mu\hat{\phi}\star\partial^\mu\hat{\phi}-\frac{m^2}{2}\hat{\phi}\star\hat{\phi}-\frac{g^2}{4!}\hat{\phi}\star\hat{\phi}\star\hat{\phi}\star\hat{\phi}\right),$$

where the ordinary product among the fields was replaced by the Moyal product. It easy to show that $\int dx\ (f\star g)(x)=\int dx\ f(x)g(x)$ so that the quadratic terms in the action are not affected by the Moyal product. This means that propagators are not modified in noncommutative theories. But the interaction vertex is modified

$$\int d^4x\ \hat{\phi}\star\hat{\phi}\star\hat{\phi}\star\hat{\phi}=-\frac{1}{3}\int dk_1dk_2dk_3dk_4\ \delta(k_1+k_2+k_3+k_4)\times[\cos(\frac{1}{2}k_1\wedge k_2)\cos(\frac{1}{2}k_3\wedge k_4)+$$

$$+\cos(\frac{1}{2}k_1\wedge k_3)\cos(\frac{1}{2}k_2\wedge k_4)+\cos(\frac{1}{2}k_1\wedge k_4)\cos(\frac{1}{2}k_2\wedge k_3)]\ \hat{\phi}(k_1)\,\hat{\phi}(k_2)\,\hat{\phi}(k_3)\,\hat{\phi}(k_4),$$

where $k\wedge p=\theta_{\mu\nu}k^\mu p^\nu$. This interaction leads to the the usual UV divergence which requires a UV regulator Λ. But it also gives rise to contributions which are singular in the IR. Symbolically the effective regulator has the form

$$\Lambda_{eff}=\frac{1}{\frac{1}{\Lambda^2}+(\theta p)^2},$$

mixing with the UV with the IR divergence. This mixing does not spoils renormalizability at one loop but it is fatal at higher orders [7]. Several proposals were made to overcome this trouble but all of them require non trivial modifications of the original theory [8, 9, 10, 11, 12]. The simplest solution, however, is the inclusion of supersymmetry which turns the noncommutative chiral multiplet renormalizable to all orders in perturbation theory [13]. Other troubles caused by the UV/IR mixing may also be healed by supersymmetry like spontaneous symmetry breaking [14] but other problems like the renormalizability of supersymmetric gauge theories [15, 16] are not.

The noncommutative abelian gauge theory is described by the action

$$S_A=-\frac{1}{4}\int d^4x\ \hat{F}^{\mu\nu}\star\hat{F}_{\mu\nu},$$

with the field strength and gauge transformation modified by the addition of the Moyal commutator $\left[\hat{A},\hat{B}\right]_\star=\hat{A}\star\hat{B}-\hat{B}\star\hat{A}$ as

$$\hat{F}_{\mu\nu}=\partial_\mu\hat{A}_\nu-\partial_\nu\hat{A}_\mu-i[\hat{A}_\mu,\hat{A}_\nu]_\star,$$
$$\delta\hat{A}_\mu=\hat{D}_\mu\hat{\lambda}=\partial_\mu\hat{\lambda}-i[\hat{A}_\mu,\hat{\lambda}]_\star.$$

The noncommutative gauge field $\hat{A}_\mu$ can be mapped to the conventional abelian gauge field A_μ, with the conventional field strength and gauge transformation, through the Seiberg-Witten map, which to first order in θ is [4]

$$\hat{A}_\mu=A_\mu-\frac{1}{2}\theta^{\alpha\beta}A_\alpha(\partial_\beta A_\mu+F_{\beta\mu}).$$

The resulting abelian gauge theory is now an interacting theory with action

$$S_A=-\frac{1}{4}\int d^4x\ \left[F^{\mu\nu}F_{\mu\nu}+2\theta^{\mu\rho}F_\rho{}^\nu\left(F_\mu{}^\sigma F_{\sigma\nu}+\frac{1}{4}\eta_{\mu\nu}F^{\alpha\beta}F_{\alpha\beta}\right)\right].$$

It gives rise to a nonrenormalizable theory due to the new interactions generated by the noncommutativity and also presents the UV/IR mixing. Its properties have been studied by several groups. Dualities among gauge theories usually do not survive after the Seiberg-Witten map [17, 18].

3. Signatures of noncommutativity

The existence of a constant antisymmetric tensor $\theta_{\mu\nu}$ means that Lorentz symmetry is broken. However, it does not mean that Lorentz symmetry is not valid at a fundamental level. We can think of $\theta_{\mu\nu}$ as a field (as in string theory) which has a constant vacuum expectation value breaking Lorentz symmetry spontaneously. So one of the manifestations of noncommutativity is through small deviations from Lorentz symmetry. The noncommutative Maxwell equations for the conventional abelian field show that the photon has a modified dispersion relation [19]

$$k^2 - 2\theta^{\mu\alpha} F_\alpha{}^\nu k_\mu k_\nu = 0.$$

This is similar to what happens to photons in an anisotropic medium with no Faraday rotation nor birefringence [20]. Many other studies of modifies dispersion relations have also been performed [21, 22, 23, 24].

There are several limits imposed on noncommutativity coming from many experiments. Observations of blazars give $|\theta| < (10^3 TeV)^{-2}$ [25], atomic transitions in noncommutative quantum mechanics and Lamb shift in noncommutative QED give $|\theta| < (10TeV)^{-2}$ [26], spin-statistics violations in noncommutative QED from Gran Sasso and Super-Kamiokande give $|\theta| < (10^5 TeV)^{-2}$ [27], noncommutative extensions of the standard model give $|\theta| < (10TeV)^{-2}$ [28] and noncommutative symplectic structure in classical mechanics and perihelium of Mercury give $|\theta| < (10^{13} TeV)^{-2}$ [29].

4. Emergent Gravity

There is a surprising connection between translations and gauge transformations. Consider a rigid translation for a noncommutative scalar field

$$\delta_T \hat{\phi} = \xi^\mu \partial_\mu \hat{\phi},$$

and compare it with a noncommutative gauge transformation

$$\delta\hat{\phi} = -i[\hat{\phi}, \hat{\lambda}]_\star = -i(\hat{\phi} \star \hat{\lambda} - \hat{\lambda} \star \hat{\phi}).$$

We can write the derivative of the field in terms of the Moyal commutator as $\partial_\mu \hat{\phi} = -i\theta^{-1}_{\mu\nu}[x^\nu, \hat{\phi}]_\star$ assuming that θ is invertible, so that a translation can be written as a gauge transformation

$$\delta_T \hat{\phi} = \delta\hat{\phi},$$

with gauge parameter $\hat{\lambda} = -\theta^{-1}_{\mu\nu}\xi^\mu x^\nu$. Therefore, translations in the noncommutative directions are equivalent to gauge transformations [30]. Of course, this has a flavor of general relativity!

Let us consider this proposal in more detail. The action for a noncommutative scalar field coupled to a noncommutative abelian gauge field is given by

$$S_\varphi = \frac{1}{2}\int d^4x \; \hat{D}^\mu \hat{\varphi} \star \hat{D}_\mu \hat{\varphi}.$$

We can now use the Seiberg-Witten map to work with the conventional scalar field φ and the conventional abelian gauge field A_μ to get

$$S_\varphi = \frac{1}{2}\int d^4x \, [\partial^\mu\varphi\partial_\mu\varphi + 2\theta^{\mu\alpha} F_\alpha{}^\nu(-\partial_\mu\varphi\partial_\nu\varphi + \frac{1}{4}\eta_{\mu\nu}\partial^\rho\varphi\partial_\rho\varphi)].$$

Notice that the tensor inside the parenthesis is traceless. Now consider the coupling of the conventional scalar field to linearized gravity

$$S_{g,\varphi} = \frac{1}{2}\int d^4x \, (\partial^\mu\varphi\partial_\mu\varphi - h^{\mu\nu}\partial_\mu\varphi\partial_\nu\varphi + h\partial^\rho\varphi\partial_\rho\varphi),$$

where $h_{\mu\nu}$ is the traceless linearized metric and h its trace. Comparing both actions we can identify the linearized gravitational field as

$$h^{\mu\nu} = \theta^{\mu\alpha} F_\alpha{}^\nu + \theta^{\nu\alpha} F_\alpha{}^\mu + \frac{1}{2}\eta^{\mu\nu}\theta^{\alpha\beta}F_{\alpha\beta},$$
$$h = 0.$$

Therefore, the effect of noncommutativity on the scalar field is similar to a field dependent gravitational field. The coupling of matter to the abelian gauge field in noncommutative theories has effects that mimic those of gravity. In this way gravity is an emergent phenomena in noncommutative theories [31]. Remarkably charged fields feel a gravitational field that is half of that felt by the uncharged field so that the gravity coupling is now charge dependent. We can also compute the geometry induced by noncommutativity: it is that of a plane wave, more precisely a pp-wave [31]. These linearized results can be extended to the full theory [32]. As remarked in the introduction matrix models also appear when we consider D-branes in string theory. Again the emergence of gravity does arise in noncommutative matrix models. The matrix-model for a noncommutative $U(N)$ gauge theory actually describes $SU(N)$ gauge theory coupled to gravity [33].

5. Noncommutative Gravity

There are many attempts to formulate a noncommutative gravity theory leading to several extensions of general relativity. I will not try review all these proposals and will concentrate only in my attempt for such a formulation [34]. We tried to implement noncommutativity in curved spacetime in a manner which is as simple as possible. In flat spacetime $\theta_{\mu\nu}$ is a constant matrix. How could we consider it in curved spacetime? Before answering this question let us see what we know in flat spacetime. To this end let us assume that $\theta_{\mu\nu}$ is a tensor and perform an infinitesimal general coordinate transformation

$$\delta\theta^{\mu\nu} = \xi^\lambda \partial_\lambda \theta^{\mu\nu} - \partial_\lambda \xi^\mu \theta^{\lambda\nu} + \partial_\lambda \xi^\nu \theta^{\lambda\mu}.$$

Let us also assume that $\delta\theta^{\mu\nu} = 0$. For a rigid translation ξ^μ is constant and we get that $\partial_\lambda \theta^{\mu\nu} = 0$, so that $\theta^{\mu\nu}$ is constant. Let us now consider a rigid Lorentz transformation $\xi^\mu = \Lambda^\mu{}_\nu x^\nu$. This time we get $\Lambda^\mu{}_\lambda \theta^{\lambda\nu} - \Lambda^\nu{}_\lambda \theta^{\lambda\mu} = 0$. Let us choose just one non-vanishing component of $\theta_{\mu\nu}$, say $\theta^{12} \neq 0$. Then Lorentz boosts in the 3-direction and rotations in the $1-2$ plane are still preserved as well as translations in any direction. We might think that these are all the solutions of $\delta\theta^{\mu\nu} = 0$ but there is still one more. It is

$$\xi^\mu = \theta^{\mu\nu}\partial_\nu \xi,$$

where ξ is a scalar function. Notice that $\partial_\mu \xi^\mu = 0$. This means that this transformation forms a symplectic subgroup of volume preserving diffeomorphisms which also preserves $\theta^{\mu\nu}$. This symmetry is not very useful in flat spacetime but will be very important in curved spacetime.

Let us now move to curved spacetime where the general coordinate transformation is

$$\delta\theta^{\mu\nu} = \xi^\lambda D_\lambda \theta^{\mu\nu} - D_\lambda \xi^\mu \theta^{\lambda\nu} + D_\lambda \xi^\nu \theta^{\lambda\mu}. \quad (2)$$

Assume again that $\delta\theta^{\mu\nu} = 0$ as in flat spacetime. A solution which generalizes the constancy of $\theta^{\mu\nu}$ in flat spacetime to curved spacetime is

$$D_\lambda \theta^{\mu\nu} = 0.$$

This implies $D_\lambda \xi^\mu \theta^{\lambda\nu} - D_\lambda \xi^\nu \theta^{\lambda\mu} = 0$ and the solution is $D_\mu \xi^\mu = 0$. Hence there is a residual symmetry similar to what happens in flat spacetime. We are left with the symplectic subgroup of the volume preserving diffeomorphisms which also preserves a covariantly constant $\theta^{\mu\nu}$. This is our extension of noncommutativity to curved spacetime. We assume that $\theta^{\mu\nu}$ is a covariantly constant tensor. We can then define all geometric objects like the Christoffel symbol, Riemann tensor and so on in the usual way.

We can now couple $\theta^{\mu\nu}$ to the Riemann tensor to build a noncommutative contribution to the usual Einstein-Hilbert action. If we go up to second order in $\theta^{\mu\nu}$ we find that the only independent combination is

$$S_{NC} = \frac{1}{16\pi} \int d^4x \sqrt{-g}\, \theta^{\mu\nu} \theta^{\alpha}{}_{\beta} R_{\mu\nu\alpha}{}^{\beta}.$$

The local symmetry is now reduced to volume preserving transformations. Gravity theories invariant under volume preserving transformations are well known and are called unimodular gravity [35]. Its main property is that the determinant of the metric is constant and can be chosen to be one, the raison d'tre of its name. Also the cosmological constant appears as an integration constant. Another interesting property is that the Einstein-Hilbert action has a finite polynomial form in the metric.

We can derive the linearized equations of motion in the noncommutative case

$$\frac{1}{G}\left[\Box h_{\mu\nu} + \partial_\mu \partial_\nu h_\rho{}^\rho - \partial^\rho \partial_\mu h_{\nu\rho} - \partial^\rho \partial_\nu h_{\mu\rho} - \eta_{\mu\nu}\left(\Box h_\rho{}^\rho - \partial^\rho \partial^\sigma h_{\rho\sigma}\right)\right]$$
$$+\theta^{\alpha\beta}\theta_{\mu\gamma}\left(\partial_\nu \partial_\alpha h_\beta{}^\gamma - \partial^\gamma \partial_\alpha h_{\beta\nu}\right) + \theta^{\alpha\beta}\theta_{\nu\gamma}\left(\partial_\mu \partial_\alpha h_\beta{}^\gamma - \partial^\gamma \partial_\alpha h_{\beta\mu}\right) + 2\eta_{\mu\nu}\theta^{\alpha\beta}\theta^\rho{}_\sigma \partial_\rho \partial_\alpha h_\beta{}^\sigma = 0,$$

to find out that

$$\begin{aligned} g_{00} &= 1 + \left(1 + 3G\vec{\theta}^2 - G(\vec{n}\cdot\vec{\theta})^2\right) h, \\ g_{0i} &= 0, \\ g_{ij} &= -\delta^{ij} + \left[n^i n^j + G\delta^{ij}\left(\vec{\theta}^2 - (\vec{n}\cdot\vec{\theta})^2\right) + G\left(\theta^i n^j + \theta^j n^i\right)\vec{n}\cdot\vec{\theta}\right] h, \end{aligned}$$

where $h = -2GM/r$. We can find the correction to the Newtonian potential felt by a test particle

$$\frac{d^2x^i}{dt^2} = -\frac{1}{2}\partial_i \left[h + G\left(3\vec{\theta}^2 - (\vec{n}\cdot\vec{\theta})^2\right) h\right].$$

The Newtonian potential has a contribution proportional to $\vec{\theta}^2$ which can be regarded as giving rise to an effective Newton constant $G(1 + \frac{3}{2}G\vec{\theta}^2)$. The angular dependent piece $(\vec{n}\cdot\vec{\theta})^2$ also contributes to the potential but the potential still goes as $1/r$. The force on a test particle is given by

$$m\frac{d^2x^i}{dt^2} = m\left[n^i + 3G\left(\vec{\theta}^2 - (\vec{n}\cdot\vec{\theta})^2\right) n^i + 2G(\vec{n}\cdot\vec{\theta})\theta^i\right]\frac{h}{2r},$$

so it is not purely radial. The radial component is modified by $3G(\vec{\theta}^2 - (\vec{n}\cdot\vec{\theta})^2$. The term proportional to $\vec{\theta}$ produces, in general, a force off the plane of the orbit. It is also periodic for closed orbits. If $\vec{\theta}$ is perpendicular to the plane of the orbit then no periodic effect due to noncommutativity is seen.

Noncommutativity can be incorporated into gravity in alternative ways [36] and also in cosmological models and there is a huge literature on the subject [37, 38, 39].

6. Conclusions

Noncommutative theories is still a broad area of research in physics and mathematics with applications ranging from condensed matter physics to particle physics and cosmology. We presented a short review on some topics of noncommutativity in field theory and gravity. Noncommutative field theories can be easily written by replacing the ordinary product among fields by the Moyal product. Other more complicated recipes also do exist. The main consequence of the Moyal product is the appearance of the UV/IR divergence which usually spoils renormalizability. Many proposal to overcome this problem were presented and several are still under study. Noncommutative gravity is a very broad area of research. Many proposals for it are available and we discussed just one of them. Also, gravity as an emergent phenomena can be realized in the noncommutative context, at the field theory level and also at the matrix model level. Many other application of noncommutativity were not mentioned here and are still an active area of research.

Acknowledgments

I would like to thank the organizers of the XIV Mexican School on Particles and Fields for the kind invitation to deliver this talk and specially to David Vergara for hospitality. The work of V.O.R. is supported by CNPq grant No. 304495/2007-7 and FAPESP grant No. 2008/05343-5.

References

[1] Doplicher S, Fredenhagen K and Roberts J E 1995 *Commun.Math.Phys.* **172** 187–220 (*Preprint* hep-th/0303037)
[2] Snyder H S 1947 *Phys.Rev.* **71** 38–14
[3] Filk T 1996 *Phys. Lett.* **B376** 53–58
[4] Seiberg N and Witten E 1999 *JHEP* **9909** 032 (*Preprint* hep-th/9908142)
[5] Douglas M R and Nekrasov N A 2001 *Rev.Mod.Phys.* **73** 977–1029 (*Preprint* hep-th/0106048)
[6] Szabo R J 2003 *Phys.Rept.* **378** 207–299 (*Preprint* hep-th/0109162)
[7] Minwalla S, Van Raamsdonk M and Seiberg N 2000 *JHEP* **0002** 020 (*Preprint* hep-th/9912072)
[8] Langmann E and Szabo R J 2002 *Phys.Lett.* **B533** 168–177 (*Preprint* hep-th/0202039)
[9] Grosse H and Wulkenhaar R 2005 *Commun.Math.Phys.* **256** 305–374 (*Preprint* hep-th/0401128)
[10] Rivasseau V, Vignes-Tourneret F and Wulkenhaar R 2006 *Commun.Math.Phys.* **262** 565–594 (*Preprint* hep-th/0501036)
[11] Disertori M, Gurau R, Magnen J and Rivasseau V 2007 *Phys.Lett.* **B649** 95–102 (*Preprint* hep-th/0612251)
[12] Langmann E, Szabo R and Zarembo K 2004 *JHEP* **0401** 017 (*Preprint* hep-th/0308043)
[13] Girotti H, Gomes M, Rivelles V O and da Silva A 2000 *Nucl.Phys.* **B587** 299–310 (*Preprint* hep-th/0005272)
[14] Girotti H, Gomes M, Petrov A, Rivelles V O and da Silva A 2003 *Phys.Rev.* **D67** 125003 (*Preprint* hep-th/0207220)
[15] Ferrari A, Girotti H, Gomes M, Petrov A, Ribeiro A *et al.* 2004 *Phys.Rev.* **D70** 085012 (*Preprint* hep-th/0407040)
[16] Ferrari A, Girotti H, Gomes M, Petrov A, Ribeiro A *et al.* 2004 *Phys.Rev.* **D69** 025008 (*Preprint* hep-th/0309154)
[17] Harikumar E and Rivelles V O 2005 *Phys.Lett.* **B625** 156–164 (*Preprint* hep-th/0506078)
[18] Ajith K, Harikumar E, Rivelles V O and Sivakumar M 2008 *Phys.Rev.* **D77** 085010 (*Preprint* 0801.2040)
[19] Guralnik Z, Jackiw R, Pi S and Polychronakos A 2001 *Phys.Lett.* **B517** 450–456 (*Preprint* hep-th/0106044)
[20] Mariz T, Nascimento J and Rivelles V O 2007 *Phys.Rev.* **D75** 025020 (*Preprint* hep-th/0609132)
[21] Agostini A, Amelino-Camelia G and D'Andrea F 2004 *Int.J.Mod.Phys.* **A19** 5187–5220 (*Preprint* hep-th/0306013)
[22] Helling R C and You J 2008 *JHEP* **0806** 067 (*Preprint* 0707.1885)
[23] Bietenholz W 2010 (*Preprint* 1008.3886)
[24] Bietenholz W, Nishimura J, Susaki Y and Volkholz J 2006 *JHEP* **0610** 042 (*Preprint* hep-th/0608072)
[25] Amelino-Camelia G 2004 *New J.Phys.* **6** 188 (*Preprint* gr-qc/0212002)
[26] Chaichian M, Sheikh-Jabbari M and Tureanu A 2001 *Phys.Rev.Lett.* **86** 2716 (*Preprint* hep-th/0010175)
[27] Balachandran A, Mangano G, Pinzul A and Vaidya S 2006 *Int.J.Mod.Phys.* **A21** 3111–3126 (*Preprint* hep-th/0508002)

[28] Carroll S M, Harvey J A, Kostelecky V, Lane C D and Okamoto T 2001 *Phys.Rev.Lett.* **87** 141601 (*Preprint* hep-th/0105082)
[29] Romero J M and Vergara J 2003 *Mod.Phys.Lett.* **A18** 1673–1680 (*Preprint* hep-th/0303064)
[30] Gross D J and Nekrasov N A 2000 *JHEP* **0010** 021 (*Preprint* hep-th/0007204)
[31] Rivelles V O 2003 *Phys.Lett.* **B558** 191–196 (*Preprint* hep-th/0212262)
[32] Banerjee R and Yang H S 2005 *Nucl.Phys.* **B708** 434–450 (*Preprint* hep-th/0404064)
[33] Steinacker H 2007 *JHEP* **0712** 049 (*Preprint* 0708.2426)
[34] Harikumar E and Rivelles V O 2006 *Class.Quant.Grav.* **23** 7551–7560 (*Preprint* hep-th/0607115)
[35] Alvarez E 2005 *JHEP* **03** 002 (*Preprint* hep-th/0501146)
[36] Nicolini P 2009 *Int.J.Mod.Phys.* **A24** 1229–1308 (*Preprint* 0807.1939)
[37] Garcia-Compean H, Obregon O and Ramirez C 2002 *Phys.Rev.Lett.* **88** 161301 (*Preprint* hep-th/0107250)
[38] Barbosa G and Pinto-Neto N 2004 *Phys.Rev.* **D70** 103512 (*Preprint* hep-th/0407111)
[39] Bastos C, Bertolami O, Dias N and Prata J 2009 *J.Phys.Conf.Ser.* **174** 012053 (*Preprint* 0812.3488)

Journal of Physics: Conference Series **287** (2011) 012013 doi:10.1088/1742-6596/287/1/012013

Radiative Corrections in the LHC and ILC Era

Doreen Wackeroth

Department of Physics, SUNY at Buffalo, Buffalo, NY 14260-1500, USA

E-mail: dow@ubpheno.physics.buffalo.edu

Abstract. In this talk I will present a brief review of the role of radiative corrections in present and future collider experiments with emphasis on the CERN Large Hadron Collider (LHC). After a general discussion of the importance, typical features and status of higher-order calculations I will discuss their impact on the example of Higgs boson production in association with heavy quarks and precisions studies of W and Z bosons at the LHC.

1. Why radiative corrections ?

With the CERN Large Hadron Collider (LHC) currently successfully colliding protons at a center-of-mass (CM) energy of 7 TeV the discovery of the Higgs boson and of physics Beyond the Standard Model (BSM) may be imminent. While some direct signals of BSM physics may be relatively easy to detect, more likely scenarios are signals overwhelmed by large Standard Model (SM) backgrounds and/or indirect signals in form of minute deviations from SM predictions. In the latter two cases, full advantage has to be taken of the ability of the LHC to not only probe the energy but also the precision frontier. Moreover, the identification of newly discovered particles will rely on high-precision measurements of their properties, such as masses and couplings. Many aspects of the interpretation of LHC data will rely on the precise knowledge of total production rates and kinematic distributions, including radiative corrections, i. e. higher-order terms beyond leading order (LO) in perturbation theory.

In the past two decades, the two pillars of precision physics, availability of high-precision collider experiments, such as the CERN LEP/SLAC SLC e^+e^- colliders, the DESY HERA ep and the Fermilab Tevatron $p\bar{p}$ collider, and the predictive power of the SM beyond LO in perturbation theory, have made it possible

- to test the SM of electroweak (EW) and strong interactions as a fully-fledged Quantum Field Theory (QFT) at the per mille and per cent level, respectively,
- to constrain the mass of the SM Higgs boson,
- to exclude or severely constrain many new physics extensions of the SM, and
- to measure SM input parameters with high precision.

Figures 1 and 2 illustrate the impressive agreement of precision measurements at the Z resonance at LEP/SLC and above the W-pair production threshold at LEP-II, respectively, with SM predictions.

Electroweak and QCD radiative corrections will also play an important role in all aspects of the physics program at the LHC or a future International Linear e^+e^- Collider (ILC), i. e.

Journal of Physics: Conference Series **287** (2011) 012013 doi:10.1088/1742-6596/287/1/012013

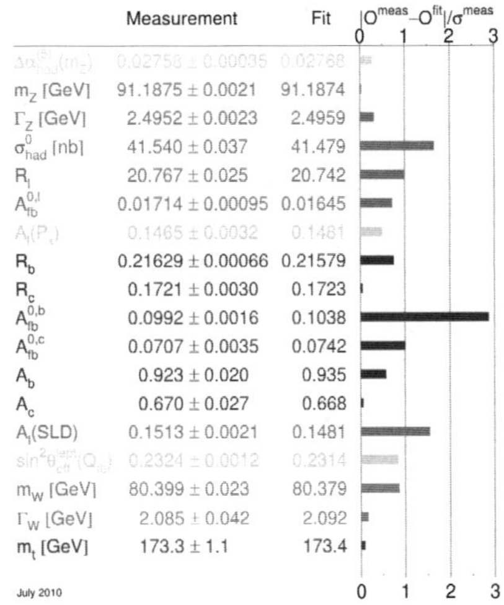

Figure 1. EW precision data are compared with SM predictions. Taken from Ref. [1].

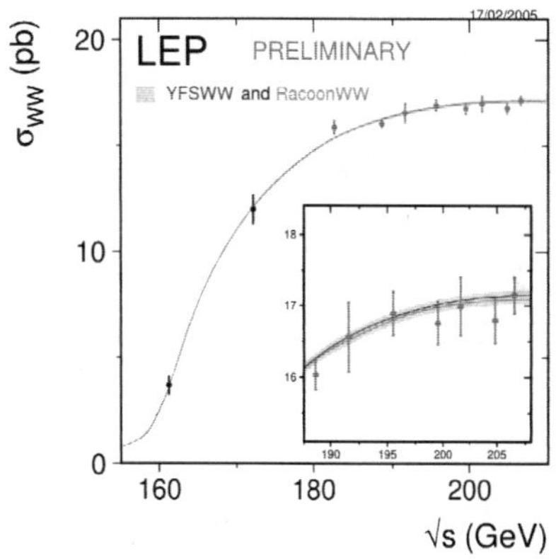

Figure 2. Precise measurements of W-pair production cross sections (σ_{WW}) at LEP-II and SM predictions including EW $\mathcal{O}(\alpha)$ corrections. A precise W mass measurement of $M_W = 80.376 \pm 0.033$ GeV [1] has been extracted from σ_{WW}. Taken from Ref. [2].

- *Discovery*: in modeling signal and background processes for Higgs and new physics searches *to distinguish the new from the known*, and in searching for *indirect* signals of BSM physics in quantum loops,
- *Identity*: in precisely extracting model parameters, such as masses and couplings of the underlying model from data,
- *Precision*: in reducing systematic errors by improving studies of effects of selection/analysis of data, by using single W and Z production cross sections as luminosity monitor, and by constraining Parton Distribution Functions (PDFs) using, e. g., the W charge asymmetry, and γ and jet production processes, and last but not least
- *Fundamentals*: in probing and exploring the QFT structure of the SM, which is interesting in its own right.

Before I present examples of the role of radiative corrections at the LHC, I will briefly review the general structure and typical features of radiative corrections. This discussion follows closely Ref. [3].

The differential hadronic cross section for a $h_1 h_2 \to N$ particles scattering process at hadron colliders can be written as:

$$d\sigma(S) = \sum_{ij} \frac{1}{1+\delta_{ij}} dx_1 dx_2 \{f_{i/h_1}(x_1,\mu_f) f_{j/h_2}(x_2,\mu_f) d\hat{\sigma}_{ij}(\hat{s},\mu_r,\mu_f) + x_1 \leftrightarrow x_2\}\mathcal{O} \tag{1}$$

where f_{i/h_1} denotes the PDF of parton i in hadron h_1 and $\mathcal{O}$ a set of kinematic cuts or jet functions. μ_r and μ_f are the renormalization and factorization scales, respectively, and $x_{1,2}$ denote the momentum fractions of the hadron momentum carried by the partons. The partonic CM energy ($\hat{s}$) is related to the hadronic CM energy by $\hat{s} = x_1 x_2 S$. Naturally, at hadron colliders one is first concerned with the impact of QCD corrections. A QCD prediction of the partonic cross section $d\hat{\sigma}_{ij}$ for the process $ij \to N$ at fixed order in the strong coupling constant α_s in perturbation theory exhibits the following general structure:

$$d\hat{\sigma}_{ij}(\mu_r, \mu_f) = \alpha_s^k(\mu_r)[d\tilde{\sigma}^{LO} + \alpha_s(\mu_r)d\tilde{\sigma}^{NLO}(\mu_r, \mu_f) + \alpha_s^2(\mu_r)d\tilde{\sigma}^{NNLO}(\mu_r, \mu_f) + \ldots ,] \quad (2)$$

where $\alpha_s^k d\tilde{\sigma}^{LO}$ constitutes the LO prediction, $\alpha_s^k[d\tilde{\sigma}^{LO} + \alpha_s d\tilde{\sigma}^{NLO}]$ the next-to-leading order (NLO) prediction of $d\hat{\sigma}_{ij}$, and so on. Radiative corrections in general consist of virtual, i. e. quantum loop, corrections and the real radiation of extra partons. For instance, the NLO contribution in Eq. (2) reads

$$d\tilde{\sigma}^{NLO} = d\sigma_{virtual}(ij \to N) + d\sigma_{real}(ij \to N+1) \ . \quad (3)$$

After employing a suitable renormalization procedure to remove UltraViolet (UV) divergences from $d\sigma_{virtual}$ and a regularization procedure for InfraRed (IR) singularities to facilitate their cancellation between $d\sigma_{virtual}$ and $d\sigma_{real}$, $d\tilde{\sigma}^{NLO}$ is finite, but may exhibit numerically large logarithms. For example, soft gluon radiation in single W and Z boson production at hadron colliders gives rise to the occurrence of large logarithms in the transverse momentum (q_T) of the W and Z bosons of the form $L = \alpha_s/\pi \ln(q_T^2/M_V^2), V = W, Z$, which can be resummed to all orders in perturbation theory. For instance, an expansion of $d\hat{\sigma}_{ij}$ in $\alpha^k L^n$

$$d\hat{\sigma}_{ij} \propto 1 + \alpha(L^2 + L + 1) + \alpha^2(L^4 + L^3 + L^2 + L + 1) + \ldots \quad (4)$$

reads after resummation

$$d\hat{\sigma}_{res} \propto C(\alpha) \exp\left[Lg_1(\alpha L) + g_2(\alpha L) + \alpha g_3(\alpha L) + \ldots\right] + R(\alpha) \ , \quad (5)$$

where $g_{1,2,\ldots}$ are computable functions.

The inclusion of higher orders in the perturbative series in $\alpha_s(\mu_r^2)$ decreases the dependence on the unknown scales μ_r and μ_f, and thus increases the reliability of the QCD prediction. Typical features of LO, NLO, NNLO and resummed QCD calculations are:

- LO: provides an order of magnitude estimate, but has large theoretical uncertainties due to the renormalization/factorization scale dependence.
- NLO: provides a reliable cross section prediction (reduced scale dependence), can considerably reduce or enhance the LO cross section and may distort the shape of kinematic distributions. As a result of the latter, reweighting of LO distributions to estimate the effect of NLO corrections with so-called K-factors, i. e. $K = \sigma_{NLO}/\sigma_{LO}$, may not be a good approximation.
- NNLO: provides reliable cross section predictions (reduced scale dependence) **and** a reliable estimate of the theoretical uncertainties due to dependence on μ_r and μ_f.
- Resummation: large logarithmic corrections are tamed, reduced scale dependence, provides an estimate of unknown fixed-order corrections.

It is quite remarkable that already at the Tevatron experimental studies of some SM processes, e. g., $p\bar{p} \to W + X$ and jet$+X$, require precise theoretical predictions even beyond NLO QCD.

Electroweak corrections also play an important role, especially in EW gauge boson production, such as Z boson production at LEP-I/SLC, W-pair production at LEP-II, and the production of one, two or three EW gauge bosons at hadron colliders. Electroweak corrections are of special interest since they exhibit a rich structure of different energy scales due to the presence of massive particles, such as the Higgs, weak gauge bosons and top and bottom quarks. Even at hadron colliders and in QCD dominated processes NLO EW corrections can be numerically at least as important as NNLO QCD corrections. At high-energies they may even be the dominant corrections due to soft/collinear virtual and real radiation of W and Z bosons resulting in so-called EW Sudakov logarithms.

2. State-of-the-art of higher-order QCD and EW corrections

In recent years impressive progress has been made in all aspects of higher-order QCD and EW calculations, i. e. in fixed-order, complete matrix element calculations of multi-leg processes, in the resummation of large logarithms in multi-scale processes and in the interface or merging of these two approaches. A recent review including an extensive list of references of the status of QCD calculations can be found, e. g., in Refs. [3, 4] and of EW calculations in Ref. [3]. The following very condensed overview of the status of higher-order calculations also includes examples of state-of-the-art calculations:

- *multi-loops* and *multi-legs*: Fixed-order calculations for $2 \to 2, 3, 4, 5$ processes at NLO QCD, $2 \to 2, 3$ processes at NNLO QCD (fully differential cross sections), and $2 \to 4$ processes at NLO EW. Examples: $pp, p\bar{p} \to W + 3, 4$ jets [5, 6, 7, 8], $pp, p\bar{p} \to t\bar{t}b\bar{b}$ [9, 10], $p\bar{p} \to 4b$ [11], $e^+e^- \to 5$ jets [12] at NLO QCD; $pp, p\bar{p} \to H$ (with decays) [13, 14], Drell-Yan (DY) production of single W or Z bosons (with decays) [15, 16], $e^+e^- \to 3$ jets [17, 18] at NNLO QCD; $e^+e^- \to 4f$ [19, 20] at NLO EW,
- *multi-scales*: Logarithmic enhanced corrections of the form $\alpha^k \ln^n(Q_1^2/Q_2^2), Q_1^2 \gg Q_2^2$ are resummed up to NNLL accuracy. Examples: q_T distributions in $pp, p\bar{p} \to H$ [21] and in DY [22]; threshold corrections in top-pair production (for a review see, e. g., [23]); EW Sudakov logarithms in 4-fermion processes (for a review see, e. g., Ref. [24]). The thrust distribution in e^+e^- collisions has even been determined at N^3LL accuracy using an effective field theory (SCET) [25].
- *hard-soft transition*: Fixed-order NLO matrix element calculations are interfaced/matched or merged with parton-shower Monte Carlo (MC) programs (e. g., MC@NLO [26] and POWHEG [27]) or multi-purpose MC programs are improved with matrix element corrections in kinematic regions where the parton shower is not sufficient.

Many of these state-of-the-art calculations required new and innovative ideas to be able to handle a large number of potentially numerically unstable one-loop diagrams, new multi-loop integrals, and a complex structure of IR singularities. These new developments include (see, e. g., Ref. [3] for more details and references)

- methods for automation and numerical evaluation of one-loop integrals include sector decomposition (IR) and contour deformation (thresholds), and new tensor reductions (to avoid spurious singularities due to vanishing Gram determinants),
- methods for dealing with IR singularities include phase space slicing and dipole or antenna subtraction, and
- methods that obtain one-loop amplitudes for multi-leg processes directly from poles and cuts reducing them to lower-leg/lower-point amplitudes (for a review see, e.g., Ref. [28]).

The so-called Les Houches wish list of higher-order calculations, recently updated in Ref. [29] and shown in Table 1, not only nicely illustrates the enormous progress in recent years in providing reliable predictions for processes that have been identified to be essential for the LHC physics program, but also that there is still much work to do.

3. Radiative corrections and the search for the Higgs boson

The Higgs boson, predicted as a direct consequence of W and Z boson mass generation in the SM via spontaneous symmetry breaking of the $\mathrm{SU}(2)_L \otimes \mathrm{U}(1)_Y$ gauge group, so far eluded direct observation. The mass of the SM Higgs boson is severely constrained (95 % C.L.): $M_H < 185$ GeV [1] by global fits to EW precision data, $M_H > 114.4$ GeV [1] and $M_H \notin [158, 175]$ GeV [30] by direct searches at LEP and the Tevatron, respectively. Moreover, the Tevatron may be able to exclude the entire mass range preferred by EW precision data. It is expected that if the SM Higgs boson exists, it will not escape detection at the LHC. Direct

Table 1. The Les Houches 2009 wish list [29] (slightly modified) for processes which require NLO or NNLO QCD precision at the LHC ($V = \gamma, Z$ or W). The NLO QCD calculations of processes 1-12 are completed or close to completion. The need for EW corrections is not reflected in this list. For references see [29].

process	relevant for
1. $pp \to V\,V$+jet	$t\bar{t}H$, new physics
2. $pp \to H + 2$ jets	H production by VBF
3. $pp \to t\bar{t}\,b\bar{b}$	$t\bar{t}H$
4. $pp \to t\bar{t} + 2$ jets	$t\bar{t}H$
5. $pp \to V\,V\,b\bar{b}$	VBF$\to H \to VV$, $t\bar{t}H$, new physics
6. $pp \to V\,V + 2$ jets	VBF$\to H \to VV$
7. $pp \to V + 3$ jets	various new physics signatures
8. $pp \to V\,V\,V$	SUSY trilepton searches
9. $pp \to t\bar{t} \to 6$ fermions	background to Higgs
10. $pp \to t\bar{t}j$	background, $t\bar{t}$@NNLO
11. $pp \to b\bar{b}b\bar{b}$	background to MSSM Higgs
12. $pp \to V + 4$ jets	$t\bar{t}$, various new physics signatures
13. $pp \to Wb\bar{b}j$	top, various new physics signatures
14. $pp \to t\bar{t}t\bar{t}$	various new physics signatures
15. NNLO $pp \to t\bar{t}$	SM benchmark
16. NNLO to VBF and Z/γ+ jet	Higgs couplings
17. NNLO QCD+NLO EW for W/Z	SM benchmark

searches for the Higgs boson require that the underlying production as well as background processes are well under theoretical control. All SM Higgs production processes at the Tevatron and the LHC and most background process are at least known at NLO QCD and for some processes, such as $gg \to H$, NNLO QCD corrections had to be included to obtain reliable predictions. In addition, for many processes and observables also the impact of EW corrections has been studied and resummed calculations of soft gluon radiation have been made available. A caveat of many of these higher-order calculations is that they did not include decays of the final-state particles, as observed in the experiments, so that much of the recent and ongoing theory effort goes now into remedying this situation.

Once the Higgs boson is observed, it will be crucial to measure its properties, such as the Yukawa couplings to fermions. The production of Higgs bosons in association with top and bottom quarks, $pp \to t\bar{t}H, b\bar{b}H$, plays a special role in this endeavour, since it directly probes the top and bottom-quark Yukawa coupling. $b\bar{b}h$ production can even be the dominant Higgs production channel in the Minimal Supersymmetric extension of the SM (MSSM) because of the possibility of enhanced Yukawa couplings, $g^{MSSM}_{bb(h^0,H^0)} = \frac{(-\sin\alpha, \cos\alpha)}{\cos\beta} g^{SM}_{bbh}$ (in the MSSM there are three neutral (h^0, H^0, A^0) and a charged Higgs boson $(H^\pm)$). For the extraction of top and bottom-quark Yukawa couplings at the LHC the predictions for the underlying Higgs production processes need to be under good theoretical control. For instance, the study of Ref. [31] finds that the top-quark Yukawa coupling can be determined with a 15% precision when LHC and ILC measurements are combined. The inclusion of NLO QCD corrections to $t\bar{t}h$ [32, 33, 34, 35] and $b\bar{b}h$ [36, 37, 38, 39, 40, 41] production at the LHC resulted in a reduced scale dependence. For instance, the remaining theoretical uncertainty from scale variation of the total $t\bar{t}h$ production rate at NLO QCD is estimated to be about $10-15\%$ at the Tevatron and $15-20\%$ at the LHC.

For $\mu_r = \mu_f = \mu$ in the range $m_t < \mu < 2m_t$ the $\mathcal{O}(\alpha_s)$ corrections reduce the LO cross section at the Tevatron, $K = \sigma_{NLO}/\sigma_{LO} = 0.7 - 0.95$, but at the LHC for $m_t + M_h/2 < \mu < 4m_t + 2M_h$ they lead to an enhancement, $K = 1.2 - 1.4$ [34, 35]. Although $t\bar{t}h$ production is considered to be experimentally very difficult due to its small production rate and large QCD background, recent studies that consider boosted Higgs bosons and top quarks [42] suggest that the situation is not as hopeless as originally thought, which triggered renewed theoretical and experimental interest in this process (see also Ref. [29]). Recently the decay $h \to b\bar{b}$ has been included in the NLO QCD prediction of $t\bar{t}H$ [29] which is the dominant decay for a light Higgs boson ($M_h < 135$ GeV). Also important background processes to $pp \to t\bar{t}h$ and $pp \to b\bar{b}h$ with $h \to b\bar{b}$, such as $pp \to t\bar{t}b\bar{b}$ and $pp \to b\bar{b}b\bar{b}$ have been made available at NLO QCD as shown in Figures 3 and 4, respectively.

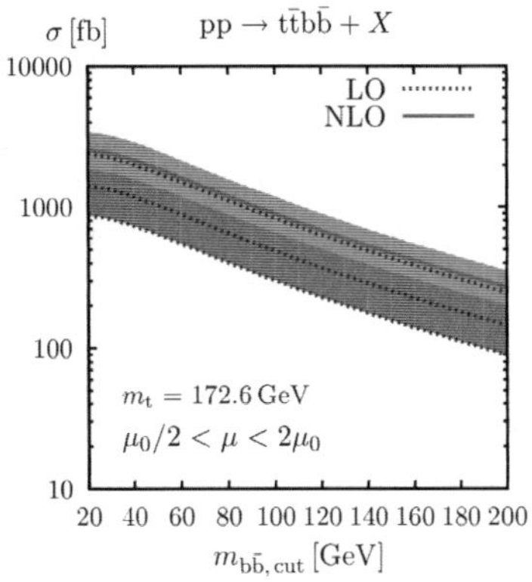

Figure 3. LO and NLO predictions for the total $t\bar{t}b\bar{b}$ cross section at the LHC in dependence of the cut on the $b\bar{b}$ invariant mass. The μ_r, μ_f scale uncertainty is represented by the bands and is considerably reduced at NLO QCD. Taken from Ref. [9]. See also Ref. [10].

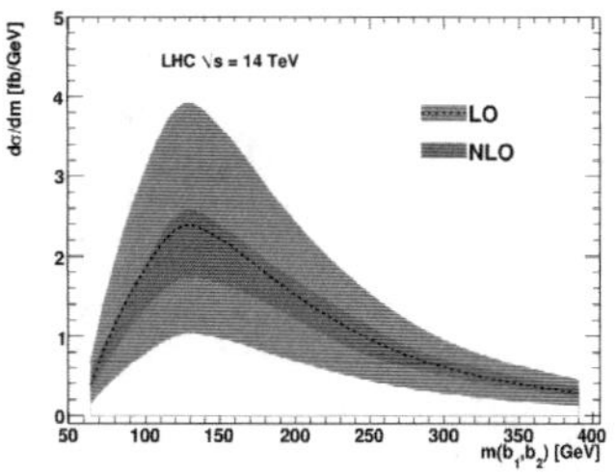

Figure 4. LO and NLO predictions for the invariant mass distribution of the two leading b-jets in $pp \to b\bar{b}b\bar{b}$ production at the LHC. The bands are obtained by varying μ_r (μ_f is fixed) and again the scale dependence is much less pronounced at NLO QCD. Taken from Ref. [11].

4. Electroweak precision physics with W and Z bosons

The W and Z production process are one of the theoretically best understood, most precise probes of the SM. As illustrated in Figs. 1 and 2 precision measurements and calculations of Z boson observables at LEP-I/SLC and of W-pair production cross sections at LEP-II probed the SM at the per mille level. For instance, for the precision measurement of the W mass at LEP-II the theoretical uncertainty of predictions for the total W-pair production cross section ($\delta\sigma_{WW}^{theory}$) had to better than 1% above the W-pair threshold, which was achieved by including EW $\mathcal{O}(\alpha)$ to $e^+e^- \to WW \to 4f$, resulting in $\delta\sigma_{WW}^{theory} = 0.4\%$ at $\sqrt{s}$ =200 GeV [43]. At a future ILC, precision measurements at the W-pair threshold require an even more dedicated theory effort, resulting in the calculation of the complete EW one-loop corrections to $e^+e^- \to 4f$ [19, 20]. Precision measurements of the W mass at the Tevatron recently surpassed the LEP-II precision and will continue to improve (for a review see, e. g., [44]). At the LHC, a measurement of M_W with $\delta M_W = 15$ MeV (or even 7 MeV) is anticipated. Besides precision measurements of SM parameters, M_W and the effective weak mixing angle $\sin^2\theta_W$, W and Z boson observables

Journal of Physics: Conference Series **287** (2011) 012013 doi:10.1088/1742-6596/287/1/012013

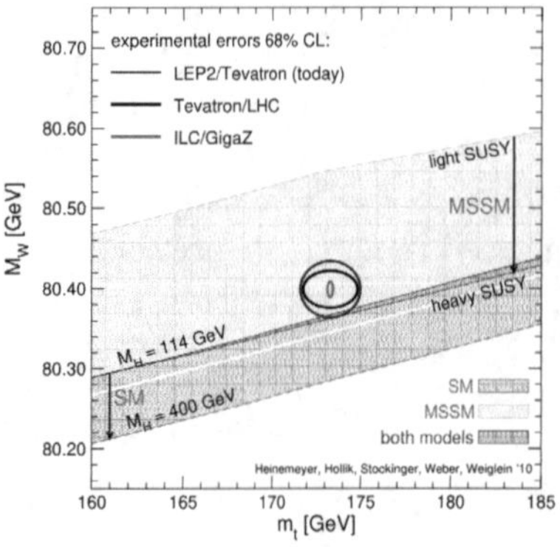

Figure 5. The predictions for the W mass in dependence of the top quark mass within the SM and MSSM in comparison with experimental M_W and m_t measurements at 68% C.L.. The bands represent theory predictions and are obtained by varying the free parameters of the underlying model. Taken from Ref. [45].

constrain the SM Higgs boson mass and are sensitive to signals of BSM physics. For example, as illustrated in Fig. 5, the virtual presence of new particles in loop corrections may lead to observable deviations from SM predictions of EW observables, so that ever more precise measurements and SM and MSSM predictions of M_W (together with a precise knowledge of m_t) could indicate that the MSSM is preferred over the SM. Moreover, BSM physics could mainfest itself as resonances of new, heavy gauge bosons in invariant mass distributions and forward-backward asymmetries at high energies, and as anomalous triple and quartic EW gauge boson self-couplings in the production of two and three EW gauge bosons.

In view of the importance and rich physics potential of EW gauge boson production processes, it is not surprising that these processes have been well studied not only experimentally but also theoretically. For hadron collider physics, for instance, NLO QCD and EW calculations are available for the production of one, two and three gauge bosons, as well as fully differential NNLO QCD predictions for the DY process. As mentioned earlier, an interesting signature of EW corrections at higher energies $\sqrt{s} \gg M_{W,Z}$ is their enhancement by EW Sudakov logarithms of the form $\alpha^l \ln^N(\frac{s}{M_V^2})$ (with $1 \leq N \leq 2l$ and $l = 1, 2, \ldots$ denotes the number of loops). They are remnants of UV singularities after renormalization and of soft/collinear emission of virtual and real W/Z bosons. EW logarithmic corrections to 4-fermion processes are known up to 2-loop N^3LL order and are available in form of compact analytical formula (for a review see, e. g, Ref. [24]). The W/Z boson masses act as a physical cut-off parameter and real W/Z radiation is usually not included, since it leads to a different initial/final state. However, as shown on the example of a number of observables in Ref. [48] it may be the case that the experimental setup requires their inclusion, which can result in a reduction of the impact of the logarithms of virtual origin. A recent study [49] discusses in more detail this partial cancellation of real and virtual corrections.

Apart from providing ever more precise predictions, theorists are also increasingly getting involved in the usage of their calculations in the experimental analysis and in determining the residual theoretical uncertainty. This is especially important when observables need to be under control at the per cent or even per mille level in the relevant kinematic regions. The latter is the case for the transverse mass distribution of the lepton pair (M_T) in single W production from which M_W is extracted at the Tevatron. Figure 6 illustrates the combined impact of EW and QCD NLO corrections on M_T. As shown in the lower panel the ambiguity in combining these two NLO calculations results into to small differences which can only be resolved by a calculation of the two-loop $\mathcal{O}(\alpha\alpha_s)$ corrections to single W production. Such a calculation will be necessary, if such small differences affect the W mass at the few per cent level. The question whether W and Z boson observables at the Tevatron and the LHC are under theoretical control at the required level of precision is presently also under investigation in the context of the

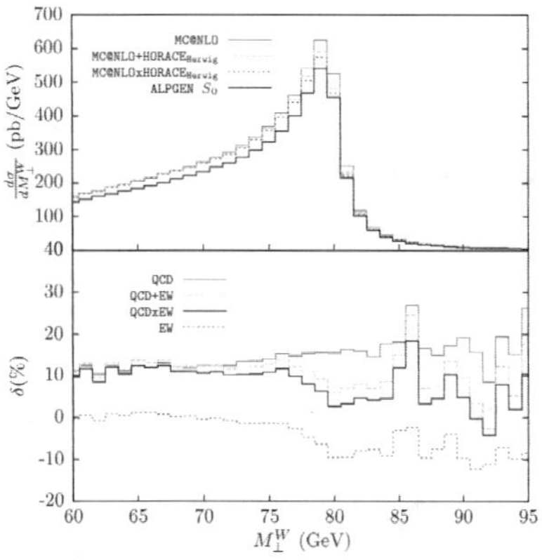

Figure 6. Impact of combined EW and QCD NLO corrections on the W transverse mass distribution using MC@NLO/HORACE. The percent difference shown in the lower panel (solid black and dashed green curves) is due to an ambiguity in performing this combination. Taken from Ref. [46].

Milano/Fermilab W mass workshop [50]. The efforts in providing a definite answer include an assessment of uncertainties of the available QCD predictions due to:

- QCD factorization/renormalization scale dependence,
- different treatments of non-perturbative QCD contributions and different implementation of subleading logarithms for $q_T \sim 20$ GeV in soft-gluon resummation,

and of EW calculations due to:

- EW input scheme dependence and different implementations of higher-order corrections,
- QED scale dependence of the PDFs, and
- unknown higher-order EW corrections.

Results of earlier studies can be found in Refs. [51, 52] and a report of the findings of the Milano/Fermilab W mass workshop is in preparation [50].

It is also important to quantify the uncertainty when using a theory prediction in the experimental analysis that does not include all known higher order corrections, for instance, as done in Ref. [53] where predictions of PHOTOS, which includes multiple photon radiation, in the experimental analysis and not HORACE, which in addition also provides the full EW one-loop corrections. The study of predictions of the total W production cross section at the LHC in Ref. [53] finds $\delta\sigma_W/\sigma_W = -0.904 \pm 0.109\%(-0.322 \pm 0.014\%)$ at 7 (14 TeV) due to missing $O(\alpha)$ EW corrections and $\delta\sigma_W/\sigma_W = 2.86 \pm 0.10\%(2.24 \pm 0.11\%)$ due to missing NNLO QCD corrections.

5. Conclusions

Radiative corrections are needed to fully exploit the potential of present and future collider experiments to discover the Higgs boson and/or signals of new physics and, once detected, to determine their properties to reconstruct the underlying theory. In view of high precision measurements at the LHC and a future ILC, SM predictions have to be under control at an unprecedented level of precision, requiring the calculation of QCD and EW radiative corrections to multi-leg processes, in some case up to the NNLO level (and NNLL accuracy). New techniques and automated tools have been and are being development to meet this challenge. Apart from checking off the Les Houches wish list, the interpretation of LHC data will require close collaboration between theorists and experimentalists to make the best use of these higher-order calculations. LHC physics results are already being compared to NLO and NNLO predictions and first steps have been made towards combining multi-purpose MCs that are needed for modeling of hadronization and underlying event (such as HERWIG and PYTHIA) and higher-order (parton-level) calculations.

Journal of Physics: Conference Series **287** (2011) 012013 doi:10.1088/1742-6596/287/1/012013

Acknowledgments
I am grateful to the organizers of the *XIV Mexican School of Particles and Fields* for the opportunity to attend and present at this very interesting conference. I especially would like to thank Maria E. Tejeda-Yeomans for her kind hospitality. The research of the author is supported in part by the National Science Foundation under grant No. NSF-PHY-0757691 and No. NSF-PHY-0547564.

Appendix
A number of tools for higher-order calculations are publicly available and a partial list is provided in the following (please see the program's webpages for references). There are many more Monte Carlo programs publicly available that calculate higher-order corrections for specific processes. The list below is intended to be a starting point for loop practitioners to-be rather than an exhaustive collection of available codes.

- Some automated tools for the calculation of tree-level amplitudes and cross sections:
  Madgraph/MadEvent: `madgraph.hep.uiuc.edu`
  SUSY-Madgraph: `www.thphys.uni-heidelberg.de/~plehn/smadgraph`
  O'Mega/Whizard: `projects.hepforge.org/whizard`
  Sherpa/Amegic++: `www.montecarlonet.org/index.php?p=Projects/sherpa`
  Helac/Phegas: `helac-phegas.web.cern.ch/helac-phegas/helac-phegas.html`
  Comphep: `comphep.sinp.msu.ru`
  Alpgen: `mlm.home.cern.ch/mlm/alpgen`
  carlomat: `kk.us.edu.pl/carlomat.html`
- Some tools for manipulating strings of Dirac spinors and γ matrices:
  FORM: `www.nikhef.nl/~form` and Tracer.m: `library.wolfram.com/infocenter/MathSource/2987`
- Some tools for NLO calculations:
  FeynArts: `www.feynarts.de`, FormCalc: `www.feynarts.de/formcalc`, and LoopTools: `www.feynarts.de/looptools`
  Cuttools: `www.ugr.es/~pittau/CutTools`
  Golem95: `lappweb.in2p3.fr/lapth/Golem/golem95.html`
  Samurai: `websupport1.citytech.cuny.edu/faculty/gossola/index2.html`
- Collections of NLO calculations:
  MCFM: `mcfm.fnal.gov` and GRACE: `minami-home.kek.jp`
- NLO QCD calculations interfaced with parton showers:
  MC@NLO: `https://www.hep.phy.cam.ac.uk/theory/webber/MCatNLO`
  POWHEG: `https://twiki.cern.ch/twiki/bin/view/Main/PowhegBOX`
- A collection of scalar one-loop integrals can be found, e. g., in Refs. [54, 55].

References

[1] J. Alcaraz *et al.*[ALEPH and CDF and D0 and DELPHI and L3 and OPAL and SLD Collaboration], arXiv:0911.2604 [hep-ex], and updates from `http://lepewwg.web.cern.ch/LEPEWWG`.
[2] J. Alcaraz *et al.* [ALEPH and DELPHI and L3 and OPAL and LEP Electroweak Working Group Collaborations], hep-ex/0612034.
[3] E. Laenen, D. Wackeroth, Ann. Rev. Nucl. Part. Sci. **59**, 367-396 (2009).
[4] T. Gehrmann, PoS D **IS2010**, 004 (2010) [arXiv:1007.2107 [hep-ph]].
[5] C. F. Berger *et al.*, arXiv:1009.2338 [hep-ph].
[6] C. F. Berger *et al.*, Phys. Rev. D **80**, 074036 (2009) [arXiv:0907.1984 [hep-ph]].
[7] K. Melnikov and G. Zanderighi, Phys. Rev. D **81**, 074025 (2010) [arXiv:0910.3671 [hep-ph]].
[8] R. K. Ellis, K. Melnikov and G. Zanderighi, Phys. Rev. D **80**, 094002 (2009) [arXiv:0906.1445 [hep-ph]].
[9] A. Bredenstein, A. Denner, S. Dittmaier and S. Pozzorini, Phys. Rev. Lett. **103**, 012002 (2009) [arXiv:0905.0110 [hep-ph]].
[10] G. Bevilacqua, M. Czakon, C. G. Papadopoulos, R. Pittau, M. Worek, JHEP **0909**, 109 (2009). [arXiv:0907.4723 [hep-ph]].

[11] T. Binoth, N. Greiner, A. Guffanti, J. Reuter, J. P. Guillet and T. Reiter, Phys. Lett. B **685**, 293 (2010) [arXiv:0910.4379 [hep-ph]].
[12] R. Frederix, S. Frixione, K. Melnikov and G. Zanderighi, JHEP **1011**, 050 (2010) [arXiv:1008.5313 [hep-ph]].
[13] C. Anastasiou, K. Melnikov, F. Petriello, Nucl. Phys. **B724**, 197-246 (2005). [hep-ph/0501130].
[14] S. Catani, M. Grazzini, Phys. Rev. Lett. **98**, 222002 (2007). [hep-ph/0703012].
[15] K. Melnikov, F. Petriello, Phys. Rev. Lett. **96**, 231803 (2006). [hep-ph/0603182].
[16] S. Catani, L. Cieri, G. Ferrera, D. de Florian, M. Grazzini, Phys. Rev. Lett. **103**, 082001 (2009). [arXiv:0903.2120 [hep-ph]].
[17] A. Gehrmann-De Ridder, T. Gehrmann, E. W. N. Glover, G. Heinrich, Phys. Rev. Lett. **100**, 172001 (2008). [arXiv:0802.0813 [hep-ph]].
[18] G. Dissertori, A. Gehrmann-De Ridder, T. Gehrmann, E. W. N. Glover, G. Heinrich and H. Stenzel, Phys. Rev. Lett. **104**, 072002 (2010) [arXiv:0910.4283 [hep-ph]].
[19] A. Denner, S. Dittmaier, M. Roth and L. H. Wieders, Phys. Lett. B **612**, 223 (2005) [arXiv:hep-ph/0502063].
[20] A. Denner, S. Dittmaier, M. Roth and L. H. Wieders, Nucl. Phys. B **724**, 247 (2005) [arXiv:hep-ph/0505042].
[21] D. de Florian and M. Grazzini, Phys. Lett. B **674**, 291 (2009) [arXiv:0901.2427 [hep-ph]].
[22] G. Bozzi, S. Catani, G. Ferrera, D. de Florian, M. Grazzini, Phys. Lett. **B696**, 207-213 (2011). [arXiv:1007.2351 [hep-ph]].
[23] N. Kidonakis, Phys. Rev. D **82**, 114030 (2010) [arXiv:1009.4935 [hep-ph]].
[24] J. H. Kuhn, Acta Phys. Polon. B **39**, 1639 (2008).
[25] T. Becher and M. D. Schwartz, JHEP **0807**, 034 (2008) [arXiv:0803.0342 [hep-ph]].
[26] S. Frixione, F. Stoeckli, P. Torrielli, B. R. Webber and C. D. White, arXiv:1010.0819 [hep-ph].
[27] S. Alioli, P. Nason, C. Oleari and E. Re, JHEP **1006**, 043 (2010) [arXiv:1002.2581 [hep-ph]].
[28] Z. Bern, L. J. Dixon and D. A. Kosower, Annals Phys. **322**, 1587 (2007) [arXiv:0704.2798 [hep-ph]].
[29] J. R. Andersen *et al.* [SM and NLO Multileg Working Group], arXiv:1003.1241 [hep-ph].
[30] The TEVNPH Working Group [CDF and D0 Collaboration], arXiv:1007.4587 [hep-ex].
[31] K. Desch and M. Schumacher, Eur. Phys. J. C **46**, 527 (2006) [arXiv:hep-ph/0407159].
[32] W. Beenakker, S. Dittmaier, M. Kramer, B. Plumper, M. Spira and P. M. Zerwas, Phys. Rev. Lett. **87**, 201805 (2001) [arXiv:hep-ph/0107081].
[33] W. Beenakker, S. Dittmaier, M. Kramer, B. Plumper, M. Spira and P. M. Zerwas, Nucl. Phys. B **653**, 151 (2003) [arXiv:hep-ph/0211352].
[34] S. Dawson, L. H. Orr, L. Reina and D. Wackeroth, Phys. Rev. D **67**, 071503 (2003) [arXiv:hep-ph/0211438].
[35] S. Dawson, C. Jackson, L. H. Orr, L. Reina and D. Wackeroth, Phys. Rev. D **68**, 034022 (2003) [arXiv:hep-ph/0305087].
[36] J. M. Campbell, R. K. Ellis, F. Maltoni and S. Willenbrock, Phys. Rev. D **67**, 095002 (2003) [arXiv:hep-ph/0204093].
[37] S. Dittmaier, M. Kramer and M. Spira, Phys. Rev. D **70**, 074010 (2004) [arXiv:hep-ph/0309204].
[38] S. Dawson, C. B. Jackson, L. Reina and D. Wackeroth, Phys. Rev. D **69**, 074027 (2004) [arXiv:hep-ph/0311067].
[39] S. Dawson, C. B. Jackson, L. Reina and D. Wackeroth, Phys. Rev. Lett. **94**, 031802 (2005) [arXiv:hep-ph/0408077].
[40] J. M. Campbell *et al.*, arXiv:hep-ph/0405302.
[41] S. Dawson, C. B. Jackson, L. Reina and D. Wackeroth, Mod. Phys. Lett. A **21**, 89 (2006) [arXiv:hep-ph/0508293].
[42] T. Plehn, G. P. Salam, M. Spannowsky, Phys. Rev. Lett. **104**, 111801 (2010). [arXiv:0910.5472 [hep-ph]].
[43] M. W. Grunewald *et al.*, arXiv:hep-ph/0005309.
[44] A. V. Kotwal and J. Stark, Ann. Rev. Nucl. Part. Sci. **58**, 147 (2008).
[45] S. Heinemeyer, W. Hollik, D. Stockinger, A. M. Weber and G. Weiglein, JHEP **0608**, 052 (2006) [arXiv:hep-ph/0604147]; and updates from `www.ifca.unican.es/~heinemey/uni/plots`.
[46] G. Balossini *et al.*, JHEP **1001**, 013 (2010) [arXiv:0907.0276 [hep-ph]].
[47] R. V. Harlander and W. B. Kilgore, Phys. Rev. D **68**, 013001 (2003) [arXiv:hep-ph/0304035].
[48] U. Baur, Phys. Rev. D **75**, 013005 (2007) [arXiv:hep-ph/0611241].
[49] G. Bell, J. H. Kuhn and J. Rittinger, Eur. Phys. J. C **70**, 659 (2010) [arXiv:1004.4117 [hep-ph]].
[50] A. Vicini *et al.*, in preparation; see also `http://wwwteor.mi.infn.it/ vicini/wmass.html`.
[51] C. E. Gerber *et al.* [TeV4LHC-Top and Electroweak Working Group], arXiv:0705.3251 [hep-ph].
[52] C. Buttar *et al.*, arXiv:0803.0678 [hep-ph].
[53] N. Adam, V. Halyo and S. A. Yost, JHEP **1011**, 074 (2010) [arXiv:1006.3766 [hep-ph]].
[54] R. K. Ellis, G. Zanderighi, JHEP **0802**, 002 (2008). [arXiv:0712.1851 [hep-ph]].
[55] A. Denner, S. Dittmaier, Nucl. Phys. **B844**, 199-242 (2011). [arXiv:1005.2076 [hep-ph]].

Journal of Physics: Conference Series **287** (2011) 012014 doi:10.1088/1742-6596/287/1/012014

Lattice QCD with dynamical quarks

Roger Horsley

School of Physics and Astronomy, University of Edinburgh, Edinburgh EH9 3JZ, UK

E-mail: rhorsley@ph.ed.ac.uk

Abstract. An overview is given of lattice QCD with dynamical quarks taking as a case study the QCDSF collaboration results for two and three flavours of dynamical quarks for hadronic stucture. We first sketch the lattice simulation and the idea of numerical simulations and why it is a difficult numerical problem. Measuring correlation functions on the generated configuration then gives access to hadronic masses and matrix elements. Several examples are given of results ranging from the mass spectrum and decay constants to the computation of low moments of structure functions.

1. Introduction and the problem of QCD

The goal of the lattice QCD programme is a numerical computation of strong interaction properties of particle physics such as the determination of the fundamental parameters of the QCD Lagrangian and physical properties of mesons/baryons such as masses, decay constants, (nucleon) matrix elements, form factors, generalised parton distributions, (GPDs), ..., all directly from the underlying theory. In this article we present a case study from the QCDSF Collaboration of what these calculations involve concentrating on the light quark sector: u, d, s. Many other groups in Europe, USA, Japan ..., cover these and other aspects of lattice gauge theories (LGTs) and a good overview of the state of the field can be found from the plenary and parallel talks in the yearly Lattice Conference, the latest in the series being Lattice 2010, [1]. A recent review of LGTs and hadronic physics is given in [2]. There are also several introductory texts on lattice gauge theories, [3, 4, 5, 6, 7].

QCD is a non-abelian gauge theory, locally invariant under the 'colour' group $SU(3)$ with Lagrangian

$$\mathcal{L} = -\tfrac{1}{4}F^a_{\mu\nu}F^{\mu\nu a} + \sum_q \overline{q}(i\gamma^\mu D_\mu - m_q)q\,, \tag{1}$$

a generalised QED with 6 quarks and 8 gluons, where the gluons unlike the photon can now interact with themselves. At high energy scales, it is found that perturbation theory is valid, with loosely bound quarks – 'asymptotic freedom' while at low energy scale the coupling constant becomes large and binds quarks into hadrons (either mesons, $q\overline{q}$ states or baryons, qqq states) – 'confinement'. QCD is thus fundamentally different to QED and possesses a rich dynamical content.

How do we get predictions from QCD? There are two classes: perturbative methods based on asymptotic freedom and non-perturbative methods such as using a chiral Lagragians (write down the most general Lagrangian compatible with QCD symmetries and expand in $m_\pi, p \ll 4\pi f_\pi$);

Journal of Physics: Conference Series **287** (2011) 012014 doi:10.1088/1742-6596/287/1/012014

heavy quark effective theories; sum rules; potential models (quarkonium) and lattice simulations. Of these it is fair to say that only the latttice approach directly tests the validity of the QCD Lagrangian directly, eq. (1). While perturbative methods are much the main connection between QCD and experiment they need additional hadronisation assumptions, lattice QCD is the only known direct bridge from theory which in principle covers all scales.

What would we like to determine? Some illustrative possibilities are

- The fundamental parameters of the QCD Lagrangian such as the running coupling constant, $\alpha_s(Q)$, and quark masses, m_q^R. Quark masses may be defined through the axial or vector currents

$$\partial_\mu \mathcal{A}_\mu^R = (m_{q1}^R + m_{q2}^R)P^R\,, \qquad \partial_\mu \mathcal{V}_\mu^R = (m_{q1}^R - m_{q2}^R)S^R\,, \tag{2}$$

(where $\mathcal{A}_\mu = \overline{q}_1\gamma_\mu\gamma_5 q_2$ etc.).

- The hadron spectrum – both meson octet pseudoscalar/vector and baryon octet and decuplet multiplets of particles. As an example the meson vector octet and baryon octet are depicted in Fig. 1.

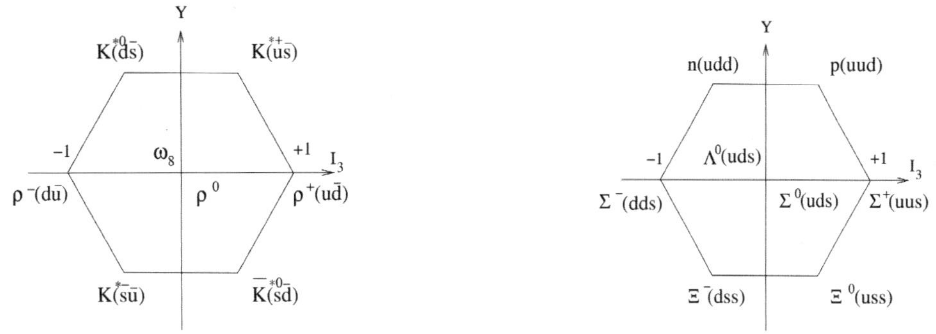

Figure 1. Vector meson (ρ, K^*, ω) and baryon (n, p, Λ, Σ, Ξ) octets, left, right panels.

- Decay constants – a typical example is the semi-leptonic decay $\pi^- \rightarrow \mu^-\overline{\nu}$ as shown in

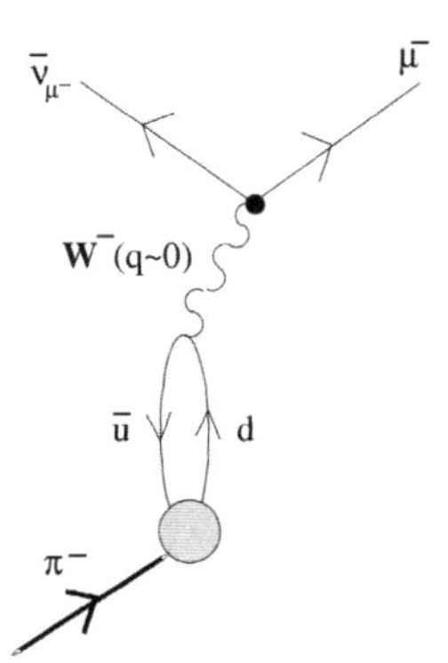

Figure 2. Decay of a pion.

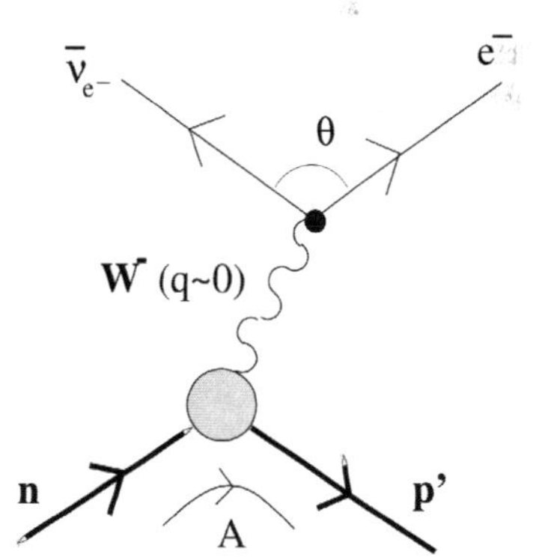

Figure 3. β-decay.

Fig. 2 with amplitude

$$T_{fi} = (-ig_W)^2 \cos\theta_c \overline{u}_\mu(\vec{k}_\mu)\gamma_\lambda(1-\gamma_5)v_\nu(\vec{k}_\nu) \times \frac{i}{m_W^2} \times \langle 0|\overline{u}\gamma_\lambda(1-\gamma_5)d|\pi^-(0)\rangle\,, \tag{3}$$

f_π being defined by $\langle 0|\overline{u}\gamma^0\gamma_5|\pi\rangle = m_\pi f_\pi$.

- β decay where a neutron decay to a proton $n \to p + e^- + \overline{\nu}_e$ (or $d \to u + e^- + \overline{\nu}_e$) as depicted in Fig. 3 with amplitude

$$T_{fi} = (-ig_W^2 \cos\theta_c)\overline{u}_e(\vec{p}_e)\gamma_\mu(1-\gamma_5)v_{\nu_e}(\vec{p}_\nu) \times \frac{1}{m_W^2} \times \langle p(\vec{p})|\overline{u}\gamma^\mu(1-\gamma_5)d|n(\vec{p})\rangle\,, \tag{4}$$

Decomposition of the blob in Fig. 3 is

$$\langle p(\vec{p})|\overline{u}\gamma^\mu(1-\gamma_5)d|n(\vec{p})\rangle = \overline{u}_p(p)\gamma^\mu(1-\gamma_5 g_A)u_n(p)\,. \tag{5}$$

Isospin invariance gives the matrix element

$$\langle \vec{p},\vec{s}|\overline{u}\gamma^\mu\gamma_5 u - \overline{d}\gamma^\mu\gamma_5 d|\vec{p},\vec{s}\rangle = 2s^\mu g_A\,, \tag{6}$$

which is the convenient expression used for a lattice computation of g_A.

- Form factors for example in electron–proton scattering as shown in Fig. 4 with

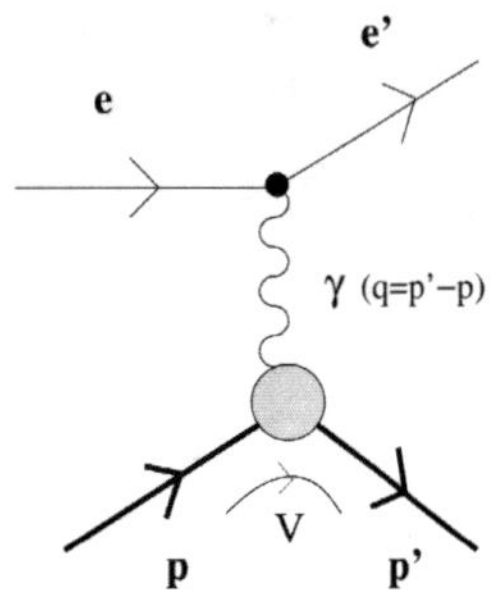

Figure 4. Electron – proton scattering.

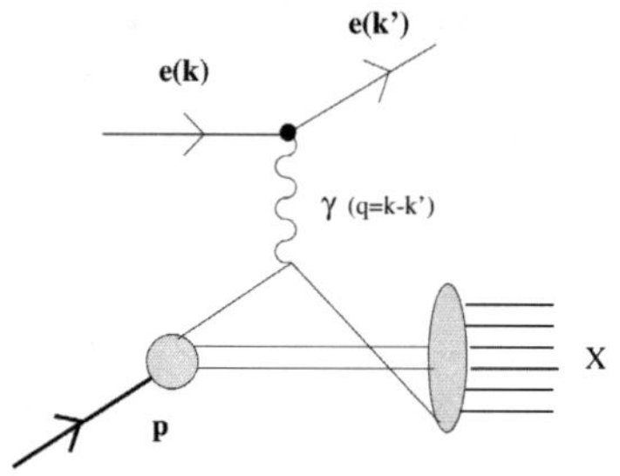

Figure 5. Electron–proton DIS.

$$T_{fi} = e^2\overline{u}_e(\vec{p}_e)\gamma^\mu u_e(\vec{p}_e')\frac{1}{q^2}\langle \vec{p}'|J^\mu(\vec{q})|\vec{p}\rangle\,, \qquad J^\mu = \tfrac{2}{3}\overline{u}\gamma^\mu u - \tfrac{1}{3}\overline{d}\gamma^\mu d + \ldots\,. \tag{7}$$

The proton blob can be decomposed into two form factors $F_1(q^2)$, $F_2(q^2)$

$$\langle \vec{p}'|J^\mu(\vec{q})|\vec{p}\rangle = \overline{u}(\vec{p}')\left[\gamma^\mu F_1(q^2) + i\sigma^{\mu\nu}\frac{q_\nu}{2m_N}F_2(q^2)\right]u(\vec{p})\,, \tag{8}$$

where $F_1^p(0) = 1$ (due to vector current conservation and $F_2^p(0) = \mu^p - 1$ the anomalous magnetic moment (in units of $e/(2m_N)$) of the nucleon. From the form factor we can find the (charge) radii of the nucleon, defined by

$$r_i^2 = -6\left.\frac{dF_i(q^2)}{dq^2}\right|_{q^2=0}\,, \qquad i = 1, 2\,. \tag{9}$$

- High Energies / Hard Processes – DIS (deep inelastic scattering) which are described by the parton model and structure functions F as shown in Fig. 5 where $ep \to eX$ when $Q^2 \equiv -q^2 \to \infty$, with $x = -q^2/2\nu$ = fixed ($\nu = p \cdot q \to \infty \sim$ energy transfer). In this case α_s is small, so there are 'quasifree' quarks (and gluons) – partons with $F \equiv F(x)$ only – Bjorken scaling. x is fraction of nucleon momentum carried by parton $p_{part} = xp_N$. With

QCD we have additional $\ln Q^2$ scaling violations. Quantitatively using the OPE (operator product expansion) for both unpolarised/polarised processes

$$\int_0^1 dx x^{n-2} F_2^{p-n}(x, Q^2) = E_{NS}^{\overline{MS}}(\alpha_s) v_{n;NS}^{\overline{MS}}(Q) , \qquad n \text{ even} , \tag{10}$$

where NS, non-singlet: $p - n$ or $u - d$

$$\begin{aligned} \langle N | \mathcal{O}_q^{\gamma;\{\mu_1 \cdots \mu_n\}} - \text{tr} | N \rangle &= 2 v_n^{(q)} \left[p^{\mu_1} \cdots p^{\mu_n} - \text{tr} \right] \\ \mathcal{O}_q^{\gamma;\mu_1 \cdots \mu_n} &= \bar{q} \gamma^{\mu_1} i \overleftrightarrow{D}^{\mu_2} \cdots i \overleftrightarrow{D}^{\mu_n} q . \end{aligned} \tag{11}$$

Alternatively the matrix elements can be given in terms of *parton densities*: q, $\overline{q}$, g with

$$\langle x^{n-1} \rangle \equiv v_n^{(q)\overline{MS}}(\mu) = \int_0^1 dx x^{n-1} \left[q^{\overline{MS}}(x, \mu) + (-1)^n \overline{q}^{\overline{MS}}(x, \mu) \right] , \tag{12}$$

where q is the probability of finding a quark with a fraction x of the nucleon momentum. Naively, we would expect for the nucleon, the u and d to be δ-functions at $x \sim \frac{1}{3}$ and it is indeed found that u, d are smeared out δ-functions around this x value.

- A more general approach connecting these various methods is given by GPDs (which we shall not consider further here).

So in general we need computation of non-perturbative quantities: such as hadron masses, m_H and matrix elements $\langle H | \widehat{\mathcal{O}} | H \rangle$.

2. The lattice approach

The lattice approach first Euclideanises and discretises space-time with a lattice spacing a. The path integral thus becomes a partition function which is a $N_S \times N_T \times d \times (n_c^2 - 1)$ integral. For example for a $\sim 32^3 \times 64 \times 4 \times 8$ lattice this means that we have a $\sim O(50,000,000)$ dimensional integral. The only known technique for such high dimensional integrals is the Monte Carlo method for finding expectation values of

$$\langle \mathcal{O} \rangle = \frac{1}{\mathcal{Z}} \int [dU][dq d\overline{q}] \mathcal{O} e^{-(S_G + S_F)} . \tag{13}$$

The continuum $a \to 0$ limit is equivalent to $\xi \to \infty$ i.e. a second order phase transition. In Fig. 6 we depict this limit. Note that $1\text{fm} = 10^{-15}\text{m}$ is a typical hadronic scale ie $\sim$ nucleon radius.

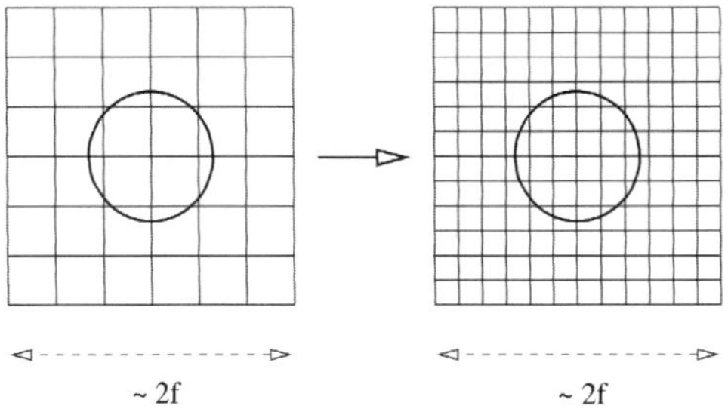

Figure 6. The lattice box.

What do we choose for the action? It should have the classical limit $a \to 0$ of eq. (1). For the gluon action the choice is relatively easy. Defining the gauge links by $U_\mu(x) = e^{igaA_\mu(x)}$ then a closed loop (or sum of closed loops) with suitable coefficient is sufficient, being gauge invariant. The simplest choice is

$$S_G = \tfrac{1}{3}\beta \sum_{plaq} tr_c(1 - U_{\mu\nu}^{plaq}) \to \int d^4x \tfrac{1}{4} F_{\mu\nu}^{a2} + O(a^2) . \tag{14}$$

For the fermion action we require

$$S_F = \sum_{q;n} \overline{q} \underbrace{\mathcal{M}}_{?} q \to \int d^4x \sum_q \overline{q}(\gamma_\mu D_\mu + m_q)q + O(a^p) . \tag{15}$$

But there is an immediate fundamental problem: a naive discretisation of the Dirac action leads to 2^4 continuum flavours (rather than just one). Several types of fermion discretisation have been proposed to ameliorate this problem, as summarised in Table 1 where the pros and cons

	chiral symmetry	flavour symmetry	numerical simulation
Wilson / $O(a)$-improved Wilson (clover)	violated recovered in the continuum	correct	medium expensive
twisted mass	violated	2 flavours a q mixing mass term	medium expensive
staggered	exact $U(1)$ from $U(4)$	4 flavours non-trivial mixing	fast
Ginsparg-Wilson: overlap / domain wall	exact at finite a	corrrect	most expensive

Table 1. Various possibilites for lattice fermions.

are illustrated. The QCDSF Collaboration has used clover fermions where each quark in the action is given by

$$S_F^{(q)} = a^4 \sum_x \left\{ \overline{q}(D_W + m_0)q + \frac{1}{4} a c_{sw} \sum_{\mu\nu} \overline{q} \sigma_{\mu\nu} F_{\mu\nu} q \right\} ,$$

with

$$D_W = \frac{1}{2} \sum_\mu \left(\gamma_\mu (\Delta_\mu + \Delta_\mu^*) - a(\Delta_\mu \Delta_\mu^*) \right) . \tag{16}$$

(Δ_μ, Δ_μ^* are the forward and backward derivatives including the gauge link.) The mass, m_0 is given by $m_0 = m_q + m_c$ with

$$m_q(c_{sw}) = \frac{1}{2a}\left(\frac{1}{\kappa} - \frac{1}{\kappa_c(c_{sw})} \right) \qquad m_c(c_{sw}) = \frac{1}{2a}\left(\frac{1}{\kappa_c(c_{sw})} - \frac{1}{1/8} \right) . \tag{17}$$

So to find the chiral limit when the quark mass vanishes we need to tune to a critical kappa, $\kappa_c(c_{sw})$ which corresponds to an additive mass renormalisation $m_c(c_{sw})$. Discretisation errors can be reduced from $O(a)$ to $O(a^2)$, if it can be arranged for one physical quantity (eg mass ratio) to have no $O(a)$ terms this fixes $c_{sw}(g^2)$, then all other masses are automatically improved to $O(a^2)$. Matrix elements, however, require further improvement operators.

Correlation functions are calculated by averaging over gauge field configurations with a weight given by the Boltzmann factor

$$\langle \mathcal{O} \rangle = \int [dU] \mathcal{O}(U) p^*(U) , \qquad p^*(U) = \frac{1}{\mathcal{Z}} e^{-S(U)} , \tag{18}$$

generated using Monte Carlo methods. The Markov Chain, with Q being the transition matrix is given by

$$p_{n+1}(U') = \int [dU] p_n(U) Q(U, U') , \qquad \int [dU'] Q(U, U') = 1 . \tag{19}$$

In 'equilibrium', $p_n \to p^*$ and the Markov chain equation is satisfied by $p^*(U)Q(U, U') = p^*(U')Q(U', U)$ or 'detailed' balance

$$\begin{aligned} Q(U, U') &= \begin{array}{c}\text{prob. of choosing a} \\ \text{Candinate config.}\, U'\end{array} \times \begin{array}{c}\text{prob. of Accepting} \\ \text{new config.}\end{array} \\ &= p_C(U \to U') \times p_A(U \to U') . \end{aligned} \tag{20}$$

For dynamical fermions the algorithm of choice is the Hybrid Monte Carlo algorithm. First (as with all algorithms) the Grassmann fermions must be integrated out (possible as we have quadratic Grassmann fields)

$$\int [dq][d\overline{q}]\, e^{\overline{q} \mathcal{M} q} = \det \mathcal{M} = \int [d\phi][d\phi^*]\, e^{-\phi^\dagger \mathcal{M}^{-1} \phi} . \tag{21}$$

Then

- p_C: candidate config (U', π')

$$p_C[(U, \pi) \to (U', \pi')] = e^{-\frac{1}{2}\pi^2} \delta[(U', \pi') - (U, \pi)] ,$$

  using discretised Hamilton's equations (leapfrog in computer time)
- p_A: accept/reject using Metroplis algorithm

$$p_A[(U, \pi) \to (U', \pi')] = \min \left(1, e^{-\Delta H}\right) , \quad \Delta H = H(U', \pi') - H(U, \pi) .$$

Monte Carlo methods are like doing an experiment (in the computer) and consideration must be taken of computing the error bars,

$$\overline{\mathcal{O}} = \frac{1}{N_{conf}} \sum_{n=1}^{N_{conf}} \mathcal{O}_n , \qquad \text{error} \propto \frac{1}{\sqrt{N_{conf}}} . \tag{22}$$

The next step is to compute the appropriate masses/matrix elements. In Figs. 7, 8 we illustrate the computation required, the masses being given by

$$\langle M(t) M^\dagger(0) \rangle = |\langle 0 | \widehat{M} | M \rangle|^2 e^{-m_M t} , \qquad \langle B(t) \overline{B}(0) \rangle \propto e^{-m_N t} . \tag{23}$$

Similarly in Figs. 9, 10 we indicate the computation of baryonic quark bilinear matrix elements,

Journal of Physics: Conference Series **287** (2011) 012014 doi:10.1088/1742-6596/287/1/012014

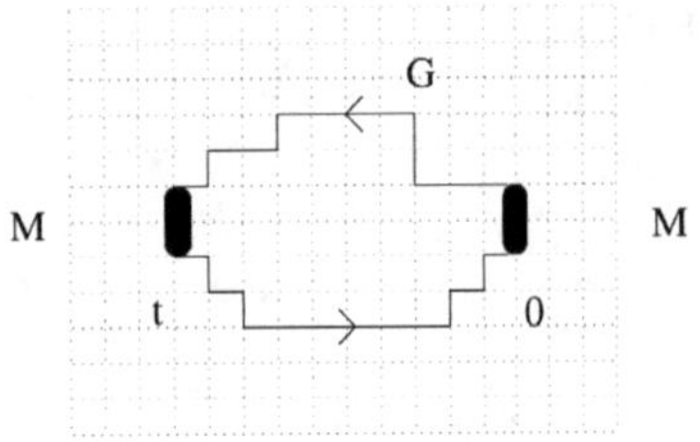

Figure 7. The meson two point correlation functions.

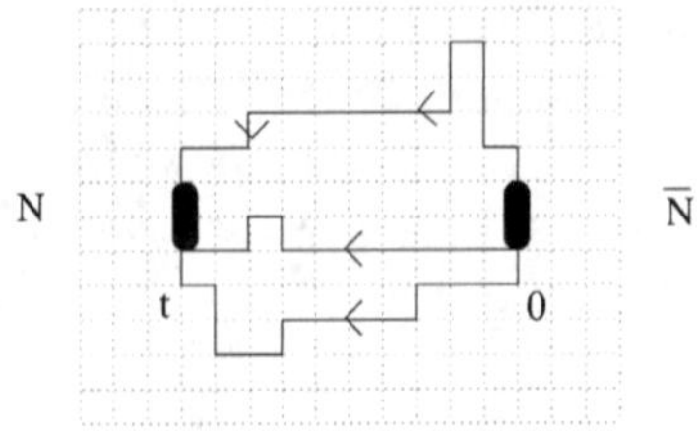

Figure 8. The baryon two point correlation functions.

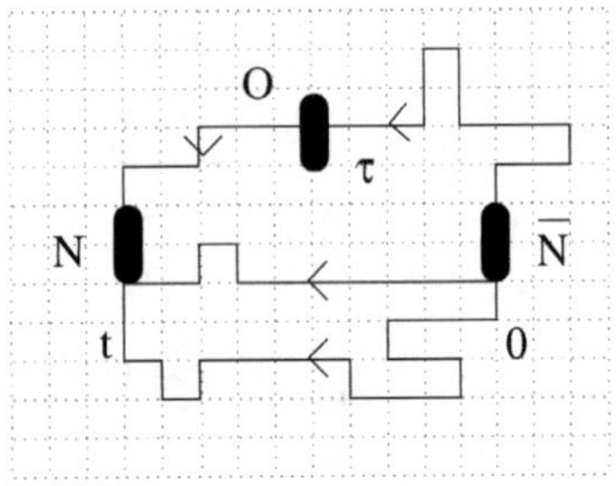

Figure 9. The quark line connected baryon three point function.

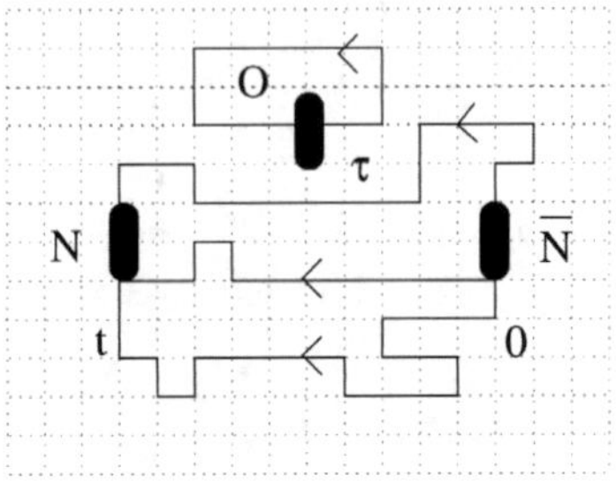

Figure 10. The quark line disconnected baryon three point function.

using

$$R(t,\tau;\vec{p}) = \frac{\langle N(t;\vec{p})O(\tau;\vec{0})\overline{N}(0;\vec{p})\rangle}{\langle N(t;\vec{p})\overline{N}(0;\vec{p})\rangle} \propto \langle N(\vec{p})|\widehat{O}|N(\vec{p})\rangle \,. \tag{24}$$

(For form factors, this result must be generalised to allow for momentum transfer.) Practically, due to the CPU cost, only the quark line connected part of the matrix element is calculated, which means that we are restricted to non-singlet matrix elements, for example $\langle N| \left[\overline{u}Ou - \overline{d}Od\right] |N\rangle$.

A lattice computation requires the limits:

- Many independent configurations ($N_{conf} \to \infty$) typically at present $N_{conf} \sim 1000$
- Large box ($L \to \infty$) typically at present $L \sim 2 - 3$fm
- Continuum extrapolation ($a \to 0$) typically at present $a \sim 0.11 - 0.08 - 0.06$ fm
- Extrapolation ($m_q^R \to m_q^{R phys}$, $q = u,\, d,\, s$) typically at present $m_\pi^2 \sim 400 - 200$ MeV.

All these limits make LGT a very difficult computation. It is easy to see that we still have a long way to go. Taking $m_\pi L = 4$ (a reasonable value to minimise finite size effects) and $p_{min} \equiv 2\pi/L = m_\pi$ then this gives Fig. 11. Either we are forced to work with a small a but with 'large' m_π (no continuum extrapolation needed but a chiral extrapolation) or 'coarse' a with $m_{ps} \sim 140$ MeV (no chiral extrapolation necessary but continuum extrapolation). It should be noted that small changes in a, m_π or L produce huge changes in cost as we expect approximately

$$\text{flops / config} \sim \left(\frac{L}{3\,\text{fm}}\right)^{\sim 5} \times \left(\frac{0.1\,\text{fm}}{a}\right)^{\sim 6} \times \left(\frac{1}{am_\pi}\right)^{\sim 1} \tag{25}$$

Currently Teraflop machines are used. It is not all gloom though: reached pion masses have been dropping rapidly in the last few years.

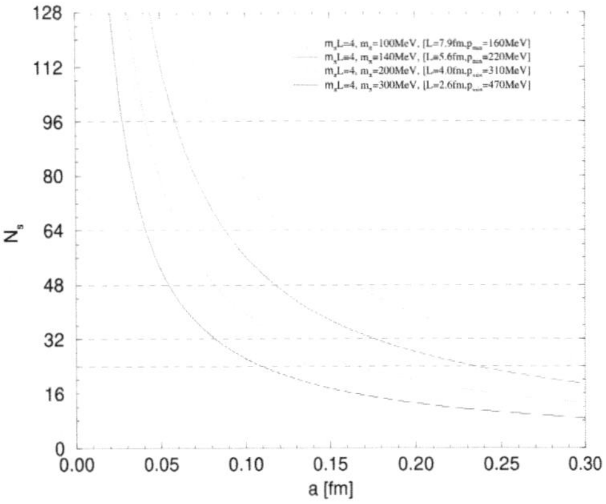

Figure 11. Plot of spatial lattice size, N_S versus lattice spacing, a (where $L = N_S a$).

Figure 12. The path to the physical point: $n_f = 2$ and $2 + 1$ flavours, left, right panels.

The path to the physical pion mass has to be chosen. While for two (degenerate) quark flavour masses, there is little to decide; as depicted in Fig. 12 for $2+1$ flavours, we have a plane, so many paths are possible. We have decided to start on a point on the $SU(3)$ flavour symmetry line and to keep the singlet quark mass constant, [8],

$$\overline{m}^R = \frac{1}{3}\,(2m_l^R + m_s^R) = \text{const.}\,, \tag{26}$$

as also sketched in Fig. 12. In this case we find results shown in Fig. 13.

To determine the scale (i.e. the value of a in femotometers), there are various possibilities, we have chosen to use a flavour singlet quantity such as the centre of mass of the baryon octet, $X_N = \frac{1}{3}(m_N + m_\Sigma + m_\Xi)$.

After the simulation we still need to renormalise the matrix elements. Several methods have been proposed, the most popular general purpose method proposed was in [9] which mimics (continuum) perturbation theory in a certain 'MOM' scheme. We sketch the procedure here. The amputated Green's function is first found (in the Landau gauge which maximizes $\sum_{x,\mu}(1 - \frac{1}{3}\mathrm{ReTr}U_\mu(x))$. Using a 'momentum source', [10], we first compute the correlation functions

$$C^{ab}_{\mathcal{O},\alpha\beta}(p) = \frac{1}{V}\sum_{xyz} e^{-ip\cdot(x-y)}\langle q^a_\alpha(x)\mathcal{O}(z)\overline{q}^b_\beta(y)\rangle_{q,U}\,, \tag{27}$$

and then find the 1PI vertex function

$$\Gamma_{\mathcal{O}}(p) = S^{-1}(p)C_{\mathcal{O}}(p)S^{-1}(p)\,, \tag{28}$$

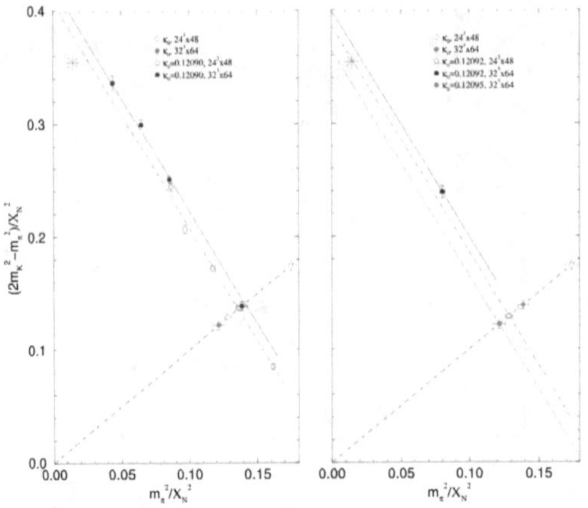

Figure 13. $(2m_K^2 - m_\pi^2)/X_N^2$ versus m_π^2/X_π^2 (with $X_\pi^2 = (2m_K^2 + m_\pi^2)/3$). The dashed black line, $y = x$ represents the $SU(3)$ flavour symmetric line. Shown are points on the flavour symmetric line (orange) followed by results with $\overline{m} = \text{const.}$. Physical values are denoted by stars.

(where S is the quark propagator). The renormalised operator (and quark wavefunction) are

$$\mathcal{O}_R = Z_\mathcal{O}\mathcal{O}\,, \qquad q_R = \sqrt{Z_q}q\,, \tag{29}$$

so the renormalised 1PI correlation is then $\Gamma_{\mathcal{O},R}(p) = S_R^{-1}(p)C_{\mathcal{O},R}(p)S_R^{-1}(p) = Z_q^{-1}Z_\mathcal{O}\Gamma_\mathcal{O}(p)$. $Z_\mathcal{O}$ is fixed from the renormalisation condition at some scale $p^2 = \mu_p^2$:

$$\tfrac{1}{12}\,\mathrm{tr}\left[\Gamma_{\mathcal{O},R}(p)\Gamma^{-1}_{\mathcal{O},\mathit{BORN}}(p)\right]\Big|_{p^2=\mu_p^2} = 1\,. \tag{30}$$

This has solved the problem, but phenomenologists use the (perturbative) $\overline{MS}$ scheme, so we have to

- Convert the MOM scheme to the renormalisation group invariant or RGI form (which is a scale and scheme independent number)
- Measure the RGI constant
- Convert back to the $\overline{MS}$ scheme

The disadvantage is that some perturbative knowledge used. The RGI form is found from

$$\mathcal{O}^{\mathit{RGI}} = Z_\mathcal{O}^{\mathit{RGI}}\mathcal{O} = \Delta Z_\mathcal{O}^\mathcal{S}(\mu)\mathcal{O}^\mathcal{S}(\mu)\,, \qquad \mathcal{S} = MOM\,, \overline{MS}\,, \tag{31}$$

with

$$[\Delta Z_\mathcal{O}^\mathcal{S}(\mu)]^{-1} = \left[2b_0 g^\mathcal{S}(\mu)^2\right]^{-\frac{d_{\mathcal{O};0}}{2b_0}} \exp\left\{\int_0^{g^\mathcal{S}(\mu)} d\xi\left[\frac{\gamma_\mathcal{O}^\mathcal{S}(\xi)}{\beta^\mathcal{S}(\xi)} + \frac{d_{\mathcal{O};0}}{b_0\xi}\right]\right\}. \tag{32}$$

This allows us to measure a plateau and then to run to 2 GeV. For $n_f = 2$ flavours typical results are shown in Figs. 14 and 15.

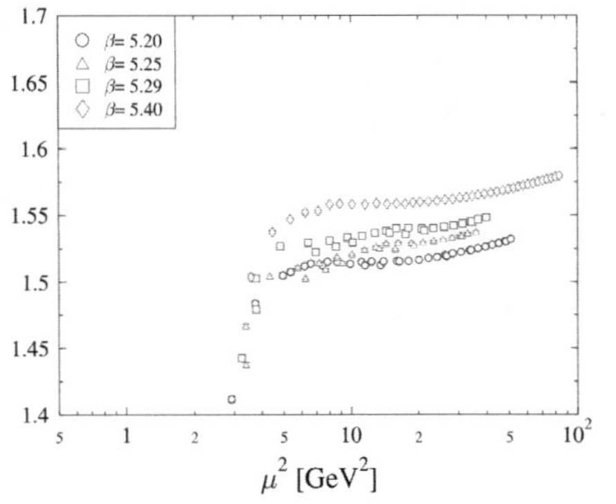

Figure 14. Plot of Z^{RGI} (for v_{2b}^{u-d} where v_2 is defined in eq. (11)) versus μ^2, [11].

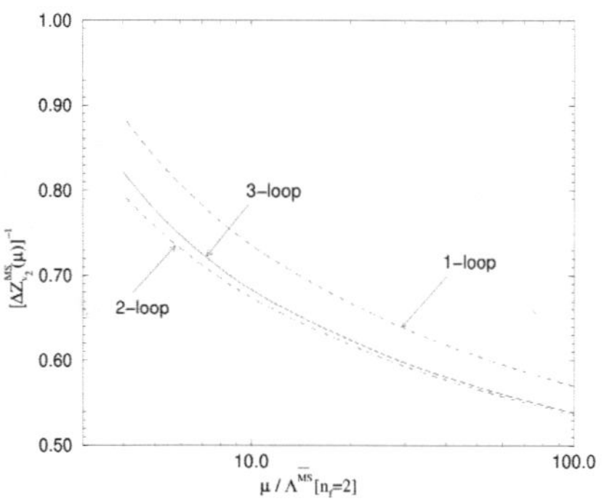

Figure 15. Plot of $\Delta Z_{v_2}(\mu)$ versus scale, $\mu/\Lambda^{\overline{MS}}$ for $n_f = 2$ flavours.

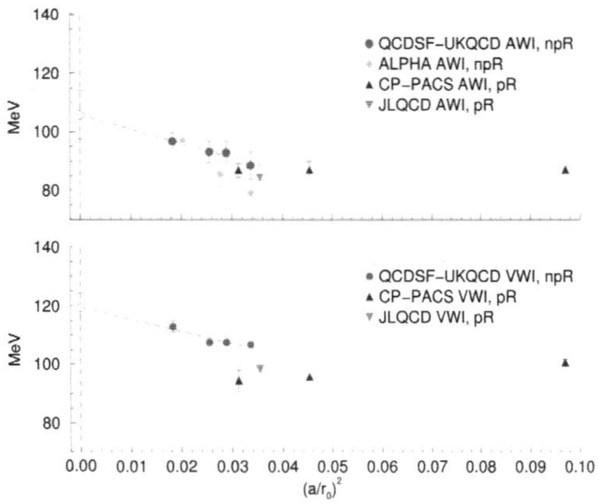

Figure 16. Plot of the continuum extrapolation of the strange quark mass.

3. Illustrative results

We first consider the strange quark mass. We can define the (bare) quark mass through either the axial or vector currents, which must then be renormalised, as in eq. (2) and section 2 and then extrapolated to the physical point. For $n_f = 2$, [12] we find the results shown in Fig. 16.

We next consider the vector meson spectrum. A flavour expansion about the symmetric point (Gell-Mann–Okubo) gives the constrained fits

$$\begin{aligned} m_\rho &= m_0 + 2\alpha\delta m_l + (\beta_0 + 2\beta_1)\delta m_l^2 \\ m_{K^*} &= m_0 - \alpha\delta m_l + (\beta_0 + 5\beta_1 + 9\beta_2)\delta m_l^2 \\ m_\phi &= m_0 - 4\alpha\delta m_l + (\beta_0 + 8\beta_1)\delta m_l^2 \, , \end{aligned} \tag{33}$$

where $\delta m_q = m_q - \overline{m}$. We see that the linear terms are described by 1 coefficient; the quadratic terms with 3 coefficients. In Fig. 17 we show the octet multiplet as a 'fan' plot. Very little curvature is seen. A similar result is obtained for the baryon spectrum. We also show in Fig. 17 the baryon octet 'fan' plot.

In Fig. 18 we show the ratio f_K/f_π, while in Fig. 19, g_A is plotted, [13].

The F_1 form factors and r_1 charge radius, (for $n_f = 2$ flavours) are given by eq. (8) and are shown in Figs. 20, 21, [14]. It is also seen that there are small $O(a^2)$ scaling violations and that

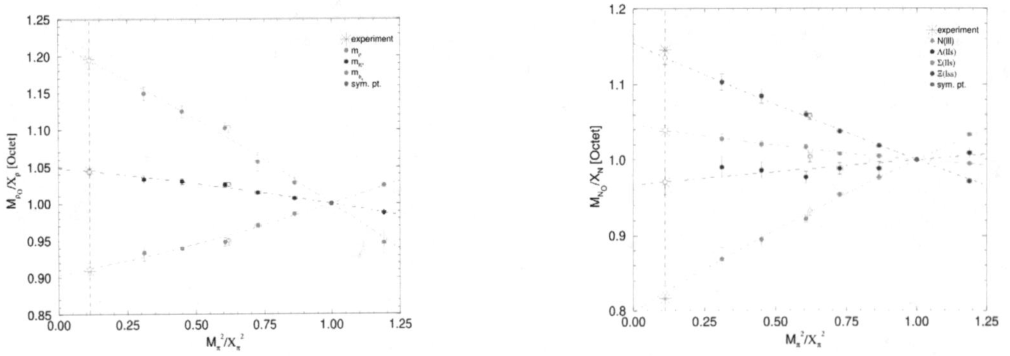

Figure 17. m_{ρ_O}/X_ρ ($\rho_O = \rho$, K^*, ϕ_s) and m_{N_O}/X_N ($N_O = N$, Λ, Σ, Ξ) against m_π^2/X_π^2 (left, right panels respectively), together with the constrained fit of e.g. eq. (33).

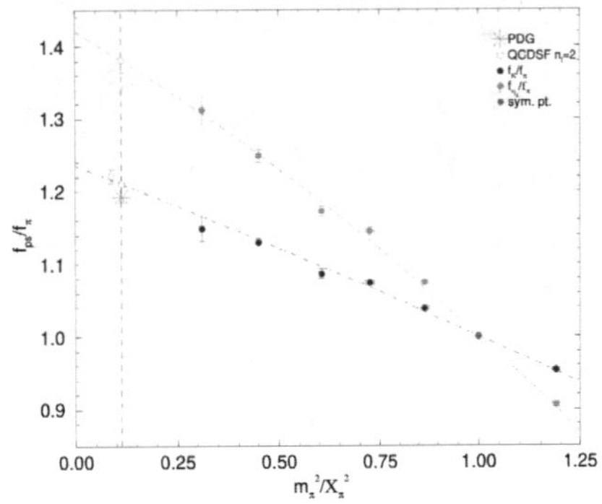

Figure 18. Plot of f_K/f_π versus m_π^2/X_π^2.

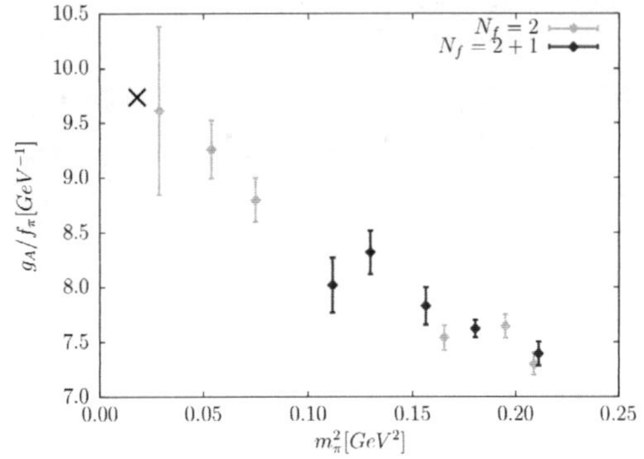

Figure 19. Plot of g_A/f_π against m_π^2.

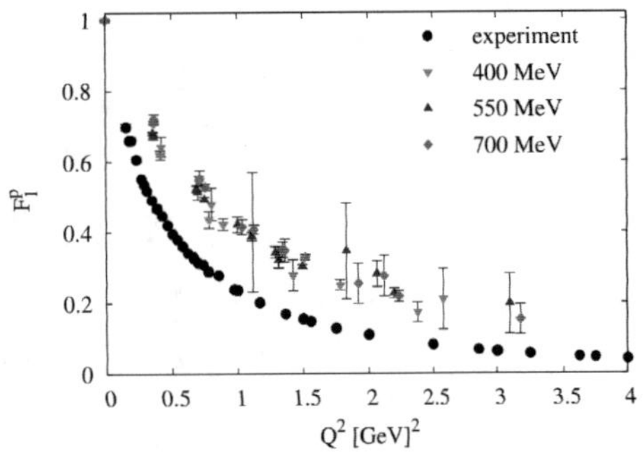

Figure 20. Plot of $F_1(Q^2)$ versus $Q^2 = -q^2$.

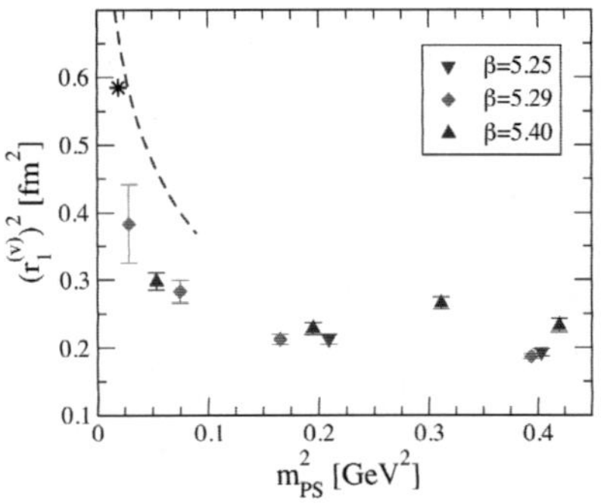

Figure 21. Plot of r_1^2 versus m_π^2.

all the 'activity' appears to be taking place close to the physical point. Form factor radii are defined in eq. (9) and r_1 (for example) determined from a fit

$$F_1(q^2) = \frac{1}{1 + q^2/m_1^2}, \tag{34}$$

is shown in Fig. 21. (The dashed line is from χ-PT.) It is seen that $r_1^{LAT} < r_1^{EXPT}$ (and this is also true for r_2) – so probably we are (still) missing dynamics due to the 'pion cloud'.

Another interesting quantity is the lowest moment of the structure function F_2^{NS} of the nucleon, as shown in Fig. 22. It is apparent that the recent $n_f = 2$ (left) results, [14] and older

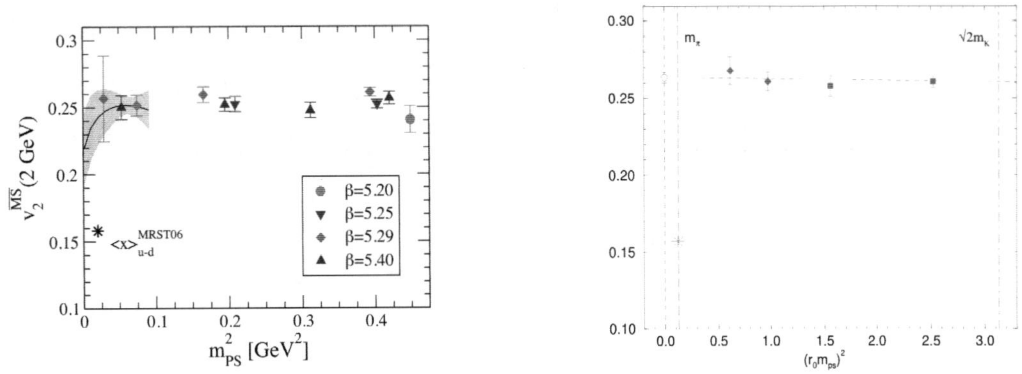

Figure 22. Plots of $\langle x \rangle^{NS}$ in the $\overline{MS}$ scheme at a scale of 2 GeV, eq. (12) for $n_f = 2$, $n_f = 2+1$ flavours, left, right panels respectively. The x-scale of these pictures are similar.

$n_f = 0$ (right) results, [15] are very similar, and that there is also a challenge to understand this and the discrepancy to the phenomenological value.

4. Conclusions

We first emphasise that the lattice approach is a fundamental test of QCD. In this preoceedings we have sketched the lattice programme (taking the QCDSF Collaboration as an example) from beginning to end. The approach is very much like an experiment – the precision can be improved by increasing the data set. Lattice gauge theory, LGT, can compute hadron masses and, more interestingly is capable of computing matrix elements. We need to take several limits however, each of which is painful, such as a large box, quark mass extrapolation to the physical point and a continuum extrapolation, $a^2 \to 0$. We have showed here some results for masses, decay constants, form factors (moments of) structure functions. Of course there are many other quantities under active consideration such as hadron decays, excited nucleon states, nucleon distribution amplitudes, GPDs, charm and b-quark physics finite temperature QCD, 'technicolor' experiments to name only a few. This will all need much faster machines for the programme to reach its goals.

References

[1] The XXVIII International Symposium on Lattice Field Theory, Lattice 2010 - Villasimius, Sardinia Italy, June 14-19, 2010, Proceedings of Science, PoS(Lattice 2010), http://pos.sissa.it/ .
[2] Ph. Hägler, *Phys. Rept.* **490** (2010) 49, [arXiv:0912.5483]].
[3] M. Creutz, *Quarks, gluons and lattices* (CUP 1983).
[4] H. J. Rothe, *Lattice Gauge theories: an introduction* (World Scientific 1992).
[5] J. Smit, *Introduction to Quantum Fields on a Lattice* (CUP 2002).
[6] T. Degrand and C. DeTar, *Lattice methods for quantum chromodynamics*, (World Scientific 2006).
[7] C. Gattringer, C.B. Lang, *Quantum Chromodynamics on the Lattice*, (Springer 2010).
[8] W. Bietenholz et. al., *Phys. Lett.* **B690** (2010) 436, [arXiv:1003.1114[hep-lat]]; arXiv:1102.5300[hep-lat].
[9] G. Martinelli et. al., *Nucl. Phys.* **B445** (1995) 81, [arXiv:hep-lat/9411010].
[10] M. Göckeler et. al., Nucl. Phys. **B544** (1999) 699, [arXiv:hep-lat/9807044].
[11] M. Göckeler et. al., *Phys. Rev.* **D82** (2010) 114511, [arXiv:1003.5756[hep-lat].
[12] M. Göckeler et. al., *Phys. Rev.* **D74** (2006) 094508, [arXiv:hep-lat/0602028].
[13] M. Göckeler et. al., Proceedings of Science PoS(Lattice 2010) 163, arXiv:1102.3407[hep-lat].
[14] S. Collins et. al., Proceedings of Science PoS(Lattice 2010) 153, arXiv:1101.2326[hep-lat].
[15] T. Bakeyev et. al., *Nucl. Phys. Proc. Suppl.* **128** (2004) 82, [arXiv:hep-lat/0311017].

The S_3 symmetry: Flavour and texture zeroes

F. González Canales[*] and A. Mondragón[**]

Instituto de Física, Universidad Nacional Autónoma de México, 04510, México D.F., México

E-mail: *ffelix@fisica.unam.mx, **mondra@fisica.unam.mx

Abstract. We use the permutational symmetry group S_3 as a symmetry of flavour, which leads to a unified treatment of masses and mixings of the quarks and leptons. In this framework all mass matrices of the fermions in the theory have the same form with four texture zeroes of class I. Also, with the help of six elements of real matrix representation of S_3 as transformation matrices of similarity classes, we make a classification of the sets of mass matrices with texture zeroes in equivalence classes. This classification reduce the number of phenomenologically viable textures for the non-singulars mass matrices of 3×3, from thirty three down to only eleven independent sets of matrices . Each of these sets of matrices has exactly the same physical content.

1. Introduction

The Standard Model (SM) can be extended by adding of three right-handed neutrino states, which would be singlets of the gauge group of SM, but coupled to matter just through the neutrino masses [1]. But in minimal extensions of the Standard Model, considering a mass term for left-handed neutrinos purely of Dirac is not theoretically favored, because it can not explain easily why neutrinos are much lighter than the charged leptons. Thus, we assume that neutrinos have Majorana masses and acquire their small masses through of the type I seesaw mechanism.

In both lepton and quark sectors of the extended Standard Model, analogous fermions in different generations, say u, c and t or d, s and b, have completely identical couplings to all gauge bosons of the strong, electroweak interactions. Prior to the introduction of the Higgs boson and mass terms, the Lagrangian is chiral and invariant with respect to any permutation of the left and right quark fields. The introduction of a Higgs boson and the Yukawa couplings give mass to the quarks and leptons when the gauge symmetry is spontaneously broken. The mass term in the Lagrangian, obtained by taking the vacuum expectation value of the Higgs field in the quark and lepton Higgs couplings, gives rise to mass matrices $\mathbf{M_d}$, $\mathbf{M_u}$, $\mathbf{M_l}$ and $\mathbf{M}_\nu$;

$$\mathcal{L}_Y = \bar{\mathbf{q}}_{d,L}\mathbf{M}_d\mathbf{q}_{d,R} + \bar{\mathbf{q}}_{u,L}\mathbf{M}_u\mathbf{q}_{u,R} + \bar{\mathbf{L}}_L\mathbf{M}_l\mathbf{L}_R + \bar{\nu}_L\mathbf{M}_\nu \left(\nu_L\right)^c + h.c. \tag{1}$$

Therefore, we propose, as well as many other authors [2, 3, 4, 5, 6], that the texture zeroes of the mass matrices of the quarks and leptons, are the result of a flavour permutational symmetry S_3 and its spontaneous or explicit breaking.

On the other hand, in the last ten years, important theoretical advances have been made in the understanding of the mechanisms for the mass fermion generation and flavour mixing. Phenomenologically, some striking progress has been made with the help of the texture zeroes and flavour symmetries in specifying the quantitative relationship between flavour mixing angles and quark or lepton mass ratios [3, 4, 7, 5] with a minimum of free parameters. In fact it, can be noted that different mass matrices with texture zeroes located in different positions may have exactly the same physical content. Therefore the question arises, is there any relation between these matrices?. We find an answer to this question through the similarity clases, recently proposed by Branco [8].

In this paper, we use the permutational symmetry S_3 as a flavour symmetry, in a unified treatment of masses and mixings of quarks and leptons. Also, with the help of a real matrix representation of the group S_3, as a basis for the transformation matrices of the similarity classes, we make a classification of the set of mass matrices with texture zeroes.

2. Flavour permutational symmetry S_3

A phenomenologically and theoretically meaningful approach for reducing the number of free parameters in the Standard Model is the imposition of texture zeroes [5, 7] or flavour symmetries. Recent flavour symmetry models are reviewed in [9]. Also, certain texture zeroes may be obtained from a flavour symmetry. In particular, a permutational S_3 flavour symmetry and its sequential explicit breaking justifies taking the same generic form for the mass matrices of all Dirac fermions, conventionally called a four texture zeroes form [3, 4]. Some reasons to propose the validity of a matrix with four texture zeroes as a universal form for the mass matrices of all Dirac fermions in the theory are the following:

(i) The idea of S_3 flavour symmetry and its explicit breaking has been succesfully realized as a mass matrix with four texture zeroes in the quark sector to interpret the strong mass hierarchy of up and down type quarks [2, 3]. Also, the numerical values of mixing matrices of the quarks determined in this framework are in good agreement with the experimental data [3].

(ii) Since the mass spectrum of the charged leptons exhibits a hierarchy similar to the quark's one, it would be natural to consider the same S_3 symmetry and its explicit breaking to justify the use of the same generic form with four texture zeroes for the charged lepton mass matrix.

(iii) As for the Dirac neutrinos, we have no direct information about the absolute values or the relative values of the Dirac neutrino masses, but the mass matrix with four texture zeroes can be obtained from an grand unified theory $SO(10)$ which describes well the data on masses and mixings of Majorana neutrinos [10]. Furthermore, from supersymmetry arguments, it would be sensible to assume that the Dirac neutrinos have a mass hierarchy similar to that of the u-quarks and it would be natural to take for the Dirac neutrino mass matrix also a matrix with four texture zeroes.

2.1. *Mass matrices from the breaking of $S_{3L} \otimes S_{3R}$*

Some authors have pointed out that realistic Dirac fermion mass matrices results from the flavour permutational symmetry $S_L(3) \otimes S_R(3)$ and its spontaneous or explicit breaking [2, 3, 4, 5, 6]. The group S_3 treats three treats three objects symmetrically, while its 3×3 representation structure $\mathbf{1} \oplus \mathbf{2}$ treats the generations differently and adapts itself readily to the hierarchical nature of the mass spectrum. Under exact $S_{3L} \otimes S_{3R}$ symmetry, the mass spectrum for either quark sector (up or down quarks) or leptonic sector (charged leptons or Dirac neutrinos) consists of one massive particle in a singlet irreducible representation and a pair of massless particles in a doublet irreducible representation, the corresponding quark mass matrix with the exact $S_{3L} \otimes S_{3R}$ symmetry will be denoted by $\mathbf{M}_{i3}$ with $i = u, d, l, \nu_D$. Here, u, d, l and ν_D denote the up quarks, down quarks, charged leptons and Dirac neutrinos, respectively. Assuming that there is only one Higgs boson in the theory, this $SU(2)_L$ doublet can be accommodated in a singlet representation of S_3.In order to generate masses for the first and second families, we add the terms $\mathbf{M}_{i2}$ and $\mathbf{M}_{i1}$ to $\mathbf{M}_{i3}$. The term $\mathbf{M}_{i2}$ breaks the permutational symmetry $S_{3L} \otimes S_{3R}$ down to $S_{2L} \otimes S_{2R}$ and mixes the singlet and doublet representation of S_3. $\mathbf{M}_{i1}$ transforms as the mixed symmetry term in the doublet complex tensorial representation of $S_3^{diag} \subset S_{3L} \otimes S_{3R}$. Putting the first family in a complex representation will allow us to have a CP violating phase in the mixing matrix. Then, in a symmetry adapted basis, $\mathbf{M}_i$ takes the form

$$M_i = m_{i3} \left[\begin{pmatrix} 0 & |A_i|e^{i\phi_i} & 0 \\ |A_i|e^{-i\phi_i} & 0 & 0 \\ 0 & 0 & 0 \end{pmatrix} + \begin{pmatrix} 0 & 0 & 0 \\ 0 & -\Delta_i + \delta_i & C_i \\ 0 & C_i & \Delta_i - \delta_i \end{pmatrix} + \begin{pmatrix} 0 & 0 & 0 \\ 0 & 0 & 0 \\ 0 & 0 & 1 - \Delta_i \end{pmatrix} \right]$$

$$M_i = m_{i3} \begin{pmatrix} 0 & A_i & 0 \\ A_i^* & B_i & C_i \\ 0 & C_i & D_i \end{pmatrix}. \tag{2}$$

where $A_i = |A_i|e^{i\phi_i}$, $B_i = -\Delta_i + \delta_i$ and $D_i = 1 - \delta_i$. From the strong hierarchy in the masses of the Dirac fermions, $m_{i3} >> m_{i2} > m_{i1}$, we expect $1 - \delta_i$ to be very close to unity. The small parameter δ_i is a function of the flavour symmetry breaking parameter $Z_i{}^{1/2}$ [3, 4]. In other words, each possible symmetry breaking pattern is now characterized by the flavour symmetry breaking parameter $Z_i{}^{1/2}$, which is defined as the ratio $Z_i{}^{1/2} = \frac{(M_i)_{23}}{(M_i)_{22}}$. This ratio measures the mixing of singlet and doublet irreducible representations of S_3. The mass matrix (2) may be written as $M_i = P_i^\dagger \bar{M}_i P_i$, where $\bar{M}_i$ is a real symmetric matrix and $P_i \equiv \text{diag}\left[1, e^{i\phi_i}, e^{i\phi_i}\right]$.

3. Classification of texture zeroes in equivalence classes

In this section we make a classification of mass matrices with texture zeroes in terms of the similarity clases. The similarity classes are defined as follows: Two matrices M and M' are similar if there exists an invertible matrix T such that

$$M' = TMT^{-1} \quad \text{or} \quad M' = T^{-1}MT. \tag{3}$$

The equivalence classes associated with a similarity transformation are called similarity classes. Another way to see the similarity classes is that the matrices that satisfy the similarity transformation, eq. (3), have the same invariants; Trace, determinant and χ:

$$\text{Tr}\{M\} = \text{Tr}\{M'\}, \quad \det\{M\} = \det\{M'\} \quad \text{and} \quad \chi' = \chi \equiv \frac{\text{Tr}\{M^2\} - \text{Tr}\{M\}^2}{2}. \tag{4}$$

In this paper we will count the texture zeroes in a matrix as follows: two texture zeroes off-diagonal counts as one zero, while one on the diagonal counts as one zero [5]. But in the literature we find that a mass matrix has double number of texture zeroes than the number that we obtain with our rule. This is so, because in the literature the total number of texture zeroes counted in a mass matrix is the sum of the texture zeroes coming from the two mass matrices in the sector of quarks (up and down quarks) or leptons (charged leptons and left-handed neutrinos). Hence, to avoid confusion in the nomenclature of the matrices, we count the number of texture zeroes in a matrix with the rule previously enunciated, but when referring to this matrix we will follow the literature's rule and name it with the double of texture zeroes it actually has.

Now, from the most general form of the mass matrices of 3×3, symmetric and Hermitian:

$$M^{\text{s}} = \begin{pmatrix} g & a & e \\ a & b & c \\ e & c & d \end{pmatrix} \quad \text{and} \quad M^{\text{h}} = \begin{pmatrix} g & a & e \\ a^* & b & c \\ e^* & c^* & d \end{pmatrix}, \tag{5}$$

we can see that only six of the nine elements of these matrices are independent of each other. Therefore, in a certain sense the similarity transformation, eq. (3), realized the permutation of the six independent elements in the nine entries of the mass matrices. But if we want to preserve the invariants (4), the elements on the diagonal can only exchanged positions on the diagonal, while the off-diagonal elements can only exchange positions outside of the diagonal. Thus we have that all these operations reduce to the permutations of three objects. So it is natural propose to as transformation matrices T in the similarity clases, see eq.(3), the six elements of the real representation of the group of permutations S_3, which are:

$$\begin{array}{c} T\left(A_0\right)=\begin{pmatrix}1&0&0\\0&1&0\\0&0&1\end{pmatrix}, T\left(A_1\right)=\begin{pmatrix}0&1&0\\1&0&0\\0&0&1\end{pmatrix}, T\left(A_2\right)=\begin{pmatrix}0&0&1\\0&1&0\\1&0&0\end{pmatrix}, \\ T\left(A_3\right)=\begin{pmatrix}1&0&0\\0&0&1\\0&1&0\end{pmatrix}, T\left(A_4\right)=\begin{pmatrix}0&1&0\\0&0&1\\1&0&0\end{pmatrix}, T\left(A_5\right)=\begin{pmatrix}0&0&1\\1&0&0\\0&1&0\end{pmatrix}. \end{array} \tag{6}$$

Then, applying the transformations (3) and taking the matrices (6) as matrices of transformation, we get the classification of mass matrices with texture zeroes, which is shown in the tables 1, 2, 3 and 4. In this tables, the "$\star$" and "$\times$" denote an arbitrary non-vanishing matrix element on the diagonal and off-diagonal entries, respectively.

Class	Textures	Invariants	
		Symmetric	Hermitian
I	$\begin{pmatrix}0&\times&0\\\times&0&0\\0&0&\star\end{pmatrix}\begin{pmatrix}\star&0&0\\0&0&\times\\0&\times&0\end{pmatrix}\begin{pmatrix}0&0&\times\\0&\star&0\\\times&0&0\end{pmatrix}$	$\mathrm{Tr}=d$ $\det=-a^2d$ $\chi=a^2$	$\mathrm{Tr}=d$ $\det=-\lvert a\rvert^2d$ $\chi=\lvert a\rvert^2$

Table 1. Matrix with eight texture zeroes.

4. Seesaw mechanism

The left-handed Majorana neutrinos acquire their small masses through the type I seesaw mechanism $M_{\nu_L} = M_{\nu_D} M_{\nu_R}^{-1} M_{\nu_D}^T$, where M_{ν_D} and M_{ν_R} denote the Dirac and right handed Majorana neutrino mass matrices, respectively. The form of M_{ν_D} is given in eq. (2); which is a matrix with four texture zeroes class I, Hermitian, and from our conjecture of a universal S_3 flavour symmetry in a unified treatment of all fermions, it is natural to take for M_{ν_R} also a matrix with four texture zeroes of class I, but symmetric. Let us further assume that the phases in the entries of the M_{ν_R} may be factorized out as $M_{\nu_R} = R\bar{M}_{\nu_R}R$, where $R \equiv \mathrm{diag}\left[e^{-i\phi_c}, e^{i\phi_c}, 1\right]$ with $\phi_c \equiv \arg\left\{c_{\nu_R}\right\}$ and

$$\bar{M}_{\nu_R} = \begin{pmatrix} 0 & a_{\nu_R} & 0 \\ a_{\nu_R} & |b_{\nu_R}| & |c_{\nu_R}| \\ 0 & |c_{\nu_R}| & d_{\nu_R} \end{pmatrix}, \tag{7}$$

Then, the type I seesaw mechanism takes the form $M_{\nu_L} = P_D^\dagger \bar{M}_{\nu_D} P_D R^\dagger \bar{M}_{\nu_R}^{-1} R^\dagger P_D \bar{M}_{\nu_D} P_D^\dagger$ and the mass matrix of the left-handed Majorana neutrinos has the following form, with four texture zeroes of class I:

$$M_{\nu_L} = \begin{pmatrix} 0 & a_{\nu_L} & 0 \\ a_{\nu_L} & b_{\nu_L} & c_{\nu_L} \\ 0 & c_{\nu_L} & d_{\nu_L} \end{pmatrix}, \tag{8}$$

Class	Textures	Invariants	
		Symmetric	Hermitian
I	$\begin{pmatrix} \star & 0 & 0 \\ 0 & 0 & \times \\ 0 & \times & \star \end{pmatrix} \begin{pmatrix} 0 & 0 & \times \\ 0 & \star & 0 \\ \times & 0 & \star \end{pmatrix} \begin{pmatrix} \star & \times & 0 \\ \times & 0 & 0 \\ 0 & 0 & \star \end{pmatrix}$ $\begin{pmatrix} \star & 0 & 0 \\ 0 & \star & \times \\ 0 & \times & 0 \end{pmatrix} \begin{pmatrix} 0 & \times & 0 \\ \times & \star & 0 \\ 0 & 0 & \star \end{pmatrix} \begin{pmatrix} \star & 0 & \times \\ 0 & \star & 0 \\ \times & 0 & 0 \end{pmatrix}$	$\mathrm{Tr} = d + g$ $\det = -c^2 g$ $\chi = c^2 - gd$	$\mathrm{Tr} = d + g$ $\det = -\|c\|^2 g$ $\chi = \|c\|^2 - gd$
II	$\begin{pmatrix} 0 & \times & 0 \\ \times & 0 & \times \\ 0 & \times & \star \end{pmatrix} \begin{pmatrix} 0 & \times & \times \\ \times & 0 & 0 \\ \times & 0 & \star \end{pmatrix} \begin{pmatrix} \star & \times & 0 \\ \times & 0 & \times \\ 0 & \times & 0 \end{pmatrix}$ $\begin{pmatrix} 0 & 0 & \times \\ 0 & \star & \times \\ \times & \times & 0 \end{pmatrix} \begin{pmatrix} 0 & \times & \times \\ \times & \star & 0 \\ \times & 0 & 0 \end{pmatrix} \begin{pmatrix} \star & 0 & \times \\ 0 & 0 & \times \\ \times & \times & 0 \end{pmatrix}$	$\mathrm{Tr} = d$ $\det = -a^2 d$ $\chi = a^2 + c^2$	$\mathrm{Tr} = d$ $\det = -\|a\|^2 d$ $\chi = \|a\|^2 + \|c\|^2$
III	$\begin{pmatrix} \star & 0 & 0 \\ 0 & \star & 0 \\ 0 & 0 & \star \end{pmatrix}$	$\mathrm{Tr} = g + b + d$ $\det = gbd$ $\chi = -gb - gd$ $-bd$	$\mathrm{Tr} = g + b + d$ $\det = gbd$ $\chi = -gb - gd$ $-bd$
IV	$\begin{pmatrix} 0 & \times & \times \\ \times & 0 & \times \\ \times & \times & 0 \end{pmatrix}$	$\mathrm{Tr} = 0$ $\det = 2ace$ $\chi = a^2 + e^2$ $+c^2$	$\mathrm{Tr} = 0$ $\det = a^* c^* e$ $+ace^*$ $\chi = \|a\|^2 + \|e\|^2$ $+\|c\|^2$

Table 2. Matrix with six texture zeroes.

Class	Textures	Invariants	
		Symmetric	Hermitian
I	$\begin{pmatrix} 0 & \times & 0 \\ \times & \star & \times \\ 0 & \times & \star \end{pmatrix} \begin{pmatrix} 0 & 0 & \times \\ 0 & \star & \times \\ \times & \times & \star \end{pmatrix} \begin{pmatrix} \star & 0 & \times \\ 0 & 0 & \times \\ \times & \times & \star \end{pmatrix}$ $\begin{pmatrix} \star & \times & \times \\ \times & 0 & 0 \\ \times & 0 & \star \end{pmatrix} \begin{pmatrix} \star & \times & 0 \\ \times & \star & \times \\ 0 & \times & 0 \end{pmatrix} \begin{pmatrix} \star & \times & \times \\ \times & \star & 0 \\ \times & 0 & 0 \end{pmatrix}$	$\mathrm{Tr} = b + d$ $\det = -a^2 d$ $\chi = a^2 + c^2$ $-bd$	$\mathrm{Tr} = b + d$ $\det = -\|a\|^2 d$ $\chi = \|a\|^2 + \|c\|^2$ $-bd$
II	$\begin{pmatrix} 0 & \times & \times \\ \times & \star & 0 \\ \times & 0 & \star \end{pmatrix} \begin{pmatrix} \star & \times & 0 \\ \times & 0 & \times \\ 0 & \times & \star \end{pmatrix} \begin{pmatrix} \star & 0 & \times \\ 0 & \star & \times \\ \times & \times & 0 \end{pmatrix}$	$\mathrm{Tr} = b + d$ $\det = -a^2 d$ $-e^2 b$ $\chi = a^2 + e^2$ $-bd$	$\mathrm{Tr} = b + d$ $\det = -\|a\|^2 d$ $-\|e\|^2 b$ $\chi = \|a\|^2 + \|e\|^2$ $-bd$
III	$\begin{pmatrix} 0 & \times & \times \\ \times & 0 & \times \\ \times & \times & \star \end{pmatrix} \begin{pmatrix} 0 & \times & \times \\ \times & \star & \times \\ \times & \times & 0 \end{pmatrix} \begin{pmatrix} \star & \times & \times \\ \times & 0 & \times \\ \times & \times & 0 \end{pmatrix}$	$\mathrm{Tr} = d$ $\det = 2ace - a^2 d$ $\chi = a^2 + c^2$ $+e^2$	$\mathrm{Tr} = d$ $\det = a^* c^* e$ $+ace^* - \|a\|^2 d$ $\chi = \|a\|^2 + \|c\|^2$ $+\|e\|^2$
IV	$\begin{pmatrix} \star & 0 & 0 \\ 0 & \star & \times \\ 0 & \times & \star \end{pmatrix} \begin{pmatrix} \star & 0 & \times \\ 0 & \star & 0 \\ \times & 0 & \star \end{pmatrix} \begin{pmatrix} \star & \times & 0 \\ \times & \star & 0 \\ 0 & 0 & \star \end{pmatrix}$	$\mathrm{Tr} = g + b + d$ $\det = -gc^2 + gbd$ $\chi = c^2 - gb$ $-gd - bd$	$\mathrm{Tr} = g + b + d$ $\det = -g\|c\|^2 - gbd$ $\chi = \|c\|^2 - gb$ $-gd - bd$

Table 3. Matrix with four texture zeroes.

where

$$\begin{aligned} a_{\nu_L} &= \frac{|a_{\nu_D}|^2}{a_{\nu_R}}, \quad d_{\nu_L} = \frac{d^2_{\nu_D}}{d_{\nu_R}}, \quad c_{\nu_L} = \frac{c_{\nu_D} d_{\nu_D}}{d_{\nu_R}} + \frac{|a_{\nu_D}|}{|a_{\nu_R}|}\left(c_{\nu_D} e^{-i\phi_{\nu_D}} - \frac{|c_{\nu_R}| d_{\nu_D}}{d_{\nu_R}} e^{i(\phi_c - \phi_{\nu_D})}\right) \\ b_{\nu_L} &= \frac{c^2_{\nu_D}}{d_{\nu_R}} + \frac{|c_{\nu_R}|^2 - |b_{\nu_R}| d_{\nu_R}}{d_{\nu_R}} \frac{|a_{\nu_D}|^2}{a^2_{\nu_R}} e^{i2(\phi_c - \phi_{\nu_D})} + 2\frac{|a_{\nu_D}|}{|a_{\nu_R}|}\left(b_{\nu_D} e^{-i\phi_{\nu_D}} - \frac{c_{\nu_D} |c_{\nu_R}|}{d_{\nu_R}} e^{i(\phi_c - \phi_{\nu_D})}\right). \end{aligned} \tag{9}$$

Class	Textures	Invariants	
		Symmetric	Hermitian
I	$\begin{pmatrix} 0 & \times & \times \\ \times & \star & \times \\ \times & \times & \star \end{pmatrix} \begin{pmatrix} \star & \times & \times \\ \times & 0 & \times \\ \times & \times & \star \end{pmatrix} \begin{pmatrix} \star & \times & \times \\ \times & \star & \times \\ \times & \times & 0 \end{pmatrix}$	$\mathrm{Tr} = b + d$ $\det = 2ace - a^2 d$ $-be^2$ $\chi = a^2 + c^2$ $+e^2 - bd$	$\mathrm{Tr} = b + d$ $\det = -\lvert a\rvert^2 d - b\lvert e\rvert^2$ $+ace^* + a^* c^* e$ $\chi = \lvert a\rvert^2 + \lvert c\rvert^2 + \lvert e\rvert^2$ $-bd$
II	$\begin{pmatrix} \star & 0 & \times \\ 0 & \star & \times \\ \times & \times & \star \end{pmatrix} \begin{pmatrix} \star & \times & \times \\ \times & \star & 0 \\ \times & 0 & \star \end{pmatrix} \begin{pmatrix} \star & \times & 0 \\ \times & \star & \times \\ 0 & \times & \star \end{pmatrix}$	$\mathrm{Tr} = b + d + g$ $\det = gbd$ $-c^2 g - e^2 b$ $\chi = e^2 + c^2 - gb$ $-gd - bd$	$\mathrm{Tr} = b + d + g$ $\det = gbd$ $-\lvert c\rvert^2 g - \lvert e\rvert^2 b$ $\chi = \lvert e\rvert^2 + \lvert c\rvert^2 - gb$ $-gd - bd$

Table 4. Matrix with two texture zeroes.

The elements a_{ν_L} and d_{ν_L} are real , while b_{ν_L} and c_{ν_L} are complex. It may also be noticed that M_{ν_L} may also have four texture zeroes of class I when M_{ν_R} has four texture zeroes of class I, six texture zeroes of class II, six texture zeroes of class I and eight texture zeroes of class I. From ecs. (9) we conclude that the information of the number of texture zeroes in M_{ν_R} is found in elements (2,2) and (2,3) of the matrix M_{ν_L}. In this case, without loss of generality, 2we may choose $\arg\left\{b_{\nu_L}\right\} = 2\arg\left\{c_{\nu_L}\right\} = 2\varphi$, the analysis simplifies since the phases in M_{ν_L} may be factorized as $M_{\nu_L} = Q\bar{M}_{\nu_L}Q$, where Q is a diagonal matrix of phases $Q \equiv \mathrm{diag}\left[e^{-i\varphi}, e^{i\varphi}, 1\right]$ and $\bar{M}_{\nu_L}$ is a real symetric matrix.

5. Mass matrix with four texture zeroes as function of the fermion masses

Now, computing the invariants (4) of the real symetric matrix $\bar{M}_{\nu_L}$, we may express the parameters A_i, B_i, C_i and D_i occuring in (2) in terms of the mass eigenvalues. In this way, we get the $\bar{M}_i$ matrix ($i = u, d, l, \nu_L$), reparametrized in terms of its eigenvalues and the parameter δ_i is

$$\bar{M}_i = \begin{pmatrix} 0 & \sqrt{\frac{\widetilde{m}_{i1}\widetilde{m}_{i2}}{1-\delta_i}} & 0 \\ \sqrt{\frac{\widetilde{m}_{i1}\widetilde{m}_{i2}}{1-\delta_i}} & \widetilde{m}_{i1} - \widetilde{m}_{i2} + \delta_i & \sqrt{\frac{\delta_i}{(1-\delta_i)} f_{i1} f_{i2}} \\ 0 & \sqrt{\frac{\delta_i}{(1-\delta_i)} f_{i1} f_{i2}} & 1 - \delta_i \end{pmatrix}, \tag{10}$$

where $\widetilde{m}_{i1} = \frac{m_{i1}}{m_{i3}}$, $\widetilde{m}_{i2} = \frac{|m_{i2}|}{m_{i3}}$, $f_{i1} = 1 - \widetilde{m}_{i1} - \delta_i$ and $f_{i2} = 1 + \widetilde{m}_{i2} - \delta_i$. The small parameters δ_i are also functions of the mass ratios and the flavour symmetry breaking parameter $Z_i^{1/2}$. The small parameter δ_i is obtained as the solution of the cubic equation $(1-\delta_i)(\widetilde{m}_{i1} - \widetilde{m}_{i2} + \delta_i)^2 Z - \delta_i f_{i1} f_{i2} = 0$, and may be written as

$$\delta_i = \frac{Z_i}{Z_i + 1} \frac{\left(\widetilde{m}_{i2} - \widetilde{m}_{i1}\right)^2}{W_i(Z)} \tag{11}$$

where $W_i(Z)$ is the product of the two roots that do not vanish when Z_i vanishes

$$\begin{aligned} W_i(Z) = {} & \left[p_i^3 + 2q_i^2 + 2q\sqrt{p_i^3 + q_i^2}\right]^{\frac{1}{3}} - |p_i| + \left[p_i^3 + 2q_i^2 - 2q_i\sqrt{p_i^3 + q_i^2}\right]^{\frac{1}{3}} + \\ & + \tfrac{1}{9}\left(Z_i\left(2\left(\widetilde{m}_{i2} - \widetilde{m}_{i1}\right) + 1\right) + \left(\widetilde{m}_{i2} - \widetilde{m}_{i1}\right) + 2\right)^2 - \tfrac{1}{3}\left(\left[q_i + \sqrt{p_i^3 + q_i^2}\right]^{\frac{1}{3}} + \left[q_i - \sqrt{p_i^3 + q_i^2}\right]^{\frac{1}{3}}\right) \times \\ & \times \left(Z_i\left(2\left(\widetilde{m}_{i2} - \widetilde{m}_{i1}\right) + 1\right) + \left(\widetilde{m}_{i2} - \widetilde{m}_{i1}\right) + 2\right) \end{aligned}$$

with $p_i = -\frac{1}{3}\frac{Z_i\left(Z_i\left(2\left(\widetilde{m}_{i2}-\widetilde{m}_{i1}\right)+1\right)+\widetilde{m}_{i2}-\widetilde{m}_{i1}+2\right)^2}{Z_i+1} + \frac{\left[Z_i\left(\widetilde{m}_{i2}-\widetilde{m}_{i1}\right)\left(\widetilde{m}_{i2}-\widetilde{m}_{i1}+2\right)\left(1+\widetilde{m}_{i2}\right)\left(1-\widetilde{m}_{i1}\right)\right]}{Z_i+1}$, and $q_i = \frac{1}{6}\frac{\left[Z_i\left(\widetilde{m}_{i2}-\widetilde{m}_{i1}\right)\left(\widetilde{m}_{i2}-\widetilde{m}_{i1}+2\right)\left(1+\widetilde{m}_{i2}\right)\left(1-\widetilde{m}_{i1}\right)\right]\left(Z_i\left(2\left(m_{i2}-m_{i1}\right)+1\right)+m_{i2}-\widetilde{m}_{i1}+2\right)}{\left(Z_i+1\right)^2} - \frac{1}{27}\frac{\left(Z_i\left(2\left(\widetilde{m}_{i2}-\widetilde{m}_{i1}\right)+1\right)+\widetilde{m}_{i2}-\widetilde{m}_{i1}+2\right)^3}{\left(Z_i+1\right)^3}$.

Journal of Physics: Conference Series **287** (2011) 012015 doi:10.1088/1742-6596/287/1/012015

Also, the values allowed for the parameters δ_i are in the following range $0 < \delta_i < 1 - \widetilde{m}_{i1}$. Now, the entries in the real orthogonal matrix $\mathbf{O}$ that diagonalize the matrix $\bar{M}_i$, may also be expressed as

$$\mathbf{O_i} = \begin{pmatrix} \left[\frac{\widetilde{m}_{i2} f_{i1}}{\mathcal{D}_{i1}}\right]^{\frac{1}{2}} & -\left[\frac{\widetilde{m}_{i1} f_{i2}}{\mathcal{D}_{i2}}\right]^{\frac{1}{2}} & \left[\frac{\widetilde{m}_{i1}\widetilde{m}_{i2}\delta_i}{\mathcal{D}_{i3}}\right]^{\frac{1}{2}} \\ \left[\frac{\widetilde{m}_{i1}(1-\delta_i) f_{i1}}{\mathcal{D}_{i1}}\right]^{\frac{1}{2}} & \left[\frac{\widetilde{m}_{i2}(1-\delta_i) f_{i2}}{\mathcal{D}_{i2}}\right]^{\frac{1}{2}} & \left[\frac{(1-\delta_i)\delta_i}{\mathcal{D}_{i3}}\right]^{\frac{1}{2}} \\ -\left[\frac{\widetilde{m}_{i1} f_{i2}\delta_i}{\mathcal{D}_{i1}}\right]^{\frac{1}{2}} & -\left[\frac{\widetilde{m}_{i2} f_{i1}\delta_i}{\mathcal{D}_{i2}}\right]^{\frac{1}{2}} & \left[\frac{f_{i1} f_{i2}}{\mathcal{D}_{i3}}\right]^{\frac{1}{2}} \end{pmatrix}, \qquad (12)$$

where, $\mathcal{D}_{i1} = (1-\delta_i)(\widetilde{m}_{i1} + \widetilde{m}_{i2})(1-\widetilde{m}_{i1})$, $\mathcal{D}_{i2} = (1-\delta_i)(\widetilde{m}_{i1} + \widetilde{m}_{i2})(1+\widetilde{m}_{i2})$ and $\mathcal{D}_{i3} = (1-\delta_i)(1-\widetilde{m}_{i1})(1+\widetilde{m}_{i2})$.

6. Mixing Matrices as Functions of the Fermion Masses

The quark mixing matrix $V_{CKM} = U_u U_d^\dagger$ takes the form

$$V_{CKM}^{th} = \mathbf{O_u}^T P^{(u-d)} \mathbf{O}_d, \qquad (13)$$

where $P^{(u-d)} = \text{diag}\left[1, e^{i\phi}, e^{i\phi}\right]$ with $\phi = \phi_u - \phi_d$, and $\mathbf{O}_{u,d}$ are the real orthogonal matrices (12) that diagonalize the real symmetric mass matrices $\bar{M}_{u,d}$. In a similar way, the lepton mixing matrix $U_{PMNS} = U_l^\dagger U_\nu$ may be written as

$$U_{PMNS}^{th} = \mathbf{O}_l^T P^{(\nu-l)} \mathbf{O}_\nu K, \qquad (14)$$

where $P^{(\nu-l)} = \text{diag}\left[1, e^{i\Phi_1}, e^{i\Phi_2}\right]$ is the diagonal matrix of the Dirac phases, with $\Phi_1 = 2\varphi - \phi_l$ and $\Phi_2 = \varphi - \phi_l$. The real orthogonal matrices $\mathbf{O}_{\nu,l}$ are defined in eq. (12). Exact explicit expressions for the unitary matices (13) and (14) are given by J. Barranco, F. Gonzalez Canales and A. Mondragon [4].

6.1. *The χ^2 fit for the Quark Mixing Matrix*

We made a χ^2 fit of the theoretical expressions for the modulii of the entries of the quark mixing matrix $|(V_{CKM}^{th})_{ij}|$, eq.(13) to the experimental values $|(V_{CKM}^{exp})_{ij}|$ [11]. We computed the modulii of the entries of the quark mixing matrix and the inner angles of the unitarity triangle from the theoretical expresion (13) with the following numerical values of the quark mass ratios [11]:

$$\widetilde{m}_u = 2.5469 \times 10^{-5},\ \widetilde{m}_c = 3.9918 \times 10^{-3},\ \widetilde{m}_d = 1.5261 \times 10^{-3},\ \widetilde{m}_s = 3.2319 \times 10^{-2}. \qquad (15)$$

The resulting best values of the parameters $\delta_u = 3.829 \times 10^{-3}$, $\delta_d = 4.08 \times 10^{-4}$ and the Dirac CP violating phase is $\phi = 90^o$. The best values for the moduli of the entries of the CKM mixing matrix are given in the following expresion

$$\left|V_{CKM}^{th}\right| = \begin{pmatrix} 0.97421 & 0.22560 & 0.003369 \\ 0.22545 & 0.97335 & 0.041736 \\ 0.008754 & 0.04094 & 0.99912 \end{pmatrix} \qquad (16)$$

and the inner angles of the unitary triangle $\alpha^{th} = 91.24^o$, $\beta^{th} = 20.41^o$ and $\gamma^{th} = 68.33^o$. The Jarlskog invariant takes the value $J_q^{th} = 2.9 \times 10^{-5}$. All these results are in very good agreement with the experimental values. The minimum value of χ^2 obtained in this fit is 4.6 and the resulting value of χ^2 for degree of freedom is $\frac{\chi^2_{min}}{d.o.f.} = 0.77$. In this way, we obtain the following numerical values for the mixing angles: $\theta_{12}^{q^{th}} = 13^o$, $\theta_{23}^{q^{th}} = 2.38^o$ and $\theta_{13}^{q^{th}} = 0.19^o$. Which are also in very good agreement with the latest analysis of the experimental data [11].

Journal of Physics: Conference Series **287** (2011) 012015 doi:10.1088/1742-6596/287/1/012015

6.2. The χ^2 fit for the Lepton Mixing Matrix

In the case of the lepton mixing matrix, we made a χ^2 fit of the theoretical expressions for the modulii of the entries of the lepton mixing matrix $|(U^{th}_{PMNS})_{ij}|$ to the values extracted from experiment as given by Gonzalez-Garcia [12].The computation was made using the following values for the charged lepton masses [11]:

$$m_e = 0.5109 \text{ MeV}, \quad m_\mu = 105.685 \text{ MeV}, \quad m_\tau = 1776.99 \text{ MeV}. \tag{17}$$

We took for the masses of the left-handed Majorana neutrinos a normal hierarchy. From the best values obtained for m_{ν_3} and the experimental values of the Δm^2_{32} and Δm^2_{21}, we obtained the following best values for the neutrino masses

$$m_{\nu_1} = 2.7 \times 10^{-3}\text{eV}, \quad m_{\nu_2} = 9.1 \times 10^{-3}\text{eV}, \quad m_{\nu_3} = 4.7 \times 10^{-2}\text{eV}. \tag{18}$$

The resulting best values of the parameters $\delta_l = 0.06$, $\delta_\nu = 0.522$ and the Dirac CP violating phases are $\Phi_1 = \pi$ and $\Phi_2 = 3\pi/2$. The best values for the modulii of the entries of the $PMNS$ mixing matrix are given in the following expresion

$$\left|U^{th}_{PMNS}\right| = \begin{pmatrix} 0.820421 & 0.568408 & 0.061817 \\ 0.385027 & 0.613436 & 0.689529 \\ 0.422689 & 0.548277 & 0.721615 \end{pmatrix}. \tag{19}$$

The value of the rephasing invariant related to the Dirac phase is $J^{th}_l = 8.8 \times 10^{-3}$. In this numerical analysis, the minimum value of the χ^2, corresponding to the best fit, is $\chi^2 = 0.288$ and the resulting value of χ^2 degree of freedom is $\frac{\chi^2_{min}}{d.o.f.} = 0.075$. All numerical results of the fit are in very good agreement with the values of the moduli of the entries in the matrix U^{exp}_{PMNS} as given in Gonzalez-Garcia [12]. We obtain the following numerical values for the mixing angles: $\theta^{lth}_{12} = 34.7^o$, $\theta^{lth}_{23} = 43.6^o$ and $\theta^{lth}_{13} = 3.5^o$. Which are also in very good agreement with the latest experimental data [12].

7. Conclusions

In this work we have shown that, starting from the flavour permutational symmetry S_3, a simple and explicit ansatz about the pattern of symmetry breaking leads to a unified treatment of masses of quarks and leptons, in which the left-handed Majorana neutrinos acquire their masses via the type I seesaw mechanism. The mass matrices of all Dirac fermions have a similar form with four texture zeroes of class I and a normal hierarchy of masses. Then, the mass matrix of the left-handed Majorana neutrinos also has a four texture zeroes class I and a normal hierarchy of masses. In this scheme, we have a parametrization of the $CKM(PMNS)$ mixing matrix in terms of four quark (lepton) mass ratios and one (two) CP violating phase in very good agreement with all available experimental data. Also, with help of the symmetry group S_3 we made a classification of mass matrices with texture zeroes in equivalence classes. In this classification we use the similarity transformation $M' = TMT^{-1}$ or $M' = T^{-1}MT$, in which we take the transformation matrices T as six elements of real representation of S_3. With this classification we reduce the number of phenomenologically viable textures for non-singulars mass matrices of 3×3, from thirty three down to only eleven independent sets of matrices . Each of these sets of matrices has exactly the same physical content.

Acknowledgments

We thank Dr. J. Barranco for many inspiring discussions on this problem. This work was partially supported by CONACyT Mexico under Contract No. 82291, and DGAPA-UNAM Contract No. PAPIIT IN112709.

Journal of Physics: Conference Series **287** (2011) 012015 doi:10.1088/1742-6596/287/1/012015

References

[1] Xing Z z 2010 *Nucl. Phys. Proc. Suppl.* **203-204** 82–117, Hernandez P 2010 (*Preprint* 1010.4131)
[2] Fritzsch H 1977 *Phys. Lett.* **B70** 436, Pakvasa S and Sugawara H 1978 *Phys. Lett.* **B73** 61, Harari H, Haut H and Weyers J 1978 *Phys. Lett.* **B78** 459, Fritzsch H 1979 *Nucl. Phys.* **B155** 189.
[3] Mondragon A and Rodriguez-Jauregui E 1999 *Phys. Rev.* **D59** 093009 (*Preprint* hep-ph/9807214), Mondragon A and Rodriguez-Jauregui E 2000 *Phys. Rev.* **D61** 113002 (*Preprint* hep-ph/9906429)
[4] Barranco J, Gonzalez Canales F and Mondragon A 2010 *Phys. Rev.* **D82** 073010 (*Preprint* 1004.3781)
[5] Fritzsch H and Xing Z z 2000 *Prog. Part. Nucl. Phys.* **45** 1–81 (*Preprint* hep-ph/9912358)
[6] Kubo J, Mondragon A, Mondragon M and Rodriguez-Jauregui E 2003 *Prog. Theor. Phys.* **109** 795–807 (*Preprint* hep-ph/0302196), Xing Z z, Yang D and Zhou S 2010 *Phys. Lett.* **B690** 304–310
[7] Xing Z z and Zhang H 2003 *Phys. Lett.* **B569** 30–40 (*Preprint* hep-ph/0304234)
[8] Branco G C, Emmanuel-Costa D, Gonzalez Felipe R and Serodio H 2009 *Phys. Lett.* **B670** 340–349
[9] Altarelli G, Feruglio F and Merlo L 2009 *JHEP* **05** 020, de Adelhart Toorop R, Bazzocchi F and Merlo L 2010 *JHEP* **08** 001, Plentinger F, Seidl G and Winter W 2008 *JHEP* **04** 077, Plentinger F and Seidl G 2008 *Phys. Rev.* **D78** 045004 (*Preprint* 0803.2889)
[10] Buchmuller W and Wyler D 2001 *Phys. Lett.* **B521** 291–298, Bando M and Obara M 2003 *Prog. Theor. Phys.* **109** 995–1015 Bando M, Kaneko S, Obara M and Tanimoto M 2004 (*Preprint*
[11] Nakamura K *et al.* (Particle Data Group) 2010 *J. Phys.* **G37** 075021
[12] Gonzalez-Garcia M C and Maltoni M 2008 *Phys. Rept.* **460** 1–129 (*Preprint* 0704.1800), Gonzalez-Garcia M C, Maltoni M and Salvado J 2010 *JHEP* **04** 056 (*Preprint* 1001.4524)

QCD in the δ-Regime

W. Bietenholz[a], N. Cundy[b], M. Göckeler[c], R. Horsley[d], Y. Nakamura[e], D. Pleiter[f], P.E.L. Rakow[g], G. Schierholz[c,h] and J.M. Zanotti[d] (QCDSF Collaboration)

[a] Instituto de Ciencias Nucleares, Universidad Nacional Autónoma de México, A.P. 70-543, C.P. 04510 México, Distrito Federal, Mexico
[b] Lattice Gauge Theory Research Center, FPRD, and CTP, Department of Physics & Astronomy, Seoul National University, Seoul, 151-747, South Korea
[c] Institut für Theoretische Physik, Universität Regensburg, 93040 Regensburg, Germany
[d] School of Physics, University of Edinburgh, Edinburgh EH9 3JZ, United Kingdom
[e] Center for Computational Sciences, University of Tsukuba, Tsukuba, Ibaraki 305-8577, Japan
[f] Deutsches Elektronen-Synchrotron DESY, 15738 Zeuthen, Germany
[g] Theoretical Physics Division, Dept. of Mathematical Sciences, University of Liverpool, Liverpool, L69 3BX, United Kingdom
[h] Deutsches Elektronen-Synchrotron DESY, 22603 Hamburg, Germany

E-mail: wolbi@nucleares.unam.mx

Abstract. The δ-regime of QCD is characterised by light quarks in a small spatial box, but a large extent in (Euclidean) time. In this setting a specific variant of chiral perturbation theory — the δ-expansion — applies, based on a quantum mechanical treatment of the quasi one-dimensional system. In particular, for vanishing quark masses one obtains a residual pion mass M_π^R, which has been computed to the third order in the δ-expansion. A comparison with numerical measurements of this residual mass allows for a new determination of some Low Energy Constants, which appear in the chiral Lagrangian. We first review the attempts to simulate 2-flavour QCD directly in the δ-regime. This is very tedious, but results compatible with the predictions for M_π^R have been obtained. Then we show that an extrapolation of pion masses measured in a larger volume towards the δ-regime leads to good agreement with the theoretical predictions. From those results, we also extract a value for the (controversial) sub-leading Low Energy Constant $\bar{l}_3$.

1. QCD and Chiral Perturbation Theory

The Lagrangian of QCD is scale invariant, but its quantisation singles out an intrinsic energy $\Lambda_{\rm QCD}$, which sets the scale for the hadron spectrum. Our daily life is dominated by low energy and therefore by the lightest quark flavours, *i.e.* quarks with masses $m_{\rm q} \ll \Lambda_{\rm QCD}$. In the limit of vanishing quark masses, their left- and right-handed spinor components (Ψ_L and Ψ_R) decouple. Thus the Lagrangian takes the structure

$$\mathcal{L}_{\rm QCD} = \bar{\Psi}_L D \Psi_L + \bar{\Psi}_R D \Psi_R + \mathcal{L}_{\rm gauge} \ , \qquad (1)$$

where D is the Dirac operator. For N_f massless quark flavours, this Lagrangian has the global symmetry

$$U(N_f)_L \otimes U(N_f)_R = \underbrace{SU(N_f)_L \otimes SU(N_f)_R}_{\text{chiral flavour symmetry}} \otimes \underbrace{U(1)_V}_{\text{baryon number conservation}} \otimes \underbrace{U(1)_A}_{\text{axial symmetry}} . \qquad (2)$$

Here we split off the phases; the (vectorial) symmetry under simultaneous left- and right-handed phase rotation, $U(1)_V$, corresponds to the conservation of the baryon number. The remaining $U(1)_A$ symmetry — for opposite L and R phase rotations — is the axial symmetry. That is a symmetry of the classical theory, which breaks explicitly under quantisation, *i.e.* it is anomalous. We are interested in the remaining *chiral flavour symmetry,* which (in infinite volume) breaks spontaneously,

$$SU(N_f)_L \otimes SU(N_f)_R \longrightarrow SU(N_f)_{L+R} \, . \qquad (3)$$

Chiral Perturbation Theory (χPT) deals with an effective Lagrangian in term of fields in the coset space of this spontaneous symmetry breaking, $U(x) \in SU(N_f)$ [1]. Thus it captures the lightest degrees of freedom, which dominate low energy physics, in this case given by $N_f^2 - 1$ Nambu-Goldstone bosons. The effective chiral Lagrangian $\mathcal{L}_{\rm eff}$ embraces all terms which are compatible with the symmetries. This concept also extends to the case where small quark masses are added, $m_{\rm q} \gtrsim 0$. Then one deals with light pseudo Nambu-Goldstone bosons, which are identified with the light mesons; for $N_f = 3$ this includes the pions, the kaons and the η-meson.

Here we consider the case $N_f = 2$, so we only deal with the quark flavours u and d. We assume them to be degenerate, *i.e.* to have both the mass $m_{\rm q}$. In this case the field $U(x) \in SU(2)$ describes the pion triplet. The terms in $\mathcal{L}_{\rm eff}$ are ordered according to an energy hierarchy, which depends on the number of derivatives and powers of $m_{\rm q}$. Some of the first terms are

$$\begin{aligned} \mathcal{L}_{\rm eff} &= \frac{1}{4} F_\pi^2 \, {\rm Tr}[\partial_\mu U^\dagger \partial_\mu U] + \frac{1}{2} \Sigma m_{\rm q} \, {\rm Tr}[U + U^\dagger] \\ &\quad - \frac{1}{4} l_1 \Big({\rm Tr}[\partial_\mu U^\dagger \partial_\mu U] \Big)^2 - \frac{1}{4} l_2 \Big({\rm Tr}[\partial_\mu U^\dagger \partial_\nu U] \Big)^2 \\ &\quad - (l_3 + l_4) \Big(\frac{\Sigma m_{\rm q}}{2F_\pi} \Big)^2 \Big({\rm Tr}[U + U^\dagger] \Big)^2 + l_4 \frac{\Sigma m_{\rm q}}{4F_\pi} {\rm Tr}[\partial_\mu U^\dagger \partial_\nu U] \, {\rm Tr}[U + U^\dagger] + \dots \end{aligned} \qquad (4)$$

Each term comes with a coefficient, which is a free parameter of the effective theory. These coefficients are known as the *Low Energy Constants* (LECs). The leading LECs are the pion decay constant F_π (which was measured as $F_\pi = 92.4$ MeV) and the chiral condensate Σ (the order parameter for chiral symmetry breaking). The l_i are sub-leading LECs. In the chiral limit $m_{\rm q} = 0$, only F_π occurs in the leading order. This determines an intrinsic scale $\Lambda_\chi = 4\pi F_\pi \simeq 1.2$ GeV, which adopts the rôle of $\Lambda_{\rm QCD}$ in χPT.

The LECs are of primary importance for low energy hadron physics. To some extent they can be fixed from phenomenology. On the theoretical side, they can only be determined from QCD as the underlying fundamental theory. Since this refers to low energy, it is a non-perturbative task, and therefore a challenge for lattice simulations. If one succeeds in their determination, we arrive at a rather complete, QCD-based formalism for low energy hadron physics.

2. Pions in a finite volume: p-regime, ϵ-regime and δ-regime

For a system of pions in a finite volume, say with periodic boundary conditions and with a characteristic extent L, the low energy expansion can be formulated in terms of the dimensionless lowest non-zero momentum $p_\mu/\Lambda_\chi \sim 1/(2F_\pi L)$, and of the correlation length $\xi = M_\pi^{-1}$ (M_π is the pion mass). Both p_μ and M_π should be light compared to Λ_χ. Depending on the size and shape of the volume, one distinguishes various regimes, with different counting rules for these expansion parameters:

- p-Regime: This is the standard setting with a large volume, $L \gg \xi$, and therefore small finite size effects [2]. In this case the counting is simply $O(1/L) = O(M_\pi)$. From the Lagrangian we can read off (on tree level) the Gell-Mann–Oakes–Renner relation
$$M_\pi^2 = \frac{\Sigma}{F_\pi^2} m_q \ . \tag{5}$$
- ϵ-Regime: This regime refers to a small box in Euclidean space [3], say $V = L^4$ with $L \lesssim \xi$ (where ξ is still the would-be inverse pion mass in a large volume, with the same quark mass; in Nature $\xi \simeq 1.5$ fm). In this regime, the χPT counting rules read $O(1/L) = O(m_q) = O(M_\pi^2/\Lambda_\chi)$, unlike the p-regime counting.
Experimentally the ϵ-regime is not accessible, but QCD simulations in (or at least close to) this regime are feasible. They are of interest in particular because the LECs determined in this regime are the same that occur in large volume. Therefore we can extract physical information even from an unphysical regime. Since this can be achieved with a modest lattice size, this method is attractive from a practical perspective [4]. It has been intensively explored since 2003. It is difficult to simulate safely inside this regime, but certain properties, which are characteristic for the ϵ-regime, have been observed. Ref. [5] provides a short overview.
- δ-Regime: Here one deals with a small *spatial* volume, but a large extent T in Euclidean time [6], say
$$L^3 \times T \quad , \qquad L \lesssim \xi \ll T \ . \tag{6}$$
This relation for L and T is depicted in Fig. 1 on the left (it is exactly opposite to the setting used in studies of QCD at finite temperature). The counting rules for the corresponding *δ-expansion* are
$$\frac{1}{\Lambda_\chi L} = O(\delta) \quad , \qquad \frac{M_\pi}{\Lambda_\chi} \, , \, \frac{1}{\Lambda_\chi T} = O(\delta^3) \ . \tag{7}$$

A map of these three regimes in terms of the pion mass and the inverse extent in Euclidean time (the temperature) is shown in Fig. 1 on the right.

This article addresses the δ-regime. It is far less known and explored than the p- and the ϵ-regime, but it shares with the latter the exciting property that physical LECs can be extracted from an unphysical setting. A further motivation for studying QCD in a "δ-box" is that its shape allows (approximately) for a simplified analytical treatment in terms of 1d field theory, *i.e. quantum mechanics*. In this case one considers a quantum rotator as described by the 1d $O(4)$ model, due to the local isomorphism between the orthogonal group $O(4)$ and the chiral symmetry group $SU(2)_L \otimes SU(2)_R$. Closely related systems have applications in solid state physics, in particular regarding quantum anti-ferromagnets [7, 8].

Spontaneous symmetry breaking does not occur in a finite volume. Therefore the pions[1] — *i.e.* the pseudo Nambu-Goldstone bosons — cannot become massless in the chiral limit $m_q = 0$, in contrast to the infinite volume. We may consider a fixed box of a δ-shape and vary the quark mass: a large value of m_q implies a large pion mass, so that we enter the p-regime and the Gell-Mann–Oakes–Renner relation (5), $m_q \propto M_\pi^2$, is approximated. For small m_q the pion mass turns into a plateau, which ends in the chiral limit at a *residual pion mass* M_π^R. This behaviour is illustrated schematically in Fig. 2.

[1] One might argue if the term "pion" is adequate in the δ-regime. We find it acceptable and convenient, but readers who disagree may simply denote M_π^R (see below) as the "mass gap".

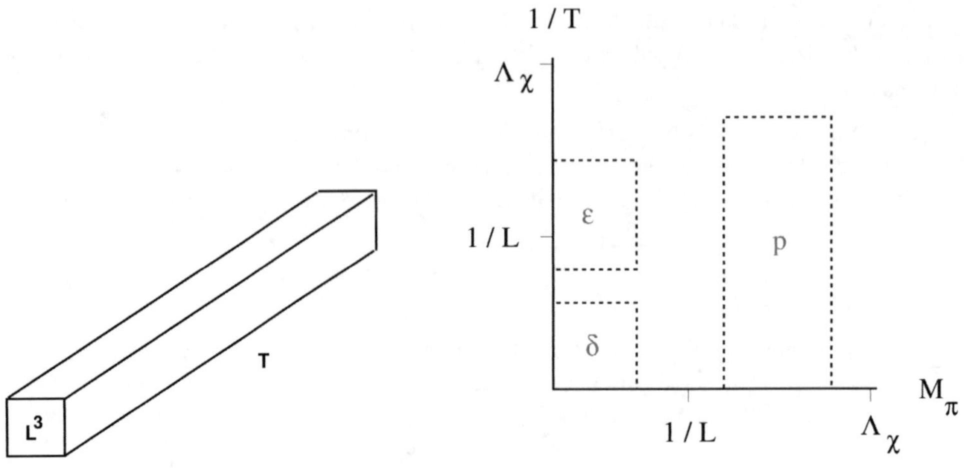

Figure 1. On the left: an illustration of a typical shape of a δ-box, *i.e.* an anisotropic finite volume where a pion gas can be treated by the δ-expansion. On the right: a schematic map of the applicability domains of three different expansion rules of χPT, namely the p- , the ϵ- and the δ-regime. The dashed lines indicate regions where clearly one expansion holds; in the transition zones between these regions various expansions could work more or less.

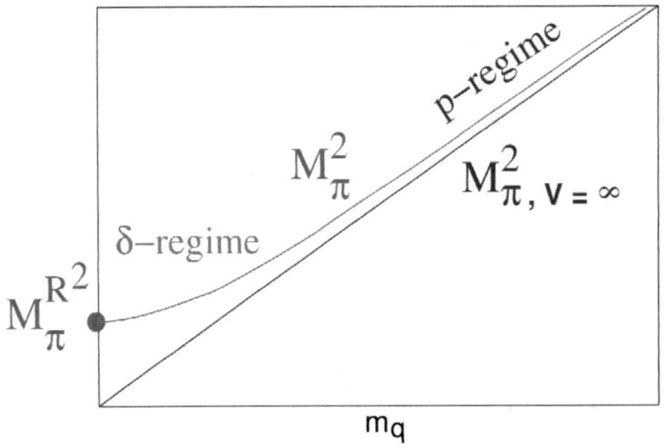

Figure 2. A qualitative picture of the expected behaviour of the pion mass squared in a δ-box. For heavy quarks and pions we approximate the p-regime relation $m_q \propto M_\pi^2$. For light quarks the pion mass attains a plateau, and finally (in the chiral limit $m_{\rm q} = 0$) the residual value M_π^R.

The value of M_π^R can be computed with the δ-expansion. The spectrum of the $O(4)$ quantum rotator (a quantum mechanical particle on the sphere S^3) is given by $E_\ell = \ell(\ell+2)/(2\Theta)$, so the mass gap amounts to $M_\pi^R = 3/(2\Theta)$. The challenge is now to compute the moment of inertia Θ. In his seminal paper on the δ-regime, H. Leutwyler gave its value to leading order (LO) as $\Theta \approx F_\pi^2 L^3$. Thus the residual pion mass can be written as

$$M_\pi^R = \frac{3}{2F_\pi^2 L^3 (1+\Delta)} \ . \tag{8}$$

The shift Δ captures higher order corrections, which are suppressed in powers of $1/(F_\pi L)^2$. They have been evaluated to next-to-leading order (NLO) in Ref. [8], and recently even to next-to-next-to-leading order (NNLO) [9], which yields

$$\Delta = \frac{0.4516\ldots}{F_\pi^2 L^2} + \frac{0.08843\ldots}{F_\pi^4 L^4}\Big[1 - 0.1599\ldots\Big(\ln(\Lambda_1 L) + 4\ln(\Lambda_2 L)\Big)\Big] . \tag{9}$$

Λ_i are scale parameters for the sub-leading LECs. The latter are given at the scale of the physical pion mass as

$$\bar{l}_i = \ln(\Lambda_i / M_\pi^{\rm phys})^2 . \tag{10}$$

Even more recent papers addressed again the NNLO of the δ-expansion [10], and the corrections due to finite $m_{\rm q}$ [11]. In the following we will discuss numerical results for M_π^R. We see that a confrontation with the analytical prediction in eqs. (8), (9) could enable a new determination of a set of LECs from first principles of QCD.

3. Attempts to simulate QCD in the δ-regime

The straight way to measure M_π^R are simulations directly *in* the δ-regime. Since the δ-box differs from the lattice shapes in usual simulations, this requires the special purpose generation of configurations. Moreover, precise chirality is vital in this regime, hence one is supposed to use a formulation of lattice quarks which preserves chiral symmetry. Such lattice fermions are known since the late 90ies, but their simulation is extremely tedious, in particular with dynamical quarks (*i.e.* keeping track of the fermion determinant in the generation of gauge configurations). We anticipate that so far there are no robust results of simulations clearly inside the δ-regime. In this section we summarise the efforts that have been carried out so far.

At the Symposium LATTICE2005 D. Hierl presented a first attempt to simulate 2-flavour QCD in the δ-regime [12]. That study used a truncated version of a chiral lattice Dirac operator, so a first question is if the quality of approximate chirality was sufficient for that purpose. Since this Dirac operator is very complicated, that simulation was performed with a non-standard algorithm, which probes the fermion determinant with a stochastic estimator. The spatial volume was $\approx (1.2\ \text{fm})^3$, and the results for M_π at small quark masses agreed well with the LO of the δ-expansion, *i.e.* eq. (8) at $\Delta = 0$ (and with the phenomenological value of F_π).

In 2007 the QCDSF Collaboration generated a new set of data, which have not been published. They were obtained with dynamical overlap quarks, which are exactly chiral. This simulation used the Hybrid Monte Carlo algorithm (*i.e.* the reliable standard algorithm). The lattice had a modest size of $8^3 \times 16$ sites, and the spatial box length was again $L \approx 1.2$ fm. At first sight the results seemed to look fine: for decreasing $m_{\rm q}$ we saw a transition from a Gell-Mann–Oakes–Renner type behaviour towards a plateau. Its value agreed with the chirally extrapolated value of Ref. [12], and therefore also with the LO of eq. (8).

Unfortunately, this is *not* the end of the story. If we proceed to the NLO correction, *i.e.* if we include the first term of Δ given in eq. (9), the predicted value for M_π^R decreases drastically in this small volume — from 782.5 MeV down to 321.7 MeV — and the agreement with the above data is gone. Considering this dramatic effect of the NLO correction one might worry that the δ-expansion could converge only very slowly in this small box, and such simulations are not instructive at all. However, adding also the NNLO correction alters the NLO result only a little — to (336.3 ± 7.6) MeV — so it is reasonable to assume the δ-expansion to be already well converged.[2]

Thus simulations in this small box *can* be useful in view of a confrontation with the analytical predictions for M_π^R. In fact a second sequence of runs by the QCDSF Collaboration (performed

[2] To obtain this theoretically predicted value, we inserted the LECs as far as they are known. In particular the sub-leading LECs $\bar{l}_1 = -0.4 \pm 0.6$ and $\bar{l}_2 = 4.1 \pm 0.1$ are taken from Ref. [13], along with their uncertainties, which imply the uncertainty in the NNLO value of M_π^R.

m_{q}	a [fm]	M_π	M_π^R (NNLO)
0.008	0.102(9)	(535.8 ± 54.5) MeV	(510.3 ± 30.7) MeV
0.01	0.104(8)	(580.5 ± 47.2) MeV	(505.8 ± 30.3) MeV
0.03	0.114(8)	(444.8 ± 31.4) MeV	(476.2 ± 33.4) MeV
0.04	0.111(8)	(476.4 ± 34.7) MeV	(486.2 ± 32.6) MeV
0.05	0.111(9)	(511.9 ± 43.9) MeV	(486.2 ± 35.0) MeV

Table 1. The results by QCDSF Collaboration of the year 2008, on a $8^3 \times 16$ lattice with 5 values of the mass m_{q} for the two degenerate dynamical overlap quark flavours. We display the measured values of the physical lattice spacing a and the pion mass, as well as the NNLO prediction for the residual pion mass M_π^R. (The errors capture the uncertainty in a, in the pion mass in lattice units, and in the LECs $\bar{l}_1$ and $\bar{l}_2$, cf. eq. (14).)

in 2008) yielded data much closer to the NNLO prediction. They involved 5 quark masses; the results for the physical lattice spacing a and the pion mass are given in Table 1. Since a varies for the different values of m_{q}, the box length $L = 8a$ and the prediction for M_π^R vary as well. In Fig. 3 we compare the numerically measured pion masses and the corresponding NNLO δ-expansion results. They are compatible, in contrast to the earlier data, which were probably not well thermalised. In these new runs thermalisation is accomplished, but the ratio $T/L = 2$ is still modest. We therefore proceeded to a $8^3 \times 32$ lattice, where our runs are ongoing. Still, the quality of agreement with the δ-expansion that we observed already on the $8^3 \times 16$ lattice is impressive.

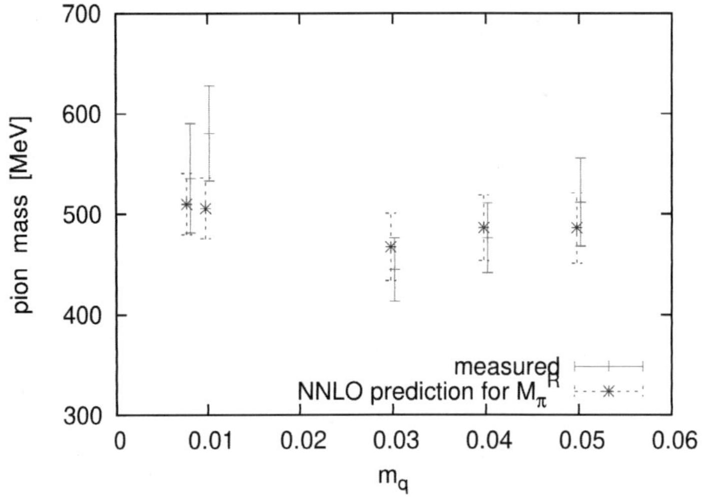

Figure 3. An illustration of our results given in Table 1: we show the numerically measured pion mass and the theoretical residual mass M_π^R, as predicted by the NNLO δ-expansion. (The latter actually refer to the chiral limit, but its value depends on L and therefore on the lattice spacing a, which varies for the simulations at different m_{q}.) We observe good agreement between the measured and the predicted values.

4. Residual pion mass by an extrapolation from the p-regime

In this section we proceed to a different approach. It is based on simulation results in the p-regime (up to the transition zone), which are then extrapolated towards the δ-regime. Details of this study are given in Ref. [14]. Also in this framework we consider it essential to use dynamical quarks, but we do not insist on exact chiral symmetry in the p-regime. Hence we used Wilson fermions, which is an established standard lattice fermion formulation, in a form which corrects $O(a)$ lattice artifacts. Thus the simulation was much faster, and we could tackle much larger lattices than those mentioned in the Section 3; our data reported below were obtained on three lattice sizes: $24^3 \times 48$, $32^3 \times 64$ and $40^3 \times 64$. On the other hand, this lattice regularisation breaks the chiral symmetry explicitly, so that additive mass renormalisation sets in. Nevertheless we were able to attain very light pion masses.

For the gauge part we used the standard plaquette lattice action. Our simulations were carried out at two values for the strong gauge coupling g_s, respectively the parameter $\beta = 6/g_s^2$. We determined the physical lattice spacing a from the measured nucleon mass, which revealed that we were dealing with fine lattices,

$$\beta = 5.29 \ \rightarrow \ a \simeq 0.075 \text{ fm} \quad , \quad \beta = 5.4 \ \rightarrow \ a \simeq 0.067 \text{ fm} \quad . \tag{11}$$

Thus the spatial size was in the range $L \simeq 1.6$ fm…3.0 fm. Due to the additive mass renormalisation, we could not refer to the bare quark mass anymore. We measured the current quark mass by means of the PCAC relation,

$$m_q = \frac{\langle \partial_4 A_4(\vec{0}, x_4) P(0) \rangle}{\langle P(\vec{0}, x_4) P(0) \rangle} , \tag{12}$$

where P is the pseudoscalar density, and A_μ is the axial current. We observed practically no finite size effects on a and on m_q, but we did see a striking L-dependence of M_π, as expected. These quantities were found in the range

$$m_q = 3.60 \text{ MeV} \dots 231 \text{ MeV} \quad , \quad M_\pi = 174 \text{ MeV} \dots 1.52 \text{ GeV} \quad , \quad M_\pi L = 2.7 \dots 9.7 \ . \tag{13}$$

The latter confirms that our data range from the deep p-regime to the transition zone. A few missing data points have been completed with a (lengthy) formula for exponentially suppressed finite size effects [15, 16] within the p-regime, which holds up to $O(p^4)$. This formula involves the renormalised sub-leading LECs $\bar{l}_i$, $i = 1 \dots 4$, see eqs. (4) and (10). Well established phenomenological values were composed in Ref. [13] ($\bar{l}_1$, $\bar{l}_2$ were anticipated in footnote 2),

$$\bar{l}_1 = -0.4 \pm 0.6 \quad , \quad \bar{l}_2 = 4.1 \pm 0.1 \quad , \quad \bar{l}_3 = 2.9 \pm 2.4 \quad , \quad \bar{l}_4 = 4.4 \pm 0.4 \quad . \tag{14}$$

They were estimated from $\pi\pi$ scattering data, and in particular the $\bar{l}_4$ value is based on the scalar pion form factor. The dark horse in this context is $\bar{l}_3$: we replaced the above value by $\bar{l}_3 \approx 4.2$, which we obtained in our study, see below.

Our measured and interpolated data are given in Ref. [14]. We extrapolated them towards the δ-regime and extracted in particular a value for M_π^R based on the relatively simple chiral extrapolation formula

$$M_\pi(L)^2 = M_\pi^{R\,2} + C_1 m_q [1 + C_2 m_q \ln(C_3 m_q)] \ , \tag{15}$$

which interpolates between the $O(p^4)$ correction formula (for large LM_π) and the chiral limit [13]. The C_i and M_π^R are treated as free parameters to be fixed by the fit. Our data and the fits for $\beta = 5.29$ and for $\beta = 5.4$ are shown in Figs. 4 and 5, respectively. These plots also illustrate the extrapolation result for M_π^R in the chiral limit.

Our main result is shown in Fig. 6. It compares the extrapolated values for M_π^R with the predictions based on the δ-expansion to LO, NLO and NNLO, as a function of L. The latter two predictions are very close to each other for the volumes under consideration, so we can again assume the expansion to be well converged.

Journal of Physics: Conference Series **287** (2011) 012016 doi:10.1088/1742-6596/287/1/012016

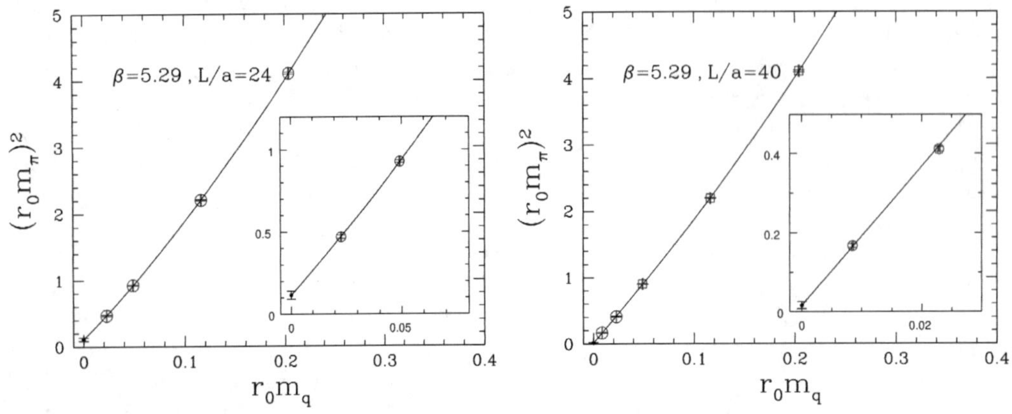

Figure 4. Our chiral extrapolation referring to the lattice spacing $a \simeq 0.075$ fm (corresponding to $\beta = 5.29$) with $L \simeq 1.8$ fm (on the left) and $L \simeq 3.0$ fm (on the right). The data points show the measured pion mass squared against the quark mass in the p-regime (the Sommer scale parameter $r_0 = 0.467$ fm is employed to convert them into dimensionless units). The curve is the fit according to eq. (15). It ends in the chiral limit, where we illustrate the extrapolated value for M_π^R and its error.

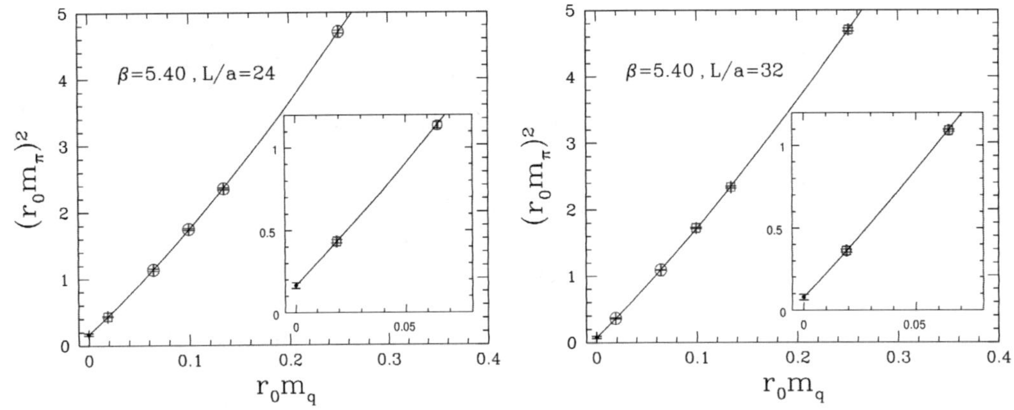

Figure 5. The same as Fig. 4, but now for the finer lattices with $a \simeq 0.067$ fm (corresponding to $\beta = 5.4$), such that $L \simeq 1.6$ fm (on the left) and $L \simeq 2.1$ fm (on the right).

The extrapolation results for M_π^R reach down to values even below the physical pion mass. The plot also shows the line where the product $M_\pi L$ decreases below 1; this can be roughly considered as the boundary of the δ-regime. Our extrapolations lead close to this boundary, and in one case *into* the δ-regime.[3]

In particular this plot shows that the extrapolated masses M_π^R match this curve remarkably well. It would not have been obvious to predict this feature, because our data were obtained in

[3] In that case, the extrapolation should actually turn into a plateau at tiny m_q, but this would hardly change the result.

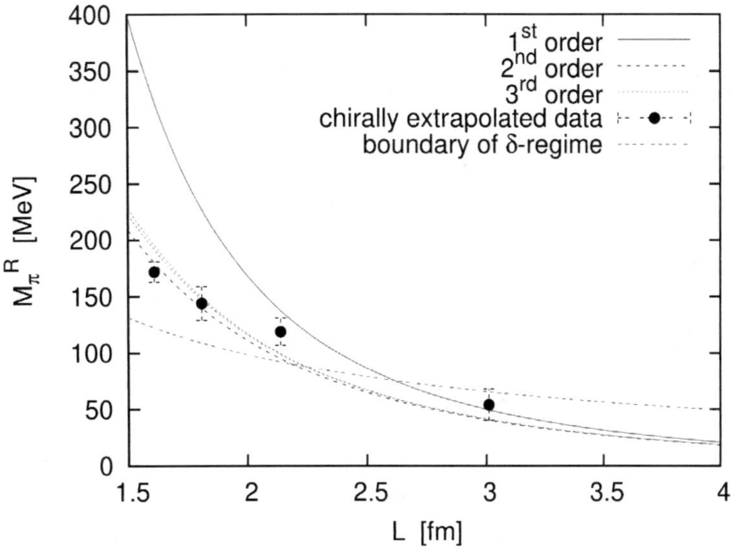

Figure 6. A comparison of our extrapolated values for the residual pion mass M_π^R with the predictions by the δ-expansion (LO, NLO, NNLO), as functions of the spatial box size L. We observe a remarkably good agreement. In our largest volume, the extrapolated M_π^R is located below the physical pion mass, and in the region $M_\pi^R L < 1$, which can be viewed as the domain of the δ-regime.

a regime where the basis of the δ-expansion (frozen spatial degrees of freedom) does not hold. This suggests that the extrapolation formula (15) applies well in a sizable domain.

5. Evaluation of Low Energy Constants

By fitting our pion mass data according to eq. (15), we obtained results for the four free fitting parameters. In Section 4 we discussed the results for M_π^R that we obtained in this way. Moreover, the results for C_1, C_2 and C_3 can be used to evaluate the notorious sub-leading LEC $\bar{l}_3$ [14]. This is how we obtained our value that we anticipated in Section 4,

$$\bar{l}_3 = 32\pi^2 F_\pi^2 \frac{C_2}{C_1} \ln \frac{C_1}{C_3 M_\pi^{\mathrm{phys}\ 2}} = 4.2 \pm 0.2 \ . \tag{16}$$

(The error given here emerges from the fits, the additional systematic error would be hard to estimate). In view of other results in the literature, this value is in the upper region. An overview has been presented in Ref. [17], in particular in Table 11 and Figure 9. That overview estimates the world average as $\bar{l}_3 = 3.3(7)$.

In the previous considerations of Sections 4 and 5, we inserted the phenomenological value of the pion decay constant, $F_\pi = 92.4$ MeV. Alternatively, we could also treat F_π as a free parameter to be determined by the fits. In particular, matching M_π^R this yields in the chiral limit [14]

$$F_\pi^{\mathrm{numerical}}\Big|_{m_q=0} = 78^{+14}_{-10} \ \mathrm{MeV} \ , \tag{17}$$

which seems a bit low. However, effective field theory considerations suggest that the value of F_π in the chiral limit should indeed be below the physical value; Ref. [17] estimates $F_\pi|_{m_q=0} \approx 86$ MeV.

6. Conclusions

The δ-regime refers to a system of pions in a finite box, typically of the shape $L^3 \times T$ with $L \lesssim M_\pi^{-1} \ll T$. It can be treated by Chiral Perturbation Theory with suitable counting rules, the δ-expansion. In particular this predicts the residual pion mass in the chiral limit, M_π^R, which has been computed recently to NNLO [9].

First attempts to simulate 2-flavour QCD in the δ-regime were confronted with technical difficulties, in particular thermalisation problems. Nevertheless we obtained good results for the pion mass with light quarks on a small lattice of size $8^3 \times 16$, with $L \simeq (0.82 \dots 0.91)$ fm.

In another pilot study we measured pion masses in the p-regime (up to the transition region) and extrapolated them towards the δ-regime. This yields numerical results for M_π^R, which agree remarkably well with the predictions by the δ-expansion. That expansion is based on assumptions, which do not hold in the regime where the data were obtained. Therefore it would not have been obvious to predict the observed agreement. This comparison can be viewed as a numerical experiment, which led to an interesting observation.

Our pion mass fits from the p- towards the δ-regime also fix some further constants, which allow for a new determination of the (mysterious) sub-leading Low Energy Constant $\bar{l}_3$; we obtained $\bar{l}_3 = 4.2(2)$. This is somewhat above the average of the phenomenological and numerical estimates in the literature. The constants $\bar{l}_1, \dots, \bar{l}_4$ are relevant in the effective description of $\pi\pi$ scattering. Finally we also considered the option to treat F_π as a free parameter. In the chiral limit the fits lead to a value somewhat below the phenomenological F_π.

Robust numerical results, measured manifestly inside the δ-regime, are still outstanding.

Acknowledgements : W.B. thanks for kind hospitality at the University of Bern, where this talk was written up. The simulations for this project were performed on the SGI Altix 4700 at LRZ (Munich), the IBM BlueGeneL and BlueGeneP at NIC (Jülich), the BlueGeneL at EPCC (Edinburgh), and the apeNEXT at NIC/DESY (Zeuthen). We also thank for financial support by the EU Integrated Infrastructure Initiative "HadronPhysics2" and by the DFG under contract SFB/TR 55 (Hadron Physics from Lattice QCD). J.M.Z. is supported through the UKs STFC Advanced Fellowship Program under contract ST/F009658/1.

References

[1] Weinberg S 1979 *Physica* A **96** 327. Gasser J and Leutwyler H 1984 *Ann. Phys.* **158** 142
[2] Gasser J and Leutwyler H 1987 *Phys. Lett.* B **184** 83
[3] Gasser J and Leutwyler H 1987 *Phys. Lett.* B **188** 477
[4] Giusti L, Hoelbling C, Lüscher M and Wittig H 2003 *Comput. Phys. Commun.* **153** 31
[5] Bietenholz W 2010 arXiv:1007.0285 [hep-lat]
[6] Leutwyler H 1987 *Phys. Lett.* B **189** 197
[7] Neuberger H and Ziman T 1989 *Phys. Rev.* **B 39** 2608. Hasenfratz P and Leutwyler H 1990 *Nucl. Phys.* **B 343** 241. Wiese U-J and Ying H-P 1994 *Z. Phys.* B **93** 147. Beard B B and Wiese U-J 1996 *Phys. Rev. Lett.* **77** 5130. Gerber U, Hofmann C P, Jiang F-J, Nyfeler M and Wiese U-J 2009 *J. Stat. Mech.* P03021. Hofmann C P 2010 *Phys. Rev.* **B 81** 014416. Jiang F-J and Wiese U-J 2010 arXiv:1011.6205 [cond-mat.str-el]
[8] Hasenfratz P and Niedermayer F 1993 *Z. Phys.* B **92** 91
[9] Hasenfratz P 2010 *Nucl. Phys.* B **828** 201
[10] Niedermayer F and Weiermann C 2011 *Nucl. Phys.* B **842** 248
[11] Weingart M 2010 arXiv:1006.5076 [hep-lat]
[12] Hasenfratz A, Hasenfratz P, Hierl D, Niedermayer F and Schäfer A 2006 *PoS(LAT2006)178*
[13] Colangelo G, Gasser J and Leutwyler H 2001 *Nucl. Phys.* B **603** 125
[14] Bietenholz W, Göckeler M, Horsley R, Nakamura Y, Pleiter D, Rakow P E L, Schierholz G and Zanotti J M (QCDSF Collaboration) 2010 *Phys. Lett.* B **687** 410
[15] Lüscher M 1986 *Commun. Math. Phys.* **104** 177, 1986 *Commun. Math. Phys.* **105** 153
[16] Colangelo G, Dürr S and Haefeli C 2005 *Nucl. Phys.* B **721** 136
[17] FLAG Working Group of FLAVIAT (Colangelo G *et al.*) 2010 arXiv:1011.4408 [hep-lat]

Journal of Physics: Conference Series **287** (2011) 012017 doi:10.1088/1742-6596/287/1/012017

Gauge invariance and radiative corrections in an extra dimensional theory

H. Novales-Sánchez[(a)] **and J. J. Toscano**[(b)]

Facultad de Ciencias Físico Matemáticas, Benemérita Universidad Autónoma de Puebla, Apartado Postal 1152, Puebla, Puebla, México.

E-mail: [(a)]`hnovales@alumnos.fcfm.buap.mx`, [(b)]`jtoscano@fcfm.buap.mx`

Abstract. The gauge structure of the four dimensional effective theory originated in a pure five dimensional Yang-Mills theory compactified on the orbifold S^1/Z_2, is discussed on the basis of the BRST symmetry. If gauge parameters propagate in the bulk, the excited Kaluza-Klein (KK) modes are gauge fields and the four dimensional theory is gauge invariant only if the compactification is carried out by using curvatures as fundamental objects. The four dimensional theory is governed by two types of gauge transformations, one determined by the KK zero modes of the gauge parameters and the other by the excited ones. Within this context, a gauge-fixing procedure to quantize the KK modes that is covariant under the first type of gauge transformations is shown and the ghost sector induced by the gauge-fixing functions is presented. If the gauge parameters are confined to the usual four dimensional space–time, the known result in the literature is reproduced with some minor variants, although it is emphasized that the excited KK modes are not gauge fields, but matter fields transforming under the adjoint representation of $SU_4(N)$. A calculation of the one-loop contributions of the excited KK modes of the $SU_L(2)$ gauge group on the off-shell W^+W^-V, with V a photon or a Z boson, is exhibited. Such contributions are free of ultraviolet divergences and well–behaved at high energies.

1. A general idea

Standard Model (SM) extensions with TeV^{-1}-sized extra dimensions [1] involve at least three key scales, the first of which is the electroweak scale, at 246GeV, where spontaneous symmetry breaking takes place. Another main scale is the compactification one, whose size should be smaller than the current experimental sensitivity in order to make this sort of models phenomenologically consistent. Finally, the third scale is beyond the former ones, and there the physics should drastically change. Such asseverations suggest that any generalization of the SM to a given version with TeV^{-1}-sized extra dimensions is, from the commencement, an effective theory defined in a larger space–time manifold. A symptom of such a non–fundamental behavior is the fact that coupling constants in extra dimensional models have dimensions, which implies that these models are nonrenormalizable. When one compactifies the extra dimensions and integrates them, one obtains a four dimensionsal (4D) effective theory where the dynamic variables are the so-called KK modes, which are infinite in number. In the 4D space–time the coupling constants are dimensionless, but the nonrenormalizable nature of the extra dimensional model is however inherited at the four dimensional level and manifests itself through the infinite KK towers. In connection with that, it has been proven [2] that, in the case of one universal extra dimension, one-loop level effects are insensitive to the new physics scale and, in fact, the KK sums of the radiative corrections are convergent. By contrast, for two or more universal extra dimensions [2] the loop effects are highly sensitive to the cut-off imposed by the fundamental scale, and the KK sums are rather divergent.

In this paper we shall start with a five dimensional (5D) Yang-Mills theory, and shall discuss a route to obtain a theory properly quantized at the 4D level. In order to reach such a goal, the fifth dimension shall be compactified on the orbifold S^1/Z_2 and then the extra coordinate shall be integrated in the action. By following this path, a 4D effective theory is obtained. Such a theory is constituted by the ordinary 4D Yang-Mills theory and other couplings among light gauge bosons and heavy ones. This 4D theory possesses an interesting gauge structure [3], which we shall discuss. We shall brefly comment on the employment of the Becchi-Rouet-Stora-Tyutin (BRST) formalism [4] to quantize this 4D effective theory. The quantum action so obtained can be utilized to perform radiative corrections on light Green functions. In particular, we shall exhibit the one-loop KK contributions to the W^+W^-V vertex. Such contributions are well behaved [5] and consistent with previous results reported in the literature. In fact, a comparison of such effects with the tree-level contributions of higher canonical dimension operators shows [5] that the former are dominant over the latter, as was already claimed in the case of one universal extra dimension [2].

2. The gauge structure of the 4D effective theory

We consider a 5D space-time manifold with a flat metric diag$(1, -1, -1, -1, -1)$, where we define the $SU_5(N)$-invariant Lagrangian

$$\mathcal{L}_{5\mathrm{YM}} = -\frac{1}{4}\mathcal{F}^a_{MN}(x,y)\mathcal{F}^{aMN}(x,y), \tag{1}$$

with x labeling the ordinary four dimensional space-time coordinates, while y is the coordinate for the fifth dimension. Capital letters indices run over 0, 1, 2, 3, 5, and greek indices shall refer to the usual four-dimensional space-time. The 5D dynamic variables in Eq.(1) are the vector bosons $(\mathcal{A}^a_\mu(x,y), \mathcal{A}^a_5(x,y))$, which compose the 5D curvatures

$$\mathcal{F}^a_{MN}(x,y) = \partial_M \mathcal{A}^a_N(x,y) - \partial_N \mathcal{A}^a_M(x,y) + g_5 f^{abc}\mathcal{A}^b_M(x,y)\mathcal{A}^c_N(x,y). \tag{2}$$

Here, g_5 is the 5D coupling constant, which has dimensions of $(\mathrm{mass})^{-1/2}$. This Lagrangian is invariant under the set of gauge transformations

$$\delta\mathcal{A}^a_M = \mathcal{D}^{ab}_M \alpha^b, \tag{3}$$

which are defined by the gauge parameters α^b. As we shall comment below, these parameters are crucial objects when deriving the 4D effective Lagrangian, as they determine two essentially different scenarios [3]. One of them occurs when the gauge parameters propagate in the fifth dimension, whereas the other is defined by gauge parameters that do not depend on the fifth dimension. If the extra dimension is compactified on the orbifold S^1/Z_2 with radius R and the gauge parameters depend on the extra dimension, such parameters can be expanded in Fourier series as

$$\alpha^a(x,y) = \frac{1}{\sqrt{2\pi R}}\alpha^{(0)a}(x) + \sum_{n=1}^{\infty}\frac{1}{\sqrt{\pi R}}\alpha^{(n)a}(x)\cos\left(\frac{ny}{R}\right). \tag{4}$$

In extra dimensional models, the zero-modes of the fields are identified with the light fields corresponding to the already known particles in the 4D space–time, while the excited KK modes are heavy new fields. By virtue of this, one should expect the zero-mode gauge parameters to define the *standard gauge transformations* (SGT) of the 4D Yang-Mills theory. On the other hand, the excited KK gauge parameters should define a different set transformations that we call *nonstandard gauge transformations* (NSGT). We shall see that under such circumstances all the zero-mode fields are gauge fields, and that some KK excited modes fields are also gauge fields. By contrast, if gauge parameters do not propagate in the extra dimension, these parameters cannot be Fourier–expanded, which in turn implies that they remain the same before and after compactification. This suggests that in four dimensions there must be only SGT.

As there are two scenarios, one should be able to derive two different 4D effective Lagrangians. In the case of gauge parameters propagating in the extra dimension the Fourier expansions must be performed over the 5D curvatures [3],

$$\mathcal{F}^a_{\mu\nu}(x,y) = \frac{1}{\sqrt{2\pi R}}\mathcal{F}^{(0)a}_{\mu\nu}(x) + \sum_{n=1}^{\infty}\frac{1}{\sqrt{\pi R}}\mathcal{F}^{(n)a}_{\mu\nu}(x)\cos\left(\frac{ny}{R}\right), \tag{5}$$

$$\mathcal{F}^a_{\mu 5}(x,y) = \sum_{n=1}^{\infty}\frac{1}{\sqrt{\pi R}}\mathcal{F}^{(n)a}_{\mu 5}(x)\sin\left(\frac{ny}{R}\right). \tag{6}$$

This procedure contrasts with the path most commonly followed in the literature, where gauge fields are expanded instead [6]. The lesson to learn is that expansion and integration of the extra dimension over covariant objects leads to 4D covariant objects [3]. In the other scenario, where gauge parameters do not depend on the extra dimension, the objects to expand and integrate over are the gauge fields [3].

2.1. First scenario: gauge parameters propagate in the extra dimension

We define the 4D Lagrangian

$$\mathcal{L}_{4\mathrm{YM}} \equiv \int_0^{2\pi R} dy \mathcal{L}_{5YM}, \tag{7}$$

and insert the expansions shown in Eqs.(5,6) into it, which results in [3]

$$\mathcal{L}_{4\mathrm{YM}} = -\frac{1}{4}\left(\mathcal{F}^{(0)a}_{\mu\nu}\mathcal{F}^{(0)a\mu\nu} + \mathcal{F}^{(m)a}_{\mu\nu}\mathcal{F}^{(m)a\mu\nu} + 2\mathcal{F}^{(m)a}_{\mu 5}\mathcal{F}^{(m)a\mu 5}\right), \tag{8}$$

where repeated indices, including the mode ones, denote a sum. This convention shall be followed during the rest of the paper. This 4D Lagrangian has an elegant structure, for it is constituted exclusively by objects that resemble the Yang-Mills curvature. In fact, the precise form of the curvatures $\mathcal{F}^{(0)a}_{\mu\nu}$, $\mathcal{F}^{(m)a}_{\mu\nu}$ and $\mathcal{F}^{(m)a}_{\mu 5}$ can be straightforwardly determined [3]. A crucial feature of the 4D Lagrangian, Eq.(8), is that it is invariant under the SGT [3]

$$\delta A^{(0)a}_\mu = \mathcal{D}^{(0)ab}_\mu \alpha^{(0)b}, \tag{9}$$

$$\delta A^{(m)a}_\mu = g f^{abc} A^{(m)b}_\mu \alpha^{(0)c} \tag{10}$$

$$\delta A^{(m)a}_5 = g f^{abc} A^{(m)b}_5 \alpha^{(0)c}, \tag{11}$$

as well as under the NSGT [3]

$$\delta A^{(0)a}_\mu = g f^{abc} A^{(m)b}_\mu \alpha^{(m)c}, \tag{12}$$

$$\delta A^{(m)a}_\mu = \mathcal{D}^{(mn)ab}_\mu \alpha^{(n)b}, \tag{13}$$

$$\delta A^{(m)a}_5 = \mathcal{D}^{(mn)ab}_5 \alpha^{(nb)}, \tag{14}$$

where $\mathcal{D}^{(0)ab}_\mu$ is the ordinary 4D Yang–Mills covariant derivative, $\mathcal{D}^{(mn)ab}_\mu$ is a sort of covariant derivative, given by [3]

$$\mathcal{D}^{(mn)ab}_\mu = \delta^{mn}\mathcal{D}^{(0)ab}_\mu - g f^{abc}\Delta^{mrn} A^{(r)c}_\mu, \tag{15}$$

and $\mathcal{D}_5^{(mn)ab}$ is [3] an object that does not involve derivatives. The factors Δ^{mrn} in the covariant drerivative Eq.(15) are sums of Kronecker deltas and their precise form is not important for the present discussion. These sets of gauge transformations can be derived at least by means of three different methods [3]: Fourier analysis, the Dirac's method [7] toghether with the Castellani's formalism [8], and the BRST formalism [9]. Note that the zero–modes $A_\mu^{(0)a}$ are gauge fields under the SGT, but under the NSGT they change in a way that resembles a transformation in the adjoint representation, although in this case there is a mixing of KK modes. On the other hand, the excited modes $A_\mu^{(m)a}$ are matter fields under the SGT, but gauge fields under the NSGT. Finally, the scalar fields $A_5^{(n)a}$ are pseudo-Goldstone bosons, as they can be eliminated from the theory [3] by performing a nonstandard gauge transformation with gauge parameters $\alpha^{(n)a} = (R/n)A_5^{(n)a}$. Concerning the Dirac's method, the only fields that generate constraints are $A_\mu^{(0)a}$ and $A_\mu^{(n)a}$, and such constraints are all first class, which implies that these fields are gauge fields, a feature that is consistent with the form of transformations (9–14). It is worth emphasizing that the 4D curvatures $\mathcal{F}_{\mu\nu}^{(0)a}$, $\mathcal{F}_{\mu\nu}^{(m)a}$ and $\mathcal{F}_{\mu 5}^{(m)a}$ transform covariantly [3] under the SGT and the NSGT.

2.2. Second scenario: gauge parameters do not propagate in the extra dimension

If gauge parameters are not allowed to propagate in the extra dimension, they remain the same after compactification and integration of the extra dimension, and instead of an infinite number of them at the 4D level, there are only $N^2 - 1$ gauge parameters in four dimensions. With this in mind, one should only expect SGT-invariance after the fifth dimension is compactified and intagrated. In this case, the objects to expand in Fourier series are the gauge fields instead of the curvatures [3]. After compactifying and integrating the fifth dimension, one obtains the 4D Lagrangian [3]

$$\hat{\mathcal{L}}_{4\text{YM}} \equiv \int_0^{2\pi R} dy \mathcal{L}_{5\text{YM}} = \mathcal{L}_{4\text{YM}} + \Delta\mathcal{L}. \tag{16}$$

Here, the $\mathcal{L}_{4\text{YM}}$ Lagrangian is the one exhibited in Eq.(8), which was obtained within the context of the first scenario. The term $\Delta\mathcal{L}$ is hence a difference between the Lagrangians derived in the two scenarios, and it is given by [3]

$$\Delta\mathcal{L} = -\frac{1}{4} g^2 f^{abc} f^{ade} \left(\Delta^{rnpq} A_\nu^{(n)c} A^{(q)e\nu} - \Delta'^{rnpq} A_5^{(n)c} A_5^{(q)e} \right) A_\mu^{(r)b} A^{(p)d\mu}, \tag{17}$$

where the factors Δ^{rnpq} and Δ'^{rnpq} are sums and differences of Kronecker deltas. The $\hat{\mathcal{L}}_{4\text{YM}}$ Lagrangian is invariant under the SGT, Eqs.(9,10,11), but not under the NSGT, Eqs.(12,13,14), consistently with the fact that there are no KK excitations for the gauge parameters that define the latter sort of gauge transformations. In this scenario the nature of the fields involved in the 4D theory is quite different from the one found in the first scenario, as the zero-mode fields, $A_\mu^{(0)a}$, are still gauge fields, but the vector KK excited modes, $A_\mu^{(m)a}$, are matter fields and are not gauge fields. An important difference among the two scenarios is that, in this second context, the scalar KK fields, $A_5^{(n)a}$, are not Pseudo-Goldstone bosons, but matter fields, which cannot be removed from the theory [3]. The last comment concerning this scenario is that the Dirac's method leads [3], under these circumstances, to first class as well as second class constraints, suggesting that there are matter fields, which is in agreement with the above discussion.

3. Quantization

Through this section, the quantization of the 4D effective theory, in the context of the first scenario, will be briefly discussed. Such a quantization, which is based on the BRST

Journal of Physics: Conference Series **287** (2011) 012017 doi:10.1088/1742-6596/287/1/012017

formalism [9], leads [3] to the most general Faddeev-Popov ghost sector. An interesting feature of this quantization scheme is that the quantization of the zero-modes and the excited ones can be performed independently of each other [3]. As we are mainly interested in the one-loop contributions to light Green functions with KK excited modes, exclusively, inside the loops [5], we shall quantize only the KK excitations, leaving gauge invariance with respect to the SGT, but removing the degeneracy associated with the NSGT. This does not mean that the invariance under the SGT cannot be eliminated. In fact, the gauge with respect to such transformations can be fixed as in the ordinary Yang-Mills theory or by a nonstandard quantization scheme, such as the background field method [10].

According to the BRST formalism [9], one starts with a configuration space that comprehends the ghost fields from the beginning. Such ghost fields coincide with the gauge parameters, although they have opposite statistics. As a next step, the configuration space is enriched by the introduction of an antifield per each field of the theory. Also a set of auxiliary fields is inserted. In this larger configuration space a symplectic structure, called the *antibracket*, is defined and the master equation imposed. Then a proper solution to the master equation must be obtained. Particularly, the solution to the master equation in the case of the ordinary 4D Yang-Mills theory is already known, and its generalization to the 5D version is straightforward. The obtainment of the proper solution for the 4D effective theory can be accomplished [3] by compactifying the extra dimension and performing a Fourier analysis on the 5D proper solution. As mentioned above, one can introduce a gauge-fixing procedure that maintains gauge invariance under the SGT. Such a procedure depends crucially on the employing of an appropriate set of gauge-fixing functions, which shall appear later in the quantized version of the Lagrangian and shall impact directly on the structure of the Faddeev-Popov ghost term. The following SGT–covariant set of gauge-fixing functions is introduced [3]:

$$f^{(m)a} = \mathcal{D}_\mu^{(0)ba} A^{(m)b\mu} - \xi \frac{m}{R} A_5^{(m)a}. \tag{18}$$

These functions are rather similar to other introduced [11] some years ago, in the context of the so-called *331 model* [12]. Once the proper solution to the master equation of the effective 4D theory is known, one can derive [3] the gauge-fixed action and consequently the effective Lagrangian

$$\mathcal{L}_{\text{eff}} = \mathcal{L}_{4\text{YM}} + \mathcal{L}_{\text{GF}} + \mathcal{L}_{\text{FPG}}. \tag{19}$$

This expression comprehends some very important results, such as the Lagrangian $\mathcal{L}_{4\text{YM}}$, which was shown in Eq.(8). The second term is the gauge-fixing part, $\mathcal{L}_{\text{GF}}$, which is consituted by the gauge-fixing functions, Eq.(18), and is given by

$$\mathcal{L}_{\text{GF}} = -\frac{1}{2\xi} f^{(m)a} f^{(m)a}. \tag{20}$$

where ξ is the gauge parameter. The last part, $\mathcal{L}_{\text{FPG}}$, is the most general Faddeev-Popov ghost term, as it includes up to quartic ghost couplings. Its precise expression is [3]

$$\begin{aligned}\mathcal{L}_{\text{FPG}} = {} & \bar{C}^{(m)c}\left(\frac{\partial f^{(m)c}}{\partial A_\mu^{(n)a}}\mathcal{D}^{(nr)ab\mu} - \frac{\partial f^{(m)c}}{\partial A_5^{(n)a}}\mathcal{D}_5^{(nr)ab}\right) C^{(r)b} - \frac{1}{\xi} g f^{abc} \Delta^{mrn} f^{(m)a} \bar{C}^{(r)b} C^{(n)c} \\ & + \frac{1}{2} g^2 f^{abc} f^{cde} \Delta^{mpq} \bar{C}^{(p)d} \bar{C}^{(q)e} \left(C^{(0)b} C^{(m)a} + C^{(0)a} C^{(m)b} + \Delta^{mrn} C^{(r)b} C^{(n)a}\right), \end{aligned} \tag{21}$$

with $C^{(n)b}$ and $\bar{C}^{(n)b}$ denoting the ghost and antighost fields, respectively. The effective Lagrangian, Eq.(19), contains all the couplings among light and heavy fields, which are crucial pieces to perform radiative corrections on light Green functions.

4. Radiative corrections: The W^+W^-V vertex

The principal prediction of extra dimensional models is the existence of new particles, which correspond to the KK excited modes. If there is enough energy available at the Large Hadron Collider, the lightest of such new particles could be directly produced there. However, if no signals are observed in that collider, the International Linear Collider could detect the KK excited modes through their virtual effects on low energy observables. In connection with that, an interesting process is the W^+W^- production, which could be studied at the one-loop level by inserting KK excited modes into the loops of the light W^+W^-V vertex, with V an off–shell photon or Z boson. It is worth mentioning that the one–loop contributions involving KK excited modes, exclusively, to any light Green function are [3] renormalizable, as, under such circumstances, the only divergencies that arise are those already present in the standard Yang-Mills theory and can therefore be [3] absorbed by the parameters of the light theory.

Consider the one-loop correction to the W^+W^-V vertex at the one-loop level. Suppose that the loops are formed by KK excited virtual fields, exclusively, and that V is off–shell. The calculation of this vertex leads [5] to the well-known parametrization [13] of such an interaction, at the one-loop level, when only CP-even quantities are involved. In this case, there arise three form factors, given by [5]

$$\begin{aligned} A &= \frac{g^2}{96\pi^2(4x_W-1)^3}\sum_{n=1}^{\infty}\Big(-24x_n(1-4x_W)^2B_0(1)+48x_W[5(x_W-1)x_W+4x_n(4x_W-1)B_0(2) \\ &\quad -6[-4x_W(6x_W^2-2x_W+3)+x_n(64x_W^2-4)+1]B_0(3)+36Q^2x_W[-4x_W^3+2x_W^2+x_W \\ &\quad +4x_n(x_W-1)(4x_W-1)]C_0-4\{1-2x_W[x_W(20x_W-33)+9]\}\Big), \end{aligned} \tag{22}$$

$$\begin{aligned} \Delta\kappa_V &= \frac{g^2}{96\pi^2(4x_W-1)^3}\sum_{n=1}^{\infty}\{48x_n[B_0(1)-B_0(2)]+12x_W(26x_W+1)[B_0(2)-B_0(3)] \\ &\quad -384x_nx_W[B_0(1)-2B_0(2)+B_0(3)]+768x_nx_W^2[B_0(1)-3B_0(2)+2B_0(3)] \\ &\quad -72Q^2x_W(x_W^2-12x_nx_W+x_W+3x_n)C_0+12x_W(4x_W-1)(8x_W+3)\Big\}, \end{aligned} \tag{23}$$

$$\begin{aligned} \Delta Q_V &= \frac{g^2}{96\pi^2(4x_W-1)^3}\sum_{n=1}^{\infty}\Big\{-1536x_nx_W^3[B_0(1)-B_0(3)]+96(x_W-2)x_W^2(6x_W+1)[B_0(2)-B_0(3)] \\ &\quad +768x_nx_W^2[B_0(1)+B_0(2)-2B_0(3)]-96x_nx_W[B_0(1)+2B_0(2)-3B_0(3)] \\ &\quad +144Q^2x_W[4x_W^4-4(4x_n+1)x_W^3+2(6x_n+1)x_W^2-6x_nx_W+x_n]C_0 \\ &\quad -24x_W(4x_W-1)[2x_W(6x_W-1)+1]\}, \end{aligned} \tag{24}$$

where Q is the momentum of the off–shell neutral boson, $x_W \equiv m_W^2/Q^2$ and $x_n \equiv m_n^2/Q^2$, with $m_n = n/R$, $n = 1, 2, 3, \ldots$. The following short–hand notation for the Passarino-Veltman scalar functions has been also introduced: $B_0(1) \equiv B_0(0, m_n^2, m_n^2)$, $B_0(2) \equiv B_0(m_W^2, m_n^2, m_n^2)$, $B_0(3) \equiv B_0(Q^2, m_n^2, m_n^2)$ and $C_0 \equiv C_0(m_W^2, m_W^2, Q^2, m_n^2, m_n^2 m_n^2)$. It is worth emphasizing that these form factors were calculated by employing the Feynman-'t Hooft gauge, as it has shown to be an election that leads to well–behaved results in nonconventional quantization approaches such as the background field method [10]. The form factors $\Delta\kappa_V$ and ΔQ_V are interesting as they define the magnetic dipole moment and the electric quadrupole moment of the W boson.

It is a remarkable feature of the form factors (23,24) that they are free of UV divergencies [5]. Such an insensitivity to the cut–off scale is in agreement with the results reported in the context of universal extra dimensions [2]. Altough these form factors are gauge dependent, it turns out that [5] they are well behaved at high energies and decouple in the limit of a very large compactification scale. For 0.5TeV $< R^{-1} <$ 1.0TeV, both $\Delta\kappa_V$ and ΔQ_V range [5] from approximately 10^{-5} to 10^{-6}, at best. The magnitude of the form factors in this range of energies is of the same order than those generated by new gauge bosons in the 331 model [11]. For a size of the fifth dimension of $R^{-1} \sim$ 1TeV, the one-loop contribution of the KK modes to these vertices is about one order of magnitude lower than the corresponding Standard Model radiative correction. If the compactification scale is taken as $R^{-1} \sim$ 300TeV, as it was realized in the case of one universal extra dimension [2], the magnitudes of $\Delta\kappa_V$ and ΔQ_V is [5] of $\mathcal{O}(10^{-4})$, which should be at the reach of the International Linear Collider.

Consider the operator of higher canonical dimension

$$\mathcal{O} = \frac{g_5 \alpha_W}{M_S^2} \frac{\epsilon^{abc}}{3!} \mathcal{A}^a_{\lambda\rho} \mathcal{A}^{a\rho\nu} \mathcal{A}^{c\lambda}_\nu , \tag{25}$$

where α_W is an unknown parameter that can be determined in terms of fundamental constants if the theory lying beyond the cut–off scale is known. Another ingredient of this operator is the cut–off scale, denoted by M_S. In connection with the form factors (22,23,24), this operator generates the term

$$\frac{g\alpha_W}{M_S^2} \frac{\epsilon^{abc}}{3!} A^{(0)a}_{\lambda\rho} A^{(0)b\rho\nu} A^{(0)c\lambda}_\nu , \tag{26}$$

which produces tree-level corrections to the W^+W^-V vertex. The magnitude of such corrections for a compactification scale ranging from 0.5TeV to 1.0TeV is [5] lower by an order of magnitude than the contributions introduced by the one-loop corrections discussed in the last section.

5. Conclusions

The structure of the 4D effective Lagrangian that emerges from a 5D Yang-Mills theory after compactification and integration of the extra dimension was discussed. It was emphasized that the election concerning whether the 5D gauge parameters can propagate in the fifth dimension or not is a crucial step, as it defines two essentially different contexts. If gauge parameters depend on the extra dimension there is an infinite number of KK excitations, for such parameters, that determine an infinite number of gauge transformations, which can be divided into two types: the standard gauge transformations and the nonstandard gauge transformations. It was asseverated that, in order to obtain a Lagrangian that is invariant under both types of gauge transformations, the objects to Fourier-expand, after compactification, are the curvatures, as such a procedure generates 4D covariant objects after the integration of the fifth dimension. Contrastingly, if the 5D gauge parameters are not supposed to propagate in the fifth dimension, there are only SGT at the 4D level and the Fourier expansions, allowed by the compactification of the extra dimension, must be performed on the 5D gauge fields. The 4D effective Lagrangians defined by each of these approaches are quite different. The quantization of the 4D effective theory within the first context (5D gauge parameters propagate in the fifth dimension) was discussed on the basis of the BRST formalism, and a SGT–covariant gauge–fixing procedure was introduced. The gauge–fixing functions defined within such an approach are similar to another scheme proposed some years ago for the so-called 331 model. The quantized Lagrangian, which includes a SGT–invariant gauge–fixing term and the most general ghost sector, was shown. From such an expression, the Lagrangian linking the light and heavy physics can be extracted, which is crucial to perform one-loop corrections to light Green's functions. It occurs that the one-loop level radiative corrections to light Green's functions are renormalizable, which is not necessarily

Journal of Physics: Conference Series **287** (2011) 012017 doi:10.1088/1742-6596/287/1/012017

true for the case of higher order corrections. The one-loop corrections to the W^+W^-V vertex, involving only KK excitations inside the loops, were calculated and three form factors emerged. The ones that define the CP–even electromagnetic properties of the W boson are independent of the cut-off scale, well behaved at high energies and they decouple for a large compactification scale. Their effects are dominant over those induced at the three–level by higher canonical dimension operators, consistently with the results reported in the literature.

References

[1] I. Antoniadis, Phys. Lett. B **246**, 377 (1990).

[2] T. Appelquist, H. -C. Cheng, B. A. Dobrescu, Phys. Rev. D **64**, 035002 (2001); T. Appelquist, H-U. Yee, Phys. Rev. D **67**, 055002 (2003).

[3] H. Novales-Sánchez and J. J. Toscano, Phys. Rev. D **82**, 116012 (2010).

[4] C. Becchi, A Rouet and R. Stora, Commun. Math. Phys. **42**, 127 (1975); Ann. Phys. (NY) **98**, 287 (1976); I. V. Tyutin: Preprint FIAN (P N: Lebedev Physical Institute of the USSR Academy of Science), No. 39 (1975).

[5] A. Flores-Tlalpa, J. Montaño, H. Novales-Sánchez, F. Ramírez-Zavaleta and J. J. Toscano, e-Print: arXiv:1009.0063 (to appear in Phys. Rev. D).

[6] K. R. Dienes, E. Dudas, T. Gherghetta, Nucl. Phys. B **537**, 47 (1999); C. Hills, S. Pokorski, and J. Wang, Phys. Rev. D **64**, 105005 (2001); T. Appelquist, H. -C. Cheng, and B. A. Dobrescu, Phys. Rev. D **64**, 035002 (2001); A. Mück, A. Pilaftsis and R. Rückl, Phys. Rev. D **65**, 085037 (2002); R. S. Chivukula, D. A. Dicus, H-J. He, Phys. Lett. B **525**, 175 (2002)

[7] P. A. M. Dirac, Lectures on quantum mechanics, (Yeshiva University, New York, 1964); K. Sundermeyer, Constrained Dynamics, (Springer-Verlag, Berlin, Heidelberg, New York, 1982); D. M. Gitman and I. V. Tyutin, Quantization of Fields with Constraints, (Springer, Berlin, 1990); M. Henneaux and C. Teitelboim, Quntization of Gauge Systems, (Princeton University Press, Princeton, 1992).

[8] L. Castellani, Annals Phys. **143**, 357 (1982).

[9] For a review of the BRST symmetry at both the classical and quantum level within the context of the field–antifield formalism, see J. Gomis, J. Paris and S. Samuel, Phys. Rep. **259**, 1 (1995).

[10] B.S. DeWitt, Phys. Rev. 162, 1195 (1967); Dynamical Theory of Groups and Fields (Gordon and Breach, New York, 1965); C. J. Isham, R. Penrose, and D. W. Sciama, Quantum Gravity 2 (Oxford University Press, Oxford, 1981); G. t Hooft, Acta Univ. Wratislavensis 368, 345 PHYSICAL REVIEW D 00 (1976); M. T. Grisaru, P. van Nieuwenhuizen, and C. C. Wu, Phys. Rev. D 12, 3203 (1975); H. Kluberg-Stern and J. B. Zuber, Phys. Rev. D 12, 482 (1975); D. G. Boulware, Phys. Rev. D 23, 389 (1981); C. F. Hart, Phys. Rev. D 28, 1993 (1983); L. F. Abbott, Nucl. Phys. B185, 189 (1981); Acta Phys. Pol. B 13, 33 (1982); L. F. Abbott, M. T. Grisaru and R. K. Schaefer, Nucl. Phys. B229, 372 (1983).

[11] J. Montaño, F. Ramírez-Zavaleta, G. Tavares-Velasco, J. J. Toscano, Phys. Rev. D **72**, 055023 (2005); F. Ramírez-Zavaleta, G. Tavares-Velasco, J. J. Toscano, Phys. Rev. D **75**, 075008 (2007).

[12] F. Pisano and V. Pleitez, Phys. Rev. D **46**, 410 (1992); P. H. Frampton, Phys. Rev. Lett. **69**, 2889 (1992).

[13]] K. J. F. Gaemers and G. J. Gounaris, Z. Phys. C **1**, 259 (1979); K. Hagiwara, R. D. Peccei, D. Zeppenfeld, and K. Hikasa, Nucl. Phys. B **282**, 253 (1987); U. Baur and D. Zeppenfeld, *ibid.* B **308**, 127 (1988); B **325**, 253 (1989).

Journal of Physics: Conference Series **287** (2011) 012018 doi:10.1088/1742-6596/287/1/012018

The partition function of a ferromagnet up to three loops

C P Hofmann

Facultad de Ciencias, Universidad de Colima, Bernal Díaz del Castillo 340, Colima, Colima 28045, México.

E-mail: `christoph@ucol.mx`

Abstract. The low-temperature behavior of ferromagnets with a spontaneously broken symmetry O(3) → O(2) is analyzed within the perspective of effective Lagrangians. The leading coefficients of the low-temperature expansion for the partition function are calculated up to three loops and the manifestation of the spin-wave interaction in this series is discussed. The effective field theory method has the virtue of being completely systematic and model-independent.

1. Motivation

The question of how the low-temperature expansion of the partition function of a ferromagnet is affected by the spin-wave interaction, has a long history. After a series of erroneous attempts, Dyson, in his pioneering paper on the thermodynamic behavior of an ideal ferromagnet [1], showed that the spin-wave interaction in the free energy density starts manifesting itself only at order T^5. Here we go beyond Dyson's analysis and explicitly calculate the effect of the spin-wave interaction beyond T^5 in the free energy density of a ferromagnet. To the best of our knowledge, this is the first time that the general structure of this power series and the explicit calculation of the corresponding coefficients is given. Here, we just present some preliminary results - a detailed exposition of the evaluation and, in particular, an extensive discussion of the low-temperature series for the spontaneous magnetization of an ideal ferromagnet, will be given elsewhere [2].

In the present work, we will make use of an approach which has the virtue of being completely systematic and model-independent: the method of effective Lagrangians. Within the effective Lagrangian framework, the structure of the low-temperature expansion of the partition function for an O(3) ferromagnet was analyzed in Ref. [3] up to order T^5 and Dyson's series was reproduced in a straightforward manner. In the effective language, this corresponds to including Feynman diagrams for the partition function up to two loops. Here we consider Feynman diagrams up to three-loop order in the perturbative expansion of the partition function. As it turns out, in the free energy density of a ferromagnet, the next-to-leading interaction term already sets in at order $T^{\frac{11}{2}}$ – remarkably, this term is completely determined by the two low-energy coupling constants of the leading-order effective Lagrangian $\mathcal{L}^2_{eff}$ and does not involve any higher-order effective constants from $\mathcal{L}^4_{eff}$ related to the anisotropies of the cubic lattice.

Journal of Physics: Conference Series **287** (2011) 012018 doi:10.1088/1742-6596/287/1/012018

2. Effective field theory evaluation

The effective Lagrangian method relies on an analysis of the symmetry properties of the underlying theory, i.e., the Heisenberg model in our case. In particular, it can universally be applied to systems with a spontaneously broken symmetry and is formulated in terms of Goldstone boson fields which are the relevant degrees of freedom at low energies [4]. Microscopic details of the system, such as the structure of the lattice, are taken into account through a few low-energy coupling constants in the effective Lagrangian. Symmetry does not fix the actual numerical values of these couplings – these have to be determined experimentally or in a numerical simulation of the underlying model. Symmetry, however, does unambiguously determine the derivative structure of the terms in the effective Lagrangian.

Whereas the Heisenberg model is invariant under global O(3) spin rotations, the ground state of the ferromagnet is invariant under the subgroup O(2) only. We then have one type of spin-wave excitation – or one magnon particle – in the low-energy spectrum of the ferromagnet which obeys a quadratic dispersion relation.

The leading-order effective Lagrangian (see Ref. [5]) is of order p^2 and takes the form

$$\mathcal{L}^2_{eff} = \Sigma \frac{\epsilon_{ab} \partial_0 U^a U^b}{1+U^3} + \Sigma \mu H U^3 - \frac{1}{2} F^2 \partial_r U^i \partial_r U^i \, . \tag{1}$$

The two real components of the magnon field, $U^a (a = 1, 2)$ are the first two components of the three-dimensional unit vector $U^i = (U^a, U^3)$, which transforms with the vector representation of the rotation group. The quantity H is the third component of the external magnetic field $\vec{H} = (0, 0, H), H > 0$. While the structure of the above terms is unambiguously determined by the symmetries of the underlying theory, at this order, we have two a priori unknown low-energy constants: the spontaneous magnetization Σ and the constant F. The above Lagrangian implies a quadratic dispersion relation

$$\omega(\vec{k}) = \gamma \vec{k}^2 + \mathcal{O}(|\vec{k}|^4) \, , \quad \gamma \equiv \frac{F^2}{\Sigma} \, , \tag{2}$$

characteristic of ferromagnetic magnons. It is important to note that – in the power counting scheme – one temporal derivative is on the same footing as two spatial derivatives – in the derivative expansion, two powers of momentum thus count as only one power of energy or temperature: $k^2 \propto \omega, T$.

The next-to-leading order Lagrangian is of order p^4 and takes the form [3]

$$\mathcal{L}^4_{eff} = l_1 (\partial_r U^i \partial_r U^i)^2 + l_2 (\partial_r U^i \partial_s U^i)^2 + l_3 \Delta U^i \Delta U^i \, . \tag{3}$$

It involves the three effective coupling constants l_1, l_2 and l_3. Higher order pieces which are also relevant for our calculation are

$$\mathcal{L}^6_{eff} = c_1 U^i \Delta^3 U^i \, , \qquad \mathcal{L}^8_{eff} = d_1 U^i \Delta^4 U^i \, . \tag{4}$$

We conclude this section with a remark concerning effects induced by the anisotropy of the lattice. For a cubic lattice, as shown in [6], the anisotropies start manifesting themselves only at the four-derivative level: indeed, the pieces $\mathcal{L}^4_{eff}$, $\mathcal{L}^6_{eff}$ and $\mathcal{L}^8_{eff}$ contain additional terms – not displayed in Eqs.(3) and (4) – which are not invariant under space rotations, such as

$$\sum_{s=1,2,3} \partial_s \partial_s U^i \, \partial_s \partial_s U^i \, . \tag{5}$$

In the present analysis, however, we do not take care of these extra terms and assume space rotation symmetry up to order p^8. The conclusions of the present paper regarding the manifestation of the spin-wave interaction in the partition function are not affected by this idealization.

In finite-temperature field theory the partition function is represented as a Euclidean functional integral

$$\mathrm{Tr}\left[\exp(-\mathcal{H}/T)\right] = \int [dU]\, \exp\Big(- \int_T d^4x\, \mathcal{L}_{eff}\Big) . \tag{6}$$

The integration is performed over all field configurations which are periodic in the Euclidean time direction $U(\vec{x}, x_4 + \beta) = U(\vec{x}, x_4)$, with $\beta \equiv 1/T$. The periodicity condition imposed on the magnon fields also reflects itself in the thermal propagator

$$G(x) = \sum_{n=-\infty}^{\infty} \Delta(\vec{x}, x_4 + n\beta) , \tag{7}$$

where $\Delta(x)$ is the Euclidean propagator at zero temperature,

$$\Delta(x) = \int \frac{dk_4 d^3k}{(2\pi)^4} \frac{e^{i\vec{k}\vec{x} - ik_4x_4}}{\gamma\vec{k}^2 - ik_4 + \mu H} . \tag{8}$$

At finite temperature the derivative expansion in powers of the momenta amounts to an expansion in powers of the temperature.

3. Results

While a detailed account of the perturbative evaluation can be found in Ref. [2], here we just present some of the main results. The free energy density of the O(3) ferromagnet up to order p^{11} is given by

$$\begin{aligned} z = & - \Sigma\mu H - \frac{1}{8\pi^{\frac{3}{2}}\gamma^{\frac{3}{2}}} T^{\frac{5}{2}} \sum_{n=1}^{\infty} \frac{e^{-\mu H n\beta}}{n^{\frac{5}{2}}} - \frac{15\, l_3}{16\pi^{\frac{3}{2}}\Sigma\gamma^{\frac{7}{2}}} T^{\frac{7}{2}} \sum_{n=1}^{\infty} \frac{e^{-\mu H n\beta}}{n^{\frac{7}{2}}} \\ & - \frac{105}{32\pi^{\frac{3}{2}}\Sigma\gamma^{\frac{9}{2}}} \left(\frac{9l_3^2}{2\gamma\Sigma} - c_1 \right) T^{\frac{9}{2}} \sum_{n=1}^{\infty} \frac{e^{-\mu H n\beta}}{n^{\frac{9}{2}}} \\ & - \frac{3(8l_1 + 6l_2 + 5l_3)}{128\pi^3\Sigma^2\gamma^5} T^5 \Big\{ \sum_{n=1}^{\infty} \frac{e^{-\mu H n\beta}}{n^{\frac{5}{2}}} \Big\}^2 \\ & - \frac{945\, d_1}{64\pi^{\frac{3}{2}}\Sigma\gamma^{\frac{11}{2}}} T^{\frac{11}{2}} \sum_{n=1}^{\infty} \frac{e^{-\mu H n\beta}}{n^{\frac{11}{2}}} + \frac{10395\, l_3 c_1}{64\pi^{\frac{3}{2}}\Sigma^2\gamma^{\frac{13}{2}}} T^{\frac{11}{2}} \sum_{n=1}^{\infty} \frac{e^{-\mu H n\beta}}{n^{\frac{11}{2}}} \\ & - \frac{\Sigma^{\frac{5}{2}}}{2F^9} \bar{\jmath}\, T^{\frac{11}{2}} + \mathcal{O}(T^6) . \end{aligned} \tag{9}$$

The first term is temperature-independent. Contributions which involve half integer powers of the temperature – $T^{\frac{5}{2}}, T^{\frac{7}{2}}$, $T^{\frac{9}{2}}$ and the first two terms of order $T^{\frac{11}{2}}$ – arise from one-loop graphs and are thus all related to the free energy density of noninteracting magnons. Remarkably, up to order p^{10} there is only one term in the above series – the order T^5 contribution, originating from a two-loop graph – which is due to the spin-wave interaction and represents the Dyson term. Note that there is no term of order T^4 in the above series of the free energy density.

The last term of order $T^{\frac{11}{2}}$ is related to a three-loop graph: The dimensionless function $\bar{\jmath}$,

$$\bar{\jmath} = \bar{\jmath}(\sigma) , \quad \sigma = \frac{\mu H}{T} , \tag{10}$$

in the limit $\sigma \to 0$, may be parametrized by

$$\bar{j}(\sigma) = j_1 + \mathcal{O}(\sigma) , \tag{11}$$

where the coefficient j_1 is pure number given by

$$j_1 = 0.0000091 . \tag{12}$$

With the above results, the low-temperature series for the pressure takes the form

$$P \;=\; h_0 T^{\frac{5}{2}} \,+\, h_1 T^{\frac{7}{2}} \,+\, h_2 T^{\frac{9}{2}} \,+\, h_3 T^5 \,+\, h_4 T^{\frac{11}{2}} + \mathcal{O}(T^6) , \tag{13}$$

with coefficients h_i given by

$$\begin{aligned}
h_0 &= \frac{1}{8\pi^{\frac{3}{2}}\gamma^{\frac{3}{2}}} \sum_{n=1}^{\infty} \frac{e^{-\mu H n\beta}}{n^{\frac{5}{2}}} , \\
h_1 &= \frac{15\, l_3}{16\pi^{\frac{3}{2}}\Sigma\gamma^{\frac{7}{2}}} \sum_{n=1}^{\infty} \frac{e^{-\mu H n\beta}}{n^{\frac{7}{2}}} , \\
h_2 &= \frac{105}{32\pi^{\frac{3}{2}}\Sigma\gamma^{\frac{9}{2}}} \left(\frac{9 l_3^2}{2\gamma\Sigma} - c_1 \right) \sum_{n=1}^{\infty} \frac{e^{-\mu H n\beta}}{n^{\frac{9}{2}}} , \\
h_3 &= \frac{3(8l_1 + 6l_2 + 5l_3)}{128\pi^3\Sigma^2\gamma^5} \left(\sum_{n=1}^{\infty} \frac{e^{-\mu H n\beta}}{n^{\frac{5}{2}}} \right)^2 , \\
h_4 &= \frac{945}{64\pi^{\frac{3}{2}}\Sigma\gamma^{\frac{11}{2}}} \left(d_1 - 11 l_3 c_1 F^{-2} \right) \sum_{n=1}^{\infty} \frac{e^{-\mu H n\beta}}{n^{\frac{11}{2}}} + \frac{\Sigma^{\frac{5}{2}}}{2F^9} \bar{j} .
\end{aligned} \tag{14}$$

In the limit $\sigma = \frac{\mu H}{T} \to 0$, these coefficients become temperature independent and the sums reduce to Riemann zeta functions,

$$\begin{aligned}
\tilde{h}_0 &= \frac{1}{8\pi^{\frac{3}{2}}\gamma^{\frac{3}{2}}} \zeta(\tfrac{5}{2}) , \\
\tilde{h}_1 &= \frac{15\, l_3}{16\pi^{\frac{3}{2}}\Sigma\gamma^{\frac{7}{2}}} \zeta(\tfrac{7}{2}) , \\
\tilde{h}_2 &= \frac{105}{32\pi^{\frac{3}{2}}\Sigma\gamma^{\frac{9}{2}}} \left(\frac{9 l_3^2}{2\gamma\Sigma} - c_1 \right) \zeta(\tfrac{9}{2}) , \\
\tilde{h}_3 &= \frac{3(8l_1 + 6l_2 + 5l_3)}{128\pi^3\Sigma^2\gamma^5} \zeta^2(\tfrac{5}{2}) , \\
\tilde{h}_4 &= \frac{945}{64\pi^{\frac{3}{2}}\Sigma\gamma^{\frac{11}{2}}} \left(d_1 - 11 l_3 c_1 F^{-2} \right) \zeta(\tfrac{11}{2}) + \frac{\Sigma^{\frac{5}{2}}}{2F^9} j_1 .
\end{aligned} \tag{15}$$

Of particular interest are the last two terms involving the coefficients h_3 and h_4, as they contain the spin-wave interaction part. The contribution displaying five powers of the temperature is the famous Dyson term, while the interaction contribution contained in $T^{\frac{11}{2}}$ – the last term in the coefficient h_4 – is our main new result. Note that all other contributions to the pressure originate from one-loop graphs – hence, those graphs describe noninteracting magnons and merely modify the dispersion relation.

Up to order T^5, we thus reproduce Dyson's series. In the effective Lagrangian framework, the famous interaction term of order T^5 in the free energy density originates from a two-loop graph with an insertion from the next-to-leading order Lagrangian $\mathcal{L}^4_{eff}$. Remarkably, the next-to-leading term resulting from the spin-wave interaction already sets in at order $T^{\frac{11}{2}}$. Note again that the coefficient of this three-loop interaction term, contained in h_4, does not involve any higher-order low-energy coupling constants. It only involves Σ and F, as well as the quantity $\bar{j}$, which is a dimensionless function determined by the symmetries of the underlying Heisenberg model.

4. Conclusions

The effective Lagrangian method, which is based on symmetry considerations, has the virtue of being completely systematic and model-independent. It allowed us to go beyond the results of Dyson in a rigorous way and thereby solve the long-standing question of how the spin-wave interaction manifests itself in the low-temperature expansion of the partition function of an ideal ferromagnet. In particular, we have identified the correct temperature-power beyond the Dyson term due to the spin-wave interaction: it is of order $T^{\frac{11}{2}}$ in the free energy density and does not depend on the anisotropies of the cubic lattice.

A detailed exposition of the perturbative evaluation of the partition function within the effective theory framework will be presented elsewhere [2]. In the same article, more results will be provided. In particular, a detailed account of the low-temperature series for the spontaneous magnetization of an ideal ferromagnet and its comparison with the condensed matter literature will be given there.

Acknowledgments

Support by CONACYT grant No. 50744-F is gratefully acknowledged.

References

[1] Dyson F J 1956 *Phys. Rev.* **102** 1217; **102** 1230
[2] Hofmann C P 2011 to be published
[3] Hofmann C P 2002 *Phys. Rev.* B **65** 094430
[4] Leutwyler H 1994 *Ann. Phys.* **235** 165
[5] Leutwyler H 1994 *Phys. Rev.* D **49** 3033
[6] Hasenfratz P and Niedermayer F 1993 *Z. Phys. B: Condens. Matter* **92** 91

Journal of Physics: Conference Series **287** (2011) 012019 doi:10.1088/1742-6596/287/1/012019

Conformal Symmetry Patterns in Baryon Spectra

Mariana Kirchbach[1], **Cliffor B. Compean**[2]

[1]Institute of Physics, Autonomous University of San Luis Potosi, Manuel Nava 6, San Luis Potosi, SLP 98290, Mexico

[2]Institute of Physics and Mathematics, University of Michoacan San Nicolas de Hidalgo, Morelia Michoacan 58040, Mexico

E-mail: mariana@ifisica.uaslp.mx

Abstract. Attention is drawn to the fact that the spectra of the baryons of the lightest flavors, the nucleon and the Δ, carry quantum numbers characteristic for an unitary representation of the conformal group. We show that the above phenomenon is well explained for baryons whose internal structure is dominated by a quark-diquark configuration that resides in a conformally compactified Minkowski space time, $\mathbf{R}^1 \otimes S^3$, and is described by means of the conformal scale equation there. The $\mathbf{R}^1 \otimes S^3$ space-time represents the boundary of the conformally compactified AdS_5, on which one expects to encounter a conformal theory in accord with the gauge-gravity duality. Within this context, our model is congruent with AdS_5/CFT_4.

1. The enigma of the baryon excitations-introductory remarks

The baryon spectra represent themselves enigmatic despite the long history of the respective studies [1], [2]. Specifically for the N and Δ particles, the great unknown is the number of the resonances with masses below 2500 MeV. Compared to the predictions of the traditional quark model based upon the full Hilbert space of three quark degrees of freedom and the non-relativistic $SU(6)_{SF} \otimes O(3)_L$ classification scheme, the experimental observations reported so far [3] identify much lesser a number of states. The respective resonance deficit, termed to as "missing" states, is still awaiting for explanation. Quark-diquark (q–(qq)) models [4] based on a diquark with limited angular momentum values, carry reduced spin-flavor degrees of freedom, and are obvious candidates for providing a lesser number of "missing" states, an option taken into consideration by several authors. Additional restrictions on the quantum numbers of the q-(qq) excitations can come from imposing on the spatial wavefunctions a symmetry higher than $O(3)_L$. Natural candidates are the exact Lorentz and the approximate conformal symmetries, the latter having relevance for the lightest flavors. It is the goal of the present work to examine consequences of these symmetries for the systematic of the N and Δ spectra and construct corresponding radial wavefunctions of a q–(qq) system. We implement conformal symmetry into a quark Hamiltonian describing quarks in a position space of a finite volume [5]. This is achieved in placing the q-(qq) system directly on the AdS_5 boundary, which is the AdS_5 cone, considered as conformally compactified to $S^1 \otimes S^3$ [6], or, $\mathbf{R}^1 \otimes S^3$, at a microscopic scale [7]. According to [7], correlation functions of CFT_4 on regular Minkowski spacetime, $\mathcal{M}=\mathbf{R}^{1+3}$, can be analytically continued to the full Einstein universe because $\mathbf{R}^{1+3}$ can be conformally mapped on $\mathbf{R}^1 \otimes S^3$. The implication of this important observation is that each CFT_4 state on $\mathbf{R}^1 \otimes S^3$ can be brought into unique correspondence with a state of the brane theory on $AdS_5 \otimes S^5$. Consequences on thermal states

Journal of Physics: Conference Series **287** (2011) 012019 doi:10.1088/1742-6596/287/1/012019

have been worked out in ref. [8]. We here examine consequences for the systematics of the N and Δ spectra.
The paper is organized as follows. In the next section we briefly review the space time symmetries of QCD and relate them to the observed degeneracy phenomenon in the N and Δ. In section 3 we present the conformal equation of ref. [9] and apply it to a quark-diquark (q-qq) model of light baryons. There, we also review calculations of the N and Δ spectra and the proton electric charge form-factor. The paper closes with concise conclusions.

2. Degeneracy phenomena in N and Δ spectra from space-time symmetries

The exact fundamental internal and external (space-time) symmetries of the QCD Lagrangian are the color-gauge, and the global relativistic Lorentz symmetries, respectively. They are mathematically expressed in terms of the invariance of the Lagrangian under local $SU(3)_c$, and global $SO(1,3)$ group transformations. In addition, in the light flavor sector, one considers some approximate internal and external symmetries, the most important being the three flavor $SU(3)_F$, and the conformal $SO(2,4)$ symmetries. Relativity requires fermionic high-spin states to emerge as part of Lorentz group representations, not necessarily single spin valued. The most important representations are described by means of totally symmetric Lorentz tensors of rank K with Dirac spinor components and are generated in colliding bosonic projectiles with fermionic targets. Consider for example the excitation of a proton target, a Dirac spinor Ψ, by a photon beam, a four-vector A_μ. In effect of the (γ, p) collision, the two incoming states described by the above two Lorentz representations merge to their direct product, $A_\mu \otimes \Psi$, which gives rise to a four-vector spinor,

$$A_\mu \otimes \Psi = \Psi_\mu. \tag{1}$$

The spin- and parity content of Ψ_μ is given by,

$$\frac{1}{2}^+ \oplus \left(\frac{1}{2}^-, \frac{3}{2}^-\right) \in \Psi_\mu. \tag{2}$$

The Ψ_μ representation is reducible and its irreducible parts in standard notations [10] read,

$$\begin{aligned} \left(\frac{1}{2}^-, \frac{3}{2}^-\right) &\in \{1, \frac{1}{2}\} \oplus \{\frac{1}{2}, 1\}, \\ \frac{1}{2}^+ &\in \{\frac{1}{2}, 0\} \oplus \{0, \frac{1}{2}\}. \end{aligned} \tag{3}$$

Were the beam to be described by means of the electromagnetic field strength tensor, $F_{\mu\nu}$, the outgoing state would be the totally anti-symmetric Lorentz tensor of rank 2 with Dirac spinor components, $\Psi_{[\mu\nu]}$, which is reducible too,

$$F_{\mu\nu} \otimes \Psi = \Psi_{[\mu\nu]} = \Psi_\eta \oplus \{\frac{3}{2}, 0\} \oplus \{0, \frac{3}{2}\}. \tag{4}$$

The spin-parity content of $\Psi_{[\mu\nu]}$ is,

$$\frac{1}{2}^+ \oplus \left(\frac{1}{2}^-, \frac{3}{2}^-\right) \oplus \frac{3}{2}^+ \in \Psi_{[\mu\nu]}. \tag{5}$$

The example shows that $\Psi_{[\mu\nu]}$ can serve as a vehicle for the single-spin valued $\{\frac{3}{2}, 0\} \oplus \{0, \frac{3}{2}\}$, provided, one would be able to pro ect out covariantly. Following this logic, it is not difficult to show that a spin-$\left(K+\frac{1}{2}\right)^P$ field (with K=0,1,2,3,...) of parity $P=\pm$, naturally emerges as the highest (maximal) spin in,

$$J^P_{\mathrm{max}} = \left[K+\frac{1}{2}\right]^P \in \Psi_{\mu_1\ldots\mu_K}. \tag{6}$$

The spin and parity content of $\Psi_{\mu_1\ldots\mu_K}$ is,

$$\left(\frac{1}{2}^{\pm}, \frac{3}{2}^{\pm}, \left[K-\frac{1}{2}\right]^{\pm}, \left[K+\frac{1}{2}\right]^{P}\right) \in \Psi_{\mu_1\ldots\mu_K}, \tag{7}$$

with the sign of $P=(-1)^K$, or $P=(-1)^{K+1}$, depending whether $\Psi_{\mu_1\ldots\mu_K}$ is a tensor, or pseudo-tensor. Take as an example $K=3$. The pseudo-tensor of rank-3 is reducible as,

$$\begin{aligned}\Psi_{\mu_1\mu_2\mu_3} &= \left[\{1,\frac{3}{2}\}\oplus\{\frac{3}{2},1\}\right]\oplus\left[\{\frac{3}{2},2\}\oplus\{2,\frac{3}{2}\}\right] \\ &\longrightarrow \left(\frac{1}{2}^{+},\frac{3}{2}^{-},\frac{5}{2}^{+}\right)\oplus\left(\frac{1}{2}^{-},\frac{3}{2}^{+},\frac{5}{2}^{-},\frac{7}{2}^{+}\right).\end{aligned} \tag{8}$$

The latter equation implies that the states belonging to the natural parity spin cascade, $\left(\frac{1}{2}^{+},\frac{3}{2}^{-},\frac{5}{2}^{+}\right)$, being part of the same irreducible Lorentz representation, will have same mass, say, M_1. Same is valid for the states of the accompanying unnatural parity cascade, $\left(\frac{1}{2}^{-},\frac{3}{2}^{+},\frac{5}{2}^{-},\frac{7}{2}^{+}\right)$, which will have in general an other mass, say, M_2. In this way, $\Psi_{\mu_1\mu_2\mu_3}$ will be a two mass state,

$$M_1: \left(\frac{1}{2}^{+},\frac{3}{2}^{-},\frac{5}{2}^{+}\right), \qquad M_2: \left(\frac{1}{2}^{-},\frac{3}{2}^{+},\frac{5}{2}^{-},\frac{7}{2}^{+}\right). \tag{9}$$

In consequence, relativity of boson-fermion collisions favors higher spin nucleon excitations to emerge in groups of $(2K+1)$, and as part of $\Psi_{\mu_1\ldots\mu_K}$, rather than as single spins. Were higher spins to be fundamental, Lorentz symmetry would require exact degeneracy among the members of anyone of the irreducible pieces of $\Psi_{\mu_1\ldots\mu_K}$. For example, the negative parity states in eq. (2) had to have equal masses, while the positive parity state would be allowed to have a different mass. The $\Psi_{\mu_1\ldots\mu_K}$ tensors are referred to as Rarita-Schwinger fields. As shown above, the highest spins always appear accompanied by lower-spin states. For example, the lowest fractional spin $\frac{3}{2}^{-}$ appears accompanied by just one spin $\frac{1}{2}^{-}$ state, as illustrated by eqs. (2)-(3). It would be natural to describe this group of states as a whole, a path taken by ref. [11]. Instead, the canonical opinion is that one has to find a method to restrict $\{1,\frac{1}{2}\}\oplus\{\frac{1}{2},1\}$ to a single spin-$\frac{3}{2}^{-}$. As long as the representation under consideration is irreducible, finding such methods is by no means an easy task and if not done with care, employing irreducible multi-spin parity representations of the Lorentz group in the description of single spins may result inconsistent. Precisely this happens when the restriction of $\Psi_{\mu_1\ldots\mu_K}$ to a single spin-$\left[K+\frac{1}{2}\right]^P$ is executed along the line of the Rarita-Schwinger prescription [12]. According to the latter, it is conjectured that the highest spin is described by means of a Dirac equation for any Lorentz index, μ_i, supplemented by auxiliary conditions,

$$(\not p - m)\Psi_{\mu_1..\mu_i..\mu_K}=0, \qquad p^{\mu_i}\Psi_{\mu_1..\mu_i..\mu_K}=0, \quad \gamma^{\mu_i}\Psi_{\mu_1..\mu_i..\mu_K}=0. \tag{10}$$

It has been known for a long time that this prescription is plagued by several inconsistencies. Specifically, upon electromagnetic gauging, a superluminal propagation can occur (so called

Velo-Zwanziger problem). Because of this and other inconsistencies, the requirement of the relativity that higher spins should emerge as $\Psi_{\mu_1...\mu_K}$, i.e. as groups of K pairs of states of opposite parities and one unpaired state of maximal spin (c.f. eq. (7)), has rarely been taken into account. In so doing, the important point has been missed that Lorentz symmetry prescribes the type of degeneracy between resonances with rising spins and alternating parities illustrated by eq. (9).

However, in the recent work [13] significant progress in the consistent description of spin-$\frac{3}{2}$ in terms of the four-vector spinor has been achieved, and the venue hit toward a liberation of $\Psi_{\mu_1...\mu_K}$ from the stigma of being plagued by pathologies. The study in ref. [13] executes the idea of ref. [14] that the highest spin in $\Psi_{\mu_1...\mu_K}$ has to be pinned down by a projector consistent with the Poincaré symmetry (i.e. the Lorentz symmetry extended by translations). Such covariant projectors are built up from the two Casimir operators of the Poincaré group, the squared operators of the linear momentum, P^2, and the Pauli-Lubanski vector, $\mathcal{W}^2$, whose eigenvalue upon a state, $|MJ\rangle$ of given mass M and spin J equals, $\mathcal{W}^2|MJ\rangle = -M^2J(J+1)|MJ\rangle$. Specifically for a spin-$\frac{3}{2}$ residing in Ψ_μ such projectors have been designed in [14] and shown to lead to fermion wave equations quadratic in the momenta. The spin-$\frac{3}{2}$ wave equation worked out along this line in ref. [13] has been proved to be free from the Velo-Zwanziger pathology provided the gyromagnetic factor takes the universal value of $g = 2$ and not as in the Rarita-Schwinger framework, the inverse spin value, $g = 1/J$. In this manner the consistent description of spin-$\frac{3}{2}^-$ in terms of Ψ_μ has been taken under control at least at the classical level. Later on, the method of the covariant Poincaré projectors has been successfully applied also to the spin-1 case in [15]. With that the venue toward a consistent description of high spins has been hit and the confidence in the observability of mass degenerate states belonging to a $\Psi_{\mu_1..\mu_K}$ in accord with relativity, regained.

Now it is quite instructive to check whether data on the N and Δ spectra are compatible with the degeneracies predicted in eq. (9). We here focus for the sake of concreteness on the Δ spectrum. We observe that the resonances with masses around 1900 MeV fit perfectly well in the $\Psi_{\mu_1\mu_2\mu_3}$ field from eq. (9). Specifically, the three states from $\{1, \frac{3}{2}\}\oplus\{\frac{3}{2}, 1\}$ are identified with the $\Delta(1910)$, $\Delta(1940)$, and $\Delta(1905)$ resonances of an averaged mass $M_1 = 1918$ MeV. The remaining states belong to $\{\frac{3}{2}, 2\}\oplus\{2, \frac{3}{2}\}$, and they can be identified with the $\Delta(1900)$, $\Delta(1920)$, $\Delta(1930)$, and $\Delta(1950)$ states of an averaged mass $M_2 = 1925$ MeV. Though the degeneracy phenomenon by itself is well pronounced, the mass separation between the two irreducible representations constituting $\Psi_{\mu_1\mu_2\mu_3}$ appears insignificant. Rather, it seems that $\Psi_{\mu_1\mu_2\mu_3}$ as a whole can be characterized by the mass of the underlying $\left(\frac{K}{2}, \frac{K}{2}\right)$ irrep around which the physical masses of the states then spread due to relativistic fine-splitting effects. This type of degeneracy can also be interpreted in terms of a symmetry group larger than $SO(1,3)$ and specifically in terms of the conformal symmetry. The conformal group, $SO(2,4)$ contains the Lorentz group as a subgroup, and has an unitary representation which is an infinite tower of levels which carry precisely all the quantum numbers of $\Psi_{\mu_1...\mu_K}$ at their rest frames. Within this context, all the states belonging to same $\Psi_{\mu_1...\mu_K}$ appear to leading order mass degenerate modulo the relativistic effect of the fine splitting. Stated differently, according to eq. (7), and modulo the fine-splitting, one expects to observe degeneracy among parity pairs with spins rising from $\frac{1}{2}^\pm$ to $\left[K-\frac{1}{2}\right]^\pm$, and the unpaired state of maximal spin $\left[K+\frac{1}{2}\right]^P$. The values of K have to be determined by comparison to data.

The nucleon and the Δ spectra have been analyzed already in terms of $\Psi_{\mu_1..\mu_K}$ in ref. [16]. There, it has been found that the reported N and Δ spectra can be organized each in three $\Psi_{\mu_1...\mu_K}$ states with K=1,3, and 5, from which a total of solely five resonances is "missing".

In Fig. 1 the Δ levels with $K = 3$ and 5 are depicted for illustrative purposes. On the (J, M) plane, with the states of positive and negative parity being placed below and above the

Journal of Physics: Conference Series **287** (2011) 012019 doi:10.1088/1742-6596/287/1/012019

dashed line indicating the averaged mass, the $\Psi_{\mu_1\ldots\mu_K}$'s appear shaped after fern-grass, a reason for which we here coin the notion of a "F E R N E O N" for them. As long as the $\Psi_{\mu_1..\mu_k}$ representations are uniquely identified by their maximal spins, $J_{\rm max}$, defined in eq. (6), the specification " spin-$J_{\rm max}$ ferneons" will be used occasionally. The figure shows that all the Δ resonances reported in that region so far by the Particle Data Group in [3] fit amazingly well into spin-7/2 and spin-11/2 ferneons, levels of a conformal representation. We here challenge the above degeneracy phenomenon and search for reasons for its occurrence. In the next section we shall motivate relevance of conformal symmetry for the spectra of the lightest flavor baryons and develop a quantitative model for its description.

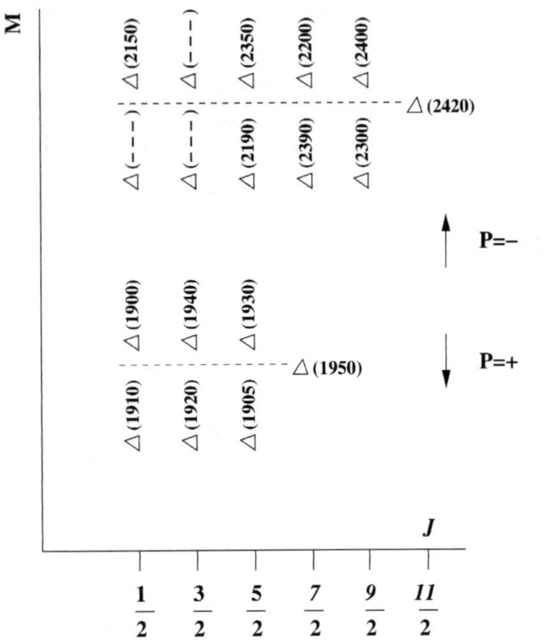

Figure 1. Schematic presentation of the spin-parity degeneracy phenomenon in the Δ spectrum on the spin (J)-mass (M) plane. The positive parity states have been placed below, the negative parity– above the dashed line marking the averaged mass. The graphical shape of the spectrum designed in this fashion resembles of fern, a reason for which we hereafter shall referrer to $\Psi_{\mu_1..\mu_K}$ as a "*spin-*$\left(K+\frac{1}{2}\right)$ *ferneon*". Resonances marked by $\Delta(- - -)$ are "missing" from the spin-11/2 ferneon. The spin-7/2 ferneon is complete. The ferneons carry same quantum numbers as the levels of the $SO(2,4)$ unitary infinite tower known from the H atom. In this way one identifies conformal patterns in the Δ spectrum. Same applies to the N spectrum (shown elsewhere [5]).

The conformal symmetry is to a good approximation a symmetry of QCD in the lightest flavors, u and d, sector and in the ultraviolet, i.e. in the regime of the asymptotic freedom where the strong coupling constant starts approaching zero, $\alpha_s(q^2) \stackrel{q^2\to\infty}{\longrightarrow} 0$. It is well known that several sum rules for "hard" processes can be derived from conformal symmetry arguments. In the infrared, the $SO(2,4)$ invariance of the QCD Lagrangian would require $\alpha_s(q^2)$ to approach a fixed point, so called "conformal window", $\alpha_s(q^2) \stackrel{q^2\to 0}{\longrightarrow} \alpha_s^*$=const. Such a behavior has been predicted in theories with many flavors [17]. However, ref. [18] has also admitted for the possibility of having a conformal window for a number a flavors between 2 and 3, just as is the case in QCD. Recent measurements of the (e,p) scattering performed at the Jefferson Laboratory are strongly indicative of the opening of a conformal window in the infrared [19], and it is justified to expect from the conformal symmetry to leave a footprint in the spectra of the lightest flavor baryons, the nucleon and the Δ and in support of the ferneon patterns in Fig. 1, advocated here. Moreover, according to the gauge-gravity duality [20], QCD is expected to share several features in common with a conformal string theory residing at the AdS_5 boundary. In the next section we shall present a quark model approach to QCD that captures the essential aspects of a conformal theory on the AdS_5 boundary and explains the degeneracy phenomenon in the light flavor baryon spectra.

3. A conformal quark model on a finite volume

The goal is to develop a potential quark model that describes the conformal patterns in the nucleon and Δ spectra and at same time captures correctly the aspects of the quark-gluon dynamic in all three regimes. As long as in the perturbative regime the $q-q$ interaction is well described by an inverse distance potential, and the non-perturbative flux-tube regime requires a linear potential, while the regime of the asymptotic freedom is adequately described by means of an infinite well, the potential of interest has to interpolate between a Coulombic potential and the infinite well while passing through a region of a linear growth. The cotangent potential, $(-\cot r)$, is an interaction of the required properties. Indeed, the first terms in its Taylor decomposition, $-\cot r = -1/r + r/3 + ...$ coincide in form with the Cornell potential predicted by Lattice QCD [21]. The higher terms can be viewed as a phenomenological parametrization of non-perturbative corrections [22]. The cotangent potential is better known under the name of the "curved" Coulomb potential on the closed Einsteinian Universe, $\mathbf{R}^1 \otimes S^3$, a notion indicating that the cotangent function of the second polar angle, χ, parameterizing the three dimensional surface of a constant positive curvature, the hypersphere S^3, is a harmonic function there. According to potential theory it can serve as an interaction on S^3, in a way similar as $1/r$ serves as a potential in flat three space, $\mathbf{R}^3$. The above considerations indicate that the most adequate geometry for employing the cotangent interaction is the closed Einstein Universe. The latter emerges naturally within the context of the conformally compactified AdS_5 boundary. Indeed, the AdS_5 boundary, defined by the light cone in 2+4 dimensions,

$$u^2 + v^2 - x^2 - y^2 - z^2 - x_4^2 = 0, \tag{11}$$

can be compactified to $S^1 \otimes S^3$ as,

$$S^1 \otimes S^3: \quad u^2 + v^2 = R^2 = x^2 + y^2 + z^2 + x_4^2, \tag{12}$$

locally equivalent to $\mathbf{R}^1 \otimes S^3$ at a microscopic scale. The latter manifold is a conformally compactified Minkowski space time. According to [7], correlation functions of CFT_4 on regular Minkowski spacetime, $\mathcal{M}=\mathbf{R}^{1+3}$, can be analytically continued to the full Einstein universe, this because $\mathbf{R}^{1+3}$ can be conformally mapped on $\mathbf{R}^1 \otimes S^3$. The implication of this important observation is that each CFT_4 state on $\mathbf{R}^1 \otimes S^3$ can be brought into unique correspondence with a state of the brane theory on $AdS_5 \otimes S^5$. We here wish to explore consequences of this for the spectra of the lightest flavors baryons, the nucleon and the Δ, assumed as two-body quark-diquark systems for the reasons explained in the introduction. We place this system in a compactified Minkowski space time, a manifold of a finite volume, and therefore well suited for confinement description [23]. Finally, we use the scale conformal equation in $\mathbf{R}^1 \otimes S^3$ gauged by the cotangent interaction. The scale conformal equation can be found in [9], [24] and reads:

$$\frac{\hbar^2}{R^3}\left(\left(\frac{d}{d\tau}\right)^2 - (\mathcal{K}^2+1)\right) R\Phi\left(\frac{x}{R}\right) = 0, \ \tau = \ln R, \qquad \mathcal{K}^2 = -\frac{1}{\sin^2\chi}\frac{\partial}{\partial\chi}\sin^2\chi\frac{\partial}{\partial\chi} - \frac{L^2}{\sin^2\chi}.$$

Here, $\ln R$ behaves as time coordinate, and the scale equation describes free geodesic motion on S^3. Replacing $d/dlnR$ by E, transforms the wave equation for $\Phi(x)$ into the eigenvalue problem of the squared 4D angular momentum, $\mathcal{K}^2$,

$$\hbar^2\mathcal{K}^2\psi_{Klm} = E^2\psi_K \quad , \qquad E^2 = \hbar^2[K(K+2)+1]. \tag{13}$$

Noticing that $K(K+2) = K(K+D-2)|_{D=4}$, and $1=\left(\frac{D-2}{2}\right)^2|_{D=4}$, one realizes that the canonical dimension of the field, (D-2)/2, has become "mass", while the "anomalous dimensions", K(K+D-2), have become $\mathcal{K}^2$ eigenvalues. The $SO(2,4)$ generators on $\mathbf{R}^1 \times S^3$ have been constructed by [25]. The interaction on S^3 is then concluded from the Green functions obtained in [26] as,

$$\mathcal{K}^2 G(\chi,\chi') = \delta(\chi-\chi') - \frac{1}{2\pi^2},$$
$$G(0,\chi) \equiv G_0 = \frac{1}{4\pi^2}(\pi-\chi)\cot\chi + c', \qquad G(\pi,\chi) \equiv G_\pi = -\frac{1}{4\pi^2}\chi\cot\chi + c''. \tag{14}$$

Here, G_π and G_0 refer to sources placed on the respective North, and South poles. They can be equivalently obtained from one-cusp Wilson loops giving rise to cusp anomalous dimensions [27]. As a potential in the quantum-mechanical eigenvalue problem on S^3 one employs the "electric" one-form,

$$V(\chi) = -4\pi A(G_\pi - G_0) = -A\cot\chi, \tag{15}$$

with $\mathcal{K}^2\cot\chi = 0$, because of its property to make the Dirichlet energy functional on S^3 stationary. Its derivative, $\mathcal{D} = \frac{d}{d\chi}V(\chi) = A\frac{1}{\sin^2\chi}$, gives $A\csc^2\chi$ the meaning of a dipole potential on S^3, while $\cot\chi$ acts as the "curved" Coulomb potential there. We consider a q-(qq) configuration on S^3 bound within a cotangent potential, introduced as a gauge interaction, $i\hbar\partial_\tau \longrightarrow i\hbar\partial_\tau - [(-2G\sqrt{\kappa})\cot\chi + \bar{a}]$. The resulting exactly solvable equation is:

$$[-\kappa\hbar^2\frac{d^2}{d\chi^2} + \mathcal{V}(\chi)]\Psi(\chi) = ((E-\bar{a})^2 - c_0)\Psi(\chi),$$
$$\mathcal{V}(\chi) = -2G\sqrt{\kappa}(E-\bar{a})\cot\chi + \hbar^2\kappa\alpha(l)(\alpha(l)+1)\csc^2\chi, \qquad c_0 = \mu^2 - \hbar^2\kappa + \bar{a}^2 + (2G\sqrt{\kappa})^2,$$
$$\mu^2 = \frac{1}{6R^2}, \quad \alpha(l) = -\frac{1}{2} + \sqrt{\left(l+\frac{1}{2}\right)^2 - \frac{(2G)^2}{\hbar^2}} = l + \Delta l, \qquad \Delta l \approx -\frac{1}{2}\frac{(2G)^2}{\hbar^2}\frac{1}{l+\frac{1}{2}}, \tag{16}$$

with Δl standing for the kinematic $(P_{2I,1} - S_{2I,1})$ fine splitting. The wavefunctions are:

$$\Psi_{Kl}(\chi) = N_{Kl}e^{-a\chi}(\sin\chi)^{K+1+\Delta l}R_n^{(a,b)}(\cot\chi),$$
$$a = \frac{2G(E-\bar{a})}{\sqrt{\kappa}\hbar^2(K+1+\Delta l)}, \qquad b = -(K+1+\Delta l), \quad K = n + l, \tag{17}$$

Here, $R_n^{(a,b)}(\cot\chi)$ are the non-classical orthogonal polynomials of Routh-Romanovski, rediscovered in [28], and reviewed in [29]. The mass formula is obtained as,

$$(E-\bar{a})^2 = \frac{c_0 + \hbar^2\kappa(K+1)^2}{1+\frac{4G^2}{\hbar^2(K+1)^2}} + 2\Delta l\left(\frac{\hbar^2\kappa(K+1)}{1+\frac{4G^2}{\hbar^2(K+1)^2}} - \frac{4G^2\left(\frac{c_0}{(K+1)^3} + \frac{\hbar^2\kappa}{K+1}\right)}{\hbar^2\left(1+\frac{4G^2}{\hbar^2(K+1)^2}\right)^2}\right). \tag{18}$$

The Δ spectrum of conformal patterns has been generated by a best fit to eq. (18) and presented in Fig. 2 next to the calculation of the proton electric charge form-factor as an illustration of the quality of the wave functions.

4. Conclusions

We developed a model suitable for the description of the conformal symmetry patterns in the light baryon spectra.The partial $F_{35}(1905) - F_{37}(1950) - P_{33}(1920) - P_{31}(1910)$ degeneracy of the positive parity states from the spin-7/2 ferneon has been noticed independently in LF-QCD [30], [31]. We conclude relevance of conformal symmetry for the light baryon spectra.

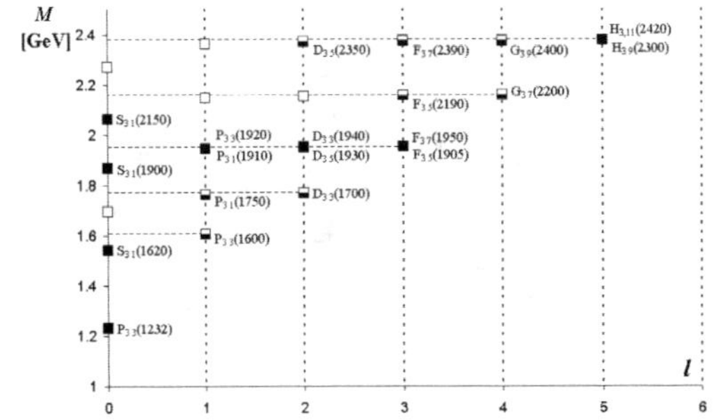

Figure 2. Δ spectrum from the Finite Volume Conformal Quark Model (FV-CQM). For the N spectrum and more details see [5].

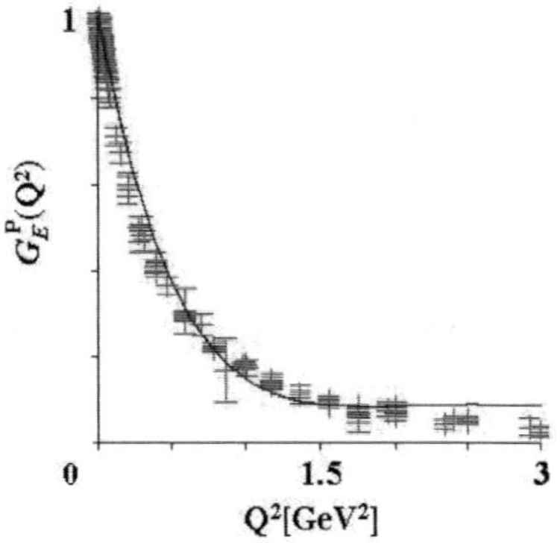

Figure 3. The proton electric charge form factor from the FV-CQM in comparison to data (for details, see [5]).

4.1. Acknowledgments

Work partly supported by CONACyT-Mexico under grant number CB-2006-01/61286.

[1] Burkert V D and Lee T S H 2004, *Int. J. Mod. Phys.* E **13** 1035
[2] Afonin S S 2007 *Int. J. Mod. Phys.* A **22**, 4537
[3] Amsler S G *et al.* 2008 *Phys. Lett.* B **667** 1
[4] Jaffe R L and Wilczek F 2003 *Phys. Rev. Lett.* **91** 232003
[5] Kirchbach M and Compean C B 2010 *Phys. Rev.* D **82** 034008
[6] Nicolai H, Sezgin E and Tanii Y 1998 *Nucl. Phys.* B **305** 483
[7] Lüscher M and Mack G 1975 *Commun. Math. Phys.* **41** 203
[8] Horowitz G T and Ooguri H 1998 *Phys. Rev. Lett.* **80** 4116.
[9] Gibbons G W and Steif A R 1995 *Phys. Lett.* B **346** 255
[10] Wybourne B G 1974 *Classical groups for physicists* (New York: Wiley-Inter-science)
[11] Rico L M and Kirchbach M 2006 *Mod. Phys. Lett.* A **21** 2961
[12] Rarita W and Schwinger J 1941 *Phys. Rev.* **60** 61
[13] Napsuciale M, Kirchbach M and Rodriguez S 2006 *Eur. Phys. J.* A **29** 289
[14] Kirchbach M 2003 *Found. Phys.* **33** 781
[15] Napsuciale M, Rodriguez S, Delgado-Acosta G and Kirchbach M 2008 *Phys. Rev.* D **77** 014009
[16] Kirchbach M 1997 *Mod. Phys. Lett.* A **12**, 2373
[17] Banks T and Zacks A 1982 *Nucl. Phys.* B **196** 189
[18] Seiberg N 1994 *Phys. Rev.* D **49** 6857
[19] Deur A, Burkert V, Chen J P and Korsch W 2008 *Phys. Lett.* B **665** 349
[20] Maldacena J 1998 *Phys. Rev. Lett.* **80**, 4859
[21] Takahashi T T, Suganuma H, Nemoto Y and Matsufuru H 2002 *Phys. Rev.* D **65** 114509
[22] Brambilla N, Vairo A, Rosch T 2005 *Phys.Rev.* D **72** 034021
[23] Witten E 1998 *Adv. Theor. Math. Phys.* **2** 505
[24] Fubini S, Hanson A J and Jackiw R 1973 *Phys. Rev.* D **7** 1732
[25] Antoniadis I, Mazur P O and Motola E 1997 *Phys. Rev.* D **55** 4756
[26] Alertz B 1990 *Annales de l' I.H.P.* A **53** 319
[27] Belitsky A, Gorsky A S and Korchemski G P 2003 *Nucl. Phys.* B **667** 3
[28] Compean C B and Kirchbach M 2006 *J. Phys. A:Math.Gen.* **39** 547
[29] Raposo A, Weber H J, Alvarez-Castillo D E and Kirchbach M 2007 *C. Eur. J. Phys.* **5**, 253
[30] de Téramond G F and Brodsky S J 2005 *Phys. Rev. Lett.* **94** 201601
[31] Brodsky S J and de Téramond G F 2006 *Phys. Rev. Lett.* **96** 201601

Journal of Physics: Conference Series **287** (2011) 012020 doi:10.1088/1742-6596/287/1/012020

New Forward and Diffractive Physics at CMS

Alberto Santoro

Departamento de Física Nuclear e Altas Energias
Instituto de Física Universidade do Estado do Rio de Janeiro
Rua São Francisco Xavier, 524 - Maracanã
20559-900 - Rio de Janeiro - RJ - Brazil

E-mail: alberto.santoro@cern.ch

Abstract. Forward and Diffractive Physics (FWP) in LHC is a new open window to understand this type of strong interactions. We will present a didactic description of the topics being developed at CMS. As we know there still is no new results to present for FWP. We are accumulating data to have soon new results. We will show a number of topics and the detectors properties to do the observation of several topologies. We expect to give an optimistic view of the area.

1. Introduction

High Energy Physics (HEP) is the science of the Fundamental Interactions of Nature and of its most elementary particles. HEP, as a frontier science, is permanently changing its facilities with new Instrumentation, new computing, new analysis methodology and so on. One region of the spectrum of the phase space available for interactions is the Diffraction situated on the Forward direction of the scattering. Diffraction is perhaps one of the oldest subjects of physics. Just for curiosity, the Diffraction in Optics was first observed by Leonardo Da Vinci [1] (1452-1519). [1] Historical information can be found in the references [2, 3].

The Large Hadron Collider (LHC) is the last new machine which represent one of the most important challenge scientific and technologically. We never did an experiment at 14 TeV. It is the great opportunity to make advancements in Science. We are working on the limits of the present technology. In particular, the Compact Muon Solenoid (CMS) is one of the 4 main detectors of LHC. (There are three other experiments, ALICE, LHCb and ATLAS (figs.1, 2). Each one of these experiments is dedicated to one main subject. ALICE for heavy Ions interactions; LHCb for b-Physics, and ATLAS, like CMS are the exceptions, they are general purpose experiments.

The main parameters of LHC are:

- It is a proton x proton collider;
- it has 27 km of circumference;

[1] An interesting reference: "Leonardo Da Vinci - An Artabras Book" - Reynal and Company in Association with William Marrow and Company N.Y. Chapter: Leonardo's Optics, Domenico Argentieri p. 405, says the following: "I did a great discovery founding the transcription of Gian Battista Venturini which was in the Reggio Emilia Library. The text shown clearly that Leonardo observed the diffraction phenomena but made a wrong nterpretation".

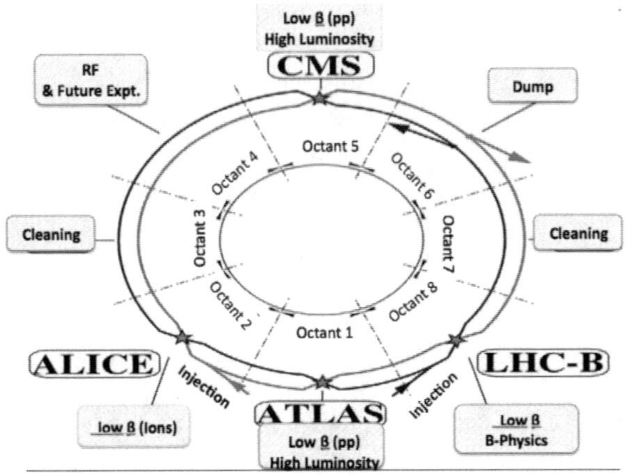

Figure 1. The LHC and the Interaction Points for each Experiment/Detector

.

Figure 2. This is part of the 27 km of the tunel of the LHC

.

- It is at 100 m under surface;
- The beam Energy of each is 7 TeV;
- Energy at SCM (Center of Mass) is 14 TeV;
- It has 1800 Superconductor Magnets;
- It has 7000 km of Superconductor wires;
- It has 120 Tones of Liquid Helium ($\tilde{7}00$ K litters);
- The temperature of the supercondutor is 1,9 Kelvin;
- It is built to work with 2808 bunches per Beam;
- And $1,15$ x 10^{11} Particles per Bunch.

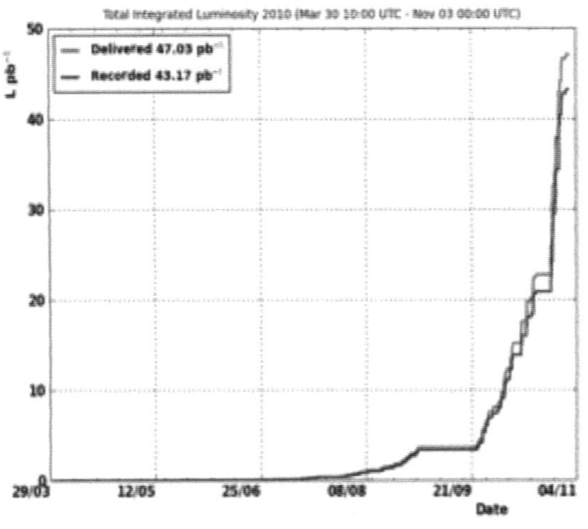

Figure 3. The Luminosity between March and November 2010

The present status of LHC is working in a very good performance. Only for curiosity, after to start in 20/09/2008 an unexpected accident heppened on the tunel. A small device of the cryogen system stopped to work causing problem on the process of the Superconducting Magnets. 53 Dipoles and 14 Quadripoles was removed and repaired. In November of 2009 LHC started to work again up till now. We had a very rich run period and stop to prepare the Heavy Ions run. However the figure 3 show a plot of Luminosity which represent the sucess of the LHC. That allow us to expect for the near future good results and even descoveries.

Journal of Physics: Conference Series **287** (2011) 012020 doi:10.1088/1742-6596/287/1/012020

The CMS Detector as we show in figure 4 is all parts of the Central Detector: Pixels Detector, Tracker Detector, the Electromagnetic and Hadronic Calorimeters, the Solenoid, the Muons Detector and the steel Yokes. These are the main parts constituting the main CMS Detector. To complete the view of CMS Detectors we show in figure 5 the set of detectors of Forward Physics. These detectors open a new window for CMS. Finally we would like to warn the reader that we will not show results from LHC since they are still in a very preliminar studies but we expect soon to have new results to present.

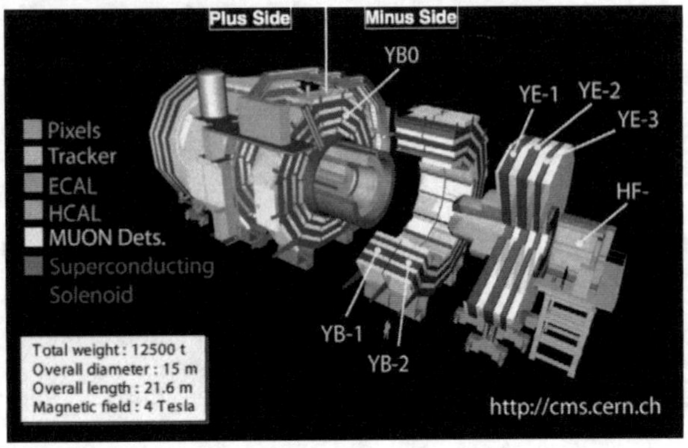

Figure 4. CMS Detectors and its main parts

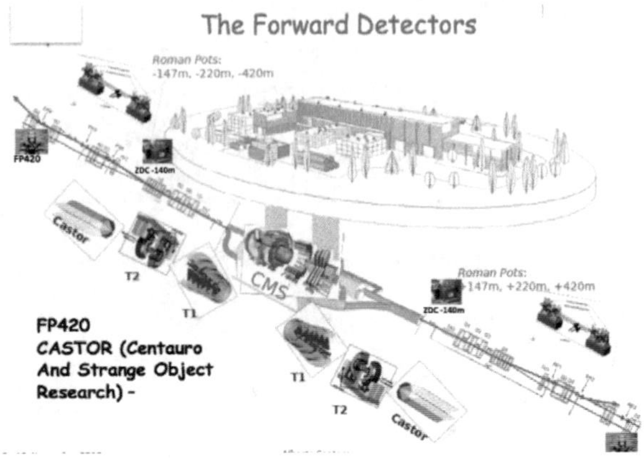

Figure 5. The Forward Detectors of CMS

2. Forward Physics and Diffraction

In general CMS Detector will study a large number of subjects like: Higgs (the most popular subject), Dark Matter, Matter and Antimatter, Heavy Ions Physics, Supersymmetry, Extra Dimensions, Leptoquarks, Jets and Multijets, New Physics or Beyond the Standard Model topics, Centaurus and Strange Objects and Forward and Diffractive Physics. But in this talk we are interested mainly on Diffractive Physics.

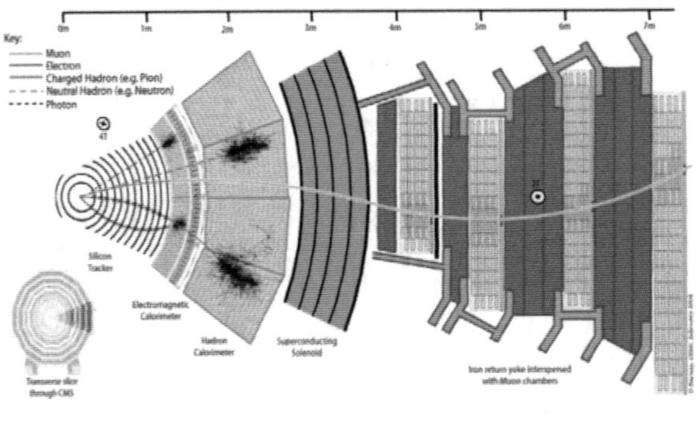

Figure 6. A slice of the Detector showing the particles detected by each part of the spectrometer

In figure 6 we show the particles that can be detected by the Central Detector. There are 5 particles (γ, e^-, Neutral Particles (e.g. neutron), Charged Particles (e.g. pions), muons) and its path crossing the sub-detectors. The neutral particles deposit the energy mainly in the Electromagnetic Calorimeter and the charged particles go to the Hadronic Calorimeters. Muons cross all sub-detectors leaving their signatures on the Muon detectors. Photons and electrons go to the Electromagnetic Calorimeter (EM). All charged particles suffer deviation with the magnetic field.

Many topologies of Forward Physics and Diffraction, as shown in fig. 7, need to have the two proton scattered triggered with the event in the Central Detector well defined. But in the present we have to work with the rapidity gap. We still do not have triggered events. There are many subtopics in Forward and Diffractive Physics: Elastic Scattering and Total cross section after the installation of Roman Pots of the TOTEM detectors; Single and Double Hard Diffraction; Hard Double Pomeron exchange, Higgs production, Glueballs, Diffractive W and Z production and so on.

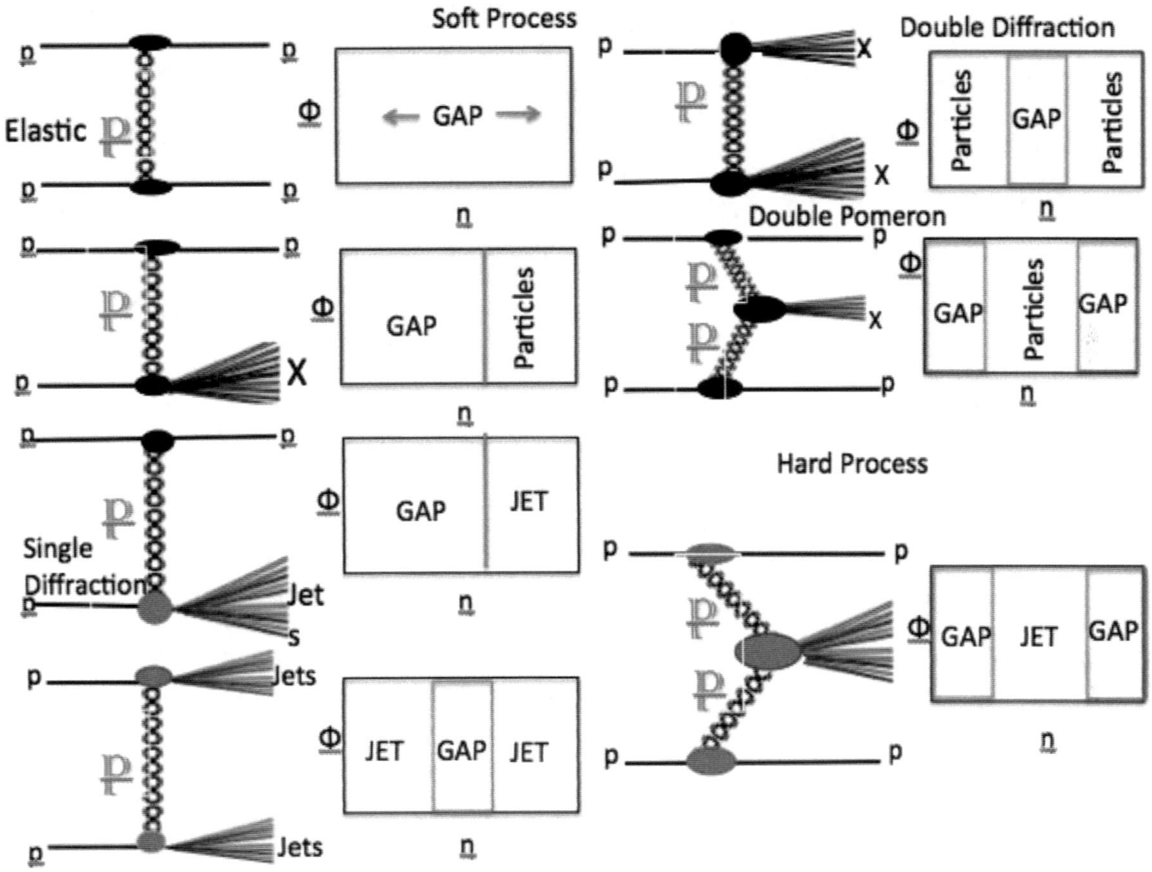

Figure 7. Diffraction topologies and the Gap on φ versus η

The plots and distributions are always presented with the Kinematics Variables appropriated. They are the:

- $t = (p_f - p_i)^2 \rightarrow$ Transfer momentum
- $x_{\mathbb{P}} = p_f/p_{Beam} \rightarrow$ Scattered Proton Momentum Fraction
- $\zeta = 1 - x_{\mathbb{P}} \rightarrow$ Momentum Fraction
- $M_X = \sqrt{s\zeta} \rightarrow$ Diffractive Mass
- $\eta = -\ln(\tan(\theta/2)) \rightarrow$ Pseudorapidity

Journal of Physics: Conference Series **287** (2011) 012020 doi:10.1088/1742-6596/287/1/012020

2.1. Elastic Scattering and Soft Diffraction

The Elastic Scattering is an important diffraction component. It is with the TOTEM Detectors, the Roman Pots, that will be possible to do measurement of Elastics Scattering and Total Cross Section for proton proton at 14 TeV. These future results has an important aspect for the asymptotic Theorems [4]. Soft component of Diffraction is very well described by Regge Theory [5]. The main characteristic of Diffraction is the exchange of the vacuum quantum number ($J^{PC} = 1^{++}; I^G = 0^+; Q = B = S = 0$) and a good experimental signature is the rapidity gap found when we do the 3-dimensional plot of the energy deposit on forward hadron calorimeter cells.

2.2. Diffraction Topologies and Associated Gap Expected

As we know, 1/3 of the Inelastic cross section at the LHC is diffractive. This give us a good optimistic view for the next years. The gaps and the topologies of the diffraction scattering are show in figure 7.

At the partonic interaction level, Diffraction is a good way for probing proton and measure the Diffractive Parton Distribution which give us information about the structure of the proton and also give us the probability of the event type being studied as we can see in the figure 8.

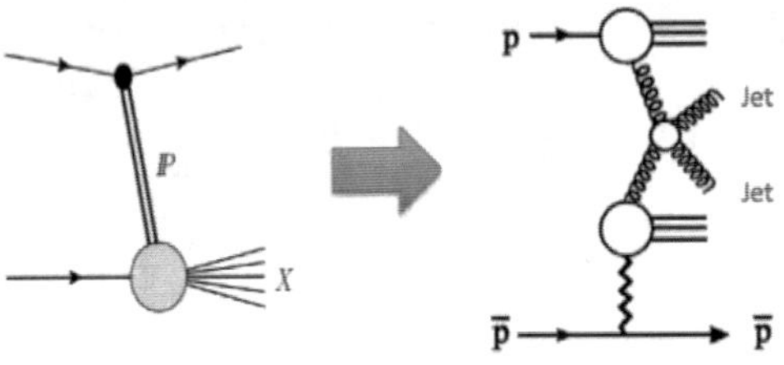

On the left side we see the Single Hard Diffraction with a Pomeron exchanged. The X represent the jets produced by the diffractive interaction between the Pomeron and the Proton. At Partonic level of the right side is represented by the Jets produced by the interaction.

Figure 8. This is the representation of the partonic interaction or the Pomeron probing the proton.

When we face the partonic approach for diffraction scattering using QCD to calculate the cross section it appear a number of difficulties and some of them are not definitely solved.The first one is the Factorization theorem. The calculation of the differential cross-section of Diffractive (DIS) Deep Inelastic Scattering is Structure Function dependent as

$$F_2^{D(4)}(\beta, x_P, t; Q^2) = \sum_i \int_\beta^1 \frac{dz}{z} C_i\left(\frac{\beta}{z}\right) f_i^D(\beta, x_P, t; Q^2)$$

where, $f_i^D(\beta, x_P, t; Q^2)$ is the universal Diffractive Parton Distribution Function (DPDF). Using the Data from HERA, for DIS the structure function can be obtained,

$$\sigma^D(\gamma^* p \to Xp) = \sum_i \hat{\sigma} \otimes \int_i^D (x_{Pom}, t, z, Q2)$$

This is the QCD factorization theorem proven by J. Collins [10, 11].

While the Factorization theorem works for HERA [12] for p $\bar{p}$ the factorization theorem breakdown. The single diffraction rates of W and Z [13, 14], Dijet [15, 16], b-quark [17], J/ψ show a factor 10 lower than this one expected using DPFs from HERA. This is a strong disagreement for hard diffraction. A possible explanation is based on the difference between the initial states. $p\bar{p}$ has an important number of partons then the gap could be filled by the multiple interactions and the diffractive events not appear properly. The way to have a quantitative measurement of this is calculate the survive probability.

It is very important to test again on the LHC, this time on proton proton scattering.

Another difficulty for Particle Diffraction it will be the pile up produced for high luminosity. [6] We have to pay attention to this physical effect consequence of high energy and high luminosity. Then it is very important for the future of the Diffraction to have all Forward detectors installed at CMS. These detectors, as we said above are:

- The spectrometer TOTEM with its Telescopes and Roman Potes [7]
- The CASTOR, a Forward Calorimeter [20]
- The ZDC, a Zero Degree Calorimeter [8]
- The FP420/HPS, a fundamental detector to get scattered protons [9]

2.3. Forward Detectors

CMS is by far the most complete experiment for Forward Physics, since there is an important number of Forward Detectors dedicated to Forward Physics. It is important to know that it is the first experiment to consider a so significative number of subdetectors which will open the number of possibilities for CMS. One very interesting exemple is this one of the possible diffractive observation of Higgs with the Central Production Topology. Let us see a bit about each one of the most important but not exhaustive list of detectors.

2.3.1. TOTEM - TOTEM is a set of sub-detectors to measure the Elastic and Total Cross-Section [7]. The spectrometer is constituted by Roman Pots at 147 m and 220 m near to the beam. The rapidity gap range fill the space between 3.1 and 6.5. This experiment is important for studies of elastic scattering and it is also important to get a new number for total cross-section in view of the values found by the E811 and CDF at Tevatron/Fermilab [19]. We need to compare to the predictions of the asymptotic extrapolation of the two know measures.

2.3.2. CASTOR - Centauro And STrange Objects Research - Castor is a Forward Calorimeter with two components, one Eletromagnetic and another Hadronic. This was designed to attend the request for an eventual Centauro event. The Eletromagnetic sector has 16 segments in ϕ and 12 for hadronic part. The active part is made of Tungsten sandwiched with plates of Quartz. A air-core lightguides transmit the Cerenkov light to the PMTs. These PMTs have special treatment for radiation resistence. Its position is at the end of the HF-Hadronic Calorimeter. The range of pseudo-rapidity η is: $5.2 < \eta < 6.6$. Centauros are the phenomena observed in Cosmic Ray Physics [20]. The main characteristic of Centauros are the large multiplicity of hadrons with a low activity electromagnetic or almost no γ no π^0 production. In Cosmic Ray Physics, as is shown in the refereces given above, there are different level of mass for Centauros: 200 to 300 GeV/c^2 those called Centauros, for mini-centauros they can have 20 - 30 GeV/c^2, for Chiron 200 GeV/c^2 and for Geminion 15 - 30 GeV/c^2. These value have to be a guide/hint to our experiment at LHC/CMS.

2.3.3. ZDC-Zero Degree Calorimeter - The (ZDC) is a subdetector located 140 m down the tunnel on each side of the central detector CMS interaction point. This Calorimeter is part of the complex of calorimeters like Castor, with five eletctromagnetic channels and four hadronic, as a sub-system of the CMS Forward HCAL hadron Calorimeters. The physics associated to this calorimeter are the luminosity, then the LHC itself, the heavy ion physics, cosmic ray physics and diffraction [8].

Journal of Physics: Conference Series **287** (2011) 012020 doi:10.1088/1742-6596/287/1/012020

2.3.4. FP420/HPS - The FP420 is a project [9] for a new Forward sub-detector for CMS. He was conceived to be a silicon detector and having the possibility to detect scattered protons at high precision. He will be on the both side, of the tunnel of LHC at 420 m from the interaction point of CMS. The physics associated to this sub-detector include several important topics connected to the diffrative central exclusive production of the interaction of $pp \to p + \phi + p$ where both outgoing protons remain intact ϕ can be a Higgs boson decay in the $b\bar{b}$ or any other heavy flavor. The reader will find many interesting details on the reference above.

3. Conclusion

Forward Physics and Diffraction are very important to complete the understanding of Strong Interactions. Many comparisons between Theory/Models and Experiment data does not consider the diffractive component. Then it is interesting to know how much the diffractive part of the cross section contribute. This is being studying calcuting the rate of the diffractive and no diffractive or diffractive and all production. This can give us a better understand of the mechanisms of production and eventually make the eventual corrections on the calculations. On the other hand these detectors for measurement of diffraction, helps in different ways other topics like the calculation of the luminosity. There are many topics under investigation on the Forward Physics Group of CMS. The rate of diffractive/non diffractive of Heavy Flavor, the topology Jet Gap Jet in which we can have some light about the production of a number of states; Single Diffraction Z and W production, Dijets and multijets in diffraction, Centauros, and so on. We expect for next years by the introduction of new detectors like FP420 we could continue to do this physics even with the high rate of pile-up. We could not extend each subtopic of this talk due to limitation of space. Then we did a very short overview of the physics of Diffraction on CMS.

3.1. Acknowledgments

I would like to thank the organizing committee of the XIV Mexican School on Particles and Fields, for having invited me to give this talk, giving me the opportunity to participate to this interesting Meeting. We thank to FAPERJ and CNPq for partial financial support.

References

[1] Max Born and Emil Wolf "Principles of Optics Electromagnetic Theory of Light", Pergamon Press - USA (1975) Leonardo Da Vinci

[2]
- (a)Alberto Santoro - "Diffractive Physics in the Near Future" talk on the LISHEP DIffractive Workshop \http:\\www.lishep.uerj.br
- (b)Alberto Santoro - "Problemas Atuais da Fsica Difrativa" (in portuguese) - Thesis presented to UERJ Concours (2001).

[3] V. Barone and Enrico Predazzi - High-Energy Particle Diffraction - Springer (2002)

[4] These 3 references below are excellent to get update with Soft Diffraction.
- (i)Uri Maor, arXiv:1010.5323[hep-ph] (to be appear on JHEP)
- (ii)Uri Maor et al., (arXiv:0903.0247[hep-ph] and
- (iii)Uri Maor e at., arXiv:0910.1196[hep-ph])

[5]
- E. Feinberg, I. Ya. Pomeranchuk, Suppl. N. Cim. 4, 652(1956);
- M.L. Good and W. D. Walker, P.R. 120, 1855, 1857 (1960);
- A. Santoro - The Future of Diffaction at Tevatron also see many other references in the Proceedings of the International Workshop on Diffraction in High Energy Physics - Cetraro -Italy 2-7 September (2000) - Edited by Roberto Fiori, Maxim I, Kotsky, Alessandro Papa, Enrico Predazzi, Giancarlo Susinno -North Holland - Nuclear Physics B (Proc. Suppl) 99A (2001) April 2001.

[6] Eliza Melo da Costa - Master Thesis "Pile-up impact in Diffraction at CMS/LHC"-UERJ- 2009.

[7] For a more complete information see: The TOTEM Experiment at the CERN Large Hadron Collider The TOTEM Collaboration et al 2008 JINST 3 S08007.

[8] O. Grachov, M. Murray, J. Wood, (Kansas U.) , Y. Onel, S. Sen, T. Yetkin, (Iowa U.) Commissioning of the CMS zero degree calorimeter using LHC beam. .Presented at 14th International Conference on Calorimetry in High Energy Physics (CALOR 2010), Beijing, China, 10-14 May 2010; e-Print: arXiv:1008.1157.
[9] M.G. Albrow, et al. FP420 R&D collaboration, "The FP420 R&D project: Higgs and New Physics with forward protons at the LHC" JINST 4 T10001(2009).
[10] J. Collins P.R.D57(1998) 3051.
[11] A. Solano, Talk on the Diffraction 2010 conference. This is a very clear talk about DPDFs and Factorization at HERA.
[12]
- (i) CERN/LHC 2006-039/G-124, CMS Note 2007/002, TOTEM Note 06-5 2006.
- (ii)ZEUS Collaboration: N.P. B816 (2009)1; N.P.B672 (2003), 3; EPJ C52 (2007), 873.
- (iii)H1 Collaboration: EPJ C50 (2007), 1; JHEP 0710:042 (2007).
- (iv) A.B. Kaidalov, V.A. Khoze, A.D. Martin ad M.G. Riskin, hep-ph: arXiv:0911.3716-v1 (2009).

[13] F. Abe et al., CDF collaboration, Phys. Rev. Lett. 78 (1997) 2698-2703, arXiv:hep-ex/9703010.
[14] V.M. Abazov et tal., Dzero Collaboration, Phys. Lett. B574 (2003) 169-179; arXiv: hep-ex/0308032.
[15] CF. Abe et al., DF collaboration, Phys. Lett. B531 (2002) 52-60.
[16] V.M. Abazov B. Abbott tal., Dzero Collaboration, Phys. Lett. B574 (2003) 169-179; arXiv: hep-ex/9912061.
[17] A. A. Affolder et al., CDF collaboration, Phys. Rev. Lett. 84 (2000) 232-237.
[18] A. A. Affolder et al., CDF collaboration, Phys. Rev. Lett. 87 (2001) 2241802, arXiv: hep-ex/0107071.
[19] M. Delle et al. Diffraction and total cross-section at the Tevatron and the LHC- arXiv:hep-ex/0602021 v1 (2006).
[20]
- a) F. Halzen - Felix home Page.
- b) Brazil-Japan Collab. -Proceed. Of the 21st. Int. Conf.- Adelaide Australia-vol.8,259,1990.
- c) idem idem, Prog. Theo.Phys.Suppl. 47,1,1971.
- d) C.M.G.Lattes, Y.Fujimoto and S. Hasegawa, Phys.Rep. 65,151 (1980).
- e) C.E.Navia et al. Phys. Rev. D40,2898 (1989).
- f) C.M.G. Lattes et all. Phys.Rep.,65,151,(1980)].
- g) A. Santoro, Cesar Lattes, a Descoberta do Meson Pi e outras historiaspg.167-169 (1999) edited by F.Caruso, A. Marques e A. Troper CBPF/MCT.
- h) Castor: Centauro and Strange Object Research in nucleus-nucleus ast LHCArXiv: hep-ex/029008v2, 2002.
- i) Technical Proposal submitted by CMS-HI-CASTOR-Group 2004.

Moduli fixing in semirealistic string compactifications

Saúl Ramos-Sánchez

Instituto de Física, Universidad Nacional Autónoma de México, Apdo. Postal 20-364, México 01000, México

E-mail: ramos@fisica.unam.mx

Abstract. Heterotic orbifold compactifications yield a myriad of models that reproduce many properties of the supersymmetric extension of the standard model and provide potential solutions to persisting problems of high energy physics, such as the origin of the neutrino masses and the strong CP problem. However, the details of the phenomenology in these scenarios rely on the assumption of a stable vacuum, characterized by moduli fields. In this note, we drop this assumption and address the problem of moduli stabilization in realistic orbifold models. We study their qualities and their 4D effective action, and discuss how nonperturbative effects indeed lift all bulk moduli directions. The resulting vacua are typically de Sitter and there are generically some quasi-flat directions which can help to deal with cosmological challenges, such as inflation.

1. Introduction

There has been a great effort in deriving models of elementary particle physics from string theory. Promising models have arisen from intersecting D-branes in type IIB string theory, F-theory, Calabi-Yau and orbifold compactifications of the heterotic string. In particular, heterotic orbifolds have been extensively studied and successful models reproducing several features of the minimal supersymmetric extension of the standard model (the MSSM) have been found [1, 2, 3, 4, 5, 6, 7]. Unfortunately, contact between all these different constructions and particle phenomenology requires a full understanding of the details of the vacua that each model offers. These details are associated with the dynamics of the *moduli* fields that describe all geometrical features of the particular compactification. For instance, the masses of quarks and leptons as well as the dynamics of supersymmetry breakdown depend on the vacuum expectation values (VEVs) of the moduli, which are arbitrary at tree level in many string constructions. Therefore, in order to achieve a prediction from string theory and simultaneously to avoid severe cosmological constraints, a mechanism that fixes the VEVs of the moduli and gives them large masses is needed. This mechanism is usually called *moduli fixing* or *stabilization.*

Besides constructions that potentially explain particle physics, moduli stabilization has been independently studied in several scenarios. In type IIB string theory, it has been shown that conjugating fluxes, nonperturbative effects and/or α' corrections could yield stable vacua [8, 9, 10]. There has also been some progress in fixing Kähler and complex structure moduli in heterotic compactifications on Calabi-Yau [11] and (generalized) half-flat [12] manifolds. In heterotic orbifold compactifications [13, 14], which is the focus of the present note, it is frequently argued that the absence of fluxes (other than the NS flux) can render much more difficult –if not impossible– the search for a stable vacuum. However, it is known that the inherent target

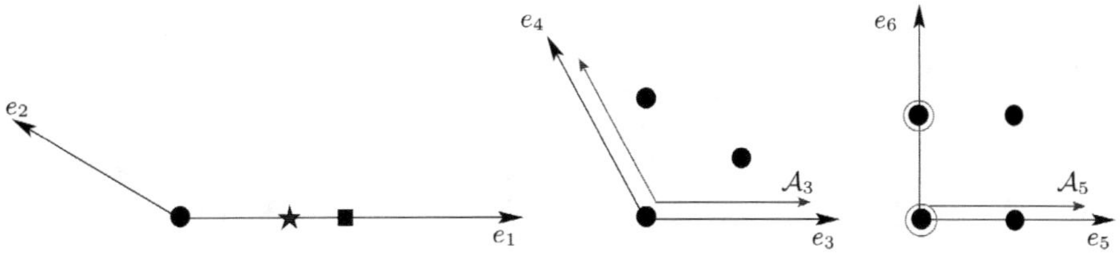

Figure 1: Geometry of the 6D compact space of a $\mathbb{Z}_6$–II orbifold. The 6D space is chosen to be the torus $T^6 = T^2 \times T^2 \times T^2$ spanned by the roots e_i of the Lie algebra $G_2 \times SU(3) \times SO(4)$. The smaller/blue arrows denote the noncontractible loops that are embedded as nontrivial Wilson lines $\mathcal{A}_i$ in the gauge degrees of freedom. Modding out a $\mathbb{Z}_6$ symmetry in this geometry leaves the points $\bullet$, $\star$ and ■ untouched.

space modular symmetries [15, 16, 17] together with nonperturbative effects such as gaugino condensation and string worldsheet instantons can provide a scalar potential with metastable minima, at least in toy models [18, 19, 20, 21, 22].

The purpose of the present note is to review the main results of [23]. We go beyond the usual claims and tackle explicitly the problem of moduli fixing in MSSM candidates arising from heterotic orbifolds. We base our study on a subset of the $\mathcal{O}(300)$ realistic $\mathbb{Z}_6$–II orbifold models of the minilandscape [24, 25], in which gaugino condensation can successfully yield an acceptable scale of supersymmetry breakdown [26]. The main phenomenological properties of these models shall be discussed in sec. 2, where the details of a sample model are given. We consider all bulk moduli fields, i.e. the dilaton S and those fields describing the deformations in size T_i and shape U_i of the compact dimensions. We analyze the 4D effective action of these models, which is well understood [27, 28, 29]. In particular, the scalar potential consists of various computable, perturbative and nonperturbative contributions, which, as it turns out, lift all bulk moduli directions. This will be the topic of sec. 3. Once moduli stabilization is achieved in these setups, it results immediate to explore some cosmological consequences. We speculate in sec. 4 that inflation might emerge readily, since some of the moduli directions exhibit very small curvature, rendering the associated fields natural inflaton/curvaton candidates.

In the following, we shall work in Planck units, specifically we set $M_{Planck} = 1$.

2. An MSSM candidate

The so-called minilandscape is perhaps the most fertile search of string-derived MSSM candidates. There have been found about 300 models with the exact spectrum of the MSSM (or NMSSM [30]) and no exotics[1] at low energies. Besides, these models exhibit other appealing properties: approximate gauge coupling unification [31, 32], see-saw neutrino masses [33], predominantly TeV gravitino mass [26], gauge-top unification [34], R-symmetries [6, 35] and other discrete symmetries [36] that prevent rapid proton decay and other lepton/baryon-number violating processes.

The minilandscape is based on $\mathbb{Z}_6$–II orbifold compactifications of the $E_8 \times E_8$ heterotic string. The 6D torus is chosen to be the product of three 2-tori, $T^6 = T^2 \times T^2 \times T^2$, which are spanned by the simple roots of the Lie algebra $G_2 \times SU(3) \times SO(4)$, as depicted in fig. 1. In this basis, the action of the $\mathbb{Z}_6$–II symmetry acts identifying the points related by a rotation (or repeated rotations) by $\pi/3$ in the G_2 2-torus, a rotation $2\pi/3$ in the SU(3) 2-torus, and a reflection through the origin in the last 2-torus. The special points that are left unaffected by this identification, depicted by $\bullet$, $\star$ and ■ in fig. 1, are called fixed points and are of crucial relevance in the minilandscape, as we shall see shortly. The intrinsic modular invariance of the heterotic

[1] The term *exotic* refers to any field charged under the SM gauge group *and* which does not appear in the MSSM matter spectrum.

Table 1: The matter spectrum of a realistic $\mathbb{Z}_6$–II orbifold model. It corresponds to the spectrum of the MSSM plus standard model singlets and vectorlike exotics, which acquire masses through a Higgs-like effect after some singlets attain VEVs. The bulk moduli are uncharged under the gauge group. Quantum numbers w.r.t. the gauge group $\mathrm{SU}(3)_c \times \mathrm{SU}(2)_L \times [\mathrm{SO}(8) \times \mathrm{SU}(3)]_{hidden}$ are given in parenthesis. The hypercharge $\mathrm{U}(1)_Y$ appears as subindex.

Bulk moduli: S, T_1, T_2, T_3, U					
3 (net) generations					
3	$(\mathbf{3},\mathbf{2};\mathbf{1},\mathbf{1})_{1/6}$	q_i			
3	$(\overline{\mathbf{3}},\mathbf{1};\mathbf{1},\mathbf{1})_{-2/3}$	$\bar{u}_i$			
3	$(\mathbf{1},\mathbf{1};\mathbf{1},\mathbf{1})_{1}$	$\bar{e}_i$			
3+4	$(\overline{\mathbf{3}},\mathbf{1};\mathbf{1},\mathbf{1})_{1/3}$	$\bar{d}_i$	4	$(\mathbf{3},\mathbf{1};\mathbf{1},\mathbf{1})_{-1/3}$	d_i
3+4	$(\mathbf{1},\mathbf{2};\mathbf{1},\mathbf{1})_{-1/2}$	ℓ_i	4	$(\mathbf{1},\mathbf{2};\mathbf{1},\mathbf{1})_{1/2}$	$\bar{\ell}_i$
Higgses					
1	$(\mathbf{1},\mathbf{2};\mathbf{1},\mathbf{1})_{-1/2}$	h_d	1	$(\mathbf{1},\mathbf{2};\mathbf{1},\mathbf{1})_{1/2}$	h_u
Standard model singlets					
43	$(\mathbf{1},\mathbf{1};\mathbf{1},\mathbf{1})_{0}$	n_i	8	$(\mathbf{1},\mathbf{1};\mathbf{8},\mathbf{1})_{0}$	$\widetilde{N}_i$
10	$(\mathbf{1},\mathbf{1};\mathbf{1},\mathbf{3})_{0}$	N_i	10	$(\mathbf{1},\mathbf{1};\mathbf{1},\overline{\mathbf{3}})_{0}$	$\bar{N}_i$
Exotics					
8	$(\mathbf{3},\mathbf{1};\mathbf{1},\mathbf{1})_{1/6}$	δ_i	8	$(\overline{\mathbf{3}},\mathbf{1};\mathbf{1},\mathbf{1})_{-1/6}$	$\bar{\delta}_i$
8	$(\mathbf{1},\mathbf{1};\mathbf{1},\mathbf{1})_{1/2}$	s_i^+	8	$(\mathbf{1},\mathbf{1};\mathbf{1},\mathbf{1})_{-1/2}$	s_i^-
16	$(\mathbf{1},\mathbf{2};\mathbf{1},\mathbf{1})_{0}$	m_i			

string demands the orbifold action to be embedded in the gauge degrees of freedom, producing thereby the breakdown of the original $\mathrm{E}_8\times\mathrm{E}_8$ gauge group to a subgroup thereof, which in the minilandscape models takes the generic form $\mathrm{SU}(3)_c \times \mathrm{SU}(2)_L \times \mathrm{U}(1)_Y \times G_{hidden} \times \mathrm{U}(1)^n$. All standard model (SM) fields are uncharged under the nonabelian gauge factor G_{hidden}, but some other "hidden" fields form nontrivial representations under this group.

The guiding principle of the minilandscape search is local grand unification, which refers to the possibility that certain fixed points of an orbifold be endowed with the gauge symmetry of a grand unified theory (GUT) like SU(5) or SO(10), while the 4D gauge group be the one of the SM. This indeed happens in the small moduli-space region of the minilandscape and leads to scenarios that preserve the appealing features of GUTs such as gauge coupling unification while solving their common phenomenological drawbacks such as problematic quark/lepton mass relations. The crucial advantage is that matter fields living at points endowed e.g. with an SO(10) local symmetry can transform as **16**-plets locally. Since the **16** spinor of SO(10) contains a SM family of quarks and leptons, if ideally there were three of such special local GUTs, one would find a geometric origin of the three SM generations. Moreover, the SM Higgs doublets need not arise from such local GUTs, but from the bulk or a fixed point with no local GUT. Therefore, no doublet-triplet splitting must be enforced in this situation. Unfortunately, the scenario with three degenerate SM families coming from local GUTs is not favored by orbifold constructions [24]. Instead, only two generations arise from equivalent local SO(10) GUTs (see the encircled/red • in fig. 1), what is not necessarily bad news, for we might prefer a setting in which a distinction between the two lightest and the heaviest SM generation emerges naturally.

Let us focus now on one particular minilandscape model. The resulting matter spectrum is presented in tab. 1. The 4D gauge group is $\mathrm{SU}(3)_c \times \mathrm{SU}(2)_L \times \mathrm{U}(1)_Y \times [\mathrm{SO}(8) \times \mathrm{SU}(3)]_{hidden} \times \mathrm{U}(1)^6$. Note that no SM field is charged under $[\mathrm{SO}(8) \times \mathrm{SU}(3)]_{hidden}$, what "hides" this sector, in the sense that it only interacts gravitationally with the observed matter. In addition to the 4D $\mathcal{N} = 1$ supergravity multiplet (which is not displayed in tab. 1), one obtains a net number of three SM generations, the up and down Higgses required in the MSSM, some

$SU(3)_c \times SU(2)_L \times U(1)_Y$ singlets and a bunch of vectorlike exotics that are decoupled at a scale M_d without breaking supersymmetry.[2] The SM singlets charged under the hidden group can also acquire masses of order M_d, breaking spontaneously the additional $U(1)^6$ and leaving a pure Yang-Mills $SO(8)\times SU(3)$ hidden sector at lower energies. This situation leads to gaugino condensation in this sector, which renders the scale of supersymmetry breakdown as low as ~ 100 TeV [26], not far from the admissible/expected value.

In the absence of matter parity in this model[3], some of the SM singlets (those without VEVs) can be considered right-handed neutrinos. We have verified that they become massive when the exotics decouple, i.e. they get masses of order $M_d \sim M_{GUT}$ too. This large right-handed neutrino mass enables the see-saw mechanism to produce left-handed neutrino masses of order $10^{-3} - 10^{-2}$ eV, as suggested by neutrino-oscillation probes [38].

As shown in tab. 1, this model has five bulk moduli. The dilaton S arising from the supergravity multiplet, and four deformation parameters. The latter are obtained from demanding invariance of the metric $g_{\alpha\beta} = e_\alpha \cdot e_\beta$ under the $\mathbb{Z}_6$–II orbifold action. The free parameters in this case are the "radius" of the cycle e_1 in the G_2 torus, e_3 in the SU(3) torus, and the cycle radii and angle of the last torus, i.e. the magnitude of e_5 and e_6 and the angle between them. These deformations are usually collected in three Kähler moduli T_1, T_2, T_3 for the radii of each of the tori, and a complex structure modulus U for the angle in the last torus.

As a last remark, we would like to mention that this model possesses some useful symmetries. Like many minilandscape models, it presents an SO(2) family symmetry [39] which is the result of the vertical symmetry between the fixed points of the last torus (see e.g. the encircled/red • in fig. 1). This symmetry is particularly useful for particle phenomenology, as it could shed light on the structure of the quark and lepton mass matrices. Besides, all orbifold models exhibit target-space modular symmetries [40, 41] that act as discrete transformations on the moduli, winding numbers and momenta. In the present model, they are $[SL(2,\mathbb{Z}) \times \Gamma_1(3) \times \Gamma_0(4)]_T \times [\Gamma^0(4)]_U$ transforming the three Kähler T_i and one complex structure U moduli, respectively.[4] These discrete transformations do not only ensure (and help to verify) the consistency of the theory, but are a great tool for moduli fixing, as we shall see.

3. Fixing the moduli

In order to stabilize the moduli, a full knowledge of the effective supergravity theory is required. Remarkably, heterotic orbifold compactifications are the only scenario in which all relevant quantities can be computed explicitly from string theory. Perturbatively, the effective theory receives contributions from nonvanishing couplings between matter states that can be found by applying the so-called selection rules [27, 28], based on the string symmetries that are left unbroken by the orbifold action. The coupling strengths have been worked out using conformal field theory techniques [42, 43, 44] finding that they are exponentially suppressed by the Kähler moduli. Perturbative α' corrections do not contribute to the tree-level superpotential [45]. At nonperturbative level, the main contribution to the effective theory comes from gaugino condensation, which is affected by threshold corrections due to the decoupling of exotic matter and string excitations [46, 47].

To simplify somewhat our discussion, we adopt the following assumptions:

- all (vectorlike) exotics decouple consistently with supersymmetry at a unique scale M_d, such that $(M_{GUT} \sim) M_d \lesssim M_{string} \lesssim 1$;[5]

[2] It has been checked that e.g. operators $n_i^x\ \ell_j\ \bar{\ell}_k$ exist, implying that the additional four (exotic) pairs $\ell_i - \bar{\ell}_i$ attain large masses once some n_i develop VEVs preserving $\mathcal{N}=1$ supersymmetry and the three desirable lepton doublets. All other exotic fields ($\bar{d}_i, d_i, \bar{\delta}_i, \delta_i, s_i^{\pm}$ and m_i) are decoupled analogously.

[3] We point out that there are many other models which do exhibit matter parity [6],[37, app. E].

[4] Details on these symmetries can be found in [23, app. A.4].

[5] In fact, this assumption has been confirmed in several cases [6, 35].

- at some scale close to M_d, the charged hidden matter $\widetilde{N}_i, N, \bar{N}$ are decoupled too, yielding hidden-sector gaugino condensation at the scale $\Lambda_{gc} \lesssim M_d$;
- nonrenormalizable couplings among matter fields are negligible;
- some SM singlets (those present in admissible renormalizable couplings) n_i do not enter in the process of decoupling of unwanted matter and attain VEVs $A_i \ll 1$. It follows that matter kinetic terms proportional to $|A_i|^2$ in the Kähler potential can be safely ignored;
- no twisted moduli nor blow-up modes appear, so that we can remain at the orbifold point;
- D-term contributions to the scalar potential are negligible.

Under these conditions, the (large-volume) Kähler potential of the model introduced in sec. 2 is computable [29, 41, 48] and given by

$$K = -\log\left[S + \bar{S} - \frac{19}{24\pi^2}\log(T_1 + \bar{T}_1) + \frac{7}{24\pi^2}\log(T_2 + \bar{T}_2)\right] - \sum_j \log(\phi_j + \bar{\phi}_j)\,, \tag{1}$$

where we have included all loop effects and denoted all bulk moduli by $\phi_j = \{T_1, T_2, T_3, U, S\}$.

The superpotential is split in the perturbative bit coming from matter couplings W_{yuk} and the gaugino condensation part W_{gc}, so that we obtain for our model

$$W_{yuk} = 2\,N_{255}\,A_1^2\,e^{-2\pi\,T_2/3}\left(A_2 + A_3 e^{-2\pi\,T_1/3}\right) + \ldots\,, \tag{2a}$$

$$\begin{aligned} W_{gc} = & -\frac{c}{e}\frac{3}{16\pi^2}\,e^{-4\pi^2 S/3}\,M_d^{3/2}\,\eta(3\,T_2)^{-16/9}\,\left[\eta(4\,T_3)\,\eta(U/4)\right]^{1/3} \\ & -\frac{c}{e}\frac{3}{32\pi^2}\,e^{-8\pi^2 S/3}\,M_d^{10/3}\,\eta(3\,T_2)^{4/9}\,\left[\eta(4\,T_3)\,\eta(U/4)\right]^{2}\,, \end{aligned} \tag{2b}$$

where c is an unknown constant arising from integrating out the condensate, $\eta(p_i\phi_i)$ is the well-known Dedekind function, and $N_{255} \approx 1.6$ comes from the explicit computation of string worldsheet instantons. Two remarks are in order: 1) the ellipsis in W_{yuk} contains terms with higher powers in e^{-T_i} which we neglect; and 2) modular symmetries control the moduli dependence of the eta functions, e.g. only $\eta(4\,T_3)$ transforms covariantly under the unbroken $\Gamma_0(4)$ modular symmetry. The two contributions to W_{gc} come from the two hidden groups, SO(8) and SU(3) respectively.

The structure of the superpotential is very rich. First, $W(S)$ takes a racetrack form which can indeed lead to dilaton stabilization [20, 21] (although it must be noticed that the SU(3) contribution is subleading mainly due to the powers of M_d). Second, $W(T_1)$ is KKLT-like which is known to yield potentially stable vacua with no or little fine-tuning. Third, $W(T_2, T_3, U)$ is a complicated combination of cusp forms, which exhibits several critical points. We would like to point out that the inclusion of matter fields in W could be crucial for arriving to a de Sitter vacuum [49]. It is noticeable that, contrary to what happens in toy models in which there are gazillions of tunable parameters, the only parameters left free to tune in this promising model are A_i, M_d and c, connected to the decoupling of unwanted states and the gaugino condensates. Therefore, the next task is just to compute the scalar potential and locate a minimum by varying these few free parameters around admissible values: $c \sim 1$, $A_i \sim M_d \ll 1$.

Applying eqs. (1) and (2) to the general expression of the F-term potential, $V_F = e^K\left(K^{A\bar{B}} D_A W D_{\bar{B}}\overline{W} - 3|W|^2\right)$ (the indices A, B label all chiral supermultiplets present and $D_A W = \partial_A W + W\,\partial_A K$), we have computed the scalar potential of the model. In the limits of our assumptions and with $c = 1/10$, $20A_1 = 100A_2 = A_3 = 1/10$, $M_d = 1/65$, we have found numerically dozens of vacua out of which we chose the one shown in tab. 2. The scalar potential is plotted in fig. 2. Unfortunately, all vacua are unstable for they turn out to have at least one

Table 2: An unstable de Sitter solution of an orbifold MSSM candidate. The dominant contribution to the mass eigenstate ρ_i arises from the modulus $\phi_i \in \{T_1, T_2, T_3, U, S\}$. as can also be appreciated in fig. 2. We give the solution to 5 significant figures, but we have computed it to a precision of 1000. The value of Λ is given in Planck units. The tachyonic eigenstate is given by $\text{Re}\rho_2 \sim 0.9\text{Re}T_2 + 0.4\text{Re}S$.

Parameters			Cosmological constant $\Lambda = 6.45 \times 10^{-19}$		
$c = 1/10$	$20A_1 = 100A_2 = A_3 = 1/10$	$M_d = 1/65$			
Moduli VEVs			Mass eigenstates		
$\langle\phi_i\rangle$	$\text{Re}\langle\phi_i\rangle$	$\text{Im}\langle\phi_i\rangle$	$\rho_i \sim \phi_i$	$m^2_{\text{Re}\rho_i}$	$m^2_{\text{Im}\rho_i}$
T_1	3.51166	3/2	ρ_1	1.87×10^{-18}	3.57×10^{-18}
T_2	0.24201	$-1/3$	ρ_2	-4.76×10^{-17}	1.2×10^{-17}
T_3	7.05981	35/8	ρ_3	3.15×10^{-20}	3.29×10^{-33}
U	112.95695	-506	ρ_U	9.84×10^{-23}	1.51×10^{-94}
S	0.09385	$1079/144\pi$	ρ_S	1.14×10^{-16}	2.16×10^{-16}

tachyon. In the current model, the unstable direction is dominated by $\text{Re}T_2$. The problem relies most likely in the fact that there are plenty of vacua. Exploring all of them becomes rapidly a numerical issue because of the complicated structure of the potential the models. There are, however, reasons to expect that a stable solution will emerge in a more refined search that will be discussed elsewhere [50].

On the bright side, the vacua we find have the following properties:

- all vacua are de Sitter with small cosmological constant;
- our setting favors anisotropic compactifications, which is one alternative to achieve precision gauge coupling unification and to solve the infamous hierarchy between the string scale and the unification scale M_{GUT};
- the overall volume turns out to be always large, in consistency with the Kähler approximation;
- although difficult, it is possible to achieve larger values of S, close to the expected value of the gauge coupling at M_{GUT}.

We expect these features to be generic consequences of moduli stabilization via gaugino condensation in heterotic orbifolds.

4. Some speculation: cosmological prospects

If a metastable vacuum can be found in these models, there are further questions connected to the dynamics of the moduli that must be answered. A successful string model should be able not only to reproduce the particle physics of our universe, but also to explain the structure and dynamics of the cosmos. One persisting problem is to explain the origin and dynamics of the rapid period of expansion of the observable dimensions, known as inflation (for a nice review, see [51]). Typically, it is assumed that there is some field, the inflaton, whose potential is almost flat and that for some value of the inflaton has a minimum. Many single and multi-field inflationary models have been concocted and lead to acceptable e-folds, primordial fluctuations and nongaussianities.

In the context of string theory, there has been an effort mostly in the type IIB string [52, 53, 54, 55] to derive inflation from first principles. Remarkably, it has been found that, in certain scenarios, admissible fluctuations and nongaussianities are more natural in a universe dominated by multi-field dynamics [56], as is always the case in string constructions. However, the heterotic string has not been explored sufficiently in this regard.

It seems plausible that the experience gathered in type IIB will be enriched by the structures found in heterotic orbifolds. In particular, the present study produces generically at least two

Journal of Physics: Conference Series **287** (2011) 012021 doi:10.1088/1742-6596/287/1/012021

almost flat directions. In the model investigated in this note, they correspond to the directions $\mathrm{Im}T_3$ and $\mathrm{Im}U$. Two fields of this kind seem to accommodate perfectly the scenario introduced in [56], yielding more accessible values for cosmological observables. Therefore, together with the question of moduli fixing, the feasibility of a heterotic inflationary scenario must be investigated.

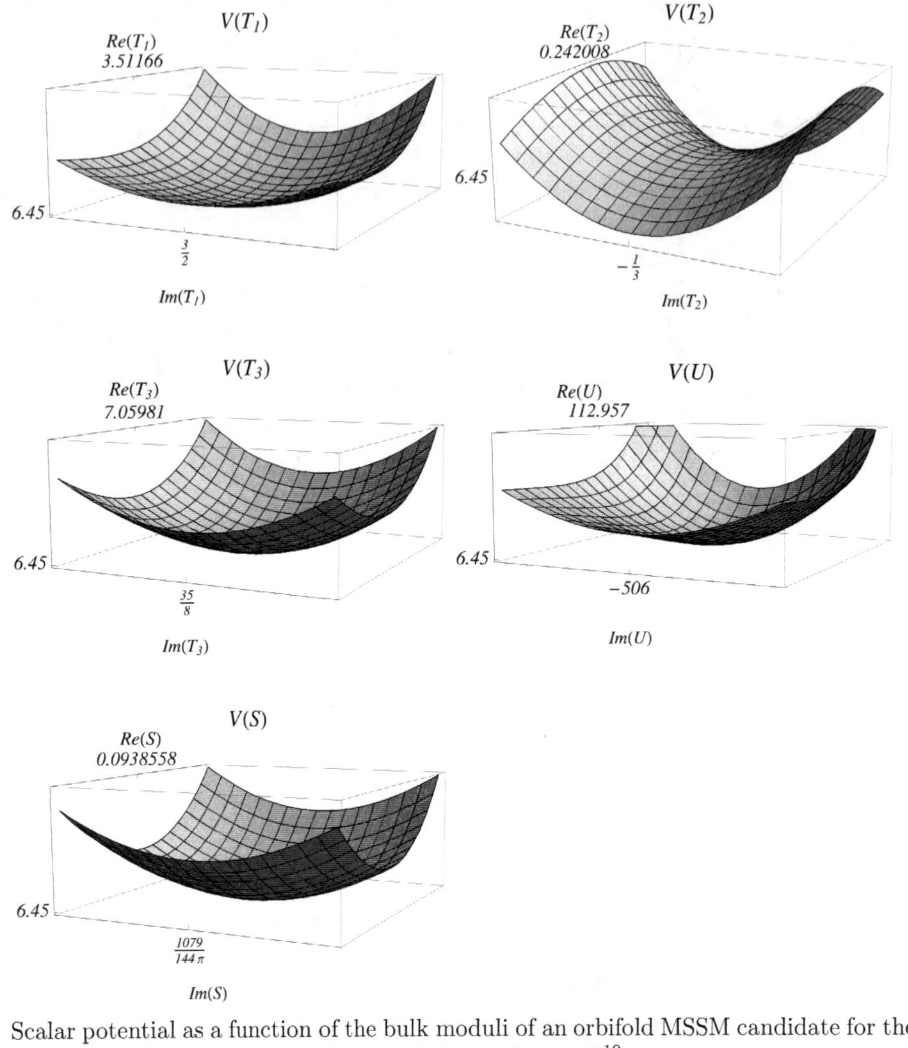

Figure 2: Scalar potential as a function of the bulk moduli of an orbifold MSSM candidate for the vacuum presented in tab. 2. The scalar potential is rescaled by a factor 10^{19}.

References

[1] Ibáñez L E, Kim J E, Nilles H P and Quevedo F 1987 *Phys. Lett.* **B191** 282–286
[2] Faraggi A E, Nanopoulos D V and Yuan K j 1990 *Nucl. Phys.* **B335** 347
[3] Kobayashi T, Raby S and Zhang R J 2005 *Nucl. Phys.* **B704** 3–55 (*Preprint* `hep-ph/0409098`)
[4] Buchmüller W, Hamaguchi K, Lebedev O and Ratz M 2006 *Phys. Rev. Lett.* **96** 121602 (*Preprint* `hep-ph/0511035`)
[5] Buchmüller W, Hamaguchi K, Lebedev O and Ratz M 2007 *Nucl. Phys.* **B785** 149–209 (*Preprint* `hep-th/0606187`)
[6] Lebedev O *et al.* 2008 *Phys. Rev.* **D77** 046013 (*Preprint* `0708.2691`)
[7] Blaszczyk M *et al.* 2010 *Phys. Lett.* **B683** 340–348 (*Preprint* `0911.4905`)

[8] Giddings S B, Kachru S and Polchinski J 2002 *Phys. Rev.* **D66** 106006 (*Preprint* hep-th/0105097)
[9] Becker K, Becker M, Haack M and Louis J 2002 *JHEP* **06** 060 (*Preprint* hep-th/0204254)
[10] Kachru S, Kallosh R, Linde A D and Trivedi S P 2003 *Phys. Rev.* **D68** 046005 (*Preprint* hep-th/0301240)
[11] Anderson L B, Gray J, Lukas A and Ovrut B 2010 (*Preprint* 1010.0255)
[12] de Carlos B, Gurrieri S, Lukas A and Micu A 2006 *JHEP* **03** 005 (*Preprint* hep-th/0507173)
[13] Dixon L J, Harvey J A, Vafa C and Witten E 1985 *Nucl. Phys.* **B261** 678–686
[14] Dixon L J, Harvey J A, Vafa C and Witten E 1986 *Nucl. Phys.* **B274** 285–314
[15] Ferrara S, Lust D, Shapere A D and Theisen S 1989 *Phys. Lett.* **B225** 363
[16] Lauer J, Mas J and Nilles H P 1989 *Phys. Lett.* **B226** 251
[17] Lauer J, Mas J and Nilles H P 1991 *Nucl. Phys.* **B351** 353–424
[18] Cvetič M, Font A, Ibáñez L E, Lüst D and Quevedo F 1991 *Nucl. Phys.* **B361** 194–232
[19] Dundee B, Raby S and Westphal A 2010 *Phys. Rev.* **D82** 126002 (*Preprint* 1002.1081)
[20] de Carlos B, Casas J A and Muñoz C 1991 *Phys. Lett.* **B263** 248–254
[21] de Carlos B, Casas J A and Muñoz C 1993 *Nucl. Phys.* **B399** 623–653 (*Preprint* hep-th/9204012)
[22] Bailin D, Love A, Sabra W A and Thomas S 1994 *Mod. Phys. Lett.* **A9** 2543–2555 (*Preprint* hep-th/9405031)
[23] Parameswaran S L, Ramos-Sánchez S and Zavala I 2011 *accepted in JHEP* (*Preprint* 1009.3931)
[24] Lebedev O *et al.* 2007 *Phys. Lett.* **B645** 88–94 (*Preprint* hep-th/0611095)
[25] Lebedev O, Nilles H P, Ramos-Sánchez S, Ratz M and Vaudrevange P K S 2008 *Phys. Lett.* **B668** 331–335 (*Preprint* 0807.4384)
[26] Lebedev O *et al.* 2007 *Phys. Rev. Lett.* **98** 181602 (*Preprint* hep-th/0611203)
[27] Dixon L J, Friedan D, Martinec E J and Shenker S H 1987 *Nucl. Phys.* **B282** 13–73
[28] Hamidi S and Vafa C 1987 *Nucl. Phys.* **B279** 465
[29] Dixon L J, Kaplunovsky V and Louis J 1990 *Nucl. Phys.* **B329** 27–82
[30] Lebedev O and Ramos-Sánchez S 2010 *Phys. Lett.* **B684** 48–51 (*Preprint* 0912.0477)
[31] Dundee B, Raby S and Wingerter A 2008 *Phys. Rev.* **D78** 066006 (*Preprint* 0805.4186)
[32] Anandakrishnan A and Raby S 2011 (*Preprint* 1101.1976)
[33] Buchmüller W, Hamaguchi K, Lebedev O, Ramos-Sánchez S and Ratz M 2007 *Phys. Rev. Lett.* **99** 021601 (*Preprint* hep-ph/0703078)
[34] Hosteins P, Kappl R, Ratz M and Schmidt-Hoberg K 2009 *JHEP* **07** 029 (*Preprint* 0905.3323)
[35] Lee H M *et al.* 2011 *Phys. Lett.* **B694** 491–495 (*Preprint* 1009.0905)
[36] Förste S, Nilles H P, Ramos-Sánchez S and Vaudrevange P K S 2010 *Phys. Lett.* **B693** 386–392 (*Preprint* 1007.3915)
[37] Ramos-Sánchez S 2009 *Fortsch. Phys.* **10** 907–1036 (*Preprint* 0812.3560)
[38] Schwetz T, Tortola M A and Valle J W F 2008 *New J. Phys.* **10** 113011 (*Preprint* 0808.2016)
[39] Raby S and Ramos-Sánchez S 2011 *work in progress*
[40] Ferrara S, Lüst D and Theisen S 1989 *Phys. Lett.* **B233** 147
[41] Ibáñez L E and Lüst D 1992 *Nucl. Phys.* **B382** 305–364 (*Preprint* hep-th/9202046)
[42] Burwick T T, Kaiser R K and Müller H F 1991 *Nucl. Phys.* **B355** 689–711
[43] Erler J, Jungnickel D, Spalinski M and Stieberger S 1993 *Nucl. Phys.* **B397** 379–416 (*Preprint* hep-th/9207049)
[44] Choi K S and Kobayashi T 2008 *Nucl. Phys.* **B797** 295–321 (*Preprint* 0711.4894)
[45] Dine M and Seiberg N 1985 *Phys. Rev. Lett.* **55** 366
[46] Dixon L J, Kaplunovsky V and Louis J 1991 *Nucl. Phys.* **B355** 649–688
[47] Mayr P and Stieberger S 1993 *Nucl. Phys.* **B407** 725–748 (*Preprint* hep-th/9303017)
[48] Lüst D and Muñoz C 1992 *Phys. Lett.* **B279** 272–280 (*Preprint* hep-th/9201047)
[49] Lebedev O, Nilles H P and Ratz M 2006 *Phys. Lett.* **B636** 126–131 (*Preprint* hep-th/0603047)
[50] Akrami Y *et al.* 2011 *work in progress*
[51] Mazumdar A and Rocher J 2011 *Phys. Rept.* **497** 85–215 (*Preprint* 1001.0993)
[52] Kallosh R and Linde A D 2004 *JHEP* **12** 004 (*Preprint* hep-th/0411011)
[53] Lalak Z, Ross G G and Sarkar S 2007 *Nucl. Phys.* **B766** 1–20 (*Preprint* hep-th/0503178)
[54] Covi L *et al.* 2008 *JHEP* **08** 055 (*Preprint* 0805.3290)
[55] Conlon J P, Kallosh R, Linde A D and Quevedo F 2008 *JCAP* **0809** 011 (*Preprint* 0806.0809)
[56] Burgess C P *et al.* 2010 *JHEP* **08** 045 (*Preprint* 1005.4840)

Journal of Physics: Conference Series **287** (2011) 012022 doi:10.1088/1742-6596/287/1/012022

MINERνA Project (FNAL e938)'s Mexico Contribution

J. Félix [1], G. Zavala [2], Z. Urrutia [1], A. Higuera [1], E. Valencia[1], and MINERνA[3] Collaboration

[1] Laboratorio de Partículas. Departamento de Física. DCeI, Campus León. Universidad de Guanajuato. México.

E-mail: felix@fisica.ugto.mx

[2] Departamento de Ciencias administrativas, Campus Guanajuato. Universidad de Guanajuato. México.

E-mail: gzg@quijote.ugto.mx

[3] http://minerva.fnal.gov.

Abstract. The MINERνA Project (http://minerva.fnal.gov) (Main INjector ExpeRimentνA) is an experiment that uses Fermilab NuMI line. Its main goals are measure the interactions neutrino (antineutrino)-Nucleon at low energies, improve neutrino oscillation studies, study the strong dynamics between nucleons and between nuclei (nucleons) and neutrinos, and between nuclei (nucleons) and anti-neutrinos. I report on the current status of MINERνA experiment, studies currently under way, studies that can be done, and the Mexican (Universidad de Guanajuato) participation in MINERνA experiment.

1. Introduction

Mexican group, at Universidad de Guanajuato, in November 2007 started MINERνA collaboration (http://minerva.fnal.gov). Universidad de Guanajuato MINERνA collaborators in the past three years are: Professors Julian Felix and Gerardo Zavala; PhD students Zaida Urrutia, Aaron Higuera and Ranferi Gutierrez; Msc students Edgar Valencia, Aaron Higuera and Jorge Castorena; and Bsc students Cesar Capetillo, Luis Balcazar and Maria Cristina Zarazua. Our group, including our students, contributed to installation and commissioning of the MINERνA detector, which was completed in March 2010 and to construction and operation of the MINERνA test beam experiment which completed its data taking in July 2010. Many of our students have already obtained degrees from the work on MINERνA: Aaron Higuera, Jorge Castorena graduated getting their MSc degrees in physics based on commissioning of the test beamline and the MINERνA tracking prototype detector, respectively. Julio Cesar Capetillo and Luis Alfonso Balcazar had graduated with BSc degrees in mechanical engineering with projects on the MINERνA test beamline. Edgar Valencia will graduate in March 2011 with an MSc thesis on the three dimensional simulation of the dipole MINERνA Magnets. Maria Cristina Zarazua will receive a BSc in 2011 in part based on work comparing MINERνA detector simulations and data.

A key part of the experience of each of these students has been a period of residence at Fermilab which is realized through generous support for the students from Fermilab. Professors

Journal of Physics: Conference Series **287** (2011) 012022 doi:10.1088/1742-6596/287/1/012022

Felix and Zavala also are working to improve the experience for the students on campus at Universidad de Guanajuato. We are working towards installation of small cluster of computers for data analysis for MINERνA collaboration. A more ambitious future plan is to develop a high energy detector laboratory for teaching and research, which includes a nearly completed laboratory for high performance computation. Part of the student work in this lab will be analysis of MINERνA data and MINERνA Monte Carlo simulations.

Since neutrino interaction first measurement in 1956, the field on neutrino physics has increased in both experimental and theoretical efforts and in interest. These are some of the actual problems in neutrino physics: The nature of neutrino -Dirac or Majorana-: Is the neutrino exactly equal to the anti-neutrino? The mass of the neutrino. The CP violation in the lepton sector. The mixing parameter. The oscillation of neutrinos.

Some of the proposed experiments in the world, to study neutrinos, are as follows: MINOS(http://www-numi.fnal.gov/); BooNE(http://www-boone.fnal.gov/); K2K(http://neutrino.kek.jp/); T2K(http://jnusrv01.kek.jp/public/t2k/); NOνA(http://www-nova.fnal.gov/); OPERA(http://www.nu.to.infn.it/exp/all/opera/); LBNE(http://lbne.fnal.gov/); and MINERνA(http://minerva.fnal.gov/). More accurate, and technical, information about these experiments are in their respective web page. MINERνA experiment is an international collaboration, it comprehend 21 institutions and 80 physicists from USA, Russia, Mexico, Peru, Chile, Brazil.

In the next sections, I describe MINERνA collaboration and emphasize on Mexican contribution accomplishments.

2. MINERνA Project

The project MINERνA (Main INjector ExpeRimentνA) is an experiment that uses the Fermilab NuMI line. The main goals of MINERνA experiments are: To measure neutrino (antineutrino)-nucleon cross sections, at low energies, with high statistics, to improve neutrino oscillation measurements. To measure interactions neutrino-nucleus -and to measure interactions anti neutrino-nucleus.- as a whole. This process is called coherent neutrino-nucleus scattering. Figure 1 shows MINERνA Main detector. Additionally a test beam detector was constructed and commissioning to obtain calibration parameters for MINERνA detector. Figure 2 shows MINERνA test beam line.

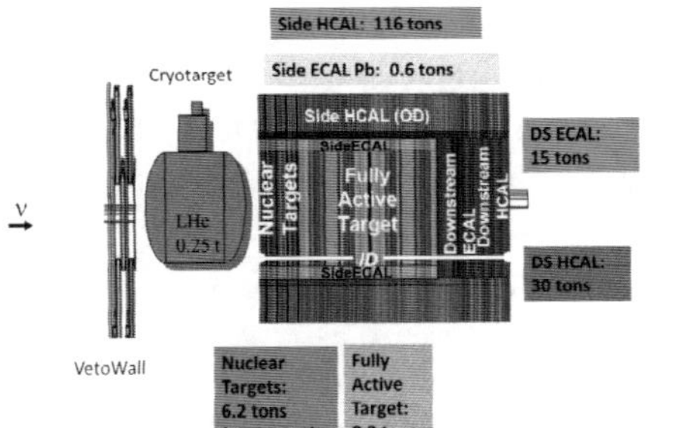

Figure 1. Layout of Minerva main detector.

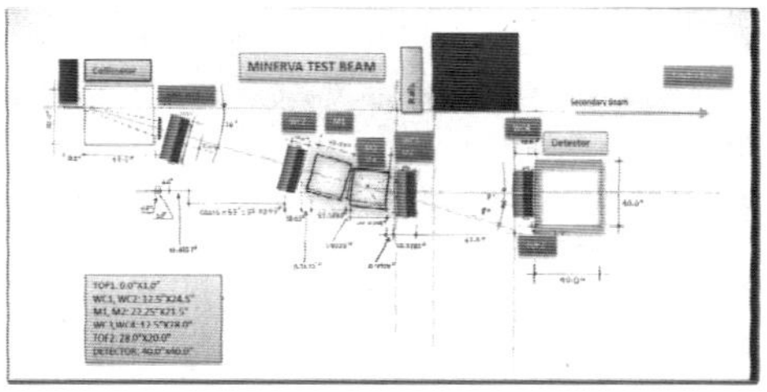

Figure 2. Layout of Minerva test beam line.

2.1. MINERνA Current Status

Both detectors are assembled and taking data. By March, 2010, the MINERνA main detector was completed and commissioned; the MINERνA test beam detector was completed and commissioned by July, 2010. MINERνA Monte Carlo is under current development and improvement. Figure 3 shows MINERνA main detector completely installed; and Figure 4 shows MINERνA test beam line completely installed.

Figure 3. Photo of MINERνA main detector installed.

Figure 4. Photo of MINERνA test beam line completely installed.

The main MINERνA detector is 105 meters under ground in the line of Fermilab NUMI, upstream of MINOS detector.

The main characteristics of MINERνA detector are as follows: It is fine grained scintillator, 120 segmented planes, time resolution is about 4 ns, spatial resolution about 3 mm, covers a wide atomic numbers -He, C, H2O, Fe, Pb-. It has recorded 1.37×10^20 POT; and can perform $\frac{d\sigma}{dE}$ measurements in QE, DIS, and $s\pi$ production modes.

MINERνA detector is running in two different modes: antineutrino beam, and neutrino beam. And at three different energies: Low energy configuration, the most probable energy is around 3 GeV, and the rate of production data is about 60 000 events/tone-10^20 POT. Medium Energy configuration, the most probable energy is around 7.0 GeV, and the rate of production data is about 230 000 events/tone-10^20 POT. High Energy configuration, the most probable energy is around 12.0 GeV, and the rate of production data is about 525 000 events/tone-10^20 POT. The neutrino flux is known with 5% of uncertainty.

MINOS detector gives muon momentum and charge.

Figure 5 and Figure 6 give another views of MINERν detector.

2.2. MINERνA Physics

The physics topics, among others, that MINERνA can address are the following: Rock muons, Michel electrons. Neutrino oscillations. Neutrino physical properties. Quasi elastic interactions. Resonances. Baryon polarization. Coherent production of pions. Interactions neutrino

Journal of Physics: Conference Series **287** (2011) 012022 doi:10.1088/1742-6596/287/1/012022

Figure 5. Another view of MINERνA main detector installed.

Figure 6. Another view of MINERνA main detector installed.

e-nucleon. Interactions neutrino nucleon (H20, He, Pb, C, Fe). Neutral currents. Charged Currents.

Figure 7 and Figure 8 show some real data.

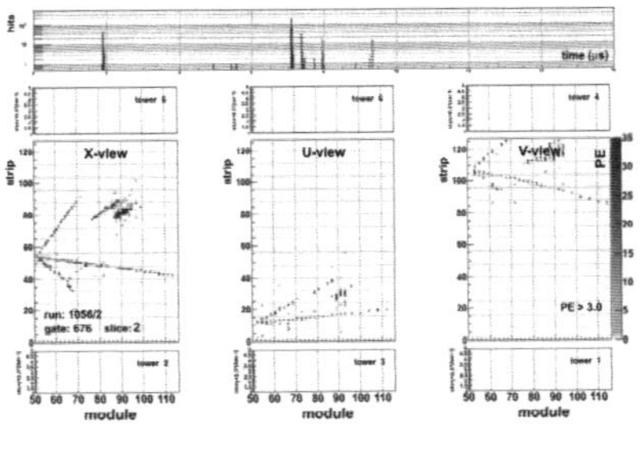

Figure 7. MINERνA data.

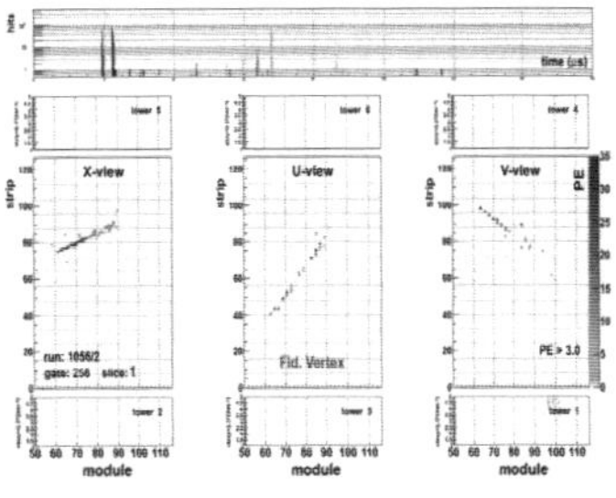

Figure 8. MINERνA data.

Figure 9 and Figure 10 show some real data.

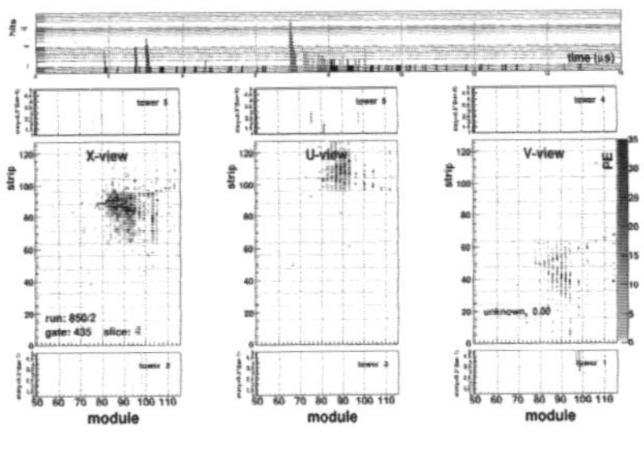

Figure 9. MINERνA data.

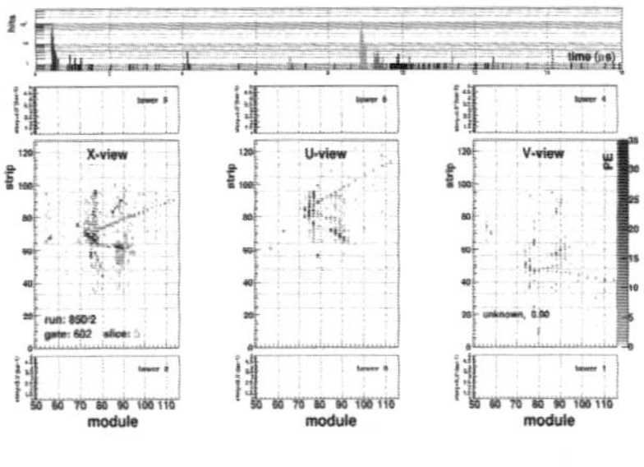

Figure 10. MINERνA data.

Figure 11 and Figure 12 show some real data.

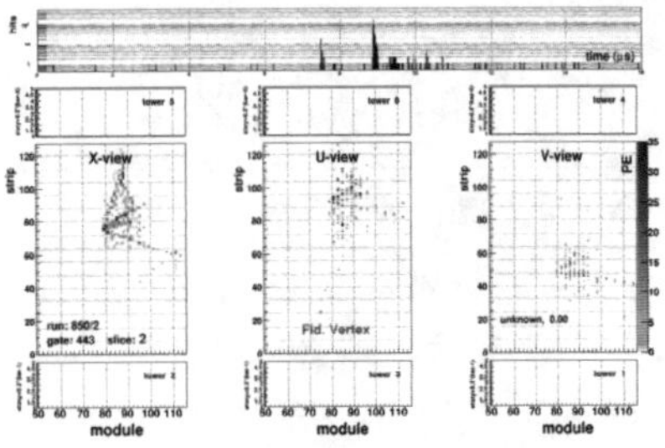

Figure 11. MINERνA data.

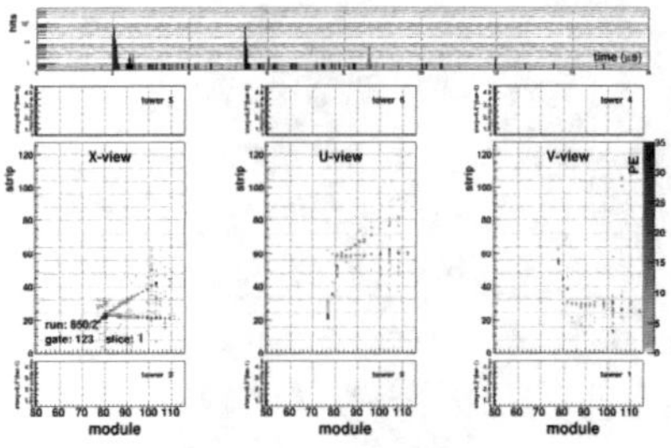

Figure 12. MINERνA data.

Figures 13 and 14 show some Monte Carlo data.

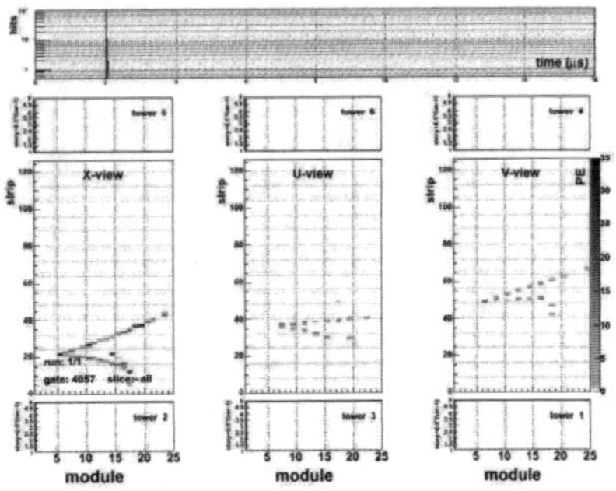

Figure 13. MINERνA Monte Carlo data.

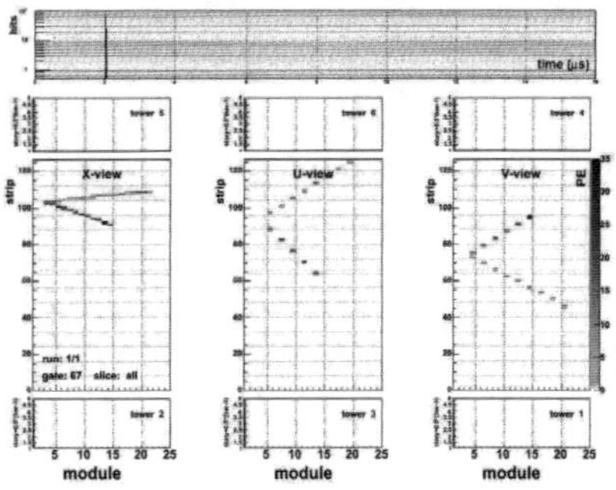

Figure 14. MINERνA Monte Carlo data.

3. Mexican Participation

The Mexican group participating in MINERνA collaboration, so far, is from the Universidad de Guanajuato. We participate since 2007, and currently this group includes two professors -Gerardo Zavala and Julian Felix-, two Ph. D students, one MsC student and five Bachelor students.

Figure 15 shows some students, and professor, at Fermilab.

3.1. MINERνA Detector

Mexican group participated in assembling the MINERνA detector (one MsC student and one Ph. D student), in the plane detector (two BsC students), in data taking (the whole group) and in the analysis of data (the whole group).

3.2. MINERνA Test Beam Line

The group from Guanajuato has participated in the Test Beam Line doing the following: planning (Julian Felix), designing (Julian Felix), installing (Julian Felix), studying the time of flight

Figure 15. Students and Professors at Fermilab.

(Zaida Urrutia, Julian Felix), designing a Cerekov detector (Luis Balcazar, Julio Capetillo, Julian Felix), simulating the magnets (Julian Felix, Edgar Valencia), taking data and analyzing data (the whole group).

More details about MINERνA dipole magnets simulations are given by Edgar Valencia paper, in this proceedings.

3.3. Data Processing

The infrastructure we are creating for data processing, both, real data and Monte Carlo data. The cluster assembled in Leon, in Guanajuato, and the small one at Leon.

To process and analyze data we have created a prototype of cluster, formed by 5 laptops, see Figure. We have installed Condor (http://www.cs.wisc.edu/condor/) and we are testing it. It has 9 processors with different speed, 9 GB of Ram, 1.5 TB hard disk space, and a local network of 10/100 MHz transferred speed.

Figures 16 and 17 show a prototype of cluster operated with condor.

Also we have created two clusters, almost similar, to connect as a GRID -260 processors,

Journal of Physics: Conference Series **287** (2011) 012022 doi:10.1088/1742-6596/287/1/012022

Figure 16. Universidad de Guanajuato cluster prototype, one view.

Figure 17. Universidad de Guanajuato cluster prototype, another view.

300 GB of memory, and 10 TB of storage. With these, we are planning process the data and analyzed data.

3.4. Data Analysis

Lambda polarization studies are possible in MINERνA data. Neutrino (nucleon) going to Lambda0 X; anti neutrino (nucleon) going to Lambda0 X. The polarization is measured by fitting to straight line the cosine of the angle between proton momentum from Lambda 0 and the normal of the production plane, in the coordinate system where Lambda 0 is at rest.

Aaron Higuera, Ph. D. Student, is working on neutrino-nucleus targets interactions. He is measuring cross section of the neutrino target interactions as function of neutrino energy and atomic mass (He, H2O, Pb, C, Fe).

Zaida Urrutia is working on vertex activity. What happens on the vertex of interaction of the neutrino and the target?

4. Conclusions

Mexican group has participated on MINERνA detector installation, on Test beam line, on Data analysis, on data processing.

These activities have impacted on the creation of infrastructure -a laboratory for particle physics, and two clusters and one GRID for processing and analyzing data- and human resources -BSc, MSc, and PhD students-. This collaboration will continue for some years ahead.

5. References

[1] A. Higuera et al. MINERνA test beam commissioning. By MINERνA Collaboration. 2009. 4pp. Prepared for CIPANP 2009: 10th Conference on the Intersections of Particle and Nuclear Physics, San Diego, California, 26-31 May 2009. Published in AIP Conf.Proc.1182:112-115,2009.

[2] J. Felix et al. Simulation and characterization of the MINERνA dipole magnets. By MINERνA Collaboration. 2009. 4pp. Prepared for CIPANP 2009: 10th Conference on the Intersections of Particle and Nuclear Physics, San Diego, California, 26-31 May 2009. Published in AIP Conf.Proc.1182:108-111,2009.

[3] J. Castorena et al. Commissioning of the MINERνA tracking prototype. By MINERνA Collaboration. 2009. 4pp. Prepared for CIPANP 2009: 10th Conference on the Intersections of Particle and Nuclear Physics, San Diego, California, 26-31 May 2009. Published in AIP Conf.Proc.1182:104-107,2009.

[4] J. Felix et al. The FNAL E938 experiment: The Mexican contribution to the MINERνA Collaboration. By MINERνA Collaboration. 2009. 4pp. Published in AIP Conf.Proc.1116:223-226,2009.

[5] Communication using neutrinos. By MINERνA Collaboration. To be send to Nature Letters.

6. Acknowledgements

CONACyT, Mexico, Grant No. 80249, Ciencia Básica, 2008.
MINERνA Collaboration (http://minerva.fnal.gov).

Journal of Physics: Conference Series **287** (2011) 012023 doi:10.1088/1742-6596/287/1/012023

Functional approaches to infrared Yang-Mills theory in the Coulomb gauge

A Weber[1,5], M Leder[2], J M Pawlowski[3,4] and H Reinhardt[2]

[1] Instituto de Física y Matemáticas, Universidad Michoacana de San Nicolás de Hidalgo, Edificio C-3, Ciudad Universitaria, 58040 Morelia, Michoacán, Mexico
[2] Institut für Theoretische Physik, Universität Tübingen, Auf der Morgenstelle 14, 72076 Tübingen, Germany
[3] Institut für Theoretische Physik, Universität Heidelberg, Philosophenweg 16, 69120 Heidelberg, Germany
[4] ExtreMe Matter Institute EMMI, GSI Helmholtzzentrum für Schwerionenforschung, Planckstr. 1, 64291 Darmstadt, Germany

E-mail: `axel@ifm.umich.mx`

Abstract. We present the current status of ongoing efforts to use functional methods, Dyson-Schwinger equations and functional renormalization group equations, for the description of the infrared regime of nonabelian (pure) gauge theories. In particular, we present a new determination of the color-Coulomb potential with the help of the functional renormalization group that results in an almost linearly rising potential between static color charges at large spatial distances.

1. Introduction

Important progress has been achieved over the last decade in the description of the deep infrared region of nonabelian gauge theories with the help of functional methods, employing Coulomb gauge fixing. By functional methods we refer to semi-analytical tools that do *not* make use of the discretization of space-time as does lattice gauge theory. Specifically, equations of Dyson-Schwinger-type arising from a variational principle have been used, and more recently functional renormalization group equations. In this contribution, we will report on the current status of these investigations. We will focus exclusively on pure gauge theories, more specifically SU(N) Yang-Mills theory, but include static color charges so as to obtain a description of the heavy quark potential, as in quenched approximations.

We will start by briefly describing the general theoretical setup: the Hamiltonian framework is used, where the Weyl *and* Coulomb gauge conditions, $A_0^a(\mathbf{x}) = 0$ and $\nabla \cdot \mathbf{A}^a(\mathbf{x}) = 0$, are imposed on the SU(N) gauge fields. Physical states are described by wave functionals of $\mathbf{A}^a(\mathbf{x})$ with scalar product

$$\langle \phi | \psi \rangle = \int D[\mathbf{A}]\, J[\mathbf{A}]\, \phi^*[\mathbf{A}]\, \psi[\mathbf{A}] \,. \qquad (1)$$

Here, $J[\mathbf{A}]$ stands for the Faddeev-Popov (FP) determinant $J[\mathbf{A}] = \mathrm{Det}\,(-\nabla \cdot \mathbf{D})$ with the spatial covariant derivative $\mathbf{D}^{ab} = \delta^{ab}\nabla + g f^{abc}\mathbf{A}^c(\mathbf{x})$. The functional integral in (1) is understood to be

[5] Speaker.

restricted to spatially transverse gauge fields, i.e. those that fulfill the gauge fixing conditions.

The dynamics is defined by the Christ-Lee Hamiltonian H [1] that we do not write out. In the presence of a static color charge density $\rho_q^a(\mathbf{x})$, H contains the interaction term

$$H_q = \frac{1}{2} \int d^3x\, d^3y\, \rho_q^a(\mathbf{x}) F^{ab}(\mathbf{x}, \mathbf{y}) \rho_q^b(\mathbf{y}) \tag{2}$$

with the integral kernel

$$F^{ab}(\mathbf{x}, \mathbf{y}) = \langle \mathbf{x}, a | (-\nabla \cdot \mathbf{D})^{-1} (-\nabla^2) (-\nabla \cdot \mathbf{D})^{-1} | \mathbf{y}, b \rangle \,. \tag{3}$$

The vacuum expectation value $\langle F^{ab}(\mathbf{x}, \mathbf{y}) \rangle$ is called the color-Coulomb potential. It is suppposed to give the dominant contribution to the confining interaction between color charges. More precisely, for large spatial distances the color-Coulomb potential provides an upper bound for the Wilson potential [2].

For the following, it will be convenient to write the FP determinant in a local form by introducing ghost fields,

$$J[\mathbf{A}] = \mathrm{Det}\,(-\nabla \cdot \mathbf{D}) = \int D[\bar{c}, c] \, \exp\left(- \int d^3x\, \bar{c}^a(\mathbf{x}) (-\nabla \cdot \mathbf{D}^{ab}) c^b(\mathbf{x}) \right) . \tag{4}$$

In our analysis, we will focus on the equal-time correlation functions, i.e. the vacuum expectation values of products of the field operators $\mathbf{A}^a(\mathbf{x})$ (transverse), $c^a(\mathbf{x})$ and $\bar{c}^a(\mathbf{x})$. We can easily write down an expression for the generating functional of these correlation functions,

$$\begin{aligned} Z[\mathbf{J}, \eta, \bar{\eta}] = \int D[\bar{c}, c, \mathbf{A}] \, e^{-\int d^3x\, \bar{c}(-\nabla \cdot \mathbf{D}) c} \, |\psi[\mathbf{A}]|^2 \\ \times \exp\left(\int d^3x \left[\mathbf{J}^a(\mathbf{x}) \cdot \mathbf{A}^a(\mathbf{x}) + \bar{c}^a(\mathbf{x}) \eta^a(\mathbf{x}) + \bar{\eta}^a(\mathbf{x}) c^a(\mathbf{x}) \right] \right) , \end{aligned} \tag{5}$$

where $\psi[\mathbf{A}]$ is the vacuum wave functional. If we formally define an "action" $S[\mathbf{A}]$ through $|\psi[\mathbf{A}]|^2 = e^{-S[\mathbf{A}]}$, (5) looks like the usual generating functional of Euclidean Green's functions in the covariant Lagrangian formulation of the theory, only in three instead of four dimensions. Of course, $S[\mathbf{A}]$ is a complicated and a priori unknown functional of $\mathbf{A}^a(\mathbf{x})$. We will parametrize the "propagators", the equal-time two-point correlation functions of the theory, in the most general way (restricted by symmetries) as follows:

$$\left\langle A_i^a(\mathbf{p})\, A_j^b(-\mathbf{q}) \right\rangle = \frac{1}{2\omega(p)} \delta^{ab} \left(\delta_{ij} - \frac{p_i p_j}{p^2} \right) (2\pi)^3 \delta(\mathbf{p} - \mathbf{q}) \,, \tag{6}$$

$$\left\langle c^a(\mathbf{p})\, \bar{c}^b(-\mathbf{q}) \right\rangle = \langle \langle \mathbf{p}, a | (-\nabla \cdot \mathbf{D})^{-1} | \mathbf{q}, b \rangle \rangle = \frac{d(p)}{p^2} \delta^{ab} \, (2\pi)^3 \delta(\mathbf{p} - \mathbf{q}) \,. \tag{7}$$

Here and in the following, we use the notation $p = |\mathbf{p}|$. The functions $\omega(p)$ and $d(p)$ will be of central interest in the rest of this contribution. Notice that the ghost propagator (7) is just the vacuum expectation value of the inverse FP operator (or rather, its integral kernel).

2. Variational principle: Dyson-Schwinger equations

A set of equations of Dyson-Schwinger-type for the equal-time correlation functions of the theory was obtained in Ref. [3] from the variational principle, using a Gaussian ansatz for the vacuum wave functional. The contribution of the FP determinant was fully taken into account in [4, 5]. We write the ansatz for the vacuum functional as

$$|\psi[\mathbf{A}]|^2 = e^{-\widetilde{S}[\mathbf{A}]}\,, \qquad \widetilde{S}[\mathbf{A}] = \frac{1}{2}\int \frac{d^3p}{(2\pi)^3}\, A_i^a(-\mathbf{p})\, 2\widetilde{\omega}(p)\, A_i^a(\mathbf{p})\,. \tag{8}$$

Then the variational principle with respect to the unknown function $\widetilde{\omega}(p)$,

$$\frac{\delta}{\delta\widetilde{\omega}(p)}\langle H\rangle = 0\,, \tag{9}$$

leads to a gap equation for the equal-time gluon propagator. The detailed form of the equation as well as the approximations involved in its derivation can be found in [4, 5].

The gap equation involves, apart from the gluon propagator, the ghost propagator and the color-Coulomb potential, hence further input is needed in order to arrive at a closed system of equations. The generating functional (5) can be used to derive a Dyson-Schwinger (DS) equation for the ghost propagator in the usual way:

$p^2 d^{-1}(p) \equiv \Big($ 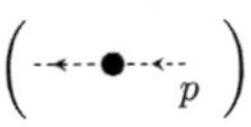 $\Big)^{-1} = Z_c p^2 -$ 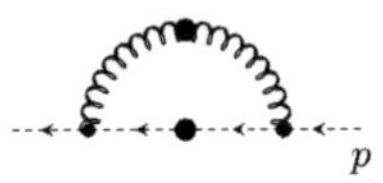 . (10)

In the diagrams, we represent the full equal-time ghost propagator by a dashed line and the gluon propagator by a curly line, with a dot on the lines. By extending Taylor's non-renormalization theorem [6] to the present situation, we have replaced in (10) the full ghost-gluon vertex (one of the vertices on the right-hand side) with the bare one. This replacement is also used in the gap equation.

For the color-Coulomb potential,

$$\left\langle F^{ab}(\mathbf{p},-\mathbf{q})\right\rangle = \langle\langle \mathbf{p},a|(-\nabla\cdot\mathbf{D})^{-1}(-\nabla^2)(-\nabla\cdot\mathbf{D})^{-1}|\mathbf{q},b\rangle\rangle = V_c(p)\,\delta^{ab}\,(2\pi)^3\delta(\mathbf{p}-\mathbf{q})\,, \tag{11}$$

we use the following parameterization and diagrammatic representation motivated by the appearance of the inverse FP operator [cf. (7)]:

$V_c(p) = \dfrac{d(p)}{p^2}\, p^2 f(p)\, \dfrac{d(p)}{p^2} =$  , (12)

thereby defining the Coulomb form factor $f(p)$. Of course, the function $f(p)$ is itself unknown, and before discussing the possibility of determining it in terms of the gluon and ghost propagators, we will resort to the factorization hypothesis [7]

$$\langle(-\nabla\cdot\mathbf{D})^{-1}(-\nabla^2)(-\nabla\cdot\mathbf{D})^{-1}\rangle = \langle(-\nabla\cdot\mathbf{D})^{-1}\rangle\,(-\nabla^2)\,\langle(-\nabla\cdot\mathbf{D})^{-1}\rangle\,, \tag{13}$$

which is equivalent to

$$V_c(p) = \frac{d(p)}{p^2}\, p^2\, \frac{d(p)}{p^2}\,, \tag{14}$$

or $f(p) = 1$. Adopting this (so far unjustified) assumption, we obtain a closed system of equations.

Before disussing the numerical solutions of the equations, we have to comment on the Gribov-Zwanziger confinement scenario [8, 9]. In brief, the idea is that the existence of Gribov copies (gauge-equivalent but different transverse gauge field configurations) forces one to restrict the functional integral over the gauge field to the first Gribov region where the FP operator is positive definite.[6] By a statistical argument, in the infrared (IR) regime the dominant contribution to the functional integral comes from the region close to the Gribov horizon where the FP operator has zero modes. Since the ghost propagator is the vacuum expectation value of the inverse FP operator, it can be argued that the ghost propagator $d(p)/p^2$ should be more singular in the IR than p^{-2}, thus $d^{-1}(p=0)=0$, the "horizon condition" [10]. Hence, one should look for solutions that fulfill the horizon condition.

It turns out that there are two different solutions of this type [4, 12]. Both show scaling behavior in the IR, i.e. the equal-time propagators obey power laws in this kinematical regime. Furthermore, among the different contributions to the gluon propagator in the gap equation, the ghost loop diagram dominates completely the IR behavior, a property known as ghost dominance. These facts make it possible to even obtain analytical solutions for the propagators in the IR region [7, 11]. With the notations

$$\omega(p) = A\, p^{-\alpha}\,, \qquad d(p) = B\, p^{-\beta}\,, \tag{15}$$

one obtains in this way a general sum rule for the IR exponents:

$$\alpha = 2\beta - 1\,. \tag{16}$$

Consistent solutions exist for the values $(\alpha = 0.592, \beta = 0.796)$ and $(\alpha = 1, \beta = 1)$. One may also define a running coupling constant from the ghost-gluon vertex as

$$\alpha(p) = \frac{8}{3}\frac{g^2(p)}{4\pi}\,, \qquad g^2(p) = g_B^2\, \frac{p}{\omega(p)}\, d^2(p) \tag{17}$$

(g_B is the bare coupling constant). In the ultraviolet (UV), the solutions show asymptotic freedom (although not with the correct power of $\ln p$ due to the approximations to the gap equation), while $\alpha(p)$ saturates at a constant value in the IR. Analytically, one obtains $N_c\,\alpha(0) = 11.99$ for the solution with $\beta = 0.796$, and $N_c\,\alpha(0) = 16\pi/3$ for $\beta = 1$ [11].

To close this section, we comment on a possible drastic simplification of the equations: if one uses, instead of the general Gaussian ansatz (8), the lowest-order perturbative vacuum wave functional

$$|\psi[\mathbf{A}]|^2 = e^{-S_0[\mathbf{A}]} = \exp\left(-\frac{1}{2}\int \frac{d^3p}{(2\pi)^3}\, A_i^a(-\mathbf{p})\, 2p\, A_i^a(\mathbf{p})\right) \tag{18}$$

in the generating functional (5), the complicated gap equation may be replaced by the following DS equation for the gluon propagator:

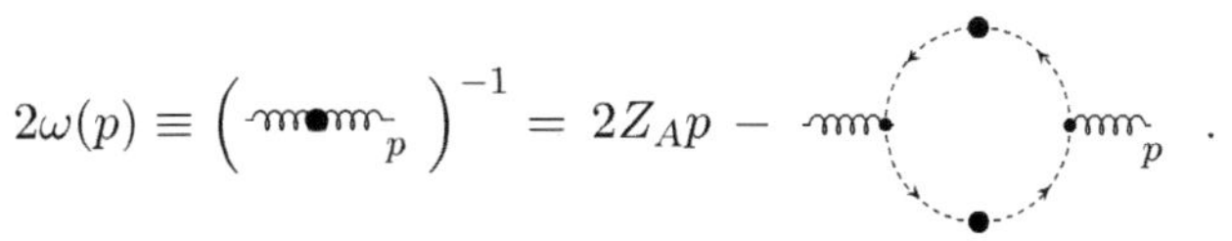

(19)

In particular, (19) does not make use of the factorization hypothesis. Due to ghost dominance, the gap equation approaches (19) in the IR. Somewhat surprisingly, we have found numerically that the solutions obtained with the two different sets of equations coincide over the whole momentum range from the IR to the UV to good numerical precision (for $\beta = 0.796$) [13]. In Fig. 1, ghost and gluon propagators are represented in a double-logarithmic plot for the two different sets of equations. The small discrepancy in the IR is due to the lower numerical precision of the earlier calculation ("DSE") in [4].

6 Actually, the first Gribov region contains Gribov copies itself, and it is necessary to further restrict the integral to the so-called fundamental modular region. It is likely that this additional complication does not affect the following argument (in the case of the Coulomb gauge).

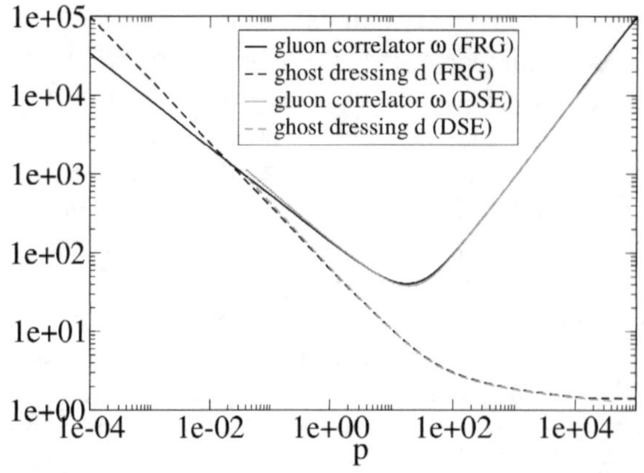

Figure 1. Comparison of the results for the propagators from the two different sets of equations [13]. The label "DSE" stands for the set that contains the gap equation and "FRG" for the set that uses (19).

3. Color-Coulomb potential and factorization hypothesis

The color-Coulomb potential is more directly related to physically observable quantities than the gluon and ghost propagators. For the solution with $\beta = 1$, the factorization hypothesis (14) leads to $V_c(p) \propto p^{-2-2\beta} = p^{-4}$ in the IR which corresponds to a potential in position space that rises exactly linearly for large distances. Unfortunately, the approximation used in the UV [see our remark following (17)] does not permit to relate the (Coulomb) string tension to the scale Λ_{QCD}.

We will now turn to the question of whether the factorization hypothesis is actually justified. To this end, it is convenient to represent the color-Coulomb potential with the help of a composite operator K,

$$\langle\langle \mathbf{x}, a|(-\nabla\cdot\mathbf{D})^{-1}(-\nabla^2)(-\nabla\cdot\mathbf{D})^{-1}|\mathbf{y}, b\rangle\rangle = \left\langle c^a(\mathbf{x}) K \bar{c}^b(\mathbf{y})\right\rangle_{\mathrm{GI}},$$
$$K = \int d^3z\, \bar{c}^d(\mathbf{z})(-\nabla^2_{\mathbf{z}})c^d(\mathbf{z}), \qquad (20)$$

where the index GI (gluon-irreducible) on the vacuum expectation value means that one has to restrict the contibuting diagrams to those where the operator K remains connected to the external points when all gluon lines are cut. The Coulomb form factor $f(p)$ is then precisely the form factor of the composite operator K. Introducing K in the standard way into the generating functional (5), one may derive a DS equation for $f(p)$. After suitable approximations, one arrives at (see also [4])

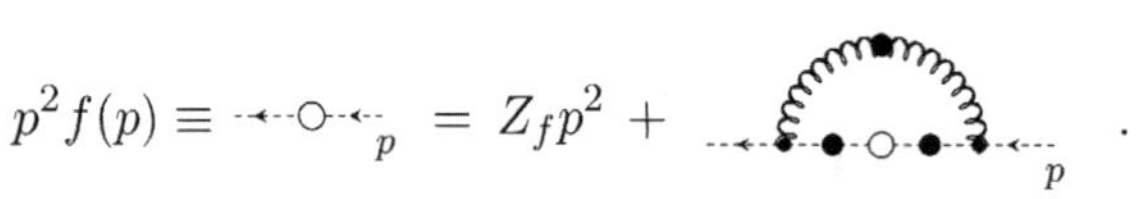

(21)

Now (21) can be used to close the system of equations instead of invoking the factorization hypothesis. The result is disappointing: no solution that fulfills the horizon condition could be found neither numerically nor analytically [14]. Numerically, solutions of the complete system of equations are found to exist only for $d^{-1}(0) \gtrsim 0.02$. For the latter solutions, $f(p)$ tends to a constant for $p \to 0$, so that $V_c(p) \propto p^{-2}$ and the color-Coulomb potential is not confining.

Let us briefly comment on the most recent results for the equal-time two-point correlation functions obtained in calculations on space-time lattices in the Coulomb gauge. They are still

Journal of Physics: Conference Series **287** (2011) 012023 doi:10.1088/1742-6596/287/1/012023

somewhat controversial. In particular, different results have been obtained for the UV-behavior of the gluon propagator in [15] and [16]. Even worse, both results are at odds with recent perturbative one-loop calculations [17]. On the other hand, [15] finds an IR behavior consistent with the ($\beta = 1$)-solution of the DS equations. As for the Coulomb potential, although no conclusion has been reached regarding its IR-behavior, it seems clear that the factorization hypothesis is violated [18].

4. The functional renormalization group

We will now turn to a different functional method, the functional (or Wilsonian) renormalization group. In order to adapt it to the case at hand, one starts with the generating functional (5) and introduces an IR cutoff k in the following way [13]:

$$Z_k[\mathbf{J}, \eta, \bar{\eta}] = \int D[\bar{c}, c, \mathbf{A}] \exp\left(-\int \frac{d^3p}{(2\pi)^3}\, \bar{c}^a(-\mathbf{p}) R_k^c(p)\, c^a(\mathbf{p})\right)$$
$$\times \exp\left(-\frac{1}{2}\int \frac{d^3p}{(2\pi)^3}\, A_i^a(-\mathbf{p}) R_k^A(p)\, A_i^a(\mathbf{p})\right) e^{-\int d^3x\, \bar{c}(-\nabla\cdot\mathbf{D})c}\, |\psi[\mathbf{A}]|^2\, e^{\int d^3x\, [\mathbf{J}\cdot\mathbf{A} + \bar{c}\eta + \bar{\eta}c]}\,. \quad (22)$$

The cutoff functions $R_k^c(p)$ and $R_k^A(p)$ have the properties

$$R_k^{c,A}(p) \to \infty \quad \text{for} \quad p \ll k\,, \qquad R_k^{c,A}(p) \to 0 \quad \text{for} \quad p \gg k\,. \quad (23)$$

This means that the IR modes $p \ll k$ in the functional integral (22) are heavily suppressed, while in the limit $k \to 0$, $R_k^{c,A}(p) \to 0$ and $Z_k[\mathbf{J}, \eta, \bar{\eta}]$ tends toward the full generating functional $Z[\mathbf{J}, \eta, \bar{\eta}]$. In the actual calculations, we have used an exponential suppression of the IR modes,

$$R_k^c(p) = p^2 r_k(p)\,, \qquad R_k^A(p) = 2p\, r_k(p)\,, \qquad r_k(p) = \exp\left(\frac{k^2}{p^2} - \frac{p^2}{k^2}\right)\,. \quad (24)$$

From (22), flow equations for the k-dependent equal-time correlation functions can be derived in the standard way [19]. They read for the propagators

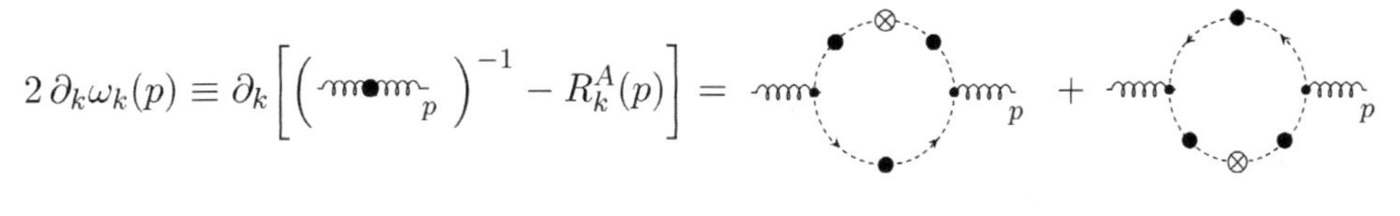
(25)

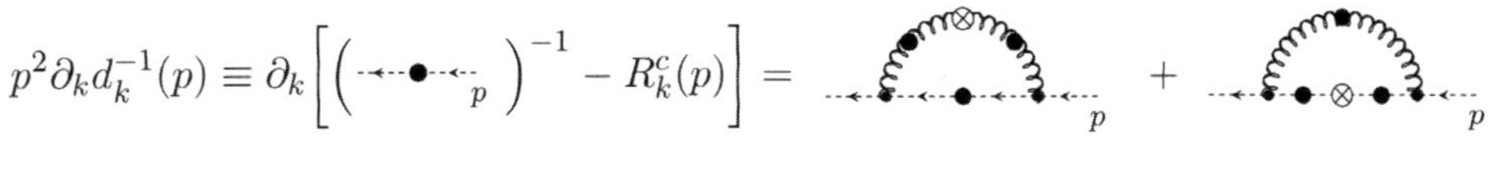
(26)

Here, the symbol $\otimes$ stands for the insertion of $\partial_k R_k^{c,A}$. The non-renormalization theorem for the ghost-gluon vertex has been used in both equations. Furthermore, we have neglected diagrams that involve three- and four-gluon vertices, which is justified for the description of the IR regime if ghost dominance is assumed. Finally, we have omitted tadpole diagrams in order to be able to close the system of differential equations. Partial inclusion of the tadpole diagrams is argued in [13] to lead back to the DS equations (10) and (19) (after integrating over k).

The general strategy is to start integrating the flow equations at a large value of k where due to asymptotic freedom the "action" $S[\mathbf{A}]$ can be replaced with $S_0[\mathbf{A}]$ from (18) and the

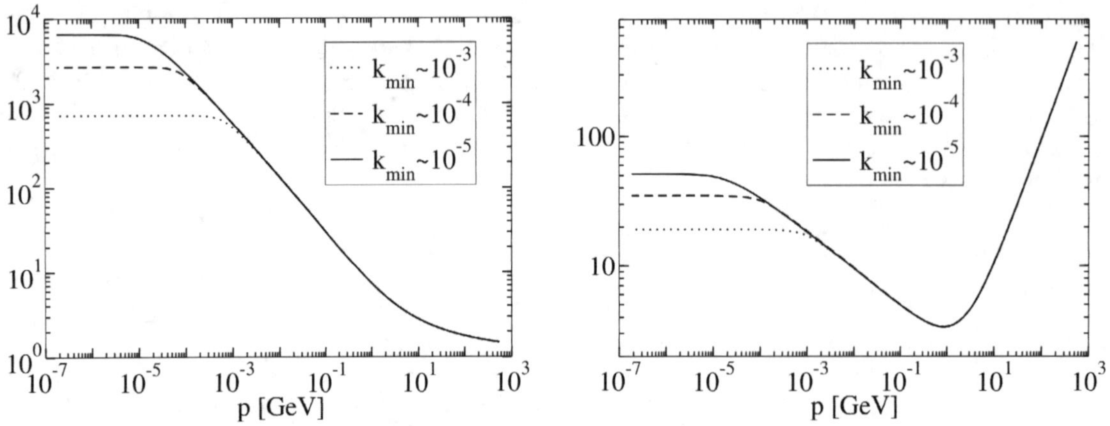

Figure 2. The results for $d(p)$ (left) and $\omega(p)$ (right) from the flow equations, for three different values of $k_{\rm min}$ [13].

coupling constant is small, so that the initial values of the flow are known. Then the flow equations are numerically integrated toward $k = 0$, where $\omega(p) = \omega_{k=0}(p)$ and $d(p) = d_{k=0}(p)$ are read off. Technically, it is important to convert the differential equations (25), (26) to integral equations first, so that the horizon condition and a renormalization condition for $\omega(p)$ can be conveniently incorporated. The results are presented in Fig. 2, again as double-logarithmic plots. For technical reasons, the integration of the flow equations stops at a minimum value $k_{\rm min} > 0$. Then $\omega(p) = \omega_{k=k_{\rm min}}(p)$ for $p \gg k_{\rm min}$, and similarly for $d(p)$. From Fig. 2 it is clear that the power-law behavior of the propagators extends toward smaller momenta p as $k_{\rm min}$ is lowered.

The exponents found numerically are ($\alpha = 0.28, \beta = 0.64$), smaller than for both solutions of the DS equations. They obey the sum rule (16). The fact that the exponents come out smaller than the ones from the DS equations is not entirely unexpected, since a similar behavior was found in analogous calculations in the Landau gauge [20]. Generally, the results for the exponents will vary slightly with the choice of the cutoff functions due to the approximations made in the system of flow equations. An "optimized" choice is expected to give exponents identical to the ones from the DS equations [13]. For the running coupling constant (17) one also finds saturation in the IR at a slightly smaller value than for the DS solutions.

By incorporating the composite operator K in the functional integral (22), one derives (after suitable approximations) the following flow equation for the Coulomb form factor:

$p^2 \partial_k f_k(p) \equiv$ 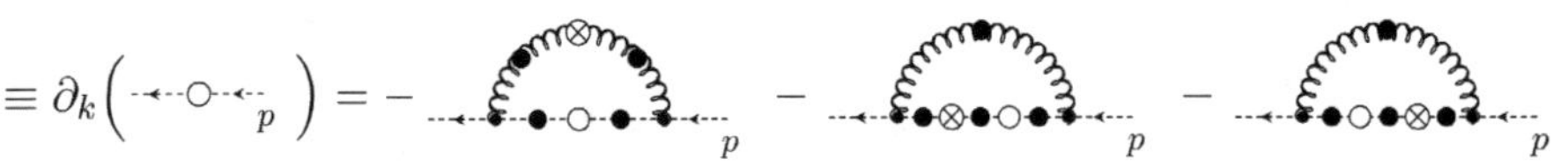 (27)

Making use of the results for $\omega_k(p)$ and $d_k(p)$, this equation can be integrated. Contrary to the DS equations, (27) has a solution that is represented in Fig. 3. The IR-behavior is determined numerically to

$$f(p) \propto p^{-\gamma} , \qquad \gamma = 0.57 . \tag{28}$$

In particular, of course, $f(p) \neq 1$, and the factorization hypothesis is violated. With the values for the exponents from the flow equations one obtains in the IR

$$V_c(p) = \frac{d(p)}{p^2}\, p^2 f(p)\, \frac{d(p)}{p^2} = \frac{1}{p^{2+2\beta+\gamma}} = \frac{1}{p^{3.85}} , \tag{29}$$

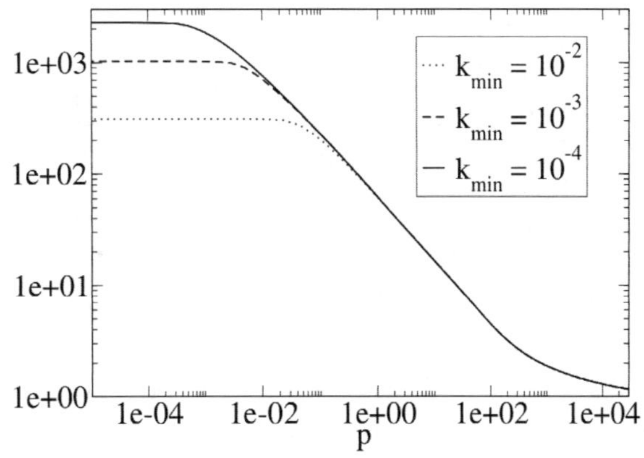

Figure 3. The Coulomb form factor $f(p)$ from (27).

close to $V_c(p) \propto p^{-4}$ which would correspond to a linearly rising potential in position space.

In summary, we find that functional methods are a powerful tool for the description of the nonperturbative infrared regime of nonabelian gauge theories. The formulation of these theories in the Coulomb gauge is particularly convenient, mainly because it gives direct access to the color-Coulomb potential. The Gribov-Zwanziger confinement scenario provides a conceptual framework to understand the confinement mechanism. It can be conveniently implemented via the horizon condition. In particular, we have seen that an almost linearly rising color-Coulomb potential is obtained from the functional renormalization group equations (and the factorization hypothesis is violated). It has also become clear that the approximations employed still have to be improved in order to achieve a quantitatively reliable description of the infrared region.

Acknowledgments

A.W. gratefully acknowledges support by CIC-UMSNH. M.L. and H.R. were supported by DFG-Re856/6-3.

References

[1] Christ N H and Lee T D 1980 *Phys. Rev.* D **22** 939
[2] Zwanziger D 1998 *Nucl. Phys.* B **518** 237
[3] Szczepaniak A P and Swanson E S 2001 *Phys. Rev.* D **65** 025012
[4] Feuchter C and Reinhardt H 2004 *Phys. Rev.* D **70** 105021
[5] Reinhardt H and Feuchter C 2005 *Phys. Rev.* D **71** 105002
[6] Taylor J C 1971 *Nucl. Phys.* B **33** 436
[7] Zwanziger D 2004 *Phys. Rev.* D **70** 094034
[8] Gribov V N 1978 *Nucl. Phys.* B **139** 1
[9] Zwanziger D 1992 *Nucl. Phys.* B **378** 525
[10] Zwanziger D 1991 *Nucl. Phys.* B **364** 127
[11] Schleifenbaum W, Leder M and Reinhardt H 2006 *Phys. Rev.* D **73** 125019
[12] Epple D, Reinhardt H and Schleifenbaum W 2007 *Phys. Rev.* D **75** 045011
[13] Leder M, Pawlowski J M, Reinhardt H and Weber A 2011 *Phys. Rev.* D **83** 025010
[14] Epple D, Reinhardt H, Schleifenbaum W and Szczepaniak A P 2008 *Phys. Rev.* D **77** 085007
[15] Burgio G, Quandt M and Reinhardt H 2009 *Phys. Rev. Lett.* **102** 032002
[16] Nakagawa Y *et al* 2009 *Phys. Rev.* D **79** 114504
[17] Campagnari D, Weber A, Reinhardt H, Astorga F and Schleifenbaum W 2010 *Nucl. Phys.* B doi:10.1016/j.nuclphys.2010.09.013
[18] Voigt A, Ilgenfritz E M, Müller-Preussker M and Sternbeck A 2008 *Phys. Rev.* D **78** 014501
[19] Wetterich C 1993 *Phys. Lett.* B **301** 90
[20] Pawlowski J M, Litim D F, Nedelko S and von Smekal L 2004 *Phys. Rev. Lett.* **93** 152002

Magnus approximation in neutrino oscillations

Mario A. Acero, Alexis A. Aguilar-Arevalo and J. C. D'Olivo

Instituto de Ciencias Nucleares, Departamento de Física de Altas Energías, Universidad Nacional Autónoma de México (ICN-UNAM).
Apdo. Postal 70-543, México, D.F. 04510, México.

E-mail: mario.acero@nucleares.unam.mx

Abstract. Oscillations between active and sterile neutrinos remain as an open possibility to explain some anomalous experimental observations. In a four-neutrino (three active plus one sterile) mixing scheme, we use the Magnus expansion of the evolution operator to study the evolution of neutrino flavor amplitudes within the Earth. We apply this formalism to calculate the transition probabilities from active to sterile neutrinos with energies of the order of a few GeV, taking into account the matter effect for a varying terrestrial density.

1. Introduction

Experimental results from solar, atmospheric, reactor and accelerator neutrino experiments, have proved the existence of neutrino mixing, showing also that neutrinos are massive particles. The dominant behavior of neutrino oscillations is well described by a three–neutrino picture with two mass splitting parameters and three mixing angles, plus a CP–violating phase.

Analytical treatments of the neutrino oscillation phenomenon have been implemented, in the framework of a two– or three–neutrino system, considering the matter potential felt by neutrinos traveling through the Earth [1, 2, 3, 4, 5, 6, 7]. It has been proved that the Magnus expansion [8] is a useful tool to obtain an approximate analytical description of matter effects on neutrino oscillations considering two or three neutrinos [9, 10, 11, 12].

On the other hand, some anomalous experimental results from the Gallium radioactive source experiments, LSND, MiniBooNE and MINOS collaborations, can be interpreted as an indication of the existence of additional (sterile) neutrinos (see for example Refs. [13, 14, 15, 16, 17, 18, 19, 20]).

Here we apply the Magnus expansion formalism to describe the propagation of neutrinos through the Earth, in the case of four neutrinos: the three standard active neutrinos plus and additional sterile neutrino.

2. Evolution of the four–neutrino system

The flavor amplitudes of the four–neutrino system at time t can be written as

$$\begin{pmatrix} \psi_e(t) \\ \psi_\mu(t) \\ \psi_\tau(t) \\ \psi_s(t) \end{pmatrix} = \mathcal{U}(t,t_0) \begin{pmatrix} \psi_e(t_0) \\ \psi_\mu(t_0) \\ \psi_\tau(t_0) \\ \psi_s(t_0) \end{pmatrix} . \tag{1}$$

where the evolution operator $\mathcal{U}(t,t_0)$ satisfies the Schrödinger-like equation ($\hbar = c = 1$)

$$i\frac{d\mathcal{U}}{dt}(t,t_0) = H(t)\mathcal{U}(t,t_0), \tag{2}$$

with the initial condition $\mathcal{U}(t_0,t_0) = I$. In the flavor basis $\{|\nu_\alpha\rangle, \alpha = e,\mu,\tau,s\}$, the Hamiltonian $H(t)$ takes the form

$$H(t) = UH_0U^\dagger + V(t), \tag{3}$$

where $V(t) = \mathrm{diag}(V_{CC}(t), 0, 0, -V_{NC}(t))$ is the matter potential, $H_0 = \mathrm{diag}(0,0,\Delta_{31},\Delta_{41})$ and U, is the mixing matrix that relates the neutrino flavor states to the states with definite masses m_i, $i = 1,\ldots,4$: $|\nu_\alpha\rangle = \sum_{i=1}^4 U^*_{\alpha i}|\nu_i\rangle$. Here, $\Delta_{ij} = \Delta m^2_{ij}/2E = (m_i^2 - m_j^2)/2E$, and V_{CC} and V_{NC} are the charge current and neutral current contributions to the matter effective potential for the active neutrinos. Since we are interested in neutrinos with an energy E of a few GeV and higher, the effect of the Δm^2_{21} mass splitting can be safely discarded and we put $\Delta m^2_{21} = 0$. We also subtracted a common term $V_{NC}(t)$, which has no observable effect on the oscillations.

Hereafter, we assume that neutrinos are of Dirac type and parametrize the mixing matrix as follows:

$$U = R_{23}\,\Gamma_3\,R_{34}\,\Gamma_3^*\,R_{14}\,\Gamma_2\,R_{24}\,\Gamma_2^*\,\Gamma_1\,R_{13}\,\Gamma_1^*\,R_{12}, \tag{4}$$

which helps to simplify the calculations. In this formula, R_{ij} are 4×4 rotation matrices expressed in terms of the vacuum mixing angles θ_{ij} and the diagonal matrices Γ_i include the three CP–violating phases δ_i $(i = 1,2,3)$ that appear in the present scheme. For $\theta_{i4} = 0$, U reduces to the standard 3–neutrino mixing matrix [21], with δ_1 identified to the corresponding CP–violating phase for this case.

It is interesting to see the important effect that the neutral current potential V_{NC} has on the four–neutrino system. If there were four active neutrinos (assuming that the sterile neutrino interacts as the ν_μ and ν_τ), one could expect a behavior similar to the three–neutrino system, given that the contribution of V_{NC} could be subtracted of from Eq. (3). This can be seen in Fig. 1, where the (numerical) evolution of the eigenvalues for a system composed by four active neutrinos is shown. In this case (where for completeness we used $\Delta m^2_{21} \neq 0$), there three *resonance* regions which are associated with the mixing angles θ_{12}, θ_{13}, θ_{14}. After each of these regions, one of the neutrinos decouples from the rest.

On the other hand, a more complex evolution is observed when one of the four neutrinos is sterile (see Fig. 2). In this case, V_{NC} is not a common factor for all the neutrinos and the potential matrix $V(t)$ in the Hamiltonian is the one given after Eq. (3). The effect of the neutral current potential is evident. The three resonances corresponding to those of Fig. 1 are also present here, but now the neutrinos do not decouple after each of them. On the contrary, as a direct effect of V_{NC} there appear other resonance regions related to the other mixing angles involved in the neutrino system.

For our analytical treatment, it is convenient to work within a new basis, by introducing the new vector $\hat{\Psi} \equiv \hat{U}^\dagger\Psi$, with $\hat{U} = R_{23}\Gamma_3R_{34}\Gamma_3^*R_{14}$. Its evolution is driven by the Hamiltonian $\hat{H} = \hat{U}^\dagger H\hat{U}$ which, under the assumption $\Delta m^2_{41} \gg \Delta m^2_{31}, V_{CC}, V_{NC}$, becomes

$$\hat{H}(t) \cong \begin{pmatrix} D_1 & 0 & Z_1^* & 0 \\ 0 & D_2 & 0 & Z_2^* \\ Z_1 & 0 & D_3 & 0 \\ 0 & Z_2 & 0 & D_4 \end{pmatrix}, \tag{5}$$

with

$$\begin{aligned} D_1 &= \Delta_{31}s_{13}^2 + V_{CC}c_{14}^2 - V_{NC}s_{14}^2c_{34}^2, & D_2 &= \Delta_{41}s_{24}^2, \\ D_4 &= \Delta_{41}c_{24}^2 + V_{CC}s_{14}^2 - V_{NC}c_{14}^2c_{34}^2, & D_3 &= \Delta_{31}c_{13}^2 - V_{NC}s_{34}^2, \\ Z_1 &= \Delta_{31}c_{13}s_{13}e^{i\delta_1} - V_{NC}s_{14}c_{34}s_{34}e^{i\delta_3}, & Z_2 &= \Delta_{41}c_{24}s_{24}e^{i\delta_2}. \end{aligned} \tag{6}$$

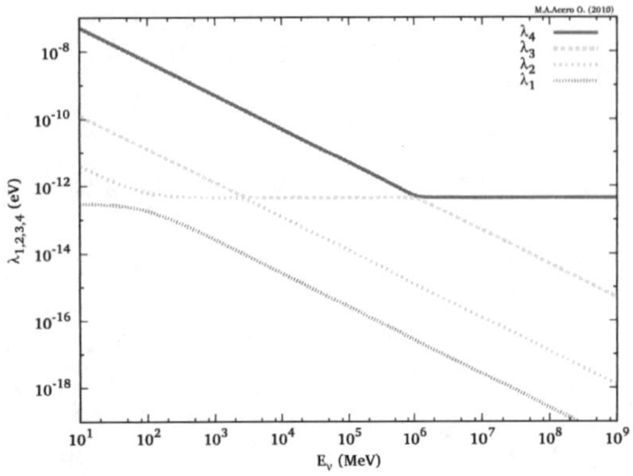

Figure 1. Eigenvalues as a function of neutrino energy, assuming four active neutrinos. Oscillation parameters are the same as in Fig. 5, plus $\Delta m^2_{21} = 7.59 \times 10^{-5}$ eV^2.

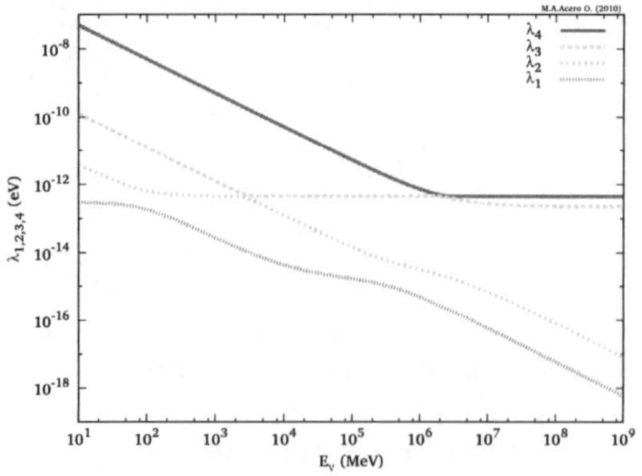

Figure 2. Eigenvalues as a function of neutrino energy, assuming three active plus one sterile neutrinos. Oscillation parameters as in Fig. 5, plus $\Delta m^2_{21} = 7.59 \times 10^{-5}$ eV^2.

For brevity we have defined $c_{ij} \equiv \cos\theta_{ij}$ and $s_{ij} \equiv \sin\theta_{ij}$.

The matrix in Eq. (5) can be diagonalized by a time-dependent unitary transformation

$$U_m = \Gamma_2\, R_{24}(\theta^m_{24})\, \Gamma^*_2 \Gamma_1\, R_{13}(\theta^m_{13})\, \Gamma^*_1\, R_{12}(\pi/2), \tag{7}$$

such that $U_m \hat{H}\, U^\dagger_m = H_D = \mathrm{diag}(\lambda_1, \lambda_2, \lambda_3, \lambda_4)$, with the energy eigenvalues in matter $\lambda_k(t)$ ordered in such a way that $\lambda_1 < \lambda_2 < \lambda_3 < \lambda_4$. The mixing angles in matter are given by formulas similar to the familiar ones for two neutrinos:

$$\cos 2\theta^m_{13} = \frac{D_3 - D_1}{\sqrt{(D_3 - D_1)^2 + 4|Z_1|^2}}, \tag{8}$$

$$\cos 2\theta^m_{24} = \frac{D_4 - D_2}{\sqrt{(D_4 - D_2)^2 + 4|Z_2|^2}}. \tag{9}$$

Figure 3 shows the mixing angles in matter and the energy eigenvalues λ_i as functions of the neutrino energy. We see that λ_2 and λ_3 are almost degenerate at around the same energy at which the angle θ^m_{13} exhibits the characteristic resonant behavior for the terrestrial density. Notice that, for the energy range under consideration, the largest eigenvalue λ_4 decreases as E^{-1}, while the smallest eigenvalue λ_1 remains constant as a function of the neutrino energy.

3. Solution of the evolution operator

The evolution operator in the flavor basis can be expressed as $\mathcal{U}(t,t_0) = \mathcal{O}_m(t)\mathcal{U}_A(t,t_0)\mathcal{O}^\dagger_m(t_0)$, where $\mathcal{O}_m(t) = \hat{U}U_m(t)$ is the mixing matrix that relates the flavor states with the set $\{|\nu^m_k(t)\rangle, k = 1,\ldots,4\}$ of the (instantaneous) energy eigenstates in matter. The latter define the adiabatic basis and in it the evolution operator $\mathcal{U}_A(t,t_0)$ obeys Eq. (2) with the Hamiltonian

$$H_A(t) = H_D(t) - iU^\dagger_m(t)\,\dot{U}_m(t), \tag{10}$$

where dot means differentiation with respect to time.

Discarding the second term in Eq. (10) corresponds to solving the problem in the adiabatic approximation. In any case, the time dependence generated by H_D can be integrated exactly

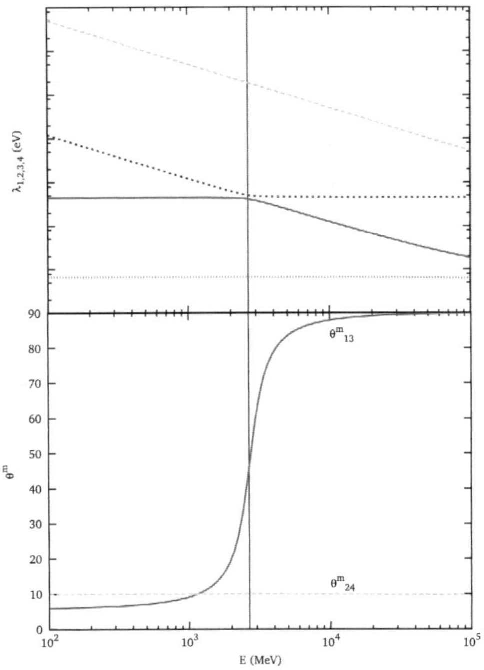

Figure 3. Approximate eigenvalues and matter mixing angles as functions of neutrino energy.

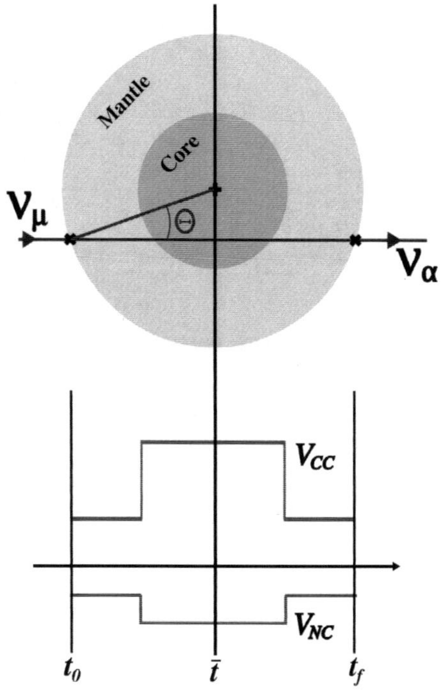

Figure 4. Mantle–Core–Mantle potential modeling the Earth's interior.

by a change of the representation accomplished by means of the unitary transformation $\mathcal{P}(t) = \exp[-\int_{t_0}^{t} dt' H_D(t')]$. For the remaining part, in general it is not possible to find an analytical solution in a close form and one has to rest on some approximation to determine it. As mentioned in the introduction, an appropriate procedure, which preserves unitarity, is based in the exponential expansion of the evolution operator. Accordingly, next we write $\mathcal{U}_A(t, t_0) = \mathcal{P}(t) \exp \Omega(t, t_0)$ and evaluate Ω in terms of the first two terms of its Magnus expansion: $\Omega \cong \Omega_1 + \Omega_2$.

Proceeding in this manner, after some algebraic manipulations we arrive at:

$$\mathcal{U}_A(t_f, t_0) = \mathcal{I}_{32}^{*} \begin{pmatrix} \mathcal{I}_{32} e^{-i\alpha_1} & 0 & 0 & 0 \\ 0 & (c_\xi - i s_\xi \frac{\xi_{(2)}}{\xi}) e^{i\phi_{\bar{t}\to t_f}} & i e^{i\delta_1} s_\xi \frac{\xi_{(1)}}{\xi} & 0 \\ 0 & i e^{-i\delta_1} s_\xi \frac{\xi_{(1)}}{\xi} & (c_\xi + i s_\xi \frac{\xi_{(2)}}{\xi}) e^{-i\phi_{\bar{t}\to t_f}} & 0 \\ 0 & 0 & 0 & \mathcal{I}_{32} e^{-i\alpha_4} \end{pmatrix} \tag{11}$$

where $\alpha_i = \int_{t_0}^{t_f} dt' \lambda_i(t')$ and

$$\mathcal{I}_{32} = \exp\Big\{ -\frac{i}{2} \big[\alpha_3(t) + \alpha_2(t)\big] \Big\}, \tag{12}$$

$$\phi_{\bar{t}\to t_f} = \int_{\bar{t}}^{t_f} dt' \big[\lambda_3(t') - \lambda_2(t')\big], \tag{13}$$

and we have introduced the notations $c_\xi = \cos\xi$, $s_\xi = \sin\xi$, and $\xi = \sqrt{\xi_{(1)}^2 + \xi_{(2)}^2}$. The quantities

$$\xi_{(1)} = 2\int_{\bar{t}}^{t_f} dt'\dot{\theta}_{13}^m(t')\sin\phi_{\bar{t}\to t'}, \tag{14}$$

$$\xi_{(2)} = \int_{t_0}^{t_f} dt' \int_{t_0}^{t'} dt''\,\dot{\theta}_{13}^m(t')\dot{\theta}_{13}^m(t'')\sin\phi_{t'\to t''}, \tag{15}$$

appear in the explicit calculation of $\Omega_1(t_f, t_0)$ and $\Omega_2(t_f, t_0)$, respectively [11], taking into account the fact that the potential is symmetric with respect to the middle point of the neutrino trajectory $\bar{t} = (t_f + t_0)/2$.

4. Transition Probabilities

We are interested the transition probabilities of muon neutrinos coming from the atmosphere and traversing the Earth. With the results of the previous section, we are able to compute the oscillation probabilities trough the relation

$$\begin{aligned} P_{\nu_\mu\to\nu_\alpha}(t) &= |\Psi_\alpha(t)|^2 \\ &= |\mathcal{U}_A(t_f, t_0)\Psi_\mu(t_0)|^2 \\ &= |\hat{U}^\dagger U_m^\dagger(t)\mathcal{U}(t_f, t_0)U_m(t)\hat{U}\Psi_\mu(t_0)|^2, \end{aligned} \tag{16}$$

which depends on the vacuum–mixing parameters through $\hat{U}$ and on the matter potential by means of U_m and $\mathcal{U}(t_f, t_0)$.

In Figures 5 and 6, we show the behavior of the transition probabilities as a function of the neutrino energy for different values of the parameters involved in the analysis. We consider a simplified model for the electron density inside the Earth, the so-called mantle-core-mantle [22] (Fig. 4). One can see that the Magnus approximation agrees very well with the numerical calculation for $E \gtrsim 7$ GeV. As the plots shows, the effect of the inclusion of the sterile neutrino generates noticeable changes on the $\nu_\mu \to \nu_\alpha$ ($\alpha = e, s$) transition probabilities. Moreover, the $\nu_\mu \to \nu_s$ transition remains as an open possibility. These effects increase for larger values of the three additional mixing angles θ_{i4} ($i = 1, 2, 3$). This is particularly relevant for θ_{24} and θ_{34}, on which no stringent limits have been placed directly, contrary to the case of θ_{14} that has been constrained by the Bugey experiment [23].

Finally, taking into account that neutrinos produced in the atmosphere reach the Earth's surface with different zenith angle Θ (see Fig. 4), in Figures 7 and 8 we plot contours of the oscillation probability $P(\nu_\mu \to \nu_{e,s})$ in vacuum and in the terrestrial matter, using our result of Eq. 16. It is important to note that we consider the initial neutrino state to be a muon neutrino, at the moment it arrives to the Earth surface, i.e., vacuum oscillations are assumed not to occurs in the atmosphere.

It is clear from the figures that Earth matter potential enhances the probability that muon neutrinos oscillate to electron neutrinos, through the MSW resonant effect [24, 25]. This effect is notorious in the high probability ($> 40\%$) regions shown in Fig. 8 at neutrino energies around 3 GeV and 6 GeV. Being aware of some subtle differences, our results are in good agreement with those presented, for instance, in Ref. [26].

5. Conclusions

We used the Magnus expansion to calculate the flavor transition of a system composed by the three active neutrinos plus a sterile neutrino under the effect of the terrestrial potential. This approximation coincides very well with the numerical computations for high neutrino energies ($E \gtrsim 7$ GeV). The inclusion of the fourth neutrino has a significative effect on the evolution of

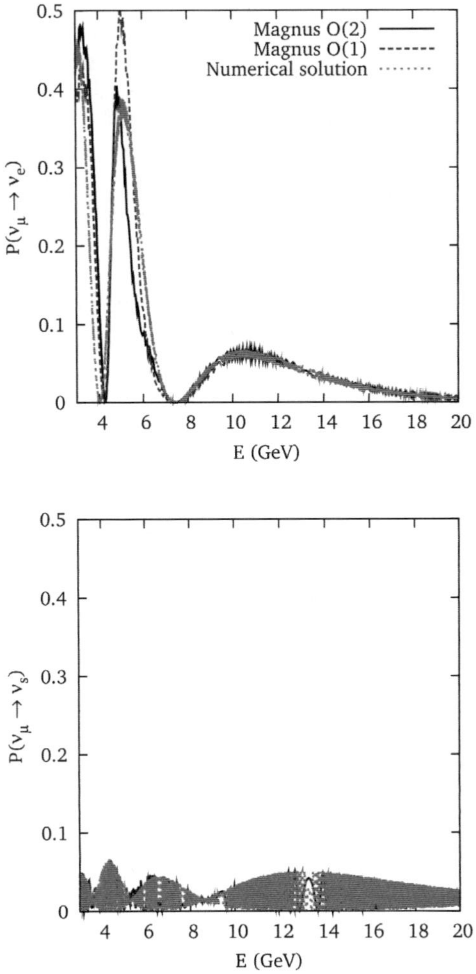

Figure 5. $P(\nu_\mu \to \nu_{e,s})$ as a function of the energy for a neutrino crossing the center of the Earth. Parameters: $\Delta m^2_{31} = 2.5 \times 10^{-3}$ eV^2, $\Delta m^2_{41} = 1.0$ eV^2, $\theta_{12} = 34°$, $\theta_{13} = 7°$, $\theta_{23} = 45°$, $\theta_{i4} = 5°$.

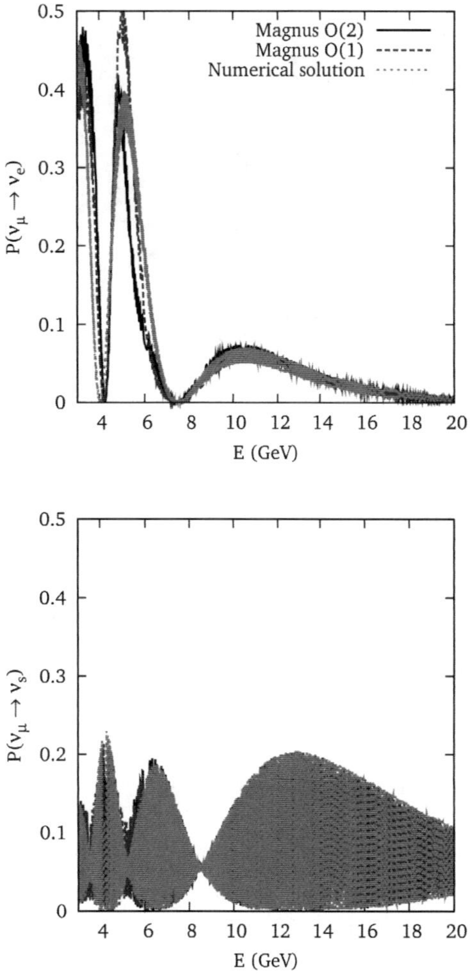

Figure 6. $P(\nu_\mu \to \nu_{e,s})$ as a function of the energy for a neutrino crossing the center of the Earth. The oscillation parameters are same as in Fig. 5, changing θ_{i4} to $\theta_{14} = 5°$, $\theta_{24} = \theta_{34} = 10°$.

matter eigenstates of the system as well as on $P(\nu_\mu \to \nu_\alpha)$ which depends strongly on θ_{i4} and exhibits a rapid oscillatory behavior superimposed to the standard oscillations for the active neutrinos.

Acknowledgments

This work has been partially supported by PAPIIT-UNAM through grant IN117210 and CONACYT through grant 83534 and Red FAE. M. A. A. acknowledges support by a postdoctoral grant from the UNAM; he also thanks the organizers for the support to participate in MSPF2010.

References

[1] E. Lisi and D. Montanino, *Phys. Rev. D* **56**, 1792 (1997).

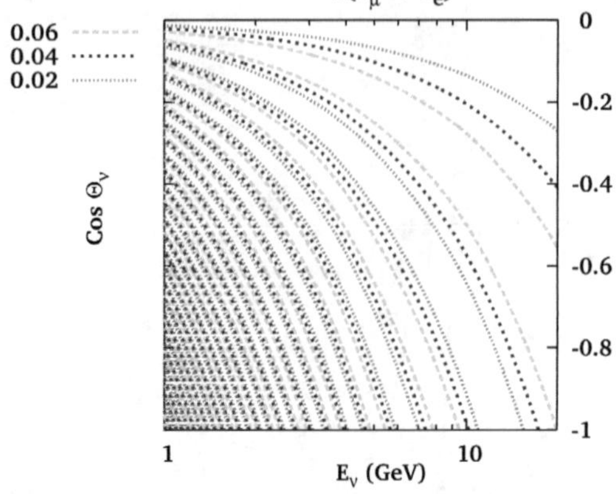

Figure 7. $P(\nu_\mu \to \nu_e)$ in vacuum as a function of the neutrino energy and the zenith angle Θ. Oscillation parameters are $\theta_{12} = 34°$, $\theta_{13} = 11.5°$, $\theta_{23} = 45°$, $\theta_{14} = 0°$, $\theta_{24} = \theta_{34} = 10°$, $\Delta m^2_{31} = 2.5 \times 10^{-3}$ eV^2, $\Delta m^2_{41} = 0.1$ eV^2

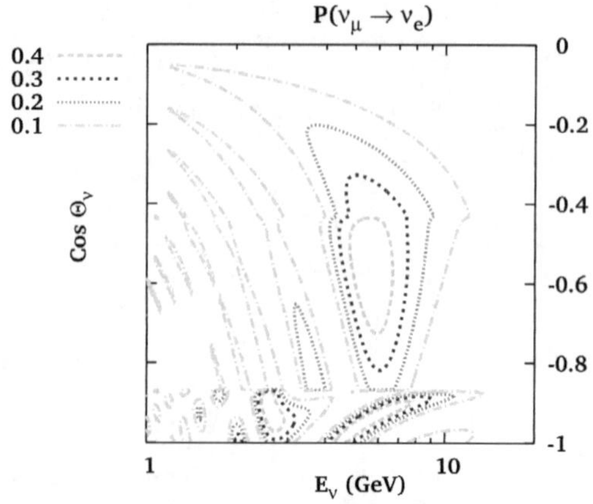

Figure 8. $P(\nu_\mu \to \nu_e)$ in matter as a function of the neutrino energy and the zenith angle Θ. Oscillation parameters are $\theta_{12} = 34°$, $\theta_{13} = 11.5°$, $\theta_{23} = 45°$, $\theta_{14} = 0°$, $\theta_{24} = \theta_{34} = 10°$, $\Delta m^2_{31} = 2.5 \times 10^{-3}$ eV^2, $\Delta m^2_{41} = 0.1$ eV^2

[2] M. Freund, *Phys. Rev. D* **64**, 053003 (2001).
[3] E. K. Akhmedov, M. A. Tortola and J. W. F. Valle, *JHEP* **0405**, 057 (2004).
[4] M. Blennow and T. Ohlsson, *J. Math. Phys.* **45**, 4053 (2004).
[5] P. C. de Holanda, W. Liao and A. Y. Smirnov, *Nucl. Phys. B* **702**, 307 (2004).
[6] A. N. Ioannisian and A. Y. Smirnov, *Phys. Rev. Lett.* **93**, 241801 (2004).
[7] A. N. Ioannisian, N. A. Kazarian, A. Y. Smirnov and D. Wyler, *Phys. Rev. D* **71**, 033006 (2005).
[8] W. Magnus, *Commun. Pure Appl. Math.*, **7**, 649 (1954).
[9] J. C. D'Olivo, *Phys. Rev. D* **45**, 924 (1992).
[10] J. C. D'Olivo and J. A. Oteo, *Phys. Rev. D* **54**, 1187 (1996).
[11] A.D. Supanitsky, J.C. D'Olivo and G. Medina-Tanco, *Phys. Rev. D* **78**, 0455024 (2008)
[12] A. Ioannisian and A.Yu. Smirnov, *Nucl. Phys.* **B916**, 94 (2009).
[13] M. Maltoni and T. Schwetz, *Phys. Rev. D* **76**, 093005 (2007).
[14] M. A. Acero, C. Giunti and M. Laveder, *Phys. Rev. D* **78**, 073009 (2008).
[15] Y. Farzan, T. Schwetz and A. Y. Smirnov, *JHEP* **0807**, 067 (2008).
[16] G. Karagiorgi, Z. Djurcic, J. M. Conrad, M. H. Shaevitz and M. Sorel, *Phys. Rev. D* **80**, 073001 (2009) [*Erratum-ibid. D* **81**, 039902 (2010)].
[17] P. Adamson *et al.* [The MINOS Collaboration], *Phys. Rev. D* **82**, 051102 (2010).
[18] P. Adamson *et al.* [The MINOS Collaboration], *Phys. Rev. D* **81**, 052004 (2010).
[19] C. Giunti and M. Laveder, *Phys. Rev. D* **82**, 053005 (2010).
[20] C. Giunti and M. Laveder, arXiv:1012.0267 [hep-ph].
[21] C. Amsler *et al.* [Particle Data Group], *Phys. Lett. B* **667**, 1 (2008).
[22] Stacey, F. D. 1977, *Physics of the earth*, John Wiley & Sons, New York, NY (USA) 414 p.
[23] Y. Declais *et al.*, *Nucl. Phys. B* **434**, 503 (1995).
[24] L. Wolfenstein, *Phys. Rev. D* **17**, 2369 (1978).
[25] S. P. Mikheyev, A. Yu. Smirnov, *Yad. Fiz.* **42**, 1441 (1985); *Sov. H. Nucl. Phys.* **42**.
[26] J. Hosaka *et al.* [Super-Kamiokande Collaboration], *Phys. Rev. D* **74**, 032002 (2006).

Journal of Physics: Conference Series **287** (2011) 012025 doi:10.1088/1742-6596/287/1/012025

Gamma-Ray Burst Physics

Shigehiro Nagataki

Yukawa Institute for Theoretical Physics, Kyoto University, Oiwake-cho Kitashirakawa Sakyo-ku, Kyoto 606-8502, Japan

E-mail: nagataki@yukawa.kyoto-u.ac.jp

Abstract. I have developed two-dimensional general relativistic magnetohydrodynamic (GRMHD) code. I have performed numerical simulations of collapsars using these codes and realistic progenitor models. In the GRMHD simulation, it is shown that a jet is launched from the center of the progenitor. We also performed simulations of collapsars with different Kerr parameters a =0, 0.5, 0.9, 0.95. It is shown that a more rapidly rotating black hole is driving a more energetic jet. No jet is seen for the case of Schwartzschild black hole case, while the total energy of the jet is as large as 10^{50} erg for a rapidly rotating Kerr black hole case ($a = 0.95$). In order to explain the high luminosity of a GRB, it is concluded that a rapidly rotating black hole is favored ('faster is better'). We also performed two-dimensional hydrodynamic simulations in the context of collapsar model to investigate the explosive nucleosynthesis happened there. It is found that the amount of ^{56}Ni is very sensitive to the energy deposition rate. This result means that the amount of synthesized ^{56}Ni can be little even if the total explosion energy is as large as 10^{52} erg. Thus, some GRBs can associate with faint supernovae. Thus we consider it is quite natural to detect no underlying supernova in some X-ray afterglows.

1. Introduction

Gamma-Ray Bursts (GRBs; in this study, we consider only long GRBs, so we refer to long GRBs as GRBs hereafter) have been mysterious phenomena since their discovery in 1969. Last decade, observational evidence for supernovae (SNe) and GRBs association has been reported. Some of the SNe that associate with GRBs were very energetic and blight. The estimated explosion energy was of the order 10^{52} ergs, and produced nickel mass was about 0.5 $M_{\odot}$. Thus they are categorized as a new type of SNe (sometimes called as hypernovae). The largeness of the explosion energy is very important, because it can not be explained by the standard core-collapse SN scenario, and other mechanism should be working at the center of the progenitors. In this paper, we investigate the dynamics of collapsars using two-dimensional general relativistic magnetohydrodynamic (GRMHD) code[1].

Also, we perform two-dimensional hydrodynamic simulations in the context of collapsar model to investigate the explosive nucleosynthesis happened there. We have to be careful to the point that where and when ^{56}Ni is synthesized in the collapsar is not unclear. This is because the explosion mechanism is not known well. One possibility is that ^{56}Ni is synthesized in the jet region like jet-like supernova explosion[2, 3]. Another possibility is that nucleosynthesis in the accretion disk, and some fraction of the accreting matter escape from the system[4]. In this paper, we consider the former possibility.

Journal of Physics: Conference Series **287** (2011) 012025 doi:10.1088/1742-6596/287/1/012025

2. GENERAL RELATIVISTIC MHD SIMULATIONS

2.1. Formulation

We have adopted a conservative, shock-capturing scheme with Harten, Lax, and van Leer (HLL) flux term with flux-interpolated constrained transport technique. We use a third-order Total Variation Diminishing (TVD) Runge-Kutta method for evolution in time, while monotonized central slope-limited linear interpolation method is used for second-order accuracy in space. 2D scheme (2-dimensional Newton-Raphson method) is usually adopted for transforming conserved variables to primitive variables.

When we perform simulations of GRMHD, Modified Kerr-Schild coordinate is basically adopted with mass of the BH (M) fixed. We use G =M = c = 1 unit. The calculated region covers from r =1.8 to 3 $\times 10^4$ (that corresponds to 5.3 $\times 10^5$ cm and 8.9 $\times 10^9$ cm in cgs units). We adopt the model 12TJ in Woosley and Heger (2006)[5]. We set the Kerr-parameter of the black hole at the center is set to be 0.5 throughout of the simulation. The initial, weak poloidal magnetic fields are put initially. The minimum plasma beta in the simulation region is greater than 100 initially.

2.2. Results

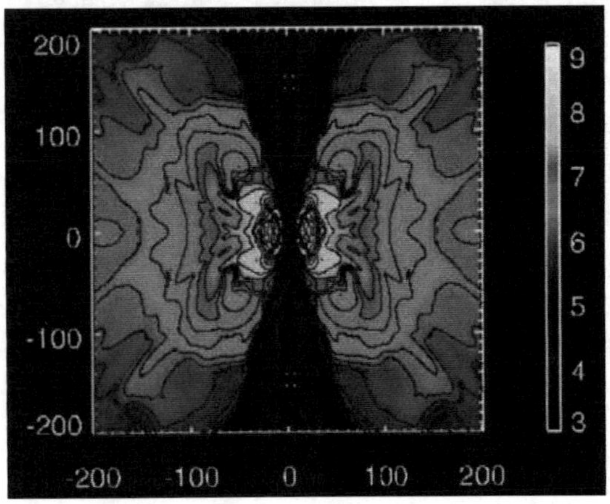

Figure 1. Contours of rest mass density at the central region in logarithmic scale at t = 180000 (that corresponds to 1.773 sec), in which cgs units are used assuming that the gravitational mass of the BH is 2 $M_{\odot}$. The length unit in the vertical/horizontal axes corresponds to 2.95 times10^5 cm.

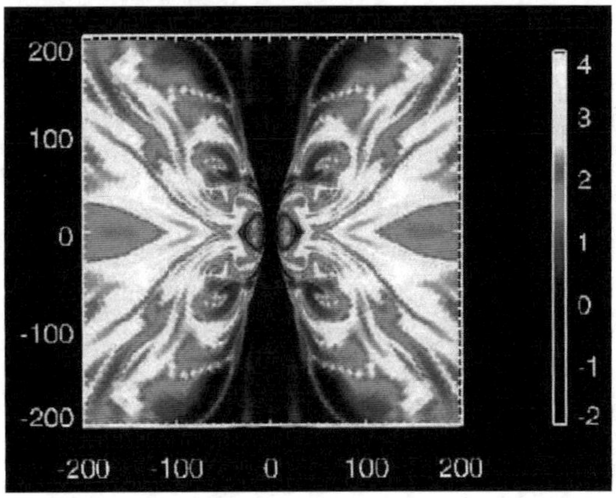

Figure 2. Contour of the plasma beta (pgas/pmag) at t = 180000 in logarithmic scale.

In Figure 1, contours of rest mass density at the central region are shown. Contours represent the density in units of g cm^{-3} in logarithmic scale. The length r = 200 corresponds to 5.9 times 10^7 cm. The time unit corresponds to 9.85 times 10^{-6} sec. Figure 2 shows contours of the plasma beta (pgas/pmag) in logarithmic scale at t = 180000. As expected, the plasma beta is low in the jet region while it is high in the accretion disk region.

3. Dependence on the Kerr Parameter

Next, we studied the dependence of dynamics on the Kerr Parameter. The initial condition is also same with the previous section, but for different Kerr Parameters. In the previous section, the Kerr parameter, a, was assumed to be 0.5, but in this study we perform simulations for $a =$ 0, 0.5, 0.9, 0.95, respectively (we name them as Model A,B,C,D, respectively).

In Figure 3, contours of rest mass density in logarismic scale for all models at the same time-slice $t = 160000$ (that corresponds to 1.5760 sec) are shown. Cgs units are used for the rest mass density, while the length in the vertical/horizontal axes is written in $G = M = c = 1$ unit. $r = 1$ and 4000 corresponds to 2.95 $\times 10^5$ cm and 1.18 $\times 10^9$ cm, respectively. These results are projected on the (r sinθ, r cos θ)-plane. Upper left panel shows the state of Model A ($a = 0$), upper right panel shows the one of Model B ($a = 0.5$), lower left panel shows the one of Model C ($a = 0.9$), and lower right panel shows the one of Model D ($a = 0.95$). It is clearly shown that the rotating black hole drives the jet (the Schwartzschild black hole cannot drive a jet (Model A), while a more rapidly rotating black hole is driving a stronger jet).

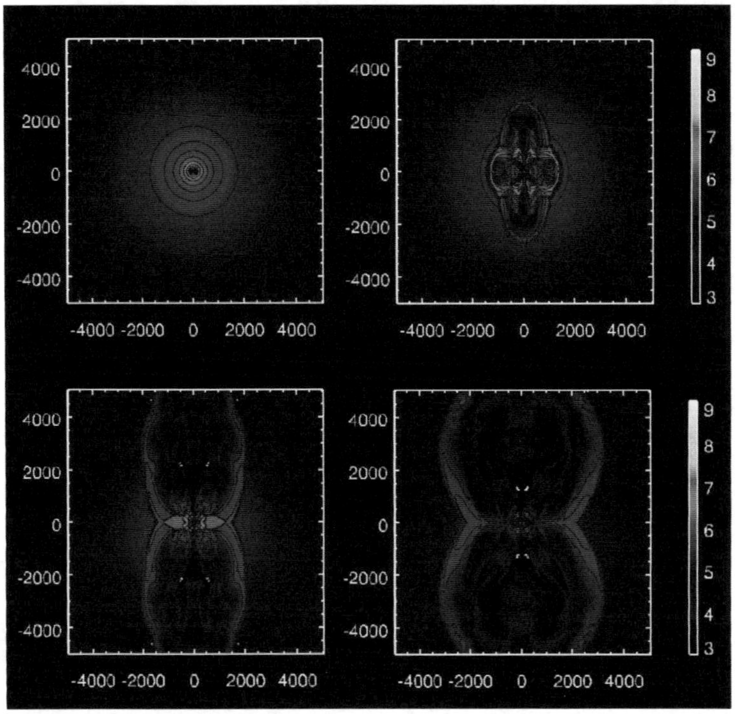

Figure 3. Contours of rest mass density in logarismic scale for all models at the same time-slice $t =$ 160000 (that corresponds to 1.5760 sec). Cgs units are used for the rest mass density, while the length in the vertical/horizontal axes is written in $G = M = c = 1$ unit. $r =$ 1 and 4000 corresponds to 2.95 $\times 10^5$ cm and 1.18 $\times 10^9$ cm, respectively. These results are projected on the (r sinθ, r cos θ)-plane. Upper left panel shows the state of Model A ($a = 0$), upper right panel shows the one of Model B ($a = 0.5$), lower left panel shows the one of Model C ($a = 0.9$), and lower right panel shows the one of Model D ($a = 0.95$).

In Figure 4, plots of the jet energy for all models at $t = 160000$ are shown. The definition of the jet energy is:

$$E_{\rm Jet} = 2 \times 2\pi \int_{r_+}^{\infty} dr \int_0^{\theta} d\theta \sqrt{-g}(T_t^t - \rho u^0 u_0), \tag{1}$$

where T_t^t is the (t,t) component of total energy-momentum tensor and integration is done only for the region where u^r (radial component of 4-velocity of fluid) is positive. It is noted that the contribution of the rest mass energy is subtracted. Factor 2 is coming from the symmetry of the system with respect to the equatorial plane. Dot curve represents the jet energy within the opening angle $\theta = 5°$, while dashed curve represents the one within $\theta = 10°$. The unit of vertical axis is 10^{48} erg. It is clearly seen that a more rapidly rotating black hole is driving a stronger jet. The total energy of the jet for Model D ($a = 0.95$) is as large as 10^{50} erg.

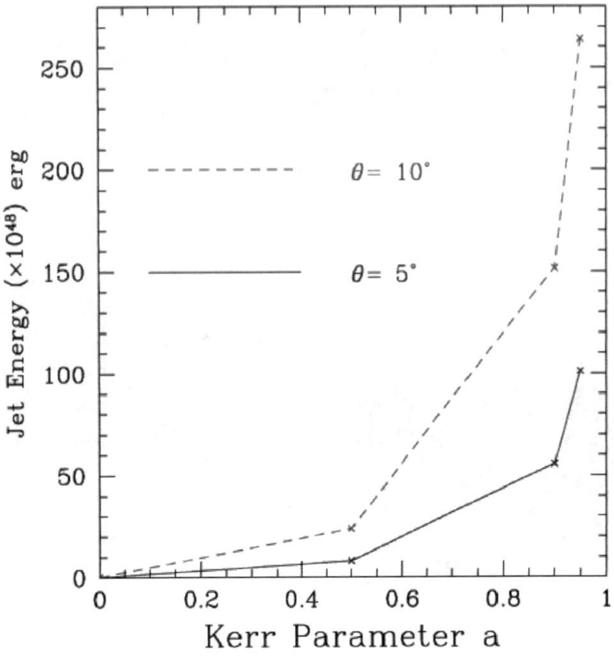

Figure 4. Plots of the jet energy (see text for the definition in detail) for all models at t = 160000 (that corresponds to 1.5760 sec). The unit of vertical axis is 10^{48} erg. Dot curve represents the jet energy within the opening angle $\theta = 5°$, while dashed curve represents the one within $\theta = 10°$.

4. EXPLOSIVE NUCLEOSYNTHESIS IN COLLAPSAR

4.1. Formulation

We have done two-dimensional hydrodynamic simulations taking into account self-gravity and the gravitational potential of the central point mass. The calculated region corresponds to a quarter of the meridian plane under the assumption of axisymmetry and equatorial symmetry. The spherical mesh with 250 by 30 grid points is used for all the computations. The radial grid is nonuniform, extending from 2 $\times 10^7$ to 3 $\times 10^{11}$ cm with finer grids near the center, while the polar grid is uniform. We adopt the collapsar model of MacFadyen & Woosley (1999)[4].

To simulate the jet-induced explosion, we deposit only thermal energy at a rate 10^{51} ergs/s homogeneously within a 30 degree cone around the rotation axis for 10 sec. We name this model E51. For comparison, we perform a calculation in which total explosion energy (10^{52} ergs) is put initially with the same deposition region as model E51. We name these models E52.

Since the chemical composition behind the shock wave is not in nuclear statistical equilibrium, the explosive nucleosynthesis has to be calculated using the time evolution of density and temperature, and a nuclear reaction network, which is called post-processing. The nuclear reaction network contains 250 species[6, 7].

4.2. Results

We deposited thermal energy to launch a jet from the central region of the collapsar. The density structure for models E51 shows a sharp, narrow jet propagates along the rotation axis in model E51. On the other hand, in the case of E52, a broad, deformed shock wave propagates in the progenitor (see also Nagataki et al. (2006)[3].

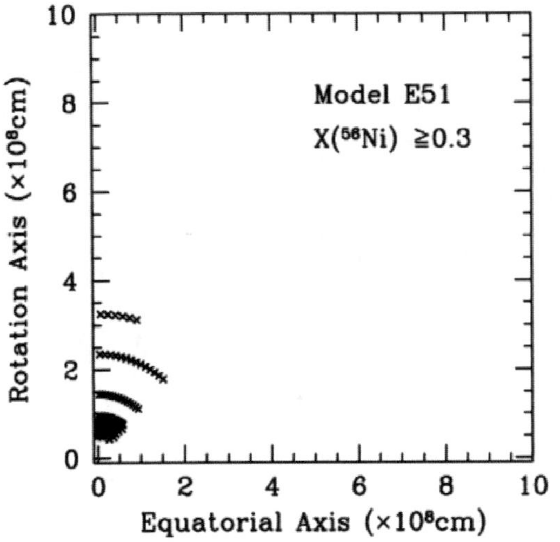

Figure 5. Positions of the ejected test particles at 0 sec that meet the condition that the mass fraction of ^{56}Ni becomes greater than 0.3 as a result of explosive nucleosynthesis for model E51. The total ejected mass of ^{56}Ni is 0.0439 $M_\odot$.

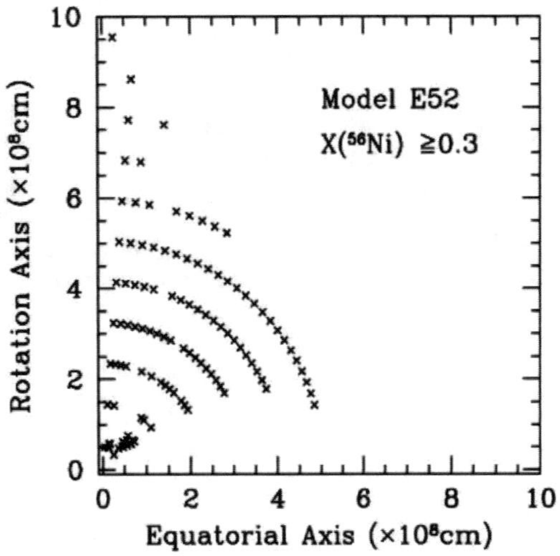

Figure 6. Same with left, but for model E52. The total ejected mass of ^{56}Ni is 0.23 solar masses.

In Figure 5, positions of the ejected test particles for model E51 at 0 sec are shown that satisfy the condition that the mass fraction of ^{56}Ni becomes greater than 0.3 as a result of explosive nucleosynthesis. The total ejected mass of ^{56}Ni becomes 0.0439 $M_\odot$, which is much smaller than the observed values of hypernovae. In Figure 6, the same values are shown as in Figure 5, but for models E52. The total ejected mass of ^{56}Ni is 0.23 $M_\odot$, which is comparable to the observed values of hypernovae.

Thus we can conclude that the resulting amount of ^{56}Ni is very sensitive to the energy deposition rate. This result means that the amount of synthesized ^{56}Ni can be little even if the total explosion energy is as large as 10^{52} erg. Thus, some GRBs can associate with faint supernovae. Thus we consider it is quite natural to detect no underlying supernova in some X-ray afterglows such as GRB060614.

5. Summary and Conclusion

I have performed numerical simulations of collapsars using these codes and realistic progenitor models. In the GRMHD simulation, it is shown that a jet is launched from the center of the progenitor. We also performed simulations of collapsars with different Kerr parameters a =0, 0.5, 0.9, 0.95. It is shown that a more rapidly rotating black hole is driving a more energetic jet. No jet is seen for the case of Schwartzschild black hole case, while the total energy of the jet is as large as 10^{50} erg for a rapidly rotating Kerr black hole case ($a = 0.95$). In order to explain the high luminosity of a GRB, it is concluded that a rapidly rotating black hole is favored ('faster is better'). We also performed two-dimensional hydrodynamic simulations in the context of collapsar model to investigate the explosive nucleosynthesis happened there. It is found that the amount of ^{56}Ni is very sensitive to the energy deposition rate. This result means that the amount of synthesized ^{56}Ni can be little even if the total explosion energy is as large as 10^{52} erg. Thus, some GRBs can associate with faint supernovae. Thus we consider it is quite natural to

Journal of Physics: Conference Series **287** (2011) 012025 doi:10.1088/1742-6596/287/1/012025

detect no underlying supernova in some X-ray afterglows such as GRB060614.

References

[1] S. Nagataki, *Astrophys. J.* **704** (2009) 937.
[2] S. Nagataki, A. Mizuta, S. Yamada, H. Takabe, K. Sato, *Astrophys. J.* **596** (2003) 401.
[3] S. Nagataki, A. Mizuta, K. Sato, *Astrophys. J.* **647** (2006) 1255.
[4] A. I. MacFadyen, S. E. Woosley, *Astrophys. J.* **524** (1999) 262.
[5] S. E. Woosley, A. Heger *Astrophys. J.* **637** (2006) 914.
[6] S. Nagataki, M. Hashimoto, K. Sato, S. Yamada, *Astrophys. J.* **486** (1997) 1026.
[7] S. Nagataki, *Astrophys. J. Suppl.* **127** (2000) 141.

Journal of Physics: Conference Series **287** (2011) 012026 doi:10.1088/1742-6596/287/1/012026

New Physics in $\Delta L = 2$ neutrino oscillations

David Delepine*, Vannia González Macías*, Shaaban Khalil† and Gabriel López Castro**

**División de Ciencias e Ingenierías de la Universidad de Guanajuato, C.P. 37150, León, Guanajuato, México.*
†Centre for Theoretical Physics, The British University in Egypt, El Sherouk City, Postal No, 11837, P.O. Box 43, Egypt
***Departamento de Fisica, Cinvestav, Apartado Postal 14-740, 07000 Mexico D.F., Mexico*

Abstract. We propose a general framework to constrain $\Delta L = 2$ processes by measuring observables associated with neutrino-antineutrino oscillations in $\pi^{\pm}$ decays. First, we use this formalism as a new strategy for detecting the CP-violating phases and the effective mass of muon Majorana neutrinos. Within the generic framework of quantum field theory, we compute the non-factorizable probability for producing a pair of same-charged muons in $\pi^{\pm}$ decays as a distinctive signature of $\nu_{\mu} - \bar{\nu_{\mu}}$ oscillations. Using the neutrino-antineutrino oscillation probability reported by MINOS collaboration, a new stringent bound on the effective muon-neutrino mass is derived. Secondly, we interpret the production of the pair of same-charged muons as a result of lepton number violating (LNV) interactions at the neutrino source, which allow us to constrain New Physics.

Keywords: lepton number violation, new physics, neutrino oscillations
PACS: 11.30.Fs,12.60.cn, 14.60Pq,14.60.St

INTRODUCTION

As is well known, the Majorana nature of neutrinos can be established via the observation of $\Delta L = 2$ processes [1]. The parameter characterizing the rate of such transitions, the effective neutrino mass $\langle m_{ll} \rangle \equiv \sum_i U_{li}^2 m_{\nu_i}$, involves a combination of neutrino masses, mixings and phases. It turns out that the only way to access the values of Majorana phases is through observables associated to $\Delta L = 2$ transitions [2]. Note, however, that the measurement of the effective electron-neutrino mass in the neutrinoless double beta decay $(0\nu\beta\beta)$ experiments can not restrict the two Majorana CP violating phases present in the PMNS mixing matrix [3, 4, 2]. This may be expected since in the $(0\nu\beta\beta)$ one measures the lifetime of the decay of two neutrons in a nucleus into two protons and two electrons, which is a CP conserving quantity. Other proposals aiming to gain access to CP-violating phases of Majorana neutrinos using neutrino-antineutrino oscillations were first discussed in [5, 6, 7, 8, 9, 10, 11].

On the other hand, direct bounds on other effective neutrino mass parameters $\langle m_{ll} \rangle$ from present experimental data are very poor. Currently, the strongest bound for the muon-neutrino case from the $K^+ \to \pi^- \mu^+ \mu^+$ branching fraction [12] is only $|\langle m_{\mu\mu} \rangle| \leq$ 0.04 TeV [13], which leads to a negligible constraint on the neutrino masses and CP violating phases.

In section I we describe the mechanism, based on neutrino-antineutrino oscillation, which would allow to derive a strong bound on the effective Majorana mass of the

Journal of Physics: Conference Series **287** (2011) 012026 doi:10.1088/1742-6596/287/1/012026

muon-neutrino $\langle m_{\mu\mu} \rangle$. In addition, it provides a method for detecting the Majorana neutrino CP violating phases through measuring the CP asymmetry of the $\pi^{\pm}$ decay where neutrino-antineutrino oscillation take place. Using the preliminary bound on the neutrino-antineutrino oscillation probability reported by the MINOS Collaboration [14], we derive a bound on $\langle m_{\mu\mu} \rangle$ which improves existing bounds by several orders of magnitude.

In section II we describe the mechanism that would allow us to constrain lepton number violating interactions. In this case we interpret the observation of the final states, same-charged muons at the production and detection of neutrinos, as a result of lepton number violating interactions in pion decays at the neutrino source. Such interactions appear for example in SUSY models with R-Parity violating terms and leptoquark models. In particular in a SUSY model without R-Parity conservation, the radiative contributions are proportional to the R parity couplings λ and λ', which in general are complex. We attempt to impose constraints over these couplings λ and λ' from the current bound on neutrino-antineutrino transitions obtained by the MINOS collaboration [14].

It is worth noting that the probability of a process associated to neutrino oscillation is usually assumed to be factorized into three independent parts: the production process, the oscillation probability and the detection cross section. Here, we adopt the S-matrix amplitude method described in [15], in order to avoid the usual factorization scheme.

NEUTRINO-ANTINEUTRINO OSCILLATION

Let us start by considering a positive charged pion which decays into a virtual neutrino at the space-time location (x,t) together with a positive charged muon. After propagating, the neutrino can be converted into an antineutrino which produces a positive charged muon at the point $(x^{'},t^{'})$ where it interacts with a target, as shown in Fig. 1. For definiteness, we illustrate this process with the production of the neutrino in π^+ decay and its later detection via its weak interaction with a target nucleon N

$$\pi^+(p_1) \to \mu^+(p_2) + \nu^s_\mu(p) \hookrightarrow \overline{\nu_l}^d(p) + N(p_N) \to N'(p_{N'}) + \mu^+(p_l)$$

where the superscript $s(d)$ refers to the virtual neutrino (antineutrino) at the source (detection) vertex. This is a $|\Delta L| = 2$ process. Notice that if two identical anti-muons ($\mu^+(p_2)$ and $\mu^+(p_l)$) are produced at very different space-time locations, well separated in distance and each identified in different detectors, then the total amplitude does not require to be anti-symmetrized.

If one ignores other flavors, the decay amplitude becomes:

$$\begin{aligned} T_{\nu_\mu-\bar{\nu}_\mu}(t) &= (2\pi)^4\delta^4(p_l - p_N + p_{N'} + p_2 - p_1)(G_F V_{ud})^2 (J_{NN'})_\mu f_\pi \\ &\times \sum_i \bar{\nu}_\mu(p_l)\gamma^\mu(1+\gamma_5)\, \not{p}_1 v(p_2) \times U_{\mu i} U_{\mu i}(m_{\nu_i}) \frac{e^{-itE_{\nu_i}}}{2E_{\nu_i}}, \end{aligned} \quad (1)$$

where the relation $\nu_k = \sum U_{k\alpha}\nu_\alpha$ between flavor k and mass α neutrino eigenstates has been used, $f_\pi = 130.4$ MeV is the $\pi^{\pm}$ decay constant, and $J_{NN'}$ parametrizes

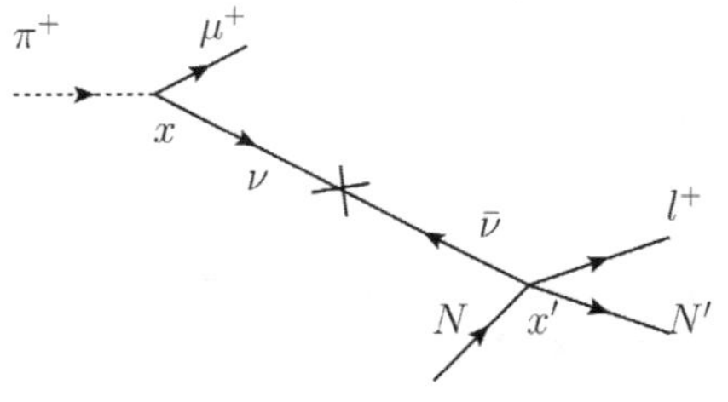

Figure 1. Feynman diagram of the process $\pi(p_1) \to \mu^+(p_2) + \nu_\mu(p)$ followed by the detection process:$\overline{\nu_l}(p) + N(p_N) \to N'(p_{N'}) + \mu^+(p_l)$.

the interaction with the nucleon. Note that, contrary to the case of neutrino-neutrino oscillations [15], only the neutrino mass term survives in this case. For simplicity, one assumes that

$$(J_{NN'})_\mu = \overline{u}_{N'}(p_{N'})\gamma_\mu[g_V(q^2) + g_A(q^2)\gamma_5]u_N(p_N) \tag{2}$$

where we keep only the contributions of leading vector $g_V(q^2)$ and axial-vector $g_A(q^2)$ form factors, with $q = p_{N'} - p_N$.

If we neglect terms of $O(m_\mu/m_{N,N'})$, one obtains

$$\begin{aligned} \left|T_{\nu_\mu-\bar{\nu}_\mu}(t)\right|^2 &= (2\pi)^4\delta^4(p_l - p_N + p_{N'} + p_2 - p_1)(G_F V_{ud})^4|f_\pi|^2 64(g_A-1)^2 m_N m_\mu^2 \\ &\times \sum_{i,j} U_{\mu i}U^*_{\mu j}U_{\mu i}U^*_{\mu j}e^{-it\triangle E_{\nu_{ij}}}\frac{m_{\nu_i}m_{\nu_j}}{4E_{\nu_i}E_{\nu_j}}(E_2 - E_p)\left(\frac{1}{2}(m_\pi^2 - m_\mu^2) + p_l\cdot p_2\right. \\ &- \left.\frac{m_N}{(E_2 - E_p)}G(g_A)p_l\cdot p_2 - 2m_N F(g_A)\left[E_2 - E_l\left(1 + \frac{m_\pi^2}{m_\mu^2}\right)\right]\right) \end{aligned} \tag{3}$$

where $E_2(E_l), E_p$ are, respectively, the initial (final) muon and the pion energies and $\triangle E_{\nu_{ij}} = E_{\nu_i} - E_{\nu_j}$. The functions $F(g_A)$ and $G(g_A)$ are given by: $F(g_A) = \frac{g_A^2+1}{(g_A-1)^2}$, $G(g_A) = \frac{g_A+1}{g_A-1}$. One can easily check that Eq. (3) is not factorizable into (production)×(propagation)×(detection) subprocesses due to the terms proportional to $p_l \cdot p_2 = E_l E_2 - |\vec{p}_l||\vec{p}_2|\cos\alpha$, where α is the angle between the directions of μ^+ particles. This is an important difference with respect to the case of neutrino-neutrino ($\Delta L = 0$) oscillations where it was shown in Ref.[15] that the S-matrix formalism reproduces the hypothesis of factorization of the probabilities.

In the following, we shall neglect the q^2-dependence of the nucleon form factors (namely, we take $g_V = g_V(q^2 = 0) = 1$ and $g_A = g_A(q^2 = 0) \approx -1.27$ [2]). As is well known [16], the cross section of charged current neutrino-nucleon quasielastic scattering is sensitive to the q^2-dependence of these form factors. However, as long as we confine to the CP rate asymmetry for neutrino↔antineutrino oscillations (see below) we expect that the effects of the momentum-transfer dependence of $g_{V,A}$ will partially cancel in the ratio of oscillation rates. Thus, after integration over kinematical variables, it is possible to write the rate of the complete process as

$$\Gamma_{\nu_\mu - \bar{\nu}_\mu} = \left| \sum_i U_{\mu i}^2 \frac{m_{\nu_i}}{2E_{\nu_i}} e^{itE_{\nu_i}} \right|^2 \times F(M,\phi), \tag{4}$$

where $F(M,\phi)$ denotes the kinematical function

$$\begin{aligned} F(M,\phi) &= \frac{32\pi}{E_p}(G_F V_{ud})^4 |f_\pi|^2 (g_A - 1)^2 \left(\left[I_4 - m_N G(g_A) I_5 - \frac{1}{2}(m_\mu^2 - m_\pi^2) I_1 \right] \right. \\ &\times \left. m_\mu^2 - 2m_N F(g_A) \left[m_\mu^2 I_2 - (m_\mu^2 + m_\pi^2) I_3 \right] \right). \end{aligned} \tag{5}$$

The functions I_a for $a = 1,..,4$ can be obtained from the following integral:

$$I_a = \int \frac{d^3p_2}{2E_2} \frac{d^3p_l}{2E_l} (E_2 - E_p) f_a \delta(E_p + E_N - E_{N'} - E_l - E_2), \tag{6}$$

with $f_1 = 1$, $f_2 = E_2$, $f_3 = E_l$, and $f_4 = (p_l \cdot p_2)$ and $f_5 = (p_l \cdot p_2)/(E_2 - E_p)$.

There are two interesting limits for this process. At very short times, (as in short-baseline neutrino experiments), $\Gamma_{\nu_\mu - \bar{\nu}_\mu} \simeq \frac{|\langle m_{\mu\mu} \rangle|^2}{E_\nu^2} \times F(M,\phi)$, where $\langle m_{\mu\mu} \rangle$ is the effective Majorana mass for the muon neutrino. However in the long time limit (long-baseline neutrino experiments) the oscillation terms cannot be neglected. In the limit of $\theta_{13} = 0$, the Majorana phases $\alpha_{1,2}$ are the only sources of CP violation and hence

$$a_{CP} \simeq \tan[2(\alpha_2 - \alpha_3)] \sin\gamma \qquad \text{where } \gamma = \frac{\Delta m_{23}^2 L(km)}{2E_\nu(GeV)} \tag{7}$$

Thus, in the case of LBL neutrino experiment like MINOS where the distance L is given by $L = 735$ km and the energy E_ν is typically around $2-3$ GeV, one finds that $\sin\gamma \sim \mathcal{O}(1)$ [17],[18]. Thus, measuring CP asymmetry will be unavoidable indication for large CP violating Majorana phases.

Application to MINOS results on neutrino-antineutrino oscillations

Recently, MINOS [18] has measured the spectrum of ν_μ events which are missing after travelling 735 km. It is these missing events which are the potential source of $\bar{\nu}_\mu$ appearance. In their preliminary analysis, they were able to put a limit on the fraction of muon neutrinos transition to muon anti-neutrinos [14]: $P(\nu_\mu \to \bar{\nu}_\mu) < 0.026$ (90% c.l.). Assuming CPT, this limit can be written as $\frac{\Gamma_{\nu_\mu - \bar{\nu}_\mu}}{\Gamma_{\nu_\mu - \nu_\mu}} < 0.026$. In the limit of ultrarelativistic neutrinos, $E_{\nu_i} \simeq E_\nu(1 + m_{\nu_i}^2/2E_\nu)$, keeping only the leading terms in the m_{ν_i}/E_ν terms, and using our expresion for the total rate, we get

$$\left| \sum_i U_{\mu i}^2 m_{\nu_i} e^{it\frac{m_{\nu_i}^2}{2E_\nu}} \right|^2 \lesssim 0.001 \times E_\nu^2 \tag{8}$$

Journal of Physics: Conference Series **287** (2011) 012026 doi:10.1088/1742-6596/287/1/012026

To illustrate the usefulness of this relation, let us consider the general case of 3 generations. In this case, one finds

$$\begin{aligned} 0.001 \quad \times \quad & E_\nu^2 \gtrsim \left|\langle m_{\mu\mu}\rangle\right|^2 - 4\sum_{i>j} \mathrm{Re}\left(U_{\mu i}^2 U_{\mu j}^{*2}\right) m_{\nu_i} m_{\nu_j} \sin^2\frac{\Delta m_{ij} L}{4E_\nu} \\ - \quad & 2\sum_{i>j} \mathrm{Im}\left(U_{\mu i}^2 U_{\mu j}^{*2}\right) m_{\nu_i} m_{\nu_j} \sin\frac{\Delta m_{ij} L}{2E_\nu} \end{aligned} \tag{9}$$

Assuming that the only phases that appear in the neutrino mixing matrix are the Majorana phases, it is possible to get a bound on the effective muon-neutrino Majorana mass, only depending on the values of the Majorana phases as the oscillation terms cannot be neglected. In such a case, Eq.(9) can be written as:

$$\begin{aligned} 0.001 \times E_\nu^2 \quad \gtrsim \quad & \left|\langle m_{\mu\mu}\rangle\right|^2 \pm 4\,\sin\left(\frac{\gamma_{32}}{2}\right) m_{\nu_2} m_{\nu_3} |U_{\mu 2}^2 U_{\mu 3}^{*2}| \sin\left(2\alpha_2 - 2\alpha_3 \pm \frac{\gamma_{32}}{2}\right) \\ - \quad & 4\sin\left(\frac{\gamma_{21}}{2}\right) m_{\nu_2} m_{\nu_1} |U_{\mu 1}^2||U_{\mu 2}^{*2}| \sin\left(\alpha_1 + \frac{\gamma_{21}}{2}\right) \\ \pm \quad & 4\sin\left(\frac{\gamma_{31}}{2}\right) m_{\nu_1} m_{\nu_3} |U_{\mu 1}^2||U_{\mu 3}^{*2}| \sin\left(\alpha_2 \pm \frac{\gamma_{31}}{2}\right) \end{aligned} \tag{10}$$

where $\gamma_{ij} = \frac{\Delta m_{ij} L(km)}{2E_\nu(GeV)}$, and the positive and negative signs refer to normal and inverted hierarchies, respectively.

We can further neglect $\sin\gamma_{21} \approx 0$ such that $m_{\nu_1} \approx m_{\nu_2}$ at first order in $\mathcal{O}(\frac{\Delta m_{12}^2}{4E_\nu})$ for the fixed experimental parameters in MINOS, then Eq.(10) can be written as

$$0.001 \times E_\nu^2 \gtrsim \left|\langle m_{\mu\mu}\rangle\right|^2 + A(\alpha_1, \alpha_2) m_{\nu_2} m_{\nu_3} \gtrsim \left|\langle m_{\mu\mu}\rangle\right|^2 \tag{11}$$

where the coefficient $A(\alpha_1, \alpha_2)$ is a function of the Majorana phases. In Fig. (2) we show the regions in which $A(\alpha_1, \alpha_2) > 0$ is satisfied for both cases of normal or inverted hierarchies, and therefore we can find a stringent bound on the effective Majorana mass.

Thus, using $E_\nu \approx 2$ GeV, one gets the following bound on $|\langle m_{\mu\mu}\rangle| \lesssim 64$ MeV.

Over the excluded regions it is not possible to get a conservative bound on $|\langle m_{\mu\mu}\rangle|$ without making extra assumptions on the neutrino mass matrix, however Eq.(9) can be used to bound Majorana parameters (masses and phases) which appear in $|\langle m_{\mu\mu}\rangle|$. As an example, if $0 \leq 2(\alpha_2 - \alpha_3) \leq \pi - \gamma/2$ and assuming that the effective muon-neutrino Majorana mass is dominated by m_{ν_2} and m_{ν_3} (the two flavour limit case) , it is still possible to get the following conservative bound: $|\langle m_{\mu\mu}\rangle| \lesssim 109$ MeV.

The bound obtained above for $|\langle m_{\mu\mu}\rangle|$ is a factor of 3 above the trivial kinematical bound $m_\pi - m_\mu \approx 34$ MeV that is allowed for the (on-shell) muon neutrino in pion decay. However, this kinematical bound applies only to the effective mass of a lepton-number conserving muon neutrino ($\pi^+ \to \mu^+ \nu_\mu$).

A bound on the effective Majorana mass of the muon neutrino, which is independent of the mass hierarchies and Majorana phases, can be obtained using the fluxes of ν_μ

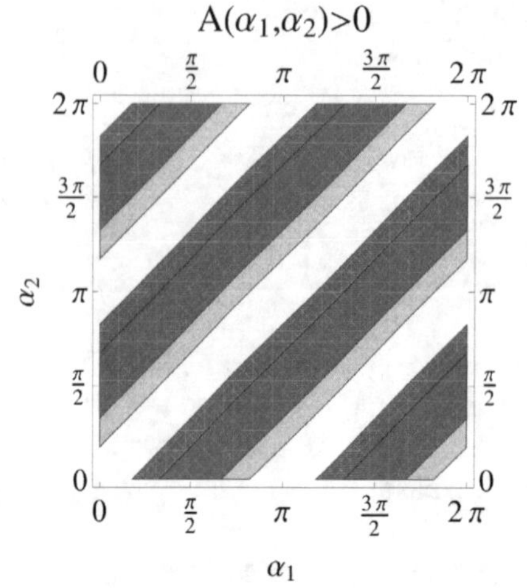

Figure 2. Region of the parametric space (α,β) for which the $A(\alpha,\beta)$ coefficient is positive. The light and dark gray zones correspond to an inverted and normal hierarchy schemes respectively.

and $\bar{\nu}_\mu$ measured with the near detector of the MINOS experiment [19]. Since the near detector is located L=1.04 km away from the target and for neutrino energies above 1 GeV, all oscillatory terms in Eq. (5) are equal to 1. Under the assumption that the excess of $\bar{\nu}_\mu$ events arises from $\nu_\mu \to \bar{\nu}_\mu$ transitions we get (note the muon-neutrino and muon-antineutrino total cross sections induced by charged currents are flat for neutrino energies above 2 GeV [2]):

$$\frac{|\langle m_{\mu\mu}\rangle|^2 \int \frac{dE_\nu}{E_\nu^2}}{\int dE_\nu} \leq \frac{\int (\Phi^{Obs}_{\bar{\nu}_\mu}(E_\nu) - \Phi^{MC}_{\bar{\nu}_\mu}(E_\nu))dE_\nu}{\int \Phi^{Obs}_{\nu_\mu}(E_\nu)dE_\nu} . \tag{12}$$

where $\Phi^{Obs(MC)}_{\nu,\bar{\nu}}$ denote the observed(expected) fluxes. Using the expected and measured integrated fluxes by the MINOS collaboration for the energy region $5 \leq E_\nu \leq 50$ GeV, we get the following bound: $|\langle m_{\mu\mu}\rangle| \leq 2.7$ GeV , which looks rather poor compared to the value reported above.

LEPTON NUMBER VIOLATION AT THE NEUTRINO SOURCE

Let us consider next a virtual neutrino(antineutrino) produced together with a positively(negatively) charged muon at the space time location (x,t), it the travels to (x',t') and is detected there because it interacts with a target producing a positively(negatively) charged lepton. We are assuming this $\Delta L = 2$ process is due to NSI interactions at the production vertex. For defitness, we illustrate this process with production of a antineutrino(neutrino) in the $\pi^+(\pi^-)$ decay and its late detection via its weak interaction with a target nucleon N.

$$\pi^+(p_1) \to \mu^+(p_2) + \nu^c_s(p) \hookrightarrow \nu^c_d(p) + N(p_N) \to N'(p_{N'}) + l^+(p_l) \tag{13}$$

This effective states are not necessarily ΔL conserving once NSI interactions are introduced. If we assume that neutrinos are left-handed, $\Delta L = 2$ semi-leptonic interactions can be described by the following effective Hamiltonian,

$$\mathscr{H} = 2\sqrt{2}G_F V_{ud}\Big\{C_1^k\left(\overline{\nu_k^c}\gamma^\alpha P_R\mu\right)\left(\overline{d}\gamma_\alpha P_R u\right) + C_2^k\left(\overline{\nu_k^c}\gamma^\alpha P_R\mu\right)\left(\overline{d}\gamma_\alpha P_L u\right) + C_{(3,4)}^k\left(\overline{\nu_k^c}P_L\mu\right)\left(\overline{d}P_{(R,L)}u\right) + C_{5(R,L)}^k\left(\overline{\nu_k^c}\sigma_{\alpha\beta}P_R\mu\right)\left(\overline{d}\sigma^{\alpha\beta}P_{(R,L)}u\right)\Big\}\,, \quad (14)$$

where ν_k denotes a neutrino with flavor k and $P_{R,L} = (1\pm\gamma_5)/2$.
In the following and for simplicity, we consider the case where lepton number violation occurs only at the π^+ decay vertex. Note that the tensor currents proportional to the $C_{5L(R)}^k$ Wilson coefficients will not contribute to π^+ decay because it is not possible to generate an antisymmetric tensor from the pion momentum alone. Thus, the only non-vanishing hadronic matrix elements at the production vertex are:

$$\langle 0|\overline{d}\gamma^\mu\gamma_5 u|\pi^+\rangle = if_\pi p_\pi^\mu\ ,\ \langle 0|\overline{d}\gamma_5 u|\pi^+\rangle = \frac{-if_\pi m_\pi^2}{m_u+m_d}\,, \quad (15)$$

where $f_\pi = 130$ MeV is the pion decay constant and $m_{u,d}$ denote the light quark masses. Again we shall neglect the q^2-dependence of the nucleon form factors (namely, we take $g_V = g_V(q^2=0) = 1$ and $g_A = g_A(q^2=0) \approx -1.27$ [2]). In this case the total rate of the $\Delta L = 2$ is factorized:

$$|T_{\nu_\mu-\nu_\mu^c}(\tau)|^2 \simeq \Gamma_\pi\frac{1}{2}(A^2m_\pi^2 + B^2m_\mu^2 - 2\Re[AB^*])P(\nu_\mu^c\to\nu_\mu^c)\sigma_\nu$$

with $\tau = (t'-t) > 0$ is the time elapsed from the production to the detection space-time locations of neutrinos. The A, B coefficients are given by,

$$B_\mu \equiv -i(C_1^{\mu*} - C_2^{\mu*})\ ,\ A_\mu = \frac{-im_\pi^2}{m_u+m_d}(C_3^{\mu*} - C_4^{\mu*}) \quad (16)$$

We shall attempt to constrain the LNV parameters with the MINOS preliminary results [14]: $P(\nu_\mu \to \bar\nu_\mu) < 0.026$ (90% c.l.)., this is,

$$(A^2m_\pi^2 + B^2m_\mu^2 - 2\Re[AB^*]) \lesssim 0.05 \quad (17)$$

As an example let us apply the previous formalism to the MSSM without R-Parity conservation, but requiring the conservation of baryon number to ensure the proton is stable. The must superpotential that conserves bayon number is

$$W_{\Delta L=1} = \lambda_{ijk}L_i\cdot L_j\bar e_k + \lambda'_{ijk}L_i\cdot Q_j\bar d_k + \mu_L^iL_i\cdot H_u \quad (18)$$

here the "·" is a $SU(2)$ product, therefore λ_{ijk} is antisymmetric in i, j. The summation over the generation indices $i, j, k = 1,2,3$ is understood. The bar symbol represents an antiparticle and not the Dirac conjugation. We can expand the Yukawa terms in the

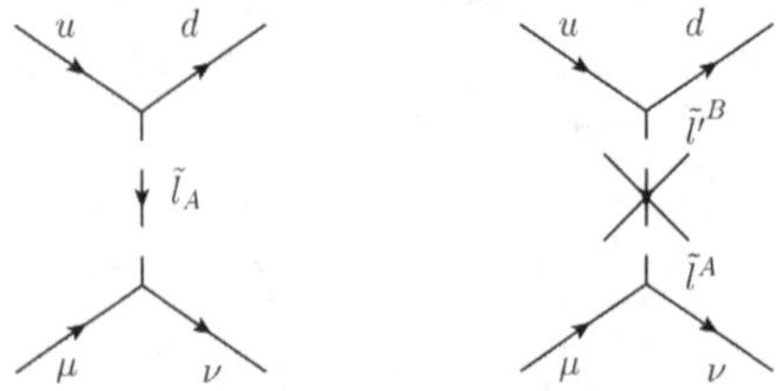

Figure 3. $\Delta L = 2$ SUSY contribution to the $\pi^+ \to \mu^+ + \nu^c$

Lagrangian as

$$\begin{aligned} L &= \lambda_{ijk}\left[\tilde{\nu}^i{}_L \overline{e^k_R} e^j_L + \tilde{e}^j{}_L \overline{e^k_R} \nu^i_L + \tilde{e}^{*k}{}_R \overline{(\nu^i_L)^c} e^j_L - (i \leftrightarrow j)\right] \\ &+ \lambda'_{ijk}\left[\tilde{\nu}^i_L \overline{d^k_R} d^j_L + \tilde{d}^j_L \overline{d^k_R} \nu^i_L + \tilde{d}^{*k}{}_R \overline{(\nu^i_L)^c} d^j_L - \tilde{e}^i_L \overline{d^k_R} u^j_L \right. \\ &- \left. \tilde{u}^j_L \overline{d^k_R} e^i_L - \tilde{d}^{*k}{}_R \overline{(e^i_L)^c} u^i_L\right] + h.c. \end{aligned} \tag{19}$$

The corresponding effective hamiltonian (Fig.3) is therefore,

$$H_{eff} = \frac{G_F}{\sqrt{2}} V_{ud} \sum_i \left((\bar{\nu}\gamma^\rho P_L \mu)(\bar{d}\gamma_\rho P_L u) + C_2 (\bar{\nu} P_R \mu)(\bar{d} P_L u) + C_3 (\overline{\nu^c} P_L \mu)(\bar{d} P_L u) \right) \tag{20}$$

The Wilson coefficients can be expressed as $C_i = C_i^{SM} + C_i^{SUSY}$. For $i = 2, 3$ the C_i^{SM} vanish identically, and $C_1 = 1$. In this respect, the Wilson coefficients C_i are given by

$$C_2 = \frac{\sqrt{2}}{G_F V_{ud}} \left(\frac{1}{\tilde{m}^2}\right) (\lambda^*_{122} \lambda'_{211} + \lambda_{232} \lambda_{311}) \ , \ C_3 = \frac{\sqrt{2}}{G_F V_{ud}} \left(\frac{1}{\tilde{m}^4}\right) \sum_{A,B} \lambda_{12A} \lambda'_{B11} (\delta^l_{RL})_{AB} \tag{21}$$

Using our previous results (Eq. 17) for typical slepton masses of 200 GeV, we obtain $\sum_{A,B} \lambda_{12A} \lambda'_{B11} (\delta^l_{RL})_{AB} \lesssim 10^{-4}$.

CONCLUSIONS

The production of leptons with same charges at the production and detection vertices of neutrinos will be a clear manifestation of $|\Delta L| = 2$ processes. One interesting result is that the time evolution probability of the whole process is not factorizable into production, oscillation and detection probabilities, as is the case in neutrino oscillations [15]. We find that, for very short times of propagation of neutrinos, the observation of $\mu^+\mu^+$ events would lead to a direct bound on the effective mass of muon Majorana neutrinos. In the case of long-baseline neutrino experiments, the CP rate asymmetry for production of $\mu^+\mu^+/\mu^-\mu^-$ events would lead to direct bounds on the difference of CP-violating Majorana phases. Finally, using the current bound on muon neutrino-antineutrino oscillations reported by the MINOS Collaboration we are able to set the bound $\langle m_{\mu\mu} \rangle \lesssim 64$ MeV, which is the first direct limit on the neutrino masses, although it is still several orders of magnitude below current indirect bounds reported in the literature. Future results from MINOS are expected from the analysis of twice the data set used to get the bound reported so far [14] and quoted in Eq. (12) above. Since current uncertainties in the observed and expected number of $\bar{\nu}_\mu$ events are dominated by statistical errors [14], we could expect only a slight improvement by a factor $1/\sqrt{2}$ on the effective Majorana mass of the muon neutrino. Neutrino factories may improve this bound by more than one order of magnitude.

We also show the possibility to constrain LNV interactions, such as the λ, λ' terms in the SUSY superpotential with R-Parity violation, for the case in which the observation of the final states, same-charged muons at the production and detection of neutrinos, is a result of non standard lepton number violating interactions at the neutrino source.

ACKNOWLEDGEMENT

This work was also partially supported by Conacyt (Mexico) and by PROMEP project. The work of S.K. is supported in part by the Science and Technology Development Fund (STDF) Project ID 437, ICTP project 30 and the Egyptian Academy of Scientific Research and Technology. The authors would like to thank M. Bishai, A. Aguilar-Arevalo and A. Himmel for useful discussions.

REFERENCES

1. M. Doi, T. Kotani, H. Nishiura, K. Okuda and E. Takasugi, Phys. Lett. B **102**, 323 (1981); J. Schechter and J.W.F. Valle, Phys. Rev. D 22, 2227 (1980); S. Bilenky, J. Hosek, and S. Petcov, Phys. Lett. 94B, > 495 (1980).
2. C. Amsler *et al.* [Particle Data Group], Phys. Lett. B **667**, 1 (2008).
3. F. T. . Avignone, S. R. Elliott and J. Engel, Rev. Mod. Phys. **80**, 481 (2008).
4. M. C. Gonzalez-Garcia and M. Maltoni, Phys. Rept. **460**, 1 (2008).
5. J. Schechter and J. F. Valle, Phys. Rev. D **23**, 1666 (1981).
6. L. F. Li and F. Wilczek, Phys. Rev. D **25**, 143 (1982).
7. J. Bernabeu and P. Pascual, Nucl. Phys. B **228**, 21 (1983).
8. B. Kayser, Phys. Rev. D **30**, 1023 (1984).
9. G. Barenboim, J. Bernabeu and O. Vives, Phys. Rev. Lett. **77**, 3299 (1996).
10. J. N. Bahcall and H. Primakoff, Phys. Rev. D **18**, 3463 (1978).
11. P. Langacker and J. Wang, Phys. Rev. D **58**, 093004 (1998).
12. R. Appel et al, Phys. Rev. Lett. 85, 2877 (2000).
13. M. Flanz, W. Rodejohann, and K. Zuber, Phys. Lett. B473, 324 (2000); 480, 418(E) (2000); W. Rodejohann, J. Phys. G28, 1477 (2002).
14. A. I. Himmel [MINOS Collaboration], AIP Conf. Proc. **1182**, 56 (2009); I. Danko [MINOS Collaboration], arXiv:0910.3439 [hep-ex].
15. D. Delepine, V. G. Macias, S. Khalil and G. Lopez Castro, Phys. Rev. D **79**, 093003 (2009).
16. M. Fukugita and T. Yanagida, *Physics of Neutrinos and Applications to Astrophysics*, Springer-Verlag, Berlin (2003).
17. K. E. Arms [MINOS and NOνA Collaborations], Nucl. Phys. Proc. Suppl. **189**, 271 (2009).
18. A. Blake [MINOS Collaboration], J. Phys. Conf. Ser. **120**, 052041 (2008).
19. P. Adamson et al. [MINOS Collaboration], Phys. Rev. **D81**, 072002 (2010).

Journal of Physics: Conference Series **287** (2011) 012027 doi:10.1088/1742-6596/287/1/012027

Spinning particle approach to higher spin field theory

Olindo Corradini

Centro de Estudios en Física y Matemáticas Basicas y Aplicadas
Universidad Autónoma de Chiapas, Tuxtla Gutiérrez, Chiapas, México

Dipartimento di Fisica, Università di Bologna
via Irnerio 46, I-40126 Bologna, Italy

INFN, Sezione di Bologna
via Irnerio 46, I-40126 Bologna, Italy

E-mail: Olindo.Corradini@bo.infn.it

Abstract. We shortly review on the connection between higher-spin gauge field theories and supersymmetric spinning particle models. In such approach the higher spin equations of motion are linked to the first-class constraint algebra associated with the quantization of particle models. Here we consider a class of spinning particle models characterized by local $O(N)$-extended supersymmetry since these models are known to provide an alternative approach to the *geometric* formulation of higher spin field theory. We describe the canonical quantization of the models in curved target space and discuss the obstructions that appear in presence of an arbitrarily curved background. We then point out the special role that conformally flat spaces appear to have in such models and present a derivation of the higher-spin curvatures for maximally symmetric spaces.

1. Introduction

Particle models –through the worldline formalism– provide an efficient alternative method, to conventional second-quantized field theories, for the computation of quantum field theory Feynman diagrams and effective actions. Moreover some spinning particle models can also be applied to the first quantization of higher spin fields. In particular, from the canonical quantization of $O(N)$-extended locally super-symmetric particle models, the first-class constraints yield the equations of motions for free spin-$\frac{N}{2}$ fields in terms of the corresponding linearized curvatures, that in this approach are the wave functions of the particle models. In other words the quantization of $O(N)$-extended susy particle models is related to the so-called "geometric formulation" of higher spin field theory. Such formulation was proposed in [1, 2] (using the the linearized higher spin geometry developed in [3, 4]) in order to relax algebraic constraints present in the conventional formulation given in terms of higher spin potentials, and investigate alternatives to introduce interactions in the higher spin fields, a problem that still constitutes an open field of research [5, 6].

Toward an alternative approach to the geometric formulation we studied the quantization of $O(N)$ spinning particle models [7, 8, 9, 10]

$$S = \int dt\Big[p_\mu \dot{x}^\mu + \frac{i}{2}\psi_{i\mu}\dot{\psi}_i^\mu - e\underbrace{\Big(\frac{1}{2}p_\mu p^\mu\Big)}_{H} - i\chi_i\underbrace{\Big(p_\mu \psi_i^\mu\Big)}_{Q_i} - \frac{i}{2}a_{ij}\underbrace{\Big(\psi_i^\mu \psi_{j\mu}\Big)}_{J_{ij}}\Big], \quad i = 1, \ldots, N \qquad (1)$$

both using canonical quantization [11, 12] and path integrals [13]. Above (x^μ, p_μ) are bosonic canonical coordinates, ψ^i_μ their fermionic partners and (e, χ_i, a_{ij}) are the gauge fields of the local worldline-Poincaré symmetry, local extended supersymmetry and local $SO(N)$ gauge symmetry respectively. The canonical quantization of the flat target space models described by (1) and their curved deformations is analyzed below. In particular for conformally flat spaces it is shown that the constraint algebra closes, although being non linear and for a further restricted class of backgrounds, characterized by maximal symmetry, the quadratically deformed constraint algebra is solved and linearized higher-spin curvatures are found, for an arbitrary N and an arbitrary (even) space-time dimension $D = 2d$; for $D = 2d + 1$ and $N > 2$ these models are quantum mechanically trivial due to a global anomaly [8, 9]. The path integral quantization is very interesting in that the worldline on the circle of the above action yields the effective action of quantum higher spin fields coupled to external backgrounds. In [13] we computed the worldline path integral on the circle with flat background and we found that it correctly reproduces the number of degrees of freedom of the propagating higher spin field, that is the information included in the flat space one-loop effective action. Extensions to curved space backgrounds are under study and will be published shortly.

In the present report we restrict our study to $O(N)$ spinning particles. However it is worth mentioning other particle models relevant to higher spin field physics, such as twistor-like superparticles [14, 15], $U(N)$ spinning particle [16, 17], and $Sp(2r)$ models [18].

2. Symmetries of the models

The above flat space action (1) is invariant under the local symmetry generated by the first class constraints (H, Q_i, J_{ij}) associated to the aforementioned gauge fields. Transformations of the canonical variables can be obtained with the help of elementary Poisson brackets $\{x^\mu, p_\nu\}_{PB} = \delta^\mu_\nu$ and $\{\psi^i_\mu, \psi^j_\nu\}_{PB} = -i\delta^{ij}\eta_{\mu\nu}$, whereas transformations of the gauge fields can be obtained using the first class algebra

$$\begin{aligned}
\{Q_i, Q_j\}_{PB} &= -i2\delta_{ij}H \\
\{J_{ij}, Q_k\}_{PB} &= \delta_{jk}Q_i - \delta_{ik}Q_j \\
\{J_{ij}, J_{kl}\}_{PB} &= \delta_{jk}J_{il} - \delta_{ik}J_{jl} + \delta_{jl}J_{ik} - \delta_{il}J_{jk}
\end{aligned} \tag{2}$$

and imposing the invariance of the action under the local transformation generated by $G = \xi H + i\epsilon_i Q_i + \frac{1}{2}\alpha_{ij}J_{ij}$. One thus gets

$$\begin{aligned}
&\delta x^\mu = \{x^\mu, G\}_{PB} = \xi p^\mu + i\epsilon_i\psi^\mu_i\,, \quad \delta p_\mu = \{p_\mu, G\}_{PB} = 0 \\
&\delta\psi^\mu_i = \{\psi^\mu_i, G\}_{PB} = -\epsilon_i p^\mu + \alpha_{ij}\psi^\mu_j \\
&\delta e = \dot{\xi} + 2i\chi_i\epsilon_i\,, \quad \delta\chi_i = \dot{\epsilon}_i - a_{ij}\epsilon_j + \alpha_{ij}\chi_j \\
&\delta a_{ij} = \dot{\alpha}_{ij} + \alpha_{im}a_{mj} + \alpha_{jm}a_{im}\,.
\end{aligned} \tag{3}$$

The rigid symmetries involve transformations under the target space Poincaré group that ensures the relativistic invariance of the model. In addition the model is conformally invariant. To prove conformal invariance we can first obtain the background symmetries of the model where background fields, such as the metric, transform as well. In order to do so we rewrite the flat space action (1) using arbitrary coordinates

$$S = \int dt\Big[p_\mu\dot{x}^\mu + \frac{i}{2}\psi_{i\mu}\dot{\psi}^\mu_i - e\underbrace{\left(\frac{1}{2}\pi_\mu\pi^\mu\right)}_{H_0} - i\chi_i\underbrace{\left(\psi^a_i e^\mu_a\pi_\mu\right)}_{Q_i} - \frac{i}{2}a_{ij}\underbrace{\left(\psi^a_i\psi_{ja}\right)}_{J_{ij}}\Big] \tag{4}$$

where tangent space indices are used for the fermions and a vielbein is accordingly introduced, $\psi^a_i = e^a_\mu(x)\psi^\mu_i$, and for later convenience the hamiltonian has been renamed H_0. A spin

Journal of Physics: Conference Series **287** (2011) 012027 doi:10.1088/1742-6596/287/1/012027

connection $\omega_{\mu ab}$ can be uniquely constructed and in turn the covariant momenta read

$$\pi_\mu = p_\mu - \frac{i}{2}\omega_{\mu ab}\psi_i^a\psi_i^b \tag{5}$$

that satisfy the bracket relations

$$\{\pi_\mu, \pi_\nu\}_{PB} = \frac{i}{2}R_{\mu\nu ab}\psi_i^a\psi_i^n \; . \tag{6}$$

It is thus now easy to check that action (4) is invariant under (i) target space diffeomorphisms, (ii) local Lorentz transformations and (iii) Weyl rescalings. As a consequence of the latter background symmetries there exist conformal Killing vectors that leave the metric invariant. Hence the model (4) is invariant under rigid symmetries belonging to the conformal group $SO(D,2)$ [19, 20]. Moreover background Weyl symmetry guarantees that the spinning particle is consistent in any conformally flat space.

2.1. First class algebra on conformally flat spaces

It is instructive to compute the constraint algebra on an arbitrarily curved background. With the only help of the elementary Poisson brackets discussed above, for the nonvanishing brackets one obtains (the $SO(N)$ subalgebra generated by J_{ij} holds unchanged)

$$\begin{aligned} \{Q_i, Q_j\}_{PB} &= -2i\delta_{ij}H_0 + \frac{i}{2}R_{abcd}\psi_i^a\psi_j^b\psi^c\cdot\psi^d \\ \{Q_i, H_0\}_{PB} &= -\frac{i}{2}\pi^a R_{abcd}\psi_i^b\psi^c\cdot\psi^d \end{aligned} \tag{7}$$

that generically fails to be first class. A possible option, as suggested by the results obtained in the previous section, is to restrict attention to conformally flat spaces. In such a case, the Weyl tensor identically vanishes and the Riemann tensor can be written as combination of Ricci and scalar tensor, so that the above algebra reduces to

$$\begin{aligned} \{Q_i, Q_j\}_{PB} &= -2i\delta_{ij}H_0 - \frac{iR}{(D-1)(D-2)}J_{ik}J_{jk} - \frac{R_{ab}}{D-2}\Big(\psi_i^a\psi_k^b J_{jk} + (i\leftrightarrow j)\Big) \\ \{Q_i, H_0\}_{PB} &= \frac{R}{(D-1)(D-2)}Q_k J_{ki} + \frac{R_{ab}}{D-2}\Big(\pi^a\psi_k^b J_{ik} + i\psi_i^a\psi_k^b Q_k\Big) \end{aligned} \tag{8}$$

which becomes first class, though with structure *functions* rather than structure *constants*.

The above algebra further simplifies if we consider maximally symmetric spaces –that are a subclass of conformally flat spaces– for which the Riemann tensor is uniquely fixed in terms of the curvature scalar $R_{abcd} = b(\eta_{ac}\eta_{bd} - \eta_{ad}\eta_{bc})$ with $b \equiv \frac{R}{D(D-1)}$. Hence (8) reduces to

$$\{Q_i, Q_j\}_{PB} = -2i\delta_{ij}H + ib\Big(J_{ik}J_{jk} - \frac{1}{2}\delta_{ij}J_{kl}J_{kl}\Big)\,, \quad \{Q_i, H\}_{PB} = 0 \tag{9}$$

where we have introduced the improved hamiltonian

$$H = H_0 + \Delta H = \frac{1}{2}\pi^a\pi_a - \frac{b}{4}J_{kl}J_{kl} \; . \tag{10}$$

In the remaining of this report we will consider the canonical quantization of the quadratically deformed fist-class constraint algebra given in (9).

3. Canonical quantization

In order to canonically quantize the models we need to promote (bosonic and fermionic) phase space variables to operators: the fundamental (anti) commutations relations read

$$[x^\mu, p_\nu] = i\delta^\mu_\nu\,, \qquad \{\psi_i^a, \psi_j^b\} = \eta^{ab}\delta_{ij} \tag{11}$$

so that the fermionic variables give rise to a (multi)-Clifford algebra and can be represented in terms of (tensor products of) gamma matrices and in turn the wave function can be written as a multispinor $R_{\alpha_1,\ldots,\alpha_N}$. We also need to define constraints with suitable ordering, so that the algebra remains first class at the quantum level. The ordering is again immediate for the $SO(N)$ subalgebra, $J_{ij} = \frac{i}{2}[\psi_i^a, \psi_{aj}]$. The ordering in the susy operators is uniquely fixed by replacing the covariant momentum with covariant derivatives acting on multispinorial wave functions, $Q_i = \psi_i^a e_a^\mu \Big(p_\mu - \frac{i}{2}\omega_{\mu ab}\psi_k^a\psi_k^b\Big)$. The non vanishing part of the algebra thus reads

$$[J_{ij}, J_{kl}] = \ SO(N) \text{ algebra} \tag{12}$$

$$\{Q_i, Q_j\} = 2\delta_{ij}H - \frac{b}{2}\Big(J_{ik}J_{jk} + J_{jk}J_{ik} - \delta_{ij}J_{kl}J_{kl}\Big) \tag{13}$$

$$[J_{ij}, Q_k] = i\delta_{jk}Q_i - i\delta_{ik}Q_j \tag{14}$$

where $H = \frac{1}{2}(\pi^a\pi_a - i\omega^a{}_{ab}\pi^b) - \frac{b}{4}J_{kl}J_{kl} - bA(D)$, with $A(D) = (2-N)\frac{D}{8} - \frac{D^2}{8}$. These constraints are to be imposed à la Dirac-Gupta-Bleuler onto the wave function. Hence, in the above Clifford basis, the susy constraints are nothing but N Dirac equations and along with the $SO(N)$ constraints give rise (in flat 4d space) to the so-called Bargmann-Wigner equations for the spin-$\frac{N}{2}$ field. Below we make use of a different basis for the Grassmann variables in which the higher spin geometry emerges in a more direct way.

3.1. Integer higher-spin fields in (A)dS space

We consider the case when the number of susy charges is even $N = 2S$, $S \in \mathbf{N}$. In this case we can complexify the fermions

$$\psi_I^a = \frac{1}{\sqrt{2}}(\psi_i^a + i\psi_{i+S})\,, \quad \psi^{aI} = \frac{1}{\sqrt{2}}(\psi_i^a - i\psi_{i+S})\,, \quad i, I = 1, \ldots, S \tag{15}$$

so that

$$\{\psi_I^a, \psi^{bJ}\} = \eta^{ab}\delta_I^J \tag{16}$$

and we can represent them using a coherent state basis $\langle\psi|$ where ψ_I^a act multiplicatively whereas $\psi^{aI} = \frac{\partial}{\partial\psi_{aI}}$ act as left derivatives. Therefore the generic wave function will be a combination of antisymmetric (multi) forms

$$\langle\psi| \otimes \langle x|R\rangle = \sum_{A_i=0}^{D} R_{\mu_1\cdots\mu_{A_1},\cdots,\nu_1\cdots\nu_{A_S}}(x)\psi_1^{\mu_1}\cdots\psi_1^{\mu_{A_1}}\cdots\psi_S^{\nu_1}\cdots\psi_S^{\mu_{A_S}}\,. \tag{17}$$

All other operators involving fermions get complexified accordingly, $Q_i \to (Q_I, Q^I)$, $J_{ij} \to (J_I{}^J, K_{IJ}, K^{IJ})$, with

$$Q_I = \psi_I^a e_a^\mu \pi_\mu\,, \quad Q^I = \psi^{aI} e_a^\mu \pi_\mu \tag{18}$$

$$K_{IJ} = \psi_I \cdot \psi_J \quad K^{IJ} = \psi^I \cdot \psi^J\,, \quad J_I{}^J = \psi_I \cdot \psi^J - d\,\delta_I^J \tag{19}$$

and we can select as set of independent constraints $(J_I{}^J, Q_I, K^{IJ})$. In the above complex basis once this set of constraints is satisfied the remaining constraints are satisfied automatically thanks to the first class algebra; here we will not report the whole algebra written in the complex basis and address the interested reader to the original papers where it is explicitly worked out [11, 12].

The $J_I{}^J$ constraints are *irreducibility* constraints in that they select out of the above sum a tensor that transforms irreducibly under $GL(D)$. Such a tensor is described by a Young tableau with d rows (note that when $I = J$ the operator selects tensors that have d indices in the I group) and S columns

$R \sim d$ 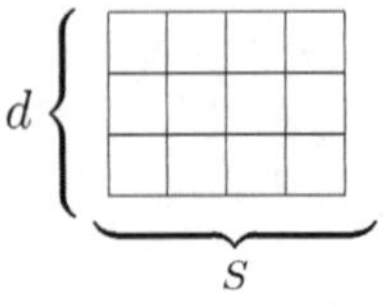 (20)

that is the linearized higher spin curvature; note that for $S = 2$ and $D = 4$ ($d = 2$) it has the symmetry of the Riemann tensor. For generic S and D it is a higher spin curvature.

The Q_I constraints are *integrability* constraints: their solution leads to write the curvature in terms of a potential (and the integrability conditions convert into differential Bianchi identities)

$$Q_I|R\rangle = 0 \quad \longrightarrow \quad |R\rangle = q|\phi\rangle \tag{21}$$

where q is an S-th order differential operator constructed out of Q_I's and K_{IJ}'s

$$q = \sum_{n=0}^{[S/2]} (-b)^n r_n(S) q_n(S)\,, \quad q_n(S) = \frac{1}{S!}\epsilon^{I_1 \cdots I_S} K_{I_1 I_2} \cdots K_{I_{2n-1} I_{2n}} Q_{I_{2n+1}} \cdots Q_{I_S} \tag{22}$$

and the numerical coefficients $r_n(S)$ are uniquely fixed [11] by imposing conditions (21) and read

$$r_n(S) = \frac{1}{2n} \sum_{k=1}^{n} r_{n-k}(S)\, a_{2k}(S - 2(n-k) + 1)\,, \qquad r_0(S) \equiv 1 \tag{23}$$

where $a_{2k}(S) = f_k\ P(S, 2k)$, with $P(S, m) = S(S-1)\cdots(S-m)$ being the Pochhammer function and f_k the coefficients generated by the Taylor expansion of the tangent function $\tan(z) = \sum_k f_k z^{2k+1}$. Above the relevant part of the algebra is

$$\{Q_I, Q_J\} = b(K_{IL} J_J{}^L + K_{JL} J_I{}^L) \tag{24}$$

and the commutators among J_I^J and other operators, that can be summarized as $[J_I{}^J, U_K^L] = \delta_K^J U_I^L - \delta_I^L U_K^J$. Note that in flat space Q_I's anticommute among themselves and the above operator simply reduces to $q_0 = Q_1 \cdots Q_S$, that correctly reproduces the flat space curvature whereas in (A)dS, again due to (24), corrections appear as powers of the curvature b. The potential is also an irreducible representation of $GL(D)$, that satisfies $J_I{}^J|\phi\rangle = -\delta_I^J|\phi\rangle$ and is described by

$\phi \sim d - 1$ 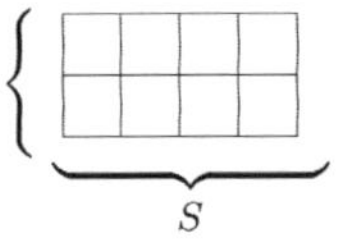 (25)

so that for $D = 4$ it describes a rank-S totally symmetric tensor, the (metric-like) spin-S irreducible representation of the Poincaré group. For $D > 4$ there is no unique massless

Journal of Physics: Conference Series **287** (2011) 012027 doi:10.1088/1742-6596/287/1/012027

irreducible representation of the Poincaré group: all mixed symmetry tensors represented by Young tableaux with S columns are spin-S irreducible representations. In the present formalism we obtain the curvature and (below) the equation of motion for the mixed-symmetry tensor described by (25).

Finally the *trace* constraints K^{IJ} lead to the equation of motion for the higher spin potential

$$0 = K^{IJ}|R\rangle = K^{IJ}q|\phi\rangle = q^{IJ}G|\phi\rangle. \tag{26}$$

Here the operator G is a second order differential operator named Fronsdal-Labastida [21, 22, 23] operator

$$G = -2H_0 + Q_I Q^I + \frac{1}{2}Q_I Q_J K^{IJ} - bK_{IJ}K^{IJ} + b\alpha_S(D) \tag{27}$$

where $\alpha_S(D) = S^2 + A_S(D)$, whereas q^{IJ} is an $(S-2)$-th order operator. The higher derivative equation of motion (26) can be cast into a second-order equation of motion by introducing a compensator field as nontrivial kernel of operator q^{IJ}

$$G|\phi\rangle = |C\rangle\,, \quad q^{IJ}|C\rangle = 0\,. \tag{28}$$

Both the higher derivative equation (26) and the compensated equation (28) are gauge invariant under the transformation $\delta|\phi\rangle = Q_K V^K|\xi\rangle$ with unconstrained gauge parameter $|\xi\rangle$; in (28) the gauge transformation of the compensator is tuned to be equal to the (generically nonvanishing) transformation of the l.h.s. and schematically reads $G\delta|\phi\rangle = \delta|C\rangle \propto K^{IJ}|\xi\rangle$. It is thus possible to perform a partial gauge fixing in order to gauge away the compensator and obtain the Fronsdal-Labastida equation of motion

$$G|\phi\rangle = 0 \tag{29}$$

that is gauge invariant provided the parameter is (for $S > 2$) traceless, $K^{IJ}|\xi\rangle = 0$, and the potential is (for $S > 3$) double-traceless, $K^{IJ}K^{MN}|\phi\rangle = 0$.

3.2. Half-integer higher-spin fields in (A)dS space

The results of the previous section can be extended to the case of half-integer spin corresponding to an odd number of supersymmetry charges, $N = 2r+1, r \in \mathbf{N}$ and $S = r + 1/2$. Now we consider the complexification of the even part of the fermionic fields and represent the Grassmann variables labelled with N as gamma matrices

$$\begin{aligned}\psi^a_I &= \frac{1}{\sqrt{2}}(\psi^a_i + i\psi_{i+r})\,, \quad \psi^{aI} = \frac{1}{\sqrt{2}}(\psi^a_i - i\psi_{i+r})\,, \quad i, I = 1, \ldots, r \\ \psi^a_N &= \frac{1}{\sqrt{2}}\gamma^a\end{aligned} \tag{30}$$

so that now the wave function will be a spinor tensor

$$\langle\psi| \otimes \langle\chi_\alpha| \otimes \langle x|R\rangle = \sum_{A_i=0}^{D} R_{\mu_1\cdots\mu_{A_1},\cdots,\nu_1\cdots\nu_{A_r};\alpha}(x)\ \psi_1^{\mu_1}\cdots\psi_1^{\mu_{A_1}}\cdots\psi_r^{\nu_1}\cdots\psi_r^{\mu_{A_r}}\,. \tag{31}$$

where $|\chi_\alpha\rangle$ is a generic state in the fermionic space $\gamma^a|\chi_\alpha\rangle = \gamma^a_{\alpha'\alpha}|\chi_{\alpha'}\rangle$. In turn the first class constraints split as $Q_i \to (Q_I, Q^I, \not{\Pi})$, $J_{ij} \to (J_I{}^J, K_{IJ}, K^{IJ}, L_I, L^I)$, where

$$\not{\Pi} = \gamma^a e^\mu_a \pi_\mu\,, \quad L^I = \gamma^a\psi^I_a\,, \quad L_I = \gamma^a\psi_{aI}\,. \tag{32}$$

Journal of Physics: Conference Series **287** (2011) 012027 doi:10.1088/1742-6596/287/1/012027

Here the fist term is the Dirac operator and the second is the "gamma trace" operator. The first class algebra allows us to choose $(Q_I, J_I{}^J, L^I)$ as an independent set of constraints, since $[L^I, L^J] = -2K^{IJ}$. Again the $J_I{}^J$ constraints are irreducibility constraints picking a single (spinor)-tensor out of the above sum, the one whose tensorial part is described by (20). The integrability constraints now satisfies a slightly more complicated algebra, namely

$$\{Q_I, Q_J\} = b(K_{IL}J_J{}^L + K_{JL}J_I{}^L) + \frac{b}{4}(L_I L_J + L_I L_J) \ . \tag{33}$$

However it is possible to rotate the irreducibility operators as

$$\mathcal{Q}_I^{(\pm)} = Q_I \pm \frac{\sqrt{b}}{2} L_I \tag{34}$$

and select $(\mathcal{Q}_I^{(-)}, J_I{}^J, L^I)$ as independent set of constraints. These operators satisfy the relations

$$\mathcal{Q}_I^{(-)}\mathcal{Q}_J^{(+)} + \mathcal{Q}_J^{(-)}\mathcal{Q}_I^{(+)} = b(K_{IL}J_J{}^L + K_{JL}J_I{}^L) = \mathcal{Q}_I^{(+)}\mathcal{Q}_J^{(-)} + \mathcal{Q}_J^{(+)}\mathcal{Q}_I^{(-)} \tag{35}$$

that is the same as (24). Hence, the integrability conditions in terms of $\mathcal{Q}_I^{(-)}$

$$\mathcal{Q}_I^{(-)}|R\rangle = 0 \quad \longrightarrow \quad |R\rangle = q|\phi\rangle \tag{36}$$

are solved by

$$q = \sum_{n=0}^{[r/2]} (-b)^n r_n(r) q_n(r) \tag{37}$$

$$q_n(r) = \frac{1}{r!}\epsilon^{I_1\cdots I_r} \times \begin{cases} K_{I_1I_2}\cdots K_{I_{2n-1}I_{2n}}\mathcal{Q}^{(+)}_{I_{2n+1}}\mathcal{Q}^{(-)}_{I_{2n+2}}\cdots\mathcal{Q}^{(-)}_{I_r}\,, & r=2p \\ K_{I_1I_2}\cdots K_{I_{2n-1}I_{2n}}\mathcal{Q}^{(+)}_{I_{2n+1}}\mathcal{Q}^{(-)}_{I_{2n+2}}\cdots\mathcal{Q}^{(+)}_{I_r}\,, & r=2p+1 \end{cases}$$

where $r_n(r)$ are the same numerical coefficients of the previous section.

Finally the *gamma trace* constraints lead to the equation of motion for the higher spin potential

$$L^I|R\rangle = L^I q|\phi\rangle = q^I \mathcal{F}_r|\phi\rangle = 0 \tag{38}$$

where

$$\mathcal{F}_r = (-)^{r-1}\left(\not{\Pi} + Q_K L^K\right) + \frac{\sqrt{b}}{2}L^K L_K \tag{39}$$

is the so-called Fang-Fronsdal operator [24, 25]. Similarly to the integer case, the higher-derivative equation (38) can be cast into a linear differential equation by introducing a compensator field that belongs to the kernel of operator q^I; namely

$$\mathcal{F}_r|\phi\rangle = |C\rangle \tag{40}$$

and again both the compensated linear equations and (38) are gauge-invariant, with unconstrained parameter, under the following transformation

$$\delta|\phi\rangle = \begin{cases} \mathcal{Q}_I^{(+)}V^I|\xi\rangle\,, & r=2p \\ \mathcal{Q}_I^{(-)}V^I|\xi\rangle\,, & r=2p+1 \end{cases} \tag{41}$$

upon which $\mathcal{F}_r\delta|\phi\rangle = \delta|C\rangle \propto L^I|\xi\rangle$. Hence, by gauging away the compensator one gets the Fang-Fronsdal equation (generalization of Dirac's and Rarita-Schwinger's) $\mathcal{F}_r|\phi\rangle = 0$ that is gauge-invariant (for $S > 3/2$) if $|\xi\rangle$ is gamma-traceless and (for $S > 5/2$) $|\phi\rangle$ is triple gamma-traceless.

4. Conclusions

We described an alternative approach to the study of higher spin fields, based on locally supersymmetric $O(N)$-extended spinning particle models. In particular, we showed that the first class constraints lead to the geometric formulation of higher spin fields and we computed linearized curvatures, in D-dimensional maximally symmetric spaces and discussed gauge invariance and linearized free equations of motion associated to the vanishing of the (gamma) trace of the curvature. The models discussed above describe a set of mixed-symmetry higher spin potentials corresponding to Young tableaux with $\frac{D}{2} - 1$ rows and S columns, thus reducing to totally symmetric tensors in $D = 4$. We also showed that the constraint algebra remains first class for the whole class of conformally flat spaces.

References

[1] Francia D and Sagnotti A 2002 *Phys. Lett.* **B543** 303–310 (*Preprint* hep-th/0207002)
[2] Francia D and Sagnotti A 2003 *Class. Quant. Grav.* **20** S473–S486 (*Preprint* hep-th/0212185)
[3] Weinberg S 1965 *Phys. Rev.* **138** B988–B1002
[4] de Wit B and Freedman D Z 1980 *Phys. Rev.* **D21** 358
[5] Vasiliev M A 2004 *Fortsch. Phys.* **52** 702–717 (*Preprint* hep-th/0401177)
[6] Sorokin D 2005 *AIP Conf. Proc.* **767** 172–202 (*Preprint* hep-th/0405069)
[7] Gershun V D and Tkach V I 1979 *JETP Lett.* **29** 288–291
[8] Howe P S, Penati S, Pernici M and Townsend P K 1988 *Phys. Lett.* **B215** 555
[9] Howe P S, Penati S, Pernici M and Townsend P K 1989 *Class. Quant. Grav.* **6** 1125
[10] Marnelius R 2009 (*Preprint* 0906.2084)
[11] Bastianelli F, Corradini O and Latini E 2008 *JHEP* **11** 054– (*Preprint* 0810.0188)
[12] Corradini O 2010 *JHEP* **09** 113 (*Preprint* 1006.4452)
[13] Bastianelli F, Corradini O and Latini E 2007 *JHEP* **02** 072– (*Preprint* hep-th/0701055)
[14] Bandos I A, Lukierski J and Sorokin D P 2000 *Phys. Rev.* **D61** 045002 (*Preprint* hep-th/9904109)
[15] Bandos I, Bekaert X, de Azcarraga J A, Sorokin D and Tsulaia M 2005 *JHEP* **05** 031 (*Preprint* hep-th/0501113)
[16] Bastianelli F and Bonezzi R 2009 *JHEP* **03** 063 (*Preprint* 0901.2311)
[17] Bastianelli F and Bonezzi R 2010 *JHEP* **05** 020 (*Preprint* 1003.1046)
[18] Bastianelli F, Corradini O and Waldron A 2009 *JHEP* **05** 017– (*Preprint* 0902.0530)
[19] Marnelius R 1979 *Phys. Rev.* **D20** 2091
[20] Siegel W 1988 *Int. J. Mod. Phys.* **A3** 2713–2718
[21] Fronsdal C 1979 *Phys. Rev.* **D20** 848–856
[22] Fronsdal C 1978 *Phys. Rev.* **D18** 3624
[23] Labastida J M F 1989 *Nucl. Phys.* **B322** 185
[24] Fang J and Fronsdal C 1978 *Phys. Rev.* **D18** 3630
[25] Fang J and Fronsdal C 1980 *Phys. Rev.* **D22** 1361

Journal of Physics: Conference Series **287** (2011) 012028 doi:10.1088/1742-6596/287/1/012028

Dynamical Mass Generation and Confinement in Maxwell-Chern-Simons Planar Quantum Electrodynamics

S Sánchez Madrigal[1], C P Hofmann[2] and A Raya[1],

[1]Instituto de Física y Matemáticas, Universidad Michoacana de San Nicolás de Hidalgo, Edificio C-3, Ciudad Universitaria, Morelia, Michoacán 58040, México.

[2]Facultad de Ciencias, Universidad de Colima, Bernal Díaz del Castillo 340, Colima, Colima 28045, México.

E-mail: saul@ifm.umich.mx, christoph@ucol.mx, raya@ifm.umich.mx,

Abstract. We study the non-perturbative phenomena of Dynamical Mass Generation and Confinement by truncating at the non-perturbative level the Schwinger-Dyson equations in Maxwell-Chern-Simons planar quantum electrodynamics. We obtain numerical solutions for the fermion propagator in Landau gauge within the so-called rainbow approximation. A comparison with the ordinary theory without the Chern-Simons term is presented.

1. Motivation

Dynamical Mass Generation (DMG) and Confinement are two emergent phenomena of Quantum Chromodynamics (QCD) responsible for the nature of the hadronic spectrum. These can be studied by means of the corresponding Schwinger-Dyson Equations (SDE), as an alternative to the formulation of QCD on the lattice. The SDEs form an infinite tower of relations among the Green's functions of a given quantum field theory, whose nature is genuinely non-perturbative. We choose to study the SDE in a simplified model that shares with QCD the phenomena of Confinement and DMG, namely, quantum electrodynamics in the plane, or QED_3. Applications of QED_3 also extend to condensed matter systems, including high-T_c superconductivity, the quantum Hall effect and, very recently, graphene. Moreover, it is interesting to include the Chern-Simons (CS) term for the gauge field, which explicitly breaks parity and induces a topological mass for the photons.

Incorporating the CS term in studies of DMG in QED_3 have found that when the CS coefficient reaches a critical value, chiral symmetry is restored in a first order phase transition [1, 2]. In this work, we are interested in the effect of the CS term regarding confinement in quenched QED_3. In Sec. 2 we present our model with a reducible representation of (2+1)-dimensional quantum electrodynamics with a CS term. Solutions to the gap equation in the rainbow approximation are presented in Sec. 3. The question of confinement in QED_3 with a CS term is addressed in Sec.4. Finally, in Sec. 5, we present our conclusions.

2. Model

We study quantum electrodynamics in (2+1)-dimensions. For fermions, there exist two irreducible representations of the Euclidean Clifford algebra $\{\gamma_\mu, \gamma_\nu\} = 2\delta_{\mu\nu}$. In any of these representations, it is impossible to define chiral symmetry [3, 4, 5]. Furthermore, the fermion mass term, regardless its origin, is parity $\mathcal{P}$ non invariant. One still can construct a parity-preserving Lagrangian considering two different species with a relative sign between their masses [6, 7]. As a result, two chiral transformations can be defined [7]. These two species are conveniently merged into a single four-component spinor, making use of a reducible representation of the Dirac matrices. As compared with its four-dimensional counterpart, only three Dirac matrices are required to describe the dynamics of planar fermions. Once we have selected a set of matrices to write down the Dirac equation, say $\{\gamma_0, \gamma_1, \gamma_2\}$, two anti-commuting gamma matrices, namely, γ_3 and γ_5 remain unused. Hence the corresponding massless Dirac Lagrangian is invariant under the chiral-like transformations $\psi \to e^{i\alpha\gamma_3}\psi$ and $\psi \to e^{i\beta\gamma_5}\psi$, that is, it is invariant under a global $U(2)$ symmetry with generators 1, γ_3, γ_5 and $[\gamma_3, \gamma_5]$, corresponding to the interchange of fermion species. This symmetry is broken by an ordinary mass term $m_e\bar{\psi}\psi$. The order parameter of the symmetry breaking is the condensate $\langle\bar{\psi}\psi\rangle_e = \langle 0|\bar{\psi}\psi|0\rangle$. There exists a second mass term which is invariant under the "chiral" transformations, referred to as the Haldane mass term $m_o\bar{\psi}\tau\psi$ [8] with $\tau = [\gamma_3, \gamma_5]/2 = diag(I, -I)$, which is associated with the condensate $\langle\bar{\psi}\psi\rangle_o = \langle 0|\bar{\psi}\tau\psi|0\rangle$. The ordinary mass term is even under parity $\mathcal{P}$ transformations, but the Haldane mass term is not. We use the subscripts e for even and o for odd in quantities of the model. The Haldane mass term radiatively induces a parity-odd contribution into the vacuum polarization, which can be traced back to an induced Chern-Simons interaction

$$\mathcal{L}_{CS} = -\frac{i\theta}{4}\varepsilon_{\mu\nu\rho}A_\mu F_{\nu\rho} \, . \tag{1}$$

Such a term is parity non invariant, and despite the fact that it is not manifestly gauge invariant, the corresponding action respects gauge symmetry. The parameter θ induces a topological mass for the photons. Thus, the model we consider is described by the Lagrangian

$$\mathcal{L} = \bar{\psi}(i\partial\!\!\!/ + eA\!\!\!/ + m_e + \tau m_o)\psi + \frac{1}{4}F_{\mu\nu}F_{\mu\nu} + \frac{1}{2\xi}(\partial_\mu A_\mu)^2 - \frac{i\theta}{4}\varepsilon_{\mu\nu\rho}A_\mu F_{\nu\rho}. \tag{2}$$

There are many condensed matter systems which can be described by this Lagrangian, for which the physical origin of the masses depends on the underlying system, including high-T_c superconductivity, the quantum Hall effect and, very recently, graphene.

Written in this form, neither m_e nor m_o correspond to poles of the fermion propagator. We introduce the chiral projectors $\chi_\pm = (1 \pm \tau)/2$ which have the properties $\chi_\pm^2 = \chi_\pm$, $\chi_+\chi_- = 0$, $\chi_+ + \chi_- = 1$. These allow us to define the right-handed ψ_+ and left-handed ψ_- fermion fields as $\psi_\pm = \chi_\pm\psi$. The $\chi_\pm$ project the upper and lower two component spinors (fermion species) out of the four-component ψ, such that the fermion sector of the Lagrangian (2) can be cast in the form

$$\mathcal{L}_F = \bar{\psi}_+(i\partial\!\!\!/ - m_+)\psi_+ + \bar{\psi}_-(i\partial\!\!\!/ - m_-)\psi_-, \tag{3}$$

with $m_\pm = m_e \pm m_o$. This Lagrangian describes two fermion species of physical masses m_+ and m_-, respectively. These masses break chiral symmetry and parity at the same time. Moreover, the effect of the parity-violating mass is seen to remove the mass degeneracy between the two species. In what follows, we will be interested in the analytical properties of the dynamically generated fermion propagator associated with these physical masses. The analytical structure of the propagator can be studied from the corresponding Schwinger-Dyson equation, which corresponds to the expression

$$S_F(p)^{-1} = S_F^{(0)}(p)^{-1} + e^2\int\frac{d^3k}{(2\pi)^3}\Gamma^\mu(k,p)S_F(k)\gamma^\nu\Delta_{\mu\nu}(k-p) \, , \tag{4}$$

where $\Gamma^\mu(k,p)$ and $\Delta_{\mu\nu}(k-p)$ are, respectively, the full fermion-photon vertex and the full photon propagator, which obey themselves their own SDE, and $S_F(p)$ and $S_F^{(0)}(p)$ stand for the full and bare fermion propagator. In QED3, the coupling e^2 has mass-dimension one. Moreover, since the theory is super-renormalizable, e^2 becomes the natural scale of massless QED3, which is directly connected to the scales of confinement and dynamical chiral symmetry breaking. In this work, we write the relevant mass scales in units of $e^2 = 1$.

From the Lagrangian (2), we observe that inverse fermion propagator has the form

$$S_F^{-1}(p) = A_e(p)\not{p} + A_o(p)\tau\not{p} - B_e(p) - B_o(p)\tau \,. \tag{5}$$

The scalar functions $A_{e,o}(p)$ and $B_{e,o}(p)$ can be expressed in terms of the fermion wavefunction renormalizations $F_{e,o}(p)$ and the mass function $M_{e,o}(p)$ as $A_{e,o}(p) = 1/F_{e,o}(p)$ and $B_{e,o}(p) = M_{e,o}(p)/F_{e,o}(p)$ both in the even and odd sectors. The bare propagator corresponds to the values $A_e^{(0)}(p) = 1$, $A_o^{(0)}(p) = 0$, $B_e^{(0)}(p) = m_e$, $B_o^{(0)}(p) = m_o$. Rather than working with parity eigenstates, we find it convenient to work with the chiral Lagrangian (3). The chiral decomposition of the fermion propagator becomes

$$\begin{aligned} S_F(p) &= -\frac{A_+(p)\not{p} + B_+(p)}{A_+^2(p)p^2 + B_+^2(p)}\chi_+ - \frac{A_-(p)\not{p} + B_-(p)}{A_-^2(p)p^2 + B_-^2(p)}\chi_- \\ &\equiv -[\sigma_+^V(p)\not{p} + \sigma_+^S(p)]\chi_+ - [\sigma_-^V(p)\not{p} + \sigma_-^S(p)]\chi_- \,, \end{aligned} \tag{6}$$

where $K_\pm = K_e \pm K_o$ for all relevant even and odd quantities of the model. The inverse transformations are simply $K_e = (K_+ + K_-)/2$ and $K_o = (K_+ - K_-)/2$. In this basis, the bare quantities are $A_\pm^{(0)}(p) = 1$ and $B_\pm^{(0)}(p) = m_\pm$. On the other hand, the photon propagator associated with the Lagrangian (2) in the Landau gauge, $\xi = 0$, takes the form

$$\Delta_{\mu\nu}^{(0)}(q) = \frac{1}{q^2+\theta^2}\left(\delta_{\mu\nu} - \frac{q_\mu q_\nu}{q^2}\right) - \frac{\varepsilon_{\mu\nu\rho}q_\rho\theta}{q^2(q^2+\theta^2)} \,. \tag{7}$$

3. Gap Equation

We consider the rainbow-ladder truncation scheme, in which one makes the replacements $\Gamma^\mu \to \gamma^\mu$ and $\Delta_{\mu\nu} \to \Delta_{\mu\nu}^{(0)}$. With the chiral decomposition of the fermion propagator, Eq. (6), both in the left- and right-handed sector, the SDE corresponds to a matrix equation which can be converted into a system of scalar equations for $A_\pm(p)$ and $B_\pm(p)$ after multiplying it by 1 and $\not{p}$ and taking the trace. In the limit $m_\pm = 0$, this leads to the decoupled system of equations

$$\begin{aligned} A_\pm(p) &= 1 + \frac{\alpha}{\pi^2 p^2}\int d^3k\, \sigma_\pm^V(k)\frac{(k\cdot q)(p\cdot q)}{q^2(q^2+\theta^2)} \mp \frac{\alpha\theta}{\pi^2 p^2}\int d^3k\, \sigma_\pm^S(k)\frac{(p\cdot q)}{q^2(q^2+\theta^2)} \,, \\ B_\pm(p) &= \frac{\alpha}{\pi^2}\int d^3k\, \sigma_\pm^S(k)\frac{1}{(q^2+\theta^2)} \mp \frac{\alpha\theta}{\pi^2}\int d^3k\, \sigma_\pm^V(k)\frac{(k\cdot q)}{q^2(q^2+\theta^2)}, \end{aligned} \tag{8}$$

with $\alpha = 1/(4\pi)$ as usual. Because of the effects of the CS term, the $A_\pm(p)$ functions become not trivial. Thus, in covariant gauges these never decouple from the equations for $B_\pm(p)$ as long as $\theta \neq 0$. We numerically solve the above systems of equations in each sector varying θ.

In Fig. 1 we show the functions $A_\pm(p)$ for different values of θ. The small deviation from the tree-level value for small momentum in each case is due to the CS term. As θ increases, both functions exhibit a plateau for small p and smoothly tend to 1 in the ultraviolet. The deviation is more pronounced in $A_-(p)$. Figure 2 shows the result for $B_\pm(p)$. We observe that the height of the plateau of $B_+(p)$ at small momentum decreases from the parity-even result as θ gets bigger, but the height of the plateau of $B_-(p)$ exhibits the opposite behavior. The effect

Journal of Physics: Conference Series **287** (2011) 012028 doi:10.1088/1742-6596/287/1/012028

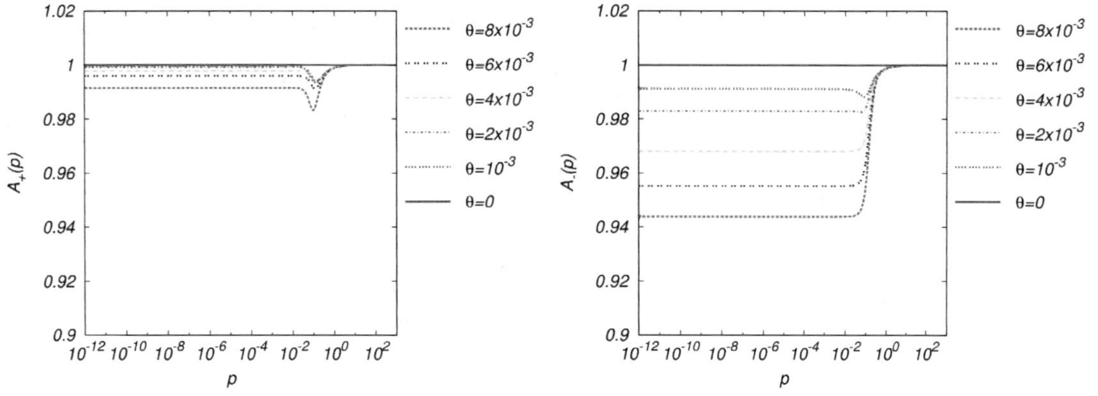

Figure 1. Functions $A_{\pm}(p)$ for various values of θ.

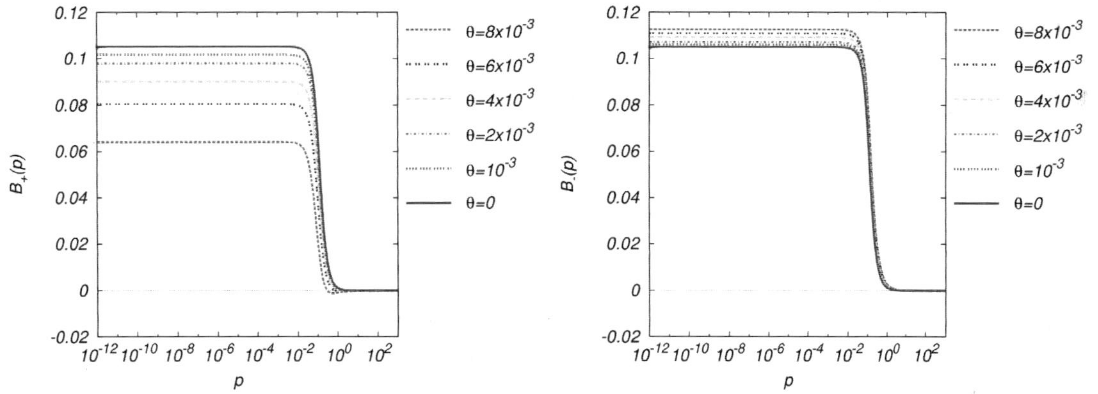

Figure 2. Functions $B_{\pm}(p)$ for various values of θ.

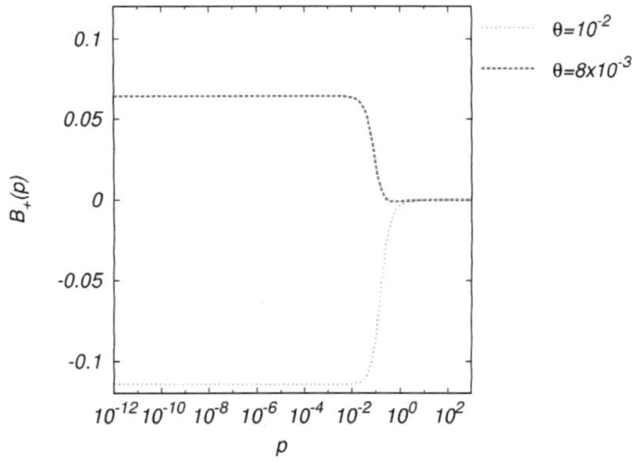

Figure 3. Critical behavior of $B_{+}(p)$ for $\theta \simeq 8 \times 10^{-3}$.

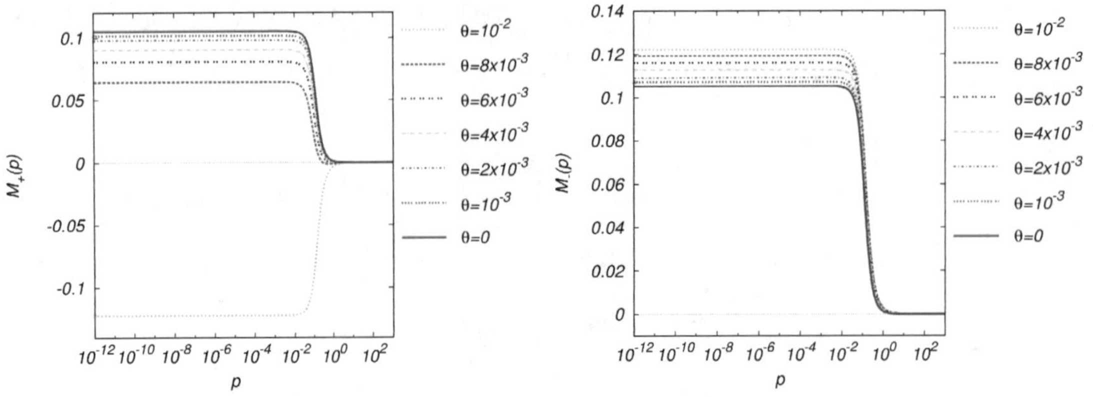

Figure 4. Mass functions $M_{\pm}(p)$ for various values of θ.

is more visible in $B_{+}(p)$. In fact, there is a critical value $\theta_c \simeq 8 \times 10^{-3}$ above which $B_{+}(p)$ strongly changes its behavior and the plateau becomes negative, as shown in Fig. 3.

This sudden drop has strong implications regarding chiral symmetry restoration in the model [1, 2]. In order to illustrate this, in Fig. 4, we show the mass functions $M_{\pm}(p) = B_{\pm}(p)/A_{\pm}(p)$ for different values of θ. At θ_c, the plateau of $M_{+}(p)$ also drops to negative values, whereas for $M_{-}(p)$, it continues to increase with θ. To ensure that these results are physically sensitive, we study the asymptotic behavior of $M_{\pm}(p)$. In the infrared, the height of the plateau can be considered as an order parameter for dynamical chiral symmetry breaking. In this connection, in Fig. 5 we draw the dependence of $\mu_{\pm} = M_{\pm}(0)$ as a function of θ below and above criticality. This parameter can be regarded as the dynamical mass of the corresponding fermion species. We observe that the role of the CS coefficient is to remove the mass degeneracy between fermion species as long as $\theta < \theta_c$. There is a light and a heavy species. At θ_c, however, there is a drastic change in this behavior, the light species develops a negative mass, which in absolute value is the same as its heavy cousin, as can be appreciated in the dotted curve in the same graph. This implies that above this θ_c, the would-be parity preserving mass $\mu_e = (\mu_{+} + \mu_{-})/2$ vanishes, and hence chiral symmetry is restored [1, 2]. Nevertheless, we want to emphasize that in this model, μ_e does not correspond to a pole in a propagator and hence to a physical mass. Physical masses $\mu_{\pm}$ are generated for arbitrarily large values of θ.

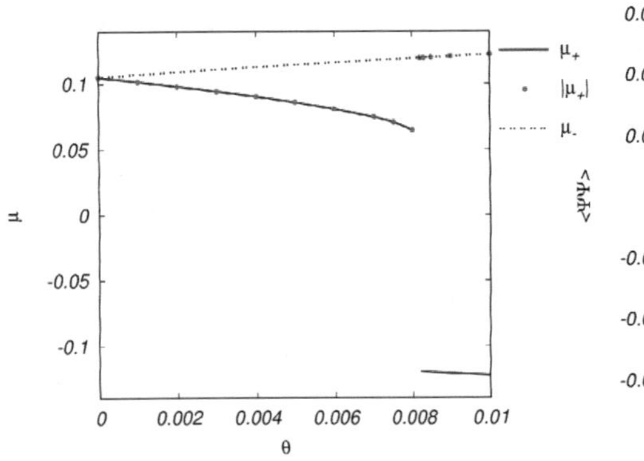

Figure 5. $\mu_{\pm}$ as a function of θ.

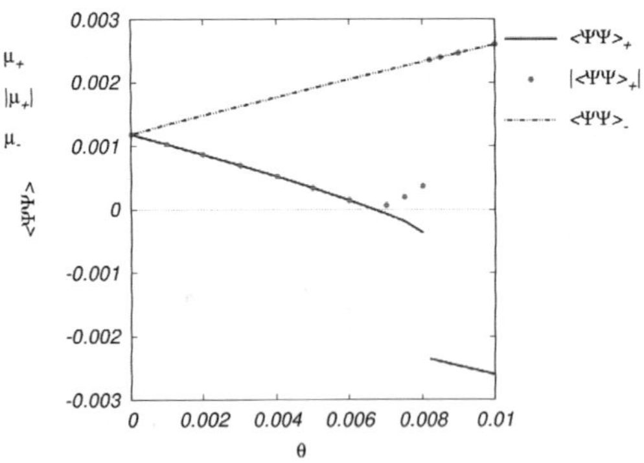

Figure 6. Chiral condensates as a function of θ.

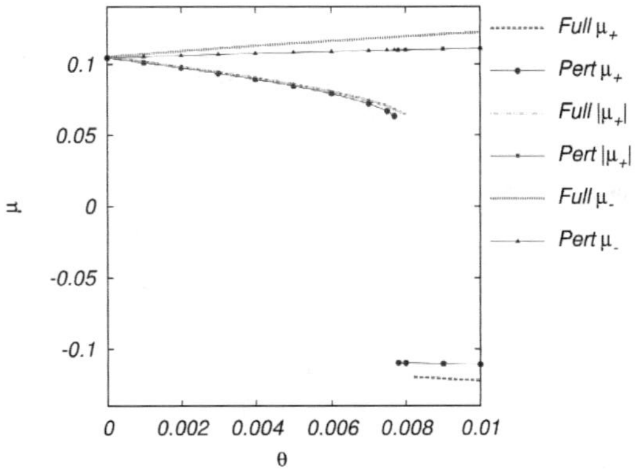

Figure 7. Comparison of the dynamically generated fermion masses for the full system (8) and the perturbatively reduced system (10).

The discontinuous behavior of the dynamically generated masses suggest a first order phase transition for the dynamical breaking of chiral symmetry. Whether there is chiral symmetry restoration can better be seen from the behavior of the condensates, which are the true order parameters. These can be extracted from the solution to the SDEs as

$$< \bar{\psi}\psi >_{\pm} = \frac{1}{\pi^2}\int_0^\infty dk k^2 \sigma_{\pm}^S(k) \; . \tag{9}$$

The results are shown in Fig. 6. Notice that in the parity preserving case, $\theta = 0$, the value of the condensate for each species is half the value of the ordinary condensate [9, 10]. The reason is that when fermion species are degenerate in mass, each one of them, corresponding to a fermion pairing mode $\bar{\psi}_\downarrow\psi_\uparrow$ or $\bar{\psi}_\uparrow\psi_\downarrow$ (arrows indicate spin orientation), contributes one half to the parity preserving condensate.

Finally, let us consider wavefunction renormalization effects. Regarding the wavefunction renormalization in the dynamically generated mass functions $M_\pm(p)$, we compare the solutions of the full systems in the previous subsection, Eq. (8), against the solution for the simplified equations resulting from the perturbation theory motivated ansatz $A_\pm(p) = 1$, namely

$$B_\pm(p) \;=\; \frac{\alpha}{\pi^2}\int d^3k \frac{B_\pm(k)}{k^2 + B_\pm^2(k)}\frac{1}{(q^2+\theta^2)} \mp \frac{\alpha\theta}{\pi^2}\int d^3k \frac{1}{k^2+B_\pm^2(k)}\frac{(k\cdot q)}{q^2(q^2+\theta^2)} \; . \tag{10}$$

The behavior of the solutions in this truncation is qualitatively similar to that of the complete system (8). The critical behavior of the dynamical masses is not washed out by the effect of wavefunction renormalization, as can be seen in Fig. 7, where a comparison of $\mu_\pm$ and $|\mu_+|$ between the full result and the perturbatively motivated one is shown. The behavior is virtually the same in both truncation schemes. The critical value θ_c^{pert} is of the same order of θ_c and thus the effects of $A_\pm(p) \approx 1$ are not significant. This means that even with the inclusion of the CS term, the rainbow-ladder truncation is still reliable in Landau gauge.

4. Confinement

In ordinary quenched QED$_3$, there is confinement. Here we are interested in the extent up to which this scenario persist in the presence of a CS term. Whether a solution to the SDE supports confinement can be tested by means of the violation of the Osterwalder-Schrader axiom

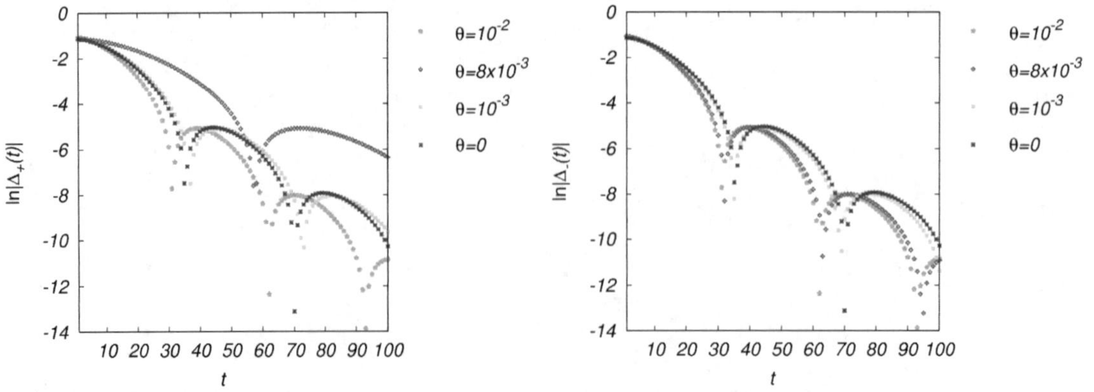

Figure 8. Confinement tests for the spatially averaged Schwinger functions.

of reflection positivity, which states that the spatially averaged Schwinger functions

$$\Delta_{\pm}(t) = \int d^2x \int \frac{d^3p}{(2\pi)^3} e^{i(tp_o + x\cdot p)} \sigma^S_{\pm}(p) \,, \tag{11}$$

should be positive definite if they are related to a stable asymptotic state. In our case, we construct the functions $\Delta_{\pm}(t)$ inserting the solutions for the SDE in the above expression, upon which we perform a confinement test. In Fig. 8 we plot the logarithm of the absolute value of each of these functions. The oscillatory behavior inferred from the pronounced peaks in the graphs reveals that even in the presence of a CS term, quenched QED_3 exhibits confinement. This behavior corresponds to two complex conjugate mass like singularities with complex masses $m = a \pm ib$ for each species, which fit to

$$\Delta_{\pm}(t) \simeq C_{\pm} e^{-a_{\pm} t} cos(b_{\pm} t + \delta_{\pm}) \,. \tag{12}$$

Thus, confinement is observed in our model.

Following Ref.[11], we perform a similar analysis on the vector parts of the propagator, $\sigma^V_{\pm}$. We define new set of spatially averaged Schwinger functions,

$$\Omega_{\pm}(t) = \int d^2x \int \frac{d^3p}{(2\pi)^3} e^{i(tp_o + x\cdot p)} \sigma^V_{\pm}(p), \tag{13}$$

and insert again the solutions found in previous sections. In Fig. 9 we draw the logarithm of the absolute values of $\Omega_{\pm}(t)$. Again, the oscillations hint confinement. Moreover, we obsverve that $\Omega_{\pm}(t)$ can be fitted according to

$$\Omega_{\pm}(t) \simeq C'_{\pm} e^{-a'_{\pm} t} cos(b'_{\pm} t + \delta'_{\pm}) \,, \tag{14}$$

which also unveils the complex pole mass structure of the propagator.

5. Conclusions

We have studied the dynamical generation of masses and confinement in Maxwell-Chern-Simons QED_3. There exists two types of fermions within the four-component spinor formalism. These species are non-degenerated in mass. The origin of these physical masses is two-fold: On the

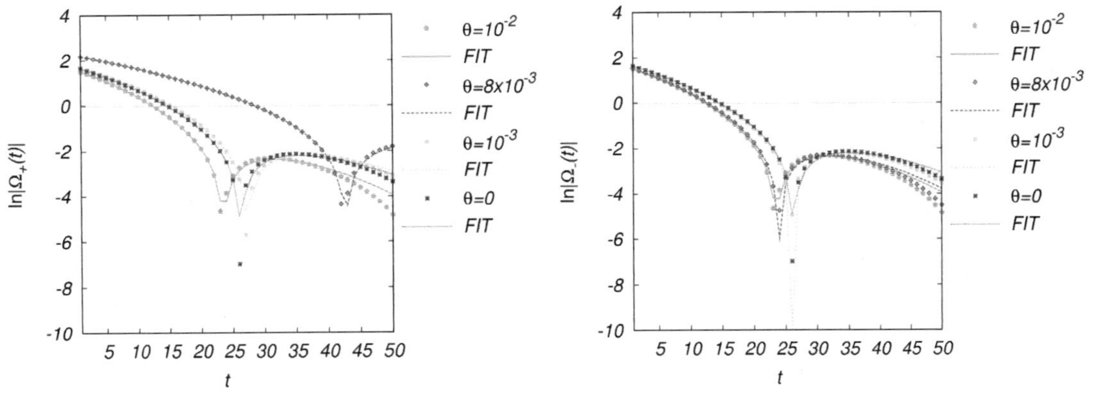

Figure 9. Spatially averaged Schwinger functions involving the vector parts of the fermion propagator.

one hand there is a parity-preserving contribution m_e coming from dynamical chiral symmetry breaking, on the other hand, there is the CS-induced Haldane mass term m_o. We have explored the consequences of confinement in our model.

For $\theta > \theta_c$ the masses for each species are equal in magnitude, $\mu+ = -\mu_-$, and we do not generate the ordinary mass $\mu_e = (\mu_+ + \mu-)/2$ that preserves parity. This is the reason why there is restoration of chiral symmetry and we only generated masses induced by the CS term. The phase transition for the mass μ_+ is of first order. Regarding confinement, according to tests performed with σ^V and σ^S, we observe that quenched QED3 with a CS term also shows this feature, regardless the value of the CS coefficient. This is in contrast to the case in which vacuum polarization effects are considered in the leading order of the $1/N$ approximation [11], where there is deconfinement for arbitrary finite values of θ.

Acknowledgments

We acknowledge support from SNI, CIC, and CONACyT grants through projects 4.22, 82230, and 50744-F, respectively.

References

[1] K.-I. Kondo and P. Maris 1995 *Phys. Rev. Lett.* **74** 18
[2] K.-I. Kondo and P. Maris 1995 *Phys. Rev.* D **52** 1212
[3] T. Appelquist, M. Bowick, D. Karabali and L.C.R. Wijewardhana 1986 *Phys. Rev.* D **33** 3704
[4] R.D. Pisarski 1984 *Phys. Rev.* D **29** 2423
[5] G.W. Semenoff and L.C.R. Wijewardhana 1988 *Phys. Rev. Lett.* **62** 2633
[6] K. Shimizu 1985 *Prog. Theor. Phys.* **74** 610
[7] Ma. de J. Anguiano and A. Bashir 2005 *Few Body Syst.* **37** 71
[8] F. D. M. Haldane 1988 *Phys. Rev. Lett.* **61** 2015
[9] C.D. Roberts and A. G. Williams 1994 *Prog. Part. Nucl. Phys.* **33** 477
[10] A. Bashir and A. Raya 2007 *Few Body Syst.* **41** 185
[11] C.P. Hofmann, A. Raya and S. Sánchez Madrigal 2010 *Phys. Rev.* D **82** 096011

Journal of Physics: Conference Series **287** (2011) 012029 doi:10.1088/1742-6596/287/1/012029

Transport and zero sound in holographic strange metals

Carlos Hoyos[1,a], **Andy O'Bannon**[2,b] **and Jackson M. S. Wu**[3,c]

[1]Department of Physics, University of Washington, Seattle, WA 98195-1560 USA
[2]DAMTP, Centre for Mathematical Sciences, Wilberforce Road, Cambridge, UK CB3 0WA
[3]Department of Physics, University of Auckland, Private Bag 92019, Auckland, New Zealand

E-mail: [a]choyos@phys.washington.edu, [b]A.OBannon@damtp.cam.ac.uk, [c]jbnwu@itp.unibe.ch

Abstract. It is a challenge to describe the non-Fermi liquid behavior that appears in strongly interacting systems of fermions using traditional field theory techniques. A proposal to model the associated "strange metal" phenomenology is to use a strongly coupled critical point with non-relativistic scale invariance, dual to a Lifshitz geometry, coupled to a finite density of probe charged carriers dual to D-branes. We use this model to study the thermodynamic and transport properties of charge carriers at finite and zero temperature. At zero temperature we show that the large wavelength behavior is dominated by a single mode whose properties are similar to the zero sound of Fermi liquids.

1. Motivation

Gauge-gravity duality [1–3] provides a novel way to compute observables in strongly-coupled, scale-invariant systems at finite density, and thus may be useful for studying condensed matter systems near quantum critical points, such as the "strange metal" phase of some heavy fermion compounds and possibly also of some high-T_c materials. In particular, gauge-gravity duality may be able to reproduce the simple properties of strange metals, such as electrical resistivity scaling linearly with temperature T, which cannot be derived in a Fermi liquid description, which assumes the appropriate degrees of freedom are weakly-interacting quasi-particles. Of course the ultimate goal is to understand why some strange metals have a high T_c.

In relativistic systems at zero temperature, holographic calculations also reveal the existence of a propagating quasi-particle excitation producing a pole in the charge density and current retarded two point functions [4–6]. In a weakly-coupled Fermi liquid, fluctuations in the shape of the Fermi surface produce a collective excitation at zero temperature, Landau's so-called "zero sound." The dispersion relation of the mode in these holographic systems was identical in form to that of Landau's zero sound, with a dispersion relation linear in momentum and a damping quadratic in momentum, hence the mode was dubbed zero sound by analogy. We will call this mode "holographic zero sound." The physical origin of this mode remains mysterious since it appears in systems with both fermions and scalars—systems that do not fit neatly into either Fermi liquid or quantum Bose liquid theory. Indeed, given the existence of the holographic zero sound, as well as unusual scaling of the heat capacity with temperature, these holographic systems may be new kinds of (strongly-coupled) quantum liquids [4].

In ref. [7] we explored the fate of holographic zero sound in the holographic strange metal models. To do so, we computed holographically the low-frequency and low-momentum forms of the retarded correlation functions of the current, for massless charge carriers. We also computed the AC conductivity associated with the charge carriers, generalizing the results of ref. [8], for two spatial field theory directions, to an arbitrary number of spatial directions. Here we give an overview of these results.

This text is organized as follows. We start reviewing the characteristics of the holographic critical point that we will use to model a strange metal in section 2. We then study the thermodynamics properties of these systems in section 3. We compute the current Green's functions in section 4, and use them to study the holographic zero sound and AC conductivity in section. We conclude with some discussion and suggestions for future research in section 5.

2. Holographic critical point

Gauge-gravity duality is holographic: it equates a weakly-coupled theory of gravity on some spacetime with a strongly-coupled field theory living on the boundary of that spacetime. The spacetime symmetries of the field theory are dual to the isometries of the bulk metric. Here we will focus on field theories invariant under the Lifshitz group, which includes scale transformations of the time coordinate t and spatial coordinates $\vec{x}$ of the form

$$t \rightarrow \lambda^z t\,, \qquad \vec{x} \rightarrow \lambda\,\vec{x}\,, \tag{1}$$

where λ is real and positive and z is called the "dynamical exponent." The dual gravity theory then lives in Lifshitz spacetime, with metric [9]

$$ds^2 = g_{\mu\nu}\,dx^\mu dx^\nu = \frac{dr^2}{r^2} - \frac{dt^2}{r^{2z}} + \frac{d\vec{x}^2}{r^2}\,, \tag{2}$$

where r is the holographic radial coordinate: $r \rightarrow \infty$ corresponds to the infrared (IR) of the field theory, while the near-boundary region $r \rightarrow 0$ corresponds to the ultraviolet (UV). Notice that when $z = 1$, the metric becomes that of anti-de Sitter space (AdS), and the dual field theory symmetry is enhanced to the relativistic conformal group.

One proposal for holographic model building of strange metals is to use probe D-branes in a Lifshitz spacetime [8]. In this approach, an appropriate configuration of fields on the probe D-brane represents a finite density of massive charge carriers [10–12], while the background spacetime represents a strongly-coupled neutral quantum critical theory. Holographic calculations then reveal that such systems can indeed reproduce the simple properties of strange metals, including electrical resistivity linear in temperature [8]. However, ingredients beyond Lifshitz spacetime, for example a nontrivial dilaton, are needed to reproduce all the measured strange metal properties [8].

In gauge-gravity duality, a theory of gravity is dual to some large-N, strongly-coupled non-Abelian gauge theory. N_f probe D-branes are dual to N_f fields in the fundamental representation of the gauge group(s) in the probe limit $N_f \ll N$, transforming under a global $U(N_f)$ group [13]. We will call these fields quarks or flavor fields, in analogy with Quantum Chromodynamics, and they will be our charged degrees of freedom. The dynamics of N_f coincident D-branes are determined by the non-Abelian Dirac-Born-Infeld (DBI) action, involving worldvolume scalars and $U(N_f)$ gauge fields. We produce a finite density of charge carriers by introducing a chemical potential for the diagonal $U(1) \subset U(N_f)$. The density operator J^t is the time component of the conserved $U(1)$ current J^μ, and is holographically dual to A_t, the time component of the $U(1)$ gauge field living on the D-branes. We will thus introduce in the bulk D-branes with only $A_t(r)$ present. The action then reduces to the Abelian DBI action.

We will only consider *massless* flavor fields. A flavor field mass operator would be dual to a scalar field on the D-branes. We omit any such scalar from our calculations. We assume our

D-branes are extended along the q spatial dimensions of the Lifshitz spacetime.The Abelian DBI action for our D-branes is then

$$\hat{S} = -N_f T_D V \int dr\, dt\, d^q x \sqrt{-\det\left[g_{ab} + (2\pi\alpha')F_{ab}\right]}\,, \tag{3}$$

where T_D is the tension, g_{ab} the induced metric, and F_{ab} the $U(1)$ field strength of the D-branes. The factor V is the volume of any internal space that the D-brane may be wrapping, and $2\pi\alpha'$ is the inverse string tension, which will typically be present in an actual string theory system. Inserting our ansatz $A_t(r)$, performing the trivial integrations over dt and $d^q x$, and dividing out the subsequent (infinite) volume factors, we obtain the action density S.

3. Thermodynamics

In this section we review the results of refs. [4,7,12,14–16] for the thermodynamics of the charge carriers in these holographic systems. The charge density in the field theory is

$$\langle J^t \rangle \equiv \widetilde{\mathcal{N}} d = \frac{\delta S}{\delta A_t'} = \mathcal{N} g_{xx}^{q/2} \frac{(2\pi\alpha')\tilde{A}_t'}{\sqrt{|g_{tt}| g_{rr} - \tilde{A}_t'^2}}\,, \tag{4}$$

where a tilde means there is a $(2\pi\alpha')$ factor. Solving for $A_t'(r)$ and plugging it back into the action, we find

$$S = -\mathcal{N} \int dr\, |g_{tt}|^{1/2} g_{rr}^{1/2} \frac{g_{xx}^q}{\sqrt{g_{xx}^q + d^2}}\,. \tag{5}$$

We obtain the field theory chemical potential and free energy by performing the integrals for $A_t'(r)$ and S. The chemical potential is

$$\mu_0 = \int_0^\infty dr\, A_t'(r) = \frac{d}{2\pi\alpha'} \int_0^\infty dr \frac{r^{-(z+1)}}{\sqrt{r^{-2q} + d^2}} = \frac{d^{z/q}}{4\pi\alpha' q} \frac{\Gamma\left(\frac{z}{2q}\right)\Gamma\left(\frac{1}{2} - \frac{z}{2q}\right)}{\Gamma\left(\frac{1}{2}\right)}\,. \tag{6}$$

The grand canonical (Gibbs) free energy density is

$$\Omega_0 = -S = \mathcal{N} \int_0^\infty dr \frac{r^{-(2q+z+1)}}{\sqrt{r^{-2q} + d^2}} = -\widetilde{\mathcal{N}} \frac{z\, d}{z+q} \mu_0\,. \tag{7}$$

These results are deceptively simple. We have actually removed various power-law divergences at the $r = 0$ endpoint of the integrations via analytic continuation. A straightforward calculation reveals that for us

$$\langle J^t \rangle = -\frac{\delta \Omega_0}{\delta \mu_0}\,, \tag{8}$$

which shows that we are working in the grand canonical ensemble. Notice that with massless flavor fields, any finite chemical potential will produce a finite density.

Analytic continuation cannot remove logarithmic divergences, which indeed occur in eqs. (6) and (7) when $z = (2n-1)q$. These divergences occur because the operators $(J^t)^{2n}$ are classically marginal at these z values [8], and if added to the action, can produce a breaking of scale invariance through quantum corrections, similarly to what happens for multi-trace scalar operators [17].

To study thermodynamics at finite temperature, we need a black hole in the bulk whose Hawking temperature is dual to the field theory temperature [18]. To capture the basic qualitative physics without committing to a specific system, we will again follow ref. [8] and write a generic Lifshitz black hole metric

$$ds^2 = \frac{dr^2}{f(r)\,r^2} - \frac{f(r)}{r^{2z}}dt^2 + \frac{d\vec{x}^2}{r^2}\,. \tag{9}$$

We do not need an exact form of $f(r)$. We only need to know that $f(r)$ has a simple zero at the horizon $r = r_H$, and that the temperature is fixed by the regularity of the Wick-rotated metric at the horizon,

$$f(r) \underset{r\to r_H}{\simeq} c_0\left(1 - \frac{r}{r_H}\right), \qquad T = c_0\frac{r_H^{-z}}{4\pi}, \tag{10}$$

where c_0 is a dimensionless constant that depends on the specific system.

The chemical potential at finite (small) temperature is

$$\mu = \int_0^{r_H} dr A_t' = \mu_0 - \frac{r_H^{-z}}{2\pi\alpha' z}\left(1 - \frac{z}{2(2q+z)}\frac{r_H^{-2q}}{d^2} + O\left(\frac{r_H^{-4q}}{d^4}\right)\right), \tag{11}$$

The finite-temperature free energy is

$$\Omega = -S = -\widetilde{\mathcal{N}}\frac{z\,d}{z+q}\,\mu - \mathcal{N}\frac{4\pi d\,T}{c_0(z+q)} + O\left(\frac{T^{2q/z+1}}{d}\right), \tag{12}$$

From here we can see that the entropy density at zero temperature is nonzero and proportional to the charge density [4, 14, 16],

$$s_0 = -\left(\frac{\partial\Omega}{\partial T}\right)_{\mu,T=0} = \frac{4\pi\mathcal{N}}{c_0(z+q)}\,d. \tag{13}$$

Indeed, in a string theory system, this entropy should be equivalent to the entropy of a single quark, represented in the bulk by a single string, times the density d of quarks [14]. The specific heat at finite temperature scales with T as

$$c_V = T\frac{\partial s}{\partial T} \sim \frac{T^{\frac{2q}{z}}}{d}\,. \tag{14}$$

A linear scaling, as in a Fermi liquid, occurs when $z = 2q$, while a scaling T^q, as in a quantum Bose liquid, occurs when $z = 2$ [16]. Notice that both the entropy and the scaling of the specific heat implies that generically the thermodynamic behavior of these models differ from that of real strange metals.

4. Transport

Zero sound should appear at zero temperature as a pole in the Fourier-transformed retarded two-point function of the density operator, $G_R^{tt}(\omega, k)$ [4]. In what follows we will also compute the density-current and current-current retarded two-point functions $G_R^{tx}(\omega, k)$ and $G_R^{xx}(\omega, k)$. The latter will give us the AC conductivity.

In gauge-gravity duality the on-shell action is the field theory generating functional [2, 3], hence we need to take two functional derivatives of the on-shell bulk action, which in turn means we need to solve the linearized equations of motion for the fields dual to the density and current operators, namely the $U(1)$ gauge fields A_t and A_x on the probe D-branes. The holographic zero sound pole will appear as a quasi-normal frequency of these fields. We thus consider fluctuations a_μ of the gauge fields, where because the field theory is invariant under rotations, we only need to consider fluctuations a_t and a_x with r, t, and x dependence. We work in a gauge where $a_r = 0$. We also expand in a plane wave basis, $a_\mu \sim e^{-i\omega t + ikx}$. Finally, we will work with the gauge-invariant electric field $E(r,\omega,k) = \omega a_x + k a_t$.

Introducing a cutoff at $r = \epsilon$ and integrating by parts, we obtain in the $\epsilon \to 0$ limit the on-shell quadratic action in terms of E

$$S_{a^2} = -\frac{\mathcal{N}}{2}\int d\omega\, dk \frac{\epsilon^{z-q+1}}{k^2}\tilde{E}(\epsilon)\tilde{E}'(\epsilon)\,. \tag{15}$$

We now need to solve the equation of motion with an ingoing boundary condition imposed at the $r \to \infty$ "horizon" [19–21], insert the solution into S_{a^2}, and then functionally differentiate to obtain the retarded correlators, for example

$$G_R^{tt}(\omega,k) = \frac{\delta^2}{\delta a_t(\epsilon)^2} S_{a^2} = \left(\frac{\delta E(\epsilon)}{\delta a_t(\epsilon)}\right)^2 \frac{\delta^2}{\delta E(\epsilon)^2} S_{a^2}\,. \tag{16}$$

Defining

$$\Pi(\omega,k) \equiv \frac{\delta^2}{\delta E(\epsilon)^2} S_{a^2}\,, \tag{17}$$

we get

$$G_R^{tt}(\omega,k) = k^2\,\Pi(\omega,k)\,, \quad G_R^{tx}(\omega,k) = \omega\, k\, \Pi(\omega,k)\,, \quad G_R^{xx}(\omega,k) = \omega^2\,\Pi(\omega,k)\,. \tag{18}$$

Our goal is to obtain (the low-frequency and low-momentum form of) the variational derivative $\Pi(\omega,k)$, which determines the retarded two-point functions that we want.

We can obtain the low-frequency behavior of $\Pi(\omega,k)$ by solving the equations of motion in two different limits and then matching the two solutions in a regime where the limits overlap, following ref. [4]. Specifically, we first obtain a solution by taking r large and then taking the small frequency and momentum limit, which means $\omega r^z \ll 1$ and $kr \ll 1$ with ωk^{-z} fixed. We then repeat the process, now taking the small frequency and momentum limit first followed by taking r large. The details are explained in ref. [7].

The boundary behavior of the solutions is given by the small r expansion of the solutions: $E \simeq C_1 + k^2 C_2 r^{q-z}/(q-z)$. When $z < q$, we have that to leading order $E(\epsilon) \simeq C_1$, while to leading order $E'(\epsilon) \simeq k^2 \epsilon^{q-z-1} C_2$, so

$$S_{a^2} = -\frac{\mathcal{N}}{2}\int d\omega\, dk \frac{\epsilon^{z-q+1}}{k^2}\tilde{E}(\epsilon)\tilde{E}'(\epsilon) = -\frac{\mathcal{N}}{2}(2\pi\alpha')^2 \int d\omega\, dk\, C_1 C_2\,. \tag{19}$$

To find $\Pi(\omega,k)$, we need to find C_2 in terms of C_1, which can be done by matching the solutions that we find in the two asymptotic regimes. After that we immediately obtain $\Pi(\omega,k)$.

Assembling all the ingredients, we have our main result, the low-frequency and low-momentum form of $\Pi(\omega,k)$:

$$\Pi(\omega,k) \propto \frac{\mathcal{N}}{\alpha_1\, k^2 - \alpha_2\, \omega^2 - \alpha_3\, G_0(\omega)}\,, \tag{20}$$

where

$$G_0(\omega) = \begin{cases} \omega^{1+2/z} & z \neq 2 \\ \omega^2 \log\left(\alpha\omega^2\right) & z = 2 \end{cases}\,, \tag{21}$$

and α_1, α_2 and α_3 are constants. For $z = 2$ another (dimensionful) constant appears inside the logarithm in $G_0(\omega)$, $\alpha = \frac{e^{2\gamma}}{16}(2d)^{-4/q}$. In the denominator of $\Pi(\omega,k)$, when $z < 2$ the ω^2 term is larger than the $G_0(\omega)$ term, whereas when $z > 2$ the $G_0(\omega)$ term is larger. The dominant term dictates the low-frequency behavior of the retarded Green's functions, and hence of the holographic zero sound and AC conductivity, as we will show in detail in what follows.

4.1. Zero sound

The denominator of $\Pi(\omega, k)$ vanishes – and hence the retarded Green's functions have a pole – whenever

$$k(\omega) = \pm\frac{1}{\sqrt{\alpha_1}}\sqrt{\alpha_2\,\omega^2 + \alpha_3\,G_0(\omega)}\,. \tag{22}$$

This determines the dispersion relation of the holographic zero sound mode: we need only expand the right-hand side of eq. (22) and then invert to find $\omega(k)$. As mentioned above, the value of z determines whether the ω^2 term or $G_0(\omega)$ term dominates at low frequencies, so we will consider two cases in turn.

4.1.1. $1 \leq z < 2$ For $1 \leq z < 2$, the $G_0(\omega)$ term under the square root has larger norm, so we can expand $k(\omega)$ and then invert to find

$$\omega(k) = \pm k\sqrt{\frac{\alpha_1}{\alpha_2} - \frac{\alpha_3}{2\alpha_2}\left(\frac{\alpha_1}{\alpha_2}\right)^{1/z} k^{2/z} + O\left(k^{-1+4/z}\right)}\,. \tag{23}$$

The mode behaves as a quasiparticle since the imaginary part goes as $k^{2/z}$, which is smaller than the real part at low momentum. As a check, if we set $z = 1$ we recover the result of ref. [4],

$$\omega(k) = \pm\frac{k}{\sqrt{q}} - i\frac{d^{-\frac{1}{q}}\Gamma(\frac{1}{2})}{\Gamma\left(\frac{1}{2}-\frac{1}{2q}\right)\Gamma\left(\frac{1}{2q}\right)}k^2 + O\left(k^3\right). \tag{24}$$

The speed of the holographic zero sound, v_0, is given by

$$v_0^2 = \frac{\alpha_1}{\alpha_2} = \frac{z}{q}\,d^{2(z-1)/q}\,\frac{\Gamma\left(\frac{z}{2q}\right)\Gamma\left(\frac{1}{2}-\frac{z}{2q}\right)}{\Gamma\left(\frac{1}{2}+\frac{z-2}{2q}\right)\Gamma\left(\frac{2-z}{2q}\right)}\,. \tag{25}$$

When $z > 1$ the speed of the holographic zero sound is dimensionful. In the relativistic case, $z = 1$, the speed of holographic zero sound coincides with the speed of normal/first sound.

4.1.2. $z \geq 2$ We consider first $z > 2$. Now the first term under the square root in eq. (22) has larger norm, so after inverting we find

$$\omega(k) = \left(\frac{\alpha_1}{i\alpha_3}\right)^{\frac{z}{z+2}} k^{\frac{2z}{z+2}} + \frac{\alpha_2}{\alpha_3}\frac{z}{z+2}\left(\frac{\alpha_1}{\alpha_3}\right)^{\frac{2(z-1)}{z+2}} k^{\frac{4(z-1)}{z+2}} + O\left(k^{\frac{2(3z-4)}{z+2}}\right). \tag{26}$$

Note that the leading term has a coefficient that is *complex*. In particular, the real and imaginary parts are of the same order, hence the excitation is not a quasi-particle.

Finally, when $z = 2$ we have

$$k = \pm\frac{\omega}{\sqrt{\alpha_1}}\sqrt{\alpha_2 + \alpha_3\log\left(\alpha\,\omega^2\right)}. \tag{27}$$

Although this mode is gapless, the dispersion relation differs from the holographic zero sound mode by logarithmic factors.

4.2. AC Conductivity

In this section we provide some physical intuition for the change in behavior of the holographic zero sound mode as a function of z by studying the low-frequency AC conductivity $\sigma(\omega)$, defined as

$$\sigma(\omega) \equiv -\frac{i}{\omega} G_R^{xx}(\omega, k=0) \,. \tag{28}$$

Recalling that the imaginary part of the retarded Green's function is proportional to the spectral function of J^x, we see that the real part of the conductivity provides a measure of the density of states that couple to J^x.

To obtain $\sigma(\omega)$ for our system, we first set $k = 0$ in our result for $\Pi(\omega, k)$ in eq. (20) and then plug into eq. (18). After this we obtain the AC conductivity at small frequency from eq. (28)

$$\sigma(\omega) = -\frac{i}{\omega} G_R^{xx}(\omega, k=0) \xrightarrow{\omega \to 0} \mathcal{N} \begin{cases} i\alpha_2^{-1}\omega^{-1} & z < 2 \\ i\alpha_3^{-1}\left(\omega \log(\alpha\omega^2)\right)^{-1} & z = 2 \\ i\alpha_3^{-1}\omega^{-2/z} & z > 2 \end{cases} . \tag{29}$$

Our result for $\sigma(\omega)$ not only generalizes that of ref. [8] to any number of spatial dimensions q, but is an independent confirmation of the result, since we derived it via a slightly different route from that of ref. [8].

When $z \neq 2$, α_2 is a real number while α_3 is complex. When $z < 2$, the conductivity is purely imaginary and has a simple pole at zero frequency. A Kramers-Kronig relation then implies that the real part of the conductivity, and hence the spectral function, consists only of a delta function at zero frequency. The system has no charged states at small frequencies. Notice that the $i\omega^{-1}$ behavior is expected in a system with free charge carriers (or with translation invariance), and indeed is formally identical to the high-frequency or "collisionless" limit of the standard Drude result. When $z > 2$ the conductivity, and hence the spectral function, has a power-law dependence, and no clean holographic zero sound quasi-particle exists.

Something analogous happens in interacting systems with a Fermi surface. Here the charge density spectral function includes a continuum of particle-hole pairs at frequencies $\omega < v_F k$, where v_F is the Fermi velocity. The zero sound gives a contribution at $\omega = v_0 k$. If $v_0 > v_F$, then the zero sound is outside the continuum, and will produce a sharp quasi-particle peak in the spectral function. However, if $v_0 < v_F$, the zero sound mode suffers Landau damping through scattering with particle-hole pairs, and hence will be reduced to a broad resonance with no good quasi-particle interpretation. A similar mechanism appears to be at work in the holographic models we have studied: whenever the spectral function is non-vanishing at small frequencies, the holographic zero sound mode becomes a broad resonance.

5. Conclusions

We determined the low-frequency and low-momentum form of the density and current retarded Green's functions for massless charge carriers in the holographic model of strange metals proposed in ref. [8]. Our main physical results are that for $z < 2$ the system's spectrum includes a quasi-particle – the holographic zero sound – whose dispersion relation we computed, whereas when $z > 2$ the quasi-particle is "washed out" by a continuum of low-frequency states. In addition, we recovered the results of ref. [8] for the AC conductivity from this single contribution to the spectral function.

Many avenues remain open for future research. As emphasized in ref. [8], a realistic analysis requires massive charge carriers, with a mass gap much larger than the temperature. In the bulk that means introducing a scalar field dual to the flavor mass operator on the probe D-brane worldvolume.

In relativistic string theory settings, the holographic zero sound quasi-particle is absent at any finite temperature, apparently as a consequence of the probe limit [6]. In these systems, a finite temperature produces an energy density of order N^2, which is enormous compared to the probe flavor's order $N_f N$ energy density. The low-frequency dynamics is thus dominated by adjoint-sector physics, which induces charge diffusion. The spectral function grows linearly with frequency in this case, so the finite-temperature width of the holographic zero sound is big and the pole is smeared into a broad resonance. Presumably the same is true for any $z < 2$. A natural question is what happens if we leave the probe limit. Does the holographic zero sound remain a sharp quasi-particle when N_f is on the order of N?

Lastly, the model building proposed in ref. [8] is worth pursuing. As noted there, some studies of the high-T_c cuprate strange metals find that the low-frequency AC conductivity scales as $\omega^{-2/3}$, which suggests that any holographic model reproducing such scaling may not exhibit a zero sound pole. A holographic system that reproduces all measured strange metal properties will be more complicated than just probe D-branes in a Lifshitz geometry however, so the question of whether a holographic zero sound quasi-particle exists in these more complicated models should be studied on a case-by-case basis.

Acknowledgments

C.H. would like to thank the organizers of the XVI Mexican School of Particles and Fields for their kind hospitality, Carlos Nunez and Amos Yarom for interesting discussions during the same and give special thanks to Elena Caceres and Alberto Guijosa for both. This work was supported in part by the U.S. Department of Energy under Grant No. DE-FG02-96ER40956.

References

[1] Maldacena J M 1998 *Adv. Theor. Math. Phys.* **2** 231–252 (*Preprint* hep-th/9711200)
[2] Gubser S S, Klebanov I R and Polyakov A M 1998 *Phys. Lett.* **B428** 105–114 (*Preprint* hep-th/9802109)
[3] Witten E 1998 *Adv. Theor. Math. Phys.* **2** 253–291 (*Preprint* hep-th/9802150)
[4] Karch A, Son D T and Starinets A O 2008 (*Preprint* 0806.3796)
[5] Kulaxizi M and Parnachev A 2008 *Phys. Rev.* **D78** 086004 (*Preprint* 0808.3953)
[6] Kim K Y and Zahed I 2008 *JHEP* **12** 075 (*Preprint* 0811.0184)
[7] Hoyos-Badajoz C, O'Bannon A and Wu J M S 2010 *JHEP* **09** 086 (*Preprint* 1007.0590)
[8] Hartnoll S A, Polchinski J, Silverstein E and Tong D 2009 (*Preprint* 0912.1061)
[9] Kachru S, Liu X and Mulligan M 2008 *Phys. Rev.* **D78** 106005 (*Preprint* 0808.1725)
[10] Kobayashi S, Mateos D, Matsuura S, Myers R C and Thomson R M 2007 *JHEP* **02** 016 (*Preprint* hep-th/0611099)
[11] Karch A and O'Bannon A 2007 *JHEP* **09** 024 (*Preprint* 0705.3870)
[12] Karch A and O'Bannon A 2007 *JHEP* **11** 074 (*Preprint* 0709.0570)
[13] Karch A and Katz E 2002 *JHEP* **06** 043 (*Preprint* hep-th/0205236)
[14] Karch A, Kulaxizi M and Parnachev A 2009 *JHEP* **11** 017 (*Preprint* 0908.3493)
[15] Benincasa P 2009 (*Preprint* 0911.0075)
[16] Lee B H and Pang D W 2010 (*Preprint* 1006.4915)
[17] Witten E 2001 (*Preprint* hep-th/0112258)
[18] Witten E 1998 *Adv. Theor. Math. Phys.* **2** 505–532 (*Preprint* hep-th/9803131)
[19] Son D T and Starinets A O 2002 *JHEP* **09** 042 (*Preprint* hep-th/0205051)
[20] Herzog C P and Son D T 2003 *JHEP* **03** 046 (*Preprint* hep-th/0212072)
[21] van Rees B C 2009 *Nucl. Phys. Proc. Suppl.* **192-193** 193–196 (*Preprint* 0902.4010)

Journal of Physics: Conference Series **287** (2011) 012030 doi:10.1088/1742-6596/287/1/012030

Unusual identities for QCD at tree-level

N E J Bjerrum-Bohr[1], Poul H Damgaard[1], Bo Feng[2,3] and Thomas Søndergaard[1]

[1]Niels Bohr International Academy and Discovery Center,
The Niels Bohr Institute, Blegdamsvej 17, DK-2100 Copenhagen, Denmark
[2]Center of Mathematical Science, Zhejiang University, Hangzhou, China
[3]Kavli Institute for Theoretical Physics China, CAS, Beijing 100190, China

E-mail:

Abstract. We discuss a set of recently discovered quadratic relations between gauge theory amplitudes. Such relations give additional structural simplifications for amplitudes in QCD. Remarkably, their origin lie in an analogous set of relations that involve also gravitons. When certain gluon helicities are flipped we obtain relations that do not involve gravitons, but which refer only to QCD.

1. Introduction

At the Large Hadron Collider at CERN intense beams of protons are already colliding at historically high energies. The amounts of data taken from these collisions are enormous and it will require massive efforts from both experimentalists and theorists to uncover signals of new physics above 'background' QCD processes. From the theoretical side the provision of precise cross sections for scattering processes is therefore crucial. Such computations are, however, tremendously complex if one employs traditional techniques. Luckily, new computational methods inspired by both string theory and twistor variables have been rapidly invented in the last decade. These novel approaches have succeeded in providing us a series of quite efficient toolboxes for computations. Amazingly, we are nevertheless still learning more about gauge theories and the surprising hidden simplifications that such theories give rise to in perturbation theory. Gauge invariance seems to be the essential ingredient: it is because different parts of calculations can look complicated or simple depending on the chosen gauge. From this perspective, the complexity of perturbative calculations in QCD and QCD-like theories is partly fake and only due to inconvenient choices of computation. It is essential to uncover as many relations among *gauge invariant observables* as possible.

In this context it came as a complete surprise to everybody a few years ago that there could be new non-trivial identities among color-ordered amplitudes in QCD, the relations conjectured by Bern, Carrasco and Johansson [1]. This set of BCJ-relations, as they are now called, can be proven most easily using string theory [2, 3], but an inductive proof based on quantum field theory alone has also recently been established [4, 5]. What is particularly remarkable about BCJ-relations is that they are sufficiently restrictive to reduce the basis of color-ordered amplitudes from size $(n-2)!$ to $(n-3)!$ independently of the helicity configurations. This follows again very naturally from string theory [2].

Recently, we have found that there are yet more relations between such color-ordered amplitudes in Yang-Mills theories (and Yang-Mills theories coupled to matter) once we fix helicities [6]. These new identities take the form of a linear sum of *products* of two color-ordered amplitudes in Yang-Mills theory that, surprisingly, vanish. Looked at from field theory alone this is a most unusual phenomenon, since quadratic relations among amplitudes have, as far as we know, never been seen before, in any context. The reason why they nevertheless appear is, amazingly, related to the perturbative relationship between gravity amplitudes and color-ordered Yang-Mills amplitudes, the so-called KLT-relations [7, 8, 9, 10, 11].

Here we discuss these quadratic relations between Yang-Mills amplitudes [6] in some greater detail. Our discussion is structured in the following way. In section 2 we review some well-known properties of tree-level gauge-theory amplitudes. In section 3 we present the quadratic relations along with several explicit examples, and in section 4 we sketch the proof of these relations. For full details we refer to the paper [9]. Finally, in section 5 we have our conclusions.

2. Gauge-theory amplitudes

To set up some notation, we begin with a brief review of the current state of tree-level gauge-theory amplitudes. This will also serve as a nice comparison between the amplitude relations that were known previosuly and the new ones we will discuss here.

It is well known that when working with tree-level gauge theory amplitudes $\mathcal{A}_n$, it is convenient to introduce the color-ordered *sub*amplitudes A_n. They are related to the full color-dressed amplitudes by

$$\mathcal{A}_n = \sum_{\sigma\in S_{n-1}} \mathrm{Tr}[T^1T^{\sigma(2)}\cdots T^{\sigma(n)}]A_n(1,\sigma(2),\ldots,\sigma(n)). \tag{1}$$

Here T^i are the generators of the gauge group, and the sum runs over all permutations of leg $2,3,\ldots,n$. In this way one has completely decoupled the color-structure from the kinematics and can now focus on the simpler objects A_n.

These color-ordered amplitudes turn out to have an immensely rich structure. In particular, they satisfy several different kind of relations, reducing the number of independent A_n's considerably. The most simple are cyclicity and reversion:

$$A_n(1,2,\ldots,n) = A_n(2,3,\ldots,n,1), \qquad A_n(1,2,\ldots,n) = (-1)^n A_n(n,n-1,\ldots,1), \tag{2}$$

The so-called photon-decoupling identity reads

$$\sum_{\sigma\in cyclic} A_n(1,\sigma(2),\sigma(3),\ldots,\sigma(n)) = 0, \tag{3}$$

with the sum running over all cyclic permutations of leg $2,3,\ldots,n$. For example, in the four-point case it reads

$$A_4(1,2,3,4) + A_4(1,3,4,2) + A_4(1,4,2,3) = 0. \tag{4}$$

The photon-decoupling identity reflects the simple fact that we can consider a Yang-Mills amplitude and replace one of the non-Abelian generators by the identity matrix. Since it commutes right through all the other generators, and since the amplitude must vanish, the identity follows.

More generally, we have the Kleiss-Kuijf (KK) relations [12]

$$A_n(1,\{\alpha\},n,\{\beta\}) = (-1)^{n_\beta} \sum_{\sigma\in \mathrm{OP}(\{\alpha\},\{\beta^T\})} A_n(1,\{\sigma\},n), \tag{5}$$

where the sum is over "ordered permutations", *i.e.* all permutations of $\{\alpha\} \cup \{\beta^T\}$ that keep the order of the individual elements belonging to each set fixed. Here n_β is the number of elements in $\{\beta\}$, and $\{\beta^T\}$ is the $\{\beta\}$ set with the ordering reversed. For example,

$$A_6(1,2,3,6,4,5) = A_6(1,2,3,5,4,6) + A_6(1,2,5,3,4,6) + A_6(1,2,5,4,3,6) + A_6(1,5,2,4,3,6) + A_6(1,5,4,2,3,6) + A_6(1,5,2,3,4,6), \tag{6}$$

where we have chosen $\{\alpha\} = \{2,3\}$ and $\{\beta\} = \{4,5\}$ (and therefore $n_\beta = 2$). Altogether these relations reduce the number of independent subamplitudes to $(n-2)!$.

More recently additional relations were discovered: the Bern-Carrasco-Johansson (BCJ) relations [1]. There are several ways of presenting them, *e.g.*

$$0 = s_{12}A_n(1,2,3,\ldots,n) + (s_{12}+s_{23})A_n(1,3,2,4,\ldots,n) + (s_{12}+s_{23}+s_{24})A_n(1,3,4,2,5,\ldots,n) + \cdots + (s_{12}+s_{23}+s_{24}+\cdots+s_{2(n-1)})A_n(1,3,4,\ldots,n-1,2,n), \tag{7}$$

along with all the relations obtained by permutation of $1,2,\ldots,n$ in the above equation. Exploiting these additional relations, the number of independent subamplitudes reduces to $(n-3)!$. What is new here is the fact that the identities involve the external momenta. Nevertheless, they are universal and hold for all choices of helicity.

The string theory generalization of these relations has also been found [2, 3]. They follow from monodromy relations in the integral representation of string and they reduce to the BCJ-relations in the field theory limit. This was in fact the first proof of the BCJ-relations in the field theory limit. The fact that the basis of amplitudes is reduced to $(n-3)!$ from $(n-2)!$ follows beautifully from string theory as a consequence of the need to fix precise three of the n positions of the external legs in the string theory amplitude. The origin of this lies in the fact that the Moebius group is a three-parameter group. It is quite remarkable that this concept, which would appear to be totally unrelated to QCD, can have have such an important consequence for QCD amplitudes. Understanding this in terms of a different representation of the Feynman diagrams involved is one outstanding question which remains to be answered. It is possible that the world-line formalism [13] holds the clue, but the details remain to be worked out. This being said, it is nevertheless now known how to prove the BCJ-relations using field theory alone [4, 5]. In fact, that proof does not use any Lagrangian-specific representation, but relies instead on very general analyticity properties of the S-matrix.

3. Quadratic amplitude relations

We now come to the main topic of this talk: a new set of identities among gauge-theory amplitudes that have recently been discovered (and proven) in refs. [6, 8]. These new identities have a rather different structure than any of the previously mentioned relations. First of all, they are *quadratic* in the amplitudes, and second, instead of being helicity-independent, like all the relations reviewed in the last section, they relate amplitudes from different helicity sectors. The fact that the relations are not linear in amplitudes is particularly surprising; almost all of our intuition about gauge theory amplitudes come from the linear level. This is where Ward Identities live, and this is where symmetries act in a simple ways. Most unusual, these quadratic identites can be seen as being dual to relations that link QCD-amplitudes to *gravity amplitudes.* The fact that they are quadratic is therefore directly related to the fact that the graviton, being of spin-2, has polarization tensors that can be factorized into to two spin-1 (gluon) polarization vectors. Say, for positive helicity: $\epsilon_{\mu\nu}^{(+)} \sim \epsilon_\mu^{(+)}\epsilon_\nu^{(+)}$. The appearance of the outer product between the two gluon polarization vectors is what leads to the quadratic relations. Yet, QCD-amplitude identities should not involve gravity. How can this be reconciled? It turns out that there are "forbidden" gravity amplitudes that vanish. This happens when we consider mixed combinations $\epsilon_\mu^{(+)}\epsilon_\nu^{(-)}$ – what should have been the scalar component of a gravity amplitude, which vanishes.

Figuratively speaking, then, these new QCD-identities arise as follows: When we consider appropriate products of like-helicity amplitudes, we produce a gravity amplitude. When we flip the helicities of some of the gluons (which legs we flip will be described in detail below), we instead get zero. This gives the new identities.

Before presenting the quadratic relations in all their gory detail, let us first introduce a useful quantity $\mathcal{S}$, that depends on k massless momenta in the following way:

$$\mathcal{S}[i_1,\ldots,i_k|j_1,\ldots,j_k]_{p_1} \equiv \prod_{t=1}^{k}\big(s_{i_t 1} + \sum_{q>t}^{k}\theta(i_t,i_q)s_{i_t i_q}\big)\,, \tag{8}$$

where $\{i_1,\ldots,i_k\}$ and $\{j_1,\ldots,j_k\}$ are two arbitrary orderings of k momenta. We have defined as usual $s_{ij} \equiv (p_i+p_j)^2$. Furthermore, $\theta(i_a,i_b)$ is zero if i_a comes sequentially before i_b in $\{j_1,\ldots,j_k\}$ and unity if it comes after.

The definition of $\mathcal{S}$ may sound complicated, in particular dues to the curious definition of the "step function" $\theta(i_a,i_b)$. It is therefore useful to write down a few explicit examples of $\mathcal{S}$ that will help in understanding how it is constructed:

$$\mathcal{S}[2|2]_{p_1} = s_{12}, \qquad \mathcal{S}[23|23]_{p_1} = s_{12}s_{13}, \qquad \mathcal{S}[234|324]_{p_1} = (s_{12}+s_{23})s_{13}s_{14}. \tag{9}$$

This $\mathcal{S}$ function turns out to have quite a few very nice properties. We will not go through all of them here, they can be found in [11] (along with its string theory generalization, which turns out to preserve these nice properties), but we wish to stress its close connection to the BCJ-relations. This is most manifest through

$$0 = \sum_{\alpha\in S_{n-2}} \mathcal{S}[\beta(2,n-1)|\alpha(2,n-1)]A_n(1,\alpha(2,n-1),n), \tag{10}$$

where we have introduced the simplifying notation $\alpha(2,n-1) \equiv \{\alpha(2),\alpha(3),\ldots,\alpha(n-1)\}$ for permutations of leg $2,3,\ldots,n-1$, and $\beta(2,n-2)$ is just some arbitrary permutation of the same legs. The sum is over all α permutations. Eq. (10) is indeed precisely satisfied due to the BCJ-relations in eq. (7)! Conversely, the system of equations one obtains from choosing different β permutations in eq. (10) can be used to reduce the basis to $(n-3)!$ [11]. In this way we see that eq. (10) is a neat and compact reformulation of the BCJ-relations in terms of our $\mathcal{S}$ function, the momentum kernel.

Since the quadratic relations which we are soon going to present are helicity-dependent, we will also introduce the following short-hand notation. We denote A_n^k an N^{k}MHV amplitude, *i.e.* A_n^k is an n-point subamplitude with $2+k$ negative helicity gluons. Of course, to have non-vanishing amplitudes $k\in\{0,1,\ldots,n-4\}$. We do not care about the exact helicity configuration, just which helicity *sector* it belongs to.

With this we can now write down the quadratic relations. Assuming $k\neq h$ the following identity is satisfied

$$0 = \sum_{\gamma,\beta\in S_{n-3}} A_n^k(n-1,n,\gamma(2,n-2),1)\mathcal{S}[\gamma(2,n-2)|\beta(2,n-2)]_{p_1}A_n^h(1,\beta(2,n-2),n-1,n), \tag{11}$$

where the sum is over all permutations of γ and β.

Note that in each of the A_n^k (A_n^h) amplitudes in the sum it is the same $k+2$ ($h+2$) legs that have flipped helicity. These are relations among subamplitudes living in different helicity sectors. They are *not* just satisfied due to the BCJ-relations, which one might have thought from the seemingly close analogy to eq. (10). The fact that the identities have quite different content is clearly pointed out from our requirement $k \neq h$ on the helicity structure, whereas the BCJ-relations are helicity-independent. We can schematically express eq. (11) as

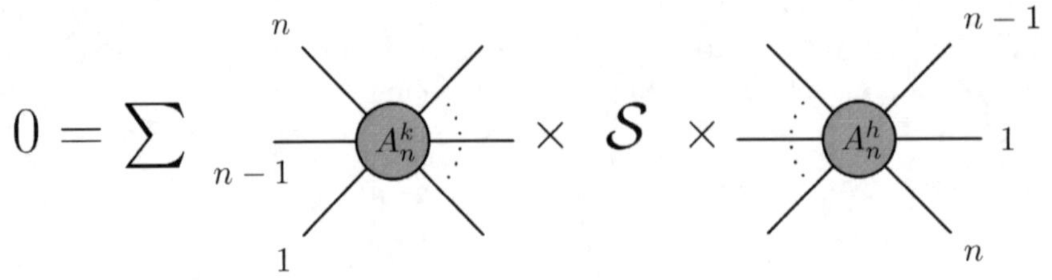

(12)

We re-emphasize that when helicities are not flipped, the same right hand side of this equation produces instead a graviton amplitude! The explicit link between the two sets of relations has been further elaborated on in ref. [14].

Although eq. (11) is probably one of the nicest forms of the general n-point case, it contains a quite huge number of terms, namely $[(n-3)!]^2$. Using BCJ-relations one can write eq. (11) in other, but completely *equivalent*, ways [9]. In general we can write the relations as

$$0 = \sum_{\sigma \in S_{n-3}} \sum_{\alpha \in S_{j-2}} \sum_{\beta \in S_{n-1-j}} A_n^k(\alpha(\sigma(2,j-1)),1,n-1,\beta(\sigma(j,n-2)),n)\mathcal{S}[\alpha(\sigma(2,j-1))|\sigma(2,j-1)]_{p_1}$$
$$\times\, \mathcal{S}[\sigma(j,n-2)|\beta(\sigma(j,n-2))]_{p_{n-1}} A_n^h(1,\sigma(2,j-1),\sigma(j,n-2),n-1,n)\,, \quad (13)$$

for *any* $j = 2, 3, \ldots, n-1$. Eq. (11) is therefore just a special case of eq. (13), namely the one with $j = n-1$. However, we could just as well choose $j = [n/2]+1$ in which case the number of terms in eq. (13) reduces to $(n-3)!([n/2]-1)!([n/2]-2)!$

Finally, we provide yet another way of writing these relations. This form has a higher degree of manifest crossing symmetry between the different external legs. However, it requires the introduction of a regularization. To this end, let us assume we make the following shift in the momentum p_1 of leg 1 and momentum p_n of leg n:

$$p_1 \to p_1' \equiv p_1 - xq, \qquad p_n \to p_n' \equiv p_n + xq, \quad (14)$$

where x is some arbitrary parameter and q a four-vector satisfying $q^2 = p_1 \cdot q = 0 \neq p_n \cdot q$. This preserves overall energy-momentum conservation, keeps p_1' on-shell, but makes ${p_n'}^2 = s_{1'2\ldots n-1} \neq 0$. After this regularization the quadractic relations can be written as

$$0 = \lim_{x\to 0} \sum_{\gamma,\beta} \frac{A_n^k(n',\gamma(2,n-1),1')\mathcal{S}[\gamma(2,n-1)|\beta(2,n-1)]_{p_1'} A_n^h(1',\beta(2,n-1),n')}{s_{1'2\ldots n-1}}. \quad (15)$$

We see that as $x \to 0$ the denominator goes to zero, but due to eq. (10) the numerator is *also* going to zero. Indeed, both the numerator and denominator vanish at the same, and eq. (15) is just another way of representing the quadratic relations [8].

Although this form may not be as practically convenient as eq. (13) it turns out to be important for the proof of the quadratic relations, even in the form of eq. (13).

3.1. Examples

To get a better feel for eq. (13) (and eq. (11)), let us write out some explicit examples. For four points the relations are trivially satisfied. Eq. (11) takes the form

$$0 = s_{12} A_4^k(3,4,2,1) A_4^h(1,2,3,4). \quad (16)$$

But if $k \neq h$ at least one of the amplitudes must have three or four of the same helicity and therefore vanish all by its own due to the MHV-rule. (Alternatively, here is a quite unusual proof of that MHV-rule). However, already at five points we start getting non-trivial cancellations among different amplitudes that do not vanish individually. In this case eq. (11) is

$$\begin{aligned} 0 = s_{12}A_5^h(1,2,3,4,5)\big[s_{13}A_5^k(4,5,2,3,1) + (s_{13}+s_{23})A_5^k(4,5,3,2,1)\big] \\ + s_{13}A_5^h(1,3,2,4,5)\big[s_{12}A_5^k(4,5,3,2,1) + (s_{12}+s_{23})A_5^k(4,5,2,3,1)\big], \end{aligned} \tag{17}$$

or using eq. (13) with $j = 3$

$$0 = s_{12}s_{34}A_5^h(1,2,3,4,5)A_5^k(4,3,5,2,1) + s_{13}s_{24}A_5^h(1,3,2,4,5)A_5^k(4,2,5,3,1). \tag{18}$$

Taking $(h,k) = (0,1)$ (or $(1,0)$) we have relations with non-vanishing subamplitudes. For instance,

$$\begin{aligned} 0 = s_{12}s_{34}A_5(1^-,2^-,3^+,4^+,5^+)A_5(4^+,3^-,5^+,2^-,1^-) \\ + s_{13}s_{24}A_5(1^-,3^+,2^-,4^+,5^+)A_5(4^+,2^-,5^+,3^-,1^-). \end{aligned} \tag{19}$$

Finally, let us give the explicit expression for six points (using eq. (13) with $j = 4$)

$$0 = s_{12}s_{45}A_6^h(1,2,3,4,5,6)\big[s_{13}A_6^k(5,4,6,2,3,1) + (s_{13}+s_{23})A_6^k(5,4,6,3,2,1)\big] + \mathcal{P}(2,3,4). \tag{20}$$

Here we start having relations between "real" NMHV and MHV amplitudes ($(h,k) = (0,1)$).

It is clear from this discussion that if we know the amplitudes for some helicity sector, we can plug them into the quadratic relations and thereby get linear relations between amplitudes in *another* helicity sector. In particular, one can choose the simple MHV amplitudes for one of them and get linear relations between the more complicated N^hMHV amplitudes. Choosing the A_n^k amplitudes to be MHV, and using a standard spinor-helicity notation, one gets the general relation [10, 15]

$$0 = \sum_{\beta\in S_{n-3}} \prod_{i=2}^{n-2} [\beta(i)|\beta(i+1)+\beta(i+2)+\cdots+\beta(n-1)|n\rangle A_n^h(1,\beta(2,n-2),n-1,n), \tag{21}$$

for $h > 0$. As an example, consider the six-point case where A_6^h is some NMHV amplitude

$$0 = [2|3+4+5|6\rangle[3|4+5|6\rangle[4|5|6\rangle A_6^h(1,2,3,4,5,6) + \mathcal{P}(2,3,4). \tag{22}$$

One might wonder from these considerations what happens when $k = h$. In this case we do not have zero on the left-hand-side of eq. (13) (or eq. (15)), but instead this is precisely where we get a tree-level gravity amplitude! That is, when the helicities in the two amplitudes are the same one has the famous Kawai-Lewellen-Tye (KLT) relations [7]. Surprisingly, the vanishing of these quadratic relations was needed as an essential input for the field theoretical proof of the KLT-relations [8, 9]. One can unify these relations by consider the KLT-relations between $\mathcal{N} = 4$ supersymmetric Yang-Mills and $\mathcal{N} = 8$ supergravity. In this picture the quadratic vanishing relations appear as a consequence of violation of R-symmetry [10, 14].

4. The proof

The reader may be more interest in the content of these QCD-identities than in their proof. Nevertheless, let us here just sketch the proof of these new quadratic relations. All we need to know are some analytical properties of subamplitudes. For full details of the proof we refer to [9].

We choose to prove the identities by induction. We thus assume that we have verified the quadratic identities up to $n-1$ points (they are easy to verify in the low-point cases) and now want to show this implies the relation at n points, *i.e.* show that

$$X_n = \sum_{\gamma,\beta\in S_{n-3}} A_n^k(n-1,n,\gamma(2,n-2),1)\mathcal{S}[\gamma(2,n-2)|\beta(2,n-2)]_{p_1} A_n^h(1,\beta(2,n-2),n-1,n) \tag{23}$$

is also zero (when $k \neq h$). Here we restrict ourselves to the form in eq. (11), since it is completely equivalent to all the forms contained in eq. (13), see [9]. Showing the new identities in this particular form is therefore equivalent to showing them all.

We now make a so-called BCFW-shift in the legs 1 and n, and consider the following contour integral [16] under the assumption that the boundary contribution is zero,

$$0 = \oint \frac{dz}{z} X_n(z) = X_n(0) + \sum(\text{residues for } z \neq 0)\,. \tag{24}$$

Since the residues are calculated from the poles of X_n, and the poles only appear in the subamplitudes, there are two different kind of terms to consider:

(A) The pole appears in only one of the amplitudes A_n^k or A_n^h.

(B) The pole is present in both amplitudes A_n^k and A_n^h.

Using the properties of the $\mathcal{S}$ function one find that the terms in category (A) always vanish due to BCJ-relations (in the form of eq. (10)).

The terms in category (B) are a bit more tricky. However, these will factorize into products of lower-point quadratic relations where at least one of them will satisfy the requirement that the amplitudes belong to different helicity sectors. For this part it is important that we know that the quadratic relations can be written in the form of eq. (15) since it will be products of quadratic relations with one in the form of eq. (11) and one in the form of eq. (15). For this step we would of course first need to show that eq. (15) is always satisfied. This can be done from the beginning by following the exact same procedure. Actually, in that case the residues will always factorize into quadratic relations in the form of eq. (15) and thereby make the induction proof direct.

With this procedure we find that *all* residues vanish and therefore also $X_n \equiv X_n(0) = 0$.

5. Conclusion

We have discussed in some detail the quadratic identites among Yang-Mills amplitudes that have recently been found [6]. Along with several explicit examples we have also sketched our proof of these new identities. What is surprising about these QCD-relation is that it seems one could not possibly have arrived at them without knowing that there is also a way to construct gravity amplitudes out of quadratic combinations of not-flipped color-orderes amplitudes. We have nevertheless succeeded in proving them directly using only self-contained quantum field theory tools, without any reference to gravity amplitudes. Further progress in this direction seems possible. It would clearly be interesting to investigate more identities at one and possibly multi-loop level. This seems a promising avenue for future studies. Some steps have already been taken in this direction [6].

References

[1] Z. Bern, J. J. M. Carrasco and H. Johansson, Phys. Rev. D **78**, 085011 (2008) [0805.3993 [hep-ph]].
[2] N. E. J. Bjerrum-Bohr, P. H. Damgaard and P. Vanhove, Phys. Rev. Lett. **103**, 161602 (2009) [0907.1425 [hep-th]].
[3] S. Stieberger, 0907.2211 [hep-th].
[4] B. Feng, R. Huang and Y. Jia, Phys. Lett. B **695** (2011) 350 [arXiv:1004.3417 [hep-th]].
[5] Y. X. Chen, Y. J. Du and B. Feng, arXiv:1101.0009 [hep-th].
[6] N. E. J. Bjerrum-Bohr, P. H. Damgaard, B. Feng and T. Sondergaard, Phys. Lett. B **691** (2010) 268 [arXiv:1006.3214 [hep-th]].
[7] H. Kawai, D. C. Lewellen and S. H. H. Tye, Nucl. Phys. B **269** (1986) 1.
[8] N. E. J. Bjerrum-Bohr, P. H. Damgaard, B. Feng and T. Sondergaard, Phys. Rev. D **82** (2010) 107702 [arXiv:1005.4367 [hep-th]].
[9] N. E. J. Bjerrum-Bohr, P. H. Damgaard, B. Feng and T. Sondergaard, JHEP **1009** (2010) 067 [arXiv:1007.3111 [hep-th]].
[10] B. Feng and S. He, JHEP **1009** (2010) 043 [arXiv:1007.0055 [hep-th]].
[11] N. E. J. Bjerrum-Bohr, P. H. Damgaard, T. Sondergaard and P. Vanhove, JHEP **1101** (2011) 001 [arXiv:1010.3933 [hep-th]].
[12] R. Kleiss and H. Kuijf, Nucl. Phys. B **312** (1989) 616.
[13] M. J. Strassler, Nucl. Phys. B **385** (1992) 145 [arXiv:hep-ph/9205205]; M. G. Schmidt and C. Schubert, Phys. Lett. B **318** (1993) 438 [hep-th/9309055].
[14] H. Tye and Y. Zhang, arXiv:1007.0597 [hep-th]; H. Elvang and M. Kiermaier, JHEP **1010** (2010) 108 [arXiv:1007.4813 [hep-th]].
[15] B. Feng, S. He, R. Huang and Y. Jia, JHEP **1010** (2010) 109 [arXiv:1008.1626 [hep-th]].
[16] R. Britto, F. Cachazo and B. Feng, Nucl. Phys. B **715** (2005) 499 [hep-th/0412308];
R. Britto, F. Cachazo, B. Feng and E. Witten, Phys. Rev. Lett. **94** (2005) 181602 [hep-th/0501052].

Journal of Physics: Conference Series **287** (2011) 012031 doi:10.1088/1742-6596/287/1/012031

Schwinger pair creation of particles and strings

Christian Schubert

Instituto de Física y Matemáticas, Universidad Michoacana de San Nicolás de Hidalgo, Edificio C-3, Apdo. Postal 2-82, C.P. 58040, Morelia, Michoacán, México.

E-mail: `schubert@ifm.umich.mx`

Abstract. I shortly review the worldline instanton method for calculating Schwinger pair production rates in (i) one-loop QED (ii) multiloop QED and (iii) one-loop open string theory.

1. History

Already in 1931 F. Sauter [1] realized that Dirac's theory of the electron predicts that an electric field of sufficient strength and extent can induce spontaneous creation of electron – positron pairs from the vacuum. Naively, a virtual pair turns real by separating out along the field and drawing a sufficient amount of energy from it to make up for both rest mass energies (fig. 1).

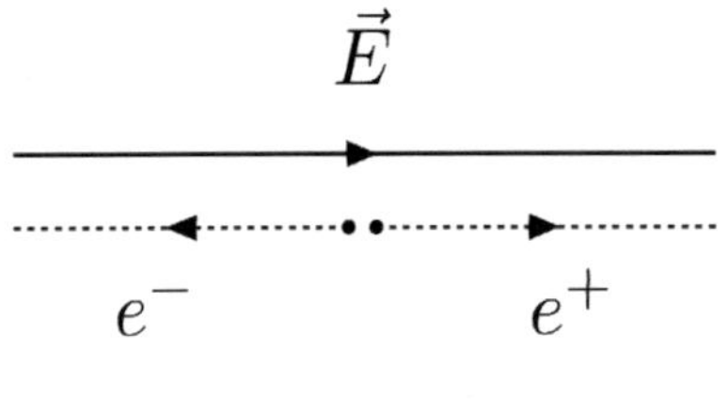

Figure 1. Pair creation by an external field.

Twenty years later, J. Schwinger [2] used the more advanced methods then available for QED to obtain his famous formulas for the imaginary parts of the one-loop effective actions in a constant homogeneous field, for both spinor and scalar QED:

$$\mathrm{Im}\mathcal{L}^{(1)}_{\mathrm{spin}}(E) = \frac{m^4}{8\pi^3}\beta^2 \sum_{n=1}^{\infty} \frac{1}{n^2} \exp\left[-\frac{\pi n}{\beta}\right] \tag{1.1}$$

$$\mathrm{Im}\mathcal{L}^{(1)}_{\mathrm{scal}}(E) = -\frac{m^4}{16\pi^3}\beta^2 \sum_{n=1}^{\infty} \frac{(-1)^n}{n^2} \exp\left[-\frac{\pi n}{\beta}\right] \tag{1.2}$$

($\beta = \frac{eE}{m^2}$; the upper index on $\mathcal{L}$ refers to the loop order). For low pair creation rates, these imaginary parts relate directly to the pair creation rate per volume w, namely one has

$w \approx 2\mathrm{Im}\mathcal{L}$. Schwinger's formulas (1.1),(1.2) are written in terms of an infinite sum of "Schwinger exponentials", where the nth term relates to the coherent production of n pairs by the field. Their dependence on the field strength is nonperturbative, which confirms Sauter's picture of the pair production as a tunneling process. Unfortunately, this also implies that the pair creation rate is exponentially small for $E \ll E_{\rm cr} := 10^{16} V/\mathrm{cm}$. Until recently, macroscopic fields of this strength were considered unattainable. However, this has changed due to the rapid evolution of laser technology. Presently there are various lasers in planning or operation, e.g. POLARIS [3], ELI [4] and XFEL [5], whose maximum field strength falls only a few orders of magnitude short of $E_{\rm cr}$ (only about one order of magnitude in the case of XFEL). This has led to a number of attempts to bring down the pair creation threshold by a superposition of various optical or X-ray beams; to mention just two recent proposals, counterpropagating linearly polarized lasers were considered in [6] and the superposition of a plane-wave X-ray beam with a strongly focused optical laser pulse in [7]. Obviously, such configurations have field strengths very far from the constant homogeneous case considered by Schwinger, and there is little hope for an exact calculation of the associated pair creation rates; approximation methods will have to be used. The tunnel effect picture suggest the use of WKB, and this is indeed the method which in the past has been almost universally used for the calculation of pair creation rates in nonconstant fields; see, e.g., [8, 9, 10, 11, 12].

The subject of this talk, the *worldline instanton method*, is related to the WKB approximation, but closer in spirit to modern relativistic quantum field theory, which leads to certain advantages. This approach was invented in 1982 by I. K. Affleck, O. Alvarez and N. S. Manton [13] ("AAM" in the following) for the case of a constant field in scalar QED, using Feynman's worldline representation of the scalar QED effective action [14]. In the quenched (single scalar loop) approximation, this representation is

$$\Gamma_{\rm scalar}[A] = \int d^4x\, \mathcal{L}(A) = \int_0^\infty \frac{dT}{T}\, \mathrm{e}^{-m^2T} \int_{x(T)=x(0)} \mathcal{D}x(\tau)\, e^{-S[x(\tau)]} \tag{1.3}$$

Here m and T are the mass and proper time of the loop scalar, and the path integral $\int \mathcal{D}x(\tau)$ is over closed trajectories in (Euclidean) spacetime with a worldline action $S[x(\tau)]$ that has three parts:

$$S = S_0 + S_{\rm ext} + S_{\rm int} \tag{1.4}$$

where

$$\begin{aligned} S_0 &= \int_0^T d\tau \frac{\dot{x}^2}{4} \\ S_{\rm ext} &= ie\int_0^T \dot{x}^\mu A_\mu(x(\tau)) \\ S_{\rm int} &= -\frac{e^2}{8\pi^2}\int_0^T d\tau_1 \int_0^T d\tau_2 \frac{\dot{x}(\tau_1)\cdot\dot{x}(\tau_2)}{(x(\tau_1)-x(\tau_2))^2} \end{aligned} \tag{1.5}$$

S_0 describes the free propagation of the scalar particle, $S_{\rm ext}$ its interaction with the background field, and $S_{\rm int}$ internal photon exchanges in the loop.

The basic idea is to calculate the path integral by a stationary phase approximation. We will carry it through first at the one-loop level.

Journal of Physics: Conference Series **287** (2011) 012031 doi:10.1088/1742-6596/287/1/012031

2. One-loop QED

The one-loop effective action is obtained from (1.3) by omitting $S_{\rm int}$,

$$\Gamma^{(1)}_{\rm scalar}[A] = \int_0^\infty \frac{dT}{T}\,{\rm e}^{-m^2T}\int \mathcal{D}x\,{\rm e}^{-\int_0^T d\tau\left(\frac{\dot x^2}{4}+ieA\cdot\dot x\right)} \tag{2.1}$$

Rescaling $\tau = Tu$, this becomes

$$\Gamma^{(1)}_{\rm scalar}[A] = \int_0^\infty \frac{dT}{T}\,{\rm e}^{-m^2T}\int \mathcal{D}x\,{\rm e}^{-\left(\frac{1}{T}\int_0^1 du\,\dot x^2+ie\int_0^1 duA\cdot\dot x\right)} \tag{2.2}$$

This makes it apparent that the T integral has a stationary point at $T_c = \sqrt{\int_0^1 du\,\dot x^2}/m$. If we are only interested in the imaginary part of the effective action at large mass, we can use this stationary point to eliminate the T integral, yielding

$${\rm Im}\Gamma^{(1)}[A] = \frac{1}{m}\sqrt{\frac{2\pi}{T_c}}\,{\rm Im}\int \mathcal{D}x\,{\rm e}^{-\left(m\sqrt{\int du\,\dot x^2}+ie\int_0^1 duA\cdot\dot x\right)} \tag{2.3}$$

The new worldline action,

$$S = m\sqrt{\int_0^1 du\,\dot x^2}+ie\int_0^1 duA\cdot\dot x \tag{2.4}$$

is stationary if

$$m\frac{\ddot x_\mu}{\sqrt{\int_0^1 du\,\dot x^2}} = ieF_{\mu\nu}\dot x_\nu \tag{2.5}$$

Contracting this equation with $\dot x^\mu$ shows that $\dot x^2 = {\rm const.} \equiv a^2$, so that $m\ddot x_\mu = ieaF_{\mu\nu}\dot x_\nu$. Thus the extremal action trajectory $x^{\rm cl}(u)$, to be called *worldline instanton*, is simply a periodic solution of the Lorentz force equation. The semiclassical (large mass) approximation becomes

$${\rm Im}\mathcal{L}(E) \sim {\rm e}^{-S[x^{\rm cl}]} \tag{2.6}$$

In the case of a constant field $\mathbf{E} = (0,0,E)$ considered by AAM, the worldline instanton turns out to be a circle in the $x_3 - x_4$ plane, of radius m/eE and winding number n:

$$x^{\rm cl}(u) = \frac{m}{eE}\Big(x_1,x_2,\cos(2n\pi u),\sin(2n\pi u)\Big) \tag{2.7}$$

$$S[x^{\rm cl}] = n\pi\frac{m^2}{eE} \tag{2.8}$$

Thus the instanton action for winding number n reproduces the nth exponent in Schwinger's representation (1.2) of ${\rm Im}\mathcal{L}^{(1)}_{\rm scal}(E)$.

Worldline instantons for more general electric fields where obtained in [15]. As an example, let us show the case of the "single-bump" field $E(x_3) = E\,{\rm sech}^2(kx_3)$. Here the nth instanton solution is

$$\begin{aligned} x_3(u) &= \frac{m}{eE}\frac{1}{\tilde\gamma}\,{\rm arcsinh}\Big(\frac{\tilde\gamma}{\sqrt{1-\tilde\gamma^2}}\sin(2n\pi u)\Big) \\ x_4(u) &= \frac{m}{eE}\frac{1}{\tilde\gamma\sqrt{1-\tilde\gamma^2}}\arcsin(\tilde\gamma\cos(2n\pi u)) \end{aligned} \tag{2.9}$$

Journal of Physics: Conference Series **287** (2011) 012031 doi:10.1088/1742-6596/287/1/012031

where the parameter $\tilde{\gamma} \equiv \frac{mk}{eE}$ is a measure for the inhomogeneity of the field. The instanton action is

$$S[x^{\rm cl}] \;=\; n\,\frac{m^2\pi}{eE}\left(\frac{2}{1+\sqrt{1-\tilde{\gamma}^2}}\right) \tag{2.10}$$

For $\tilde{\gamma} \to 0$ one recovers the constant field instanton. With increasing $\tilde{\gamma}$ the instanton action increases, which means that the associated pair creation rate goes down. For $\tilde{\gamma} > 1$ the instanton ceases to exist, in accordance with the fact that the maximal energy which can be extracted from the field by a virtual pair drops below $2m$. In fig. 2 the instanton trajectories are plotted for various values of $\tilde{\gamma}$.

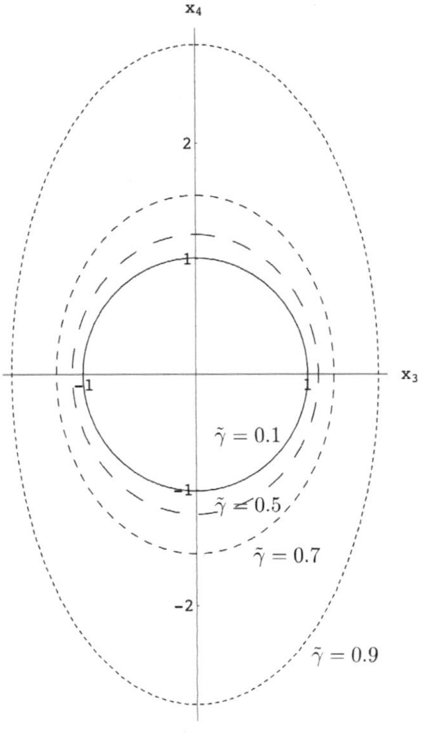

Figure 2. Plot of the instanton paths for the case $E(x) = E\mathrm{sech}^2(kx)$.

In [16] a general method was developed to also calculate the prefactor of the instanton-induced Schwinger exponential $\mathrm{e}^{-S[x^{\rm cl}]}$, that is, the fluctuation determinant of the expansion of the worldline path integral around the stationary trajectory. Including this prefactor for the single-bump case one finds excellent numerical agreement [16] with an exact result by A.I. Nikishov [17] and as well with a direct Monte Carlo integration of the worldline path integral [18].

See [16] for some other cases where the instantons can be found in closed form, [19] for more general classes of fields, and [15] for the extension to spinor QED.

3. Multiloop QED

As argued already by [13], the constant field worldline instantons (2.7) remain stationary trajectories even in the presence of photon insertions. Evaluating the photon insertion term S_{int} on the principal ($n = 1$) trajectory gives the following all-loop formula for the imaginary part of the effective Lagrangian in the large mass (= weak field) approximation:

$$\mathrm{Im}\mathcal{L}_{\mathrm{scal}}(E) = \sum_{l=1}^{\infty} \mathrm{Im}\mathcal{L}^{(l)}_{\mathrm{scal}}(E) \overset{\beta \to 0}{\sim} \frac{m^4 \beta^2}{8\pi^3} \exp\Big[-\frac{\pi}{\beta} + \alpha\pi\Big] = \mathcal{L}^{(1)}_{\mathrm{scal}}\, \mathrm{e}^{\alpha\pi} \tag{3.1}$$

This formula is rather remarkable, since in terms of standard field theory it corresponds to a summation of the infinite set of Feynman diagrams shown in fig. 3:

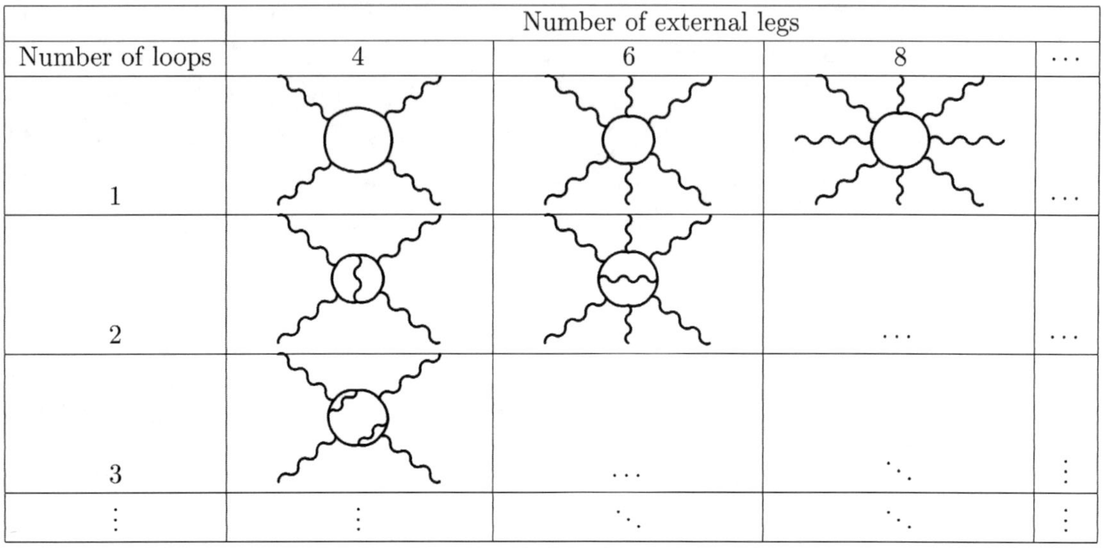

Figure 3. Feynman diagrams contributing to the AAM formula.

Moreover, the mass in (3.1) is the physical mass [13], so that fig. 3 effectively includes also an infinite set of mass renormalization counterdiagrams.

In 1984, S. L. Lebedev and V. I. Ritus [20, 21] obtained the same exponentiation by an analysis of the corrections due to photon exchanges in Sauter's tunnelling picture, for both scalar and spinor QED. They also confirmed this $\mathrm{e}^{\alpha\pi}$ factor to linear order in α by a direct calculation of $\mathcal{L}^{(2)}_{\mathrm{spin}}(E)$.

Apart from the Schwinger pair creation effect, the AAM formula (3.1) and its spinor QED equivalent hold also interesting information on the multiloop QED photon amplitudes, in the limit of low photon energy and large photon number. This information can be extracted using Borel dispersion relations [22, 23], which allows one to show, for example, that the large N behaviour of these amplitudes for $l \geq 2$ is dominated by the one-fermion loop contributions, and qualitatively different for physical and generic mass renormalization. More interestingly, it suggests that the QED perturbation series may converge for the one-fermion loop contributions to the N – photon amplitudes [23, 24, 25, 26]. If correct, this would extend an old conjecture by P. Cvitanovic [27] for the $g - 2$ factor.

4. One-loop open string theory

At the one-loop level, pair creation by a constant homogeneous electric field has also been studied for string theory, both for the open [28, 29] and the heterotic string [29]. For the open string case, C. Bachas and M. Porrati [29] obtained the following formula for the imaginary part of the effective action:

$$\mathrm{Im}\mathcal{L}^{(1)}_{\text{string}}(E) = \frac{1}{4(2\pi)^{D-1}} \sum_{\text{states S}} \frac{\beta_1+\beta_2}{\pi\epsilon} \sum_{n=1}^{\infty}(-)^{n+1}\Bigl(\frac{|\epsilon|}{n}\Bigr)^{D/2} \exp\Bigl(-\frac{\pi n}{|\epsilon|}(M_S^2+\epsilon^2)\Bigr) \tag{4.1}$$

Here the first sum is over the physical states of the bosonic string, with M_S the mass of the state. The second sum is a Schwinger-type sum. $D = 26$ is the critical dimension for the open string. We have defined

$$\beta_{1,2} = \pi q_{1,2} E \tag{4.2}$$

where $q_{1,2}$ are the $U(1)$ charges at the string endpoints, and

$$\epsilon = \frac{1}{\pi}\Bigl(\mathrm{arctanh}\beta_1 + \mathrm{arctanh}\beta_2\Bigr) \tag{4.3}$$

In the weak field limit, (4.1) reproduces the scalar Schwinger formula (1.2), as well as its generalization to arbitrary integer spin J. For stronger fields the right hand grows much faster than for the particle case, and even diverges at a critical field strength

$$E_{\text{cr}} = \frac{1}{\pi\,|\max q_i|} \tag{4.4}$$

Presumably this means that a field of this strength would break the string apart.

We will now show how to reproduce, in the large mass limit, the Schwinger exponents in the open-string result of (4.1) by a worldsheet instanton which is a natural generalization of the AAM worldline instanton (2.7) [30].

To write down a path integral for the open string at one-loop in an external Maxwell field, one has to replace the particle loop by a string annulus and insert the interaction term S_{ext} of (1.5) on both boundaries of the loop, with the charge e replaced by $q_{1,2}$ (we assume $q_1 \neq q_2$ to exclude the Möbius strip contribution to the amplitude). The effective action becomes, in conformal gauge,

$$\Gamma^{(1)}[F] = \frac{1}{2}\int_0^\infty \frac{dT}{T}(4\pi^2 T)^{-\frac{D}{2}} Z(T) \int \mathcal{D}x\, \mathrm{e}^{-S_E[x,F]} \tag{4.5}$$

where the path integral is over the embeddings of the annulus at fixed T into D - dimensional flat space. The worldsheet action is

$$S_E = \frac{1}{4\pi\alpha'}\int d\sigma d\tau\, \partial_a x^\mu \partial^a x_\mu - i\frac{q_1}{2}\int d\tau\, x^\mu \partial_\tau x^\nu F_{\mu\nu}\Big|_{\sigma=0} - i\frac{q_2}{2}\int d\tau\, x^\mu \partial_\tau x^\nu F_{\mu\nu}\Big|_{\sigma=\frac{1}{2}} \tag{4.6}$$

Here α' is the Regge slope, which will be set equal to $\frac{1}{2}$ in the following. The worldsheet is parametrized as a rectangle $\sigma \in [0, \frac{1}{2}]$ and $\tau \in [0, T]$ where $\tau = T$ is identified with $\tau = 0$. We use euclidean conventions where $\sigma^0 = -i\sigma^2 = -i\tau$, $x^0 = -ix^D$, and $A_D = -iA_0$. $Z(T)$ is the partition function,

$$Z(T) = \sum_{\text{oriented states}} \mathrm{e}^{-\pi T M_S^2} \tag{4.7}$$

Let us now consider the constant electric field case, $F_{D,D-1} = -F_{D-1,D} = iE$. We rescale $\tau = Tu$ and do the T - integral by the method of steepest descent. The stationary point is

$$T_0 = \sqrt{\frac{I_u}{I_\sigma + 2\pi^2 M_S^2}} \tag{4.8}$$

where

$$\begin{aligned} I_\sigma &:= \int_0^1 du \int_0^{\frac{1}{2}} d\sigma\, \partial_\sigma x^\mu \partial_\sigma x_\mu \\ I_u &:= \int_0^1 du \int_0^{\frac{1}{2}} d\sigma\, \partial_u x^\mu \partial_u x_\mu \end{aligned} \tag{4.9}$$

The new worldsheet action is

$$S_{\text{eff}} = \frac{1}{\pi}\sqrt{I_u}\sqrt{I_\sigma + 2\pi^2 M_S^2} - i\frac{q_1}{2}\int d\tau\, x^\mu \partial_\tau x^\nu F_{\mu\nu}\Big|_{\sigma=0} - i\frac{q_2}{2}\int d\tau\, x^\mu \partial_\tau x^\nu F_{\mu\nu}\Big|_{\sigma=\frac{1}{2}} \tag{4.10}$$

It leads to the equations of motion

$$\begin{aligned} \left[I_u \partial_\sigma^2 + (I_\sigma + 2\pi^2 M_S^2)\partial_u^2\right] x^\mu &= 0 \\ T_0 \partial_\sigma x^\mu &= i\pi q_2 F_{\mu\nu} \partial_u x^\nu \qquad (\sigma = \frac{1}{2}) \\ T_0 \partial_\sigma x^\mu &= -i\pi q_1 F_{\mu\nu} \partial_u x^\nu \qquad (\sigma = 0) \end{aligned} \tag{4.11}$$

The nth worldsheet instanton solving these equations is found by the ansatz

$$\begin{aligned} x_n^{D-1} &= \frac{2\pi M_S}{|a|} \cos(2\pi n u) \cosh(b - a\sigma) \\ x_n^D &= \frac{2\pi M_S}{|a|} \sin(2\pi n u) \cosh(b - a\sigma) \end{aligned} \tag{4.12}$$

(with the remaining coordinates constants). We take equal signs for n and a. Plugging this ansatz into (4.11) one finds that all the parameters can be determined, namely

Journal of Physics: Conference Series **287** (2011) 012031 doi:10.1088/1742-6596/287/1/012031

$$\begin{aligned} T_0 &= \frac{2\pi n}{a} \\ b &= \mathrm{arctanh}\beta_1 \\ a &= 2\Bigl(\mathrm{arctanh}\beta_1 + \mathrm{arctanh}\beta_2\Bigr) \\ N &= \frac{2\pi M_S}{|a|} \end{aligned} \tag{4.13}$$

This yields the worldsheet action

$$S[x^\mu] = 2\pi^2 M_S^2 \frac{n}{a} \tag{4.14}$$

Noting that $a = 2\pi\epsilon$ this correctly reproduces the exponents of the Bachas-Porrati formula (4.1) in the large M_S limit. The method of calculation is, however, substantially simpler than the ones applied in [28] and [29]. Thus we hope that this worldsheet instanton technique will make it feasible to obtain string pair creation rates also for certain classes of nonconstant electric fields. An interesting aspect of the constant field instanton (4.12) is, that its restriction to each boundary coincides with the worldline instanton (2.7). It is not clear whether this property may extend to more general fields, but if so it would greatly facilitate finding the corresponding worldsheet instantons.

References

[1] Sauter F 1931 *Zeit. f. Phys.* **69** 742
[2] Schwinger J 1951 *Phys. Rev.* **82** 664
[3] Ultraphotonik - Polaris, http://www.physik.uni-jena.de/inst/polaris
[4] The Extreme Light Infrastructure (ELI) project, http://www.extreme-light-infrastructure.eu/eli-home.php
[5] The European X-Ray Laser Project XFEL, http://xfel.desy.de
[6] Ruf M, Mocken G R, Müller C, Hatsagortsyan K Z and Keitel H C 2009 *Phys. Rev. Lett.* **102** 080402
[7] Dunne G V, Gies H and Schützhold R 2009 *Phys. Rev.* D **80** 111301
[8] Keldysh L V 1965 *Sov. Phys. JETP* **20** 1307
[9] Brezin E and Itzykson C 1970 *Phys. Rev.* D **2** 1191
[10] Narozhnyi N B and Nikishov A I 1970 *Sov. J. Nucl. Phys.* **11** 596
[11] Popov V S 1972 *Sov. Phys. JETP* **34** 709; Popov V S 1972 *Sov. Phys. JETP* **35** 659
[12] Popov V S and Marinov M S 1973 *Sov. J. Nucl. Phys.* **16** 449
[13] Affleck I K, Alvarez O and Manton N S 1982 *Nucl. Phys.* B **197** 509
[14] Feynman R P 1950 *Phys. Rev.* **80** 440
[15] Dunne G V and Schubert C 2005 *Phys. Rev.* D **72** 105004
[16] Dunne G V, Gies H, Schubert C and Wang Q 2006 *Phys. Rev.* D **73** 065028
[17] Nikishov A I 1970 *Nucl. Phys.* B **21** 346
[18] Gies H and Klingmüller K 2005 *Phys. Rev.* D **72** 065001
[19] Dunne G V and Wang Q 2006 *Phys. Rev.* D **74** 065015
[20] Lebedev S L and Ritus V I 1984 *JETP* **59** 237
[21] Ritus V I 1987 *Issues in Intense-Field Quantum Electrodynamics* ed V L Ginzburg (Commack: Nova Science)
[22] Dunne G V and Schubert C 2000 *Nucl. Phys.* B **564** 591
[23] Dunne G V and Schubert C 2002 *JHEP* **0206** 042
[24] Dunne G V and Schubert C 2006 *J. Phys.: Conf. Ser.* **37** 59
[25] Martin L C, Schubert C and Villanueva Sandoval V M 2003 *Nucl. Phys.* B **668** 335
[26] Huet I, McKeon D G C and Schubert C 2010 *JHEP* **1012** 036
[27] Cvitanovic P 1977 *Nucl. Phys.* B **127** 176
[28] Burgess C P 1987 *Nucl. Phys.* B **294** 427
[29] Bachas C and Porrati M 1992 *Phys. Lett.* B **296** 77
[30] Torrielli A and Schubert C 2010 *J. Phys.* A **43** 402003

XIV Mexican School on Particles and Fields IOP Publishing
Journal of Physics: Conference Series **287** (2011) 012032 doi:10.1088/1742-6596/287/1/012032

Jets structure in Pb–Pb collisions at LHC energies

E. Cuautle, I. Domínguez and G. Paić

Instituto de Ciencias Nucleares, Universidad Nacional Autónoma de México, Apartado Postal 70-543, México Distrito Federal 04510, México

E-mail: isabel@nucleares.unam.mx

Abstract.
Modifications of the structure of jets produced in ion-ion collisions represent a very sensitive tool to study the interactions of partons with the medium. Partonic energy loss manifests itself as a decrease of the number of particles carrying a high fraction of the jet energy and, the additional radiated energy, increases the number of low-energy particles. We present and analysis of jets produced in ion-ion collisions at LHC energies. Predictions with two different Monte Carlo simulation models (PYQUEN and Q-PYTHIA) for the parton energy loss, are analyzed together with modification of the jet structure.

1. Introduction

Jets are produced in hard scattering of partons from nucleons within the incoming particles or ions. The outgoing partons transverse the medium formed in the collision[1] interacting with it and losing energy before fragmenting into jets of hadrons. Evidence for jet energy loss has been reported by *RHIC* experiment. Two evidences to support this are the disappearance of away-side correlations in central $Au + Au$ compare to $p + p$ collisions and the high-p_T hadrons suppression in the nuclear modification factor R_{AB}.

In the disappearance of away-side correlations in central $Au+Au$ compare to $p+p$ collisions (left panel of figure 1), a hadron pair drawn from a single jet will generate an enhanced correlation at $\Delta\phi \approx 0$, as observed for $p+p$, $d+Au$ and $Au+Au$. A hadron pair drawn from back-to-back dijets will generate an enhanced correlation at $\Delta\phi \approx \pi$, as observed for $p+p$ and for $d+Au$ with somewhat broader width than the near-side correlation peak. However, the back-to-back dihadron correlation is absent in central $Au+Au$ collisions. If the correlation is indeed the result of jet fragmentation, the suppression is due to the final state interaction of hard-scattered partons or their fragmentation products in the dense medium generated in $Au+Au$ collisions.

The nuclear modification factor R_{AB} (right panel of figure 1) is defined as the ration of the inclusive momentum spectra of charged hadrons from $Au+Au$ events to $p+p$ reference spectrum, scaled to account for nuclear geometry. It is expected that the number of hard processes in an event scales $< N_{bin} >$, in absence of nuclear effect $R_{\mathrm{AB}} = 1$. Conventional nuclear effects, such as nuclear shadowing of the parton distribution functions and initial state multiple scattering,

[1] QCD predicts that a high temperature or density ($\epsilon_c \sim 1\mathrm{GeV}/fm^3$), matter should be close to ideal gas called quark-gluon-plasma (QGP)

cannot account for the suppression. Furthermore, the suppression is not seen in $d + Au$ but is unique to $Au + Au$ collisions, proving experimentally that it results not from nuclear effects in the initial state, but rather from the final state interaction of hard scattered partons or their fragmentation products in the dense medium generated in $Au + Au$ collisions.

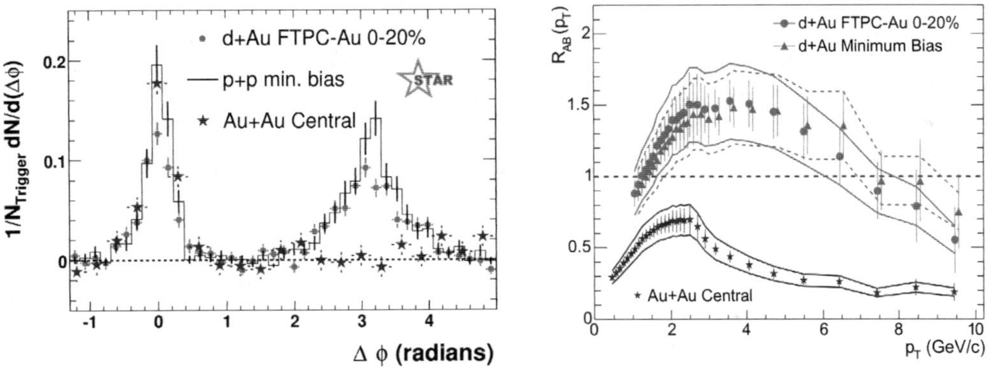

Figure 1. Dihadron azimuthal correlations at high p_T for $p + p$, central $d + Au$ and central $Au + Au$ collisions (background subtracted) from $STAR$ [1] (left). Binary-scaled ratio $R_{\rm AB}$ of charged hadron and π_0 inclusive yields from $\sqrt{s_{\rm NN}} = 200$ GeV $Au + Au$ and $d + Au$ relative to that from $p + p$ collisions, from $STAR$ [2] (right).

In this work, we studied the effects of parton energy loss in a medium under extreme conditions of temperature and density. To achieve that, we simulate high-p_T events in $p + p$ collision with PYTHIA [3], as a baseline, and events passed through two afterburner model of energy loss PYQUEN [4] and Q-PYTHIA [5]. After events simulation, jets were reconstructed with a cone jet finding algorithm and event shape analysis was done. Finally modifications in jet structure, due to energy loss, were studied.

2. Energy loss Monte-Carlo simulation

PYQUEN (PYthia QUENched) is a Monte Carlo event generator routine implemented to simulate re-scattering, radiative and collisional energy loss of hard partons in expanding quark-gluon plasma created in ultra-relativistic heavy ion-ion collisions. The internal model parameters are the initial conditions for the QGP formation expected for central ion-ion collisions (table 1).

Q-PYTHIA is a Monte Carlo implementation of medium-induced gluon radiation in the final-state branching process. Medium effects are introduced through an additive term in the vacuum splitting functions. A realistic Glauber-like collision geometry is defined, the production points of the hard scatterings are distributed in the nuclear overlapping area according to the probability of central nucleon-nucleon collisions. In this model the only free parameter is the scale of the transport coefficient $k = 6 \cdot 10^6$ fm, which corresponds to an average $\langle \hat{q} \rangle = 14$ GeV^2/fm in a 0-10% central collision.

Once particles production simulation was made, with and without energy loss, we proceed to reconstruct jets using a cone jet algorithm [6] and event shape analysis (ESA) [7].

Journal of Physics: Conference Series **287** (2011) 012032 doi:10.1088/1742-6596/287/1/012032

Table 1. PYQUEN internal model parameters.

Description	Parameter	Value
Medium-induced partonic energy loss	ienglu	0 - radiative and collisional 1 - radiative 2 - collisional
Initial temperature of quark-gluon plasma	T_0	0.2 GeV $< T_0 <$ 2 GeV
Proper time of quark-gluon plasma formation	τ_0	0.01 fm/c $< \tau_0 <$ 10 fm/c
Number of active quark flavors in quark-gluon plasma	n_f	0, 1, 2 or 3
Angular distribution of emitted gluons	ianglu	0 - small - angular 1 - wide - angular 2 - collinear

3. Cone jet algorithm

Cone jet algorithms associated particles whose trajectories lie within a circle of specific radius R in $\eta \times \phi$ space. Starting with a trial geometric center (or axis) for a cone in $\eta \times \phi$ space, the energy-weighted centroid is calculated including contributions from all particles within the cone. This new point in $\eta \times \phi$ is then used as the center for a new trial cone.

As this calculation is iterated the cone center "flows" until a "stable" solution is found, *i.e.*, until the centroid of the energy depositions within the cone is aligned with the geometric axis of the cone. The cone variables are:

$$E_T^J = \Sigma E_T^i \qquad \eta^J = \frac{1}{E_T^J}\Sigma E_T^i \eta^i \qquad \phi^J = \frac{1}{E_T^J}\Sigma E_T^i \phi^i$$

where E_T^i is the transverse energy, η^i is the rapidity and ϕ^i is the azimuthal angle for the particle i and E_T^J is the transverse energy, η^J is the rapidity and ϕ^J is the azimuthal angle for the jet J.

4. Event Shape Analysis (ESA)

The event shape analysis measures the geometrical properties of the energy flow in QCD states, its leads us a method to determine areas where one does encounter special topologies of jets in an event, allow an easy selection of events. In hadron-hadron collisions, three shapes variables are used: thrust (T), recoil (R) and thrust-minor (T_{min}).

The thrust variable (T) is defined using only transverse variables:

$$T \equiv \underbrace{max}_{\vec{n}_T} \frac{\sum_i |\vec{q}_{\perp,i} \cdot \vec{n}_T|}{\sum_i |\vec{q}_{\perp,i}|}$$

$\vec{q}_{\perp,i}$ represent the momentum components transverse to the beam and $\vec{n}_T$ is the transverse vector that maximizes the ratio. The sphericity of the event is defined as $\tau \sim 1 - T$. For events with narrow back-to-back jets, τ is close to 0 and for events with a uniform distribution of momenta τ is close to 0.3

The recoil variable R is defined as the vector sum of the transverse momentum:

$$R \equiv \frac{|\sum_i \vec{q}_{\perp i}|}{\sum_i |\vec{q}_{\perp i}|}$$

This quantity measures the balance of momenta of the event. For example for monojets events, *i.e.,* two jets in a event but only one of theme inside detector acceptance, R tends to 1, because there are no vectorial transverse momentum cancellations between particles. Otherwise, in the case of narrow back-to-back jet inside the acceptance, R tends to 0 because the vectorial transverse momentum sum is 0.

T_{min} is a measure of the momentum out of the plane formed by the transverse momentum and the beam. Given the thrust axis ($\vec{n}_t$), we can define it as:

$$T_{min} \equiv \frac{\sum_i |\vec{p}_{t,i} \; x \; \vec{n}_t|}{\sum_i |\vec{p}_{t,i}|}$$

5. Cone jet algorithm results

In table 2 are shown the parameters used in this analysis, both for particle production and for the jets reconstruction using a cone jet algorithm.

Table 2. Parameters used.

Particle production	**Cone jet algorithm**
100 000 events	Charged particles with
PYTHIA,	p_T> 0.3 GeV/c
PYQUEN and Q-PYTHIA	R = 0.7
$\sqrt{s_{\rm NN}} = 5.5$ TeV	${E_T}^J > 5$ GeV
$p_T^{hard} = 50$ GeV/c	$\lvert\eta^J\rvert < 0.5$

Taking the ration (R_{AA}) between the particles produced with PYTHIA and the both models of energy loss as a function of p_T and the transverse momentum $j_T = p_T \sin\Delta\phi$, changes in the structure of jets can be observed due to the parton energy loss.

Partonic energy loss manifests itself as a decrease of the number of particles carrying a high fraction of the jet energy and additional radiated energy increases the number of low-energy particles (left panel of figure 2). In addition, broadening of the distribution of jet-particle momenta perpendicular to the jet axis, j_T distribution was observed and clear differences between the two Monte Carlo models in the pattern of increase of low-momentum particles and depletion of high-momentum particles can be observed (right panel of figure 2).

5.1. Acoplanarity between dijets

Instead of being back-to-back in the transverse plane the partons are affected by additional transverse momentum contributions from the intrinsic k_T of the scattering partons and from NLO processes. In nucleus-nucleus collisions the interaction with the medium further increases

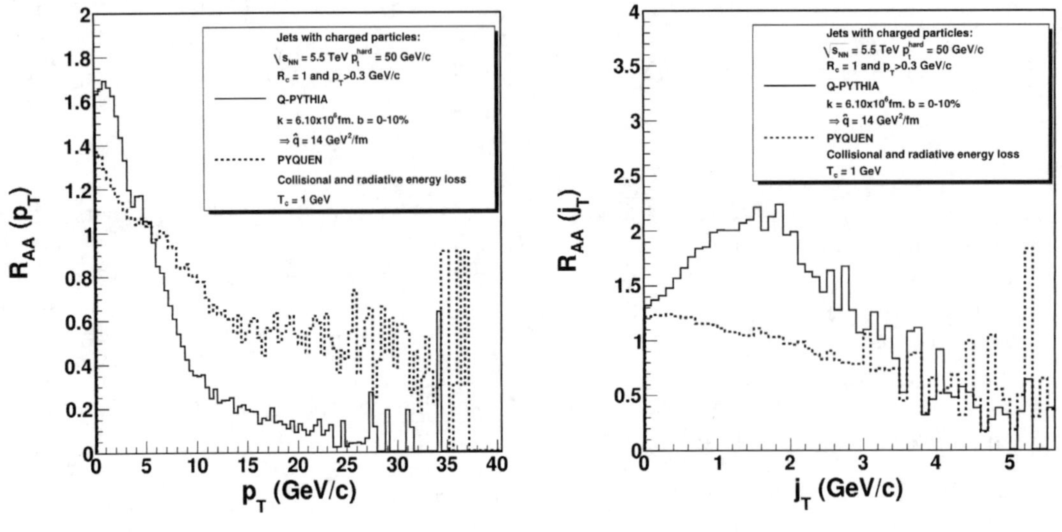

Figure 2. Ratio of particles yield (R_{AA}) between Q-PYTHIA and PYTHIA (solid line) and PYQUEN (with $T_0 = 1$ GeV) and PYTHIA (dashed line) as function of p_T (left) and j_T (right).

the so called acoplanarity of the parton pair.

Acoplanarity is defined as the average transverse momentum component of the away-side jet. $\vec{p}_{T,a}$ perpendicular to the plane formed by the beam and the near side jet with transverse momentum $\vec{p}_{T,t}$.

$$p_{\rm out} = |p_{T{\rm a}}| sin\Delta\phi = \sqrt{2}k_{T{\rm y}}$$

and the transverse momentum imbalance is the difference between the trigger jet ($\vec{p}_{T,t}$) and the away-side jet ($\vec{p}_{T,a}$) transverse momentum in the trigger jet-beam plane.

$$p_{\rm T,diff} = p_{max} - p_{proj} = |p_{T{\rm t}}| - |p_{T{\rm a}}|Proj(p_{T{\rm t}})| = \sqrt{2}k_{T{\rm y}}$$

Figure 3 show both distribution, $p_{\rm out}$ and $p_{\rm T,diff}$ comparing PYTHIA with Q-PYTHIA and PYQUEN. The increase in the acoplanarity due to the parton energy loss was calculated as:

$$\sqrt{< \mathbf{p^2_{out_{col}}} >} = \sqrt{< \mathbf{p^2_{out_{med}}} > - < \mathbf{p^2_{out_{pp}}} >}$$

and is observed for both Monte Carlo. For Q-PYTHIA we obtained $\sqrt{(8.44)^2 - (7.80)^2} =$ 3.2 GeV/c and for PYQUEN $\sqrt{(9.24)^2 - (7.80)^2} = 4.9$ GeV/c

6. Event Shape Analysis results

The event shape analysis starts by plotting a two dimensional distribution ("thrust map"), with τ $(1-T)$ in the horizontal axis and R in the vertical axis. This plot allows to identify different classes of events according with their location in the thrust map (table 3).

First of all, was observed that at low p_T^{cut}>0.3 GeV/c (left panel of figure 4), isotropics events are presents, even hard processes are simulated. Increasing p_T^{cut}>4 GeV/c (right panel of figure 4)

Journal of Physics: Conference Series **287** (2011) 012032 doi:10.1088/1742-6596/287/1/012032

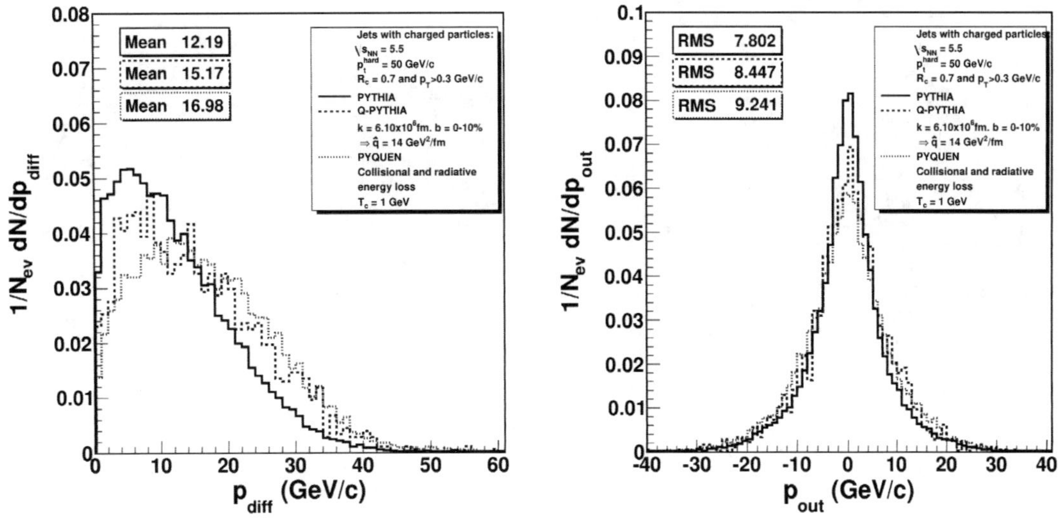

Figure 3. $p_{T,\text{diff}}$ (left) and p_{out} (right) for dijets produced with PYTHIA (solid line), Q-PYTHIA (dashed line) and PYQUEN with $T_0 = 1$ GeV (dotted line).

events with monojets and dijets are selected, decreasing background. This behavior is more notorious in the azimuthal correlations between particles (figure 5 for low (left) and high (right) p_T).

Table 3. Parameters used and cuts for events selection

Particle production	**ESA**	**Class of event**
100 000 events PYTHIA, PYQUEN and Q-PYTHIA $\sqrt{s_{\text{NN}}} = 5.5$ TeV $p_T^{hard} = 50$ GeV/c	Charged particles with $p_T^{leading}$>10 GeV/c p_T^{assoc}>0.3 GeV/c and p_T^{assoc}> 4 GeV/c $\|\eta\| < 0.9$	Dijets (A): $1 - T$ <0.05, $R \leq$0.35 Mono jet (B): $1 - T \leq$0.05, $R \geq$0.9 Isotropic (C): $1 - T \geq$0.3, $R \leq$0.4

Now, taking the ration (R_{AA}) between the particles produced with PYTHIA and the both models of energy loss as a function of event shape analysis variables, $1 - T$ and T_{min}, which provide a measure of the radiation perpendicular to that plane, shows modifications because of the parton energy loss, being higher at low p_T.

7. Summary

Two Monte Carlo models for jet quenching in $Pb + Pb$ collisions at LHC energy were studied. We observed:

- An increase of low-momentum particles and depletion of high-momentum particles.
- In Pb+Pb collisions the acoplanarity between dijets is increased due to partonic interactions with the dense medium can be observed

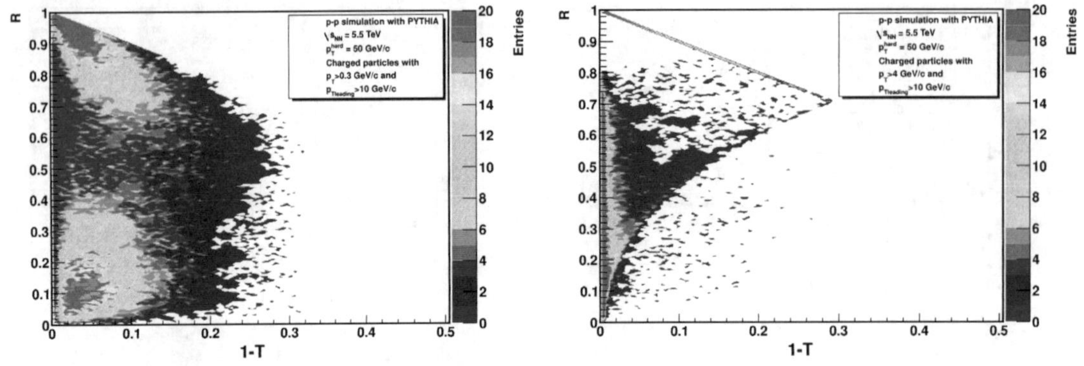

Figure 4. $1-T$ vs R for particles produced with PYTHIA. The cuts used in ESA are: $|\eta| < 0.9$, $p_{T\text{leading}}$>10 GeV/c , p_T^{cut}>0.3 GeV/c (left) and p_T^{cut}>4 GeV/c (right).

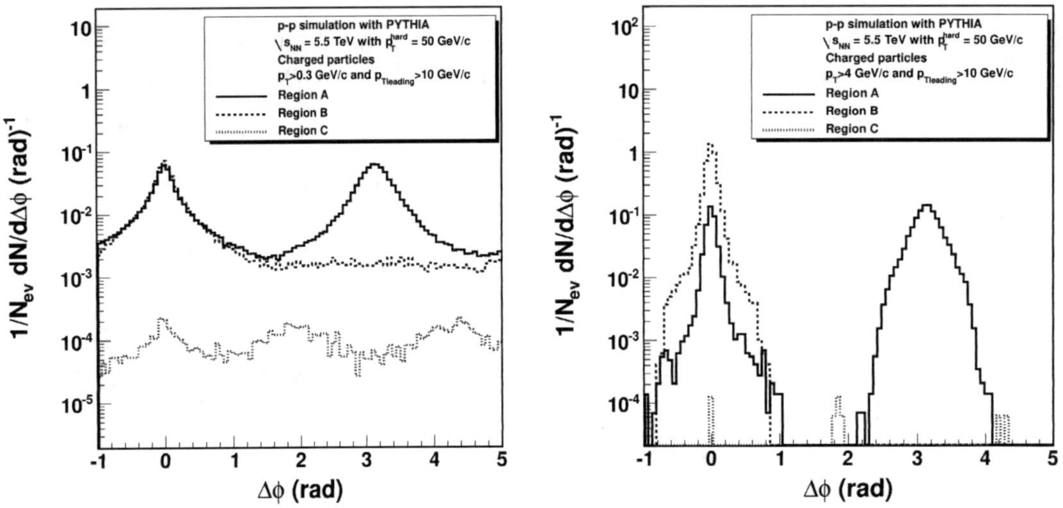

Figure 5. Azimuthal correlations produced with PYTHIA. The cuts used in ESA are: $\eta| < 0.9$, p_T^{cut}>0.3 GeV/c (left), p_T^{cut}>4 GeV/c (right) and $p_{T\text{leading}}$>10 GeV/c . Events in the different regions of the plot R vs $1-T$ are shown: events in region (A) (solid line), events in region (B) (dashed line), events of region (C) (dotted line), according to the table 3.

Event Shape Analysis (ESA) is available to describe the topology of events in $Pb + Pb$ collisions.

- Using the variables thrust and recoil and constructing the thrust is possible to study different configurations, like dijets.
- Both, thrust minor and azimuthal correlations can be used to study the acoplanarity of a pair of jets and modifications in their structure.

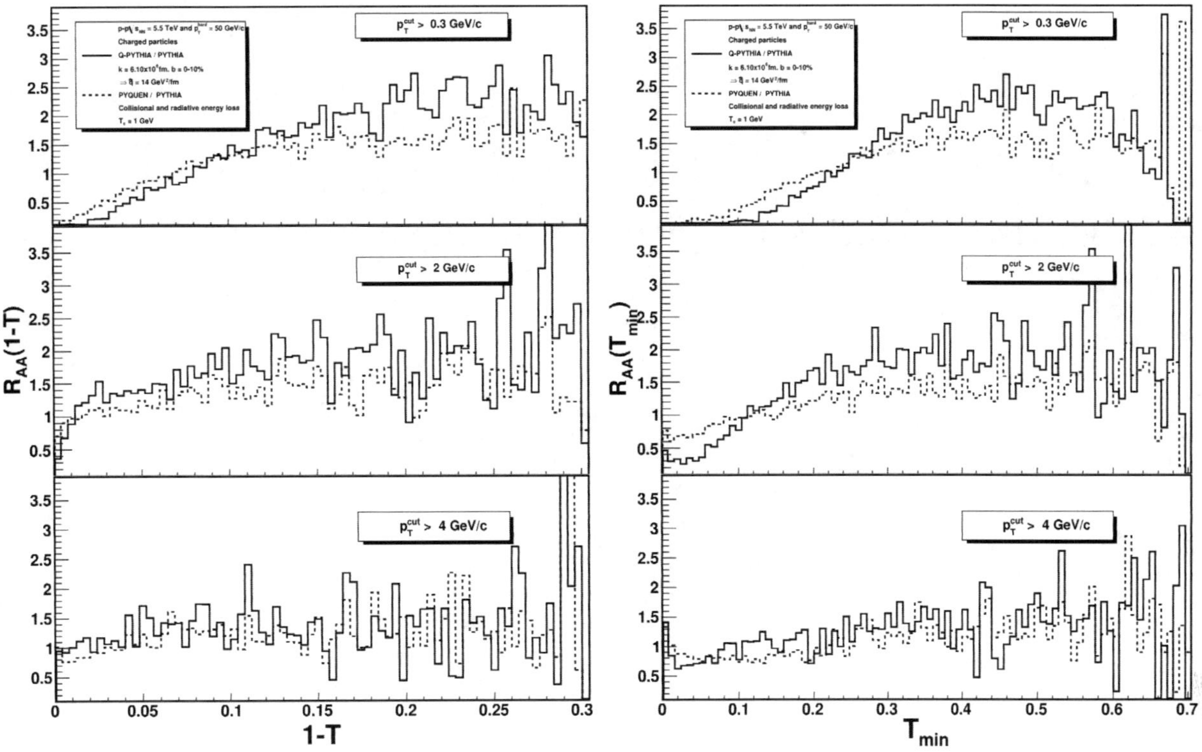

Figure 6. Ratio of particles yield (R_{AA}) between PYQUEN and PYTHIA (solid line) and Q-PYTHIA and PYTHIA (dashed line) as a function of $1-T$ (left) and R (right) with $p_{T\mathrm{l}} = 10$ GeV/c and $p_{T\mathrm{a}} = 0.3$ GeV/c (top), $p_{T\mathrm{a}} = 2$ GeV/c (middle) and $p_{T\mathrm{a}} = 4$ GeV/c (bottom).

Acknowledgments

The authors express their gratitude to Dr. Andreas Morsch, Dr. Leticia Cunqueiro and Antonio Ortiz for their valuable contributions in the present work. Also, we wishing to acknowledge to *REDFAE* and PAPIIT-IN116508 for the financial support granted for the attendance.

References

[1] J. Adams *et al.* [STAR Collaboration], Phys. Rev. Lett. **93**, 252301 (2004)
[2] J. Adams *et al.* [STAR Collaboration], Phys. Rev. Lett. **91**, 072304 (2003)
[3] T. Sjostrand, S. Mrenna and P. Skands, JHEP **0605** (2006) 026
[4] I. P. Lokhtin and A. M. Snigirev, Eur. Phys. J. C **45**, 211 (2006)
[5] N. Armesto, L. Cunqueiro and C. A. Salgado, Eur. Phys. J. C **63**, 679 (2009)
[6] S. D. Ellis and D. E. Soper, Phys. Rev. D **48**, 3160 (1993)
[7] A. O. Velasquez and G. Paic, ALICE-INT-2009-015

Journal of Physics: Conference Series **287** (2011) 012033 doi:10.1088/1742-6596/287/1/012033

Towards seeking θ_{13}: Status and initial data from T2K

Laura Lee Kormos for the T2K Collaboration

Department of Physics, Lancaster University, Lancaster LA1 1TL, UK

E-mail: l.kormos@lancaster.ac.uk

Abstract. T2K is a long-baseline neutrino oscillation experiment with the primary aim of measuring the smallest and most elusive neutrino mixing angle, θ_{13}. 2010 saw the beginning of the T2K data-taking phase. This paper gives an overview of the T2K experiment, its goals and current status, and will discuss some of the initial data. All parts of the experiment: beamline, near detectors and far detector, are functioning well. In the far detector, 23 events have been selected from the first T2K dataset. Particle identification in the far detector also is discussed in this paper. Detailed analysis has not yet been completed.

1. Introduction

Neutrino flavour-changing is well-described by the MNSP mixing matrix [1]. There are at present two parameters in this matrix which are unmeasured, θ_{13} and the CP-violating phase δ, although other parameters may exist in addition if neutrinos are Majorana particles. The measurement of these two parameters will provide values that are key to understanding not only the behaviour of neutrinos, but potentially will allow us to understand the dynamics of the early universe through which matter gained dominance over antimatter.

The T2K (Tokai to Kamioka) experiment aims to provide the first measurement of the smallest and hence most elusive of the MNSP mixing angles, θ_{13}, by searching for the appearance of ν_e in a ν_μ beam, as well as to provide a more precise measurement of θ_{23} by measuring ν_μ disappearance. In the first case, after 5 x 0.75 MW x 10^7 seconds, corresponding to 8.3 x 10^{21} protons on target (POT), T2K can measure $\sin^2 2\theta_{13}$ values as small as 0.008 at the 90% confidence limit in the region of $0.001 < \Delta\ m^2_{23} < 0.004$ assuming systematic uncertainties of 10% or less. This is more than an order of magnitude smaller than the existing best limits which come from the CHOOZ experiment [2]. In the second case, T2K aims to measure $\sin^2 2\theta_{23}$ with a precision of $\delta(\sin^2 2\theta_{23}) \sim 0.01$ and $\delta(\Delta\ m^2_{23}) < 1 \times 10^{-4}$ eV^2. This may help us to determine whether or not θ_{23} is maximal, which would have theoretical implications.

T2K consists of three main components: the ν_μ beamline; a suite of near detectors; and a far detector. The T2K collaboration includes approximately 500 members from 61 institutions in 12 countries. In order to accomplish the two main aims stated above, a ν_μ beam is produced at the Japan Proton Accelerator Research Complex (JPARC) on the east coast of Japan near Tokai and directed towards the far detector, Super Kamiokande (SK), 295 km away in the Kamioka mines. A near detector suite is positioned 280 m from the neutrino production target and consists of an on-axis detector, INGRID, and the off-axis ND280 detector. T2K uses the unique off-axis

Journal of Physics: Conference Series **287** (2011) 012033 doi:10.1088/1742-6596/287/1/012033

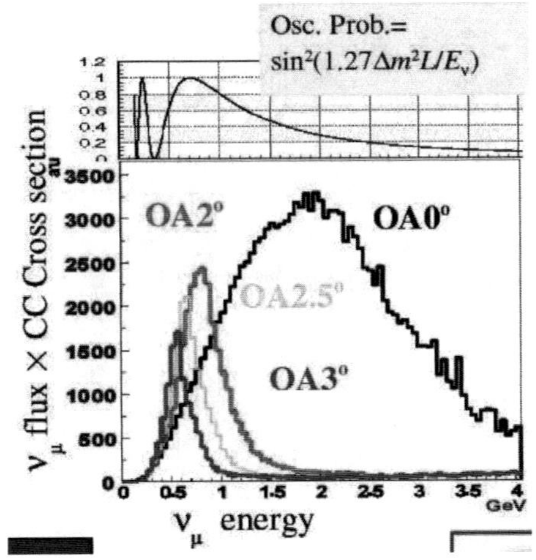

Figure 1. The normalised ν_μ flux as a function of ν_μ energy at different angles off the main axis of the beam.

beam configuration, which is discussed further in Section 2, in order to maximise the $\nu_\mu \rightarrow \nu_x$ transition probability.

2009 saw the commissioning of the T2K beam and detectors. Data for analysis was obtained during the first half of 2010. Sections 2, 3.1, 3.2 and 4 discuss the commissioning and monitoring data which has been obtained from the beamline, from each of the subdetectors within the near detector suite, and from SK. Future plans are covered in Section 5.

2. The T2K beamline

The T2K ν_μ beam is produced by bringing an intense proton beam into collision with a cylindrical, graphite target, thereby producing hadrons consisting of mainly pions with some kaons. The charged pions are focused into a Decay Volume, where they decay almost exclusively via the process $\pi \rightarrow \mu\nu_\mu$. This provides the source of ν_μ for the experiment. The beam then proceeds through a beam dump which removes any remaining hadron component, and then through approximately 150 m of rock which removes the muon component, leaving a ν_μ beam with little ν_e contamination. The beam is monitored at each stage via proton, muon and neutrino monitors. Monitoring the neutrino beam is part of the function of the near detector suite and is discussed in Sections 3.1 and 3.2. The proton and muon monitoring data is discussed below.

2.1. Off-axis beam

Contrary to intuition, the optimal neutrino beam for T2K is obtained by aiming the neutrinos 2.5° away from SK. The reason for this is elucidated in Figure 1 which shows the ν_μ flux versus ν_μ energy at different off-axis angles, where OA0° corresponds to off-axis 0°, i.e. on-axis. The further off-axis one samples the beam, the lower the neutrino energy and the narrower the energy spectrum. This allows for the beam to be tuned to the energy that corresponds to a maximum oscillation probability, and greatly reduces the number of high-energy neutrinos in the beam which are more likely to result in the production of pions which are difficult to reconstruct in the detectors, particularly in the case of neutral pions in SK, as discussed in Section 4. In addition, the off-axis beam has less ν_e contamination simply due to kinematic configurations, and actually has more ν_μ at the energy corresponding to the oscillation maximum.

The first T2K run took place from January through June 2010, during which a total of 3.23×10^{19} POT at 30 GeV were delivered for T2K analysis. The summer shutdown included installing new extraction kickers and making further improvements to the beam system,

Journal of Physics: Conference Series **287** (2011) 012033 doi:10.1088/1742-6596/287/1/012033

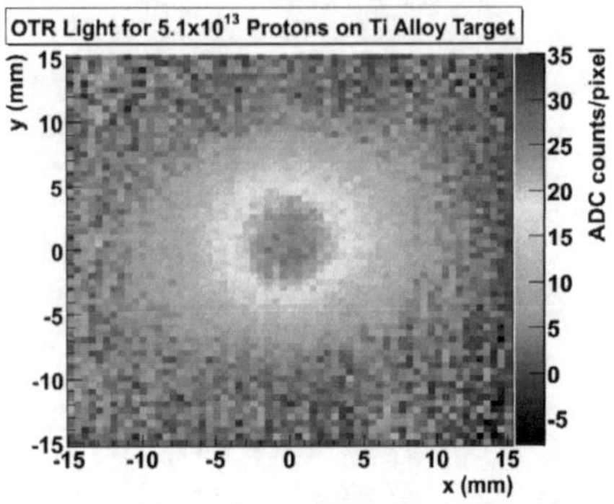

Figure 2. The transverse position of the proton beam immediately before it hits the target as measured by the OTR monitor.

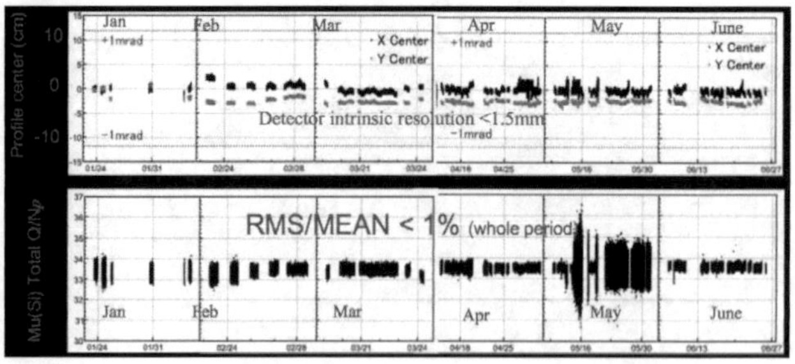

Figure 3. The position of the centre of the muon beam (top) and the beam intensity (bottom) as measured by the muon monitors immediately after the hadron beam dump.

including a change from 6 to 8 bunches per spill, a shorter spill-repetition cycle, and more protons per bunch.

2.2. Beamline monitor data

The price to pay for all of the advantages of using an off-axis beam is that the oscillation probability becomes sensitive to the off-axis angle; hence, it is important to monitor the beam position and profile well. At T2K, this is done via a series of various types of detectors placed along the beam path. The Optical Transition Radiation (OTR) Detector monitors the transverse (normal to the beam axis) position and intensity of the proton beam immediately before it impacts the target as shown in Figure 2. The beam profile is approximately Gaussian with a width of approximately 4.2 mm. The radius of the cylindrical target is 13 mm.

The position and intensity of the beam after the hadron beam dump is measured by a set of muon monitors. Figure 3 shows that the centre of the beam (top) for the entire run period was well within the target position, and the intensity of the beam (bottom) remained relatively stable except for a period of beam-testing in May. These examples are representative of the data obtained from many other monitors.

3. Near detector suite

The near detectors are situated 280 m downstream of the neutrino production target and serve the purpose of characterising the unoscillated neutrino beam. They consist of an on-axis detector, INGRID, which monitors the neutrino beam flux and profile, and an off-axis detector, ND280, which measures the beam flux, composition, neutrino interaction cross-sections and background processes at the same off-axis angle as SK. This latter detector allows for a partial cancellation of systematic uncertainties when calculating the near/far ratio, i.e. the ratio of the predicted neutrino flux between the near and far detectors, thereby reducing the uncertainties in the ν_μ normalisation used by SK in determining the oscillation parameters. The relative positions of INGRID and the off-axis ND280 are shown in Figure 4 which shows the cross-shaped INGRID below and to the left of the ND280 detector in the ND280 pit.

For all of the T2K detectors, the golden channel for identifying and measuring neutrino interactions and determining the neutrino energy is the Charged-Current Quasi-Elastic (CCQE) interaction, given by $\nu_l + \mathrm{N} \rightarrow l + \mathrm{N}'$ where $l = e$ or μ and N, N′ are nucleons within a target nucleus. The lepton in the final state is used to reconstruct the momentum of the incoming neutrino.

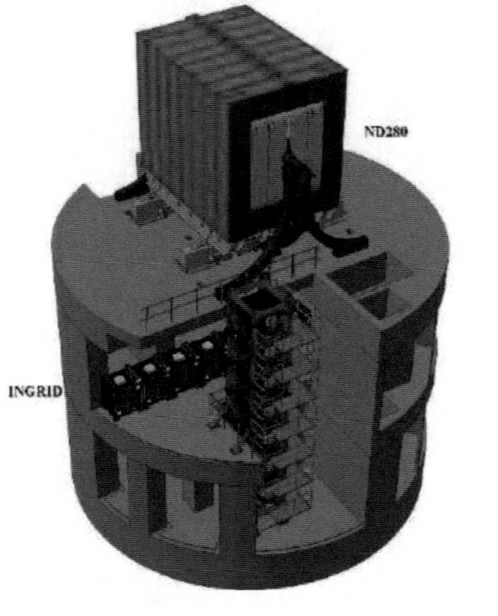

Figure 4. The relative positioning of the near detectors in the ND280 pit. The off-axis ND280 sits above and to the right of the on-axis INGRID.

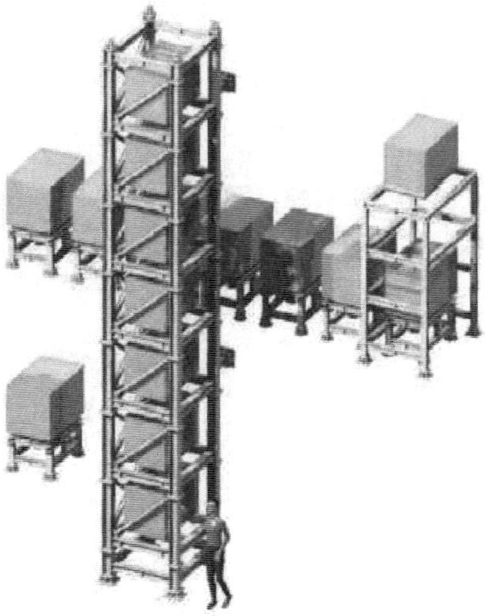

Figure 5. The INGRID detector, consisting of 16 modules.

3.1. INGRID

INGRID is composed of 16 modules, 14 forming a cross-shape and two sitting at diagonal positions, each composed of layers of plastic scintillator bars interspersed with layers of iron, as shown in Figure 5. The beam goes through the centre of the cross. The long arms of INGRID are ideal for obtaining a good measurement of the beam profile. Each scintillator bar has a central hole through which wavelength-shifting fibre is inserted to carry light from the centre to the ends of the bar, where the fibre is coupled to solid-state photosensors. For the data set presented here, INGRID included the 14 modules in the cross shape. The diagonal modules were added during the summer 2010 shutdown.

Figure 6 shows a candidate neutrino event in INGRID as seen from the side (left) and top (right) of a module. The red dots correspond to the reconstructed positions of energy depositions or "hits" within the detector, as evidenced by light collected by the photosensors at the ends of the scintillator bars, and indicate the path of the charged lepton produced by the CCQE interaction. The brown shaded layers represent the lead which functions as the interaction target for the neutrinos. The on-axis neutrino beam profile as measured by INGRID is shown in Figure 7. This horizontal profile normal to the beam axis shows the beam well-centred with the beam shape as expected. Similar results have been obtained for the vertical profile.

3.2. ND280

The off-axis ND280 detector consists of a dedicated π^0 detector (P0D), three Time Projection Chambers (TPCs) interleaved with two Fine Grain Detectors (FGDs), backed by the downstream Electromagnetic Calorimeter (ECal), all contained within a metal framed "basket" as shown in Figure 8. Surrounding the subdetectors inside the basket are the rest of the ECal modules. All of this is enclosed within the recycled UA1 magnet which produces a 0.2 T magnetic field. Although the magnet can be opened, similar to what is shown in the figure, for detector installation and maintenance, during data-taking the two halves of the magnet are brought together. The magnet yoke is instrumented for the purpose of acting as a veto for externally-produced particles, cosmic

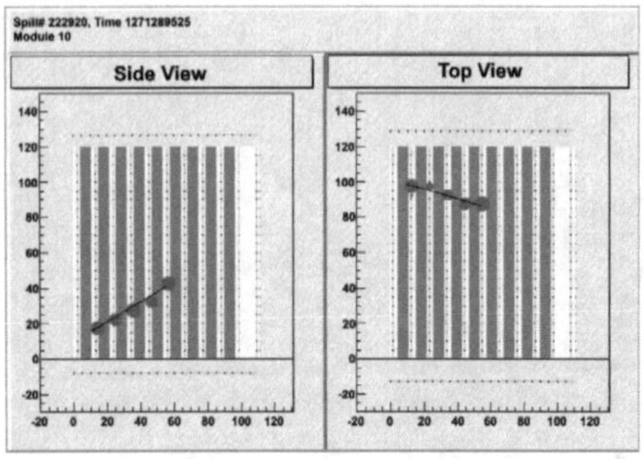

Figure 6. A candidate neutrino event detected by INGRID. The red dots represent the positions of hits in the scintillator bars.

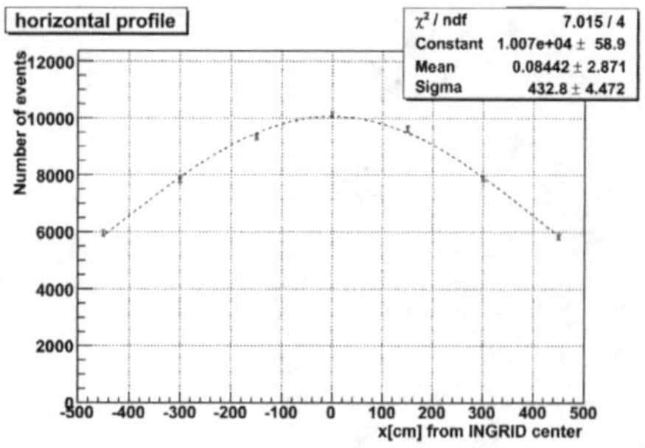

Figure 7. The horizontal profile of the ν_μ beam as measured by INGRID.

ray triggering, and muon range measurement, and is called the Side Muon Range Detector (SMRD). The TPCs provide the main momentum measurement for charged particles. The FGDs provide the main target mass for the central region of the ND280 and are optimised for detecting the proton recoil in CCQE interactions. The P0D is optimised for measuring π^0 production cross-sections. The ECal modules provide particle identification (PID) and energy measurements for neutral particles. Except for the TPCs, all of the ND280 subdetectors are composed of layers of plastic scintillator bars interspersed with layers of a target material, similar to INGRID. In the case of the P0D and the second FGD, some of the target layers are water. By comparing the interaction cross-sections measured including the water layers, with those measured excluding the water layers, the cross-sections on water will be determined, allowing for a better prediction of the events in SK. For the data set presented here, the ND280 contained all of the subdetectors except for the ECals which surround the basket, which were installed during the summer 2010 shutdown.

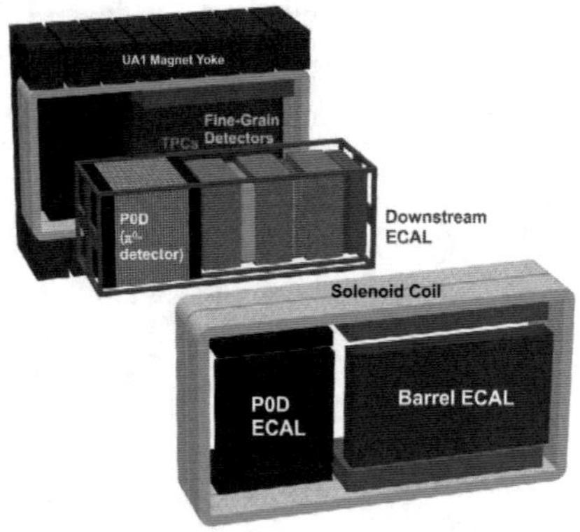

Figure 8. The ND280 off-axis detector. The ν_μ beam enters from the left of the figure. Shown are the subdetectors and the open magnet.

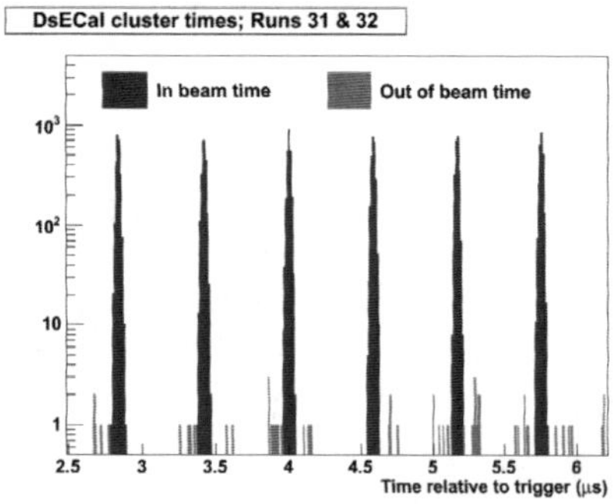

Figure 9. The timing structure of the ν_μ beam as measured by the downstream Ecal.

Each of the ND280 subdetectors has accumulated a wealth of commissioning and analysis data. A sample is presented here. Figure 9 shows the 6-bunch structure of the neutrino beam

Journal of Physics: Conference Series **287** (2011) 012033 doi:10.1088/1742-6596/287/1/012033

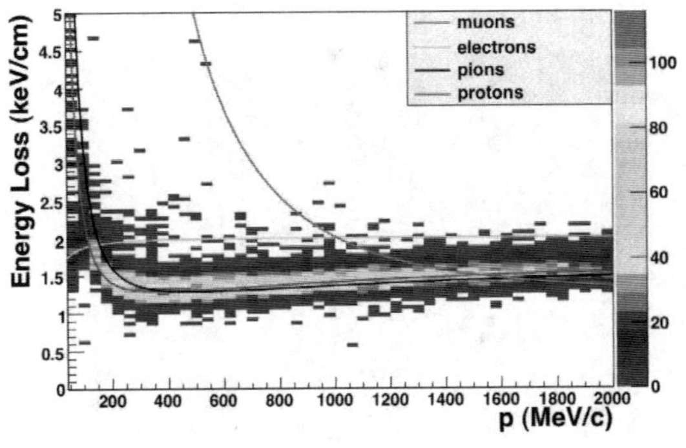

Figure 10. The rate of energy loss, dE/dx, for negative particles, as measured by the TPCs.

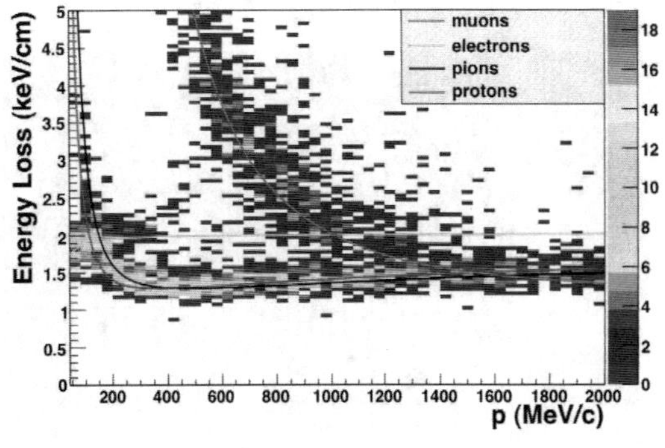

Figure 11. The rate of energy loss, dE/dx, for positive particles, as measured by the TPCs.

as measured by the identification of charge-clusters within the downstream ECal versus time. The blue peaks correspond to clusters identified during the expected bunch-crossing. The small numbers of events between the bunch-crossings are mainly due to misidentified clusters. Figures 10 and 11 show the rate of loss of energy due to ionisation for charged particles traversing the TPCs, for negative (left) and positive (right) particles. The theoretical curves for various particles overlay the data points. As expected for T2K, a large number of events for negative particles are consistent with the muon curve.

Figure 12 shows a typical selection of candidate neutrino events in the ND280 off-axis detector. In each case, the ν_μ beam is entering from the left, although due to the disperse nature of the beam the interaction vertex can occur anywhere within the ND280. The figure shows two inelastic scattering events with the vertex in FGD1 (top left and bottom right), a CCQE candidate (top right), and a candidate CC event which includes a pion (bottom left). The neutrino event in the top left plot also includes an unrelated muon (probably) traversing the detector.

4. Super Kamiokande

SK has been taking solar, atmospheric, and beam neutrino data for over 10 years and is well-described in the literature (see, for example, [3] and references therein). Neutrinos in SK are identified by the Cherenkov rings that are produced by relativistic, charged particles traversing the 50 kT ultra-pure water of the detector. As in the T2K near detectors, the CCQE interaction provides the bulk of the useful events.

In SK, the following criteria are used to select leptons produced via neutrino interactions. The timing of the event must coincide with the beam spill timing. The energy from the event must be fully-contained within the inner fiducial volume of the detector, and the reconstructed event vertex must lie within the fiducial volume. Such events are selected when they have only one reconstructed Cherenkov ring, and if they pass a threshold value for visible energy. PID is achieved by looking at specific properties of the rings. Muons leave a sharp, well-defined ring due to limited scattering within the detector, while electrons produce a more fuzzy, less well-defined ring due to scattering.

The ν_e appearance measurement will be made by counting the number of e-like events in SK and comparing it with the number expected including background ν_e inherent in the beam, other sources of electrons, and accounting for misidentified rings. The ν_μ disappearance measurement will be made by counting the number of μ-like events and comparing it with the number expected

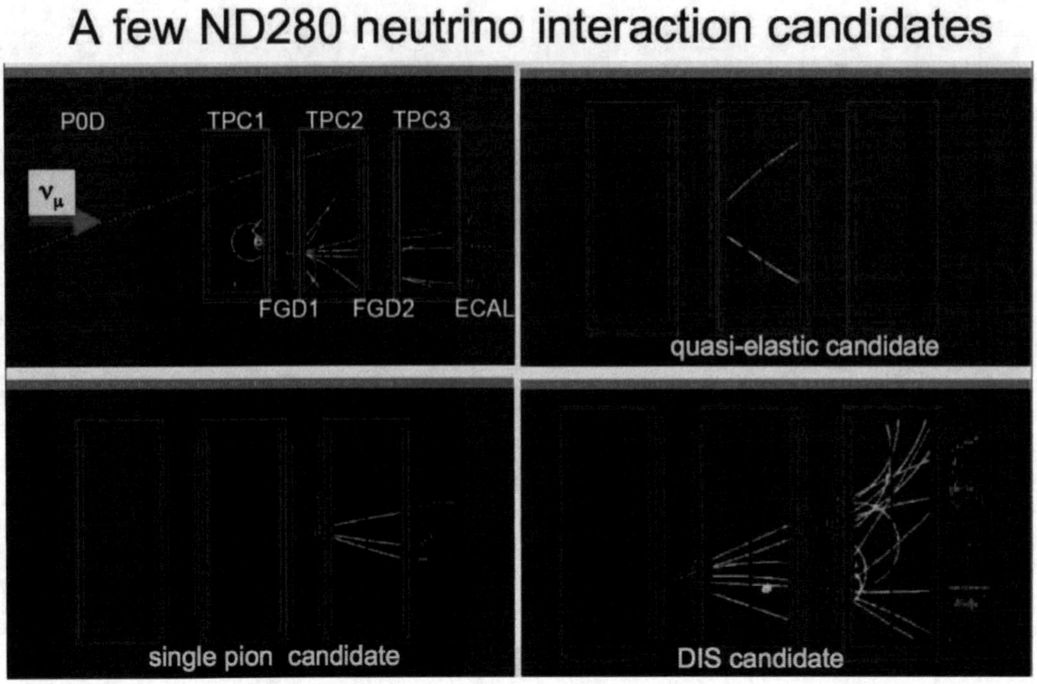

Figure 12. Typical event displays for neutrino interactions in the off-axis ND280. In all cases, the beam enters from the left of the event display.

including non-beam backgrounds. Figure 13 shows some typical muon rings in the T2K data. In these figures, the cylindrical part of the detector is mapped flat, while the top and bottom of the cylinder are shown as circles.

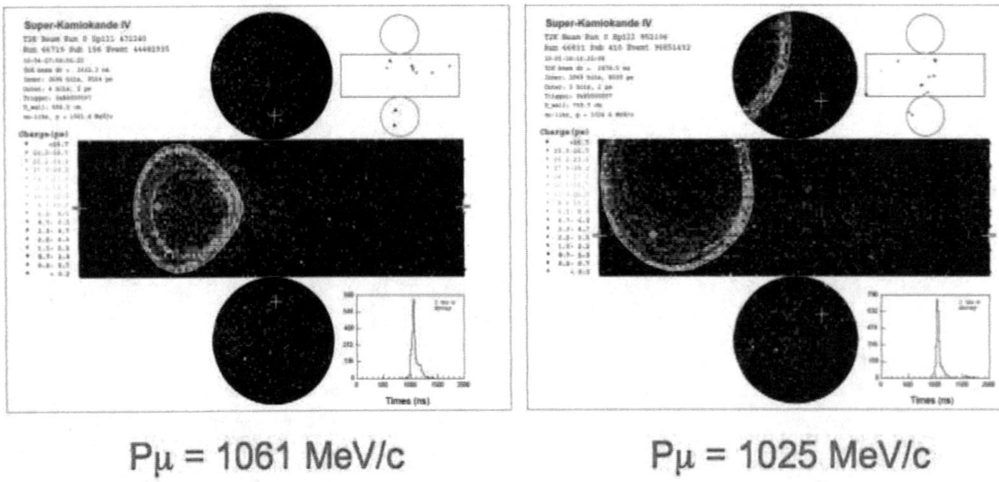

Figure 13. Typical event displays for ν_μ interactions in SK. The cylindrical detector is mapped flat, with top and bottom of the cylinder represented as circles.

One of the main backgrounds in the appearance measurement is π^0 production. π^0s decay to two photons, which produce rings very similar to those produced by electrons. These two rings may be reconstructed as one e-like ring in cases where the two photons were produced with asymmetric energy leaving only one visible photon ring, or where one ring overlays the other. Figure 14 shows the first T2K event recorded in SK. Interestingly, this event includes a π^0; however, in this case it is easy to see that there are two e-like rings rather than one.

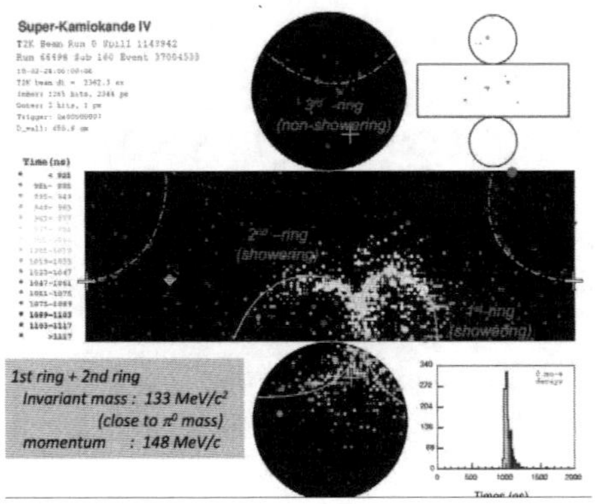

Figure 14. The first T2K event recorded in SK, an event involving a π^0. The two fuzzy rings from the decay photons are visible. The cylindrical detector is mapped flat, with the top and bottom of the cylinder represented as circles.

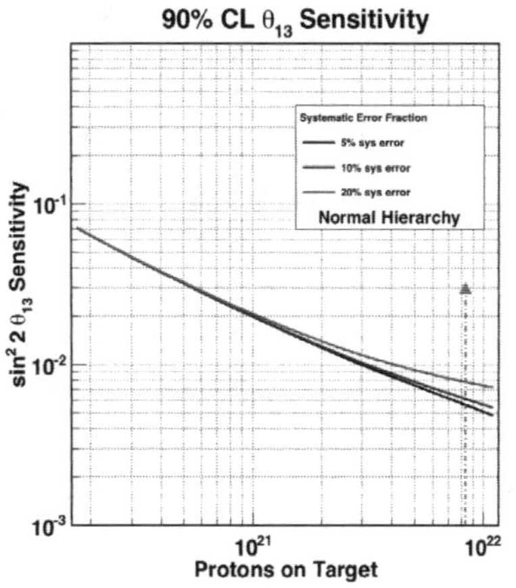

Figure 15. T2K sensitivity to $\sin^2 2\theta_{13}$ as a function of POT. The vertical line corresponds to 8.3×10^{21} POT.

SK, like the near detectors, has also observed the expected 6-bunch neutrino beam structure in their fully-contained events. In total for the January through June 2010 runs, 33 fully-contained events have been observed. Of these, 23 events meet the visible-energy criterion. Detailed analysis of these events has not yet been completed.

5. Conclusions

All modules of the ND280 detector, both on- and off-axis, now are installed and working well. The next T2K data-taking period runs from November 2010 through June 2011, during which the aim is to accumulate 150 kW x 10^7 seconds ($\sim 3 \times 10^{20}$) POT. Figure 15 shows the expected sensitivity to $\sin^2 2\theta_{13}$ at T2K as a function of number of POT. The dashed vertical line corresponds to 8.3×10^{21} POT, which is expected at the end of the first phase of T2K. The different curves correspond to different levels of systematic uncertainty. It can be seen that until 10^{21} POT have been delivered, the sensitivity is independent of the differences between systematic uncertainties that are at the level of 5, 10 or 20%. By July 2011, T2K will have increased its dataset by an order of magnitude and will be optimising the analyses.

References

[1] K. Nakamura *et al* (Particle Data Group) 2010 *J. Phys* G **37** 075021
[2] M. Apollonio *et al* (CHOOZ Collaboration) 1999 *Phys Lett B* **466** 415
[3] Y.Fukuda *et al* (Super Kamiokande Collaboration) 1998 *Phys Lett B* **433** 9; J.P.Cravens *et al* (Super Kamiokande Collaboration) 2008 *Phys Rev D* **78** 032002

Study of solar activity by measuring cosmic rays with a water Cherenkov detector

Angélica Bahena Blas[1] **and Luis Villaseñor**[2]

[1]Facultad de ciencias Físico-Matemáticas, Universidad Michoacana de San Nicolás de Hidalgo, Avenida Francisco J. Mujica S/N Ciudad Universitaria C.P. 58030 Morelia, Michoacán México
[2]Instituto de Física y Matemáticas, Universidad Michoacana de San Nicolás de Hidalgo, Avenida Francisco J. Mujica S/N Ciudad Universitaria C.P. 58030 Morelia, Michoacán México

E-mail: anbahena@ifm.umich.mx[1], villasen@ifm.umich.mx[2]

Abstract. We report on an indirect study of solar activity by using the Forbush effect which consists on the anti-correlation between the intensity of solar activity and the intensity of secondary cosmic radiation detected at ground level at the Earth. We have used a cylindrical water Cherenkov detector to measure the rate of arrival of secondary cosmic rays in Morelia Mich., Mexico, at 1950 m.a.s.l. We describe the analysis required to unfold the effect of atmospheric pressure and the search for Forbush decreases in our data, the latter correspond to more than one year of continuous data collection.

1. Introduction

Many aspects of the emergence of sunspots and coronal mass ejections on the surface of the sun are still under research as well as the details of the interaction of the solar wind with the magnetic field of the Earth. The appearance of sunspots, recorded systematically since 1700, shows an 11-year cyclic variation. The physicist Scott Ellsworth Forbush studied the secondary cosmic radiation at ground level and the rate of occurrence of sunspots in the 1930s; he discovered that there exist transient effects showing a clear of anti-correlation between the intensity of cosmic rays and solar activity, these effects are known today as Forbush decreases [1]. This phenomenon can be explained by the fact that the arrival of electrically charged cosmic rays at ground level on the Earth is directly influenced by the magnetic field carried by the solar wind as it propagates within the Solar System. In turn, the solar wind fed by coronal mass ejections is correlated with solar activity in general, and, in particular, with the rate of occurrence of sunspots.

Forbush decreases of secondary cosmic rays are detected by worldwide networks of monitors of neutron fluxes and, more recently, detectors of fluxes of secondary muons. The Forbush decrease is usually observable at the Earth within a few days after the observation of coronal mass ejection by the Sun; the decrease in the flux of secondary cosmic rays takes place over the course of a few hours.

Neutron monitors are the preferred instruments to monitor Forbush decreases because they respond to lower energy cosmic rays that suffer more heliospheric modulation compared to muon detectors. However, the advantage of muon monitors is that they are cheaper and easier to operate. Since they respond to the same phenomenon, i.e., the detection of secondary cosmic rays on the ground, the data from muon counters can be readily compared with the data from

Journal of Physics: Conference Series **287** (2011) 012034 doi:10.1088/1742-6596/287/1/012034

neutron monitors available in real time on the World Wide Web [2]. In this paper we study the sensitivity that can be achieved with a single water Cherenkov detector of small dimensions to monitor Forbush decreases by measuring the flux of secondary muons.

2. Data acquisition system

The propagation of charged particles through a medium with a higher speed than the speed of light in that medium produces the Cherenkov effect, this effect allows us to detect secondary cosmic radiation. The so called Cherenkov light can be detected by a photomultiplier located inside a water Cherenkov detector. The water Cherenkov detector used for this study, see Fig. 1, is a tank of polyethylene 1.3 m of height and 1.15 m of diameter, filled with water up to a height of 1.1 m. The water was contained in a bag of a highly reflective material called tyvek, which has a diffuse reflectivity of about 90 percent in the ultraviolet region of the EM spectrum. A single photomultiplier, EMI 9353KB, of 8 inches was located along the axis of the cylinder looking downwards from the top of the water surface. The photomultiplier signal was transmitted via RG58 coaxial cable to the electronics used to measured the rate of arrival of the secondary cosmic rays, our electronics is based on a a custom-made ADC card working at 100 MS/s controlled by a Field Programmable Gate Array (FPGA) programmed in VHDL, see Fig. 2. Our data acquisition system also has a GPS receiver to set a precise time tag on our data, an atmospheric pressure sensor and a temperature sensor. The atmospheric pressure and the temperature sensors allow us to correct our data to unfold variations in the measured rate of arrival of secondary cosmic rays due to variations in atmospheric pressure and temperature. We measured the rate of arrival of the secondary cosmic rays with four different thresholds on the PMT signals. All these dates are transmitted via a RS232 port to a computer for storage; the data are saved in binary format in one file each day; we use off-line Linux tools to converted the data into decimal format and to analyze and plot the data. It is important to note that our discrimination threshold is anchored to the baseline of the signal from the PMT; this baseline is measured every second. In summary, the tasks that our data acquisition system performs are the following:

- Get the pulse-per-second (PPS) signal from the Oncore UT+ GPS receiver.
- Measure the rates in one-second intervals from the ADC board at a threshold of 22 mV which roughly corresponds to a muon deposition energy of 22 MeV or 0.11 VEM (Vertical Equivalent Muon)

Figure 1. The water Cherenkov used in this study, in the left side, located in UMSNH.

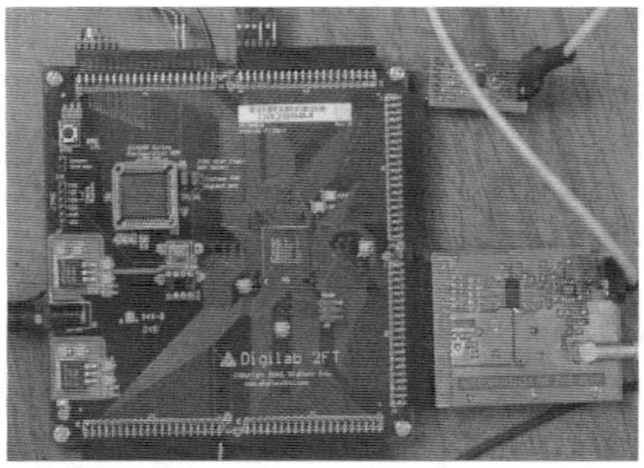

Figure 2. FPGA board and the ADC card.

Journal of Physics: Conference Series **287** (2011) 012034 doi:10.1088/1742-6596/287/1/012034

- Get the GPS time from the Oncore UT+ GPS receiver
- Read the atmospheric pressure from the HP03D sensor in synchrony with the PPS signal
- Send all the above data in binary format to the PC in synchrony with the PPS signal by using the RS232 port.

3. Analysis and results

Figure 3 shows the rate measured every second for one day (August 1st, 2010) for the second threshold; this was a typical day during a period of low solar activity. Figure 4 shows the correlation between the rates for two of the different thresholds. We have checked that the distribution of these rates is given by a Gaussian distribution with a standard deviation equal to the square root of the mean value; this means that the muon detection is a random process with a constant probability per unit time, i.e., a Poisson process. In turn this constitutes an important test of the stability of the detector.

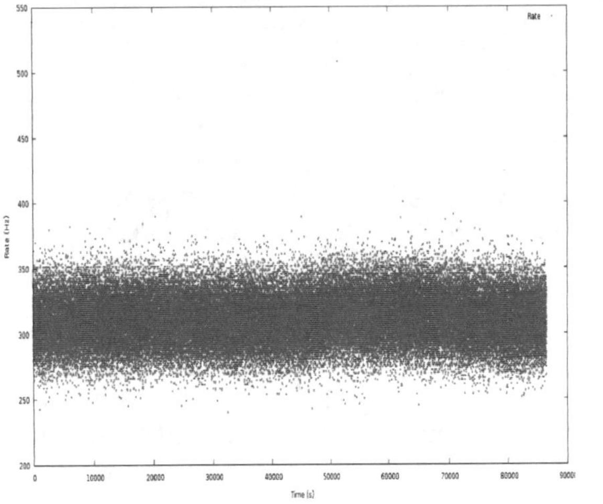

Figure 3. Rate measured per second for one day on August 1, 2010, by using the second threshold.

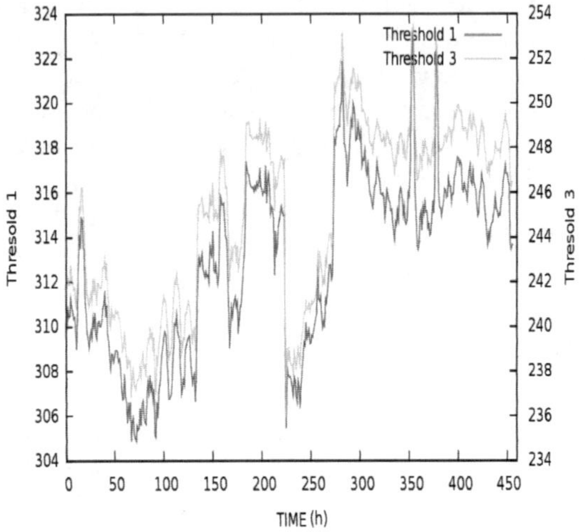

Figure 4. Hourly average rate for two different thresholds, in the period of August 1-19, 2010.

In Figure 5 we see the anti-corelacion between the average of the muon rates over periods of one hour and the atmospheric pressure. The statistical error on the average is equal to the standard deviation of the rate measured every second, roughly equal to the square root of the mean value measured every second, i.e., about 19 Hz, divided by the square root of 3600, i.e., the number of seconds in one hour. Figure 6 shows the scatter plot of the hourly average of the rate (vertical axis) versus the hourly average of the atmospheric pressure (horizontal axis). The anti-correlation between theses averages is clearly evident. The solid line is a least-squares fit, with a slope of 0.79657. This anti-correlations is the well-known barometric effect on the flux of secondary cosmic rays; it is due to the larger or smaller mass of air in the path of the secondary cosmic rays when the atmospheric pressure is higher or lower, respectively, giving rise to a higher or smaller attenuation of the secondary muons, i.e., a lower or higher flux at ground level.

Figure 7 shows two lines. In red we show the average hourly rate obtained in the period August 1-19, 2010 corrected, to remove the effect due to changes in atmospheric pressure, and normalized. In blue we show data from the McMurdo neutron monitor [3], also corrected to remove the effect due to changes in atmospheric pressure. Both lines show a Forbush decrease

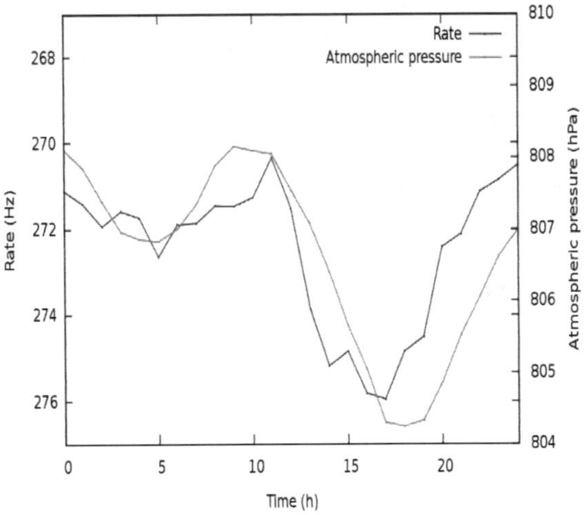

Figure 5. Hourly average of the rate of arrival of secondary muons and of atmospheric pressure, showing a clear anti-correlation between them, note that the Y axis on the left side is inverted.

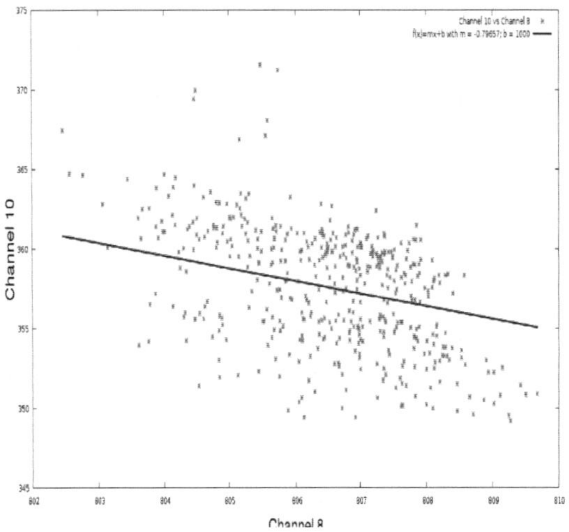

Figure 6. Hourly average of the rate (Y axis) and the hourly average of the atmospheric pressure (X axis). The solid line is the least-squares fit with a slope of -0.79657.

in coincidence during August 3rd., 2010. The second decrease shown in red in Fig. 7, at around 420 h, which does not correspond to the graph of the McMurdo data may be due to a failure in the electronics.

4. Conclusions

We conclude that a single water Cherenkov detector of 1.1 m of diameter filled with purified water up a height of 1.1 m has been shown to possess sufficient sensitivity to observe variations greater than 2% in the muon rate averaged over periods of one hour. Forbush decreases as big as 4% have been observed with muon detectors located at similar magnetic rigidity compared to Morelia. The data obtained from this work are available on line [4].

Acknowledgments

The authors thank CONACyT and the Coordinación de la Investigación Científica from UMSNH for supporting this project.

References

[1] http://astronomy.swin.edu.au/cms/astro/cosmos/F/Forbush+Decrease
[2] See for instance the on-line data from the neutron monitor of UNAM, Mexico, at http://www.geofisica.unam.mx/infra/observ.html
[3] ftp://ftp.bartol.udel.edu/pyle/BRIData/BRI2010B.txt
[4] http://www.fismat.umich.mx/~abahena/tesis/Datos_Completos_2010.

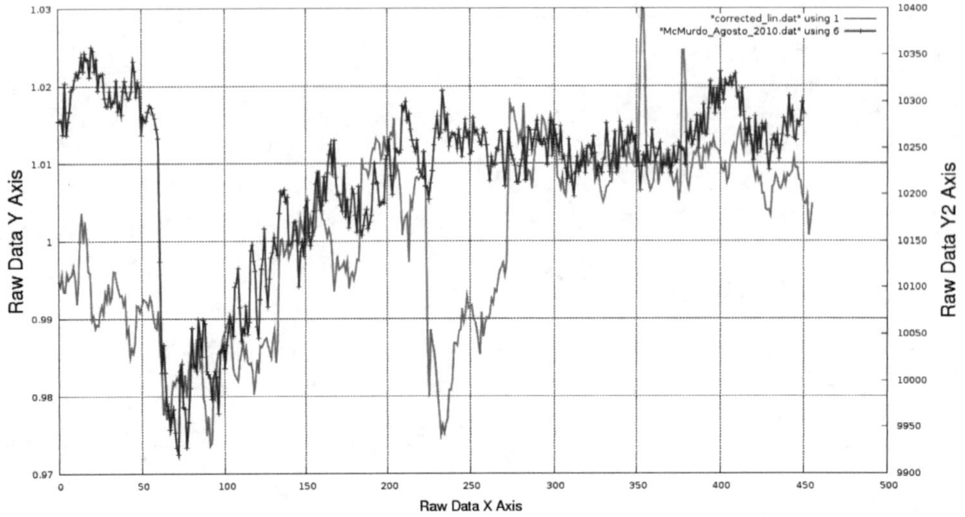

Figure 7. Comparison of our data with data from the McMurdo observatory, in the period August 1-19, 2010.

Journal of Physics: Conference Series **287** (2011) 012035 doi:10.1088/1742-6596/287/1/012035

Triple Higgs Boson Self-Coupling in a $\gamma\gamma$ Collider

A. Gutiérrez-Rodríguez

Facultad de Física, Universidad Autónoma de Zacatecas.
Apartado Postal C-580, 98060, Zacatecas, México.

Javier Peressutti and O. A. Sampayo

Instituto de Física de Mar del Plata (IFIMAR) CONICET, UNMDP. Departamento de Física, Universidad Nacional del Mar del Plata
Funes 3350, (7600) Mar del Plata, Argentina.

E-mail: alexgu@fisica.uaz.edu.mx

Abstract.
We analyzed the double production and the triple self-coupling of the standard model Higgs boson at future $\gamma\gamma$ collider energies, with the reactions $\gamma\gamma \to t\bar{t}HH$. We evaluated the total cross section for $t\bar{t}HH$ and calculated the total number of events considering the complete set of Feynman diagrams at tree-level and for different values of the triple coupling $\kappa\lambda_{HHH}$. The numerical computation was done for the energy which is expected to be available at a possible Future Linear $\gamma\gamma$ Collider with a center-of-mass energy $500-3000$ GeV and luminosities of 1000 and 5000 fb^{-1}. We found that the number of events for the process $\gamma\gamma \to t\bar{t}HH$, taking into account the decay products of both t and H, is enough to obtain relevant information about the triple Higgs boson self-coupling.

1. Introduction

The search for Higgs bosons is one of the principal missions of present and future high-energy colliders. The observation of this particle is of major importance for the present understanding of fundamental interactions. Indeed, in order to accommodate the well established electromagnetic and weak interaction phenomena, the existence of at least one isodoblete scalar field to generate fermion and weak gauge bosons masses is required. Despite repeated success in explaining the present data, the SM cannot be completely tested before this particle has been experimentally observed and its fundamental properties studied.

The triple and quartic Higgs boson couplings [1, 2, 3] λ and $\tilde{\lambda}$ are defined through the potential

$$V(\eta_H) = \frac{1}{2}M_H^2\eta_H^2 + \lambda v\eta_H^3 + \frac{1}{4}\tilde{\lambda}\eta_H^4, \tag{1}$$

where η_H is the physical Higgs field. In the SM, we obtain $M_H = \sqrt{2\lambda}v$ as the simple relationship between the Higgs boson mass M_H and the self-coupling λ where $v = 246$ GeV is the vacuum expectation value of the Higgs boson. The triple vertex of the Higgs field H is given by the coefficient $\lambda_{HHH} = \frac{3M_H^2}{M_Z^2}$. An accurate test of this relationship may reveal the extended nature of the Higgs sector. The measurement of the triple Higgs boson coupling is one of the most important goals of Higgs physics in a future e^+e^- linear collider experiment. This would provide the first direct information on the Higgs potential that is responsible for electroweak symmetry breaking.

The future e^+e^- linear collider can not only be designed to operate in e^+e^- collision mode, but can also be operated as a $\gamma\gamma$ collider. This is achieved by using Compton backscattered photons in the scattering of intense laser photons on the initial e^+e^- beams. The design of the photon linear collider and physics opportunities are described in references [4, 5] and the possibility of measuring the HHH coupling is discussed in Refs. [2, 6, 7].

The triple Higgs boson self-coupling can be measured directly in pair-production of Higgs particles at hadron and high-energy e^+e^- linear colliders. Several mechanisms that are sensitive to λ_{HHH} can be exploited for this task. Higgs pairs can be produced through double Higgs-strahlung of W or Z bosons [3, 8, 9], WW or ZZ fusion [2, 10, 11, 12] as well as through gluon-gluon fusion in pp collisions [13, 14, 15] and high-energy $\gamma\gamma$ fusion [2, 6] at photon colliders. In $\gamma\gamma$ collisions, double Higgs production is possible in several reactions at the tree level, and one process, $\gamma\gamma \to HH$, at the one-loop level:

$$\begin{aligned} WW \text{ double Higgs-fusion} &: \gamma\gamma \to WWHH, \\ \text{double Higgs-fermion} &: \gamma\gamma \to f\bar{f}HH, \\ \text{one-loop level} &: \gamma\gamma \to HH. \end{aligned} \tag{2}$$

As was studied in Ref. [2] the reaction $\gamma\gamma \to WWHH$ is free from any background except incorrect combinatorial of jets; for $M_H < 150\ GeV$, the main branching is four b-jets (with an invariant mass peak at M_H) plus jets from two additional W bosons, up to four, and/or a large missing energy. 8-jet events can be detected in total: For $M_H > 150\ GeV$, the signature includes up to twelve quark jets with the invariant mass peaking at M_W and M_H.

In the case of the process $\gamma\gamma \to f\bar{f}HH$, the important contribution happens when the considered fermion is a quark-top since the Higgs-fermion coupling is proportional to the fermion mass.

The study of the four-body processes with quark top t, $\gamma\gamma \to t\bar{t}HH$, in which the SM Higgs boson is radiated by a $t(\bar{t})$ quark at future $\gamma\gamma$ colliders [4, 5] with a c.m. energy in the range of 500 to 3000 GeV, as in the case of the DESY TeV Energy Superconducting Linear Accelerator (TESLA) machine [16], is necessary in order to know its impact on other processes and also to search for new relations that could have a clear signature of the Higgs boson production.

The Higgs coupling with top quarks, the largest coupling in the SM, is directly accessible in the process where the Higgs boson is radiated off top quarks $\gamma\gamma \to t\bar{t}HH$. Consequently, this process can be used to probe the $t-\bar{t}-H$ Yukawa coupling directly. This process depends on the Higgs boson triple self-coupling, which could lead us to obtain the first non-trivial information on the Higgs potential. We are interested in finding regions that could allow the observation of the $t\bar{t}HH$ processes at the next generation of high energy $\gamma\gamma$ linear colliders.

This paper is organized as follows: In Sec. 2, we study the triple Higgs boson self-coupling through the processes $\gamma\gamma \to t\bar{t}HH$ at next generation linear $\gamma\gamma$ colliders. Finally, we summarize our results in Sec. 3.

2. Cross-Section of the Higgs Boson Double Production with Triple Self-Coupling

To illustrate our results we present the Higgs boson number of events in Tables 1-2 (of course we have to multiplicate for the corresponding Branching Ratios to obtain the observable number of events) for several Higgs boson masses, center-of-mass energy and κ values and for a luminosity of $1000 fb^{-1}$ and $5000 fb^{-1}$.

We also include a contours plot for the number of events of the studied processes in the $(\sqrt{s},\kappa)$ plane with $M_H = 100\ GeV$ and $1000 fb^{-1}$, $5000 fb^{-1}$ in Fig. 1. These contours are obtained from Tables 1-2.

Table 1. Total production of Higgs boson pairs in the SM for $\mathcal{L} = 1000\ (5000)\ fb^{-1}$ and $\kappa = 2$.

Higgs Production	$\gamma\gamma \to t\bar{t}HH$, $\kappa = 2$					
$M_H(GeV)$	$\sqrt{s} =$ 800 GeV	$\sqrt{s} =$ 1000 GeV	$\sqrt{s} =$ 1500 GeV	$\sqrt{s} =$ 2000 GeV	$\sqrt{s} =$ 2500 GeV	$\sqrt{s} =$ 3000 GeV
100	- (-)	2 (10)	8 (40)	13 (65)	15 (75)	16 (80)
120	- (-)	- (-)	4 (20)	7 (35)	9 (45)	10 (50)
140	- (-)	- (-)	2 (10)	4 (20)	6 (30)	7 (35)

Table 2. Total production of Higgs boson pairs in the SM for $\mathcal{L} = 1000\ (5000)\ fb^{-1}$ and $\kappa = 1$.

Higgs Production	$\gamma\gamma \to t\bar{t}HH$, $\kappa = 1$					
$M_H(GeV)$	$\sqrt{s} =$ 800 GeV	$\sqrt{s} =$ 1000 GeV	$\sqrt{s} =$ 1500 GeV	$\sqrt{s} =$ 2000 GeV	$\sqrt{s} =$ 2500 GeV	$\sqrt{s} =$ 3000 GeV
100	- (-)	1 (5)	7 (35)	11 (55)	13 (65)	13 (65)
120	- (-)	- (-)	4 (20)	6 (30)	8 (40)	8 (40)
140	- (-)	- (-)	2 (10)	3 (15)	4 (20)	5 (25)

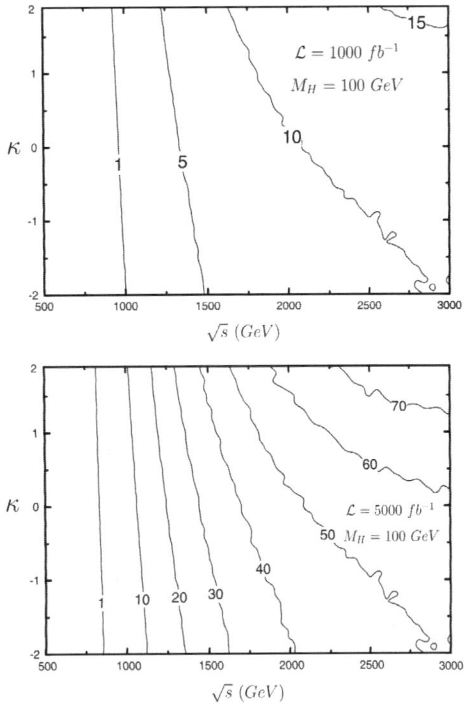

Figure 1. Contour plot for the number of events of the process $e^+e^- \to t\bar{t}HH$ as a function of $\sqrt{s}$ and κ. The variation of the number of events for modified triple couplings $\kappa\lambda_{HHH}$ is indicated for $\mathcal{L} = 1000 fb^{-1}, 5000 fb^{-1}$ and $M_H = 100\ GeV$.

3. Conclusions

As a possible option of the International Linear Collider (ILC), the feasibility of physics opportunities of high energy physics photon-photon interaction has been considered in Ref. [17]. In the high energy photon linear collider, high energy photon beams are generated by inverse Compton scattering between the electron and the laser beams. The $\gamma\gamma$ collider represents a possible opportunity for the triple Higgs boson self-coupling analysis. Therefore, we have analyzed the triple Higgs boson self-coupling at future $\gamma\gamma$ collider energies with the reaction $\gamma\gamma \to t\bar{t}HH$ and considering the complete set of Feynman diagrams at tree-level and in the frame work of the standard model.

We found that for the process $\gamma\gamma \to t\bar{t}HH$, the complete calculation at tree level gives a production cross-section of the order of a fraction of a femtobarn, i.e., 0.015 fb, 0.010 fb and 0.007 fb for $M_H = 100, 120, 140\ GeV$, respectively, and at the end of the examined energy range. These values are, however, larger than the production cross-section for $e^+e^- \to b\bar{b}HH$ and $e^+e^- \to t\bar{t}HH$ [18]. The number of events obtained considering the decay products of both t and H is enough to obtain relevant information about the triple Higgs self-coupling. Moreover, this process can be used to probe the $t - \bar{t} - H$ Yukawa coupling.

Finally, the study of this process is important and could be useful to probe the triple Higgs boson self-couplings λ_{HHH} using a $\gamma\gamma$ collider with very high luminosity and high center-of-mass energy. In addition, these results have never been reported in the literature before and could be of relevance for the scientific community.

Acknowledgments

We acknowledge support of RedFAE and CONACyT (México). O. A. Sampayo would like to thank CONICET (Argentina).

References

[1] F. Boudjema and E. Chopin, *Z. Phys.* **C73**, 85 (1996).
[2] V. A. Ilyin, A. E. Pukhov, Y. Kurihara, Y. Shimizu and T. Kaneko, *Phys. Rev.* **D54**, 6717 (1996).
[3] A. Djouadi, H. E. Haber and P. M. Zerwas, *Phys. Lett.* **B375**, 203 (1996).
[4] James, Brau, *et al.*, ILC Reference Design Report ILC-Report-2007-001 (2007).
[5] B. Badelek, *et.al.*, *Int. Jour. Mod. Phys.* **A19**, 5097 (2004).
[6] G. Jikia, *Nucl. Phys.* **B412**, 57 (1994).
[7] R. Belusevic and G. Jikia, *Phys. Rev.* **D70**, 073017 (2004).
[8] G. Gounaris, D. Schildknecht and F. Renard, *Phys. Lett.* **B83**, 191 (1979).
[9] J. Kamoshita, Y. Okada. M. Tanaka and I. Watanabe, **hep-ph/9602224**.
[10] V. Barger and T. Han, *Mod. Phys. Lett.* **A5**, 667 (1990).
[11] A. Dobrovolskaya and V. Novikov, *Z. Phys.* **C52**, 427 (1991).
[12] D. A. Dicus, K. J. Kallianpur and S. S. D. Willenbrock, *Phys. Lett.* **B200**, 187 (1988).
[13] E. W. N. Glover and J. J. van der Bij, *Nucl. Phys.* **B309**, 282 (1988).
[14] T. Plehn, M. Spira and P. M. Zerwas, *Nucl. Phys.* **B479**, 46 (1996).
[15] S. Dawson, S. Dittmaier and M. Spira, *Phys. Rev.* **D58**, 115012 (1998).
[16] TESLA Technical Desing Report, Part III, DESY-01-011C, J. Aguilar-Saavedra, *et al.*, **hep-ph/0106315**.
[17] E. Boos, *et al.* "*Gold-plated processes at photon colliders*" (2001), **arXiv:hep-ph/0103090v1**.
[18] A. Gutiérrez-Rodríguez, *et al.*, *J. of the Phys. Soc. of Jap.* **77**, 094101 (2008); A. Gutiérrez-Rodríguez, M. A. Hernández-Ruíz and O. A. Sampayo, *Mod. Phys. Lett.* **A20**, 2629 (2005); A. Gutiérrez-Rodríguez, M. A. Hernández-Ruíz and O. A. Sampayo, *Phys. Rev.* **D67**, 074018 (2003).

Bounding the Anomalous $HZ\gamma$ Coupling via the Process $e^+e^- \to \tau^+\tau^-\gamma$

A. Gutiérrez-Rodríguez[1]
Facultad de Física, Universidad Autónoma de Zacatecas.
Apartado Postal C-580, 98060, Zacatecas, México.

J. Montano and M. A. Pérez[2]
Departamento de Física, CINVESTAV.
Apartado Postal 14-740, 07000, México D.F.

E-mail: [1]alexgu@fisica.uaz.edu.mx, [2]maperez@fis.cinvestav.mx

Abstract. We obtain bounds on the anomalous coupling $HZ\gamma$ through data published by the L3 Collaboration on the process $e^+e^- \to \tau^+\tau^-\gamma$. Our analysis leads to bounds on this coupling of order 10^{-2}, for an intermediate mass Higgs boson $115 < M_H < 145$ GeV, two orders of magnitude above the Standard Model prediction.

1. Introduction

The sensitivity to the $HZ\gamma$ vertex has been studied in processes like $e^-\gamma \to e^-H$ and $e^+e^- \to H\gamma$ [1, 2, 3], rare Z and H decays [4, 5, 6], pp collisions via the basic interaction $qq \to qqH$ [6] and the annihilation process $e^+e^- \to HZ$ [3, 7, 8]. It has been found that the latter reaction with polarized beams may lead to the best sensitivity to the $HZ\gamma$ vertex [7] while an anomalous $HZ\gamma$ coupling may enhance partial Higgs decays widths by several orders of magnitude that would lead to measurable effects in Higgs signals at the LHC [6].

The general aim of the present paper is to obtain limits on the $HZ\gamma$ vertex coming from the LEP data on the reaction $e^+e^- \to \tau^+\tau^-\gamma$ [9]. We will find limits of order 10^{-2}, which are better by an order of magnitude than the bounds obtained from the known limits on the partial decay widths of the Z boson [6], but still two orders of magnitude above the SM prediction [1, 10]. The L3 collaboration has obtained also limits on the $HZ\gamma$ vertex using events with photons and a Z vector boson in the final state [11]. In this case they have used an analysis that involves the Higgs boson decay modes $H \to \gamma\gamma$, $Z\gamma$. We have found that our analysis with a tau-lepton pair in the final state induces more stringent limits on the $HZ\gamma$ vertex.

In Fig. 1, we show the Feynman diagrams which give rise to the process $e^+e^- \to \tau^+\tau^-\gamma$ in the SM at tree level and with the anomalous $HZ\gamma$ vertex when the Z vector boson is produced on mass-shell. We do not include the contribution coming from initial photon bremmstrahlung because the LEP data considered the appropriated energy cuts to eliminate this contribution.

The paper is organized as follows. In Section 2 we present the calculation of the respective cross section and in Section 3 we presented our results and conclusions.

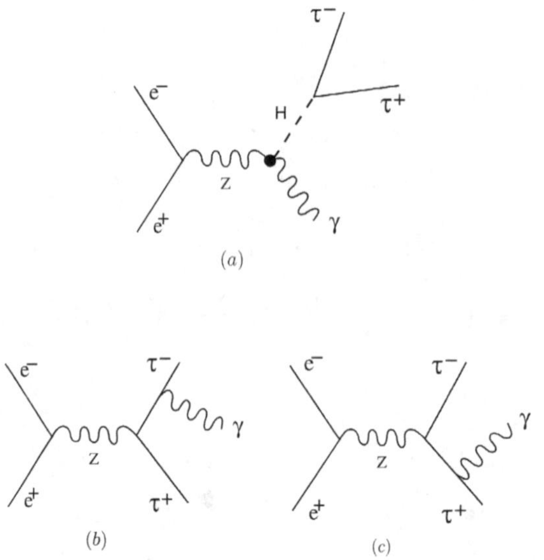

Figure 1. Feynman diagrams for the process $e^+e^- \to \tau^+\tau^-\gamma$ induced by the anomalous vertex $HZ\gamma$ (a) and the SM (b, c) when the Z vector boson is produced on mass-shell.

2. Cross-Section of the Process $e^+e^- \to \tau^+\tau^-\gamma$

The anomalous $V_1^\mu(p_1) - V_2^\nu(p_2) - H(p_H)$ vertex function is given by [3, 12]

$$\Gamma^{HV_1V_2}_{\mu\nu}(p_H, p_1, p_2) = g_Z M_Z^2 [h_1^{V_1V_2} g_{\mu\nu} + \frac{h_2^{V_1V_2}}{M_Z^2} p_{2\mu} p_{1\nu}], \tag{1}$$

where M_Z is the Z boson mass and V_1, V_2 can be $(V_1V_2) = (ZZ), (Z\gamma), (\gamma Z), (\gamma\gamma), (W^+W^-)$ or (W^+W^-).

In the present study we have considered only CP-conserving $HZ\gamma$ couplings but our results can be applied also for the CP-violating coupling.

The expression for the respective cross section, that includes the SM and the $HZ\gamma$ vertex contributions shown in Fig. 1, is given by

$$\begin{aligned} \sigma(e^+e^- \to \tau\bar{\tau}\gamma) &= \int \frac{\alpha^3}{96}[3m_\tau^2 C_1(x_W)[F_1(s, E_\gamma, \cos\theta_\gamma)(h_1^{Z\gamma})^2 + F_2(s, E_\gamma, \cos\theta_\gamma)(h_2^{Z\gamma})^2] \\ &+ m_\tau^2 C_2(x_W)[F_3(s, E_\gamma, \cos\theta_\gamma)h_1^{Z\gamma} + F_4(s, E_\gamma, \cos\theta_\gamma)h_2^{Z\gamma}] \\ &+ C_3(x_W)F_5(s, E_\gamma, \cos\theta_\gamma)]E_\gamma dE_\gamma d\cos\theta_\gamma, \end{aligned} \tag{2}$$

where E_γ and $\cos\theta_\gamma$ are the energy and scattering angle of the photon and the $C_{1,2,3}$ coefficients label the respective contributions arising from the $HZ\gamma$, SM and interference amplitudes, respectively. The kinematics is contained in the functions

$$\begin{aligned} F_1(s, E_\gamma, \cos\theta_\gamma) &\equiv \frac{(\frac{1}{2}s - \sqrt{s}E_\gamma - 2m_\tau^2)}{[(s - M_Z^2)^2 + M_Z^2\Gamma_Z^2](s + 2\sqrt{s}E_\gamma - M_H^2)^2}, \\ F_2(s, E_\gamma, \cos\theta_\gamma) &\equiv \frac{(\frac{1}{6}E_\gamma^2 - \frac{1}{3}\frac{E_\gamma^3}{\sqrt{s}} - \frac{2}{3}\frac{m_\tau^2 E_\gamma^2}{s})}{[(s - M_Z^2)^2 + M_Z^2\Gamma_Z^2](s + 2\sqrt{s}E_\gamma - M_H^2)^2}, \end{aligned}$$

$$F_3(s,E_\gamma,\cos\theta_\gamma) \equiv \frac{(-1-\frac{4}{\sin^2\theta_\gamma}+\frac{2\sqrt{s}}{E_\gamma\sin^2\theta_\gamma}+\frac{2\sqrt{s}E_\gamma}{M_Z^2}-\frac{6m_\tau^2}{\sqrt{s}E_\gamma\sin^2\theta_\gamma})}{[(s-M_Z^2)^2+M_Z^2\Gamma_Z^2](s+2\sqrt{s}E_\gamma-M_H^2)}, \tag{3}$$

$$F_4(s,E_\gamma,\cos\theta_\gamma) \equiv \frac{(-1-\frac{2}{\sin^2\theta_\gamma}-\frac{2\sqrt{s}E_\gamma}{M_Z^2}+\frac{4\sqrt{s}E_\gamma}{M_Z^2\sin^2\theta_\gamma}+\frac{2sE_\gamma^2}{M_Z^4})}{[(s-M_Z^2)^2+M_Z^2\Gamma_Z^2](s+2\sqrt{s}E_\gamma-M_H^2)},$$

$$F_5(s,E_\gamma,\cos\theta_\gamma) \equiv \frac{[(4-\sin^2\theta_\gamma)\sqrt{s}-2E_\gamma\sin^2\theta_\gamma]}{[(s-M_Z^2)^2+M_Z^2\Gamma_Z^2](\sqrt{s}\sin^2\theta_\gamma)},$$

while the coefficients $C_{1,2,3}$ are given by

$$C_1(x_W) \equiv \frac{(1-4x_W+8x_W^2)}{x_W^3(1-x_W)^3},$$

$$C_2(x_W) \equiv \frac{(1-4x_W)(1-4x_W+8x_W^2)}{x_W^{5/2}(1-x_W)^{5/2}}, \tag{4}$$

$$C_3(x_W) \equiv \frac{(1-4x_W+8x_W^2)^2}{x_W^2(1-x_W)^2},$$

where $x_W \equiv \sin^2\theta_W$.

3. Results and Conclusions

In practice, detector geometry imposes a cut on the photon polar angle with respect to the electron direction, and further cuts must be applied on the photon energy and minimum opening angle between the photon and tau in order to suppress the background from tau decay products. In order to evaluate the integral of the total cross section as a function of the parameters $h_1^{Z\gamma}$ and $h_2^{Z\gamma}$, we require cuts on the photon angle and energy to avoid divergences when the integral is evaluated at the important intervals of each experiment. We integrate over $\cos\theta_\gamma$ from -0.74 to 0.74 and E_γ from 5 GeV to 45.5 GeV for various fixed values of the Higgs boson mass M_H. For simplicity we have set the effective coupling g_Z equal to unity; using the numerical values $\sin^2\theta_W = 0.2314$, $M_{Z_1} = 91.18\ GeV$, $\Gamma_{Z_1} = 2.49\ GeV$ and $m_\tau = 1.776\ GeV$, we obtain the cross section $\sigma = \sigma(h_1^{Z\gamma}, h_2^{Z\gamma}, M_H)$. As was discussed in Ref. [9], $N \approx \sigma(h_1^{Z\gamma}, h_2^{Z\gamma}, M_H)$, and using Poisson statistics [9, 13], we require that $N \approx \sigma(h_1^{Z\gamma}, h_2^{Z\gamma}, M_H)$ be less than 1559, with $\mathcal{L} = 100$ pb^{-1}, according to the data reported by the L3 collaboration [9]. Taking this into consideration, we get limits on $h_1^{Z\gamma}$ and $h_2^{Z\gamma}$ as a function of M_H. The values obtained for these limits for several values of M_H are included in Table 1.

Table 1. Sensitivities achievable at the 95% C.L. for the $h_{1,2}^{Z\gamma}$ vertices in the process $e^+e^- \to \tau^+\tau^-\gamma$ with a luminosity of 100 pb^{-1}. We have applied the cuts used by L3 for the photon angle and energy.

M_H	$h_1^{Z\gamma}$	$h_2^{Z\gamma}$
115 GeV	[-0.042, 0.042]	[-0.045, 0.045]
130 GeV	[-0.047, 0.047]	[-0.081, 0.081]
145 GeV	[-0.11, 0.11]	[-0.19, 0.19]

We plot the total cross section in Fig. 2 as a function of the Higgs boson mass M_H for the values $h_1^{Z\gamma} = 0.047$ and $h_2^{Z\gamma} = 0.081$ given in Table 1. We observe in this figure that the cross

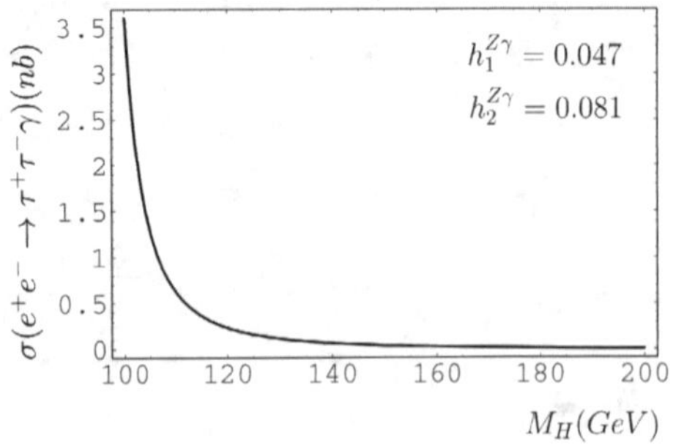

Figure 2. Cross-section for the process $e^+e^- \to \tau^+\tau^-\gamma$ as a function of M_H with $h_1^{Z\gamma} = 0.047$ and $h_2^{Z\gamma} = 0.081$.

section of the process $e^+e^- \to \tau^+\tau^-\gamma$ decreases with the increases of the Higgs bosons mass M_H.

In conclusion, we have analyzed the constraints imposed on the $HZ\gamma$ coupling from the known data for the process $e^+e^- \to \tau^+\tau^-\gamma$ obtained by the L3 Collaboration [9]. We have made similar analysis using LEP data in order to improve previous limits on the $ZZ\gamma$ and $Z\gamma\gamma$ vertices [14, 15], the magnetic and electric dipole moments of tau neutrinos [16] and the tau lepton [17], as well as some of the parameters involved in L-R symmetric and E_6 superstring model [18]. In the present case, our bounds shown in Table 1 are close to the limits expected in the annihilation process $e^+e^- \to HZ$ with polarized beams [7], and an order of magnitude better [19] than the limits obtained for the same process by the L3 Collaboration [11]. In particular, we were able to improve the bounds on the $HZ\gamma$ vertex because we did not need to use in our analysis the partial decay rates of the Higgs boson used in Ref. [11].

Acknowledgments

We acknowledge support of RedFAE and CONACyT (México).

References

[1] U. Cotti, J. L. Diaz-Cruz, J. J. Toscano, *Phys. Lett.* **B404**, 308 (1997).
[2] G. J. Gounaris, J. Layssac, F. M. Renard, *Phys. Rev.* **D67**, 013012 (2003).
[3] S. Dutta, K. Hagiwara and Y. Matsumoto, *Phys. Rev. D78*, 115016 (2008).
[4] C. S. Li, S. H. Zhu, C. F. Qiao, *Phys. Rev.* **D57**, 6928 (1998).
[5] M. A. Pérez, G. Tavares-Velasco, J. J. Toscano, *Int. J. Mod. Phys.* **A19**, 159 (2004).
[6] V. Hankele *et al.*, *Phys. Rev.* **D74**, 095001 (2006).
[7] S. D. Rindani, P. Sharma, *Phys. Rev.* **D79**, 075007 (2009).
[8] W. Killian, M. Kramer, P. M. Zerwas, *Phys. Lett.* **B381**, 243 (1996).
[9] M. Acciarri *et al.*, L3 Collab., *Phys. Lett.* **B434**, 169 (1998), and references therein .
[10] J. L. Díaz-Cruz *et al.*, *Rev. Mex. Fis.* **39**, 501 (1993); A. Djouadi *et al.*, *Eur. Phys. J.* **C1**, 163 (1998).
[11] P. Achard et al., L3 Collab., *Phys. Lett.* **B589**, 89 (2004).
[12] J. M. Hernández, M. A. Pérez, J. J. Toscano, *Phys. Rev.* **D51**, 2044 (1995).
[13] R. M. Barnett *et al. Phys. Rev.* **D54**, 166 (1996).
[14] M. A. Pérez and F. I. Ramírez-Zavaleta, *Phys. Lett.* **B609**, 68 (2005).
[15] A. Gutiérrez-Rodríguez, M. A. Hernández-Ruíz and M. A. Pérez, *Phys. Rev.* **D80**, 017301 (2009).
[16] M. Maya *et al.*, *Phys. Lett.* **B434**, 354 (1998); A. Gutiérrez-Rodríguez *et al.*, *Phys. Rev.* **D74**, 053002 (2006).
[17] A. Gutiérrez-Rodríguez, M. A. Hernández-Ruíz and M. A. Pérez, *Int. J. Mod. Phys.* **A22**, 3493 (2007).
[18] A. Gutiérrez-Rodríguez *et al.*, *Mod. Phys. Lett.* **A24**, 135 (2009).
[19] A. Gutiérrez-Rodríguez, J. Montano and M. A. Pérez, arXiv: 1009.4354 [hep-ph].

Journal of Physics: Conference Series **287** (2011) 012037 doi:10.1088/1742-6596/287/1/012037

Graded Fock–like representations for a system of algebraically interacting paraparticles

Konstantinos Kanakoglou[1,2], Alfredo Herrera–Aguilar[1]

[1] Instituto de Física y Matemáticas (IFM), Universidad Michoacana de San Nicolás de Hidalgo (UMSNH), Edificio C-3, Cd. Universitaria, CP 58040, Morelia, Michoacán, MEXICO

[2] School of Physics, Nuclear and Elementary Particle Physics Department, Aristotle University of Thessaloniki (AUTH), 54124, Thessaloniki, GREECE

E-mail: kanakoglou@ifm.umich.mx, herrera@ifm.umich.mx

Abstract. We will present and study an algebra describing a mixed paraparticle model, known in the bibliography as "The Relative Parabose Set (RPBS)". Focusing in the special case of a single parabosonic and a single parafermionic degree of freedom $P_{BF}^{(1,1)}$, we will construct a class of Fock–like representations of this algebra, dependent on a positive parameter p a kind of *generalized parastatistics order*. Mathematical properties of the Fock–like modules will be investigated for all values of p and constructions such as ladder operators, irreducibility (for the carrier spaces) and $(\mathbb{Z}_2 \times \mathbb{Z}_2)$–gradings (for both the carrier spaces and the algebra itself) will be established.

1. Introduction

Our central object of study in this short letter, will be the *Relative Parabose Set algebra* $P_{BF}^{(1,1)}$ in a single parabosonic and a single parafermionic degree of freedom. It belongs to the general family of paraparticle algebras intimately related to the Wigner quantization scheme [1]. The "*free*" paraparticle algebras (parabosonic and parafermionic algebras) have been -implicitly- introduced in the early '50's [2] while the "*mixed*" paraparticle models such as P_{BF} (and a couple of others as well) have been first introduced in [3].

The algebra $P_{BF}^{(1,1)}$ is generated (as an algebra over $\mathbb{C}$) by the four generators b^+, b^-, f^+, f^- subject to the usual trilinear relations of the *free parabosonic* and the *free parafermionic* algebras which can be compactly summarized as

$$\left[\{b^\xi, b^\eta\}, b^\epsilon\right] = (\epsilon - \eta)b^\xi + (\epsilon - \xi)b^\eta \quad , \quad \left[[f^\xi, f^\eta], f^\epsilon\right] = \tfrac{1}{2}(\epsilon - \eta)^2 f^\xi - \tfrac{1}{2}(\epsilon - \xi)^2 f^\eta \tag{1}$$

for all values $\xi, \eta, \epsilon = \pm$, together with the *mixed trilinear relations*

$$\left[\{b^\xi, b^\eta\}, f^\epsilon\right] = \left[[f^\xi, f^\eta], b^\epsilon\right] = 0, \quad \left[\{f^\xi, b^\eta\}, b^\epsilon\right] = (\epsilon - \eta)f^\xi, \quad \left\{\{b^\xi, f^\eta\}, f^\epsilon\right\} = \tfrac{1}{2}(\epsilon - \eta)^2 b^\xi \tag{2}$$

for all values $\xi, \eta, \epsilon = \pm$, which represent a kind of algebraically established interaction between parabosonic and parafermionic degrees of freedom and *characterize the relative parabose set.*

It is easy for one to see that when all combinations of the $\xi, \eta, \epsilon = \pm$ are taken into account, the first of equations (1) produces 6 relations, the second of eq. (1) produces 2 relations and equations (2) produce 24 relations. One can easily see (see also the discussion in [4]) that not all of these $24+6+2=32$ are algebraically independent. However, we keep the relations in the form given in equations (1), (2) and we do not proceed in further simplifying them, because of the compact notational and computational advantages of this presentation. (See also [4, 5, 6]).

The purpose of this letter, will be to present the construction of a class of representations for the relations (1), (2). We will present an infinite class of irreducible, $(\mathbb{Z}_2 \times \mathbb{Z}_2)$–graded representations of these relations. This class will be parametrized by the values of a positive integer p (similarly to the way that the free parabosonic and parafermionic algebra Fock–like spaces are parametrized by the values of a positive integer). The parameter p will be called *generalized parastatistics order*. It will be shown that the $b^{\pm}, f^{\pm}$ generators act as "ladder" operators (*Creation/Annihilation* operators) in a kind of two–dimensional "ladder", generalizing thus the way that the usual Canonical Commutation–Anticommutation relations (CCR–CAR) are represented in their Fock spaces and also the way that the single degree of freedom free parabosonic and free parafermionic algebras act in their own Fock–spaces [7] respectively.

2. Structure of the Fock–like spaces

In this section we are going to briefly review the results of [8] on the construction of the Fock–like spaces, which will serve as carrier spaces for the Fock–like representations of $P_{BF}^{(1,1)}$.

An old conjecture [3] on the study of the representations of the paraparticle algebras, states that if we consider representations of $P_{BF}^{(1,1)}$, satisfying the adjointness conditions $(b^-)^\dagger = b^+$ and $(f^-)^\dagger = f^+$, on a complex, infinite dimensional, pre–Hilbert [1] space, possessing a unique vacuum vector $|0\rangle$ satisfying $b^-|0\rangle = f^-|0\rangle = b^-f^+|0\rangle = f^-b^+|0\rangle = 0$ and $\langle 0|0\rangle = 1$ then the following additional conditions (where p may be an arbitrary positive integer)

$$b^-b^+|0\rangle = f^-f^+|0\rangle = p|0\rangle \tag{3}$$

single out an irreducible representation which is unique up to unitary equivalence. In other words the above statement, tells us that for any positive integer p there is an irreducible representation of $P_{BF}^{(1,1)}$ uniquely specified (up to unitary equivalence) by the above relations.

We now proceed in summarizing the results of [8]: The carrier spaces of the Fock–like representations of $P_{BF}^{(1,1)}$ are $\bigoplus_{n=0}^{p} \bigoplus_{m=0}^{\infty} \mathcal{V}_{m,n}$ where p is an arbitrary (but fixed) positive integer. The subspaces $\mathcal{V}_{m,n}$ are 2–dim except for the cases $m=0$, $n=0,p$ i.e. except the subspaces $\mathcal{V}_{0,n}$, $\mathcal{V}_{m,0}$, $\mathcal{V}_{m,p}$ which are 1–dim. Let us see how the corresponding vectors look like:

- If $0 < m$ and $0 < n < p$, then the subspace $\mathcal{V}_{m,n}$ is spanned by all vectors of the form

$$\left| \begin{array}{c} m_1, m_2, ..., m_l \\ n_0, n_1, n_2, ..., n_l \end{array} \right\rangle \equiv (f^+)^{n_0}(b^+)^{m_1}(f^+)^{n_1}(b^+)^{m_2}(f^+)^{n_2}...(b^+)^{m_l}(f^+)^{n_l}|0\rangle \tag{4}$$

where $m_1+m_2+...+m_l = m$, $n_0+n_1+n_2+...+n_l = n$ and $m_i \geq 1$ (for $i=1,2,...,l$), $n_i \geq 1$ (for $i=1,2,...,l-1$) and $n_0, n_l \geq 0$.

For any specific combination of values (m,n) the corresponding subspace $\mathcal{V}_{m,n}$ has a basis consisting of the two vectors ($R^\eta = \frac{1}{2}\{b^\eta, f^\eta\}$ for $\eta = \pm$)

$$|m,n,\alpha\rangle \equiv (f^+)^n(b^+)^m|0\rangle = \left| \begin{array}{c} m \\ n,0 \end{array} \right\rangle, \quad |m,n,\beta\rangle \equiv (f^+)^{(n-1)}(b^+)^{(m-1)}R^+|0\rangle = \frac{1}{2}\left| \begin{array}{c} m \\ n-1,1 \end{array} \right\rangle + \frac{1}{2}\left| \begin{array}{c} m-1,1 \\ n-1,1,0 \end{array} \right\rangle \tag{5}$$

- If $m=0$ or $n=0,p$, the vectors $|0,n,\beta\rangle$ and $|m,0,\beta\rangle$ are (by definition) zero and furthermore the vector $|m,p,\beta\rangle$ becomes parallel to $|m,p,\alpha\rangle$: $|m,p,\beta\rangle = \frac{1}{p}|m,p,\alpha\rangle$
- If $n \geq p+1$, all basis vectors of (5) vanish.

[1] in the sense that it is an inner product space, but not necessarily complete with respect to the inner product.

Note 1: The above described subspaces of $\bigoplus_{n=0}^{p}\bigoplus_{m=0}^{\infty}\mathcal{V}_{m,n}$ can be visualized as follows:

$$\begin{array}{cccccccccc}
\mathcal{V}_{0,0} & \mathcal{V}_{0,1} & \cdots & \mathcal{V}_{0,n} & & \cdots & \cdots & \mathcal{V}_{0,p-1} & \mathcal{V}_{0,p} \\
\mathcal{V}_{1,0} & \mathcal{V}_{1,1} & \cdots & \mathcal{V}_{1,n} & & \cdots & \cdots & \mathcal{V}_{1,p-1} & \mathcal{V}_{1,p} \\
\vdots & \vdots & \cdots & \vdots & & \cdots & \cdots & \vdots & \vdots \\
\mathcal{V}_{m,0} & \mathcal{V}_{m,1} & \cdots & \mathcal{V}_{m,n} & \mathcal{V}_{m,n+1} & \cdots & \cdots & & \mathcal{V}_{m,p} \\
\vdots & \vdots & \cdots & \mathcal{V}_{m+1,n} & & \cdots & \cdots & \vdots & \vdots \\
\vdots & \vdots & \cdots & \vdots & & \cdots & \cdots & \vdots & \vdots
\end{array}$$

• Inside each one of the above presented 2–dim subspaces $\mathcal{V}_{m,n}$ (with $m \neq 0,\ n \neq 0, p$) the vectors of (5) are linearly independent and constitute a basis. However, these vectors are neither orthogonal nor normalized. We can show that an orthonormal set of basis vectors can be obtained as

$$|m,n,+\rangle = c_+|m,n,\alpha\rangle \qquad |m,n,-\rangle = -c_-\Big(|m,n,\alpha\rangle - p|m,n,\beta\rangle\Big)$$

where $c_\pm$ are suitable normalization factors [8]. Now we can show orthonormalization $\langle m,n,s|m',n',s'\rangle = \delta_{m,m'}\delta_{n,n'}\delta_{s,s'}$ and completeness $\sum_{m=0}^{\infty}\sum_{n=0}^{p}\sum_{s=\pm}|m,n,s\rangle\langle m,n,s| = 1$.

Note 2: If we consider the Hermitian operators $N_s = \frac{1}{p}(N_f^2 - (p+1)N_f + f^+f^- + \frac{p}{2})$, $N_f = \frac{1}{2}[f^+,f^-] + \frac{p}{2}$ and $N_b = \frac{1}{2}\{b^+,b^-\} - \frac{p}{2}$ we can show that they constitute a *Complete Set of Commuting Observables* (*C.S.C.O.*): We have $[N_b,N_f] = [N_b,N_s] = [N_f,N_s] = 0$; their common eigenvectors are exactly the elements of the orthonormal basis formerly described. Any vector $|m,n,s\rangle$ is uniquely determined as an eigenvector of N_b, N_f, N_s by its eigenvalues $0 \leq m$, $0 \leq n \leq p$ and $s = \pm\frac{1}{2}$ respectively.

3. Main results: Construction of the Fock–like representations

For detailed computations and proofs of the results presented in this section see [4].

3.1. Construction of ladder operators

We now present the formulae describing explicitly the action of the generators (and hence of the whole algebra) of $P_{BF}^{(1,1)}$ on the carrier spaces $\bigoplus_{n=0}^{p}\bigoplus_{m=0}^{\infty}\mathcal{V}_{m,n}$ for any positive integer p:

$$\blacktriangledown b^-\cdot|m,n,\alpha\rangle = \begin{cases} (-1)^n m|m-1,n,\alpha\rangle - 2(-1)^n nm|m-1,n,\beta\rangle, & \underline{m: even} \\ \\ -(-1)^n\big(2n-m-(p-1)\big)|m-1,n,\alpha\rangle - 2(-1)^n n(m-1)|m-1,n,\beta\rangle, & \underline{m: odd} \end{cases}$$

$$\blacktriangledown b^-\cdot|m,n,\beta\rangle = \begin{cases} -(-1)^n|m-1,n,\alpha\rangle + (-1)^n\big(2n-m-p\big)|m-1,n,\beta\rangle, & \underline{m: even} \\ \\ -(-1)^n|m-1,n,\alpha\rangle - (-1)^n(m-1)|m-1,n,\beta\rangle, & \underline{m: odd} \end{cases}$$

$$\blacktriangledown f^-\cdot|m,n,\alpha\rangle = n(p+1-n)|m,n-1,\alpha\rangle, \blacktriangledown b^+\cdot|m,n,\alpha\rangle = (-1)^n|m+1,n,\alpha\rangle - (-1)^n 2n|m+1,n,\beta\rangle$$
$$\blacktriangledown f^-\cdot|m,n,\beta\rangle = |m,n-1,\alpha\rangle + (n-1)(p-n)|m,n-1,\beta\rangle, \quad \blacktriangledown b^+\cdot|m,n,\beta\rangle = -(-1)^n|m+1,n,\beta\rangle$$

$$\blacktriangledown f^+\cdot|m,n,\alpha\rangle = \begin{cases} |m,n+1,\alpha\rangle, \underline{if\ n\leq p-1} \\ 0, \qquad \underline{if\ n\geq p} \end{cases} \qquad \blacktriangledown f^+\cdot|m,n,\beta\rangle = \begin{cases} |m,n+1,\beta\rangle, \underline{if\ n\leq p-1} \\ 0, \qquad \underline{if\ n\geq p} \end{cases} \tag{6}$$

for all integers $0 \leq m$, $0 \leq n \leq p$. The direct proof [4] of these formulae involves lengthy "normal–ordering" algebraic computations inside $P_{BF}^{(1,1)}$. We must take into account: (*a*). the relations (1), (2) of $P_{BF}^{(1,1)}$, (*b*). the relations (3) together with $b^-|0\rangle = f^-|0\rangle = b^-f^+|0\rangle = f^-b^+|0\rangle = 0$ and (*c*). the structure and properties of the corresponding carrier space as described in Sect. 2.

Apart from the direct proof, eq. (6) have also been checked and verified via the Quantum [9] add–on for Mathematica 7.0, which is an add-on for performing symbolic algebraic computations, including the use of generalized Dirac notation. What we have actually verified via the use of this package, is that the actions formulae (6) preserve all of the relations (1), (2) of $P_{BF}^{(1,1)}$.

3.2. Irreducibility

In the following diagram we provide a "visual" interpretation of relations (6) i.e. of the action of the generators of $P_{BF}^{(1,1)}$ on the direct summand subspaces of the Fock-like carrier space $\bigoplus_{n=0}^{p}\bigoplus_{m=0}^{\infty}\mathcal{V}_{m,n}$. We can easily figure out that we have a kind of *generalized creation–annihilation operators* acting on a two dimensional ladder of subspaces:

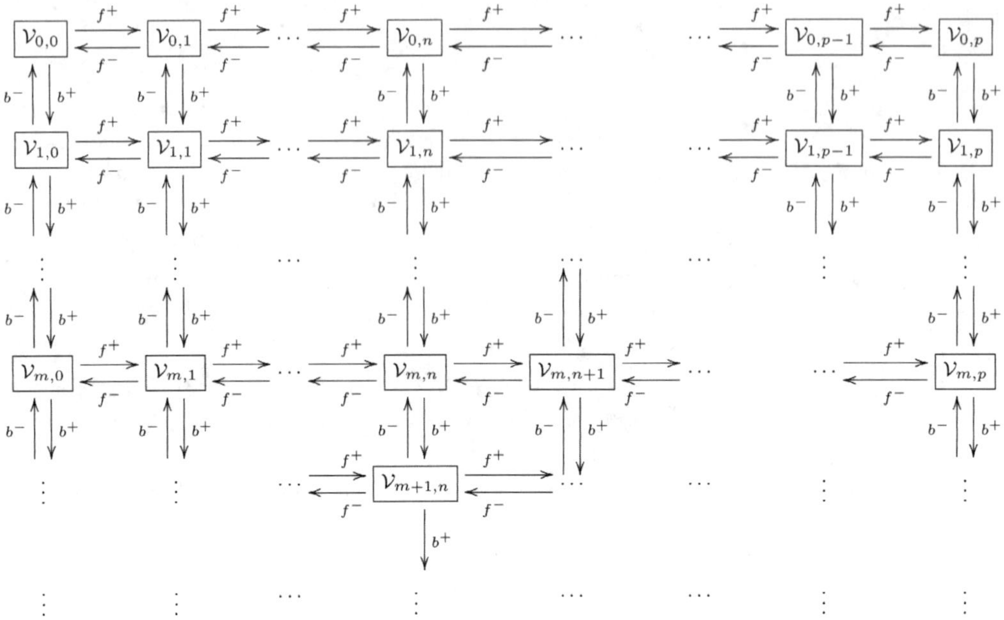

Initiating from the remark that we are dealing with a cyclic module which moreover can be generated by any of its elements, we can prove [4] that the Fock–like representations, explicitly given by (6) and visually represented in the above diagram, are irreducible representations (or: simple $P_{BF}^{(1,1)}$-modules) for any $p \in \mathbb{N}^*$.

3.3. Klein–group Grading of the representation

Defining $\mathsf{deg}|m,n,\alpha\rangle = \mathsf{deg}|m,n,\beta\rangle = \big(m \text{ mod } 2,\ n \text{ mod } 2\big) \in \mathbb{Z}_2 \times \mathbb{Z}_2$ for the carrier spaces and $\mathsf{deg}b^{\pm} = (1,0)$, $\mathsf{deg}f^{\pm} = (0,1)$ for the algebra, the Fock–like representations of $P_{BF}^{(1,1)}$ over $\bigoplus_{n=0}^{p}\bigoplus_{m=0}^{\infty}\mathcal{V}_{m,n}$, become $(\mathbb{Z}_2 \times \mathbb{Z}_2)$–graded modules, $\forall p \in \mathbb{N}^*$. (For proof and details see [4]).

Acknowledgments

KK would like to thank the whole staff of IFM, UMSNH for providing a challenging and stimulating atmosphere while preparing this article. His work was supported by the research project CONACYT/No. J60060. AHA was supported by CIC 4.16 and CONACYT/No. J60060; he is also grateful to SNI.

References

[1] Wigner E P 1950 *Physical Review* **77** 5 711-12
[2] Green H S 1953 *Phys. Rev.* **90** 2 270
[3] Greenberg O W and Messiah A M L 1965 *Phys. Rev.* **138** 5B 1155-67
[4] Kanakoglou K and Herrera–Aguilar A 2010 Ladder operators, Fock spaces, irreducibility and group gradings for the Relative Parabose Set algebra *e-print* arXiv:1006.4120v2 [math.RT]
[5] Kanakoglou K, Daskaloyannis C and Herrera–Aguilar A 2010 *AIP Conf. Proc.* **1256** 193-200
[6] Yang W and Jing S 2001 *Science in China (Series A)* **44** 9 1167-73 (*e-print* arXiv:math-ph/0212004v1)
[7] Lievens S, Stoilova N I and Van der Jeugt 2008 *Commun. Math. Phys.* **281** 805-26
[8] Yang W and Jing S 2001 *Mod. Phys. Letters* A **16** 15 963-71 (*e-print* arXiv:math-ph/0212003v1)
[9] Gómez-Muñoz J L and Delgado-Cepeda F http://homepage.cem.itesm.mx/lgomez/quantum/ Tec de Monterrey

The Effects of the Little Higgs Model on the Decay Width $Z \to e^+e^-$

A. Gutiérrez-Rodríguez

Facultad de Física, Universidad Autónoma de Zacatecas.
Apartado Postal C-580, 98060, Zacatecas, México.

E-mail: alexgu@fisica.uaz.edu.mx

Abstract.
In the framework of the Little Higgs Model (LHM), we calculate the decay width $\Gamma(Z_1 \to e^+e^-)$ with corrections of QED and QCD. We analyze this with recent data from LEP and compute the contribution of the model. We find that the deviations of the decay width of reaction $Z_1 \to e^+e^-$ from its SM value are relatively large in the parameter space preferred by the electroweak precision data. Furthermore, with reasonable free parameter values, the absolute value of the relative correction parameter $\delta\Gamma/\Gamma_{SM}$ is of $15\% - 50\%$. The experimental measurement values might generate constraints on the free parameters of the LHM.

1. Introduction

There are a number of scenarios for new physics beyond the Standard Model (SM) [1]. The most famous is the supersymmetric scenario. One of the major motivations for physics beyond the Standard Model is to resolve the hierarchy and fine-tuning problems between the electroweak scale and the Planck scale. Supersymmetric theories introduce an extended space-time symmetry and quadratically divegent quantum corrections are canceled due to the symmetry between the bosonic and fermionic partners. Technicolor theories introduce new strong dynamics at scales not much higher than the electroweak scale, thus defer the hierarchy problem. Theories with TeV scale quantum gravity reinterpret the problem completely by lowering the fundamental Planck scale. Recently, an alternative called the Little Higgs Model (LHM) [2, 3, 4, 5] has been proposed. The little Higgs model offers an alternative route to the solution of the hierarchy problem of the SM, reviving the idea that the Higgs doublet is a pseudo Goldstone boson of some global symmetry which is spontaneously broken at a TeV scale. The key feature of this kind of model is that the Higgs boson is a pseudo-Goldstone boson of an approximate global symmetry which is spontaneously broken by a vev at a scale of a few TeV and is thus naturally light. In the LHM, a set of new heavy gauge bosons (A_2, Z_2, W_2) and a new heavy vector-like quark (T) are introduced which cancel the quadratic divergence induced by SM gauge boson loops and the top quark loop, respectively. The distinguishing features of this model are the existence of these new particles and their couplings to the light Higgs. The measurement of these couplings and new particle effects might prove the existence of the little Higgs mechanism [6]. The global symmetry breaking scale is expected to be $\leq 10\ TeV$ so the little Higgs model will be relevant for the hierarchy.

In this work, we calculate the decay width of the process $Z_1 \to e^+e^-$ in the LHM. When

Journal of Physics: Conference Series **287** (2011) 012038 doi:10.1088/1742-6596/287/1/012038

compared to the process $Z_1 \to e^+e^-$ in the SM, the process in the LHM receives the additional contribution arising from the vector and axial-vector couplings as well as the parameters of the LHM [7]. We find that, with reasonable values of the free parameters, the deviation of the decay width $\delta\Gamma/\Gamma_{SM}$ from its SM is of $15\% - 50\%$.

We also study the effects of the little Higgs model in the reaction $Z_1 \to e^+e^-$. The leptonic Z_1 decays are free from the long distance effects and are thus clean.

Processes measured near the resonance have served to set bounds on the parameters of the model. This partial decay occurs in the resonance zone. As a consequence, the process is independent of the mass of the additional Z heavy gauge boson which appears in these kind of models. They also carry considerable information about the free parameters of the model used. Therefore, it is worthwhile to analyze these decay processes in the context of the new physics models.

This paper is organized as follows: In Sec. 2 we present the expressions for the decay width $Z_1 \to e^+e^-$ in the LHM, and in Sec. 3 we present our results and conclusions.

2. Width of $Z_1 \to e^+e^-$ in the Little Higgs Model

The partial Z_1 decay width $\Gamma(Z_1 \to e^+e^-)$ including QED and QCD corrections is given by:

$$\begin{aligned}\Gamma(Z_1 \to e^+e^-) &= \frac{G_F M_{Z_1}^3}{6\pi\sqrt{2}}\Bigg[(\bar{g}_V^e)^2 + (\bar{g}_A^e)^2 - 2(\bar{g}_V^e)\frac{v^2}{f^2}\left(-c_W x_Z^{W'} c/2s + \frac{s_W x_Z^{B'}}{s'c'}(2y_e - \frac{9}{5} + \frac{3}{2}c'^2)\right) \\ &- 2(\bar{g}_A^e)\frac{v^2}{f^2}\left(c_W x_Z^{W'} c/2s + \frac{s_W x_Z^{B'}}{s'c'}(-\frac{1}{5} + \frac{1}{2}c'^2)\right)\Bigg](1 + \delta\rho + \delta\rho_e + \delta_{QED}). \qquad (1)\end{aligned}$$

The vector and axial-vector $Z_1 l\bar{l}$ couplings $\bar{g}_V^e$ and $\bar{g}_A^e$ comparise one-loop and higher electroweak and internal QCD corrections through the form factors $\delta\rho_l$ and k_l, which can be written as:

$$\bar{g}_V^l = \sqrt{\rho_l}(\frac{1}{2} - 2sin^2\theta_{eff}^l), \quad \bar{g}_A^l = \sqrt{\rho_l}(\frac{1}{2}), \qquad (2)$$

with $sin^2\theta_{eff}^l = k_l sin^2\theta_W$. The term $\delta\rho$ is the deviation from the SM prediction for the ρ parameter $\rho = M_Z \cos\theta / M_W = 1 + \delta\rho$, taking into account contributions of the gauge group structure of LHM only. The contribution to $\delta\rho$ is given by

$$\delta\rho \approx -\frac{v^2}{8f^2}\left[1 + 5(c'^2 - s'^2)^2\right]. \qquad (3)$$

Also, δ_{QED} accounts for the final state photon radiation,

$$\delta_{QED} = \frac{3\alpha(s)}{4\pi}Q^2, \qquad (4)$$

where α is the QED coupling computed at the energy scale s, while Q is the lepton charge.

In order to obtain a prediction for the standard model partial Z_1 decay width into e^+e^-, we take the input parameters [8], $M_Z = 91.187\,GeV$, $G = 1.16637\times10^{-5}\,GeV^{-2}$, $\alpha(M_Z) = 1/128.95$ and $\sin^2\theta_W = 0.22335$. These parameters can be used to obtain the form factors for the decay $Z_1 \to e^+e^-$ using the Zfitter package [9], yielding

$$\delta\rho_e = 0.00531, \qquad (5)$$

and

Journal of Physics: Conference Series **287** (2011) 012038 doi:10.1088/1742-6596/287/1/012038

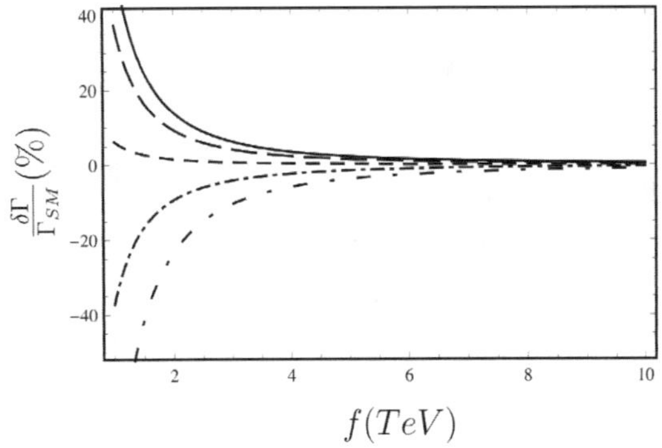

Figure 1. The relative correction $\delta\Gamma/\Gamma_{SM}$ as a function of the scale of energy f for $c = 0.5$ and different values of the mixing parameter s. Starting from the top, the curves are for $s = 0.2, 0.4, 0.6, 0.8, 1$.

$$sin^2\theta_{W(eff)} = 0.2315, \tag{6}$$

which translates into $\kappa_e = 1.0367$. Plugging these parameters into Eq. (1) together with the limit $f \to \infty$ and $\delta\rho \to 0$, we obtain the standard model prediction $\Gamma(Z_1 \to e^+e^-) = 83.99$ MeV.

3. Results and Conclusions

Our numerical results are summarized in Figs. 1-3. To see the dependence of relative correction on the parameter f, in Fig. 1 we plot $\delta\Gamma/\Gamma_{SM}$ as a function of the scale of energy f for $c = 0.5$ and different values of the mixing parameter $s = 0.2, 0.4, 0.6, 0.8, 1$. We can see that the absolute value of the relative correction decreases as f increases. The curves also demonstrate that the effect of the LHM is not sensitive to f in the range of $f \geq 6.5\ TeV$. This is because, in general, the extra contribution of the LHM to the decay width $\Gamma(Z_1 \to e^+e^-)$ is proportional to a factor of $1/f^2$. In this case, the absolute value of $\delta\Gamma/\Gamma_{SM}$ is in the range of 40% in most of the parameter space.

The relative correction $\delta\Gamma/\Gamma_{SM}$ as a function of the scale of energy f for $s = 0.5$ and different values of $c = 0.2, 0.4, 0.6, 0.8, 1$ is presented in Fig. 2. In this figure we can see that the absolute value of the relative correction $\delta\Gamma/\Gamma_{SM}$ decreases when the scale energy f increases and is sensitive to the mixing parameter c. For $c = 0.2$, the absolute value of $\delta\Gamma/\Gamma_{SM}$ is in the range of $2\% - 50\%$.

Finally, the graphic in Fig. 3 shows the allowed values for f, s and c that can be developed allowed by the decay width $\Gamma^{exp} = (83.984 \pm 0.086)\ MeV$ with 95% C.L.. We see that the values for f, s and c which can develop are in complete agreement with those reported in the literature.

The SM gauge boson Z_1 is now abundantly produced at the LHC and will be as well at the future high energy linear e^+e^- collider experiments. It is possible to examine its properties with unprecedented precision. We calculate the decay width correction of the little Higgs model of the process $\Gamma(Z_1 \to e^+e^-)$. We find that the correction is significant even when we consider the constraint of electroweak precision data on the parameters. In the favorable parameter space, the absolute value of the relative correction parameter $\delta\Gamma/\Gamma_{SM}$ is of $15\% - 50\%$. We conclude that the future experiments at the ILC could determine the effects on the $\Gamma(Z_1 \to e^+e^-)$ decay width contributed by the LHM in some parameter space or put more stringent constraints on the LHM parameters.

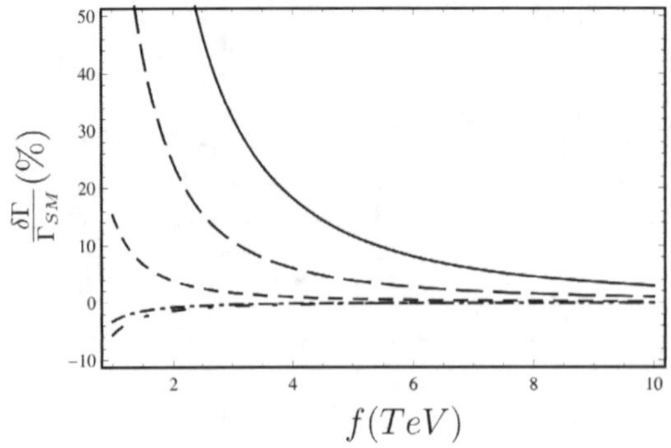

Figure 2. The relative correction $\delta\Gamma/\Gamma_{SM}$ as a function of the scale of energy f for $s = 0.5$ and different values of the mixing parameter c. Starting from the top, the curves are for $c = 0.2, 0.4, 0.6, 0.8, 1$.

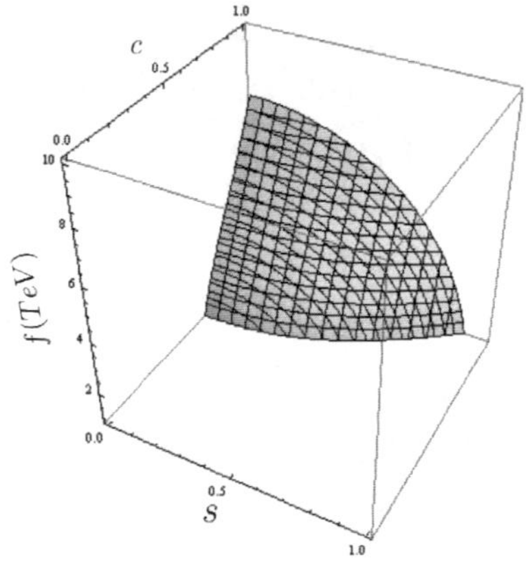

Figure 3. Possible values for s, c and f that can be developed allowed by the decay width $\Gamma^{exp}(Z_1 \to e^+e^-)$ with 95% C.L..

Acknowledgments

We acknowledge support of RedFAE and CONACyT (México).

References

[1] S. L. Glashow, *Nucl. Phys.* **22**, 579 (1961); S. Weinberg, *Phys. Rev. Lett.* **19**, 1264 (1967); A. Salam, in *Elementary Particle Theory*, Ed. N. Svartholm (Almquist and Wiskell, Stockholm, 1968) 367.
[2] N. Arkani-Hamed *et al.*, *Phys. Rev. Lett.* **86**, 4757 (2001).
[3] N. Arkani-Hamed *et al.*, *Phys. Lett.* **B513**, 232 (2001).
[4] N. Arkani-Hamed *et al.*, *JHEP* **0207**, 034 (2002).
[5] N. Arkani-Hamed *et al.*, *JHEP* **0208**, 021 (2002).
[6] Wolfgang Kilian, Jurgen Reuter Paul, *Phys. Rev.* **D70**, 015004 (2004); Cesaba Cesaki, *et al. Phys. Rev.* **D68**, 035009 (2003); Cesaba Cesaki, *et al. Phys. Rev.* **D67**, 1150029 (2003).
[7] Tao Han, Heather E. Logan, Bob McElrath and Lian-Tao Wang, *Phys. Rev.* **D67**, 095004 (2003) and references therein.
[8] Particle Data Group, C. Amsler, *et al.*, *Phys. Lett.* **B667**, 1 (2008).
[9] D. Bardin, *et al.* arXiv: hep-ph/9908433.

Journal of Physics: Conference Series **287** (2011) 012039 doi:10.1088/1742-6596/287/1/012039

Study of capabilities of the HAWC observatory to detect GRBs

M A Castillo[1], H Salazar[1] and L Villaseñor[2]

[1]FCFM-BUAP, Avenida San Claudio y 18 Sur, Colonia San Manuel, Ciudad Universitaria, Puebla, México.
[2]IFM-UMSNH. Edificio C-3, Ciudad Universitaria, Morelia, México.

E-mail: mcastillomaldonado@yahoo.com

Abstract. We describe the simulation of atmospheric air showers originated by gamma rays with energies in the 0.1-1 TeV range. We study the properties of the secondary particles at a height above level of 4100 m corresponding to the height of the array of water Cherenkov detectors (WCDs) of the HAWC (High Altitude Water Cherenkov) Observatory. In particular we study the pile-up effect of the secondary particles as they arrive at the HAWC Observatory to discern the way in which the PMTs of the WCDs of HAWC can distinguish isolated secondary particles as a function of their signal thresholds. This study is relevant to the application of the single-particle counting technique often used to try to detect gamma ray bursts (GRBs) with ground-based experiments.

1. Introduction

Gamma ray bursts (GRBs) in the keV to MeV energy range have been studied in great detail by satellite-borne experiments. Recent experiments like the LAT detector onboard the Fermi satellite are detecting GRBs with much higher energies, up to several GeVs. Up to now there are no GRBs detected with ground based experiments, except for the fact that the Milagrito experiment, a predecessor to Milagro, observed evidence for TeV emission from GRB970417a coincident with the BATSE detector [2]. In this paper we describe several properties of extensive air showers produced by high energy gamma rays at a detection altitude of 4100 m a.s.l. These results illustrate the capabilities of a ground-based experiment like the HAWC Observatory to detect the high energy components of GRBs.

2. HAWC observatory

The High Altitude Water Cerenkov detector observatory is under construction by a collaboration of scientists from México and USA at Sierra La Negra volcano, México, at an altitude of 4100 m a.s.l. (latitude: 18°59'41'' N and longitude: 97°18'28'' W). It will make use of a compact array of water Cerenkov detectors (WCDs) to survey the sky in the search of steady and transient gamma-ray sources in the 0.1-100 TeV energy range [1]. When completed, HAWC will consist of 300 WCDs of 7.3 m in diameter and 5 m in height. Each tank will hold three photomultipliers tubes (PMTs) that detect Cerenkov radiation from particles that get the ground level. HAWC's low energy sensitivity and continuous operation are unique and essential to measure the prompt emission from GRBs. HAWC can detect GRBs out to z~1 if, as predicted, their TeV fluence is comparable to their keV fluence, while for closer GRBs lower fluences can be detected.

3. Single particle technique

This technique is very simple, it is based on the measurement of the numbers of single pulses on the PMTs of the ground detectors during fixed periods (i.e., each 5 ms intervals, or less frequently depending on the temporal accuracy desired) and the study of the behavior of these rates vs. time. During the arrival of GRBs one expects significant increments on the rates of arrival of single particles (or single pulses when the particles pile up) at the ground detectors. Observation of one of such excesses in coincidence with a GRB as seen by a satellite would be an unbiased signature for GRBs with a component of high energy photons [2].

4. Simulations

We have used CORSIKA V6970 to simulate 10 000 extended atmospheric showers (EASs) originated by vertical gammas for each energy in the energy range .001 – 1 TeV in steps of 10 GeV, for a total of 1 000 000 simulated EASs, all of them corresponding to vertical gamma rays. The reported properties of these EASs correspond to an altitude of 4100 m a.s.l., i.e., the location of the HAWC Observatory. Figure 1 shows the mean number of gammas, electrons, positrons, muons and anti-muons contained in these EASs at 4100 m a.s.l. as a function of the energy of the primary gamma ray.

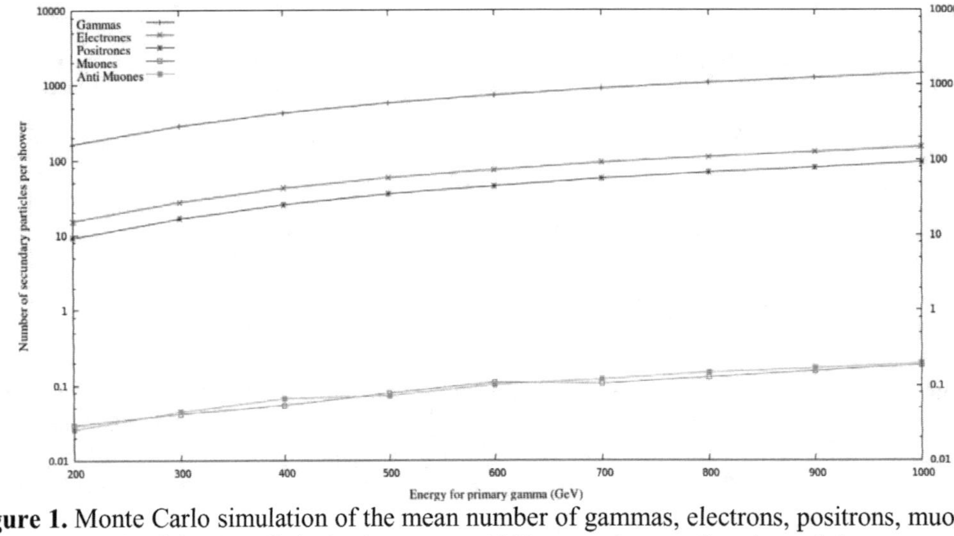

Figure 1. Monte Carlo simulation of the mean number of gammas, electrons, positrons, muons and anti-muons contained in extended air showers at 4100 m a.s.l. as a function of the energy of the primary gamma ray. We used CORSIKA V6970 to obtain these results.

What one immediately can see that about 90% of the particles that reach the ground are photons. This means that one can study the properties of EASs by focusing only on secondary gammas. Another important property is the lateral extension of the secondary gammas as they reach the ground, see figure 2. This lateral distribution has been calculated with the distances measured from the center of the shower (core) to each of the positions where the secondary gammas intercept the ground plane. The lateral distribution shown in figure 2 corresponds to vertical primary gammas with an energy of 1 TeV. Note that most of the secondary gamma particles that reach the ground are located within a distance of 1 m from the core center.

Journal of Physics: Conference Series **287** (2011) 012039 doi:10.1088/1742-6596/287/1/012039

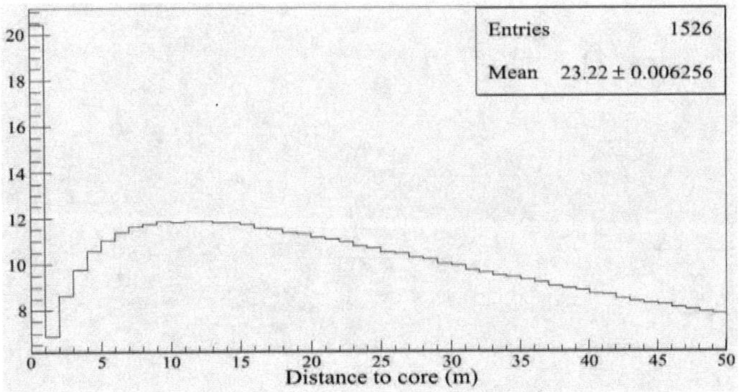

Figure 2. Lateral distribution for secondary gammas from average vertical showers simulated at the HAWC site with primary energies of 1 TeV.

Note also in the distribution of secondary gamma particles there is a second maximum around 10 m. Other important piece of information is the average number of secondary gamma particles that reach a given HAWC detector as a function of the distance from the shower core (along an axis parallel to the magnetic north pole).

Figure 3 shows the average number of secondary gamma particles that reach a given HAWC detector as a function of the energy of the primary gamma ray for different distances between the center of the HAWC detector and the shower core. For example, for showers with energy of 1 TeV there is twice the number of gammas that reach a HAWC tank localized at the origin than one localized 7.2 m away.

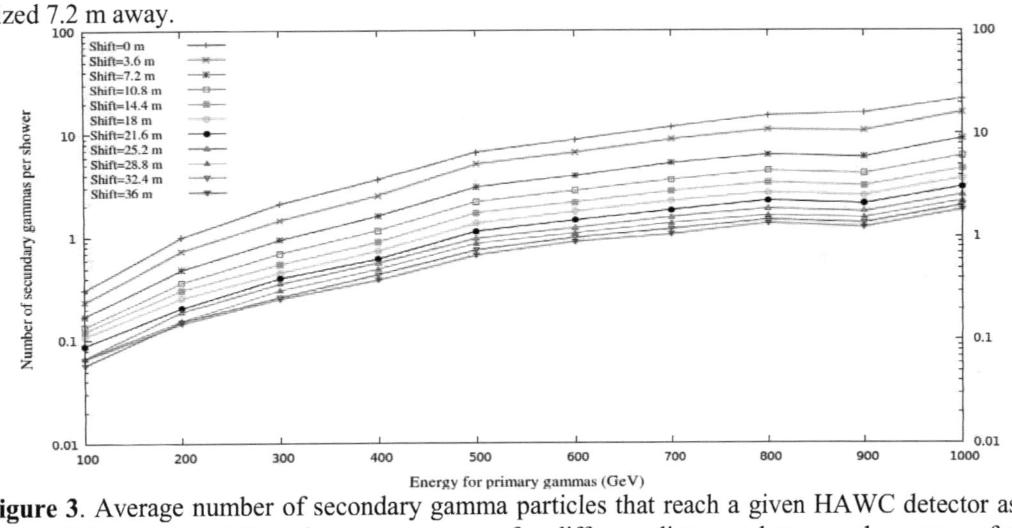

Figure 3. Average number of secondary gamma particles that reach a given HAWC detector as a function of the energy of the primary gamma ray for different distances between the center of the HAWC detector and the shower core. This simulated data correspond to EASs generated by vertical gamma rays.

Finally, we studied the structure of the arrival times of the secondary gamma particles that reach the HAWC detectors. This effect is important in order to take into account the cases where the PMT pulses do not correspond to single particles but, due to this pile-up effect, correspond to groups of particles arriving packed together with time separations lower than the typical width of single-particle pulses, which is of the order of 10 ns. Figure 4 shows the distribution of the arrival times and energies of the secondary gamma particles that pile-up within 10 ns for showers generated with energies of 1 TeV, with their vertical axes in coincidence with the detector axis of one HAWC detector.

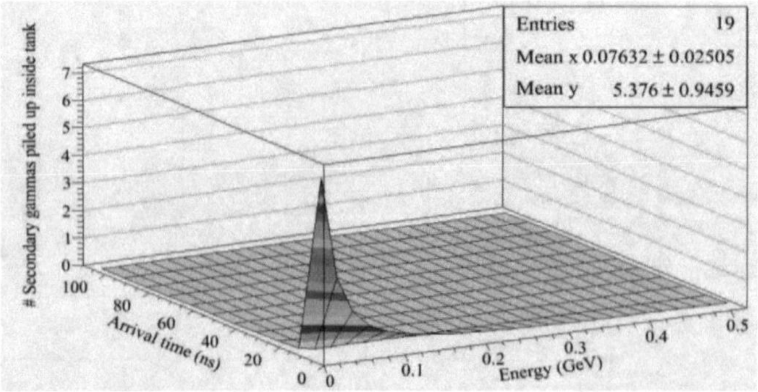

Figure 4. Energy and time distribution for secondary gammas within a given HAWC tank for an average vertical showers of 1 TeV.

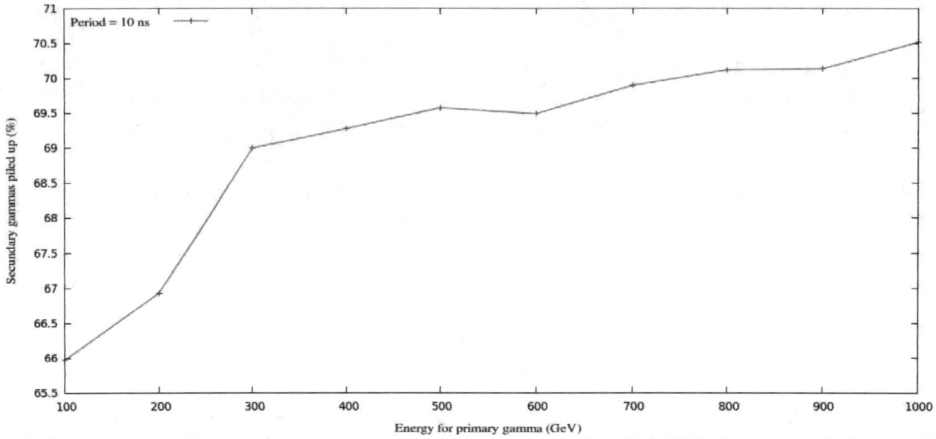

Figure 5. Percentage of secondary gamma particles that reach a HAWC detector piled up within 10 ns as a function of the energy of the primary gamma rays. The simulated showers are generated with their vertical axes in coincidence with the detector axis of one HAWC detector.

5. Conclusions

We have described the Monte Carlo simulation of extensive air showers originated by gamma rays with energies in the 0.1-1 TeV range. We studied the properties of the secondary particles at a height above see level of 4100 m, corresponding to the height of the array of water Cherenkov detectors of the HAWC Observatory. In particular we studied the pile-up effect of the secondary particles as they arrive at the HAWC detectors to discern the way in which the PMTs of the WCDs of HAWC can distinguish isolated secondary particles. The percentage of piled-up secondary gamma particles varies between 66% and 71% for primary energies in the 0.1-1 TeV range (see figure 5). We will extend this study to the pile-up effect of inclined showers generated at different distances from the HAWC detectors.

Acknowledgments

This work has been partially supported by REDFAE of CONACyT.

References

[1] Smith A J *Proceedings of the 31st ICRC 2009*
[2] Bertou X *Nuc. Inst. And Meth.* A **553** 299-303 2005
[3] Vernetto S *Astroparticle physics* **13** 75-86 2000

Journal of Physics: Conference Series **287** (2011) 012040 doi:10.1088/1742-6596/287/1/012040

Bounding the flavor-violating Hbs vertex from the $B \to X_s\gamma$ decay

J. I. Aranda[a], J. Montaño[b], F. Ramírez-Zavaleta[a], J. J. Toscano[c] and E. S. Tututi[a]

[a]Facultad de Ciencias Físico Matemáticas, Universidad Michoacana de San Nicolás de Hidalgo, Avenida Francisco J. Mújica S/N, 58060, Morelia, Michoacán, México.
[b]Departamento de Física, Universidad de Guanajuato, Campus Leon, C.P. 37150, León, Guanajuato, México.
[c]Facultad de Ciencias Físico Matemáticas, Benemérita Universidad Autónoma de Puebla, Apartado Postal 1152, Puebla, Puebla, México.

E-mail: feryuphy@fismat.umich.mx

Abstract. The nondiagonal Hbs coupling within the context of an effective Yukawa sector that comprises $SU_L(2) \times U_Y(1)$-invariant operators of up to dimension six is studied. The recent experimental result on $B \to X_s\gamma$ with hard photons is employed to constrain the Hbs vertex, with which the branching ratio for the $B_s \to \gamma\gamma$ decay is estimated. It is found that the $B_s \to \gamma\gamma$ decay can reach a branching ratio of the order of 4×10^{-8}.

1. Introduction

Suppressed observables such as $B \to X_s\gamma$, has been measured with good accuracy, showing no deviations from the standard model (SM) [1]. This means that this observable can provide stringent constraints on physics beyond the electroweak scale. We are interested in studying the flavor violating transitions $b \to s\gamma$ and $b \to s\gamma\gamma$ mediated by a SM-like Higgs boson within the context of extended Yukawa sectors that incorporates $SU_L(2) \times U_Y(1)$ invariants of up to dimension six, wich is enough to induce, in a model independent-manner, the presence of flavor and CP violation. Our main goal is to use the experimental data on the $B \to X_s\gamma$ decay to constrain the flavor violating Hbs vertex. Then we will use these results to predict the branching ratio for the $B_s \to \gamma\gamma$ transition.

2. The effective Yukawa sector

An effective Yukawa sector that generates flavor violating effects in the quark sector is:

$$\mathcal{L}^Y_{eff} = -Y^d_{ij}(\bar{Q}_i\Phi d_j) - \frac{\alpha^d_{ij}}{\Lambda^2}(\Phi^\dagger\Phi)(\bar{Q}_i\Phi d_j) - Y^u_{ij}(\bar{Q}_i\tilde{\Phi}u_j) - \frac{\alpha^u_{ij}}{\Lambda^2}(\Phi^\dagger\Phi)(\bar{Q}_i\tilde{\Phi}u_j) + H.c., \quad (1)$$

where Y_{ij}, Q_i, Φ, d_i and u_i stand for the components of the Yukawa matrix, the left-handed quark doublet, the Higgs doublet and the right-handed quark singlets of down and up type, respectively. The α_{ij} are the components of a 3×3 general matrix, which parametrize the details of the underlying physics and Λ is the new physics scale.

After spontaneous symmetry breaking, in the unitary gauge, the diagonalized Lagrangian is:

$$\mathcal{L}^{Y}_{eff} = -\Big(1+\frac{g}{2m_W}H\Big)\Big(\bar{D}M_dD+\bar{U}M_uU\Big) - H\Big(1+\frac{g}{4m_W}H\Big(3+\frac{g}{2m_W}H\Big)\Big) \times \Big(\bar{D}\Omega^d P_R D + \bar{U}\Omega^u P_R U + H.c.\Big), \quad (2)$$

where the M_a $(a = d, u)$ are the diagonal mass matrix and $\bar{D} = (\bar{d}, \bar{s}, \bar{b})$ and $\bar{U} = (\bar{u}, \bar{c}, \bar{t})$ are vectors in the flavor space. The Ω^a are matrices defined in the flavor space through the relation: $\Omega^a = (1/\sqrt{2})(v/\Lambda)^2 V^a_L \alpha^a V^{a\dagger}_R$. In general, $\Omega^{a\dagger} \neq \Omega^a$ and the Higgs boson couples to fermions through both scalar and pseudoscalar components. As a consequence, the flavor violating coupling $H\bar{q}_iq_j$ has the most general renormalizable structure of scalar and pseudoscalar type given by $-i(\Omega_{ij}P_R + \Omega^*_{ij}P_L)$.

3. Constraint on Hbs from $B \to X_s\gamma$

The leading contribution to $B \to X_s\gamma$ decay with a hard photon is dominated by the $b \to s\gamma$ process [1, 2]. We calculate the contribution of the flavor violating Hbs coupling to the $b \to s\gamma$ and $b \to sg$ decays (see Fig. 1) and study their implications for the $B \to X_s\gamma$ process.

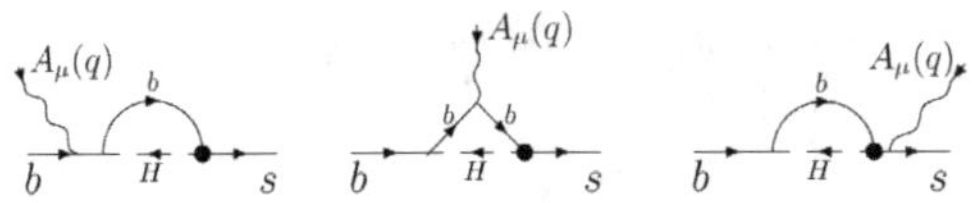

Figure 1. Diagrams contributing to the $b \to s\gamma$ transition. The $b \to sg$ process occurs via the same type of diagrams.

The total theoretical contribution to the $b - s$ transition is given by the sum of the SM contribution and the new physics effect induced by the Hbs vertex: $\mathcal{M}_T = \mathcal{M}_{SM} + \mathcal{M}_{NP}$. To get a bound for the Ω_{bs} parameter, we use the discrepancy between the theoretical prediction within the SM and the experimental measurement [3]:

$$R_{EXP-SM} \equiv \frac{\Gamma_{EXP} - \Gamma_{SM}}{\Gamma_{SM}} = \frac{Br_{EXP}(B \to X_s\gamma)}{Br_{SM}(B \to X_s\gamma)} - 1, \quad (3)$$

where Γ_{EXP} is the experimental decay width of the $B \to X_s\gamma$ transition and Γ_{SM} is the theoretical prediction of the SM. Explicitly, $R_{EXP-SM} = 0.117 \pm 0.113$. To constrain the Hbs vertex, we will assume that the SM prediction plus the Hbs contribution, coincides with the experimental value. Working out at leading order, the SM contribution is:

$$\mathcal{M}_{SM}(b \to s\gamma) = -V_{tb}V^*_{ts}\frac{\alpha^{\frac{3}{2}}}{4\sqrt{\pi}s^2_W m^2_W}C^{eff}_7(m_b)\,\bar{s}(p_s)\sigma_{\mu\nu}\epsilon^{*\mu}(q,\lambda)q^\nu(m_sP_L + m_bP_R)b(p_b), \quad (4)$$

with an effective Wilson coefficient $C^{eff}_7(m_b) = 0.689C_7(m_W) + 0.087C_8(m_W)$, which already contains the QCD contribution at the m_b scale [2].

The new physics contribution is:

$$\mathcal{M}_{NP}(b \to s\gamma) = -\frac{Q_b\,\alpha\,\mathcal{F}}{16\pi s_W m_W}\left(0.689 + \frac{0.087}{Q_b}\right)\bar{s}(p_s)\sigma_{\mu\nu}\epsilon^{*\mu}(q,\lambda)q^\nu(\Omega^*_{bs}P_L + \Omega_{bs}P_R)b(p_b), \quad (5)$$

where Q_b is the electric charge of b, s_W is the sine of the weak angle and $\mathcal{F}$ is the loop function given by $\mathcal{F} = \frac{3}{2} + x\sqrt{x^2 - 4x}\,\mathrm{sech}^{-1}\left(\frac{2}{\sqrt{x}}\right) + \frac{(2\,(2-3\,x+x^2)+(3\,x^2-x^3)\,\ln(x))}{2\,(x-1)}$, where $x = m^2_H/m^2_b$.

The problem of finding a bound for the Ω_{bs} parameter reduces now to solve a quadratic equation. The physical solution corresponds to that for which the allowed values for Ω_{bs} satisfy the $|A_{SM}|^2 > |A_{NP}|^2$ condition, which implies that $|\Omega_{bs}|^2 < (0.7 - 6.8) \times 10^{-3}$ for a Higgs mass in the range 115 GeV$< m_H <$ 200 GeV [1].

Journal of Physics: Conference Series **287** (2011) 012040 doi:10.1088/1742-6596/287/1/012040

4. The $B_s \to \gamma\gamma$ decay

The Hbs effective vertex induces the flavor violating process $b \to s\gamma\gamma$ at the one-loop level (see Fig. 2). The contribution to $b \to s\gamma\gamma$ occurs through two sets of Feynman diagrams, each given a finite and gauge invariant contribution. The first set of diagrams (see Fig. 2-a) includes box diagrams, reducible diagrams characterized by the one-loop $bs\gamma$ coupling and reducible diagrams composed by the one-loop $b-s$ bilinear coupling. Henceforth we will refer to this set of graphs as box-reducible diagrams. The second set of diagrams is characterized by the SM one-loop $H^*\gamma\gamma$ coupling, where H^* represents a virtual Higgs boson (see Fig. 2-b). These type of graphs will be named Higgs-reducible diagrams.

The amplitude for the $b \to s\gamma\gamma$ decay is:

$$\mathcal{M}^{\mu\nu} = \frac{\alpha\, g}{8\pi m_W} F_0\, \bar{u}_s(p_s)(\Omega_{bs}P_R + \Omega^*_{bs}P_L)\frac{k_2^\mu k_1^\nu - k_1\cdot k_2 g^{\mu\nu}}{2k_1\cdot k_2 - m_H^2 + im_H\Gamma_H} u_b(p_b), \tag{6}$$

with

$$\begin{aligned} F_0 =& \frac{8\,m_W^2}{2\,k_1\cdot k_2}\left(3 + \frac{2\,k_1\cdot k_2}{2\,m_W^2} + 6\,m_W^2\left(1 - \frac{2\,k_1\cdot k_2}{2\,m_W^2}\right) C_0(1)\right) \\ & - Q_t^2\, N_{c_t}\, \frac{8\,m_t^2}{2\,k_1\cdot k_2}\left(2 + (4\,m_t^2 - 2\,k_1\cdot k_2)\, C_0(2)\right), \end{aligned} \tag{7}$$

where $C_0(1) = C_0(0,0,2\,k_1\cdot k_2, m_W^2, m_W^2, m_W^2)$ and $C_0(2) = C_0(0,0,2\,k_1\cdot k_2, m_t^2, m_t^2, m_t^2)$ are the Passarino-Veltman scalar functions, m_t is the top quark mass, Q_t is the top quark charge and $N_{c_t} = 3$ is the color factor.

According to the static quark approximation [4], we can compute the decay width $\Gamma(B_s \to \gamma\gamma)$ starting from $\Gamma(b \to s\gamma\gamma)$, where it is assumed that the three-momenta of the b and s quarks vanish in the rest frame of the B_s meson. In this approximation, the B_s meson decays into two photons emitted with energies $m_{B_s}/2$ and the product $k_1\cdot k_2 = m_{B_s}^2/2$, where $m_{B_s} = m_b + m_s$ [1] is the B_s-meson mass. The decay width for the $B_s \to \gamma\gamma$ process arising from the new physics effects encoding in B_{NP} has the following form

$$\Gamma(B_s \to \gamma\gamma) = f_{B_s}^2 \frac{m_{B_s}^3}{16\pi}|B_{NP}|^2, \tag{8}$$

where

$$B_{NP} = \frac{\alpha^{\frac{3}{2}}\Omega_{bs}}{4\pi^{\frac{1}{2}} s_W}\frac{m_{B_s}}{m_W m_H^2} F_0. \tag{9}$$

We show in Fig. 3 the branching ratio for the $B_s \to \gamma\gamma$ process. From this figure, it can be appreciated that the contribution induced by the Higgs-reducible graphs is approximately 2 orders of magnitude larger than those generated by the box-reducible graphs in the range of a Higgs mass of 115 GeV$< m_H <$ 200 GeV.

5. Conclusions

We have estimated the Hbs coupling strength from the branching ratio for the $B \to X_s\gamma$ process. The effective parameter Ω_{bs} was bounded by using the discrepancy between the respective theoretical and experimental central values of the branching ratios. This constraint was used to bound the Higgs-mediated flavor violating $B_s \to \gamma\gamma$ decay and we found that its branching ratio is less than 10^{-8} in the Higgs mass interval ranging from 115 GeV to 200 GeV. Our results for the branching ratio are 2 orders of magnitude smaller than the current experimental bound imposed by the Belle Collaboration.

[1] As in Refs. [4], we will use the constituent mass for the strange quark $m_s = m_K = 0.497$ GeV.

Journal of Physics: Conference Series **287** (2011) 012040 doi:10.1088/1742-6596/287/1/012040

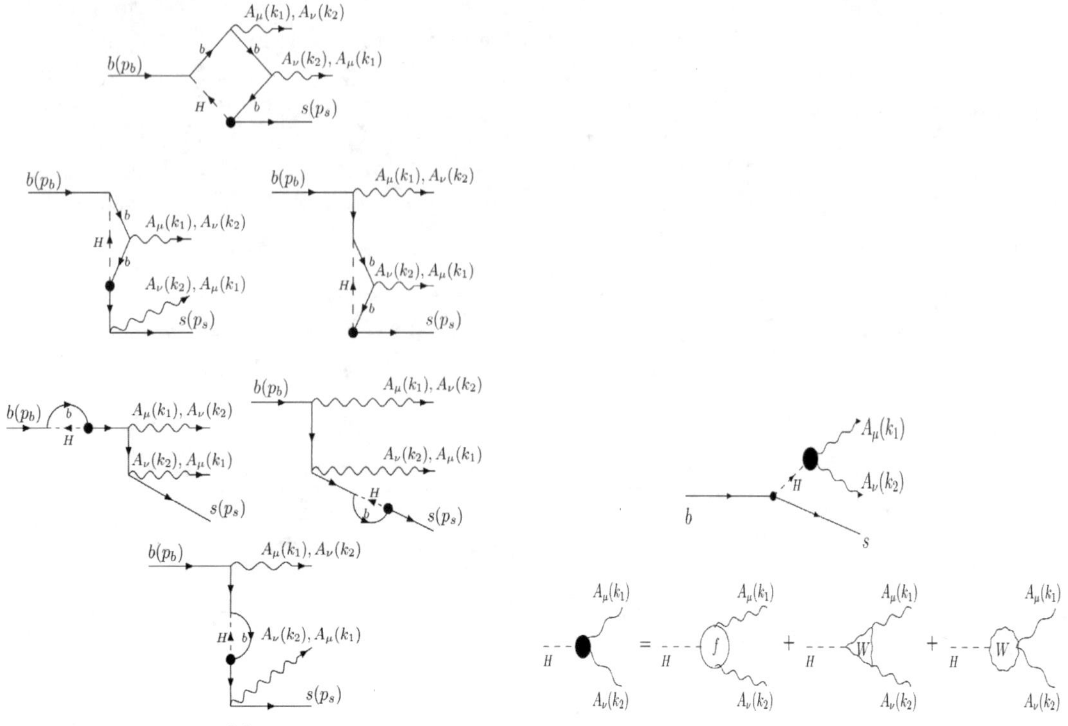

(a) (b)

Figure 2. (a) Contribution of the box and reducible diagrams to the $b \to s\gamma\gamma$ decay. (b) Contribution of the SM one-loop induced $H^*\gamma\gamma$ vertex to the $b \to s\gamma\gamma$ decay.

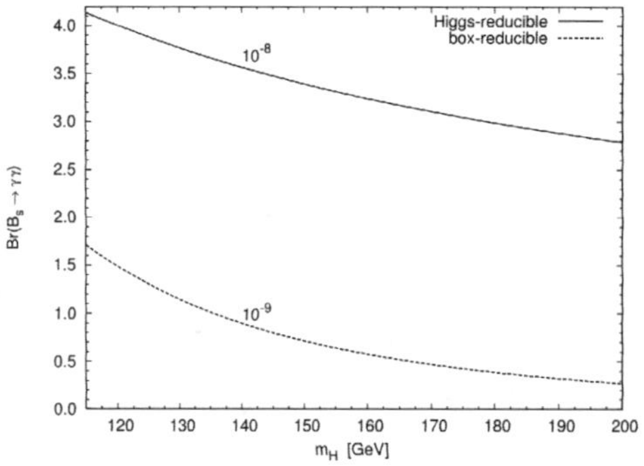

Figure 3. The branching ratio of the $B_s \to \gamma\gamma$ decay for the Higgs-reducible contribution (solid line) and box-reducible contribution (dashed line) as a function of the Higgs mass.

[1] J. I. Aranda, J. Montaño, F. Ramírez-Zavaleta, J. J. Toscano and E. S. Tututi, Phys. Rev. **D82**, 054002 (2010).

[2] B. Grinstein, R. Springer and M. B. Wise, Nucl. Phys. **B339**, 269 (1990); A. J. Buras *et al.*, Nucl. Phys. **B424**, 374 (1994).

[3] Xiao-Gang He and Lu-Hsing Tsai, JHEP **0806**, 074 (2008).

[4] S. Herrlich and J. Kalinowski, Nucl. Phys. **B381**, 501 (1992); A. Gemintern, S. Bar-Shalom and G. Eilam, Phys. Rev. **D70**, 035008 (2004).

Journal of Physics: Conference Series **287** (2011) 012041 doi:10.1088/1742-6596/287/1/012041

Quantization of Second Order Fermions

René Ángeles[1], Mauro Napsuciale[2]

Departamento de Física , Universidad de Guanajuato, Lomas del Bosque 103, Fraccionamiento Lomas del Campestre, León Guanajuato, 37150, México

E-mail: [1]`rene@fisica.ugto.mx`, [2]`mauro@fisica.ugto.mx`

Abstract. We review how second order equations for fields arise just by using projectors over Poincaré invariant subspaces. We focus in the case of fields describing massive spin $\frac{1}{2}$ particles, we propose a particular second order Lagrangian and present preliminary results in its quantization.

1. Introduction

A second order equation to describe massive spin $\frac{1}{2}$ particles was proposed in 1956 by Feynman and Gell-Mann [2] aiming to describe Fermi interactions. Further studies of second order QED and non-abelian gauge theories were done in [3]-[6]. These formalisms present some advantages in practical calculations versus the conventional approach [6].

Recently, a second order formalism for general fields was developed in [7] [8] [9] where the fundamental properties of theories for spin 1 and 3/2 and the corresponding electrodynamics were worked out. The spin 1/2 case has been recently presented in [10].

In this work we present preliminary results on the quantization of the spin 1/2 case.

2. Second Order Equations of Motion from the Poincaré Algebra

Second order equations arise naturally when one uses the Poincaré Casimirs to construct equations of motion for a field transforming under a given Homogeneous Lorentz Group (HLG) representation, to show this claim we turn to first principles. A local, Poincaré covariant, perturbative quantum theory is constructed out of fields $\psi(x)$ of the form

$$\psi(x) = \int d\lambda \{u(\lambda, x)a(\lambda) + v(\lambda, x)a^{\dagger}(\lambda)\}, \tag{1}$$

transforming as a finite dimensional representation of the HLG, where $a(\lambda)$ and $a^{\dagger}(\lambda)$ are the annihilation and creation operators of states in a unitary irreducible representation of the Poincaré group, so the quantum numbers λ are the two casimirs P^2 and spin W^2, the momentum P^{μ}, the spin projection, the discrete symmetries and internal symmetries quantum numbers. In the case of fields transforming in representations of the HLG which contain a single irreducible representation of the Poincaré group it is possible to construct the projector over subspaces with well defined mass m and spin s using the Casimir operators P^2 and W^2 as

$$\Big(\frac{P^2}{m^2}\Big)\Big(\frac{W^2}{-s(s+1)P^2}\Big)\psi = \psi, \tag{2}$$

Journal of Physics: Conference Series **287** (2011) 012041 doi:10.1088/1742-6596/287/1/012041

where $W_\alpha = \frac{1}{2}\epsilon_{\alpha\mu\nu\beta}M^{\mu\nu}P^\beta$ and therefore W^2 is quadratic in the momentum, so we can rewrite it as

$$[W^2]_{ab} = [W_\nu]_{ac}[W^\nu]_{cb} = -s(s+1)T^{(G)\alpha\beta}_{ab}P_\alpha P_\beta. \tag{3}$$

Here, the field $[\psi^{G,m,s}(x)]_a$ transforms under a representation G of the HLG and the tensor $T^{(G)\alpha\beta}_{ab}$ depends only on the generators of this representation whose internal indices are denoted by latin letters. Rearranging (2) and using (3) we get

$$\left(T^{(G)\alpha\beta}_{ab}\partial_\alpha\partial_\beta + \delta_{ab}m^2\right)\psi^{(G)}_b(x) = 0. \tag{4}$$

From this equation one can reconstruct the following Lagrangian from which the equation of motion is derived:

$$\mathcal{L} = \partial_\alpha\bar{\psi}(T^{(G)\,\alpha\beta} + T^{(A)\,\alpha\beta})\partial_\beta\psi^{(G)} - m^2\bar{\psi}\psi^{(G)}, \tag{5}$$

where $\bar{\psi}$ is some linear combination of $\psi^\dagger$ such that the action is real and Poincaré invariant, here $T^{(A)\,\alpha\beta}$ is any antisymmetric constant matrix in the HLG representation space, but this matrix plays no role for the *free theory*. Note that in the derivation above only the Poincaré algebra is used, so this equation does not have restrictions related to discrete symmetries such as parity or charge conjugation.

The method described above for obtaining equations of motion using casimirs as projectors over Poincaré subspaces forms part of the NKR formalism, actually this method has been used for the description of massive spin $\frac{3}{2}$ [7], spin 1 [8] and spin $\frac{1}{2}$ [10].

3. Second Order Fermions

Next we focus on particles with mass m and spin $\frac{1}{2}$, for the Lorentz representations $G_1 = (\frac{1}{2},0)$, $G_2 = (0,\frac{1}{2})$ couple minimally to an electromagnetic field the expresion (4) gives

$$[(i\partial_\mu - A_\mu)^2 + \vec{\sigma}\cdot(\vec{B} \pm i\vec{E})]\psi^{G_{1,2}} = m^2\psi^{G_{1,2}}, \tag{6}$$

these are just the Feynman and Gell-Mann equations [2]. The problem with these fields $\psi^{G_{1,2}}$ is that it is not possible to create a second order Lagrangian which is both *hermitian* and Poincaré scalar using only one of these two representations $\{(\frac{1}{2},0),\ (0,\frac{1}{2})\}$. To form a *hermitian and Poincaré invariant action* one should use instead the representation $G = (\frac{1}{2},0)\oplus(0,\frac{1}{2})$, then the simplest second order Lagrangian density for this representation is found to be (here $\bar{\psi} = \psi^\dagger\beta$)

$$\mathcal{L} = \frac{1}{2}\partial_\alpha\bar{\psi}(\eta^{\alpha\beta} + \frac{g}{2}[\gamma^\alpha,\gamma^\beta])\partial_\beta\psi - \frac{m^2}{2}\bar{\psi}\psi, \tag{7}$$

where we skipped the superindex G for the sake of simplicity. The details of the calculations can be found in [10].

4. Quantization of Second Order Fermions

In the literature there are two main roads for the quantization of second order theories:

- Quantize the Dirac theory and relate the Dirac fields to those obeying the second order equations (see [4] and [5]),
- Quantize a second order Poincaré invariant *non real* action [13],

$$\mathcal{L} = \partial_\mu\psi^{\dagger(\frac{1}{2},0)}\partial^\mu\psi^{(0,\frac{1}{2})} - m^2\psi^{\dagger(\frac{1}{2},0)}\psi^{(0,\frac{1}{2})} \tag{8}$$

additionally a subsidiary condition that is not contained in the Lagrangian is imposed.

Journal of Physics: Conference Series **287** (2011) 012041 doi:10.1088/1742-6596/287/1/012041

Both approaches reproduce QED results. Of course it would be nicer to quantize departing from a real action, without invoking Dirac's formalism, and using subsidiary conditions only if it is physically or mathematically necessary.

In what follows we present our preliminary results quantizing the *hermitian and Poincaré covariant theory given by* (7). If we take ψ and $\psi^\dagger$ as the independent Grassmann coordinates their associated moments are

$$\pi_\psi = -\frac{1}{2}(\gamma^\mu \partial_\mu \psi)^\dagger, \qquad \pi_{\psi^\dagger} = \frac{1}{2}(\gamma^\mu \partial_\mu \psi), \tag{9}$$

here we are using the convention of right derivatives like in [11], then the Lagrangian doesn't have constraints and we can immediately write down the Hamiltonian

$$H = \int d^3x \{\pi_\psi \gamma^0 \gamma^i \partial_i \psi - (\gamma^i \partial_i \psi)^\dagger \gamma^0 \pi_{\psi^\dagger} - 2\pi_\psi \gamma^0 \pi_{\psi^\dagger} + \frac{1}{2} m^2 \psi^\dagger \gamma^0 \psi)\}, \tag{10}$$

this would be that of Dirac if we set $\pi_\psi = i\frac{m}{2}\psi^\dagger$!

The next step in the canonical quantization is to impose the canonical anticommutation relations, the non zero relations are

$$[\psi(\vec{x},t), \pi_\psi(\vec{y},t)]_+ = [\psi^\dagger(\vec{x},t), \pi_{\psi^\dagger}(\vec{y},t)]_+ = i\delta(\vec{x}-\vec{y}), \tag{11}$$

these non vanishing anticommutators coincide with those of the Dirac theory. Since ψ and $\psi^\dagger$ are independent coordinates they must anticommute at equal times, i.e.

$$[\psi(\vec{x},t), \psi^\dagger(\vec{y},t)]_+ = 0, \tag{12}$$

this relation is not consistent since it can be proved that a massive spin $\frac{1}{2}$ field transforming under the representation $(\frac{1}{2},0)\oplus(0,\frac{1}{2})$ cannot anti-commute with its adjoint at equal times.

We have proved this claim using the Weinberg construction of fields [1], the anticommutator of the corresponding massive spin $\frac{1}{2}$ field transforming under the $(\frac{1}{2},0)\oplus(0,\frac{1}{2})$ representation with its adjoint is never zero, here Weinberg uses the Parity operator but the result holds without imposing this symmetry.

Then, how can the theory given by (7) be quantized consistently? we are studying the following possibilities:

- To consider ψ and $\psi^\dagger$ as independent variables and not establishing its relation is causing the problem in the quantization. In the Dirac formalism this works because this is a constrained theory [9].
- Another possibility is that in order to quantize the theory a constraint is necessary. In fact if we impose $\pi_\psi = i\frac{m}{2}\psi^\dagger$ as a second class constraint in the Hamiltonian, the resulting theory is equivalent to that of Dirac, but we haven't proved that this condition is indeed necessary, unique, or if there exist some condition less restrictive.

The last possibility is interesting from a theoretical point of view, because if a constraint is needed in order to quantize the theory then the necessity of this constraint would imply that a consistent quantization of massive spin $\frac{1}{2}$ needs additional symmetry (remember that constraints arises naturally from symmetries) apart from the Poincaré algebra from which these theories have been shown to arise.

This constraint could be related with the discrete symmetries of parity or charge conjugation, actually this relation appear already in the conventional formalism, the Dirac equation is closely related to parity (this relation is presented in [10]), also when passing from Dirac to Majorana charge conjugation symmetry is needed *even in the free theory.*

5. Conclusions

We have shown how the second order theories arise from the Poincaré symmetry of massive spin $\frac{1}{2}$ particles. Then we performed the canonical quantization of the theory (7), we can obtain exactly the Dirac theory adding a constraint, but it is still not clear if this condition is indeed necessary. More work is needed in order to understand the general case.

Acknowledgments

This work was partially supported by the RedFAE project of Conacyt, we thank Selim Gomez and Carlos Vaquera for useful discussions.

References

[1] S. Weinberg, The Quantum Theory of Felds, Vol. (Cambridge University Press, 1995), Chapter 5.
[2] R.P. Feynman and M. Gell-Mann, Phys. Rev. 109(1958)193.
[3] Laurie M. Brown Phys.Rev.111:957-964,1958.
[4] Levere C. Hostler, J. Math. Phys. 26, 1348 (1985).
[5] Levere C. Hostler, J. Math. Phys. 27, 2423 (1986).
[6] Morgan, A. G. Physics Letters B, v. 351, iss. 1-3, p. 249-256.
[7] M. Napsuciale, M. Kirchbach, S. Rodriguez, Eur. Phys. J. A29 (2006) 289-306. arXiv:hep-ph/0407179v3.
[8] M. Napsuciale , S. Rodriguez , E.G. Delgado-Acosta , M. Kirchbach, Phys. Rev. D77: 014009, 2008.
[9] E. G. Delgado-Acosta and M. Napsuciale, Phys. Rev. D **80**, 054002 (2009) [arXiv:0907.1124 [hep-th]].
[10] E.G. Delgado-Acosta, Mauro Napsuciale, Simon Rodriguez, Second order formalism for spin 1/2 fermions and Compton scattering. arXiv:1012.4130
[11] K. Sundermeyer, Constrained Dynamics (LNP0169, Springer, 1982).
[12] R. E. Peierls, Proc. Roy. Soc. Lond. A 214 (1952) 143-157.
[13] M. Tonin, Nuovo Cimento, 14, 1108, 1958.

Journal of Physics: Conference Series **287** (2011) 012042 doi:10.1088/1742-6596/287/1/012042

Gravity localization in non-minimally coupled scalar thick braneworlds with a Gauss-Bonnet term

D. Malagón–Morejón[1], A. Herrera–Aguilar[2] and I. Quiros[3]

[(1,2)]Instituto de Física y Matemáticas, Universidad Michoacana de San Nicolás de Hidalgo. Edificio C–3, Ciudad Universitaria, C.P. 58040, Morelia, Michoacán, México

[3]División de Ciencias e Ingeniería de la Universidad de Guanajuato, A.P. 150, 37150, León, Guanajuato, México

E-mail: malagon@ifm.umich.mx[1], herrera@ifm.umich.mx[2], iquiros@fisica.ugto.mx[3]

Abstract. We consider a warped five-dimensional thick braneworld with a four-dimensional Poincaré invariant space-time in the framework of scalar matter non-minimally coupled to gravity plus a Gauss-Bonnet term in the bulk. Scalar field and higher curvature corrections to the background equations as well as the perturbed equations are shown. A relationship between 4-dimensional and 5-dimensional Planck masses is studied in general terms. By imposing finiteness of the 4-dimensional Planck mass and regularity of the geometry, the localization properties of the tensor modes of the first order perturbed geometry are analized for an important class of solutions motivated by models with scalar fields which are minimally coupled to gravity. In order to study the gravity localization properties for this model, the normalizability condition for the lowest level of the tensor fluctuations is analized. We see that for the class of solutions examined, gravity in 4 dimensions is recovered if the curvature invariants are regular and Planck masses are finite.

1. Introduction

In constrast to Kaluza-Klein picture where extra dimensions are compactified, the braneworld models [1, 2, 3] allow an infinite extra dimension. These models requires that Standard Model fields are trapped on a 4-dimensional surface, called a 3-brane. Since gravity can propagate through all dimensions, the first important question concerning to braneworld models is to check whether they give back standard 4-dimensional gravity on the brane.

The thin braneworld models [2] have the disadvantage that the curvature is singular at the location of the wall. Two ways of smoothing out the singular character of the brane configurations are: by replacing the delta functions in the action of the system by a self-interacting scalar field [4, 5, 6, 7] or by modifying the Einstein-Hilbert action itself [8, 9].

In this work we study a 5-dimensional thick braneworld modeled by a smooth scalar domain wall non-minimally coupled to gravity with a Gauss-Bonnet term on the bulk, in the framework of conformally flat geometries. We analyze for an important class of solutions the relationship among: localization properties of the zero tensor mode, finiteness of 4-dimensional Planck mass and smoothness of the geometry.

2. The model

Consider a thick braneworld described by the following 5-dimensional action

$$S=\int d^5x\sqrt{|g|}\Big\{-\tfrac{L(\varphi)R}{2\kappa}-\alpha'\mathcal{R}^2_{\rm GB}+\tfrac{1}{2}(\nabla\varphi)^2-V(\varphi)\Big\},\ \text{where } A,B=0,....,3,5. \tag{1}$$

The constant $\alpha'>0$ and $\kappa\simeq 1/M^3$, where M is the 5-dimensional Planck mass, $L(\varphi)$ is the coupling between scalar field φ and gravity and the term $\mathcal{R}^2_{\rm GB}$ is the Gauss-Bonnet term in 5 dimensions

$$\mathcal{R}^2_{\rm GB}=R^{ABCD}R_{ABCD}-4R^{AB}R_{AB}+R^2. \tag{2}$$

Einstein equations for this model are

$$L\,R_{AB}=\kappa\,\tau_{AB}+\nabla_A\nabla_B L+\tfrac{1}{3}\,g_{AB}\,\Box L-\epsilon\,\mathcal{Q}_{AB},\quad \text{where}\quad \epsilon=2\alpha'\kappa. \tag{3}$$

The reduced energy-momentum tensor τ_{AB} corresponds to the scalar matter content on the bulk and takes the form:

$$\tau_{AB}=\partial_A\varphi\partial_B\varphi-\frac{2}{3}\,g_{AB}\,V(\varphi), \tag{4}$$

on the other hand, the term $\mathcal{Q}_{AB}$ is called Lanczos tensor and represents the corrections of Gauss-Bonnet term to Eintein equations, which can be written in the form

$$\mathcal{Q}_{AB}=\tfrac{1}{3}\,g_{AB}\,\mathcal{R}^2_{\rm GB}-2\,R\,R_{AB}+4\,R_{AC}\,R^C{}_B+4R^{CD}R_{ACBD}-2\,R_{ACDE}\,R_B{}^{CDE}.$$

The field equation for the scalar field is

$$\Box\varphi+\frac{1}{2\kappa}\,R\,L_\varphi+\frac{\partial V}{\partial\varphi}=0,\quad \text{where}\quad L_\varphi=\frac{dL}{d\varphi}. \tag{5}$$

As in [10] let us consider a warped metric in conformally flat coordinates

$$ds^2=a^2(w)[\eta_{\mu\nu}dx^\mu dx^\nu-dw^2], \tag{6}$$

the variable w is the bulk coordinate and $\eta_{\mu\nu}$ is the 4-dimensional Minkowski metric. In our work we consider the simplest case where the scalar field depends only on the bulk coordinate w and $a(w)=a(-w)$.

3. Localization of gravity

In order to study the localization of gravity let us consider the linear tensor fluctuations of the background metric (6)

$$ds^2_p=[a^2(w)\eta_{AB}+H_{AB}]dx^A dx^B,\quad \text{where}\quad H_{AB}=a^2(w)\begin{pmatrix}2h_{\mu\nu} & 0\\ 0 & 0\end{pmatrix}. \tag{7}$$

The tensor $h_{\mu\nu}$ is a divergence-less and trace-less rank-two tensor in the 4-dimensional Poincaré invariant space-time [10, 11], the tensor fluctuation equation is

$$qh_\mu^{\nu\,\prime\prime}+(3\mathcal{H}q+q')h_\mu^{\nu\,\prime}-\left[q+\frac{(q-L)'}{2\mathcal{H}}\right]\Box^\eta h^\nu_\mu=0, \tag{8}$$

where $'=\frac{d}{dw}$, $\mathcal{H}=\frac{a'}{a}$ and $q=L-\frac{4\epsilon}{a^2}\mathcal{H}^2$. If we redefine the field $h_{\mu\nu}$ as $\Psi_{\mu\nu}=\sqrt{s(w)}h_{\mu\nu}$ the equation (8) changes to:

$$\Psi''_{\mu\nu}-\frac{(\sqrt{s})''}{\sqrt{s}}\Psi_{\mu\nu}-\frac{r}{s}\Box^\eta\Psi_{\mu\nu}=0, \tag{9}$$

Journal of Physics: Conference Series **287** (2011) 012042 doi:10.1088/1742-6596/287/1/012042

where $s(w) = a^3 q$ and $r(w) = a^3\left(q + \frac{(q-L)'}{2\mathcal{H}}\right)$.

With the aim to study the spectrum of the metric fluctuations we propose a separation of variables given by $\Psi_{\mu\nu} = \psi(w)\chi_{\mu\nu}(x)$, then equation (9) is separated in two equations:

$$\Box^\eta \chi_{\mu\nu} + m^2 \chi_{\mu\nu} = 0, \quad (10)$$

$$\psi'' - \frac{(\sqrt{s})''}{\sqrt{s}}\psi + m^2 \frac{r}{s}\psi = 0. \quad (11)$$

Equation (10) describes the 4-dimensional massive tensorial modes and (11) defines the localization properties of $\Psi_{\mu\nu}$. On the other hand, a sector of gravity is localized on the brane if the associated zero mode fluctuation is normalizable, in other words, the norm for the tensor zero mode fluctuations has to be finite

$$\langle \psi_0 | \psi_0 \rangle = \int_{-\infty}^{\infty} \frac{r}{s}\, \psi_0^2\, dw < \infty. \quad (12)$$

The lowest mass eigenstate of (11) is $\psi_0 = \sqrt{s}$, then we obtain:

$$\langle \psi_0 | \psi_0 \rangle = \int_{-\infty}^{\infty} a^3\, q\, dw - 4\,\epsilon\, [a']_{-\infty}^{\infty} + 8\,\epsilon \int_{-\infty}^{\infty} \frac{a'^2}{a}\, dw. \quad (13)$$

It is interesting to ask: what is the relationship between the 4-dimensional Planck mass and the normalizability condition (12)?

4. Planck masses

For a space-time that take the following form:

$$ds^2 = a^2(w)[g_{\mu\nu}(x) dx^\mu dx^\nu - dw^2]. \quad (14)$$

where $g_{\mu\nu}(x)$ is an arbitrary 4-dimensional metric, the relationship between 4-dimensional Planck mass and 5-dimensional Planck mass can be obtained if we make a dimensional reduction by integrating (1) with respect to the w coordinate. Thus, the 5-dimensional theory (1) is reduced to a 4-dimensional Einstein-Hilbert theory plus the corrections that comes from the scalar matter and higher curvature terms of the bulk

$$S_4 \simeq M_{\rm Pl}^2 \int d^4x \sqrt{|g_4|} R_4 + \cdots, \quad (15)$$

yielding the following expression for $M_{\rm Pl}$

$$M_{\rm Pl}^2 \simeq M^3 \int_{-\infty}^{\infty} a^3(w) \left[L(\varphi) + \frac{4\epsilon}{a^2}(\mathcal{H}^2 + 2\mathcal{H}') \right] dw = M^3 \int_{-\infty}^{\infty} a^3(w)\, q\, dw + 8\, M^3\, \epsilon\, [a']_{-\infty}^{\infty}.$$

Planck mass $M_{\rm Pl}$ is closely related with $\langle \psi_0 | \psi_0 \rangle$, but a extra term $\int_{-\infty}^{\infty} \frac{a'^2}{a}\, dw$ arises in (13). A clearer relation between $M_{\rm Pl}$ and $\langle \psi_0 | \psi_0 \rangle$ can be obtained if we analyze the smoothness conditions of the curvature invariants.

5. Smoothness of geometry

A realistic thick braneworld model should not have singularities in the geometry. Let us assume that the curvature invariants and warp factor a are regular in the whole space-time and consider the class of solutions where asymptotically

$$a(w \to \infty) \simeq \frac{1}{w^\gamma}. \quad (16)$$

For (6) the curvature invariants are:

$$\begin{aligned} R &= \frac{4}{a^2}\left[2\mathcal{H}' + 3\mathcal{H}^2\right] \\ R^{AB}R_{AB} &= \frac{4}{a^4}\left[5\mathcal{H}'^2 + 9\mathcal{H}^4 + 6\mathcal{H}'\mathcal{H}^2\right], \\ R^{ABCD}R_{ABCD} &= \frac{4}{a^4}\left[4\mathcal{H}'^2 + 6\mathcal{H}^4\right]. \end{aligned}$$

Calculating the asymptotic behavior of the previous expressions at $w \to \infty$ we get:

$$R \simeq w^{2(\gamma-1)}, \quad R^{AB}R_{AB} \simeq R^{ABCD}R_{ABCD} \simeq w^{4(\gamma-1)}.$$

Thus, by imposing smoothness at infinity, the values of the exponent γ are restricted to the interval $0 < \gamma \leq 1$. Under this restriction for γ we obtain

$$M_P^2 \simeq M^3 \int_{w_\infty}^{\infty} a^3\, L\, dw + \cdots \qquad \text{and} \qquad \langle\psi_0|\psi_0\rangle \simeq \int_{w_\infty}^{\infty} a^3\, L\, dw + \cdots,$$

where $\cdots$ denotes finite terms and the interval (w_∞, ∞) formally represents the range where the approximation (16) is valid.

6. Conclusions

We have presented a thick braneworld model where the matter field is coupled non-minimally to the Einstein-Hilbert term, in addition to this, there is a Gauss-Bonnet term on the bulk. We compute the 4-dimensional Planck mass in terms of 5-dimensional Planck mass and quantities related with the matter and the geometry. In contrast with theories where the matter field is minimally coupled to a Einstein-Hilbert term, in our model M_{Pl} depend explicitly on the matter content of the bulk. A relation among smoothness of geometry, finiteness of 4-dimensional Planck mass and localization of tensorial modes was studied for an important class of solutions. We obtained that if the geometry is regular, finite 4-dimensional Planck mass implies localization of gravity on the brane.

7. Acknowledgements

This research was supported by grants CIC–4.16 and CONACYT 60060–J. DMM acknowledge a PhD grant from CONACYT. AHA and IQ thanks SNI for support.

References

[1] Gogberashvili M 2002 *Int. J. Mod. Phys.* D **11** 1635-38
[2] Randall L and Sundrum R 1999 *Phys. Rev. Lett.* **83** 3370-73
[3] Randall L and Sundrum R 1999 *Phys. Rev. Lett.* **83** 4690-93
[4] Gremm M 2000 *Phys. Lett.* B **478** 434-38
[5] DeWolfe O, Freedman D Z, Gubser S S and Karch A 2000 *Phys. Rev.* D **62** 046008
[6] Barbosa-Cendejas N, Herrera-Aguilar A, Reyes Santos M A and Schubert C 2008 *Phys. Rev.* D **77** 126013
[7] Barbosa-Cendejas N, Herrera-Aguilar A, Nucamendi U and Quiros I 2007 Mass hierarchy and mass gap on thick branes with Poincare symmetry *Preprint* arXiv:0712.3098 [hep-th]
[8] Giovannini M 2001 *Phys. Rev.* D **64** 124004
[9] Corradini O and Kakushadze Z 2000 *Phys. Lett.* B **494** 302-10
[10] Giovannini M 2001 *Phys. Rev.* D **64** 064023
[11] Csaki C, Erlich J, Hollowood T J, Shirman Y 2000 *Nucl. Phys.* B **581** 309-38

Journal of Physics: Conference Series **287** (2011) 012043 doi:10.1088/1742-6596/287/1/012043

Gauge systems and functions, hermitian operators and clocks as conjugate functions for the constraints

Vladimir Cuesta [1,†], José David Vergara [1,††] and Merced Montesinos [2,◇]

[1] Instituto de Ciencias Nucleares, Universidad Nacional Autónoma de México, 70-543, Ciudad de México, México,

[2] Departamento de Física, Centro de Investigación y de Estudios Avanzados del Instituto Politécnico Nacional, Instituto Politécnico Nacional 2508, San Pedro Zacatenco, 07360, Gustavo A. Madero, Ciudad de México, México

E-mail: [1,†] vladimir.cuesta@nucleares.unam.mx
[1,††] vergara@nucleares.unam.mx
[2,◇] merced@fis.cinvestav.mx

Abstract. We work with gauge systems and using gauge invariant functions we study its quantum counterpart and we find if all these operators are self adjoint or not. Our study is divided in two cases, when we choose clock or clocks that its Poisson brackets with the set of constraints is one or it is different to one. We show some transition amplitudes.

1. Introduction

The study of gauge systems or systems with first class constraints is an outstanding branch of theoretical physics and mathematics, its importance lies in the fact that a major number of physical systems have first class constraints, including parametric systems, quantum electrodynamics, the standard model, general relativity and a lot of systems with a finite number of degrees of freedom and so on. A special case of these theories are the covariant systems. In this case the canonical Hamiltonian vanishes and the system is invariant under the reparametrization of the coordinates. In consequence the time is not a priori defined.

Let us consider a phase space with coordinates $(q_1, \ldots, q_n, p^1, \ldots, p^n)$ and first class constraints γ_a, where the set of constraints obeys,

$$\{\gamma_a, \gamma_b\} = C_{ab}{}^c \gamma_c, \tag{1}$$

and $a = 1, \ldots, m$ and m is the number of first class constraints. We take the action principle,

$$S[q, p, \lambda^a] = \int_{\sigma_1}^{\sigma_2} \left(p_i \dot{q}^i - \lambda^a \gamma_a \right) d\sigma, \tag{2}$$

then if we vary the action we can obtain the equations of motion for our system. Now, we define a gauge invariant function R or a complete observable as a phase space function such that the Poisson brackets with the full set of constraints is zero, i. e. $\{R, \gamma_a\} = 0$. Then, all the phase space functions such that its Poisson bracket with all the constraints is not zero are called partial observables. If we take m partial observables as clocks $T_1, \ldots, T_m$, and a partial observable f, then we can find a complete observable F or a gauge invariant function (see [1] and [2] for instance). Now, our aim is to consider what happen to the quantum level. In that case we study two problems, the first problem is to define our set of clocks all along the real line and the second is the self adjoint character for the $\hat{F}$ operator, where F is a complete observable. If $Dim[Ker(\hat{F}+\imath)] = Dim[Ker(\hat{F}-\imath)]$, then $\hat{F}$ is self adjoint; if $Dim[Ker(\hat{F}+\imath)] \neq Dim[Ker(\hat{F}-\imath)]$, then $\hat{F}$ is not self adjoint and it has not self adjoint extensions (see [3] for instance). In the covariant systems we have two problems first for arbitrary defined clock the

$$Det\left[\begin{pmatrix} (\{T_i,T_j\}) & (\{T_i,\gamma_j\}) \\ (\{\gamma_i,T_j\}) & (\{D_i,D_j\}) \end{pmatrix}\right], \tag{3}$$

is zero for some regions at the phase space and then clocks are not globally well defined, the second problem is that $\hat{F}$ is not self adjoint, if we find one or two of the previous problems we propose that the clocks must be selected in such a way that $\{T_i,\gamma_j\} = \delta_{ij}$ (see [4] for instance) and we show in this work that this form of selection for the time solve the problems for several systems.

2. Non relativistic parametric free particle

In this case, the constraint for the one dimensional non-relativistic free particle is,

$$D = p_0 + \frac{p^2}{2m} = 0, \tag{4}$$

and the phase space coordinates are (x^0, x, p, p_0).

2.1. *Example 1*

For the first example, we can take the T clock like,

$$T = x^0 - ax, \tag{5}$$

and the partial observable like $f = x$, if we make that we will obtain the complete observable:

$$F = \frac{q + \frac{p}{m}\tau}{1 - a\frac{p}{m}}, \quad q = x - \frac{px^0}{m}. \tag{6}$$

However, we find a first problem for the present system, in this case we have

$$\begin{aligned} Det\left[\begin{pmatrix} \{T,T\} & \{T,D\} \\ \{D,T\} & \{D,D\} \end{pmatrix}\right] &= \left[\begin{pmatrix} 0 & 1-\frac{ap}{m} \\ \frac{ap}{m}-1 & 0 \end{pmatrix}\right] \\ &= \left(1-\frac{ap}{m}\right)^2, \end{aligned} \tag{7}$$

and *our choice for the time is not defined all along the real line.*

Now, we consider the pair of equations $(\hat{F}+\imath)\psi_+ = 0$ and $(\hat{F}-\imath)\psi_- = 0$ to determine if $\hat{F}$ is a self adjoint operator or it is not. In this case the solutions are:

$$\psi_\pm = r_\pm\sqrt{1-\frac{ap}{m}}\exp\left[\frac{\imath p^2\tau}{2m} \mp \left(p - \frac{ap^2}{2m}\right)\right], \tag{8}$$

where r_+ and r_- are constants, then,

$$|\psi_\pm|^2 = |r_\pm|^2|1-\frac{ap}{m}|\exp\left[\mp\left(2p - \frac{ap^2}{m}\right)\right]; \tag{9}$$

ψ_+ is not square integrable and ψ_- is square integrable, then $Dim[Ker(\hat{F}+\imath)] = 0$, $Dim[Ker(\hat{F}+\imath)] = 1$ and we have found a second problem: *the $\hat{F}$ operator is not self adjoint* (see [3] for instance).

Journal of Physics: Conference Series **287** (2011) 012043 doi:10.1088/1742-6596/287/1/012043

2.2. Example 2

We will study our previous system and we correct the problems of the previous example, we will use T such that $\{T, D\}$=1 (see [4] for instance) and the general solution for T will be:

$$T = \frac{mx}{p} + f\left(p, x^0 - \frac{mx}{p}, p_0\right), \tag{10}$$

then

$$Det\left[\begin{pmatrix} \{T,T\} & \{T,D\} \\ \{D,T\} & \{D,D\} \end{pmatrix}\right] = \left[\begin{pmatrix} 0 & 1 \\ -1 & 0 \end{pmatrix}\right] = 1,$$

and our T clock is well defined all along the real line.

To simplify we consider the clock,

$$T = \frac{mx}{p} + \alpha p + \beta\left(x^0 - \frac{mx}{p}\right) + \rho p_0, \tag{11}$$

with $\{T, D\} = 1$ and for the partial observable $f = x$ we obtain,

$$F = \frac{p\tau}{m} - \frac{\alpha p^2}{m} + \beta x - \frac{\beta p x^0}{m} - \frac{\rho p_0 p}{m}. \tag{12}$$

To determine the self adjointness for $\hat{F}$ we must find $Dim[Ker(\hat{F}+\imath)]$, $Dim[Ker(\hat{F}-\imath)]$ and compare our results, with the previous purpose we consider $(\hat{F}+\imath)\psi_+ = 0$ and $(\hat{F}-\imath)\psi_- = 0$. We take,

$$\psi_\pm(p, p_0) = \delta\left(p_0 + \frac{p^2}{2m}\right), \tag{13}$$

and the solutions for $(\hat{F} \pm \imath)\psi_\pm = 0$ are:

$$\begin{aligned} g_\pm &= r_\pm \exp\left(\mp\frac{p}{\beta} + \imath\frac{p^2\tau}{2m\beta} - \imath\frac{\alpha p^3}{3m\beta} + \imath\frac{\rho p^4}{8m^2\beta}\right. \\ &- \left.\imath\frac{p^2\rho(2mp_0 + p^2)}{4m^2\beta}\right), \end{aligned} \tag{14}$$

in this case we define,

$$\tilde{g}(p) = g\left(p, -\frac{p^2}{2m}\right) \tag{15}$$

and,

$$\langle\psi_1|\psi_2\rangle_{Fis} = \int_{-\infty}^{\infty} dp\tilde{g}_1^*(p)\tilde{g}_2(p), \tag{16}$$

$\hat{F}$ *is self adjoint* and we have not the second problem of the first example.

Now, we take the eigenvalue equation $\hat{X}\psi = x_1\psi$, then we take:

$$\psi(p, p_0) = \delta\left(p_0 + \frac{p^2}{2m}\right) g(p) \tag{17}$$

and,

$$g(p) = r_1 \exp\left(\frac{\imath\rho p^4}{8\beta m^2} - \frac{\imath\alpha p^3}{3\beta m} + \frac{\imath p^2\tau}{2m\beta} - \frac{\imath x_1 p}{\beta}\right), \tag{18}$$

using the results above we obtain:

$$\langle\psi_{x_1',\tau'}|\psi_{x,\tau}\rangle = |r_1|^2\sqrt{\frac{2m\pi\beta}{\imath(\tau-\tau')}}\exp\left(\frac{\imath m(x_1'-x_1)^2}{2\beta(\tau-\tau')}\right). \tag{19}$$

and in consequence we obtain the correct amplitude for the free particle (see [3] for instance).

3. Two constraints

Now, we will extend our results for covariant systems with more constraints,

$$\begin{aligned} D_1 &= \frac{1}{2}[-(p_1)^2 + (p_2)^2 + (p_3)^2], \\ D_2 &= -\frac{1}{2}[q_1 p_1 + q_2 p_2 + q_3 p_3], \end{aligned} \tag{20}$$

the phase space coordinates are $(q_1, q_2, q_3, p_1, p_2, p_3)$ and,

$$\{D_1, D_2\} = D_1. \tag{21}$$

We take similar restrictions like our previous example, in this way we define clocks all along the real line and the quantum operator associated with a complete observable is self adjoint, with the restrictions:

$$\{T_1, D_1\} = 1, \quad \{T_2, D_2\} = 1, \quad \{T_2, D_1\} = 0, \tag{22}$$

we have,

$$Det\left[\begin{pmatrix} \{T_1,T_1\} & \{T_1,T_2\} & \{T_1,D_1\} & \{T_1,D_2\} \\ \{T_2,T_1\} & \{T_2,T_2\} & \{T_2,D_1\} & \{T_2,D_2\} \\ \{D_1,T_1\} & \{D_1,T_2\} & \{D_1,D_1\} & \{D_1,D_2\} \\ \{D_2,T_1\} & \{D_2,T_2\} & \{D_2,D_1\} & \{D_2,D_2\} \end{pmatrix}\right] = 1,$$

and the general solutions for T_1 and T_2 are:

$$\begin{aligned} T_1 &= -\frac{q_1}{p_1} + f\left(p_1, p_2, q_2 + \frac{q_1 p_2}{p_1}, p_3, q_3 + \frac{q_1 p_3}{p_1}\right), \\ T_2 &= 2\ln(p_1) + g\left(\frac{p_2}{p_1}, \frac{p_1}{p_2}(p_1 q_2 + q_1 p_2), \frac{p_3}{p_1}, \right. \\ &\quad \left. \frac{p_1}{p_2}(-p_3 q_2 + p_2 q_3)\right), \end{aligned} \tag{23}$$

and our clocks are defined all along the real line.

3.1. Example 1

We can take as clocks,

$$T_1 = -\frac{q_1}{p_1}, \quad T_2 = 2\ln(p_1), \tag{24}$$

and $f = q_2$ as partial observable, then the complete observable will be:

$$F = (q_1 p_2 + q_2 p_1)\exp\left(-\frac{\tau_2}{2}\right) + \frac{p_2}{p_1}\tau_2 \exp\left(\frac{\tau_2}{2}\right), \tag{25}$$

we consider $\left(\hat{F} \pm \imath\right)\psi_\pm = 0$ and we obtain the solutions:

$$\psi_\pm = p_1^{\imath \exp(\tau_2)\tau_1}(p_1 + p_2)^{\mp \exp\left(\frac{\tau_2}{2}\right)}\left(p_2^2 - p_1^2\right) g_\pm(p_3), \tag{26}$$

then we have,

$$|\psi_\pm|^2 = |p_1 + p_2|^{\mp 2\exp\left(\frac{\tau_2}{2}\right)}\left(p_2^2 - p_1^2\right)^2 |g_\pm(p_3)|^2, \tag{27}$$

and we can choose $g_\pm(p_3)$ such that ψ_+ and ψ_- are square integrable; $Dim[Ker(\hat{F} + \imath)] = 1$, $Dim[Ker(\hat{F} - \imath)] = 1$ and $\hat{F}$ will be self adjoint.

3.2. Example 2

We use now,

$$T_1 = -\frac{q_1}{p_1}, \quad T_2 = \ln(p_2), \tag{28}$$

as clocks and $f = q_1$ like partial observable, then the complete observable will be:

$$F = -\frac{p_1}{p_2}\tau_1 \exp\left(\frac{\tau_2}{2}\right), \tag{29}$$

we must solve the equations $\left(\hat{F} \pm \imath\right)\psi_\pm = 0$ to determine if $\hat{F}$ is self adjoint or it is not, we find the solutions:

$$\psi_\pm = \exp\left[\mp\frac{q_1 p_2}{\tau_1}\exp\left(-\frac{\tau_2}{2}\right)\right] g_\pm(p_3), \tag{30}$$

and later we obtain,

$$|\psi_\pm|^2 = \exp\left[\mp\frac{2q_1 p_2}{\tau_1}\exp\left(-\frac{\tau_2}{2}\right)\right] |g_\pm(p_3)|^2. \tag{31}$$

We can choose $g_\pm(p_3)$ such that $Dim(Ker(\hat{F}+\imath)) = Dim(Ker(\hat{F}-\imath)) = 0$ and $\hat{F}$ will be self adjoint.

4. Conclusions and perspectives

To conclude our work we make the following discussion. In the first example, for the one dimensional non-relativistic free particle we found two problems: the first problem was that,

$$Det\left[\begin{pmatrix} \{T,T\} & \{T,D\} \\ \{D,T\} & \{D,D\} \end{pmatrix}\right] = 0,$$

for a selection for T and our clock was not defined all along the real line, the second problem was that the operator associated with our complete observable F was not self adjoint, we correct both problems when we choose our clock T like a phase space function such that the Poisson bracket with the constraint is one $\{T, D\} = 1$. Now, in this case,

$$Det\left[\begin{pmatrix} \{T,T\} & \{T,D\} \\ \{D,T\} & \{D,D\} \end{pmatrix}\right] = 1,$$

and our clock is well defined all along the real line and our operator associated with our complete observable F is self adjoint.

For the case of two constraints, we begin with clocks that are conjugate to the set of constraints and we did not find problems with the operators associated with the complete observable.

For the model with two constraints we studied some partial observables with different selections for the clocks in such a way that,

$$Det\left[\begin{pmatrix} (\{T_i,T_j\}) & (\{T_i,D_j\}) \\ (\{D_i,T_j\}) & (\{D_i,D_j\}) \end{pmatrix}\right] = 1,$$

and we found that $\hat{F}$ is self adjoint, where F is a complete observable constructed for a partial observable f and the clocks are globally well defined. However, with a different choice for the previous determinant, we could have the problem that our clocks are not defined all along the real line and the problem to the self adjointness for the operator associated with a complete observable could be present.

The following general study can be done: if you have a phase space with coordinates $(q_1, \ldots, q_n, p^1, \ldots, p^n)$ and first class constraints γ_a, in such a way that,

$$\{\gamma_a, \gamma_b\} = C_{ab}{}^c\gamma_c, \tag{32}$$

Journal of Physics: Conference Series **287** (2011) 012043 doi:10.1088/1742-6596/287/1/012043

where $a = 1, \ldots, m$ and m is the number of first class constraints. Then, we can choose m cloks in such a way that $\{T_i, \gamma_j\} = \delta_{ij}$ and a first problem is to determine if,

$$Det\left[\begin{pmatrix} (\{T_i, T_j\}) & (\{T_i, \gamma_j\}) \\ (\{\gamma_i, T_j\}) & (\{D_i, D_j\}) \end{pmatrix}\right] \neq 0, \tag{33}$$

for all the values of T_i, if you have that, the clocks are defined all along the real line.

A second general problem with our previous conditions is to determine if the operator $\hat{F}$ associated with the complete observable F is self adjoint, where F is obtained of a partial observable f. If our method fails in some cases, we must determine these reasons in future studies.

The definition of clocks and self adjoint operators associated with complete observables or gauge functions is open for complex systems, including general relativity.

References

[1] B. Dittrich, *Partial and complete observables for Hamiltonian constrained systems*, Gen. Rel. Grav. **39** (2007) 1891-1927.

[2] B. Dittrich and T. Thiemann, *Are the spectra of geometrical operators in loop quantum gravity really discrete?*, J. Math. Phys. **50**, 012503 (2009).

[3] R. Gambini and R. A. Porto, *Relational time in generally covariant quantum systems: Four Models*, Phys. Rev D, **63**, 105014, (2001).

[4] F. Calogero and F. Leyvraz, *General technique to produce isochronous Hamiltonians*, J. Phys. A: Math. Theor. **40**, (2007), 12931-12944.

$Z'tc$ coupling from $D^0 - \overline{D^0}$ mixing

J. I. Aranda[(a)], F. Ramírez-Zavaleta[(a)], J. J. Toscano[(b)] and E. S. Tututi[(a)]

[(a)]Facultad de Ciencias Físico Matemáticas, Universidad Michoacana de San Nicolás de Hidalgo, Avenida Francisco J. Mújica S/N, 58060, Morelia, Michoacán, México.
[(b)]Facultad de Ciencias Físico Matemáticas, Benemérita Universidad Autónoma de Puebla, Apartado Postal 1152, Puebla, Puebla, México.

E-mail: `jarandas@umich.mx`

Abstract. We bound the $Z'tc$ coupling using the $D^0 - \overline{D^0}$ meson mixing system. We obtained such coupling which is less than 5.75×10^{-2}. We have studied the Z' boson resonance considering single top production in the $e^+e^- \to Z' \to tc$ process. We obtained the number of events which is expected to be less than 10^7 at the International Linear Collider scenario. We get a branching ratio of the order of 10^{-2} for the $Z' \to tc$ decay.

1. Introduction.

Many extensions of the Standard Model (SM) predict the existence of an extra $U'(1)$ gauge symmetry group and its associated Z' boson which has been object of extensive phenomenological studies [1]. This boson can induce flavor changing neutral currents (FCNC) at tree level through $Z'q_iq_j$ couplings where q_i and q_j are up or down-type quarks. The flavor-violating parameters must fulfill experimental constraints on FCNC [2]. We focus on the Z' virtual effects we may analyze the impact of the FCNC through the single top quark production. We can use the mass difference ΔM_D of the $D^0 - \overline{D^0}$ mixing observed by the Babar and Belle collaborations to bound the strength of these couplings.

2. The $Z'tu_i$ couplings.

The FCNC Lagrangian contained in the $SU_C(3) \times SU_L(2) \times U_Y(1) \times U'(1)$ group is given by

$$\mathcal{L}_{NC} = -eJ^\mu_{EM} A_\mu - g_1 J^\mu_1 Z_{\mu,1} - g_2 J^\mu_2 Z_{\mu,2}. \tag{1}$$

J^μ_1 is the weak neutral current and J^μ_2 represents the new weak neutral current given as

$$J^\mu_2 = \sum_{i,j} \overline{\psi'_i}\, \gamma^\mu (\epsilon^\psi_{Lij}\, P_L + \epsilon^\psi_{Rij}\, P_R)\, \psi'_j. \tag{2}$$

Since the interaction between the bosons Z_1 and Z_2 is too weak to be considered, there is no mixing between them, consequently their mass eigenstates are Z^0 and Z' respectively. Let us consider the $\epsilon^u_{L,Rij}$ matrix for the sector of quarks type up. Some models assume this matrix as flavor diagonal and non-universal. The FCNC couplings in the mass eigenstates basis can be read off as

$$\Omega_{Lij} = g_2\, (V_L\, \epsilon^u_L\, V^\dagger_L)_{ij}, \;\; \Omega_{Rij} = g_2\, (V_R\, \epsilon^u_R\, V^\dagger_R)_{ij}. \tag{3}$$

3. Bounding the $Z'tc$ couplings from $D^0 - \overline{D^0}$.

The Lagrangian containing the relevant information is

$$\begin{aligned}\mathcal{L}_{NC}^{Z'q_iq_j} =& - \big[\overline{u}\,\gamma^\mu(\Omega_{Luc}\,P_L + \Omega_{Ruc}\,P_R)\,c + \overline{c}\,\gamma^\mu(\Omega_{Lcu}\,P_L + \Omega_{Rcu}\,P_R)\,u \\ &+ \overline{u}\,\gamma^\mu(\Omega_{Lut}\,P_L + \Omega_{Rut}\,P_R)\,t + \overline{t}\,\gamma^\mu(\Omega_{Ltu}\,P_L + \Omega_{Rtu}\,P_R)\,u \\ &+ \overline{c}\,\gamma^\mu(\Omega_{Lct}\,P_L + \Omega_{Rct}\,P_R)\,t + \overline{t}\,\gamma^\mu(\Omega_{Ltc}\,P_L + \Omega_{Rtc}\,P_R)\,c\big]\,Z'_\mu. \end{aligned} \tag{4}$$

From the unitary property of the $V_{L,R}$ matrices

$$|\Omega_{uc}| \approx |\Omega_{ut}\Omega_{ct}|, \tag{5}$$

provided that $\epsilon_{tt} \ll 1$. For simplicity we assume Ω's as real and $\Omega_{L,R\,q_iq_j} = \Omega_{L,R\,q_jq_i}$ and $\Omega_{L\,q_iq_j} = \Omega_{R\,q_iq_j} \equiv \Omega_{q_iq_j}$. The tree-level amplitude can be written as

$$\mathcal{M}_{\text{tree}} = -\frac{i\Omega_{uc}^2}{m_{Z'}^2}\,\overline{u}\gamma^\alpha c\,\overline{u}\gamma_\alpha c. \tag{6}$$

$\mathcal{M}_{\text{tree}}$ amplitude can be related to a four-quark effective vertex accounted by the effective Lagrangian:

$$\mathcal{L}_{eff}^{\text{tree}} = -\frac{\Omega_{uc}^2}{4m_{Z'}^2}\,(Q_1 + 2Q_2 + Q_6)\,, \tag{7}$$

where a $1/4$ factor has been introduced to compensate Wick contractions. The Q_i are dimension-six effective operators.

Analogously, the one-loop level amplitude is given by:

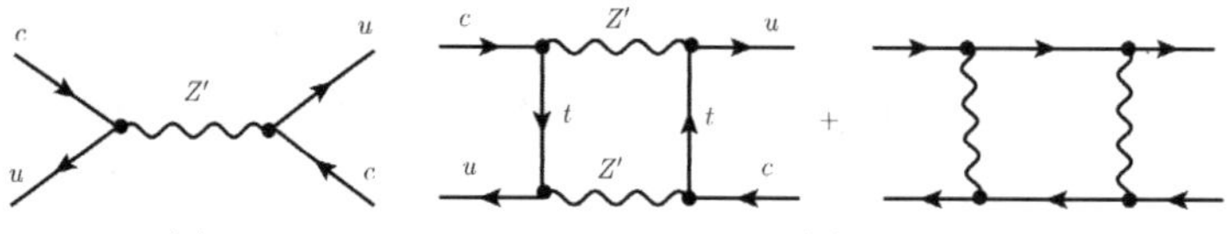

Figure 1. (a) Tree diagram;(b) Box diagrams for $D^0 - \overline{D^0}$ mixing.

$$\mathcal{M}_{\text{box}} = 2\,\Omega_{tu}^2\,\Omega_{tc}^2 \int \frac{d^4k}{(2\,\pi)^4}\,\frac{[\overline{u}\,\gamma^\lambda\,(k^\alpha\gamma_\alpha + m_t)\,\gamma^\nu\,c]\,[\overline{u}\,\gamma_\nu\,(k^\alpha\gamma_\alpha + m_t)\,\gamma_\lambda\,c]}{(k^2 - m_t^2)^2(k^2 - m_{Z'}^2)^2}. \tag{8}$$

After some algebra we arrive at the $\mathcal{M}_{\text{box}}$ amplitude which can be related to a four-quark effective vertex accounted by the effective Lagrangian:

$$\mathcal{L}_{eff}^{\text{box}} = -\frac{\Omega_{tu}^2\,\Omega_{tc}^2}{64\pi^2 m_t^2}\,\big[f(x)\,(4Q_1 + 32Q_2 + 4Q_6) + g(x)\,(8Q_3 + 4Q_4 + Q_5 + 4Q_7 + Q_8)\big], \tag{9}$$

where a $1/4$ factor has been introduced to compensate Wick contractions; $f(x)$ and $g(x)$ are loop functions given as

$$f(x) = \frac{1}{2}\,\frac{1}{(1-x)^3}\,[1 - x^2 + 2\,x\,\log x], \;\; g(x) = \frac{2}{(1-x)^3}\,[2(1-x) + (1+x)\,\log x]. \tag{10}$$

with $x = m_{Z'}^2/m_t^2$. The mass difference ΔM_D provided by the $D^0 - \overline{D^0}$ meson-mixing system is $\Delta M_D = \frac{1}{M_D} Re\langle\overline{D^0}|\mathcal{H}_{eff} = -\mathcal{L}_{eff}|D^0\rangle$. The effective Lagrangian is $\mathcal{L}_{eff} = \mathcal{L}_{eff}^{\rm tree} + \mathcal{L}_{eff}^{\rm box}$ and M_D is the D^0 meson mass. Using the modified vacuum saturation approximation [4] we have:

$$\Delta M_D = \frac{\Omega_{uc}^2 f_D^2 M_D B_D}{12 m_{Z'}^2}\left[1 + \frac{x}{8\pi^2}\big(32 f(x) - 5g(x)\big)\right], \qquad (11)$$

We used the relation in (5), B_D is the bag model parameter and f_D represents the D^0 meson decay constant. We can see from Eqs. (7), (9) and (11) that the main contribution to ΔM_D comes from the tree-level amplitude while the contribution coming from the box amplitude is of approximately 17%-19% in the range of 800 GeV $\leq m_{Z'} \leq$ 3000 GeV. Taking $B_D \sim 1$, f_D = 222.6 MeV and M_D = 1.8646 GeV and considering that ΔM_D does not exceed the experimental uncertainty

$$|\Omega_{uc}| < \frac{3.6\times 10^{-7} m_{Z'} \mathrm{GeV}^{-1}}{\sqrt{1 + \frac{x}{8\pi^2}\left(32 f(x) - 5g(x)\right)}}, \qquad (12)$$

Taking $m_{Z'}$ = 1 TeV we obtain a bound $|\Omega_{tc}\Omega_{tu}| < 3.31\times 10^{-4}$, moreover, we assume that

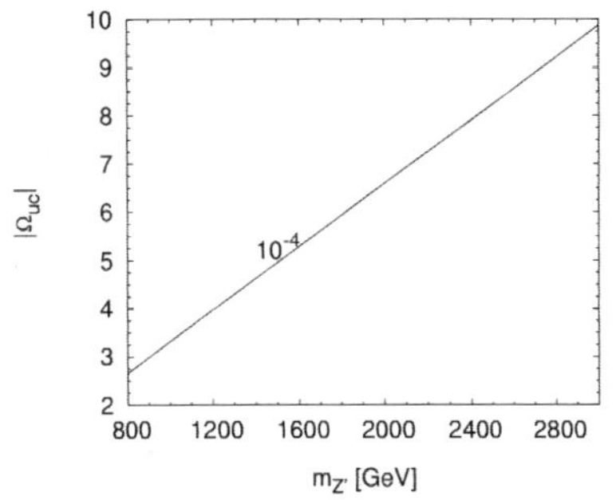

Figure 2. Behavior of $|\Omega_{uc}|$ coupling as a function of Z' boson mass.

$\Omega_{tc} = 10\,\Omega_{tu}$, as it occurs for the absolute values of U_{ts}, U_{td} elements in the CKM matrix. We found that $|\Omega_{tc}| < 5.75\times 10^{-2}$ and $|\Omega_{tu}| < 5.75\times 10^{-3}$, which are of the same order of magnitude approximately than those obtained in Ref. [3].

4. The process $e^+e^- \to Z' \to tc$ at ILC collider.

We only take the average of the chiral charges; the different values for the charges are: $Q_L^u = 0.3456$, $Q_R^u = -0.1544$, $Q_L^d = -0.4228$, $Q_R^d = 0.0772$, $Q_L^e = -0.2684$, $Q_R^e = 0.2316$ and $Q_L^\nu = 0.5$ for the Sequential Z model; $Q_L^u = \frac{1}{\sqrt{24}}$, $Q_R^u = \frac{-1}{\sqrt{24}}$, $Q_L^d = \frac{1}{\sqrt{24}}$, $Q_R^d = \frac{-1}{\sqrt{24}}$, $Q_L^e = \frac{1}{\sqrt{24}}$, $Q_R^e = \frac{-1}{\sqrt{24}}$ and $Q_L^\nu = \frac{1}{\sqrt{24}}$ for E_6 model; $Q_L^u = 0.2749$, $Q_R^u = -0.1793$, $Q_L^d = -0.1093$, $Q_R^d = -0.0635$, $Q_L^e = -0.0321$, $Q_R^e = 0.0137$ and $Q_L^\nu = 0.3521$ for Average model [4].

The Breit-Wigner resonant cross section is $\sigma(e^+e^- \to Z' \to tc) = \frac{12\pi m_{Z'}^2}{s}\frac{\Gamma(Z'\to e^+e^-)\Gamma(Z'\to tc)}{(s-m_{Z'}^2)^2 + m_{Z'}^2\Gamma_{Z'}^2}$.

For the decay width $\Gamma(Z' \to tc)$ we obtain $\Gamma(Z' \to tc) = \frac{(2m_{Z'}^4 - m_t^4 - m_{Z'}^2 m_t^2)\Omega_{tc}^2}{12\pi m_{Z'}^3}$.

We can predict around 10^7 events just at the resonance for the E_6 model. For the sequential Z model it is expected to obtain around 10^6 events. For the average of the two models, it is expected around 10^5 events. We obtain that the associated branching ratio is of the order of 10^{-2}. The production of around 10^4 tc events predicted in Ref. [3], or similar results in Ref. [5],

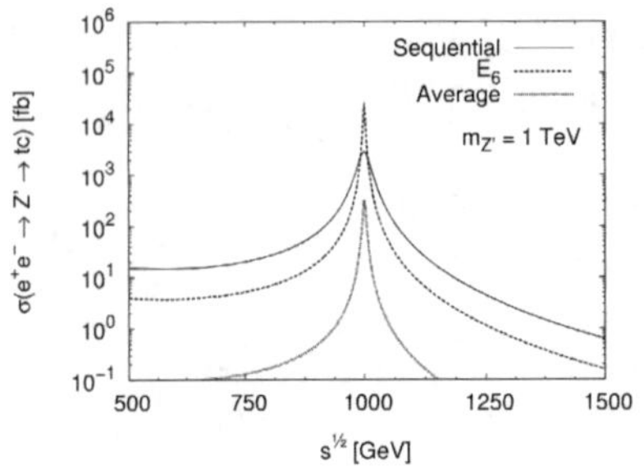

Figure 3. Cross section for $e^+e^- \to Z' \to tc$ process as a function of $\sqrt{s}$ for $m_{Z'} = 1$ TeV.

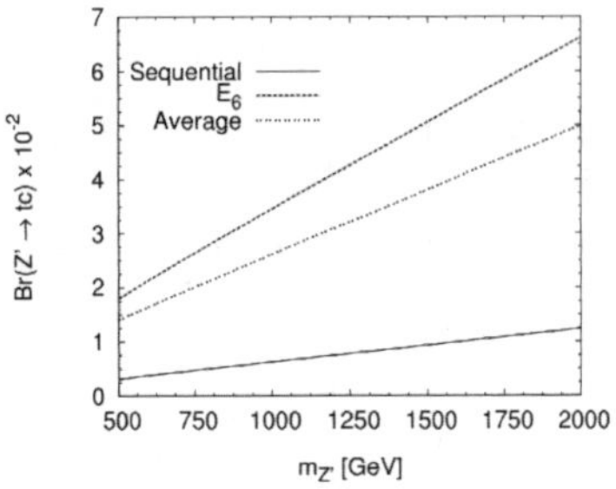

Figure 4. The branching ratio for $Z' \to tc$ decay.

predicted for the Compact Linear Collider calculated at the resonance, can be compared with our predictions and find that ours are bigger in 1 and 3 orders of magnitude for the average and the E_6 model, respectively. We found that it will be produced around 10^3 tc events for a Higgs mass of the order of top quark mass, which is two orders of magnitude less than the average prediction, calculated at the resonance. In relation to the values we have found for the branching ratios $Br(Z' \to tc) \sim 10^{-2}$ and $Br(Z' \to tu) \sim 10^{-4}$ calculated at the resonance, we can mention that these values are one order of magnitude less restrictive than corresponding branching ratios obtained in the model 3-3-1 [6].

5. Conclusions.

We have bounded the strength of the flavor-violating $Z'tc$ coupling using the experimental results coming from the $D^0 - \overline{D^0}$ meson mixing system. For a $m_{Z'} = 1$ TeV we found that $|\Omega_{tc}| < 5.75 \times 10^{-2}$. We have calculated the cross section for the $e^+e^- \to Z' \to tc$ process in the ILC collider scenario; where we found an estimation around 10^7 events for a luminosity of 500 fb^{-1} in the context of Z' boson predicted by the E_6 model. According to our results the tc flavor violation effect mediated by a Z' boson from the E_6 model is more favorable of being observed than that predicted in the sequential model one. This behavior is also repeated for the branching ratio of the $Z' \to tc$ decay.

[1] Langacker P 2008 *Rev. Mod. Phys.* **81** 1199.
[2] Aranda J I, Cordero-Cid A, Ramírez-Zavaleta F, Toscano J J and Tututi E S 2009 *Mod. Phys. Lett.* A **24** 3219; Aranda J I, Cordero-Cid A, Ramírez-Zavaleta F, Toscano J J and Tututi E S 2010 *Phys. Rev.* D **81** 077701.
[3] Arhrib A, Cheung K, Chiang C-W, Yuan T-C 2006 *Phys. Rev.* D **73** 075015.
[4] Golowich E, Hewett J, Pakvasa S and Petrov A A, 2007 *Phys. Rev.* D **76** 095009.
[5] Çakir O, Çakir I T, Senol A and Tasci AT 2010 arXiv:1003.3156 [hep-ph].
[6] Pérez M A, Tavares-Velasco G and Toscano J J 2004 *Phys. Rev.* D **69** 115004.

Journal of Physics: Conference Series **287** (2011) 012045 doi:10.1088/1742-6596/287/1/012045

Proposal for the geometrical distribution of the air cherenkov detectors for CHARM

Morales Reyes A. R., Martínez Bravo O. M.

Facultad de Ciencias Físico Matemáticas, Benemérita Universidad Autónoma de Puebla

hkanzaki@hotmail.com, omartin@fcfm.buap.mx

Abstract. In this work we propose the geometrical distribution of the air Cherenkov detectors array (ACD), who will be part of the Cosmic High Altitude Radiation Monitor Observatory (CHARM) located at Pico de Orizaba Volcano at 4300 m.a.s.l.. The proposal is based on a library of events built with photons, protons and iron nuclei as primary particles by montecarlo simulations with energies from 10^{14} eV to 10^{17} eV. The goal of this detectors will be to determinate the nature of primary cosmic radiation, through measuring the height at which the secondary particles generated reach his maximum number or X_{max}, this quantity is related with the effective cross section and finally with the atomic number A of the primary particles. In addition to this we proposed an energy estimator based on the study of the lateral distribution function of the generated events.

1. Methodology

We had constructed a database of simulations with CORSIKA software v6.735 designed to study cosmic rays and their development through the atmosphere. The CORSIKA parameters imposed to have the air Cherenkov emission corresponding to the Extensive Air Showers (EAS), implies an extensive use of cpu time and disk space, for this we were forced to configure our computational resources, building a cluster Beowulf type 1 using standard hardware and software available; in our case, the operative system used was Ubuntu 9.04 in their different versions: server and desktop.

Once we had the events database, we implement an algorithm in C language who permit us to acquire the radial particle density flux at different distances from the core of the shower for several zenithal angles, the plot generated using this results is known as the lateral distribution function who was used to determine the distance at which detectors may be located.

The reconstruction and the evaluation of the efficiency of the array is a work in progress.

2. Results

The lateral distribution function (LDF) of the EAS generated by primary particles with energies from 10^{14} eV to 10^{17} eV at Pico de Orizaba altitude will allow us to measure the secondary particles generated in the shower near the X_{max}, were the atmospheric absorption effects are minimized.

Analyzing the LDF let us determine the optimal distance at what detectors will be located at Pico de Orizaba Volcano, from figure 1 we can see this distance around 230 m.

Journal of Physics: Conference Series **287** (2011) 012045 doi:10.1088/1742-6596/287/1/012045

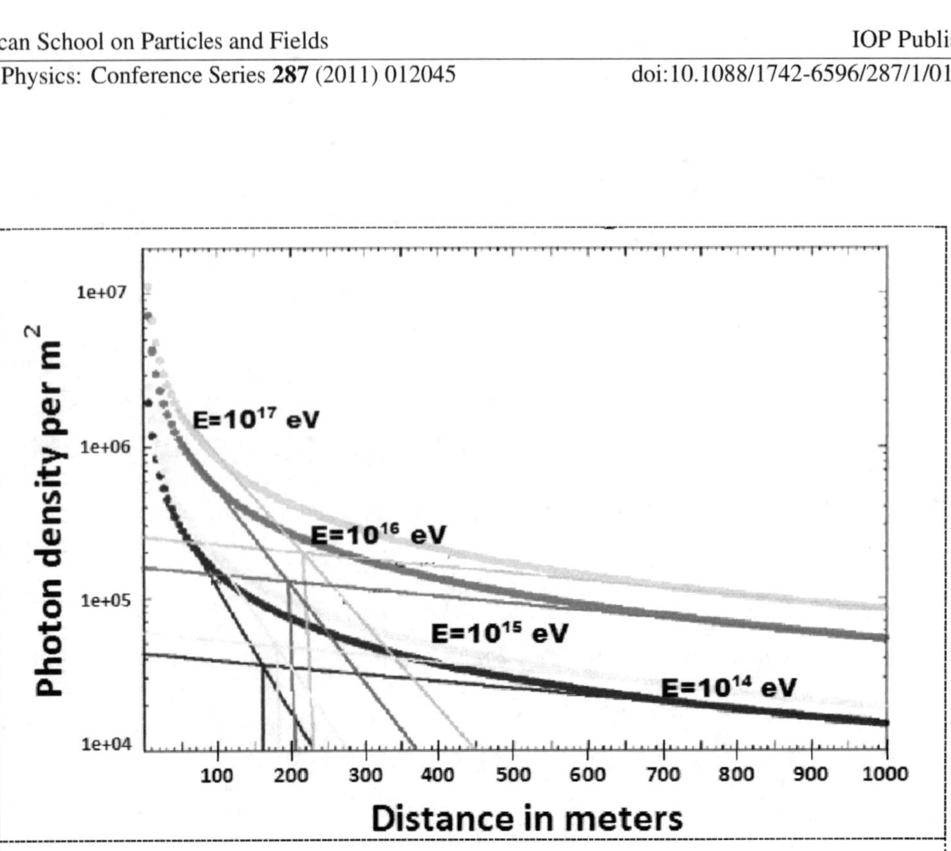

Figure 1. Determination of characteristic distance at an energy range from 10^{14} eV to 10^{18} eV for proton nuclei with zenithal angle of 0 degrees.

The combination of both detection techniques, air cherenkov detectors (ACD) to the water cherenkov detectors (WCD) in CHARM array may allow us to crosscheck the EAS parameters obtained using the advantages of each one, as we seen on figure 2, the core and arrival direction were measured by the WCD array and the effective cross section via the X_{max} altitude related with the mass of the primary particle by the ACD.

WCD detectors has a duty cycle of about 100% of time meanwhile ACD detectors need dark and moonless nights which represents a duty cycle of about 15%.

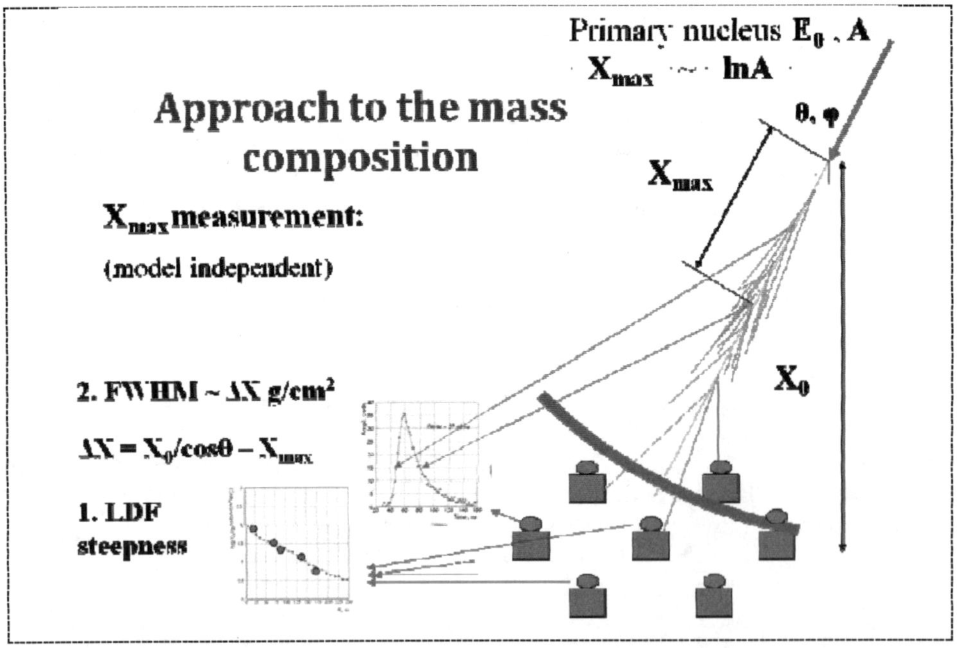

Journal of Physics: Conference Series **287** (2011) 012045 doi:10.1088/1742-6596/287/1/012045

Figure 2. Scheme of the hybrid array of WCD (squares) and ACD (circles) providing information about the X_{max} and the LDF correspondingly.

CHARM Observatory was proposed as an hexagonal array with an apothem distance of 230 meters for the first ring WCD detectors, from our results about the analysis of LDF we find a characteristic distance that correspond also with the distance at which WCD will be located.

In figure 3 we show the CHARM Observatory and the distribution of both arrays.

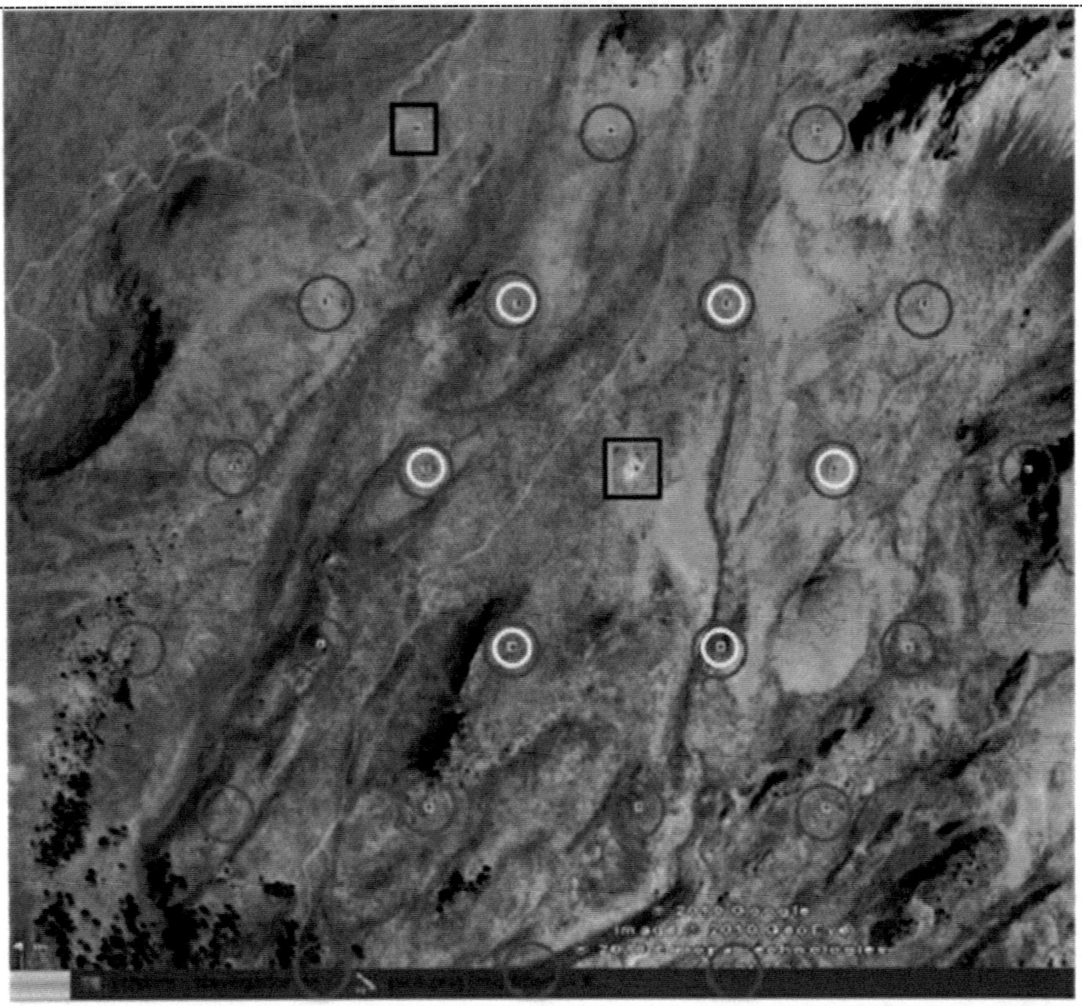

Figure 3. Satellite image form Pico de Orizaba volcano and the scheme of the Cosmic High Altitude Radiation Monitor (CHARM) observatory with the proposal location of WCD (darker circles) and ACD (lighter circles), the central square represents the acquisition data center and the outer square is the detector who located at maximum altitude of the array.

3. Conclusions

In this work we present the geometrical distribution of ACD array who will be part of CHARM Observatory. This proposed is based on the study of LDF of the events generated with CORSIKA code using the air cherenkov emission setup, in order to determine the mass composition of the cosmic primary radiation in the second knee energy range.

Journal of Physics: Conference Series **287** (2011) 012045 doi:10.1088/1742-6596/287/1/012045

As a future work we complete the array characterization by end to end simulation involving Geant4 code for the reconstruction of the CORSIKA generated events.

References

[1] O. Martínez et al, "*The Cosmic High Altitude Radiation Monitor (CHARM) at Pico de Orizaba*", Proceedings of the 31st ICRC, Lodz, 2009.
[2] J. Cotzomi et al, "*The Water Cherenkov Detector Array for Studies of Cosmic Rays at the University of Puebla*", 2004.
[3] H. Salazar et al, "*Search for Gamma Ray Burst at Sierra Negra, México*", 29th International Cosmic Ray Conference Pune, 2005.
[4] P. Bello et al, "*Air Shower array at the university of Puebla for the study of cosmic rays*", Revista Mexicana de Física, Vil. 51, 2005.
[5] A. A. Al-Rubaiee et al, "*Parametrization for Cherenkov light lateral distribution function in Extensive Air Showers*", 29th International Cosmic Ray Conference Pune, pp 249-252, 2005.
[6] M. I. Panasyuk et al, "*The Tunka-133 EAS Cherenkov array – status, first results and plans*", 29th International Cosmic Ray Conference Pune, pp 255-258, 2005.
[7] The Pierre Auger Collaboration, "*Measurement of the Depth of Maximumof extensive Air Showers above 10^{18} eV*", astro-ph.HE/1002.0699v1, 2010.
[8] The Pierre Auger Collaboration, "*Measurement of the energy of cosmic rays above 1018 eV using the Pierre Auger Observatory*", astro-ph.HE/1002.1975v1, 2010.
[9] N. M. Bundev et al, "*The Tunka Experiment: Towards a 1-km Cherenkov EAS Array in the Tunka Valley*", astro-ph.HE/1003.0089v1, 2010.
[10] N. M. Bundev et al, "*Cosmic Ray Mass Composition by the Data of Tunka EAS Cherenkov Arrays*", 21th European Cosmic Rays Symposium, Slovakia, 2008.
[11] D. Heck, J. Knapp, J. N. Capdevielle, G. Schartz, T. Thouw, "*CORSIKA: A Montecarlo Code to Simulate Extensive Air showers*", Ferschungszentrum und Universität Karlsruhe, Karlsruhe, Germany.
[12] N. N. Kalmykov et al, "*TUNKA-EAS Cherenkov Light Array*", Tunka Valley, Russia.

Journal of Physics: Conference Series **287** (2011) 012046 doi:10.1088/1742-6596/287/1/012046

Description of a4 –channel FPGA-controlled ADC-based DAQ system for general purpose PMT signals

Rubén Conde[1], Humberto Salazar[1], Oscar Martínez[1] and L. Villaseñor[2]

[1]Facultad de Ciencias Físico Matemáticas, BUAP, Puebla, México

[2]Instituto de Física y Matemáticas, Universidad Michoacana San Nicolás de Hidalgo, Morelia, México.

rbn_cnd@hotmail.com

Abstract: We describe a general purpose data acquisition system for PMT signals. Hardware-wise it consists of a 4-channel ADC daughter board, an FPGA mother board, a GPS receiver and an atmospheric pressure sensor and a temperature sensor. The four ADC channels simultaneously sample PMT input signals with a sampling rate of 100MS/s. We have evaluated the noise of our system obtaining less than -48.6dB. This DAQ system includes a firmware suitable for pulse processing in cosmic rays applications. In particular, we describe in detail the way in which this system can be used during the commissioning and early operation phases of the High Altitude Water Cherenkov Observatory (HAWC) currently under construction at Sierra Negra in Mexico.

1. INTRODUCTION

The HAWC observatory is under construction by a collaboration of scientists from the US and México at the Sierra Negra site, in México, at an altitude of 4100 m.a.s.l. It will make use of a compact array of water Cherenkov detectors (WCDs) to survey the sky in search of steady and transient gamma ray sources in the 0.1-100TeV energy range. The HAWC Observatory will reuse PMTs and electronics from the Milagro gamma Ray Observatory [1]

A prototype array for HAWC consisting of 7 cylindrical water Cherenkov detectors, with a diameter of 7.3 m and filled with water up to a height of 5 m is under construction. Each detector will have 3 PMTs looking at the water volume from the bottom of the tanks. The general purpose data acquisition system that we describe in the following sections can digitize up to 4 PMT signals with a sampling rate of 100 MS/s; it represents a low-power, portable, fast and easy-to-operate readout system that can be used along with a more powerful VME-based system during the construction phase of HAWC to test individual detectors.

2. The DAQ Hardware

Our 4-channel ADC daughter board has two independent dual analog to digital converter working at 100MS/s, with local memory and FPGA in its mother board for real-time data processing. The purpose of the pulse processing is to perform online signal processing on the digitized signals directly to minimize the data transfer size, these algorithms are implemented in the FPGA and can be reprogrammed at any time.

Each event will be tagged with precise GPS time using an embedded GPS receiver with 1 PPS (one pulse per second) synchronized with UTC within an uncertainty of 50 ns (Motorola Oncore UT+ GPS receiver). A pressure and temperature sensor (HP03D) is also connected to the FPGA board (Nexys2 from Digilent Inc.). The FPGA that we use is the Spartan 3E-500 from Xilinx.

Journal of Physics: Conference Series **287** (2011) 012046 doi:10.1088/1742-6596/287/1/012046

Figure 1 shows the block diagram of the ADC daughter board. The PMTs convert the energies deposited in the detector by charged particles into negative pulses with typical widths of 5-10 ns. These pulses are conditioned and shaped inside the ADC board into positive pulses with widths of 20-40 ns. The use of dual ADCs simplifies the firmware program written in hardware description language (VHDL) of the whole DAQ system.

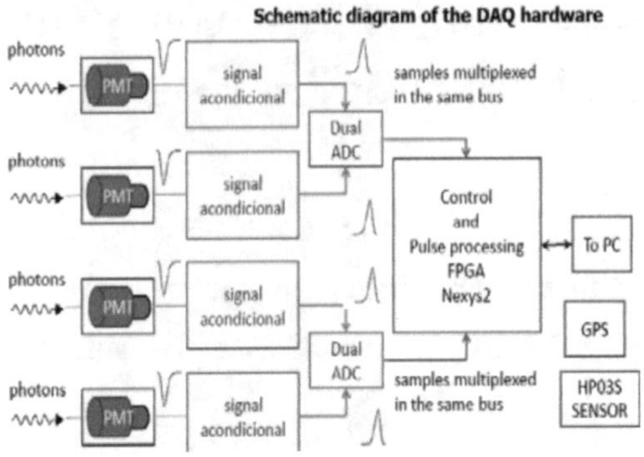

Figure 1 Block diagram of the ADC daughter board.

Hardware wise this system possess the following characteristics:

- Uses the Nexys2 board with a Spartan 3E-500 FPGA from digilent inc. [2]. The Nexys2 is a powerful digital system design platform built around a Xilinx Spartan 3E FPGA. With 16Mbytes of fast SDRAM and 16Mbytes of Flash ROM, the Nexys2 is ideally suited to embedded processors like Xilinx's 32-bit RISC Microblaze™.
- Uses a 4-channels ADC daughter board with two dual 10-bit ADC chips AD9216 [3]. The AD9216 is a dual, 3 V, 10-bit, 105 MSPS analog-to-digital converter. It features dual high performance sample and hold amplifiers and an integrated voltage reference. The AD9216 uses a multistage differential pipelined architecture with output error correction logic to provide 10-bit accuracy and guarantee no missing codes over the full operating temperature range at 105 MSPS.
- Uses one quad, 10-bit, low-power, buffered voltage-output, digital-to-analog converter (DAC), MAX5741 [4], to control the baselines of the four input channels of the ADC daughter board.
- Uses a Motorola Oncore UT+ GPS receiver [5].
- Uses an HP03S pressure and temperature sensor. It provides 16 bit word data for pressure and temperature related voltage [6].
- Uses the RS-232 port for slow communications.

Figure 2 shows a photograph of the ADC daughter board connected to the Nexys2 board. This system will be used on the first stage of the construction of the HAWC Observatory, called the VAMOS Array, to test individual detectors.

Figure 2 Photograph of a prototype of the ADC daughter board connected to the Nexyx2 board.

Journal of Physics: Conference Series **287** (2011) 012046 doi:10.1088/1742-6596/287/1/012046

3. RESULTS

We first studied the baselines of typical input pulses and measured the width of their distributions. Figure 3 shows examples of digitized baselines and their respective distributions.

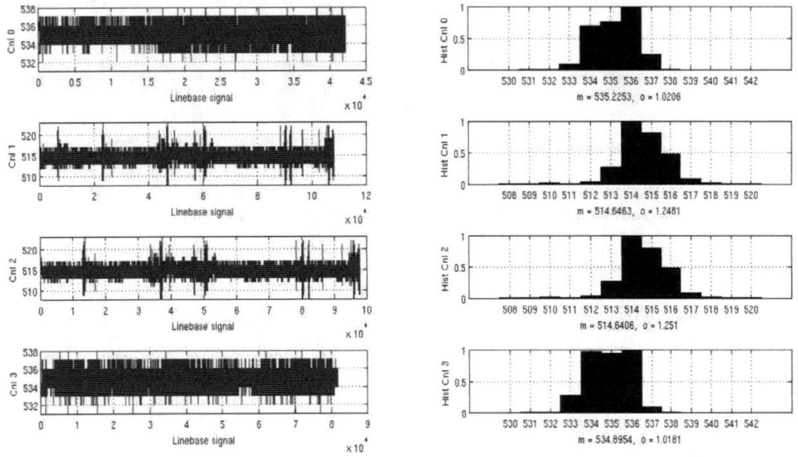

Figure 3 Examples of digitized baselines (left plots) and their respective distribution (right plots) for signals from a pulse generator.

Figure 4 shows the FFT (Fast Fourier Transform) computed for four sine signals using 4096 samples. The frequencies of these sine signals were 500 Khz, 750 Khz, 1 MHz and 2 Mhz, respectively. These four signals were sampled simultaneously at 100 MS/s by using the 4 channels of the ADC board. We used the same trigger threshold on all the four channels.

Figure 5 shows a trace firmware for a signal sampled at 100MS/s. Each trace has 16 samples, and the same threshold for all channels. Each signal corresponds to one channel with different frequency, and different pulse width.

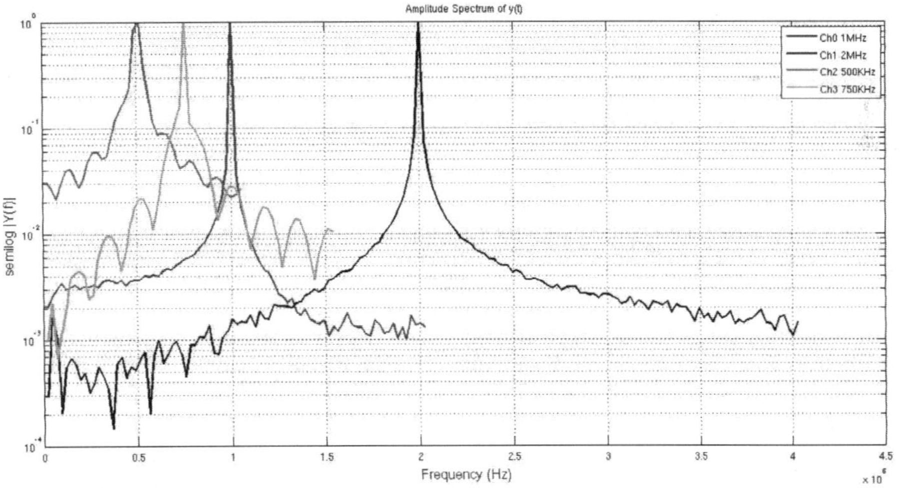

Figure 4 Fast Fourier Transforms (FFTs) computed for four sine signals 4,096 microsecond long using 4096 samples. The frequencies of these sine signals were 500KHz, 750KHz, 1MHz and 2MHz, respectively. These four signals were sampled simultaneously by using the 4 channels of the ADC board

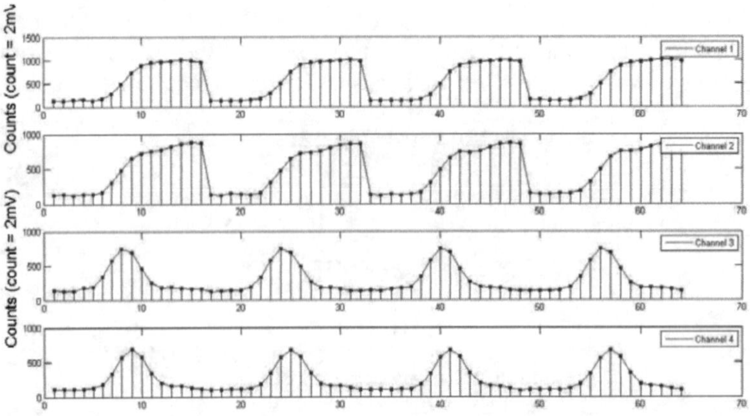

Figure 5 Examples of traces digitized at 100 MS/s with the DAQ system described in this paper. The first 16 points of each plot correspond to synchronous pulses fed into the system by using a pulse generator..

The firmware of the DAQ system described was written in hardware description language by using VHDL. On the PC side we used Perl and MATLAB under Linux to process, store and display the data acquired.

4. CONCLUSION

We have evaluated the performance of the data acquisition system to sample the PMT signals from water Cherenkov detectors at 100 MS/s and we have built a custom-made data acquisition (DAQ) too. This DAQ system includes a firmware suitable for pulse processing in cosmic rays applications.

5. ACKNOWLEDGEMENTS

The Authors acknowledge the support partially of the RedFAE of CONACYT , the coordination de la investigación científica from UMSNH and Facultad de Ciencias Físico Matemáticas BUAP.

6. REFERENCES

[1] http://hawc.umd.edu/detector.php
[2] http://www.digilentinc.com/Products/Detail.cfm?NavPath=2,400,789&Prod=NEXYS2
[3] http://www.analog.com/en/analog-to-digital-converters/ad-converters/ad9216/products/product.html
[4] http://www.maxim-ic.com/datasheet/index.mvp/id/3107/t/al
[5] http://www.tapr.org/gps_oncoreut.html
[6] http://www.hoperf.com/upfile/HP03S.pdf

Analysis of data collected by the Tatyana II satellite

Rivera Liliana[a], Martínez Oscar[b], Mendoza-Torres Eduardo[c], Salazar Humberto[d]

[a,b,d]Benemérita Universidad Autónoma de Puebla, Puebla, México
[c]Instituto Nacional de Astrofísica, Óptica y Electrónica, Puebla, México

E-mail: [a]taydarh@gmail.com, [b]moscar70@gmail.com,
[c]mend@inaoep.mx, [d]hsalazar@fismat1.fcfm.buap.mx

Abstract. The Tatyana II satellite is the second one of the University Satellite Program, which is led by the Moscow State University with the participation of the Benemérita Universidad Autónoma de Puebla. This satellite has ultraviolet, red-infrared and charged particles detectors. In this work preliminary results based on the data collected by these detectors on board the satellite over a period of ~3.5 months are presented.

1. Introduction

The study of the atmospheric background is important for studying Ultra High Energy Cosmic Rays (UHECR) from the space by satellites. The aim of this work is to analyze the ultraviolet (UV), red (R) and infrarred (IR) radiation observed at the Earth's atmosphere. The atmosphere is a "transition" region of the earth, where the dynamics of the matter is determined by the earth gravity force, geomagnetic field and external fluxes of particles. In this region space particles and space radiation penetrate to the atmosphere but also there is an opposite process of electromagnetic radiation and emission of particles from it. The most known phenomenon in this "transition" region is the aurora glow at high latitudes of the Earth where the electrons and protons of the magnetosphere penetrate to the atmosphere along the magnetic lines. Charged Particles (CP) collide with the atmosphere molecules and atoms, excite them and resulting optical radiation is observed as "aurora" [1]. Also UV emission phenomena of very short duration known as Transient Luminous Events (TLE's) thought to be related to lightnings have been observed at this region of the Earth's atmosphere [2]. In the last years a region where also charged particles penetrate the atmosphere has been observed at the South Atlantic Ocean, called the South Atlantic Anomaly (SAA). At this region the magnetic field is weaker than at similar geographic latitudes and this seems to play an important role in the appearence of SAA.

2. About Tatyana II Satellite

Tatyana II Satellite was launched from the Baikonur cosmodrome using the Soyuz2.1B carrier rocket and the Fregat upper stage on September 17, 2009 [3].

Tatyana II has a polar Sun synchronous orbit with an inclination of 98.8°, a 800 − 850 km altitude, a mass ~ 100 kg and a power of 100 W [3]. The instruments constructed by the mexican participation, with the help of students, are:

- UV and R-IR detectors, at 300−400 nanometers (nm) and 600−850 nm ranges, respectively. Each detector has an aperture of 0.024 cm ^{2} sr and a temporal resolution of 1 millisecond (ms).
- Charged Particles (CP) detector, which works with an energy threshold of 1 MeV, 350 cm^2 of effective area and temporal resolution of 1 ms

Figure 1. Tayana 2 satellite. The solar panel was installed at an angle, such that it has the maximum efficiency in energy supply and the lower part of the satellite is available to install the scientific payload.

3. Data analysis

The observations were made to the nadir satellite because in this direction the absorption of the signal in the atmosphere is minimal, the field of view is clearer and more stable than to the limb direction.

Data taken in a period corresponding to 3.5 months (October 2009-January 2010) were analized. The Interactive Data Language (IDL) software for the data analysis was used.

The coordinates of the satellite were used to identify the time and location of the satellite when it was located over the recorded events.

4. Results

The geographic coordinates for different energies of UV and R-IR emissions and for CP detections were drawn over separated world maps. For low energies the distribution is more or less homogeneously spread over all the world. At high energies the locations of the events seem to be clumped over given regions. The IR events are mainly above land. Remarkably, UV and R-IR emissions as well as CP are seen at the South Atlantic Anomaly (SAA) (Figures 2,3 and 4). It is interesting to note that among the records over the SAA there is a clear difference between the distribution of low and high energies at both UV and R-IR events (Figure 9). The less energetic events are distributed over a large region that covers a considerably part of South America and extends to South Africa through the South Atlantic Ocean while the more energetic events lie over a smaller region (around the center of the first one) and their amount is smaller than the less intense events.

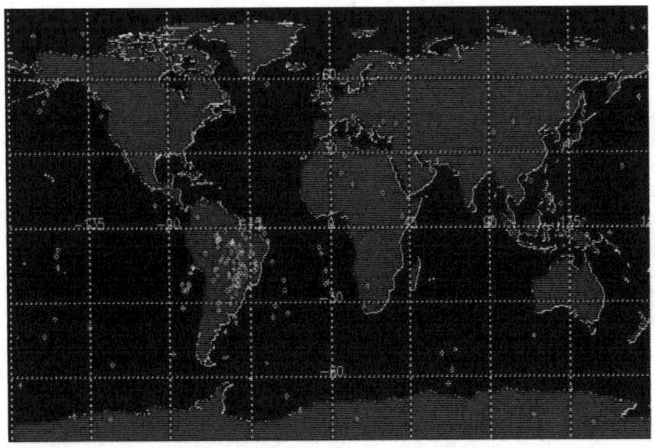

Figure 2. Distribution of UV flashes with $\sim 10^8$ photons/cm^{2}s sr. The SAA is clearly seen.

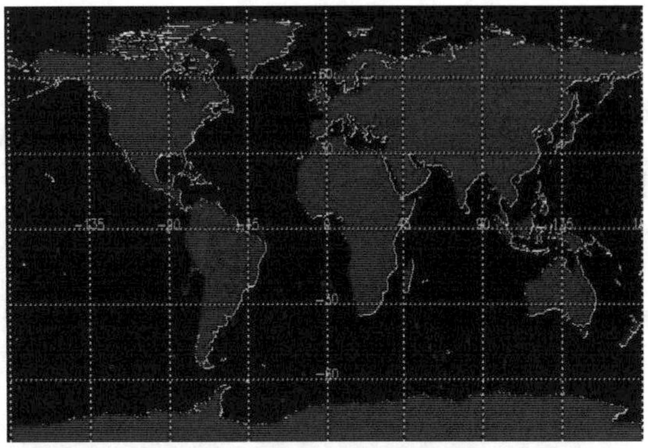

Figure 3. Distribution of R-IR flashes with $\sim 10^8$ photons/cm^{2}s sr, most events are over SAA and above land.

In the Figures 4 and 5 we have a comparison of charged particles detected zones in the Earth by Tatyana II and Along Track Scanning Radiometer (ATSR) [4]. We can see that both distributions show the South Atlantic Anomaly.

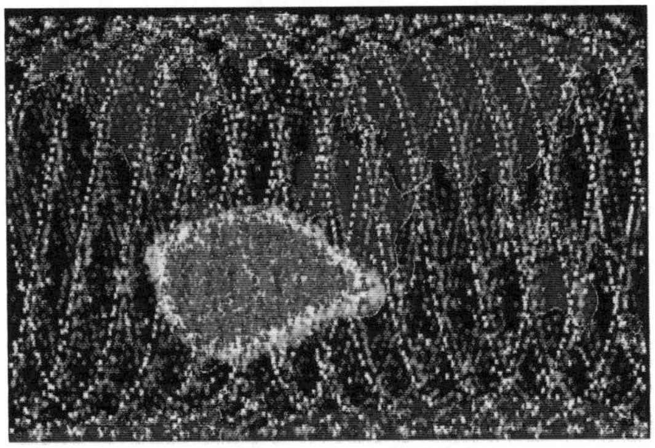

Figure 4. World Map of charged particles detected by Tatyana II satellite.

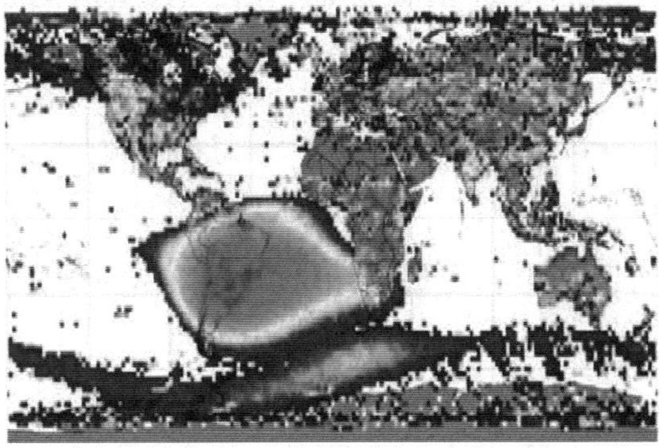

Figure 5. Charged particles distribution by Along Track Scanning Radiometer (ATSR).

The Figures 6 and 7 are compared for show the Earth's zone where the geomagnetic field intensity is lower than another regions [5]. That region corresponds at South Atlantic Anomaly.

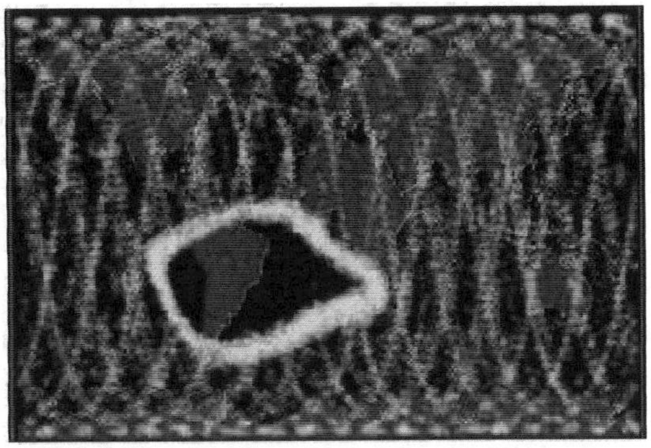

Figure 6. SAA detected by Tatyana II in the range of low energy charged particles.

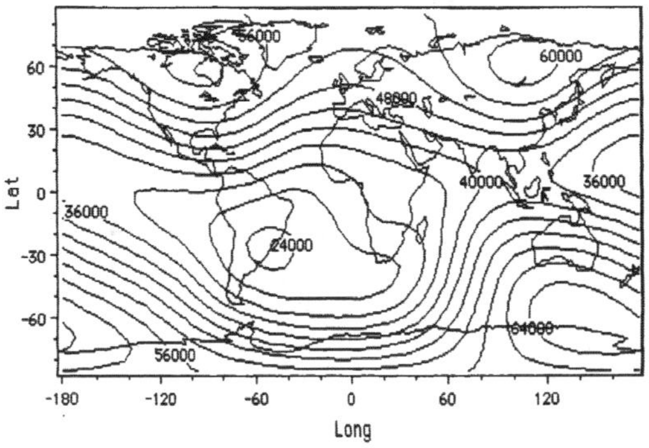

Figure 7. Contours of total field intensity at Earth's surface. It may be seen that in SAA the geomagnetic field intensity is low.

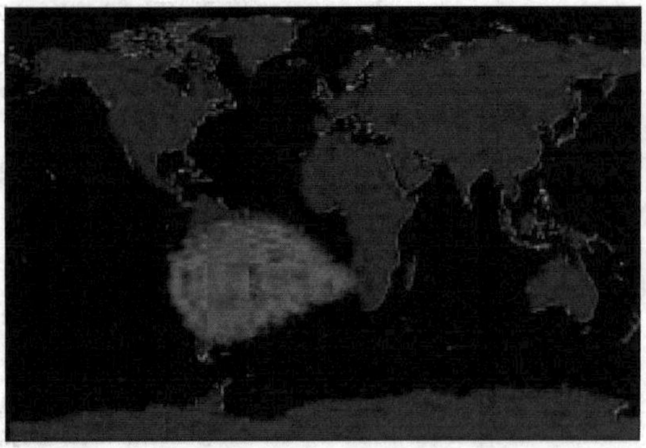

Figure 8. SAA zone in the range of high energy charged particles detected by Tatyana II satellite.

Figure 9. Distribution of UV energy events detected over SAA. The X axis is the longitude, the Y axis indicates the latitude and Z axis the amount of photons/cm^{2}s sr.

5. Conclusions

(i) We have obtained maps of UV-R-IR flashes in differents ranges of energies from Tatyana II data. The distribution of flashes shows clearly a tendency in determined regions. The IR events are mainly above land. UV and R-IR emissions are seen at the South Atlantic Anomaly (SAA).

(ii) Maps of charge particles were also obtained. The SAA was detected by Tatyana II satellite as the zone with most particle flux, so the detector works properly.

6. Acknowledgements

L.R. would like to thank the organizers of the XVI Mexican School of Particles and Fields for their support for assisting that event. This work has been supported in part by CONACyT México 106901 and VIEP, BUAP 0088.

References

[1] Program of transient uv event research at tatiana-2 satellite 2010 *Journal of Geophysical Research* **115**

[2] M. I. Panasyuk et al, Energetic particles impacting the upper atmosphere in connection with transient luminous event phenomena: Russian space experiment programs 2009 *AIP Conf. Proc.* **1118** 108-115

[3] M. I. Panasyuk, Lomonosov Moscow State University Satellite Program 2010 *Universat/LRT*

[4] S. Casadio, O. Arino and D. Serpe, Monitoring the South Atlantic Anomaly using ATSR

[5] J.R. Heirtzler, The geomagnetic field and radiation in near-Earth orbits 1999 *NASA*

AUTHOR INDEX